GCSE BIOLOGY

D. G. Mackean

John Murray

To the Student

This is a textbook to help you in studying biology for the GCSE. You will be following the GCSE syllabus of only one Examination Group, but this book contains the material needed by all the Groups. For this reason, amongst others, it is not expected that you will need to study or learn everything in the book.

Furthermore, the emphasis in GCSE is on the ability to understand and use biological information rather than on committing it all to memory. However, you will still need to use a book of this sort to find the facts and explanations before you can demonstrate your understanding or apply the biological principles.

The text is presented at two levels. *Core text* is in ordinary print and contains subject matter which occurs in nearly all the syllabuses. *Extension text* appears on a blue background and covers topics that are beyond the basic syllabus requirements or are relevant to only one syllabus.

It is therefore quite possible to concentrate only on the core material. If, however, you have a particular interest in biology or if parts of the extension material are relevant to your syllabus, you should read the extension material, *either* when you come to it *or* later after gaining a sound understanding of the core material.

The questions included in a chapter are intended to test your understanding of the text you have just read. Questions which relate to the extension material have a blue number. If you cannot answer the questions straightaway, read that section of text again with the question in mind.

There are check lists at the end of each chapter, summarising the important points covered. Points with a blue dot refer to the extension material.

The questions at the end of each section and those grouped at the end of the book are selected from the examination papers published by the GCSE Examining Groups. In many cases, they are designed to test your ability to apply your biological knowledge. The question may provide certain facts and ask you to make interpretations or suggest explanations. In such cases, the factual information may not be covered in the text.

Looking up information. To find the information you need, use the index, the contents pages and the headings at the beginning of each chapter—where blue type indicates an extension topic. If the word you want does not appear in the index, try a related word. For example, information about 'sight' might be listed under 'vision', 'eyes' or 'senses'.

Practical work. Given standard laboratory equipment, it should be possible for you to attempt any of the practical work described in the book, though you will probably not have time to do it all.

For this reason, it has been necessary to give the expected results of the experiments so that you can appreciate the design and purpose of an experiment even if you have not been able to do it yourself.

Note The first two impressions of this book erroneously stated on page 177 that late abortions may involve removal of the uterus. Unamended books still in use should be corrected to read as page 177 of this impression (second column, paragraph three).

The full-colour illustrations on pages 302–5 are by Pamela Haddon.

First published 1986
by John Murray (Publishers) Ltd
50 Albemarle Street, London W1X 4BD

Reprinted 1987, 1988, 1989 (with revisions), 1991, 1992, 1993

Printed in Hong Kong
by Colorcraft Ltd

British Library Cataloguing in Publication Data

Mackean, D. G.
GCSE biology.
1. Biology
I. Title
574 QH308.7

ISBN 0-7195-4281-2

The 1989 impression was significantly updated, and revised to include examination questions from the 1988 GCSE papers.
The opportunity was taken to amend some of the terminology to conform to the recommendations of the Institute of Biology and the Association for Science Education, in their publication *Biological Nomenclature* (Institute of Biology 1989).
The terms 'selectively permeable' and 'semi-permeable' have been replaced with 'partially permeable', and 'osmotic potential' has been replaced by 'water potential'.
In Section 4, after page 229, where the term is defined, 'allele' is used in preference to 'gene' where this is appropriate.
The word 'foetus' has been replaced by the preferred spelling 'fetus'.

Acknowledgements

I am very grateful to all the people who have provided photographs and would particularly like to thank Christine Hood and Dr C. J. Clegg, who read a substantial part of the manuscript and offered many constructive criticisms.

I also acknowledge the following Examining Groups who have kindly allowed me to reproduce questions from the 1988 GCSE papers.

W = Welsh Joint Education Committee
N = Northern Examining Association
S = Southern Examining Group
L = London and East Anglian Group
NI = Northern Ireland Schools Examinations Council

PHOTO CREDITS

Thanks are due to the following copyright owners for permission to reproduce their photographs. The abbreviations used are *t* top, *c* centre, *b* bottom, *l* left, *r* right.

ADAS Aerial Photography Unit – Crown Copyright 271*l*

Peter Addis 277*t*

Heather Angel 85*r*, 177*t*, 267, 269*r*, 299, 323*cl*, 337*l*, 343

Ardea: Liz and Tony Bomford 250; Jean-Paul Ferrero 259; Bob Gibbons 241; John Mason 72, 104, 278*c*; E. Mickleburgh 278*b*; P. Morris 265*t*, 323*r*, 334; Robert T. Smith 344*tl*; J. Swedberg 271*r*

Barnaby's Picture Library 297, 306

Dr Alan Beaumont 331

Biophoto Associates 2, 3, 5, 6*r*, 12, 13, 16, 38*l,r*, 63*t*, 64, 65*r*, 84*t*, 85*l*, 102, 138, 141, 146, 148*tr*, 152, 156*l,r*, 161, 163, 166*l,r*, 173, 190, 191*l,r*, 194, 210, 212, 213, 221, 224, 238*l*, 239*l,r*, 251, 265*b*, 266*tl,bl*, 269*t*, 279*r*, 285, 310, 311, 312*r*, 320*l,r*, 322*r*, 332, 338*r*, 342; NHPA 238*r,c*, 244

Birth Atlas, New York 176

J. v. den Brock (Biozentrum der Universität Basel) 312*l*

K. E. Carr/P. G. Toner 130

Bruce Coleman: N. G. Blake 245*b*; Jane Burton 90, 94, 105*l,r*, 337*r*, 341*b*, 344*r*; Robert P. Carr 322*bl*, 339; Alain Compost 272*l*; Eric Crichton 255, 323*tl*; Jeff Foott 279*t*; Manfred Kage 172; Gordon Langsbury 91; Hans Reinhard 57, 219, 322*tl*, 344*bl*; Frieda Sauer 328; Kim Taylor 335, 336*r*, 338*l*, 341*t*; Michael Viard 82*r*; Nicholas de Vore 272*r*; C. Zuber (WWF) 280

Central Office of Information 148*cr*

Gene Cox 4, 42*l,r*, 59, 61

Gerry Cranham 113

Engineering and Research Associates Inc, Tucson 282*t*

Finefare Ltd 318*t,c*

Glasshouse Crops Research Institute 51, 54, 55

Colin Green 245*tl,tr*

Steven Green 174

Professor W. J. Hamilton 175

Philip Harris Biological Ltd 7, 63*b*, 128, 148*tl*, 154, 232

Eric and David Hosking: D. P. Wilson 243*l,r*

ICI 314

ICI Plant Protection Division 266*tr,bl,cr*

Dr J. E. Jackson, East Malling Research Station 235

Frank Lane 282*b*

Tony Langham 118

Ian Mackean 178

Professor T. Mansfield 60

Leo Mason 168, 186, 205

Dr H. Moor 6*tl*

Nature Conservancy Council 286

Nature Photographers Ltd 268, 278*t*

Open University 6*bl*

Oxford Scientific Films 336*l*, 350

Pfizer Ltd 325

Photo Library International 273

Dr Michael Proctor 84*cl*

Rank Organisation 184*l,r*

Sir Ralph Riley 234

St Mary's Hospital Medical School 319

Swiss League Against Cancer 157

Tate and Lyle 1

Thames Water Authority 276, 277*b*

Vision International: Anthea Sieveking 177*b*

Roger Wilmshurst 294

The following photographs are by the author: 23, 41, 65*l*, 68, 81, 82*tl,cl,bl*, 84*b*, 87, 88*t,cl,cr,b*, 103*tl,tr*, 106, 107, 110, 120, 121, 249, 254, 263

Front cover picture **Bruce Coleman:** Kim Taylor

Back cover picture **Howard Jay**

Title page picture **Bruce Coleman:** Hans Reinhard

CONTENTS

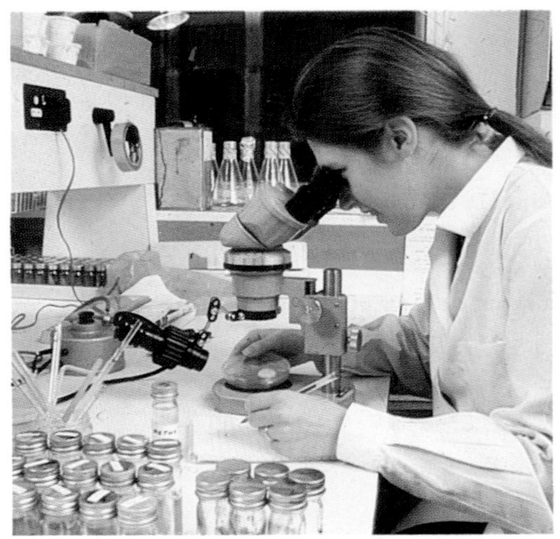

SECTION 1
SOME PRINCIPLES OF BIOLOGY

1 Cells and Tissues 2
2 The Chemicals of Living Cells 14
3 Energy from Respiration 24
4 How Substances get in and out of Cells 33
5 Photosynthesis and Nutrition in Plants 44

SECTION 2
FLOWERING PLANTS

6 Plant Structure and Function 58
7 Transport in Plants 71
8 Sexual Reproduction in Flowering Plants 80
9 Seed Germination 92
10 Vegetative Reproduction 100
11 Plant Sensitivity 105

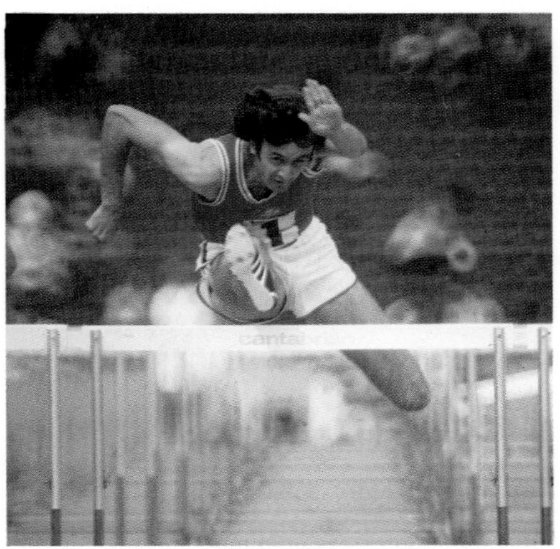

SECTION 3
HUMAN PHYSIOLOGY

12 Food and Diet 114
13 Digestion, Absorption and Use of Food 125
14 The Blood Circulatory System 137
15 Breathing 151
16 Excretion and the Kidneys 159
17 The Skin, and Temperature Control 165
18 Reproduction 170
19 The Skeleton, Muscles and Movement 182
20 Teeth 189
21 The Senses 193
22 Co-ordination 205

Louise Davies

GCSE BIOLOGY

SECTION 4
GENETICS AND HEREDITY

23 Cell Division and Chromosomes 220
24 Heredity 228
25 Variation and Selection 235

SECTION 5
ORGANISMS AND THEIR ENVIRONMENT

26 The Interdependence of Living Organisms 242
27 The Soil 254
28 The Human Impact on the Environment 264
29 Conservation and the Reduction of Pollution 275
30 Ecology 281
31 Populations 292

SECTION 6
DIVERSITY OF ORGANISMS

32 Classification 300
33 Bacteria and Viruses 310
34 Fungi 321
35 Protista and some Lower Plants 327
36 Some Invertebrates 333
37 Vertebrates 340
38 Characteristics of Living Organisms 346

Further Examination Questions 356
Appendix 1 First Aid and Emergency Treatment 360
Appendix 2 Reagents 363
Appendix 3 Book List 364
Appendix 4 Resources 365
Glossary (a) Scientific 366
(b) Biological 367
(c) Some Derivations 369
Index 370

SECTION 1
SOME PRINCIPLES OF BIOLOGY

1 Cells and Tissues

CELL STRUCTURE

How tissues are studied to see cells: the microscope; taking sections. Cell components. Plant cells. Cell structure at high magnification with the electron microscope; types of electron microscopy. Evidence for functions of cell components. Artefacts.

CELL DIVISION AND SPECIALIZATION

Cell division and growth. Specialization of cells for different functions.

TISSUES AND ORGANS

Definitions and examples of tissues, organs and systems. Tissue culture.

PRACTICAL WORK

Preparing, observing and drawing plant and animal cells.

CELL STRUCTURE

If a very thin slice of a plant stem is cut and studied under a microscope (Fig. 1), it can be seen that the stem consists of thousands of tiny, box-like structures. These structures are called **cells**. Figure 2 is a thin slice taken from the tip of a plant shoot and photographed through a microscope.

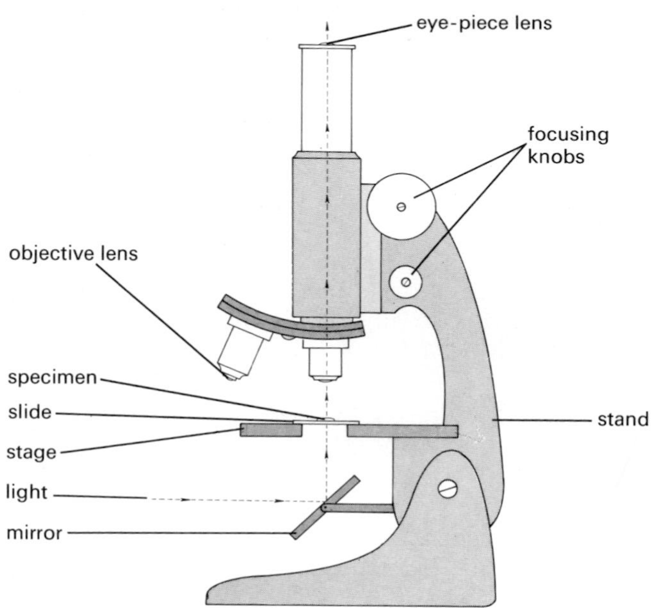

Fig. 1 The microscope. Light is reflected by the mirror and directed through the specimen into the lenses of the microscope. These lenses produce a greatly magnified image of the specimen which can be studied directly or photographed.

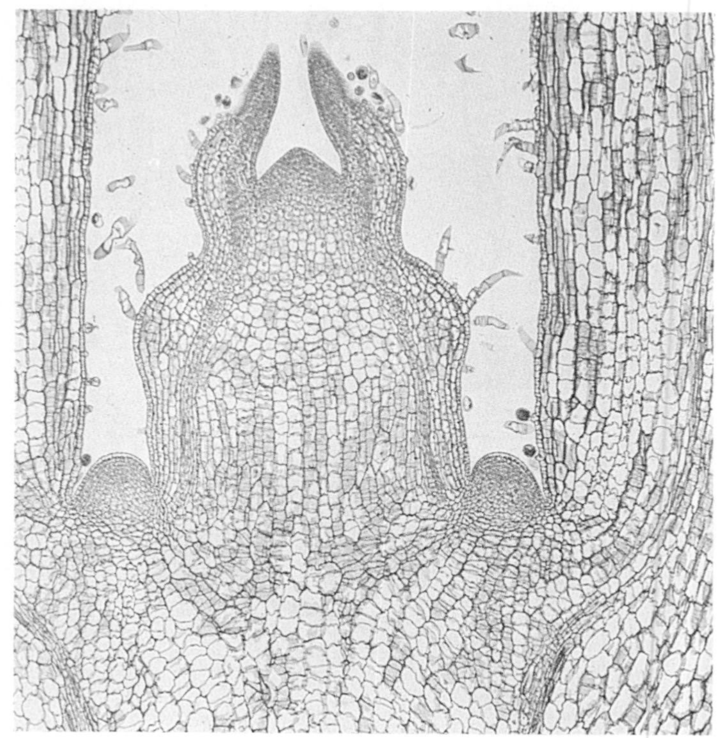

Fig. 2 Longitudinal section through the tip of a plant shoot (×60). The slice is only one cell thick, so light can pass through it and allow the cells to be seen clearly.

Photographs like this are called **photomicrographs**. The one in Fig. 2 is 60 times larger than life, so a cell which appears to be 2 mm long in the picture, is only 0.03 mm long in life.

Thin slices of this kind are called **sections**. If you cut *along the length* of the structure, you are taking a **longitudinal section**. Figure 2 is a longitudinal section which passes through two small developing leaves near the tip of the shoot, and two larger leaves below them. The leaves, buds and stem are all made up of cells. If you cut *across* the structure, you make a **transverse section** (Fig. 3).

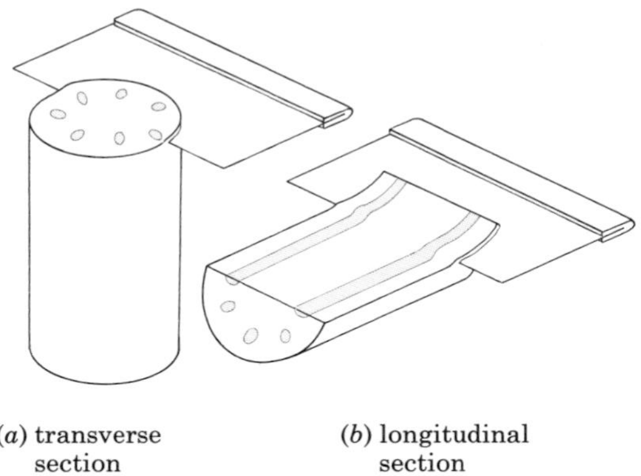

Fig. 3 Cutting sections of a plant stem

It is fairly easy to cut sections through plant structures just by using a razor blade. To make a microscopic study of animal structures is more difficult because they are mostly soft and flexible. Pieces of skin, muscle or liver, for example, first have to be soaked in melted wax. When the wax goes solid it is then possible to cut thin sections. The wax is dissolved away after making the section.

When sections of animal structures are examined under the microscope, they, too, are seen to be made up of cells but they are much smaller than plant cells and need to be magnified more. The photomicrograph of kidney tissue in Fig. 4 has been magnified 700 times to show the cells clearly. The sections are often treated with dyes, called 'stains', in order to show up the structures inside the cells more clearly.

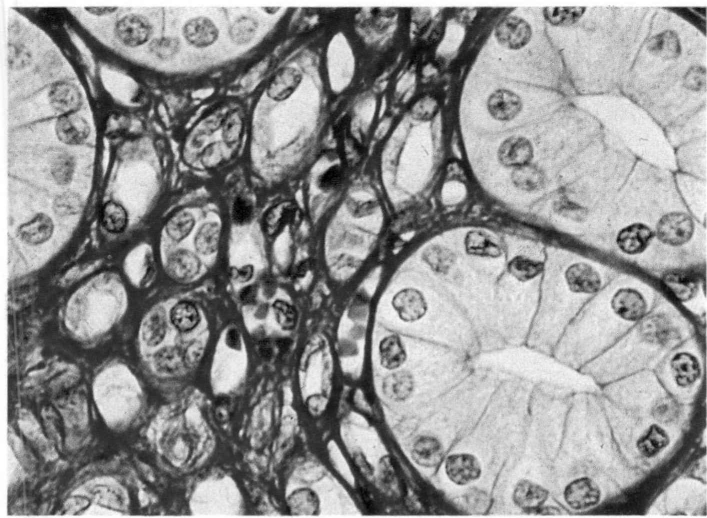

Fig. 4 Transverse section through a kidney tubule ($\times$ 700). A section through a tube will look like a ring (*see* Fig. 19*b* on p. 10). In this case, each 'ring' consists of about 10 cells.

Making sections is not the only way to study cells. Thin strips of plant tissue, only one cell thick, can be pulled off stems or leaves (Experiment 1, p. 12). Plant or animal tissue can be squashed or smeared on a microscope slide (Experiment 2, p. 13) or treated with chemicals to separate the cells before studying them.

There is no such thing as a typical plant or animal cell because cells vary a great deal in their size and shape depending on their function. Nevertheless, it is possible to make a 'generalized' drawing like Fig. 5 to show features which are present in most cells. All cells have a **cell membrane** which is a thin boundary enclosing the **cytoplasm**. Most cells have a **nucleus**.

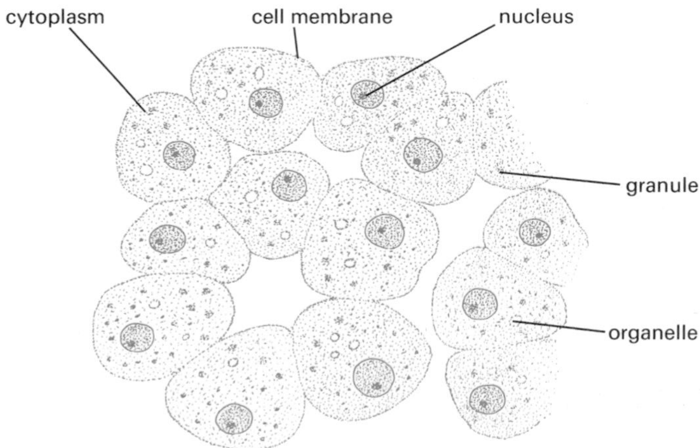

Fig. 5 A group of animal cells, e.g. cells from the lining of the cheek.

Cytoplasm Under the ordinary microscope (light microscope), cytoplasm looks like a thick liquid with particles in it. In plant cells it may be seen to be flowing about. The particles may be food reserves such as oil droplets or granules of starch. Other particles are structures which have particular functions in the cytoplasm. These structures are the **organelles**. Examples are the **ribosomes** which build up the cell's proteins (see p. 14) and the **mitochondria** which generate energy for the cell's living processes (see p. 24).

When studied at much higher magnifications with the **electron microscope** (see below), the cytoplasm no longer looks like a structureless jelly but appears to be organized into a complex system of membranes and vacuoles.

In the cytoplasm, a great many chemical reactions are taking place which keep the cell alive by providing energy and making substances that the cell needs (see pp. 14 and 24).

The liquid part of cytoplasm is about 90 per cent water with molecules of salts and sugars dissolved in it. Suspended in this solution there are larger molecules of fats (lipids) and proteins (see pp. 14–15). Lipids and proteins may be used to build up the cell structures, e.g. the membranes. Some of the proteins are **enzymes** (p. 17). Enzymes control the rate and type of chemical reactions which take place in the cells. Some enzymes are attached to the membrane systems of the cell, others float freely in the liquid part of the cytoplasm.

Cell membrane This is a thin layer of cytoplasm round the outside of the cell. It stops the cell contents from escaping and also controls the substances which are allowed to enter and leave the cell. In general, oxygen, food and water are allowed to enter; waste products are allowed to leave and harmful substances are kept out. In this way the cell membrane maintains the structure and chemical reactions of the cytoplasm.

Nucleus (plural = nuclei) Most cells contain one nucleus, usually seen as a rounded structure embedded in the cytoplasm. In drawings of cells, the nucleus may be shown darker than the cytoplasm because, in prepared sections, it takes up certain stains more strongly than the cytoplasm. The function of the nucleus is to control the type and quantity of enzymes produced by the cytoplasm. In this way it regulates the chemical changes which take place in the cell. As a result, the nucleus determines what the cell will be, e.g. a blood cell, a liver cell, a muscle cell or a nerve cell.

The nucleus also controls cell division as shown in Fig. 15. A cell without a nucleus cannot reproduce. Inside the nucleus are thread-like structures called **chromosomes** which can be seen most easily at the time when the cell is dividing. (See p. 220 for a fuller account of chromosomes.)

The term **protoplasm** is sometimes used to describe the cytoplasm, nucleus and cell membrane together.

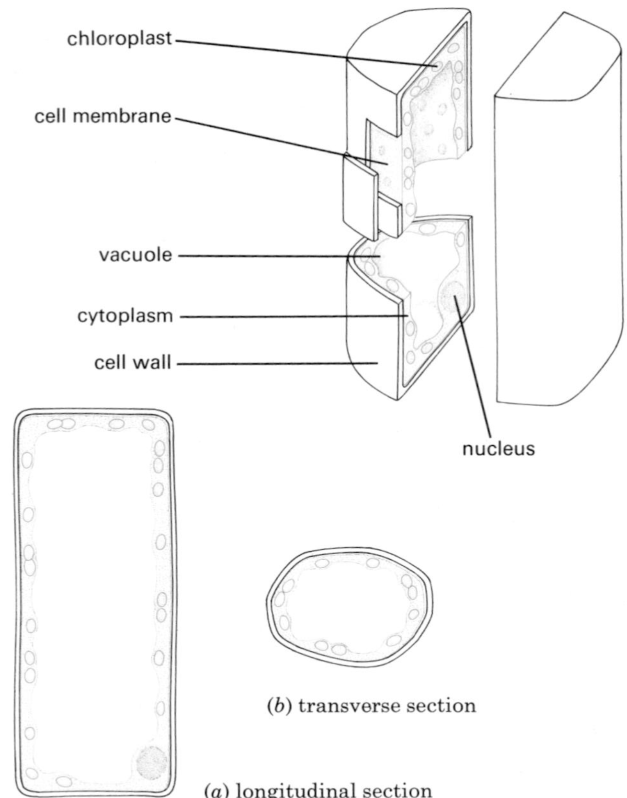

Fig. 8 Structure of a palisade cell. It is important to remember that, although cells look flat in sections or in thin strips of tissue, they are in fact three-dimensional and may seem to have different shapes according to the direction in which the section is cut. If the cell is cut across it will look like (*b*); if cut longitudinally it will look like (*a*).

Plant cells

Figure 5 represents a few generalized animal cells. Figure 6 is a photograph of plant cells in a leaf (palisade cells). Figure 7 is a simplified drawing of two of the cells.

Plant cells differ from animal cells in several ways;

1. Outside the cell membrane they all have a **cell wall** which contains cellulose and other compounds. It is non-living and allows water and dissolved substances to pass through. The cell wall is not selective like the cell membrane. (Note that plant cells *do* have a cell membrane but it is not easy to see or draw because it is pressed against the inside of the cell wall. See Fig. 8.)

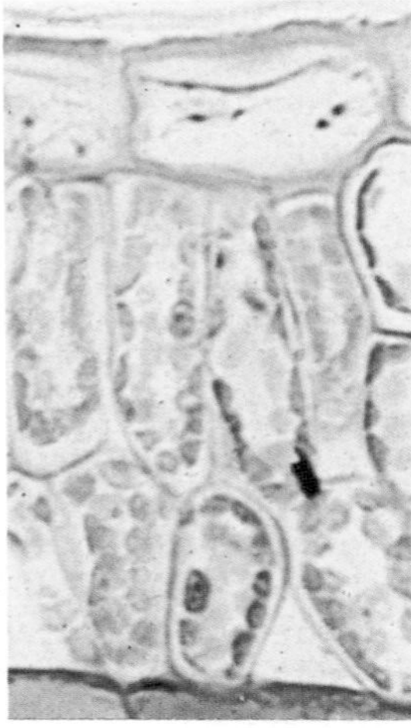

Fig. 6 Plant cells. This is a section through a leaf to show some of the tall palisade cells (×460) (*see also* p. 61).

Under the microscope, plant cells are quite distinct and easy to see because of their cell walls. In Fig. 2 it is only the cell walls (and in some cases the nuclei) which can be seen. Each plant cell has its own cell wall but the boundary between two cells side by side does not usually show up clearly. Cells next to each other therefore appear to be sharing the same cell wall.

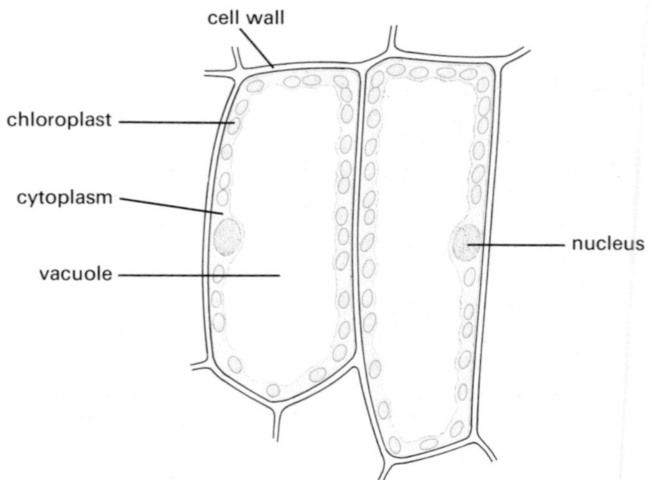

Fig. 7 Palisade cells from a leaf

2. Most mature plant cells have a large, fluid-filled space called a **vacuole**. The vacuole contains **cell sap**, a watery solution of sugars, salts and sometimes pigments. This large, central vacuole pushes the cytoplasm aside so that it forms just a thin lining inside the cell wall. It is the outward pressure of the vacuole on the cytoplasm and cell wall which makes plant cells and their tissues firm (see p. 38). Animal cells may sometimes have small vacuoles in their cytoplasm but they are usually produced to do a particular job and are not permanent.
3. In the cytoplasm of plant cells are many organelles called **plastids** which are not present in animal cells. If they contain the green substance **chlorophyll**, the organelles are called **chloroplasts** (see p. 47). Colourless plastids usually contain starch which is used as a food store.

The shape of a cell when seen in a transverse section may be quite different when the same cell is seen in a longitudinal section. Figure 8 shows why this is so. Figures 10*b* and *c* on p. 63 show the appearance of cells in a stem vein as seen in transverse and longitudinal section.

QUESTIONS

1 (a) What structures are usually present in all cells, whether they are from an animal or from a plant?
(b) What structures are present in plant cells but not in animal cells?
2 What cell structure is largely responsible for controlling the entry and exit of substances into or out of the cell?
3 In what way does the red blood cell shown in Fig. 1 on p. 137 differ from most other animal cells?
4 How does a cell membrane differ from a cell wall?
5 Why does the cell shown in Fig. 8 (*b*) appear to have no nucleus?
6 In Fig. 4, the cell membranes are not always clear. Why is it still possible to decide roughly how many cells there are in each tubule section?
7 (a) In order to see cells clearly in a section of plant tissue, would you have to magnify the tissue (i) $\times 5$, (ii) $\times 10$, (iii) $\times 100$ or (iv) $\times 1000$?
(b) What is the approximate width (in mm) of one of the largest cells in Fig. 4?
8 Make a simple drawing to show what a longitudinal section through Fig. 19*b* would look like.

The electron microscope

The structure of the cell as described so far is what might be seen with an ordinary microscope, using daylight or artificial light. A 'light' microscope such as this gives good results up to magnifications of about $\times 1000$ but is unable to distinguish structures much smaller than 0.3 microns. (A micron, μm, is a thousandth of a millimetre.)

The electron microscope passes beams of electrons instead of beams of light through the object. Because the wavelength of electron beams is very short, clear images can be produced at much greater magnifications than are possible with the light microscope. A common magnification with the electron microscope is $\times 50\,000$ but in some cases $\times 500\,000$ is used. The image can be seen by projecting it on to a fluorescent screen, as in a television tube, or by photographing it to make an electron micrograph (Fig. 9).

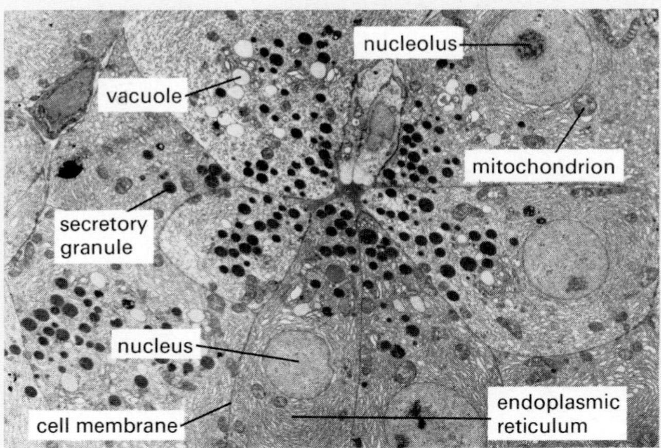

Fig. 9 Electron micrograph of cells in the pancreas ($\times 3000$).

The specimen is embedded in clear plastic and then cut into slices thin enough to allow the electron beam to pass through. The 'stains' used in this case are not dyes, but have a chemical composition which impedes the passage of electrons. This gives a greater contrast between the cell structures. Electron micrographs taken by this means show that the cytoplasm is a highly organised material, containing specialised organelles (Fig. 10).

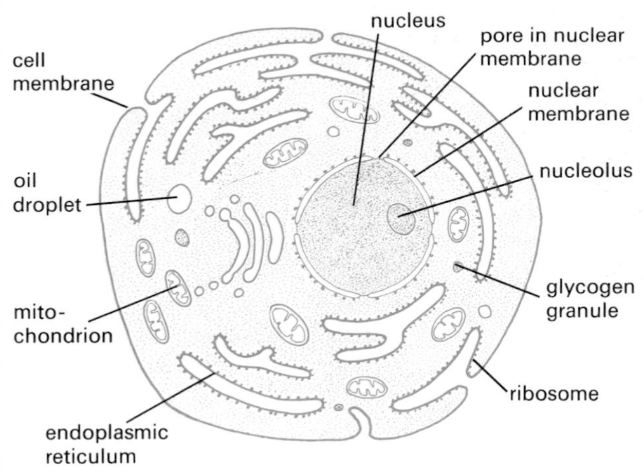

Fig. 10 General diagram of an animal cell as seen by the electron microscope. The cell has been dehydrated, stained and sectioned but it is assumed that the simplified features shown do represent structures in the living cell.

There are several other methods of electron microscopy. The tissues may be frozen in liquid nitrogen and then sectioned in various ways to expose the cell structure. Since electrons can pass straight through these 'unstained' tissues, a thin coating of carbon and platinum is applied to the surface. This makes a replica which absorbs electrons to varying extents and so produces an image of the cut surface (Fig. 11).

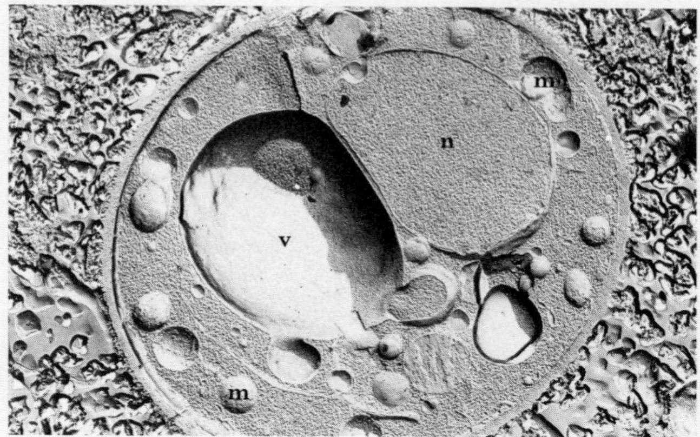

Fig. 11 Yeast cell (×12,000). Electron micrograph of a freeze-etched section. The carbon-platinum film on the exposed cut surface seems to throw the features into relief. m = mitochondrion n = nucleus v = vacuole

A structure coated in this way may also be studied by the **scanning electron microscope**. The features of the structure are revealed by electrons being scattered from the surface and redirected into a converter which produces a picture on a television screen (Fig. 12).

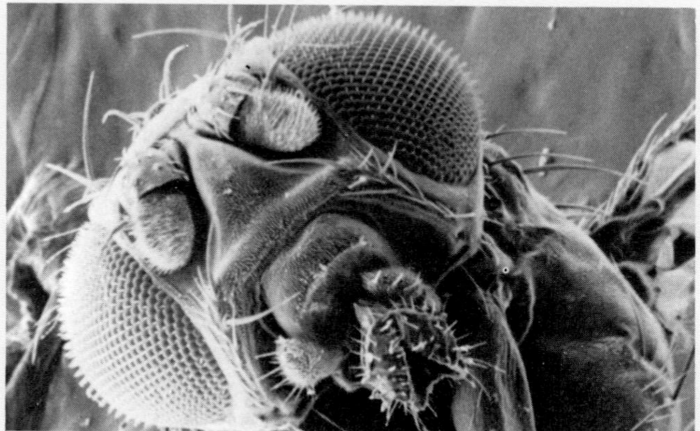

Fig. 12 Fly's head (×40). A scanning electron micrograph which shows the surface features of this insect's head showing eyes and tongue.

Organelles

Endoplasmic reticulum This is a series of tubes and flattened sacs extending throughout the cytoplasm. The channels are not permanent but change their shapes and connections. Enzyme reactions (p. 17) take place on the surface of the membranes which form the endoplasmic reticulum, and the system of channels might allow substances to move rapidly about in the cell.

Ribosomes Some of the endoplasmic reticulum membranes have particles attached to them. These are called ribosomes and are known to be the sites where proteins (p. 14) are built up.

Mitochondria (singular = mitochondrion) Mitochondria are present in the majority of cells. Their shape varies from sausage-like to spherical and they can move about in the cells. Often they collect at sites where rapid chemical activity is taking place or where oxygen is abundant. As shown in Fig. 10 their internal membranes appear to be thrown into folds which greatly increase their surface area. On these internal membranes there may be enzymes which control the process of respiration (p. 24). In respiration, molecules of food such as glucose are broken down to release energy with the aid of oxygen and enzymes. The mitochondria, therefore, are the main energy converters of the cell.

Nuclear membrane Electron micrographs show the nucleus to have a membrane round it. In this membrane there appear to be pores and it is thought that substances pass through the pores into the cytoplasm where they influence the reactions taking place.

Nucleolus The nucleolus is thought to be the region in the nucleus where a substance called RNA (ribose-nucleic acid) is made (see p. 17).

Chloroplasts These are present in plant cells and in certain single-celled organisms and algae (p. 327). Figure 13 shows a section through a chloroplast as revealed by the electron microscope. The system of membranes in the chloroplast carries molecules of chlorophyll and the enzymes which conduct the first stages of photosynthesis. The liquid material between the membranes contains enzymes which continue the process (see p. 45).

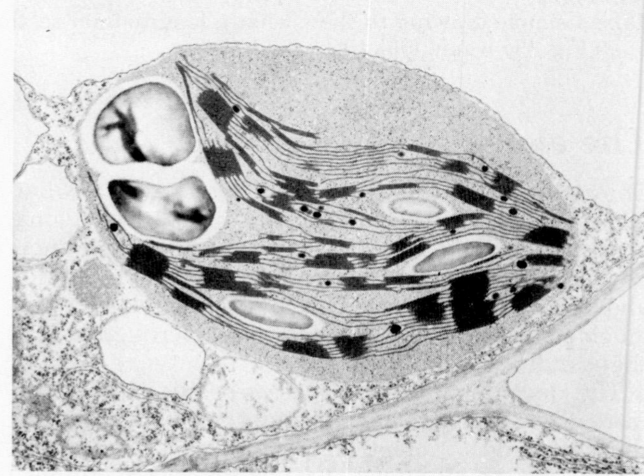

Fig. 13 Section through a chloroplast (×12,000). The system of internal membranes carries the chlorophyll and enzymes needed for photosynthesis.

Artefacts

When tissues are prepared for sectioning and microscopic study, they are often treated with chemicals called **fixatives** which harden the cytoplasm and preserve it. The tissues may then be dehydrated, e.g. by immersing them in alcohol, before embedding them in wax or plastic. Finally the sections may be 'stained' with chemicals which show up the nucleus, cell membrane, cytoplasm or cell wall more clearly. It could be argued that the cell 'structures' seen under the microscope are the results of this drastic treatment. For example, the chemical fixatives and dehydrating agents might make the cell contents shrink, become distorted and so cause the appearance of features which do not represent the structures of the living cell. Such artificial features are called **artefacts**.

It is, however, possible to study living cells and, by varying the methods of preparing tissues and the kind of illumination used, confirm that the cell membrane, nucleus, nucleolus and mitochondria, at least, are normal structures in the living cell. There is still argument about some of the details revealed by electron microscopy (e.g. what happens to the endoplasmic reticulum when the cytoplasm streams about in the cell?) but the bulk of the evidence supports the view that the visible features shown in electron micrographs do represent living structures, even if the preparation of the section has distorted them to some extent.

Nevertheless, it is best to bear in mind that 'seeing' is not always 'believing' when cells, tissues, organs and organisms are subjected to various forms of artificial treatment before observing them.

Evidence for the functions of organelles

The function of some organelles has been mentioned above. The evidence for these functions comes mainly from techniques which break open the cells and separate the different cell structures. There are many methods of breaking cells open without damaging their contents too much.

A suspension of cells is made in a suitable salt solution. The suspension may then be forced through a fine tube or put in a device similar to a food blender. The result is that the cell membranes are broken open and the organelles and granules released.

The suspension of cell fragments is then subjected to **centrifugation**. Tubes of the suspension are whirled round in a **centrifuge**. If you hold a bucket of water in your hands and then spin round on the spot, your arms and the bucket are pulled to a horizontal position. The water does not fall out because of the outward force generated by the spinning movement. A centrifuge works in a similar way with a number of special test-tubes being spun horizontally and very rapidly. The outward force causes the suspended organelles to move towards the bottom of the test-tube. The heavier organelles are forced towards the bottom faster than the light ones. So by spinning the tubes at increasing speeds, it is possible to separate out the organelles. At slow speeds, the nuclei settle out first, then chloroplasts, mitochondria, membrane fragments and ribosomes.

These suspensions of organelles can then be subjected to microscopic examination and experimentation to investigate their functions. There is still a great deal to be learned from investigations of this sort. The detailed reactions taking place in the mitochondria, ribosomes and cell membranes are far from being fully understood. It must also be remembered that the activities of isolated organelles suspended in test-tubes of solution may not be the same as their activities in a living cell.

CELL DIVISION AND CELL SPECIALIZATION

Cell division

When plants and animals grow, their cells increase in numbers by dividing. Typical growing regions are the ends of bones, layers of cells in the skin, root tips and buds (Fig. 14). Each cell divides to produce two daughter cells. Both daughter cells may divide again, but usually one of the cells grows and changes its shape and structure and becomes adapted to do one particular job—in other words, it becomes **specialized**. At the same time it loses its ability to divide any more. The other cell is still able to divide and so continue the growth of the tissue. **Growth** is, therefore, the result of cell division, followed by cell enlargement and, in many cases, cell specialization.

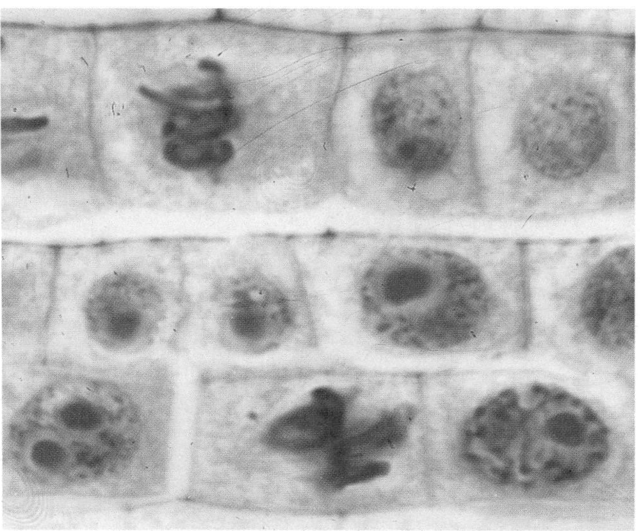

Fig. 14 Cell division in an onion root tip ($\times$ 750). The nuclei are stained pink. Most of the cells have just completed cell division.

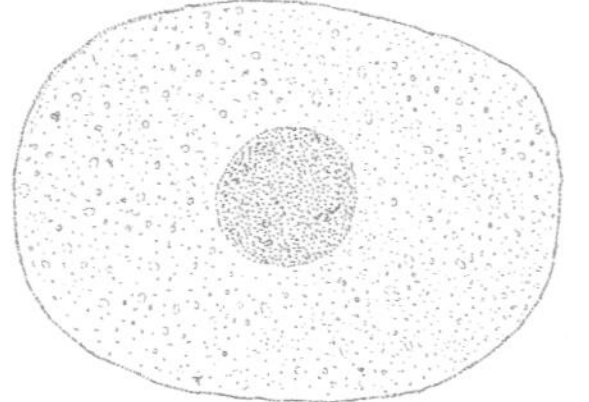

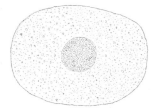

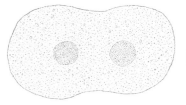

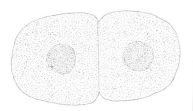

(*a*) Animal cell about to divide.

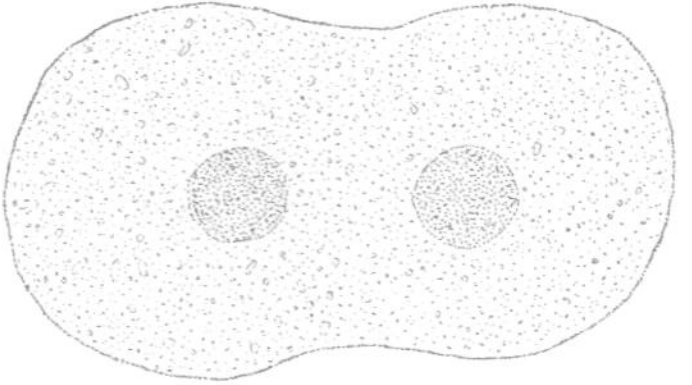

(*b*) The nucleus divides first.

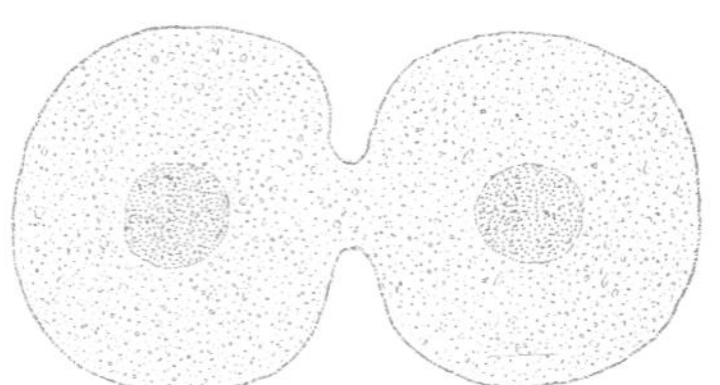

(*c*) The daughter nuclei separate and the cytoplasm pinches off between the nuclei.

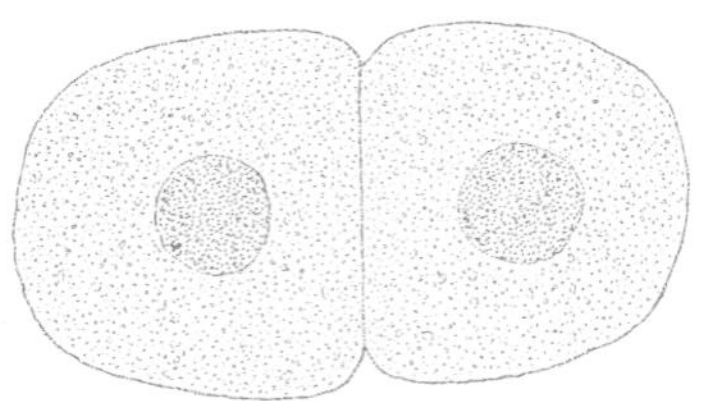

(*d*) Two cells are formed. One may keep the ability to divide, and the other may become specialized.

Fig. 15 Cell division in an animal cell

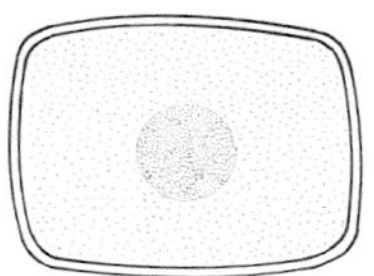

(*a*) A plant cell about to divide has a large nucleus and no vacuole.

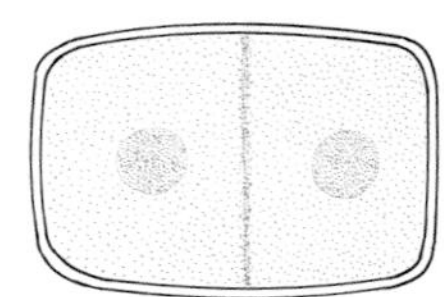

(*b*) The nucleus divides first. A new cell wall develops and separates the two cells.

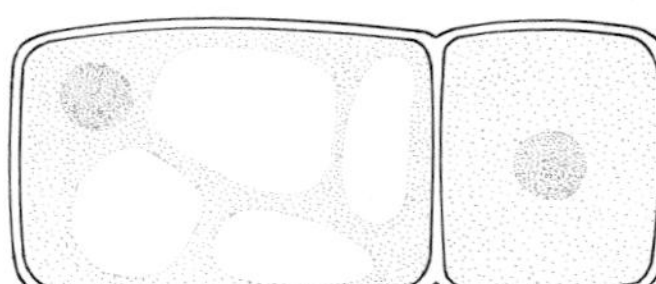

(*c*) The cytoplasm adds layers of cellulose on each side of the new cell wall. Vacuoles form in the cytoplasm of one cell.

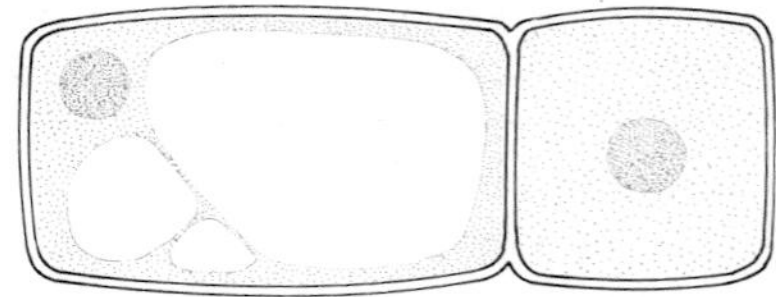

(*d*) The vacuoles join up to form one vacuole. This takes in water and makes the cell bigger. The other cell will divide again.

Fig. 16 Cell division in a plant cell

Figure 15 shows the process of cell division in an animal cell. The events in a plant cell are shown in Figs 14 and 16. Because of the cell wall, the cytoplasm cannot simply pinch off in the middle, and a new wall has to be laid down between the two daughter cells. Also a new vacuole has to form.

Organelles such as mitochondria and chloroplasts are able to divide and are shared more or less equally between the daughter cells at cell division.

Figure 17 shows the pattern of cell division that takes place at the growing point of a simple seaweed. The cells divide and expand to make the tip longer and wider. Once cells 4*a*–*d* have formed they cannot divide again but may become specialized, e.g. 4*b* and 4*c* may help carry food up to the growing point.

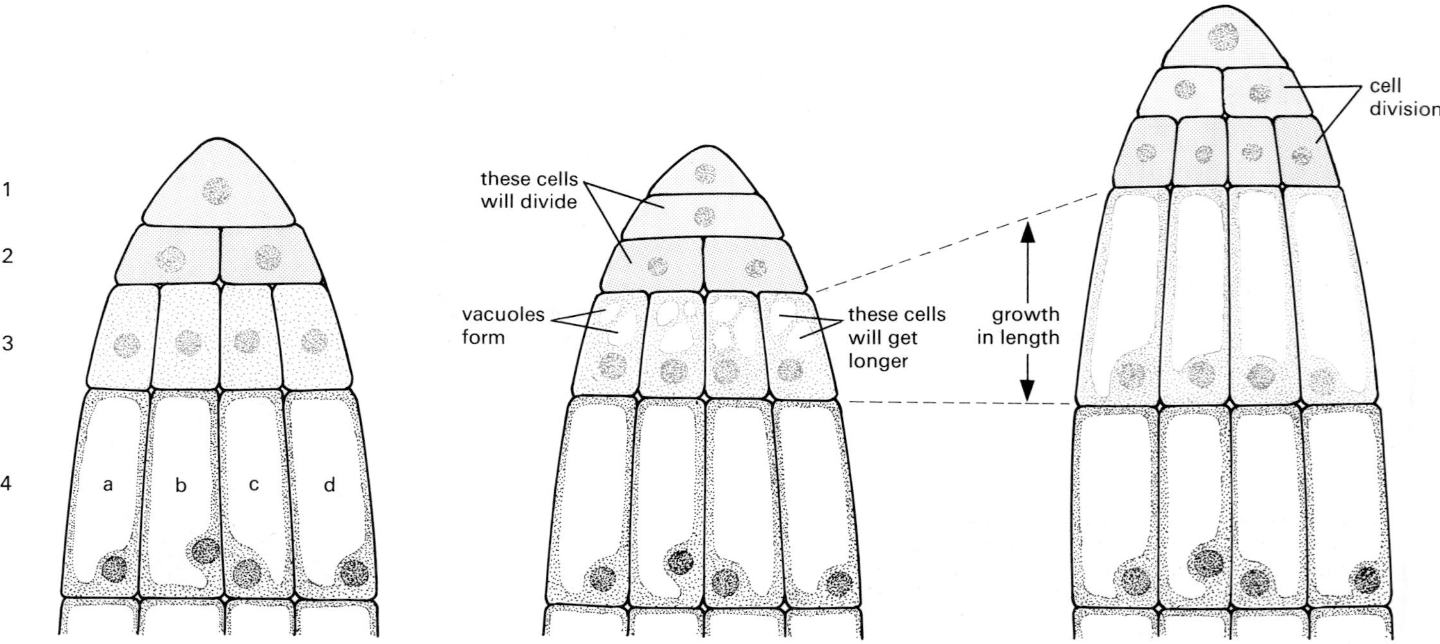

Fig. 17 Growth by cell division and cell elongation at the tip of a seaweed

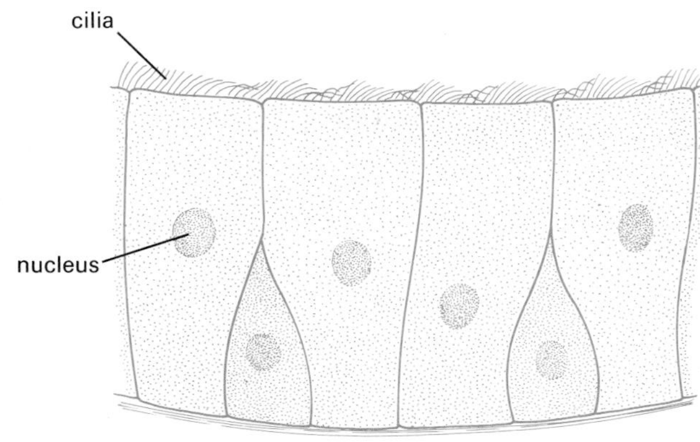

(*a*) **Ciliated cells.** These form the lining of the nose and windpipe, and the tiny cytoplasmic 'hairs', called cilia, are in continual flicking movement keeping up a stream of fluid (mucus) that carries dust and bacteria away from the lungs.

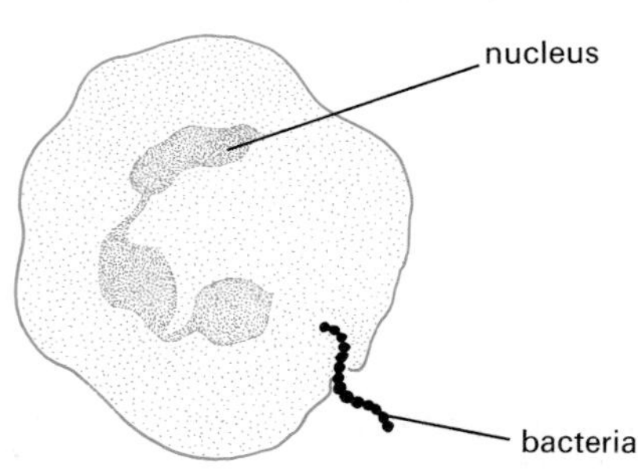

(*b*) **White blood cell.** Occurs in the blood stream and is specialized for engulfing harmful bacteria. It is able to change its shape and move about, even through the walls of blood vessels into the surrounding tissues.

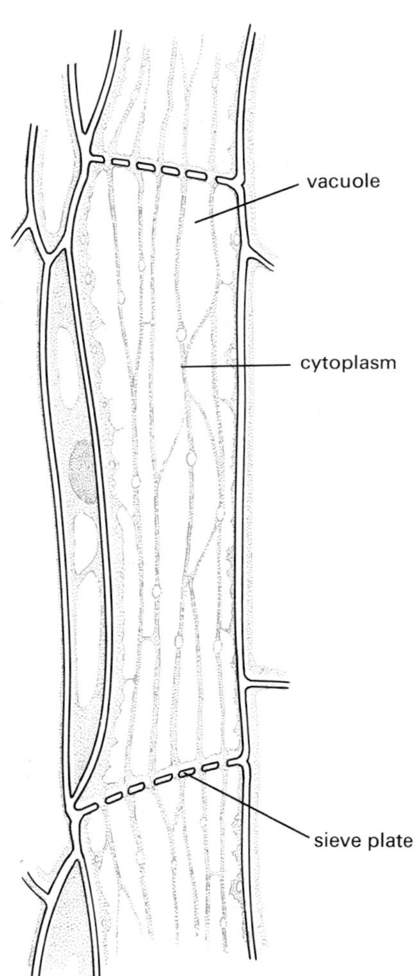

(*c*) **Food-conducting cell in a plant** (phloem cell). Long cells, joined end to end, and where they meet, perforations occur in the walls. Through these holes the cytoplasm of one cell communicates with the next. Dissolved food is thought to pass through the holes during its transport through the stem.

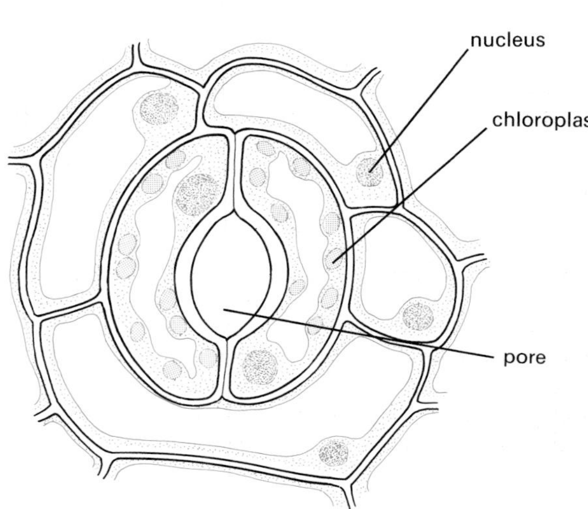

(*d*) **Guard cells of a stoma.** These two curved cells form an opening in the epidermis of a leaf and allow oxygen and carbon dioxide to pass in and out. They can change their shape and thus close the pore.

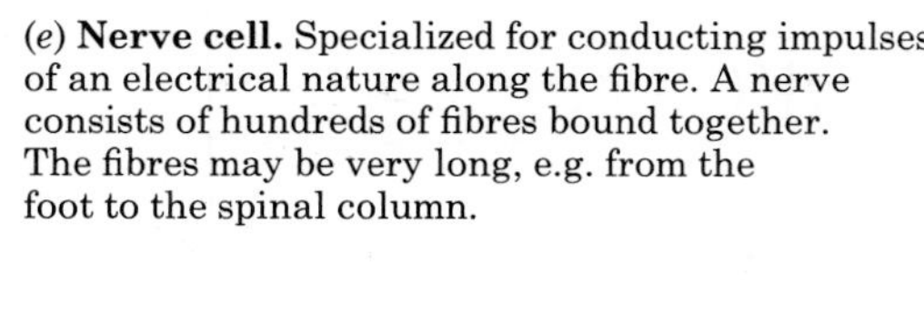

(*e*) **Nerve cell.** Specialized for conducting impulses of an electrical nature along the fibre. A nerve consists of hundreds of fibres bound together. The fibres may be very long, e.g. from the foot to the spinal column.

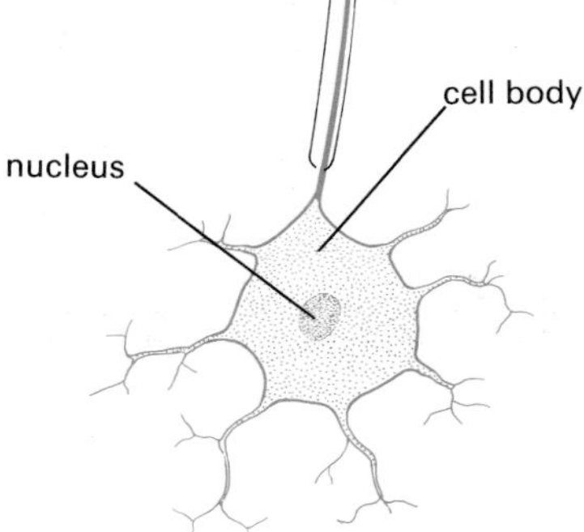

Fig. 18 Specialized cells (not to scale)

Specialization of cells

Most cells, when they have finished dividing and growing, become specialized. This means that

1. they do one particular job;
2. they develop a distinct shape;
3. special kinds of chemical change take place in their cytoplasm. The changes in shape and chemical reactions enable the cell to carry out its special function. Nerve cells and guard cells are examples of specialized cells.

Nerve cells (Fig. 18*e*):

1. conduct electrical impulses to and from the brain;
2. some of them are very long and connect distant parts of the body to the spinal cord and brain;
3. their chemical reactions cause the impulses to travel along the fibre.

Guard cells of a stoma (Fig. 18*d*):

1. open or close the stomatal pore in a leaf;
2. their curved shape and unevenly thickened cell walls causes them to change shape when the pressure in the vacuole alters;
3. changes in carbon dioxide concentration affect the chemical reactions in their cytoplasm and cell sap making the latter more or less concentrated.

The specialization of cells to carry out particular functions in an organism is sometimes referred to as **'division of labour'** within the organism. Similarly, the special functions of mitochondria, ribosomes and other cell organelles may be termed 'division of labour' within the cell.

QUESTIONS

9 Select from the following events and put them in the correct order for cell division in (i) animal cells, (ii) plant cells: (*a*) cytoplasm divides, (*b*) vacuole forms in one cell, (*c*) new cell wall separates cells, (*d*) nucleus divides.
10 Which cells in Fig. 17 make the greatest contribution to growth in length?
11 Look at Fig. 4 on page 167. When a Malpighian cell of the skin divides, which daughter cell becomes specialized and which one keeps the ability to divide again?
12 Look at Fig. 2 on page 59. (*a*) Whereabouts in a leaf are the food-carrying cells? (*b*) What other specialized cells are there in the leaf?

TISSUES AND ORGANS

There are some microscopic organisms that consist of one cell only and can carry out all the processes necessary for their survival (see p. 327). The cells of the larger plants and animals cannot survive on their own. A muscle cell could not obtain its own food and oxygen. Other specialized cells have to provide the food and oxygen needed for the muscle cell to live. Unless these cells are grouped together in large numbers and made to work together, they cannot exist for long.

Tissue A tissue such as bone, nerve or muscle in animals, and epidermis, phloem or pith (p. 62) in plants, is made up of many hundreds of cells of a few types. The cells of each type have similar structures and functions so that the tissue itself can be said to have a particular function, e.g. nerves conduct impulses, phloem carries food in plants. Figure 19 shows how some cells are arranged to form simple tissues.

Organs consist of several tissues grouped together to make a structure with a special function. For example, the stomach is an organ which contains tissues made from epithelial cells, gland cells and muscle cells. These cells are supplied with food and oxygen brought by blood vessels. The stomach also has a nerve supply. The heart, lungs, intestines, brain and eyes are further examples of organs in animals. In flowering plants, the root, stem and leaves are the organs. The tissues of the leaf are epidermis, palisade tissue, spongy tissue, xylem and phloem (see pp. 48 and 59–62).

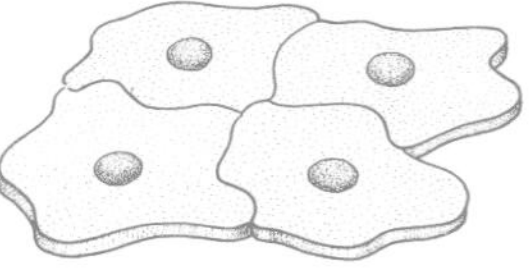

(*a*) **Cells forming an epithelium**, a thin layer of tissue, e.g. the lining of the mouth cavity. Different types of epithelium form the internal lining of the windpipe, air passages, food canal, etc., and protect these organs from physical or chemical damage.

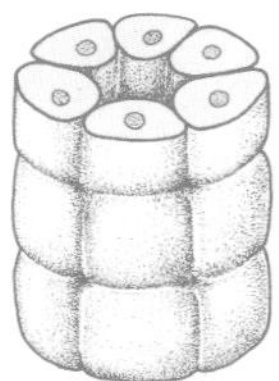

(*b*) **Cells forming a small tube**, e.g. a kidney tubule (*see* p. 160). Tubules such as this carry liquids from one part of an organ to another.

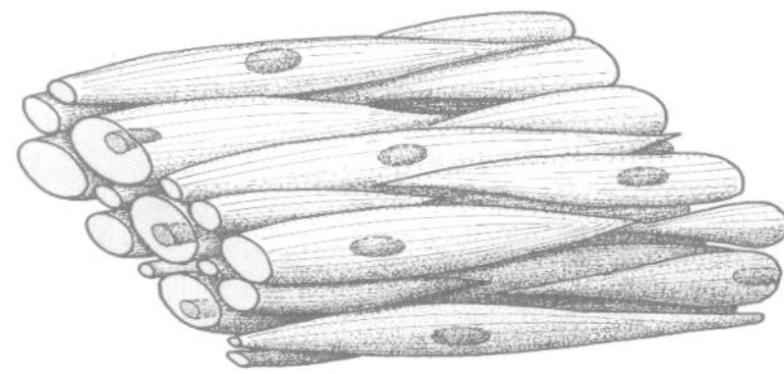

(*c*) **One kind of muscle cell** forming a sheet of muscle tissue. Blood vessels, nerve fibres and connective tissues will also be present. Contractions of this kind of muscle help to move food along the food canal or to close down small blood vessels.

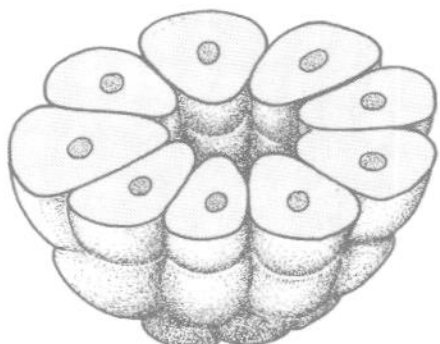

(*d*) **Cells forming part of a gland.** The cells make chemicals which are released into the central space and carried away by a tubule such as shown in (*b*). Hundreds of cell groups like this would form a gland like the salivary gland.

Fig. 19 How cells form tissues

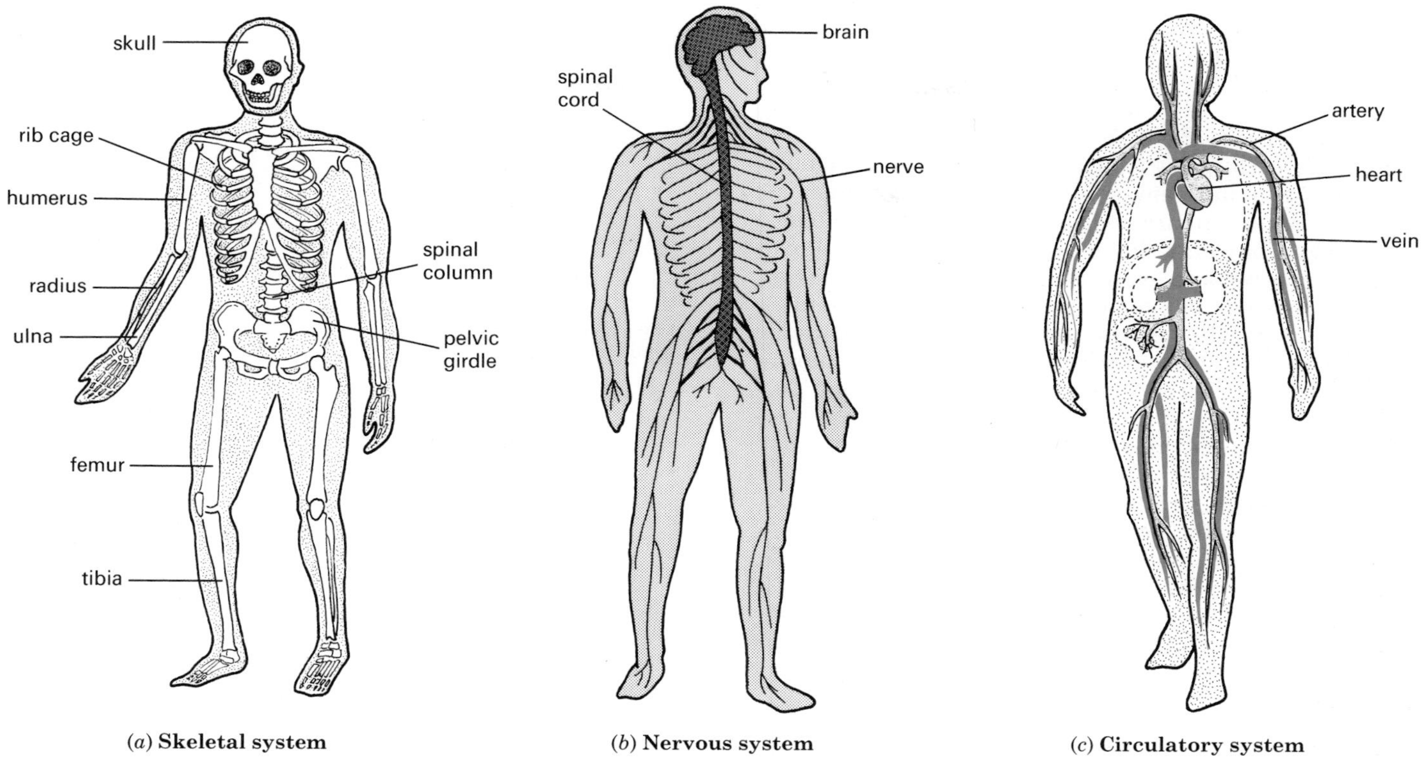

(a) **Skeletal system** (b) **Nervous system** (c) **Circulatory system**

Fig. 20 Three examples of systems in the human body

A system usually refers to a group of organs whose functions are closely related. For example, the heart and blood vessels make up the **circulatory system**; the brain, spinal cord and nerves make up the **nervous system** (Fig. 20). In a flowering plant, the stem, leaves and buds make up a system called the shoot (p. 58).

An organism is formed by the organs and systems working together to produce an independent plant or animal.

Fig. 21 An example of how cells, tissues and organs are related

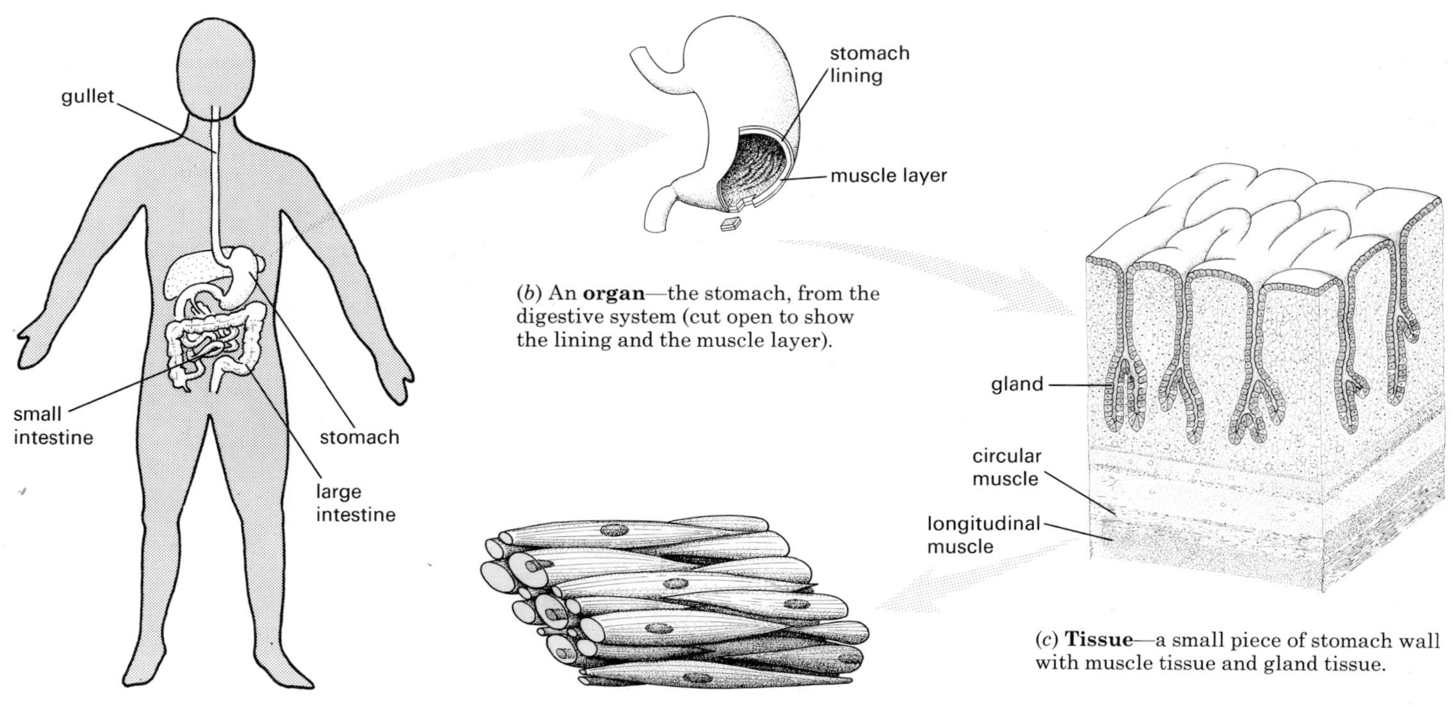

(a) **A system**—the digestive system of the human organism.

(b) An **organ**—the stomach, from the digestive system (cut open to show the lining and the muscle layer).

(c) **Tissue**—a small piece of stomach wall with muscle tissue and gland tissue.

(d) **Cells**—some muscle cells from the muscle tissue.

Tissue culture

It is possible to take samples of developing animal tissues, separate the cells by means of enzymes and make the cells grow and divide in shallow dishes containing a nutrient solution. This technique is called **tissue culture**. The cells do not become specialized but may move about, establish contact with each other and eventually cover the floor of the culture dish with a layer, one cell thick. At this point, they stop dividing unless they are separated and transferred to fresh culture vessels. Even so, most mammal cells cease to reproduce after about 20 divisions.

These tissue cultures are used to study cells and cell division, to test out new drugs and vaccines, to culture viruses or to test the effect of possible harmful chemicals. For some tests and experiments, tissue cultures are able to take the place of laboratory animals.

Large-scale tissue cultures are being developed in order to obtain cell chemicals which may be useful in medical treatments.

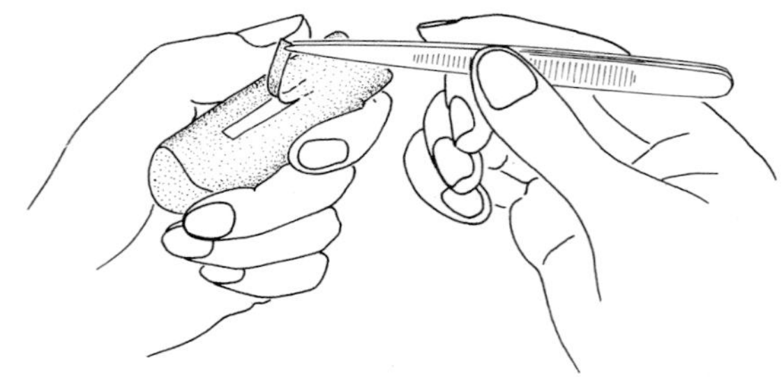

(*a*) Peel a strip of red epidermis from a piece of rhubarb stalk or . . .

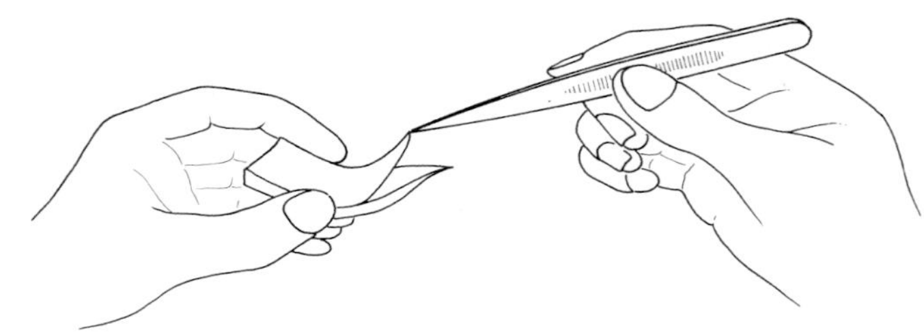

(*b*) peel the epidermis from the inside of an onion scale.

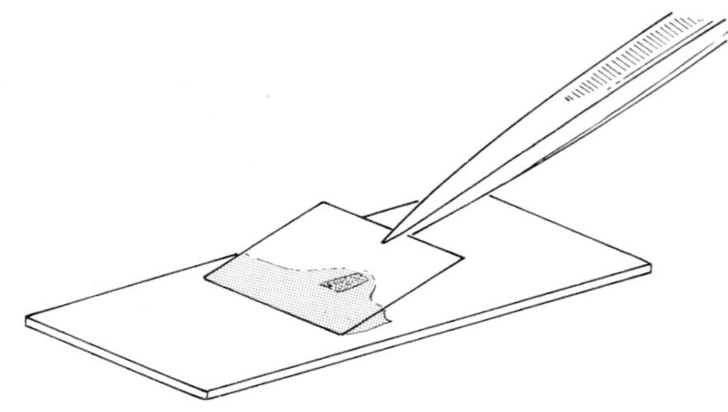

(*c*) Place the epidermis in a drop of water or weak iodine solution on a slide and carefully lower a cover slip over it.

Fig. 22 Looking at plant cells

QUESTIONS

13 Say whether you think the following are cells, tissues, organs or organisms: lungs (p. 151), skin (p. 165), root hair (p. 75), mesophyll (p. 61), multi-polar neurone (p. 206).
14 What tissues are shown in the following drawings: Fig. 10 on page 187; Fig. 8 on page 209?
15 Look at Fig 5 on page 127. Which of the structures shown do you think are organs? What system is represented by the drawing? (The first sub-heading on page 126 is a 'give-away'.)
16 (*a*) Make a list of the ways in which cells can be studied.
(*b*) For each method of study, point out one way in which the results may not represent what really happens in living cells.

PRACTICAL WORK

1. Plant cells

The outer layer of cells (epidermis) from a stem or an onion scale can be stripped off as shown in Fig. 22*a* and *b*. A piece of onion or a rhubarb stalk is particularly suitable for this. A small piece of this epidermis is placed in a drop of weak iodine solution on a microscope slide and covered with a cover-slip (Fig. 22*c*). The tissue is then studied under the microscope. The iodine will stain the cell nuclei pale yellow and the starch grains will stain blue. If red epidermis from rhubarb stalk is used, you will see the red cell sap in the vacuoles.

To see chloroplasts, use fine forceps to pull a leaf from a moss plant, mount it in water on a slide as before and examine it with the high power objective of the microscope (Fig. 23). At the base of the leaf, the chloroplasts are less densely packed and, therefore, easier to see.

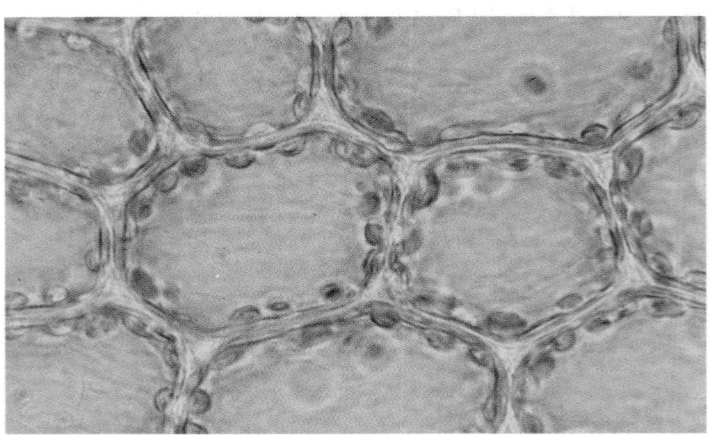

Fig. 23 Cells in a moss leaf (×500). The vacuole occupies most of the space in each cell. The chloroplasts are confined to the layer of cytoplasm lining the cell wall.

2. Animal cells

NOTE The Department of Education and Science recommends that schools no longer use the technique which involves studying the epithelial cells which appear in a smear taken from the inside of the cheek, because of the very small risk of transmitting the AIDS virus. Some Local Education Authorities may therefore forbid the use of this technique in their schools, but the Institute of Biology suggests that if the following procedure is adopted the risk is negligible. (*Biologist* 35 (4) p. 211, September 1988).

Cotton buds from a freshly opened pack are rubbed lightly on the inside of the cheek and gums. The buds are rubbed onto clean slides and then dropped into a container of absolute alcohol. The smear on the slide is covered with a few drops of methylene blue solution before being examined under the microscope. The slides are placed in laboratory disinfectant before washing.

Various alternatives to cheek epithelial smears have been put forward, e.g.

(*a*) Some Sellotape is pressed on to a 'well-washed' wrist. When the tape is removed and studied under the microscope, cells with nuclei can be seen. A few drops of methylene blue solution will stain the cells and make the nuclei more distinct.

(*b*) Microscope slides are pressed against the corneas of fresh or refrigerated bullock's eyes. Conjunctival cells stick to the slides and can be stained with methylene blue.

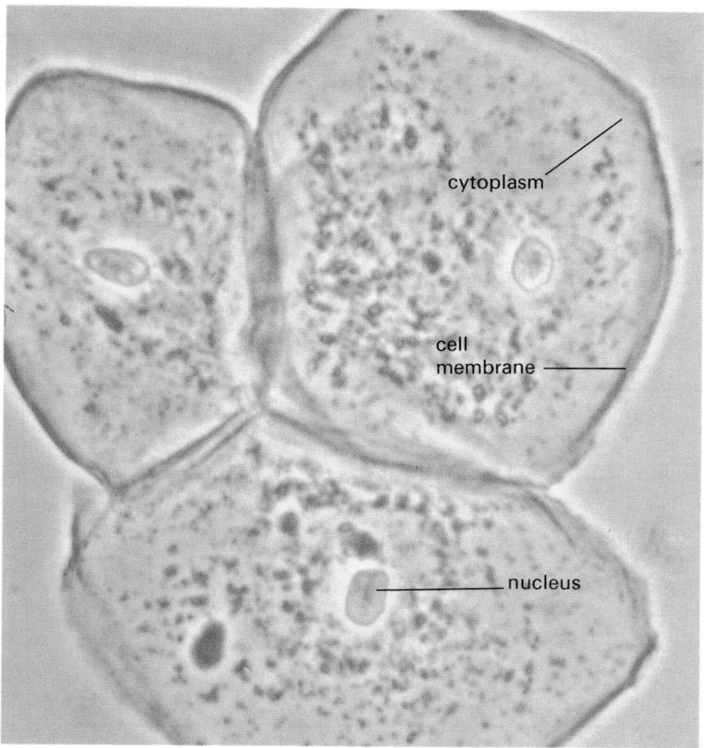

Fig. 24 Cells from the lining epithelium of the cheek (×1500).

CHECK LIST

- **Nearly all plants and animals are made up of thousands or millions of microscopic cells.**
- **All cells contain cytoplasm enclosed in a cell membrane.**
- **Most cells have a nucleus.**
  Cytoplasm contains organelles such as mitochondria, chloroplasts and ribosomes.
- **Many chemical reactions take place in the cytoplasm to keep the cell alive.**
- **The nucleus directs the chemical reactions in the cell and also controls cell division.**
- **Plant cells have a cellulose cell wall and a large central vacuole.**
- **Cells are often specialized in their shapes and activities to carry out particular jobs.**
- **Large numbers of similar cells packed together form a tissue.**
- **Different tissues arranged together form organs.**
- **A group of related organs makes up a system.**

2 The Chemicals of Living Cells

CELL PHYSIOLOGY

Description.

CHEMICAL COMPONENTS OF CELLS

Water, proteins, lipids, carbohydrates, nucleic acids, salts, ions, vitamins; their chemical structure and role in the cell. Inter-conversion of substances in cells.

ENZYMES

Definition. Methods of action. Effects of temperature and pH. Enzyme specificity. Intra- and extra-cellular enzymes.

PRACTICAL WORK

Experiments with enzymes: catalase, starch phosphorylase, pH and temperature.

Cell physiology

The term 'physiology' refers to all the normal functions that take place in a living organism. Digestion of food, circulation of blood and contraction of muscles are some aspects of human physiology. Absorption of water by roots, production of food in the leaves, and growth of shoots towards light are examples of plant physiology. The next three chapters are concerned with physiological events in individual cells.

The physiology of a whole organism is, in some ways, the sum of the physiology of its component cells. If all cells need a supply of oxygen then the whole organism must take in oxygen. Cells need chemical substances to make new cytoplasm and to produce energy. Therefore the organism must take in food to supply the cells with these substances. Of course, it is not quite as simple as this; most cells have specialized functions (p. 10) and so have differing needs. However, all cells need water, oxygen, salts and food substances and all cells consist of water, proteins, lipids, nucleic acids, carbohydrates, salts and vitamins or their derivatives.

THE CHEMICAL COMPONENTS OF A CELL

Water

Most cells contain about 75 per cent of water and will die if their water content falls much below this. Water is a good solvent and many substances move about the cell in a watery solution. Water molecules take part in a great many vital chemical reactions (e.g. photosynthesis, p. 45). In plants, it is the water pressure in the vacuoles which keeps the cells turgid (firm).

The physical and chemical properties of water differ from those of most other liquids but make it uniquely effective in supporting living activities. For example, water has a high capacity for heat (high thermal capacity). This means that it can absorb a lot of heat without its temperature rising to levels which damage the proteins in the protoplasm (see p. 15). However, because water freezes at 0 °C most cells are damaged if their temperature falls below this and ice crystals form in the protoplasm. (Oddly enough, rapid freezing of cells in liquid nitrogen at below −196 °C does not harm them.)

Proteins

Some proteins contribute to the structures of the cell, e.g. to the cell membranes, the mitochondria, ribosomes and chromosomes. These proteins are called **structural proteins**.

There is another group of proteins called **enzymes**. Enzymes are present in the membrane systems, in the mitochondria, in special vacuoles and in the fluid part of the cytoplasm. Enzymes control the chemical reactions which keep the cell alive (see p. 17).

Although there are many different types of protein, they all contain carbon, hydrogen, oxygen, nitrogen and sulphur, and their molecules are made up of long chains of simpler chemicals called **amino acids**.

One of the simplest amino acids is glycine which has the formula

H₂N–CH₂–C(=O)OH

$$\mathrm{H_2N{-}CH_2{-}C(=O){-}OH}$$ often written as $CH_2(NH_2)COOH$.

The -COOH group makes the chemical an acid and the $-NH_2$ is the **amino** group. There are about 20 different amino acids in animal proteins, including alanine, leucine, valine, glutamine, cysteine and lysine. A small protein molecule might be made up from a chain consisting of a hundred or so amino acids, e.g. glycine–valine–valine–cysteine–leucine–glutamine–, etc.

Each different protein has the amino acids arranged in its own particular order. The chemical linkage between each amino acid is called a **peptide bond**. Two amino acids joined together form a **dipeptide**, three form a **tripeptide**. More amino acids than this would make a molecule called a **polypeptide**.

A long polypeptide molecule becomes a protein when the chain of amino acids takes up a particular shape as a result of cross-linkages. Cross-linkages form between amino acids which are not neighbours, as shown in Fig. 1. The shape of a protein molecule has a very important effect on its reactions with substances as explained in 'Enzymes' below.

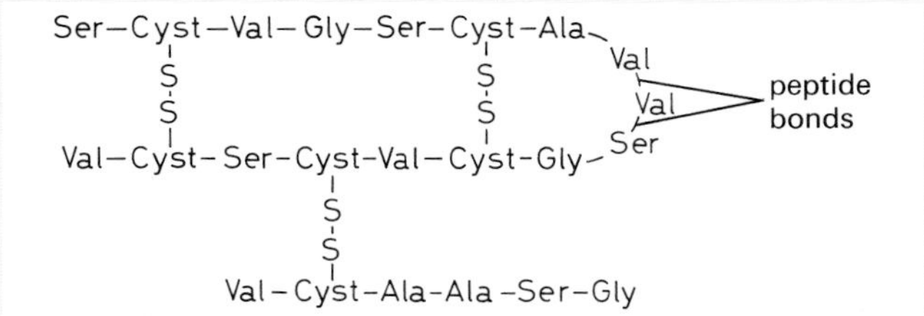

Ser–Cyst–Val–Gly–Ser–Cyst–Ala–Val
Val–Cyst–Ser–Cyst–Val–Cyst–Gly–Ser–Val
Val–Cyst–Ala–Ala–Ser–Gly
S–S
peptide bonds

Fig. 1 A small imaginary protein made from only five different kinds of amino acid. Note that cross linkage occurs between cysteine molecules with the aid of sulphur atoms.

When a protein is heated to temperatures over 50 °C, the cross-linkages in its molecule break down; the protein molecule loses its shape and will not usually regain it even when cooled. The protein is said to have been **denatured**. Because the shape of the molecule has been altered, it will have lost its original properties.

Egg-white is a protein. When it is heated, its molecules change shape and the egg-white goes from a clear, runny liquid to a white solid and cannot be changed back again. The egg-white protein, albumen, has been denatured by heat.

Proteins form enzymes and many of the structures in the cell, so if they are denatured the enzymes and the cell structures will stop working and the cell will die. Whole organisms may survive for a time above 50 °C depending on the temperature, the period of exposure and the proportion of the cells which are damaged.

Lipids

Lipids are oils or fats and substances related to or derived from them. Fats are formed from carbon, hydrogen and oxygen only. A molecule of fat is made up of three molecules of an organic acid, called a **fatty acid**, combined with one molecule of glycerol.

A commonly occurring fat is tristearin (or glyceryl tristearate).

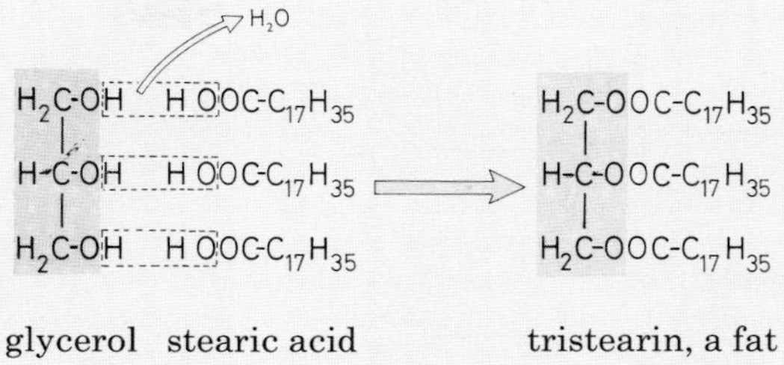

$$\begin{matrix} H_2C\text{-}OH & HOOC\text{-}C_{17}H_{35} \\ H\text{-}C\text{-}OH & HOOC\text{-}C_{17}H_{35} \\ H_2C\text{-}OH & HOOC\text{-}C_{17}H_{35} \end{matrix} \xrightarrow{-H_2O} \begin{matrix} H_2C\text{-}OOC\text{-}C_{17}H_{35} \\ H\text{-}C\text{-}OOC\text{-}C_{17}H_{35} \\ H_2C\text{-}OOC\text{-}C_{17}H_{35} \end{matrix}$$

glycerol stearic acid tristearin, a fat

Three molecules of the fatty acid, stearic acid ($C_{17}H_{35}COOH$), have combined with one molecule of glycerol (with the removal of three molecules of water) to form the **triglyceride**, tristearin. The difference between one fat and another depends on which fatty acids are combined with the glycerol. The fatty acids in any one triglyceride are not necessarily the same, e.g.

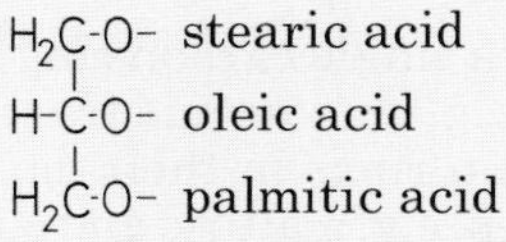

$H_2C\text{-}O\text{-}$ stearic acid
$H\text{-}C\text{-}O\text{-}$ oleic acid
$H_2C\text{-}O\text{-}$ palmitic acid

Triglycerides are often combined with proteins to form **lipoproteins**. One group of lipids, with complex molecular structures, comprises the **steroids**.

Lipids form part of the cell membrane and the internal membranes of the cell such as the nuclear membrane or the endoplasmic reticulum. Droplets of fat or oil form a source of energy when stored in the cytoplasm.

Carbohydrates

These may be simple, soluble sugars or complex materials like starch and cellulose, but all carbohydrates contain carbon, hydrogen and oxygen only. A commonly occurring simple sugar is **glucose**, whose chemical formula is $C_6H_{12}O_6$.

The glucose molecule is often in the form of a ring, represented as

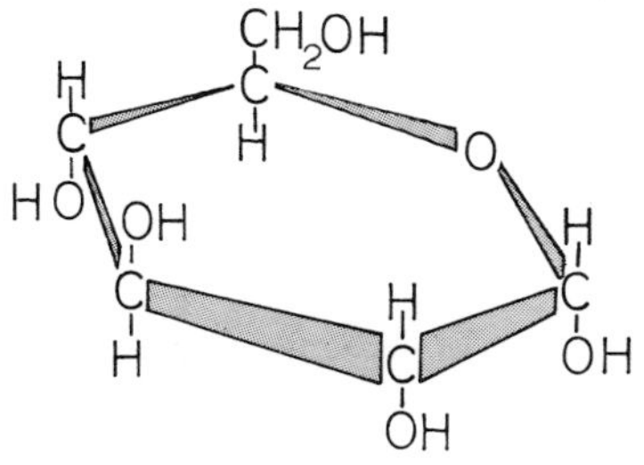

CH_2OH
H C H O
HO OH H
C H C
H OH OH
OH

or more simply as

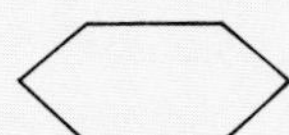

or merely

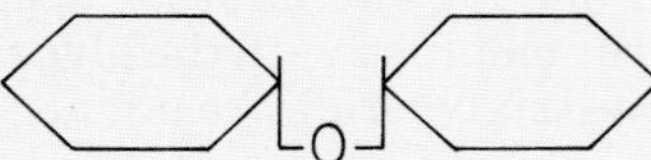

Two molecules of glucose can be combined to form a molecule of maltose $C_{12}H_{22}O_{11}$ or

Sugars with a single carbon ring are called **mono-saccharides**, e.g. glucose and fructose. Those sugars with two carbon rings in their molecules are called **di-saccharides**, e.g. maltose and sucrose. Mono- and di-saccharides are readily soluble in water.

When many glucose molecules are joined together, the carbohydrate is called a **poly-saccharide**. **Glycogen** (Fig. 2) is a poly-saccharide which forms a food storage substance in many animal cells. The **starch** molecule is made up of hundreds of glucose molecules joined together to form long chains. Starch is an important storage substance in the plastids of plant cells. **Cellulose** consists of even longer chains of glucose molecules. The chain molecules are grouped together to form microscopic fibres, which are laid down in layers to form the cell wall in plant cells (Figs 3 and 4).

Poly-saccharides are not readily soluble in water.

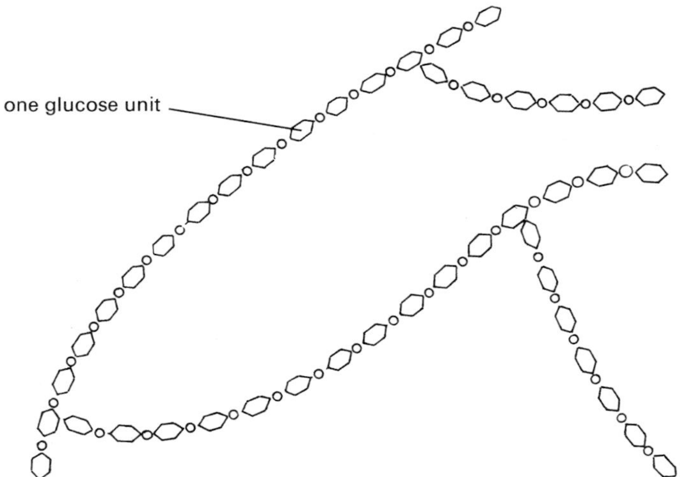

Fig. 2 Part of a glycogen molecule

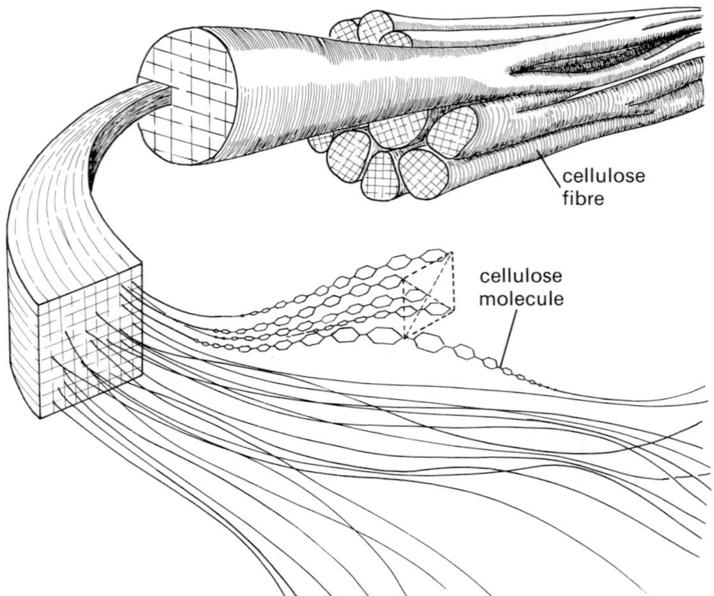

Fig. 3 Cellulose. Plant cell walls contain long, interwoven and interconnected cellulose fibres which are large enough to be seen with the electron microscope. Each fibre is made up of many long-chain cellulose molecules.
(From *Principles of Plant Physiology* by James Bonner and Arthur W. Galston. W. H. Freeman and Co. © 1952.)

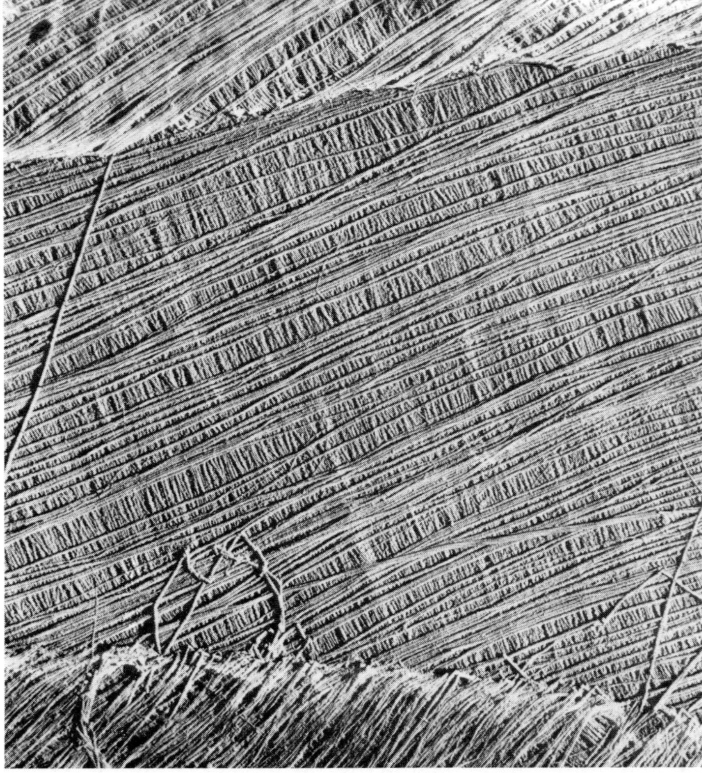

Fig. 4 Electron micrograph of a plant cell wall (× 18,000) showing the cellulose fibres.

Nucleic acids

These are large, complex molecules made up from smaller molecules called **nucleotides**. Each nucleotide is formed from a simple sugar, **ribose** or **deoxy-ribose**, combined with one or more phosphate groups, and an organic base such as **adenine** or **cytosine** (Fig. 5). Long chains of nucleotides joined together, form the nucleic acids **DNA** and **RNA**. **Deoxy-ribose-nucleic acid** (DNA) occurs in the nucleus; **ribose-nucleic acid** (RNA) is in the nucleus and the cytoplasm. DNA and RNA work together to control the sequence of amino acids which are built up into proteins by the ribosomes.

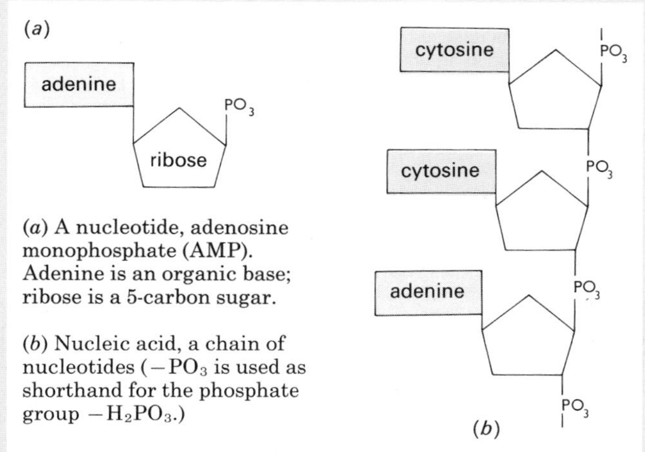

(*a*) A nucleotide, adenosine monophosphate (AMP). Adenine is an organic base; ribose is a 5-carbon sugar.

(*b*) Nucleic acid, a chain of nucleotides ($-PO_3$ is used as shorthand for the phosphate group $-H_2PO_3$.)

Fig. 5 Nucleic acid

Salts

Salts are present in cells in the form of their ions, e.g. sodium chloride in a cell will exist as sodium ions (Na^+) and chlorine ions (Cl^-). The ions may be free to move about in the water of the cell or attached to other molecules such as the proteins or lipids.

Ions, and other substances dissolved in water, usually attract water molecules around them. A sodium ion attracts three water molecules, a chloride ion attracts six. These ions are said to be **hydrated**, and it is the hydrated ion, e.g. a sodium ion surrounded by three water molecules, which moves about in solution and takes part in reactions.

Ions take part in and influence many chemical reactions in the cell, e.g. phosphate ions (PO_4^{3-}) are essential for energy transfer reactions (see 'ATP', p. 25). Ions are also involved in determining how much water enters or leaves the cell (see 'Osmosis', p. 36). Calcium, potassium and sodium ions are particularly important in chemical changes related to electrical activities of a cell, e.g. responding to stimuli or conducting nerve impulses. A shortage or excess of ions in the cells upsets their physiology and affects the normal functioning of the whole organism.

Vitamins

This is a category of substances which, in their chemical structure at least, have little in common. Plant cells can make their own vitamins. Animal cells have to be supplied with vitamins ready-made. Vitamins, or substances derived from them, play a part in chemical reactions in the cell, for example those which involve a transfer of energy from one compound to another. If cells are not supplied with vitamins or the substances needed to make them, the cell physiology is thrown out of order and the whole organism suffers.

Inter-conversion of substances in cells

Cells are able to build up and break down their proteins, lipids and carbohydrates or change one to the other. For example, animal cells build up glycogen from glucose; plant cells make starch and cellulose from glucose. All cells can make proteins from amino acids and they can build up fats from glycerol and fatty acids. Animal cells can change carbohydrates to lipids, and lipids to carbohydrates; they can also change proteins to carbohydrates but they cannot make proteins unless they are supplied with amino acids. Plant cells, on the other hand, can make their own amino acids starting from sugar and salts. The cells in the green parts of plants can even make glucose starting from only carbon dioxide and water (see pp. 44–8).

QUESTIONS

1 Why do you think it is essential for us to include proteins in our diet?
2 Which substances are particularly important in (*a*) forming the structures of the cell and (*b*) storing food in the cell?
3 Which of the following are (*a*) carbohydrates, (*b*) salts, (*c*) lipids: sugar, butter, iron sulphate, starch, sodium chloride, olive oil, cellulose?
4 What additional elements are needed to convert a carbohydrate into a protein? Where do you think plants get these elements from?
5 What is the difference between a protein and a polypeptide?
6 The fatty acid, lauric acid, has the formula $CH_3(CH_2)_{10}COOH$. Try writing the formula for its triglyceride.
7 What chemical changes can occur in a plant cell that do not take place in animal cells?

Enzymes

Enzymes are proteins that act as **catalysts**. They are made in the cells. A catalyst is a chemical substance which speeds up a reaction but does not get used up during the reaction. One enzyme can be used many times over (Fig. 6).

Figure 7*a* is a diagram of how an enzyme molecule might work to join two other molecules together and so form a more complicated substance.

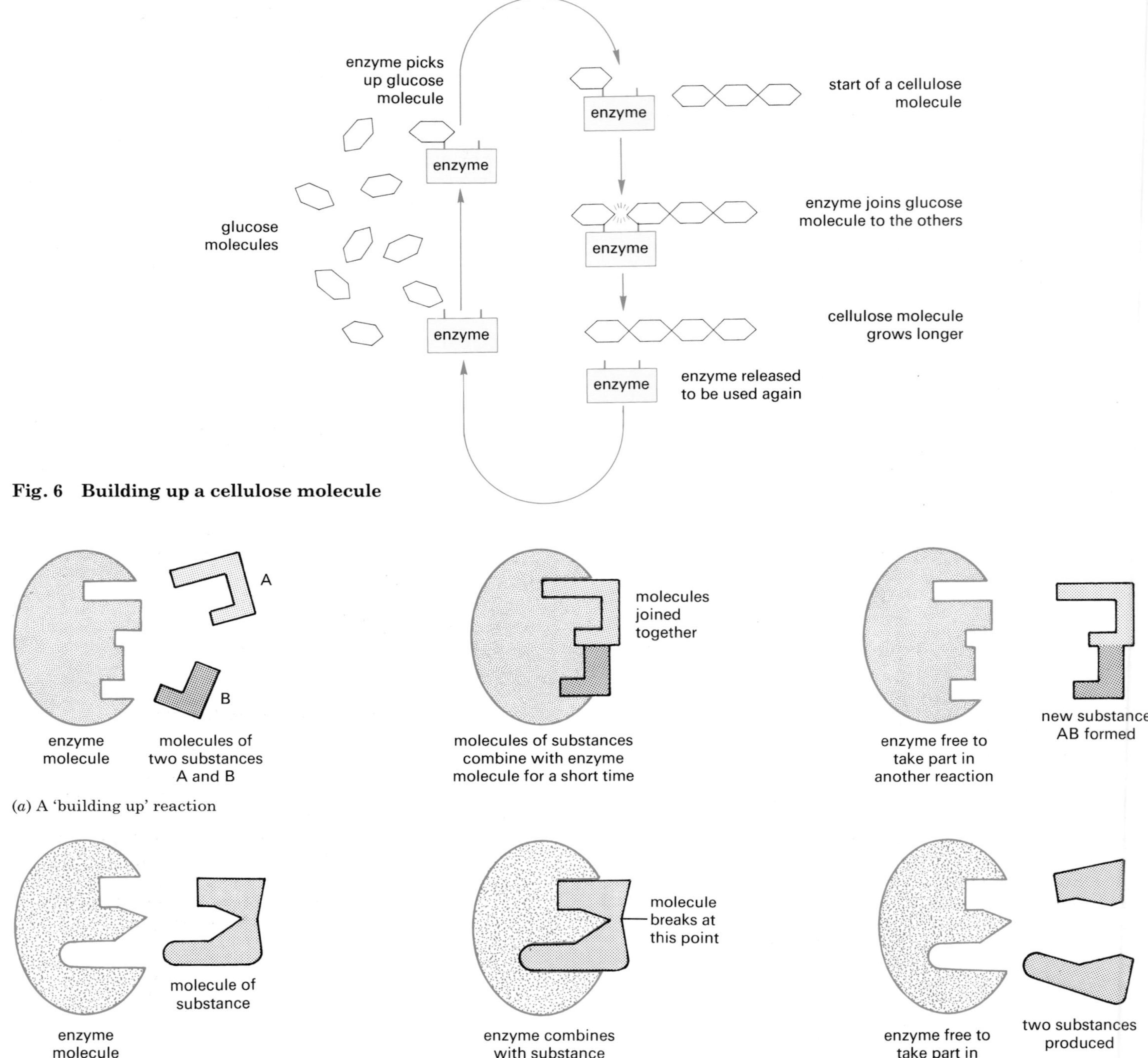

Fig. 6 Building up a cellulose molecule

(*a*) A 'building up' reaction

(*b*) A 'breaking down' reaction

Fig. 7 Possible explanation of enzyme action

An example of an enzyme-controlled reaction such as this is the joining up of two glucose molecules to form a molecule of maltose.

$$C_6H_{12}O_6 + C_6H_{12}O_6 \xrightarrow{\text{enzyme}} C_{12}H_{22}O_{11} + H_2O$$

glucose + glucose → maltose + water

In a similar way, hundreds of glucose molecules might be joined together, end to end, to form a long molecule of starch to be stored in the plastid of a plant cell. The glucose molecules might also be built up into a molecule of cellulose to be added to the cell wall. Protein molecules are built up by enzymes which join together tens or hundreds of amino acid molecules. These proteins are added to the cell membrane, to the cytoplasm or to the nucleus of the cell. They may also become the proteins which act as enzymes.

After the new substance has been formed, the enzyme is set free to start another reaction. Molecules of the two substances might have combined without the enzyme being present but they would have done so very slowly. By bringing the substances close together, the enzyme molecule makes the reaction take place much more rapidly. A chemical reaction which would take hours or days to happen on its own takes only a few seconds when the right enzyme is present.

Figure 7*b* shows an enzyme speeding up a chemical change but this time it is a reaction in which the molecule of a substance is split into smaller molecules. If starch is mixed with water it will break down very slowly to sugar, taking several years. In your saliva there is an enzyme called **amylase** which can break down starch to sugar in minutes or seconds. Inside a cell, many of the 'breaking-down' enzymes are helping to break down glucose to carbon dioxide and water in order to produce energy (see p. 24).

Enzymes and temperature

A rise in temperature increases the rate of most chemical reactions; a fall in temperature slows them down. In many cases a rise of 10 °C will double the rate of reaction in a cell. This is equally true for enzyme-controlled reactions, but above 50 °C the enzymes, being proteins, are denatured and stop working.

Figure 7 shows how the shape of an enzyme molecule could be very important if it has to fit the substances on which it acts. Above 50 °C the shapes of enzymes are changed and the enzymes can no longer combine with the substances.

This is one of the reasons why organisms may be killed by prolonged exposure to high temperatures. The enzymes in their cells are denatured and the chemical reactions proceed too slowly to maintain life.

One way to test whether a substance is an enzyme is to heat it to boiling point. If it can still carry out its reactions after this, it cannot be an enzyme. This technique is used as a 'control' (see p. 30) in enzyme experiments.

Enzymes are specific

This means simply that an enzyme which normally acts on one substance will not act on a different one. Figure 7*a* shows how the shape of an enzyme could decide what substances it combines with. The enzyme in Fig. 7*a* has a shape which exactly fits the substances on which it acts, but would not fit the substance in *b*. Thus, an enzyme which joins amino acids up to make proteins will not also join up glucose molecules to make starch. If a reaction takes place in stages, e.g.

protein ⟶ polypeptides (stage 1)
polypeptides ⟶ amino acids (stage 2)

a different enzyme is needed for each stage.

The names of enzymes usually end with **-ase** and they are named according to the substance on which they act, or the reaction which they promote. For example, an enzyme which acts on proteins may be called a **protease**; one which removes hydrogen from a substance is a **dehydrogenase**.

The substance on which an enzyme acts is called its **substrate**. Thus, the enzyme **sucrase** acts on the substrate **sucrose** to produce the mono-saccharides, glucose and fructose.

Rates of enzyme reactions

As explained above, the rate of an enzyme-controlled reaction depends on the temperature and pH. It also depends on the concentrations of the enzyme and its substrate. The more enzyme molecules produced by a cell, the faster the reaction will proceed, provided there are enough substrate molecules available. Similarly, an increase in the substrate concentration will speed up the reaction if there are enough enzyme molecules to cope with the additional substrate.

Enzymes and pH

Acid or alkaline conditions alter the chemical properties of proteins, including enzymes. Most enzymes work best at a particular level of acidity or alkalinity (pH). The protein-digesting enzyme in your stomach, for example, works well at an acidity of pH 2. At this pH, the enzyme amylase, from your saliva, cannot work at all. Inside the cells, most enzymes will work best in neutral conditions (pH 7). The pH or temperature at which an enzyme works best is often called its **optimum** pH or temperature.

Although changes in pH affect the activity of enzymes, these effects are usually reversible, i.e. an enzyme which is inactivated by a low pH will resume its normal activity when its optimum pH is restored. Extremes of pH, however, may denature some enzymes irreversibly.

Intra- and extra-cellular enzymes

All enzymes are made inside cells. Most of them remain inside the cell to speed up reactions in the cytoplasm and nucleus. These are called **intra-cellular enzymes** ('intra-' means 'inside'). In a few cases, the enzymes made in the cells are let out of the cell to do their work outside. These are **extra-cellular enzymes** ('extra-' means 'outside'). Fungi (p. 321) and bacteria (p. 310) release extra-cellular enzymes in order to digest their food. A mould growing on a piece of bread releases starch-digesting enzymes into the bread and absorbs the soluble sugars which the enzyme produces from the bread. In the digestive systems of animals (p. 125), extra-cellular enzymes are released into the stomach and intestines in order to digest the food.

PRACTICAL WORK

Tests for proteins, fats and carbohydrates are described on page 122. Experiments on the digestive enzymes amylase and pepsin are described on page 134.

Methods of preparing the reagents for the following experiments are given on page 363.

1. Extracting and testing an enzyme from living cells

In this experiment, the enzyme to be extracted and tested is **catalase**, and the substrate is hydrogen peroxide (H_2O_2). Certain reactions in the cell produce hydrogen peroxide, which is poisonous. Catalase renders the hydrogen peroxide harmless by breaking it down to water and oxygen.

$$2H_2O_2 \xrightarrow{\text{catalase}} 2H_2O + O_2$$

Grind a small piece of liver with about 20 cm^3 water and a little sand, in a mortar. This will break open the liver cells and release their contents. Filter the mixture and share it between two test-tubes, A and B (Fig. 8). The filtrate will contain a great variety of substances dissolved out from the cytoplasm of the liver cells, including many enzymes. Because enzymes are specific, however, only one of these, catalase, will act on hydrogen peroxide. Add some drops of the filtrate from test-tube A to a few cm^3 hydrogen peroxide in a test-tube. You will see a vigorous reaction as the hydrogen peroxide breaks down to produce oxygen. (The oxygen can be tested with a glowing splint.)

Now boil the filtrate in tube B for about 30 seconds. Add a few drops of the boiled filtrate to a fresh lot of hydrogen peroxide. There will be no reaction because boiling has denatured the catalase.

Next, shake a little manganese(IV) oxide powder in a test-tube with some water and pour this into some hydrogen peroxide. There will be a vigorous reaction similar to the one with the liver extract. If you now boil some manganese(IV) oxide with water and add this to hydrogen peroxide, the reaction will still occur. Manganese(IV) oxide is a catalyst but it is not an enzyme because heating has not altered its catalytic properties.

The experiment can be repeated with a piece of potato to compare its catalase content with that of the liver. The piece of potato should be about the same size as the liver sample.

2. Building starch from glucose

Grind a small piece of potato in a mortar with about 10 cm^3 distilled water and a little clean sand. Filter the mixture into a clean test-tube. On a cavity tile place 2 rows of 4 drops of a 5 per cent solution of glucose phosphate (Fig. 9). Pick up a little of the filtered potato extract in a bulb pipette and place a drop in a spare cavity on the tile. Test this drop with iodine solution. If it goes blue it contains starch and the extract must be filtered again to remove all starch grains.

If the extract is free from starch, place 1 drop of it on each of the 4 drops of glucose phosphate in the top row on the tile. Now boil the rest of the potato extract, let it cool and then place a drop on each of the glucose phosphate drops in the bottom row on the tile.

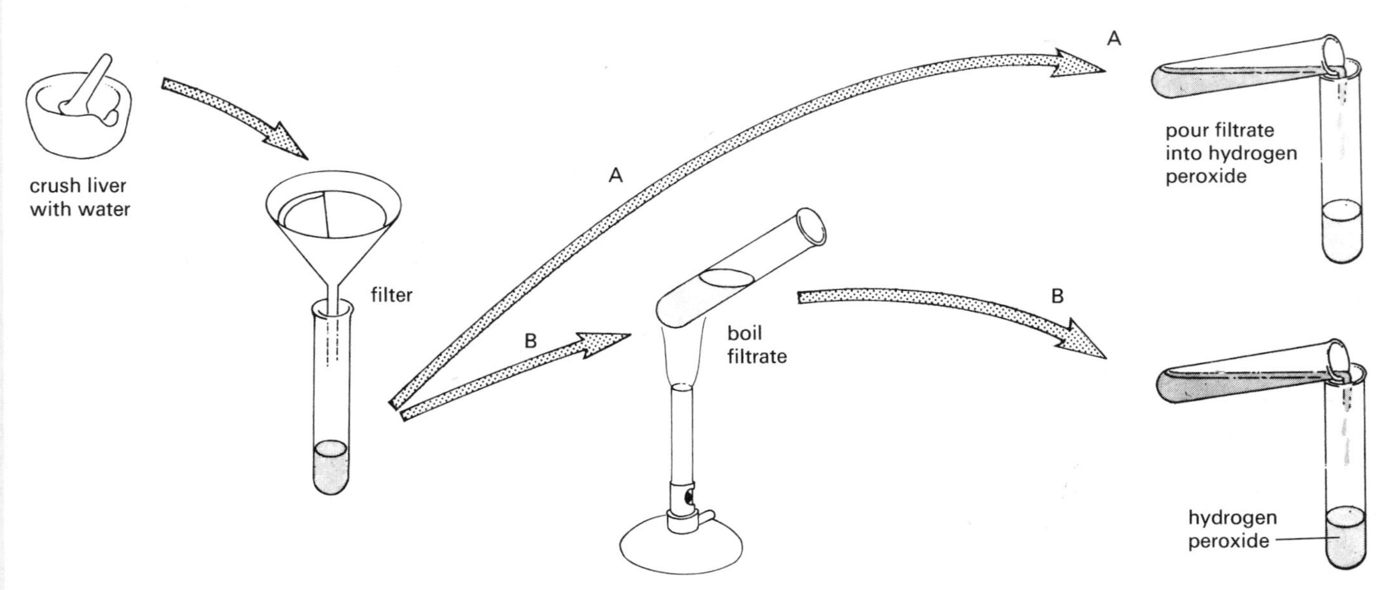

Fig. 8 Extracting and testing living cells for catalase

After 5 minutes, test the first drop in each row with iodine solution. Repeat this test at 5-minute intervals with the other 3 pairs of drops.

Finally, use a spare cavity on the tile to test a drop of the glucose phosphate solution with iodine.

Result The last 2 glucose phosphate drops in the top row should turn blue. None of the drops in the bottom row should go blue.

Interpretation A blue colour with iodine shows that starch is present. Neither the glucose phosphate solution nor the potato extract contained starch, so the starch must have formed during the course of the experiment. Presumably, an enzyme in the potato extract had joined the glucose phosphate molecules together to form starch. This is supported by the fact that the boiled extract did not produce starch, as you would expect if the reaction was brought about by an enzyme. The enzyme in question is called **starch phosphorylase**.

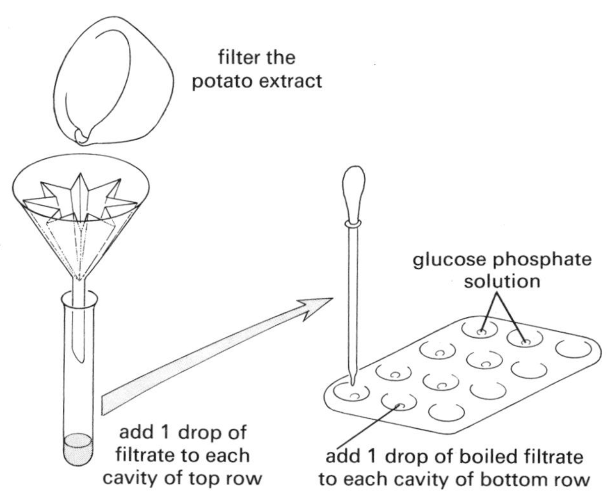

Fig. 9 Building starch from glucose

3. The effect of temperature on an enzyme reaction

Saliva contains a starch-digesting enzyme called **amylase**, which breaks down starch to a sugar (maltose).

Collect about 30 mm (depth) of saliva in a clean test-tube. Label three test-tubes A, B and C and use a graduated pipette to place 1 cm^3 saliva in each tube. Rinse the pipette.

Now label three test-tubes 1–3 and use the pipette to place 5 cm^3 of a 1 per cent starch solution in each. To each of tubes 1–3, add 6 drops only of dilute iodine solution using a dropping pipette.

Prepare three water baths by half filling beakers or jars as follows:

1. ice and water, adding ice during the experiment to keep the temperature at about 10 °C;
2. water from the cold tap at about 20 °C;
3. warm water at about 35 °C by mixing hot and cold water from the tap.

Place tubes 1 and A in the cold water bath, tubes 2 and B in the water at room temperature, and tubes 3 and C in the warm water. Leave them for 5 minutes to reach the temperature of the water. (Fig. 10.)

After 5 minutes, take the temperature of each water bath, then pour the saliva from tube A into the starch solution in tube 1 and return tube 1 to the water bath. Repeat this with tubes 2 and B, and 3 and C.

As the amylase breaks down the starch, it will cause the blue colour to disappear. Make a note of how long this takes in each case.

Answer the following questions:

(*a*) At what temperature did the salivary amylase break down starch most rapidly?

(*b*) What do you think would have been the result if a fourth water bath at 90 °C had been used?

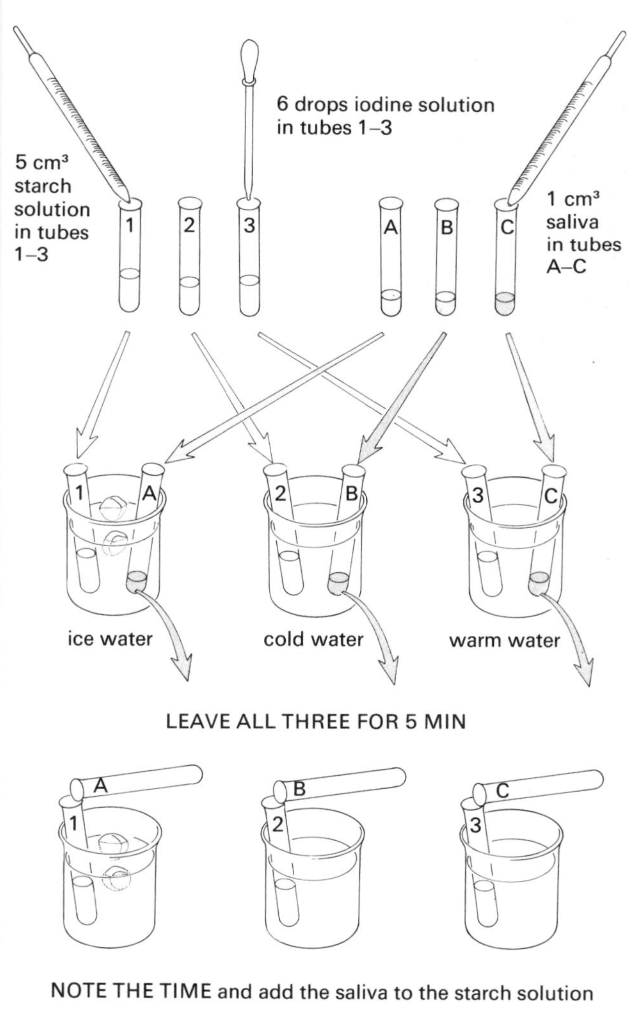

Fig. 10 The effect of temperature on an enzyme reaction

4. The effect of pH on an enzyme reaction

Label five test-tubes 1–5 and use a graduated pipette to place 5 cm^3 of a 1 per cent starch solution in each tube. Add acid or alkali to each tube as indicated below. Rinse the pipette when changing from sodium carbonate to acid.

Tube	*Chemical*	*Approximate pH*
1	1 cm^3 sodium carbonate solution (M/20)	9 (alkaline)
2	0.5 cm^3 sodium carbonate solution (M/20)	7–8 (slightly alkaline)
3	nothing	6–7 (neutral)
4	2 cm^3 ethanoic (acetic) acid (M/10)	6 (slightly acid)
5	4 cm^3 ethanoic (acetic) acid (M/10)	3 (acid)

Place several rows of iodine solution drops in a cavity tile.

Collect about 50 mm (depth) saliva in a clean test-tube and use a graduated pipette to add 1 cm^3 saliva to each tube. Shake the tubes and note the time. (Fig. 11.)

Use a clean dropping pipette to remove a small sample from each tube in turn and let one drop fall on to one of the iodine drops in the cavity tile. Rinse the pipette in a beaker of water between each sample. Keep on sampling in this way.

When any of the samples fails to give a blue colour, this means that the starch in that tube has been completely broken down to sugar by the salivary amylase. Note the time when this happens for each tube and stop taking samples from that tube. Do not continue sampling for more than about 15 minutes, but put a drop from each tube on to a piece of pH paper and compare the colour produced with a colour chart of pH values. Also test the pH of your own saliva.

Answer the following questions:

(*a*) At what pH did the enzyme, amylase, work most rapidly?
(*b*) Is this its optimum pH?
(*c*) Explain why you might have expected the result which you got.
(*d*) Your stomach pH is about 2. Would you expect starch digestion to take place in the stomach?

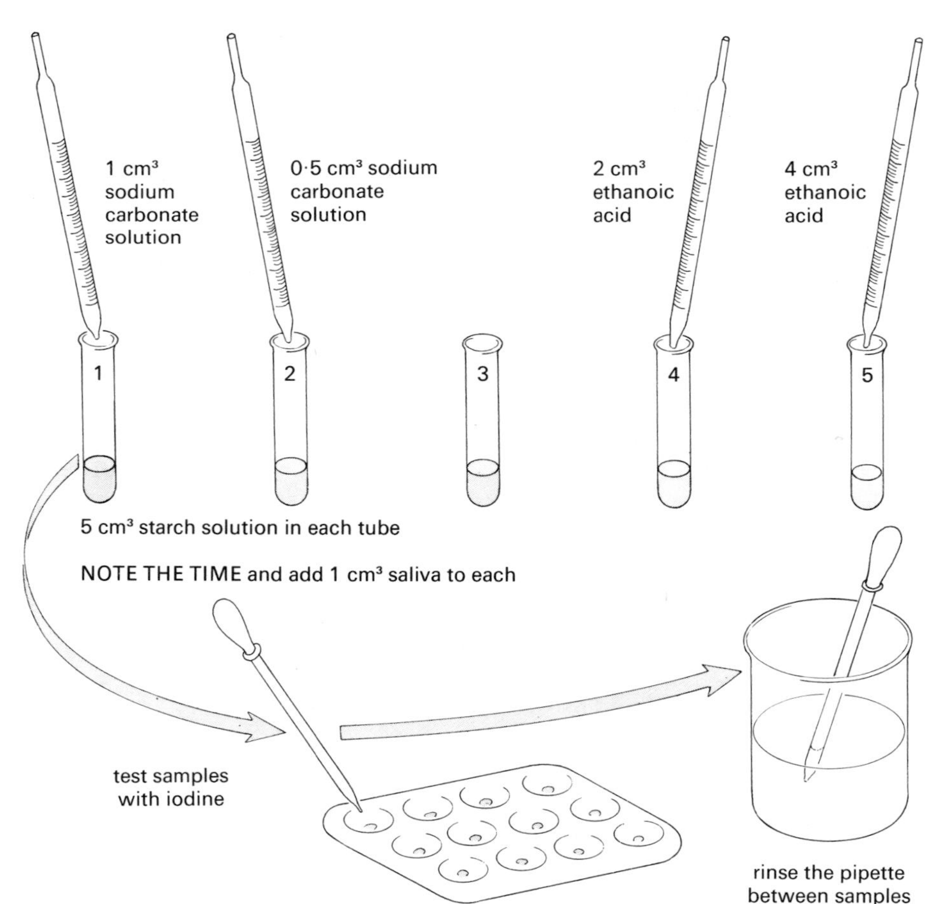

Fig. 11 The effect of pH on an enzyme reaction

Further experiments In the series *Experimental Work in Biology*, 17 experiments on food tests and 11 experiments on enzymes are described. (See p. 364.)

QUESTIONS

8 Which of the following statements apply both to enzymes and to any other catalysts:
(*a*) Their activity is stopped by high temperature,
(*b*) They speed up chemical reactions,
(*c*) They build up large molecules from small molecules,
(*d*) They are not used up during the reaction.

9 How would you expect the rate of an enzyme-controlled reaction to change if the temperature was raised (*a*) from 20 °C to 30 °C, (*b*) from 35 °C to 55 °C? Explain your answers.

10 There are cells in your salivary glands which can make an extracellular enzyme, amylase. Would you expect these cells to make intracellular enzymes as well? Explain your answer.

11 Apple cells contain an enzyme which turns the tissues brown when an apple is peeled and left for a time. Boiled apple does not go brown (Fig. 12). Explain why the boiled apple behaves differently.

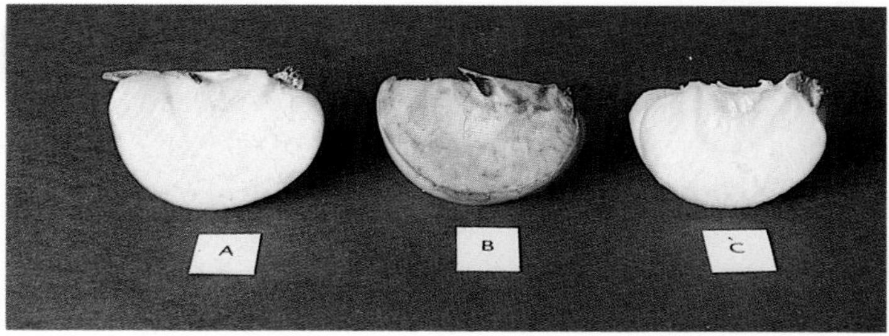

Fig. 12 Enzyme activity in an apple. Slice A has been freshly cut. B and C were cut 2 days earlier but C was dipped immediately in boiling water for one minute.

12 (*a*) What name would you give to an enzyme which converted a peptide to amino acids?
(*b*) On what kind of substance would you expect a *lipase* to act?

13 Suppose that the enzyme in Fig. 7*a* is joining a glycine molecule (p. 15) to a valine molecule. Suggest why the same enzyme will not join glycine to serine.

14 If you tried Experiment 2, suggest what might be the value in a potato tuber of an enzyme which changes glucose to starch.

CHECK LIST

- **Living matter is made up of water, proteins, lipids, nucleic acids, carbohydrates, salts and vitamins.**
- **Proteins are built up from amino acids joined together by peptide bonds.**
- **In different proteins the 20 or so amino acids are in different proportions and arranged in different sequences.**
- **Proteins are denatured by heat and some chemicals.**
- **Lipids include fats, fatty acids and steroids.**
- **Many fats are triglycerides, made from fatty acids and glycerol.**
- **Proteins and lipids form the membranes outside and inside the cell.**
- **Enzymes are proteins which catalyse chemical reactions in the cell.**
- **Enzymes are affected by pH and temperature and are denatured above 50 °C.**
- **Different enzymes may accelerate reactions which build up or break down molecules.**
- **Each enzyme acts on only one substance (breaking down), or a pair of substances (building up).**
- **The substance on which an enzyme acts is called the substrate.**

3 Energy from Respiration

RESPIRATION

Definition. Aerobic and anaerobic respiration. Energy transfer with ATP. Metabolism.

PRACTICAL WORK

Principles of the experimental designs. Output of carbon dioxide, uptake of oxygen, temperature rise. anaerobic respiration. Muscle contraction with ATP.

CONTROLLED EXPERIMENTS

Description. Hypotheses and hypothesis testing in scientific investigations. Use of carbon-14.

Most of the processes taking place in cells need energy to make them happen. Building up proteins from amino acids or making starch from glucose needs energy. When muscle cells contract, nerve cells conduct electrical impulses or plant cells form cell walls, they use energy. This energy comes from the food which cells take in. The food mainly used for energy in cells is glucose.

The process by which energy is produced from food is called **respiration**.

Respiration is a chemical process which takes place in cells. It must not be confused with the process of breathing which is also sometimes called 'respiration'. To make the difference quite clear, the chemical process in cells is sometimes called **cellular respiration**, **internal respiration** or **tissue respiration**. The use of the word 'respiration' for breathing is best avoided altogether.

Aerobic respiration

The word **aerobic** means that oxygen is needed for this chemical reaction. The food molecules are combined with oxygen. The process is called **oxidation** and the food is said to be **oxidized**. All food molecules contain carbon, hydrogen and oxygen atoms. The process of oxidation converts the carbon to carbon dioxide (CO_2) and the hydrogen to water (H_2O) and, at the same time, sets free energy which the cell can use to drive other reactions.

Aerobic respiration can be summed up by the equation

$$\underset{\text{glucose}}{C_6H_{12}O_6} + \underset{\text{oxygen}}{6O_2} \xrightarrow{\text{enzymes}} \underset{\text{carbon dioxide}}{6CO_2} + \underset{\text{water}}{6H_2O} + \underset{\text{energy}}{2830\ \text{kJ}}$$

The 2830 kilojoules (kJ) is the amount of energy you would get by completely oxidizing 180 grams of glucose to carbon dioxide and water. In the cells, the energy is not released all at once. The oxidation takes place in a series of small steps and not in one jump as the equation suggests. Each small step needs its own enzyme and at each stage a little energy is released (Fig. 1).

Although the energy is used for the processes mentioned above, some of it always appears as heat. In 'warm-blooded' animals some of this heat is retained to keep up their body temperature. In 'cold-blooded' animals the heat may build up for a time and allow the animal to move about faster. In plants the heat is lost to the surroundings (by conduction, convection and evaporation) as fast as it is produced.

Anaerobic respiration

The word **anaerobic** means 'in the absence of oxygen'. In this process, energy is still released from food by breaking it down chemically but the reactions do not use oxygen though they do often produce carbon dioxide. A common example is the action of yeast on sugar solution to produce alcohol. This process is called **fermentation**. The following equation shows what happens:

$$\underset{\text{glucose}}{C_6H_{12}O_6} \xrightarrow{\text{enzymes}} \underset{\text{alcohol}}{2C_2H_5OH} + \underset{\text{carbon dioxide}}{2CO_2} + \underset{\text{energy}}{118\ \text{kJ}}$$

As with aerobic respiration, the reaction takes place in small steps and needs several different enzymes. The yeast uses the energy for its growth and living activities, but you can see from the equation that less energy is produced by anaerobic respiration than in aerobic respiration. This is because the alcohol still contains a great deal of energy which the yeast is unable to use.

(*a*) Molecule of glucose (H and O atoms not all shown)

(*b*) The enzyme attacks and breaks the glucose molecule into two 3-carbon molecules

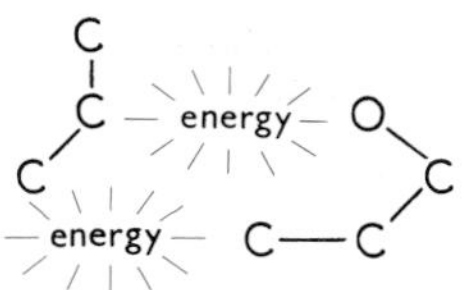

(*c*) This breakdown sets free energy

(*d*) Each 3-carbon molecule is broken down to carbon dioxide

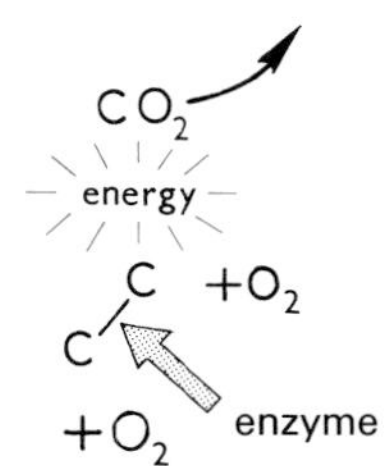

(*e*) More energy is released and CO_2 is produced

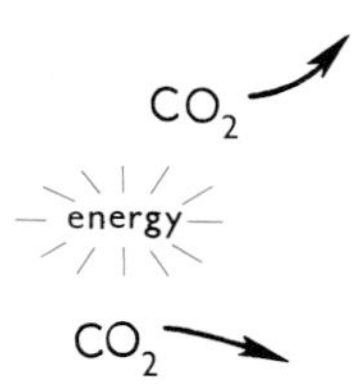

(*f*) The glucose has been completely oxidized to carbon dioxide (and water), and all the energy released

Fig. 1 Aerobic respiration

In animals, the first stages of respiration in muscle cells are anaerobic and produce **pyruvic acid** (the equivalent of the yeast's alcohol). Only later on is the pyruvic acid completely oxidized to carbon dioxide and water.

$$\text{glucose} \xrightarrow{\text{enzymes}} \text{pyruvic acid} \xrightarrow{\text{enzymes and oxygen}} CO_2 + H_2O$$

ANAEROBIC STAGE — AEROBIC STAGE

During exercise pyruvic acid may build up in a muscle faster than it can be oxidized. In this case it is turned into **lactic acid** and removed in the bloodstream. On reaching the liver, some of the lactic acid is oxidized to carbon dioxide and water, using up oxygen in the process. So, even after exercise has stopped, a high level of oxygen consumption may persist for a time as the excess of lactic acid is oxidized. This build up of lactic acid which is oxidized later is said to create an **oxygen debt**.

Energy transfer

The energy released when glucose is broken down is not used directly in the cell. Instead it is transferred to other chemicals which act as a store of readily available energy. One of these chemicals is the nucleotide **adenosine triphosphate** (ATP) (Fig. 2). At those stages in a glucose molecule's breakdown when energy is released, the energy is used to combine a phosphate ion (PO_3^{2-}) with a molecule of adenosine diphosphate (ADP) to make a molecule of adenosine triphosphate (Fig. 3).

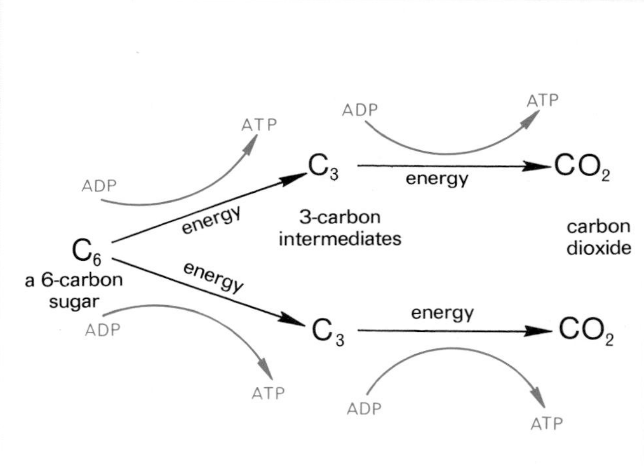

Fig. 3 Energy transfer involving ATP

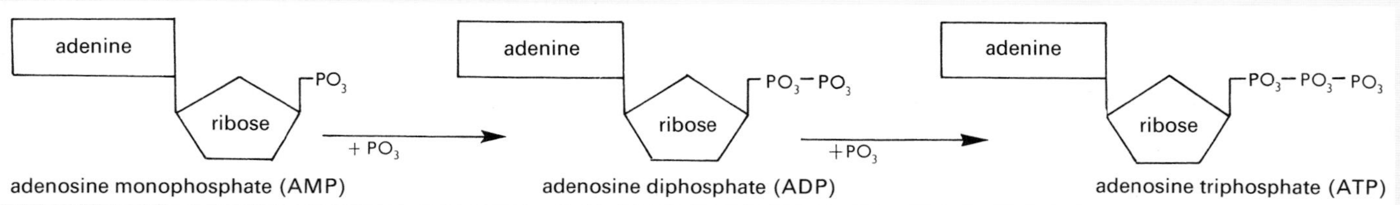

Fig. 2 The adenine nucleotides

Figure 3 seems to suggest that four molecules of ATP are made for every molecule of glucose broken down. However, many of the intermediate steps have been omitted and, in fact, 38 molecules of ATP can be built up when one molecule of glucose is completely oxidized to carbon dioxide and water.

In the presence of an appropriate enzyme, ATP readily breaks down to ADP, releasing energy and a phosphate ion. The energy can be used to drive other chemical reactions such as those producing muscle contraction.

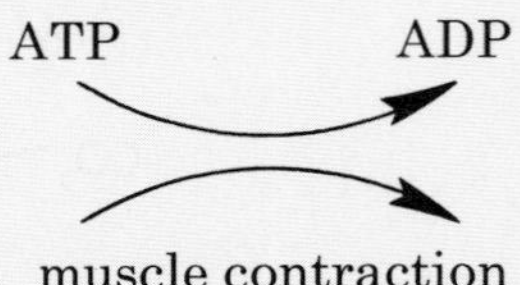

ATP is thus a kind of energy store in the cell and can be used directly for driving almost any reaction which needs energy. The breakdown of glucose to carbon dioxide and water is a comparatively slow process with many intermediate steps and enzymes. ATP breakdown is rapid and needs only a suitable ATP-ase enzyme to bring it about. This is true of all cells, plant, animal or micro-organism.

The breakdown of glucose might be compared to burning coal, in order to generate steam, which drives a turbine and so charges a battery; a fairly complex and long-winded business. The charged battery, however, is like the ATP. It can be used at a second's notice to do a wide variety of things such as light a lamp, ring a bell, drive a motor, etc.

Metabolism

All the chemical changes taking place inside a cell or a living organism are called its **metabolism**. The minimum turnover of energy needed simply to keep an organism alive, without movement or growth, is called the **basal metabolism**. Our basal metabolism maintains vital processes such as breathing, heart beat, digestion and excretion.

The processes which break substances down are sometimes called **catabolism**. Respiration is an example of catabolism in which carbohydrates are broken down to carbon dioxide and water. Chemical reactions which build up substances are called **anabolism**. Building up a protein from amino acids is an example of anabolism. The energy released by the **catabolic** process of respiration is used to drive the **anabolic** reactions which build up proteins.

You may have heard of anabolic steriods, in connection with drug-taking by athletes. These chemicals do reduce the rate of protein breakdown and may enhance the build up of certain proteins. However, their effects are complicated and not fully understood, they have undesirable side effects and their use contravenes athletics codes.

QUESTIONS

1 (*a*) If, in one word, you had to say what respiration was about, which word would you choose from this list: breathing, energy, oxygen, cells, food?
(*b*) In which parts of a living organism does respiration take place?

2 What are the two main differences between aerobic and anaerobic respiration?

3 What chemical substances (*a*) from outside the cell, (*b*) from inside the cell, must be provided for aerobic respiration to take place? (*c*) What are the products of aerobic respiration?

4 Victims of drowning who have stopped breathing are sometimes revived by a process called 'artificial respiration'. Why would a biologist object to the use of this expression? ('Resuscitation' is a better word to use.)

5 Which cell organelle would you expect to produce most ATP? (See p. 6.)

6 By comparing the equations for aerobic and anaerobic respiration on page 24, how many molecules of ATP might you expect to be produced by the anaerobic fermentation of one molecule of glucose?

PRACTICAL WORK

EXPERIMENTS ON RESPIRATION

If you look below at the chemical equation which represents aerobic respiration you will see that a tissue or an organism which is respiring should be (*a*) using up food, (*b*) using up oxygen, (*c*) giving off carbon dioxide, (*d*) giving off water and (*e*) releasing energy which can be used for other processes.

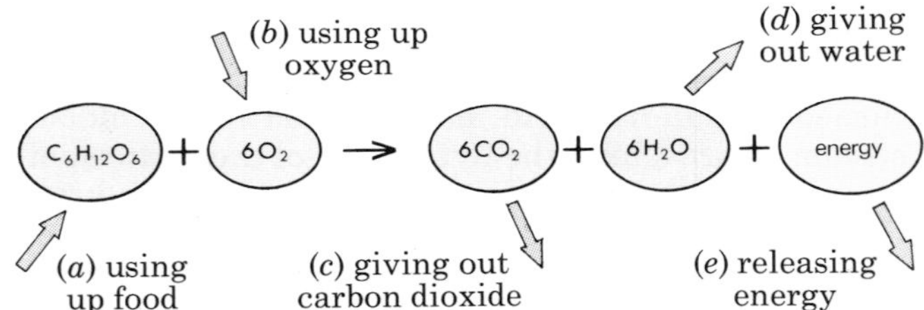

If we wish to test whether aerobic respiration is taking place:

1. '(*d*) Giving out water' is not a good test because non-living material will give off water vapour if it is wet to start with.
2. '(*a*) Using up food' can be tested by seeing if an organism loses weight. This is not so easy as it seems because most organisms lose weight as a result of evaporation of water and this may have nothing to do with respiration. It is the decrease in 'dry weight' which must be measured (p. 96).
3. (*b*), (*c*) and (*e*) are fairly easy to demonstrate either with whole organisms or pieces of living tissue, and three simple experiments will now be described.

Seeds are often used as the living organisms because when they start to grow (germinate) there is a high level of chemical activity in the cells. The seeds are easy to obtain and to handle and they fit into small-scale apparatus. In some cases blowfly maggots can be used as animal material.

1. Giving off carbon dioxide in respiration

The living organism is placed under a bell-jar which is sealed to a glass plate using a film of Vaseline (Fig. 4). Air is drawn through the apparatus with a filter pump or pushed through it with an aquarium pump. The air is led firstly through a tube of soda-lime which absorbs all the atmospheric carbon dioxide. Then the air passes through a tube of clear lime water to show that the carbon dioxide has been removed from the air. After passing through the bell-jar, the air is bubbled through a second tube of lime water. If the organism is producing carbon dioxide, the lime water should go milky. The lime water in the first tube should stay clear.

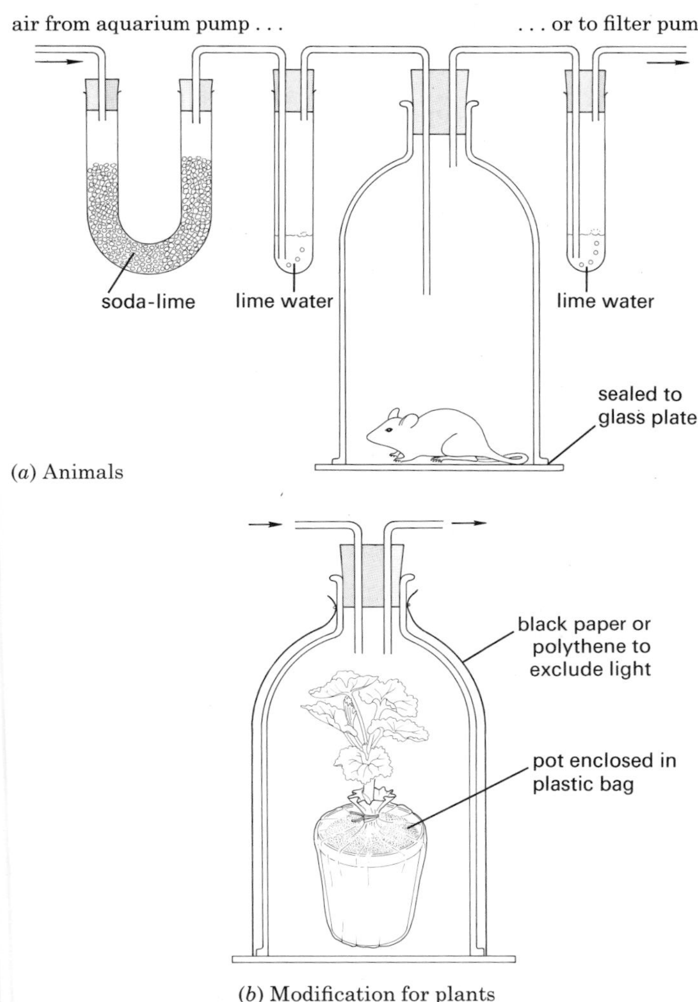

Fig. 4 Living organisms produce carbon dioxide

Modification for plants A potted plant or some leafy shoots in a jar of water can be placed under the bell-jar. However, some changes need to be made. Green plants, in daylight, use up more carbon dioxide than they produce (See 'Photosynthesis', p. 44), so the bell-jar must be covered with black paper or black polythene to keep out any light (Fig. 4*b*). If a potted plant is used, the pot and soil must be enclosed in a polythene bag, secured round the plant's stem. This is to ensure that any carbon dioxide produced comes from the plant and not from organisms in the soil.

2. Carbon dioxide from germinating seeds

Put some germinating wheat grains in a large test-tube. Cover the mouth of the tube with aluminium foil. After 15–20 minutes, take a sample of the air from the test-tube. Do this by pushing a glass tube attached to a 10 cm^3 plastic syringe through the foil and into the test-tube (Fig. 5*a*). Withdraw the syringe plunger enough to fill the syringe with air from the test-tube. Now slowly bubble this air sample through a little clear lime water in a small test-tube (Fig. 5*b*). Cover the mouth of the small test-tube and shake the lime water up.

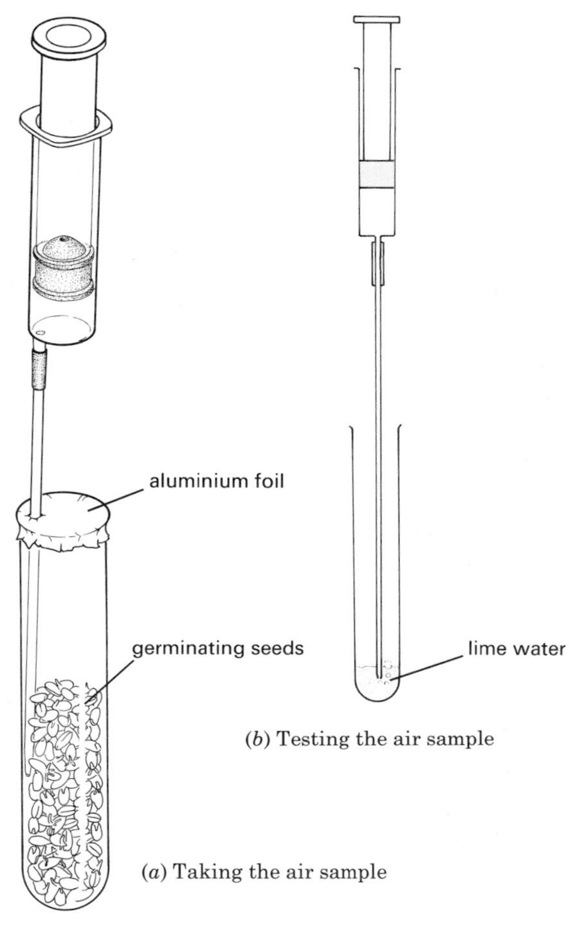

Fig. 5 Production of carbon dioxide by germinating seeds

Result The lime water will go milky.

Interpretation Lime water turning milky is evidence of carbon dioxide but it could be argued that the carbon dioxide came from the air or that the seeds give off carbon dioxide whether or not they are respiring. The only way to disprove these arguments is to do a **control experiment** (see p. 30).

Control Boil some of the germinating wheat grains before starting the experiment. When you set up the experiment, put an equal amount of boiled wheat grains in a large test-tube and cover the mouth of the tube with aluminium foil exactly as you did for the living seeds. When you test the air from the living seeds, also test the air from the dead seeds. It should not turn the lime water milky. This means that the carbon dioxide did not come from the air, nor was it given off by the dead seeds. It must be a living process in the seeds which produces carbon dioxide and this process is likely to be respiration. However, since you stopped all living processes by boiling the seeds in the control experiment, you have not been able to prove that it was respiration rather than some other chemical change which produced the carbon dioxide.

An experiment to show carbon dioxide production by plants, using hydrogencarbonate indicator, is described on page 49.

3. Using up oxygen during respiration

The apparatus in Fig. 6 is a **respirometer** (a 'respire meter'), which can measure the rate of respiration by seeing how quickly oxygen is taken up. Germinating seeds or blowfly larvae are placed in the test-tube and , as they use up the oxygen for respiration, the level of liquid in the delivery tubing will go up.

There are two drawbacks to this. One is that the organisms usually give out as much carbon dioxide as they take in oxygen. So there may be no change in the total amount of air in the test-tube and the liquid level will not move. This drawback is overcome by placing **soda-lime** in the test-tube. Soda-lime will absorb carbon dioxide as fast as the organisms give it out. So only the uptake of oxygen will affect the amount of air in the tube. The second drawback is that quite small changes in temperature will make the air in the test-tube expand or contract and so cause the liquid to rise or fall whether or not respiration is taking place. To overcome this, the test-tube is kept in a beaker of water (a water bath). The temperature of water changes far more slowly than that of air, so there will not be much change during a 30-minute experiment.

Control To show that it is a living process which uses up oxygen, a similar respirometer is prepared but containing an equal quantity of germinating seeds which have been killed by boiling. (If blowfly larvae are used, the control can consist of an equivalent volume of glass beads. This is not a very good control but is probably more acceptable than killing an equivalent number of larvae.)

The apparatus is finally set up as shown in Fig. 7 and left for 30 minutes (10 minutes if blowfly larvae are used).

The capillary tube and reservoir of liquid are called a **manometer**.

Result The level of liquid in the experiment goes up more than in the control. The level in the control may not move at all.

Interpretation The rise of liquid in the delivery tubing shows that the living seedlings have taken up part of the air. It does not prove that it is oxygen which has been taken up. Oxygen seems the most likely gas, however, because (1) there is only 0.03 per cent carbon dioxide in the air to start with and (2) the other gas, nitrogen, is known to be less active than oxygen.

If the experiment is allowed to run for a long time, the uptake of oxygen could be checked at the end by placing a lighted splint in each test-tube in turn. If some of the oxygen has been removed by the living seedlings, the flame should go out more quickly than it does in the tube with dead seedlings.

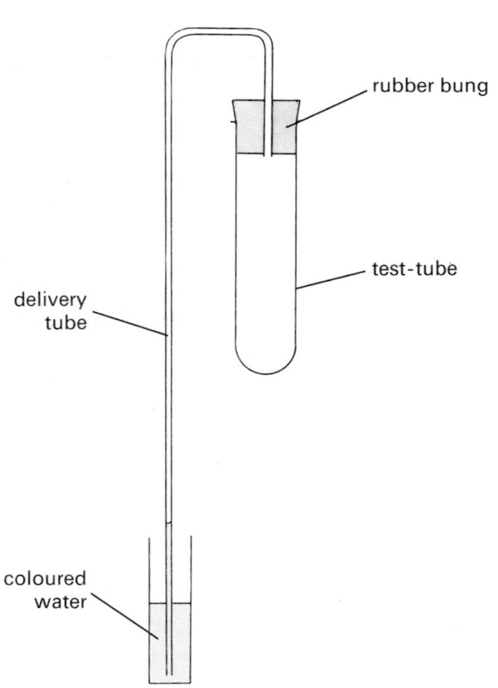

Fig. 6 A simple respirometer

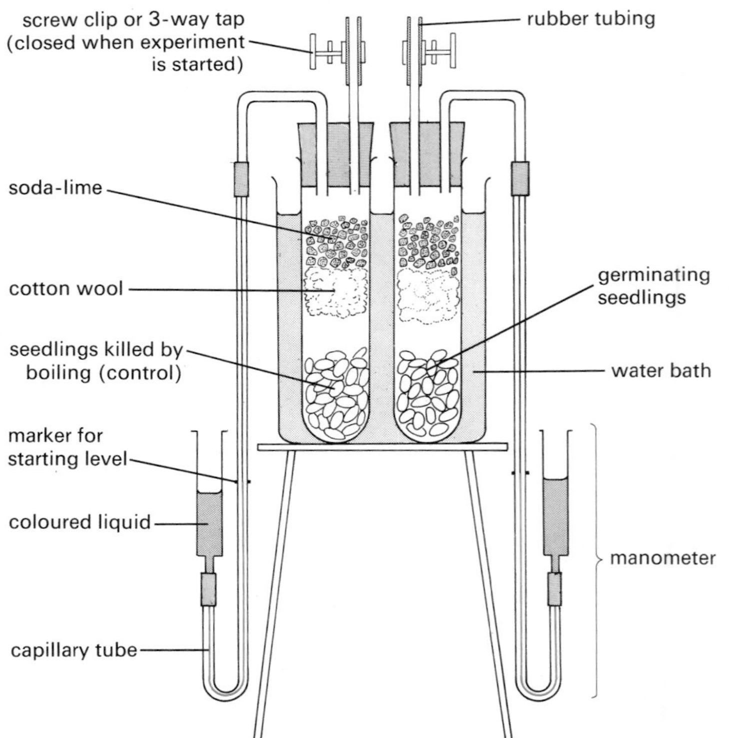

Fig. 7 To see if oxygen is taken up in respiration

4. Measuring the rate of respiration

The rate of respiration can be determined by measuring the rate of carbon dioxide production or the rate of oxygen uptake. It is fairly easy to modify Experiment 3 to measure the uptake of oxygen.

Fit a 2 cm^3 syringe to the rubber tubing above the screw clip of the test-tube with the living organisms. Better still, use a 3-way nylon tap instead of a screw clip, as in Fig. 8. Before fitting the syringe, withdraw the plunger to the 2 cm^3 mark. During the course of the experiment, the water level will rise in the manometer. From time to time, carefully press the syringe plunger to return the water level to its first position. After 20 minutes or so, take the reading on the syringe and subtract it from 2. This will be the volume of air you have had to add to make up for the oxygen absorbed by the organisms in the test-tube.

For example, if you had to add 1.3 cm^3 air in 30 minutes, the organisms are using oxygen at the rate of 2.6 cm^3 per hour. If the organisms weigh 20 g, the rate of oxygen uptake is $2.6/20 = 0.13$ cm^3 per gram per hour. This figure does not mean much on its own but can be used as the basis for comparisons, e.g. the rates of oxygen uptake at different temperatures or by different organisms.

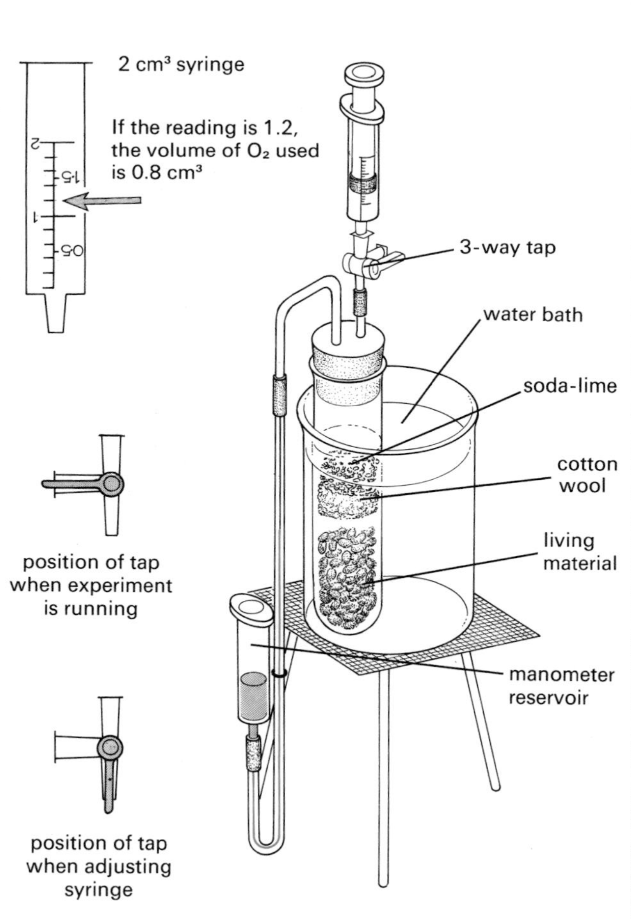

Fig. 8 Measuring the uptake of oxygen

5. Releasing energy in respiration

Fill a small vacuum flask with wheat grains which have been soaked for 24 hours and rinsed in 1 per cent formalin (or domestic bleach diluted 1+4) for 5 minutes. These solutions will kill any bacteria or fungi on the surface of the grains. Kill an equal quantity of soaked grains by boiling them for 5 minutes. Cool the boiled seeds in cold tap-water, rinse them in bleach or formalin for 5 minutes as before and then put them in a vacuum flask of the same size as the first one. This flask is the control.

Place a thermometer in each flask so that its bulb is in the middle of the seeds (Fig. 9). Plug the mouth of each flask with cotton wool and leave both flasks for 2 days, noting the thermometer readings whenever possible.

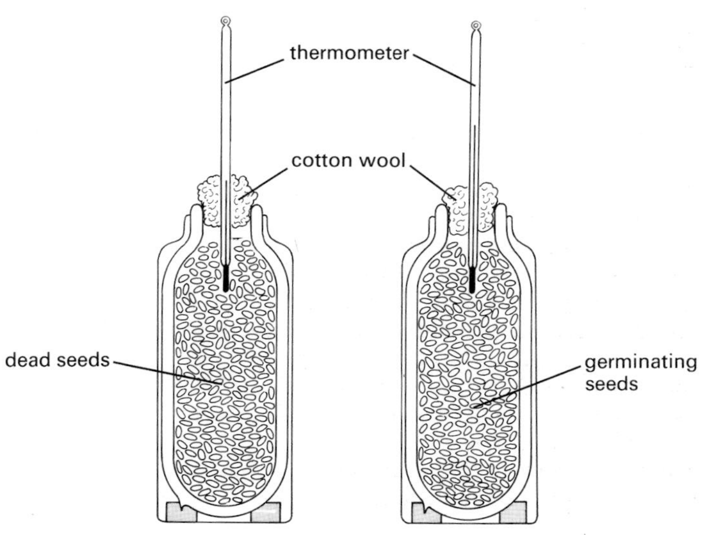

Fig. 9 Energy release in germinating seeds

Result The temperature in the flask with the living seeds will be 5–10 °C higher than that of the dead seeds.

Interpretation Provided that there are no signs of the living seeds going mouldy, the heat produced must have come from living processes in the seeds, because the dead seeds in the control did not give out any heat. There is no evidence that this process is respiration rather than any other chemical change but the result is what you would expect if respiration does produce energy.

For an experiment comparing the energy value of different food substances, see page 124.

6. Anaerobic respiration in yeast

Boil some water to expel all the dissolved oxygen. When cool, use the boiled water to make up a 5 per cent solution of glucose and a 10 per cent suspension of dried yeast. Place 5 cm^3 of the glucose solution and 1 cm^3 of the yeast suspension in a test-tube and cover the mixture with a thin layer of liquid paraffin to exclude atmospheric oxygen. Fit a delivery tube as shown in Fig. 10 and allow it to dip into clear lime water.

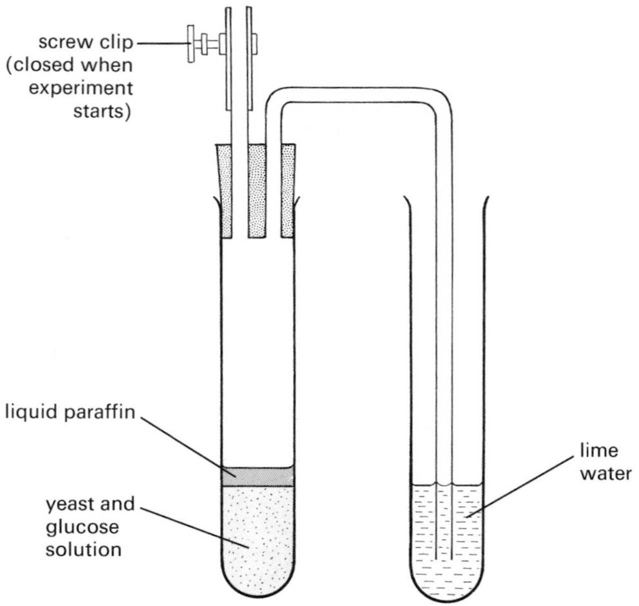

Fig. 10 Anaerobic respiration in yeast

Result After 10–15 minutes, with gentle warming if necessary, there should be signs of fermentation in the yeast-glucose mixture and the bubbles of gas escaping through the lime water should turn it milky.

Interpretation The fact that the lime water goes milky shows that the yeast-glucose mixture is producing carbon dioxide. If we assume that the production of carbon dioxide is evidence of respiration, then it looks as if the yeast is respiring. In setting up the experiment, you took care to see that oxygen was removed from the glucose solution and the yeast suspension, and the liquid paraffin excluded air (including oxygen) from the mixture. Any respiration taking place must, therefore, be anaerobic (i.e. without oxygen).

Control It might be suggested that the carbon dioxide came from a chemical reaction between yeast and glucose (as between chalk and acid) which had nothing to do with respiration or any other living process. A control should, therefore, be set up using the same procedure as before but with yeast which has been killed by boiling. The failure, in this case, to produce carbon dioxide supports the claim that it was a living process in the yeast in the first experiment which produced the carbon dioxide.

For an experiment on yeast in bread dough, see page 326.

7. Energy from adenosine triphosphate

Cut three very thin strips of fresh meat (muscle) as shown in Fig. 11. Make your cuts along the strands of muscle tissue rather than across them. Keep cutting the strips until they are only about 1 mm wide but try

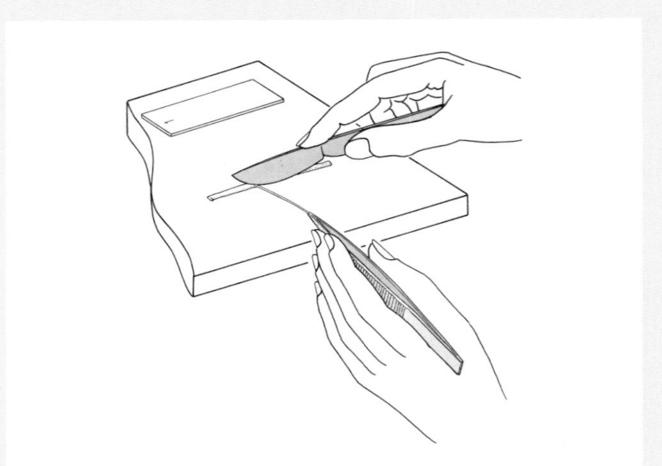

Fig. 11 Cutting thin strips of muscle

to keep them at least 20 mm long. Put each strip on a slide in a drop of Ringer's solution (see p. 363) which has a composition similar to tissue fluid. Number the slides 1–3 and measure the length of each muscle strip.

Now add a drop of 1 per cent glucose solution to slide 1, ATP solution to slide 2 and boiled (and cooled) ATP solution to slide 3. After 5 minutes, measure the muscle strips again.

Result The muscle strips on slides 2 and 3 will have contracted.

Interpretation Slide 1: although glucose contains energy, the non-living muscle cannot use it for contraction.

Slide 2: ATP produced contraction probably because the non-living muscle still contained an active enzyme for breaking down ATP to provide the necessary energy.

Slide 3: ATP is not affected by boiling and, therefore, in making muscle contract, it is not acting as an enzyme which releases energy from some other chemical.

CONTROLLED EXPERIMENTS

In most biological experiments, a second experiment called a **control** is set up. This is to make sure that the results of the first experiment are due to the conditions being studied and not to some other cause which has been overlooked.

In the experiment in Fig. 7, the liquid rising up the tube could have been the result of the test-tube cooling down, so making the air inside it contract. The identical experiment with dead seeds—the control—showed that the result was not due to a temperature change, because the level of liquid in the control did not move.

The term 'controlled experiment' refers to the fact that the experimenter (1) sets up a control and (2) controls the conditions in the experiment. In the experiment shown in Fig. 7 the seeds are enclosed in a test-tube, and soda-lime is added. This makes sure that any uptake or output of oxygen will make the liquid go up or down, and that the output of carbon dioxide will not affect the results. The experimenter had controlled both the amount and the composition of the air available to the germinating seeds.

If you did an experiment to compare the growth of plants in the house or in a greenhouse, you could not be sure whether it was the extra light or the high temperature of the greenhouse which caused better growth. This would not, therefore, be a properly controlled experiment. You must alter only one condition at a time, either the light or the temperature, and then you can compare the results with the control experiment.

A properly controlled experiment, therefore, alters only one condition at a time and includes a control which shows that it is this condition and nothing else which gave the result.

QUESTIONS

7 Which of the following statements are true? If an organism is respiring you would expect it to be (*a*) giving out carbon dioxide, (*b*) losing heat, (*c*) breaking down food, (*d*) using up oxygen, (*e*) gaining weight, (*f*) moving about?

8 What was the purpose of the soda-lime in Experiment 3 on page 28, and the lime water in Experiments 2 and 6 on pages 27 and 29?

9 In an experiment like Fig. 7 on page 28, the growing seeds took in 5 cm^3 oxygen and gave out 7 cm^3 carbon dioxide. What change in volume will take place (*a*) if no soda-lime is present, (*b*) if soda-lime is present?

10 The germinating seeds in Fig. 9 on page 29 will release the same amount of heat whether they are in a beaker or a vacuum flask. Why then is it necessary to use a vacuum flask for this experiment?

11 What is the purpose of the control in the experiment to show carbon dioxide production (Fig. 5, p. 27)?

Hypothesis testing

You will have noticed that none of the experiments described above claim to have *proved* that respiration is taking place. The most we can claim is that they have not disproved the proposal that energy is produced from respiration. There are many reactions taking place in living organisms and, for all we know at this stage, some of them may be using oxygen or giving out carbon dioxide without releasing energy, i.e. they would not fit our definition of respiration.

This inability to 'prove' that a particular proposal is 'true' is not restricted to experiments on respiration. It is a feature of many scientific experiments. One way that science makes progress is by putting forward a **hypothesis**, making predictions from the hypothesis, and then testing these predictions by experiments. A hypothesis is an attempt to explain some event or observation using the information currently available. If the results of an experiment do not confirm the predictions, the hypothesis has to be abandoned or altered.

For example, biologists observing that living organisms take up oxygen might put forward the hypothesis that 'Oxygen is used to convert food to carbon dioxide, so producing energy for movement, growth, reproduction, etc.' This hypothesis can be tested by predicting that, '*If* the oxygen is used to oxidize food *then* an organism that takes up oxygen will also produce carbon dioxide.' Experiments 2 and 3 on pp. 27 and 28 test this and fulfil this prediction and, therefore, support the hypothesis. Looking at the equation for respiration, we might also predict that an organism which is respiring will produce carbon dioxide and take up oxygen. The experiment with yeast, however, does not fulfil this prediction and so does not support the hypothesis as it stands, because here is an organism producing carbon dioxide without taking up oxygen. The hypothesis will have to be modified, e.g. 'Energy is released from food by breaking it down to carbon dioxide. Some organisms use oxygen for this process, others do not'.

There are still plenty of tests which we have not done. For example, we have not attempted to see whether it is food that is the source of energy and carbon dioxide. One way of doing this is to provide the organism with food, e.g. glucose, in which the carbon atoms are radioactive. Carbon-14 (^{14}C) is a radioactive form of carbon and can be detected by using a Geiger counter. If the organism produces radioactive carbon dioxide, it is reasonable to suppose that the carbon dioxide comes from the glucose.

$$C_6H_{12}O_6 + 6O_2 \longrightarrow 6CO_2 + 6\,H_2O + \text{energy}$$

Experiment 1 can be modifed to test this prediction. Some glucose which contains ^{14}C is dissolved in the drinking-water of the experimental animals, e.g. rats or mice (it doesn't harm them in dilute solution). If the theory of respiration is correct, the ^{14}C of the glucose molecules should be oxidized to carbon dioxide and so the carbon dioxide produced by the animals should be radioactive. This can be checked by filtering the cloudy lime water from the second tube and seeing if there is any radioactivity in the precipitated calcium carbonate.

$$Ca(OH)_2 + CO_2 \longrightarrow CaCO_3\downarrow + H_2O$$

If the calcium carbonate was not radioactive, then the carbon dioxide could not have been radioactive either and the carbon must have come from something other than the glucose. Such a result would not demolish the respiration hypothesis, but it would cast doubt on the assumption that glucose is one of the food substances which are oxidized.

Experimental Work in Biology contains 16 experiments on respiration (see p. 364).

QUESTIONS

12 The experiment with yeast on page 29 supported the claim that anaerobic respiration was taking place. The experiment was repeated using unboiled water and without the liquid paraffin. Fermentation still took place and carbon dioxide was produced. Does this mean that the design or the interpretation of the first experiment was wrong? Explain your answer.

13 Twenty seeds are placed on soaked cotton wool in a closed glass dish and after 5 days in the light 15 of the seeds had germinated. If the experiment is intended to see if light is needed for germination, which of the following would be a suitable control:

(*a*) Exactly the same set-up but with dead seeds
(*b*) The same set-up but with 50 seeds
(*c*) An identical experiment but with 20 seeds of a different species
(*d*) An identical experiment but left in darkness for 5 days?

CHECK LIST

- **Respiration is the process in cells which releases energy from food.**
- **Aerobic respiration needs oxygen, anaerobic respiration does not.**
- **The oxidation of food produces carbon dioxide as well as releasing energy.**
- **The energy from oxidation of food is transferred to ATP.**
- **ATP acts as a store of readily available energy for driving other chemical reactions or producing movement.**
- **Experiments to investigate respiration try to detect uptake of oxygen, production of carbon dioxide, release of energy as heat or a reduction in dry weight.**
- **In a controlled experiment, the scientist tries to alter only one condition at a time, and sets up a control to check this.**
- **A control is a second experiment, identical to the first experiment except for the one condition being investigated.**
- **The control is designed to show that only the condition under investigation is responsible for the results.**
- **Experiments are designed to test predictions made from hypotheses; they cannot 'prove' a hypothesis.**

4 How Substances get in and out of Cells

DIFFUSION

Explanation. Diffusion into and out of cells. Rates of diffusion. Controlled diffusion. Surface area. Endo- and exo-cytosis. Active transport.

OSMOSIS

Definition. Explanation. Artificial partially permeable membranes; the cell membrane. Water potential.

MOVEMENT OF WATER BETWEEN PLANT CELLS

PRACTICAL WORK

Experiments on diffusion of gases, diffusion in liquids, selective permeability, dialysis, turgor, osmotic flow, plasmolysis, turgor in potato tissue.

'MODELS'

Use of 'models' to explain theories.

Cells need food materials which they can oxidize for energy or which they can use to build up their cell structures. They also need salts and water which play a part in chemical reactions in the cell. On the other hand they need to get rid of substances, such as carbon dioxide, which, if they accumulated in the cell, would upset some of the chemical reactions and even poison the cell.

Substances may pass through the cell membrane either passively by diffusion, or actively by some form of active transport.

Diffusion

The molecules of a gas such as oxygen are moving about all the time. So are the molecules of a liquid, or a substance such as sugar dissolved in water. As a result of this movement, the molecules spread themselves out evenly to fill all the available space (Fig. 1). This process is called **diffusion**. One effect of diffusion is that the molecules of a gas, a liquid or a dissolved substance will move from a region where there are a lot of them (i.e. concentrated) to regions where there are few of them (i.e. less concentrated) until the concentration everywhere is the same. Figure 2*a* is a diagram of a cell with a high concentration of molecules (e.g. oxygen) outside and a low concentration inside. The effect of this difference in concentration is to make the molecules diffuse into the cell until the concentration inside and outside is the same (Fig. 2*b*).

Whether this will happen or not depends on whether the cell membrane will let the molecules through. Small molecules such as water (H_2O), carbon dioxide (CO_2) and oxygen (O_2) can pass through the cell membrane fairly easily. So diffusion tends to equalize the concentration of these molecules inside and outside the cell all the time.

molecules moving about

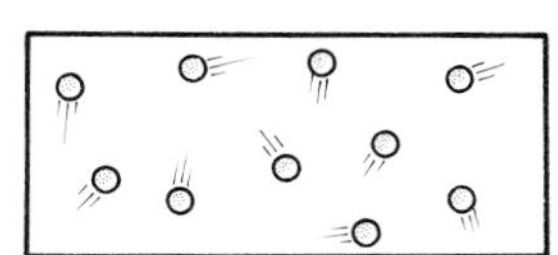

become evenly distributed

Fig. 1 Diffusion

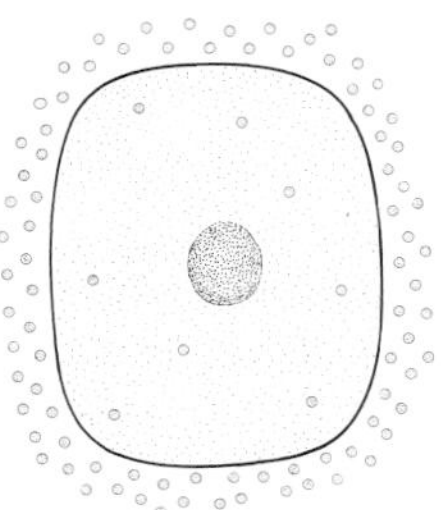

(*a*) Greater concentration outside cell

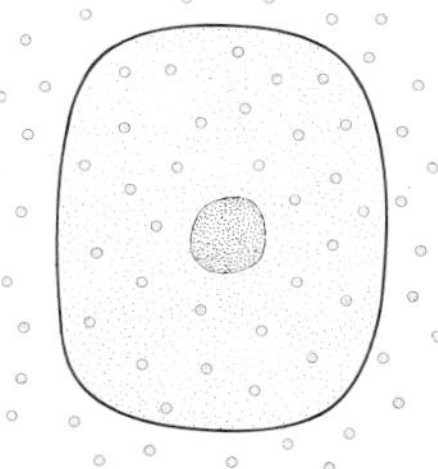

(*b*) Concentrations equal on both sides of the cell membrane

Fig. 2 Molecules entering a cell by diffusion

When a cell uses up oxygen for its aerobic respiration, the concentration of oxygen inside the cell falls and so oxygen molecules diffuse into the cell until the concentration is raised again. During tissue respiration, carbon dioxide is produced and so its concentration inside the cell goes up. Once again diffusion takes place, but this time the molecules move out of the cell. In this way, diffusion can explain how a cell takes in its oxygen and gets rid of its carbon dioxide.

Rates of diffusion The speed with which a substance diffuses through a cell wall or cell membrane will depend on temperature, pressure and many other conditions including (1) the distance it has to diffuse, (2) its concentration inside and outside the cell and (3) the size of its molecules or ions.

1. Cell membranes are all about the same thickness (about 0.007 μm) but plant cell walls vary in their thickness and permeability. Generally speaking, the thicker the wall, the slower the rate of diffusion.
2. The bigger the difference in concentration of a substance on either side of a membrane, the faster it will tend to diffuse. The difference is called a **concentration gradient** or **diffusion gradient** (Fig. 3). If a substance on one side of a membrane is steadily removed, the diffusion gradient is maintained. When oxygen molecules enter a red cell they combine with a chemical (haemoglobin) which takes them out of solution. Thus the concentration of free oxygen molecules inside the cell is kept very low and the diffusion gradient for oxygen is maintained.

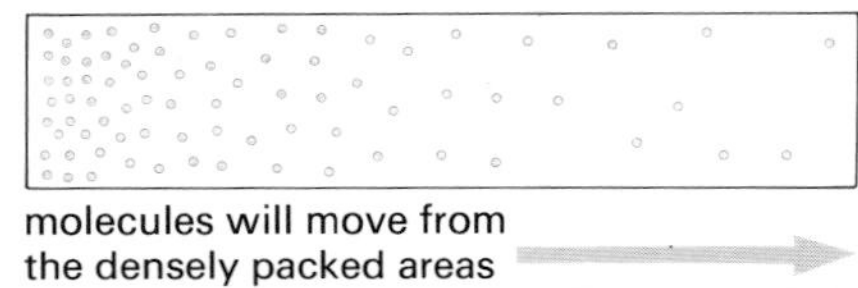

Fig. 3 Diffusion gradient

3. In general, the larger the molecules or ions, the slower they diffuse. However, many ions and molecules in solution attract water molecules around them (see p. 36) and so their effective size is greatly increased. It may not be possible to predict the rate of diffusion from the molecular size alone.

Controlled diffusion Although for any one substance, the rate of diffusion through a cell membrane depends partly on the concentration gradient, the rate is often faster or slower than expected. Water diffuses more slowly and amino acids diffuse more rapidly through a membrane than might be expected. In some cases this is thought to happen because the ions or molecules can pass through the membrane only by means of special pores. These pores may be few in number or they may be open or closed in different conditions.

In other cases, the movement of a substance may be speeded up by an enzyme working in the cell membrane. So it seems that 'simple passive' diffusion, even of water molecules, may not be so simple or so passive after all, where cell membranes are concerned.

Surface area If 100 molecules diffuse through 1 mm^2 of a membrane in 1 minute, it is reasonable to suppose that an area of 2 mm^2 will allow twice as many through in the same time. Thus the rate of diffusion into a cell will depend on the cell's surface area. The greater the surface area, the faster is the total diffusion. Cells which are involved in rapid absorption, e.g. in the kidney or the intestine, often have their 'free' surface membrane formed into hundreds of tiny projections called **microvilli** (Fig. 4) which increase the absorbing surface.

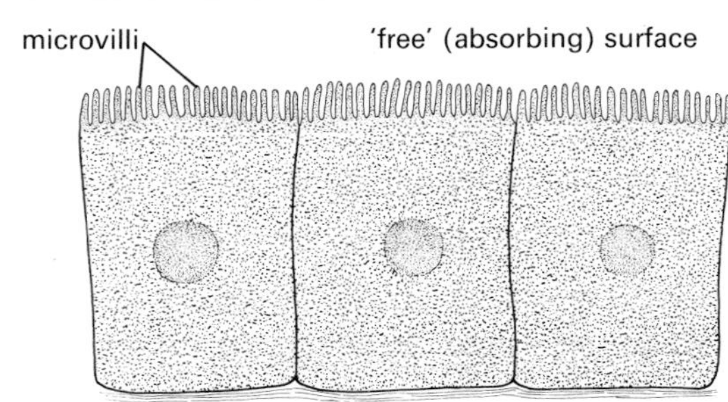

Fig. 4 Microvilli

The shape of a cell will also affect the surface area. For example, the cell in Fig. 5*a* has a greater surface area than that in Fig. 5*b*, even though they each have the same volume.

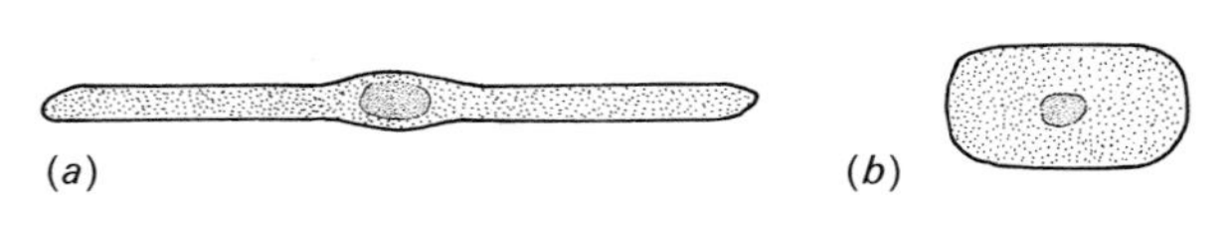

Fig. 5 Surface area. The cells have the same volume but (*a*) has a much greater surface area.

Endo- and exocytosis

Some cells can take in (**endocytosis**) or expel (**exocytosis**) solid particles or drops of fluid through the cell membrane. Endocytosis occurs in single-celled 'animals' such as *Paramecium* (p. 327) when they feed, or in certain white blood cells (phagocytes, p. 138) when they engulf bacteria, a process called **phagocytosis**. Exocytosis takes place in the cells of some glands. A secretion, e.g. a digestive enzyme, forms vacuoles or granules in the cytoplasm and these are expelled through the cell membrane to do their work outside the cell (Figs 6 and 7).

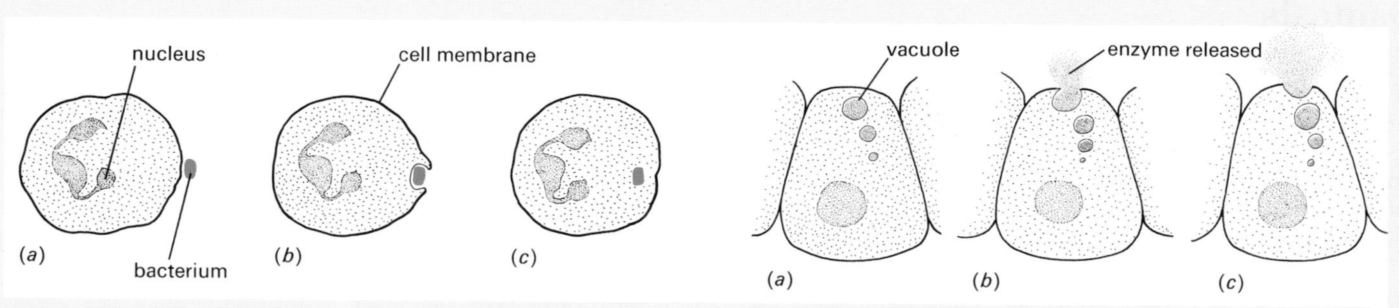

Fig. 6 Endocytosis (phagocytosis) in a white blood cell

Fig. 7 Exocytosis in a gland cell

Active transport

If diffusion were the only method by which a cell could take in substances, it would have no control over what went in or out. Anything that was more concentrated outside would diffuse into the cell whether it was harmful or not. Substances which the cell needed would diffuse out as soon as their concentration inside rose above that outside the cell. The cell membrane, however, has a great deal of control over the substances which enter and leave the cell.

In some cases, substances are taken into or expelled from the cell against the concentration gradient. For example, sodium ions may continue to pass out of a cell even though the concentration outside is greater than inside. The processes by which such reverse concentrations are produced are not properly understood and may be quite different for different substances but are all generally described as **active transport** (Fig. 8).

Anything which interferes with respiration, e.g. lack of oxygen or glucose, prevents active transport taking place. Also, during active transport, ATP is broken down to ADP (see p. 25). Thus it seems that active transport needs a supply of energy from respiration.

In some cases, a combination of active transport and controlled diffusion seems to occur. For example, sodium ions are thought to get into a cell by diffusion through special pores in the membrane and are expelled by a form of active transport. The reversed diffusion gradient for sodium ions created in this way is very important in the conduction of nerve impulses in nerve cells.

QUESTIONS

1 Look at Fig. 3 on page 138. If the symbol O_2 represents an oxygen molecule, explain why oxygen is entering the cells drawn on the left but leaving the cells on the right.

2 Look at Fig. 9 on page 154 representing one of the small air pockets (an alveolus) which form the lung.

(*a*) Suggest a reason why the oxygen and carbon dioxide are diffusing in opposite directions.

(*b*) What might happen to the rate of diffusion if the blood flow were to speed up?

3 List the ways in which a cell membrane might regulate the flow of substances into the cell.

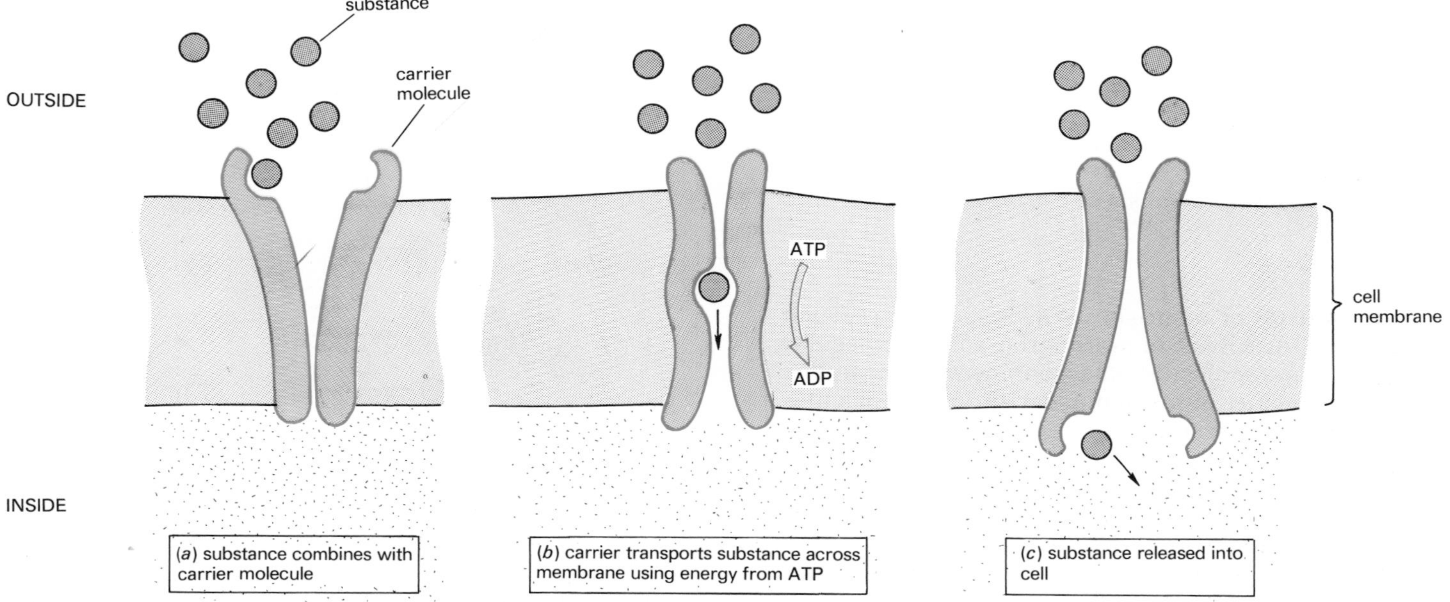

Fig. 8 Theoretical model to explain active transport

Osmosis

Osmosis is the special name used to describe the diffusion of water across a membrane, from a dilute solution to a more concentrated solution. In biology this usually means the diffusion of water into or out of cells. Osmosis is just one special kind of diffusion because it is only water molecules and their movement we are considering.

Figure 3 showed that molecules will diffuse from a region where there are a lot of them to a region where they are fewer in number; that is, from a region of highly concentrated molecules to a region of lower concentration.

Pure water has the highest possible concentration of water molecules; it is 100 per cent water molecules, all of them free to move (Fig. 10*a*).

Figure 9 shows a concentrated sugar solution, separated from a dilute solution by a membrane which allows water molecules to pass through. The dilute solution, in effect, contains more water molecules than the concentrated solution. As a result of this difference in concentration, water molecules will diffuse from the dilute to the concentrated solution.

The level of the concentrated solution will rise or, if it is confined in an enclosed space, its pressure will increase.

The membrane separating the two solutions is called **partially permeable** (sometimes 'selectively permeable' or 'semi-permeable') because it appears as if water molecules can pass through it more easily than sugar molecules can.

Osmosis, then, is the passage of water across a partially permeable membrane from a dilute to a concentrated solution.

This is all you need to know in order to understand the effects of osmosis in living organisms, but a more complete explanation is given below.

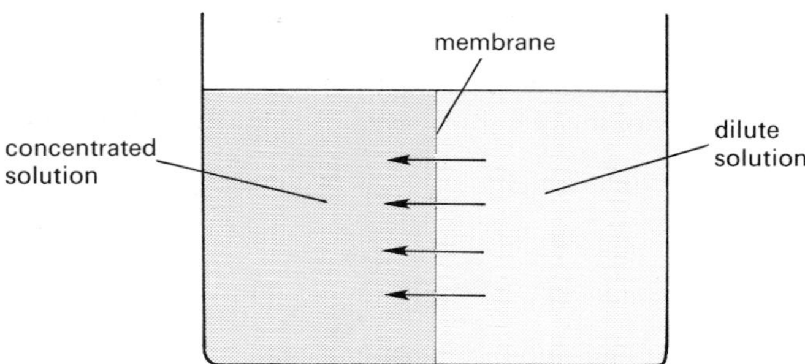

Fig. 9 The effect of osmosis

Explanation of osmosis When a substance, such as sugar, dissolves in water, the sugar molecules attract water molecules and combine with them. The water molecules now combined with sugar molecules are no longer free to move and so the effective concentration of free water molecules in the solution has been reduced (Fig. 10*b*). The more sugar molecules there are in a solution, the more water molecules will be 'tied up'. In other words, the more concentrated a solution is, the fewer free water molecules it will contain.

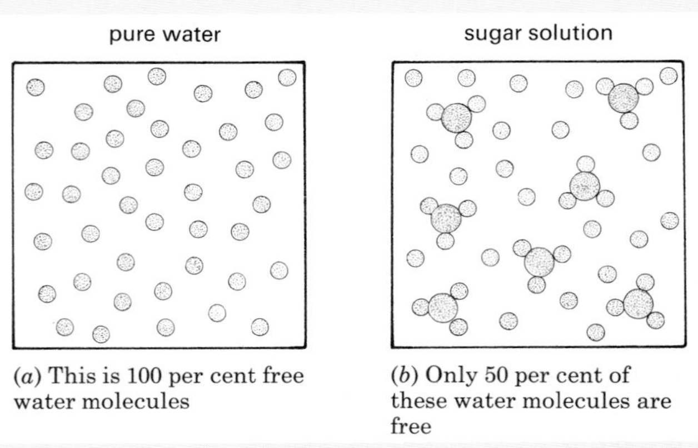

(*a*) This is 100 per cent free water molecules

(*b*) Only 50 per cent of these water molecules are free

Fig. 10 The effect of dissolved substances on water concentration

Figure 11 represents a dilute sugar solution separated from a concentrated sugar solution by a thin membrane. The membrane prevents the solutions mixing freely but allows individual water molecules and sugar molecules to pass through. There are more free water molecules on the left than on the right and so water molecules will pass more rapidly through the membrane from left to right than from right to left.

In a similar way, there will be a net movement of sugar molecules from right to left across the membrane because there are more of them on the right. However, the sugar molecules are larger and

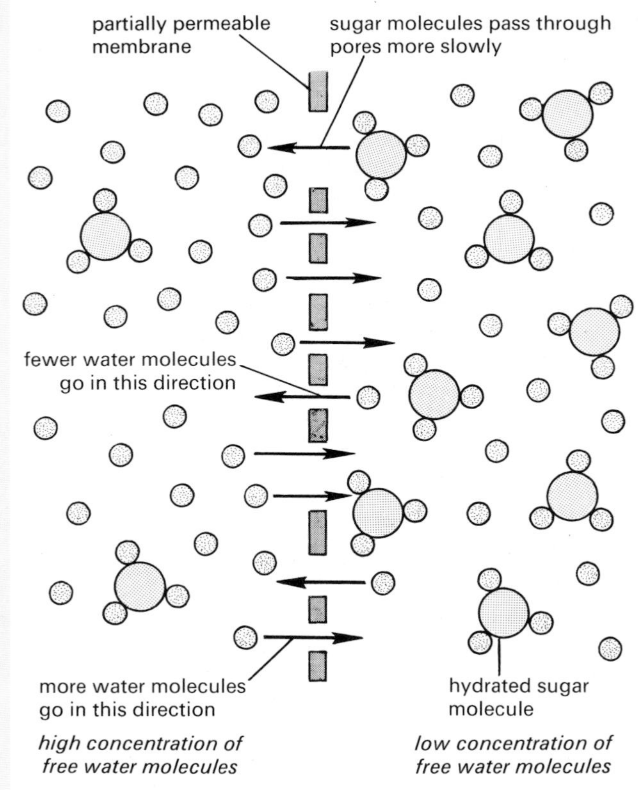

Fig. 11 The diffusion theory of osmosis

move more slowly than the water molecules. Thus the most obvious effect is the flow of water molecules from the dilute to the concentrated sugar solution.

A membrane which stops solutions from mixing freely but allows molecules below a certain size to pass through it is described as partially permeable.

The partially permeable membrane The non-living membranes used in osmosis experiments are usually made of cellulose acetate in the form of thin transparent sheets or tubes. The membranes have microscopic pores in them, too small to allow liquids to flow through freely but large enough to allow individual molecules through. Very large molecules such as starch or protein molecules cannot get through the pores, but small molecules, such as water, salts or sugars, can do so. Such a partially permeable membrane can be used for separating large molecules, such as proteins, from smaller molecules in solution (see Experiment 3, p. 39).

The pores in the artificial membrane are large enough to allow sugar molecules and water molecules through, so the membrane is not selective with respect to these two kinds of molecule. Nevertheless, in the model described above (Fig. 11) the differing concentrations of sugar and water molecules, and their relative rates of movement, cause water molecules to diffuse from the dilute to the concentrated solution, faster than the sugar molecules diffuse in the opposite direction. This makes it appear as if the membrane is selectively permeable to water molecules while resisting the passage of sugar molecules.

The cell membrane behaves like a partially permeable membrane to water and dissolved substances. The partial permeability may depend on pores in the cell membrane but the processes involved are far more complicated than in an artificial membrane and depend on the structure of the membrane and on living processes in the cytoplasm (see p. 34). The cell membrane contains lipids and proteins. Anything which denatures proteins, e.g. heat, also destroys the structure and the partially permeable properties of a cell membrane. If this happens, the cell will die as essential substances diffuse out of the cell and harmful chemicals diffuse in.

Water potential The water potential of a solution is a measure of whether it is likely to lose or gain water molecules from another solution. A dilute solution, with its high proportion of free water molecules, is said to have a higher water potential than a concentrated solution, because water will flow from the dilute to the concentrated solution (from a high potential to a low potential). Pure water has the highest possible water potential because water molecules will flow from it to any other aqueous solution, no matter how dilute.

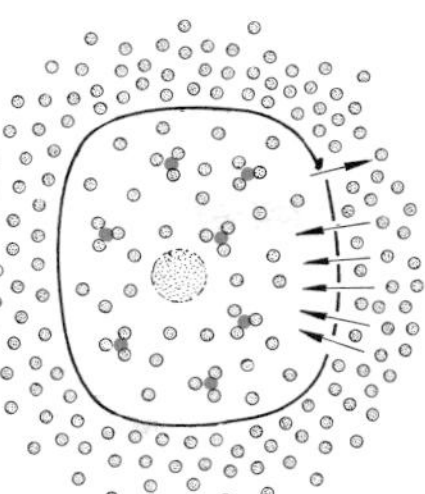

(*a*) There is a higher concentration of free water molecules outside the cell than inside, so water diffuses into the cell

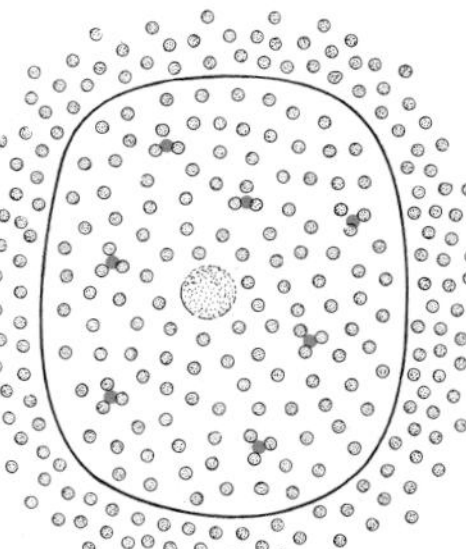

(*b*) The extra water makes the cell swell up.
(Note that molecules are really far too small to be seen at this magnification.)

Fig. 12 Osmosis in an animal cell

Animal cells

In Fig. 12 an animal cell is shown very simply. The coloured circles represent molecules in the cytoplasm. They may be sugar, salt or protein molecules. The grey circles represent water molecules.

The cell is shown surrounded by pure water. Nothing is dissolved in the water; it has 100 per cent concentration of water molecules. So the concentration of free water molecules outside the cell is greater than that inside and, therefore, water will diffuse into the cell by osmosis.

The membrane allows water to go through either way. So in our example, water can move into or out of the cell.

The cell membrane is partially permeable to most of the substances dissolved in the cytoplasm. So although the concentration of these substances inside may be high, they cannot diffuse freely out of the cell.

The water molecules move into and out of the cell, but because there are more of them on the outside, they will move in faster than they move out. The liquid outside the cell does not have to be 100 per cent pure water. As long as the concentration of water outside is higher than that inside, water will diffuse in by osmosis.

Water entering the cell will make it swell up, and unless the extra water is expelled in some way the cell will burst.

Conversely, if the cells are surrounded by a solution which is more concentrated than the cytoplasm, water will pass out of the cell by osmosis and the cell will shrink. (See Experiment 6, p. 41.) Excessive uptake or loss of water by osmosis may damage cells.

For this reason, it is very important that the cells in an animal's body are surrounded by a liquid which has the same concentration as the liquid inside the cells. The outside liquid is called 'tissue fluid' (see p. 142) and its concentration depends on the concentration of the blood. In vertebrate animals the blood's concentration is monitored by the brain and adjusted by the kidneys, as described on page 160.

By keeping the blood concentration within narrow limits, the concentration of tissue fluid remains more or less constant (see pp. 145 and 163) and the cells are not bloated by taking in too much water, or dehydrated by losing too much.

Plant cells

The cytoplasm of a plant cell and the cell sap in its vacuole contain salts, sugars and proteins which effectively reduce the concentration of free water molecules inside the cell. The cell wall is freely permeable to water and dissolved substances but the cell membrane of the cytoplasm is partially permeable. If a plant cell is surrounded by water or a solution more dilute than its contents, water will pass into the vacuole by osmosis. The vacuole will expand and press outwards on the cytoplasm and cell wall. The cell wall of a mature plant cell cannot be stretched, so there comes a time when the inflow of water is resisted by the unstretchable cell wall (Fig. 13).

This has a similar effect to inflating a soft bicycle tyre. The tyre represents the firm cell wall, the floppy inner tube is like the cytoplasm and the air inside corresponds to the vacuole. If enough air is pumped in, it pushes the inner tube against the tyre and makes the tyre hard. A plant cell with the vacuole pushing out on the cell wall is said to be **turgid** and the vacuole is exerting **turgor pressure** on the cell wall.

If all the cells in a leaf and stem are turgid, the stem will be firm and upright and the leaves held out straight. If the vacuoles lose water for any reason, the cells will lose their turgor and become **flaccid**. A leaf with flaccid cells will be limp and the stem will droop. A plant which loses water to this extent is said to be 'wilting' (Fig. 14).

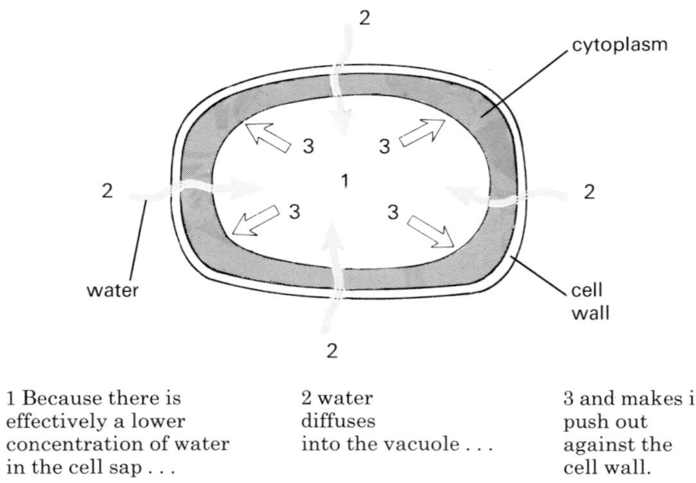

Fig. 13 Osmosis in a plant cell

(*b*) **Plant recovered after watering**

(*a*) **Plant wilting**

Fig. 14 Wilting

Water movement by osmosis Osmosis may contribute to the passage of water from one cell to another in a plant. Imagine two adjacent cells A and B (Fig. 15). A has a lower concentration of sugar in its cell sap and hence a higher water potential than B. The more dilute cell sap in A will thus force water by osmosis into cell B. The water entering cell B will dilute its cell sap and raise both its water potential and its turgor pressure, so tending to force water out into the next cell in line. Thus water passes from cell to cell down an osmotic gradient.

However, if cell B is fully turgid it is unable to expand and so can take in no more water even though it has a lower water potential than its neighbour A. In fact, if cell A is not fully turgid and still capable of expansion, the high wall pressure of cell B will force water out into cell A against the osmotic gradient. The movement of water between plant cells depends, therefore, not only on the water potential of their vacuoles but also on how turgid they are.

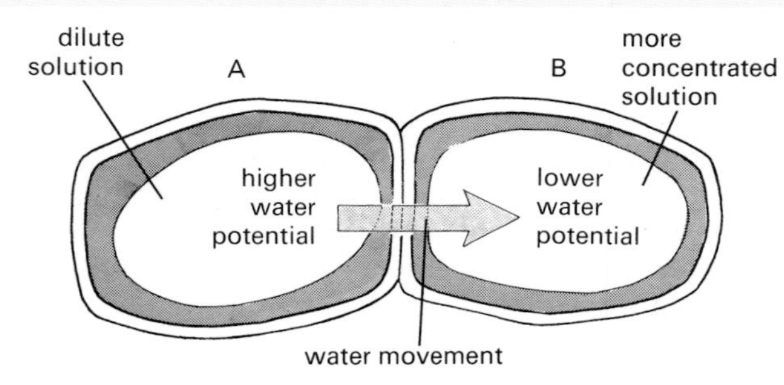

Fig. 15 Movement of water through plant cells

This kind of osmotic exchange between cells might account, to some extent, for the passage of water into the centre of a root, from a leaf vein to a palisade cell or from an epidermal cell to the guard cell of a stoma (p. 60). However, in most cases the cellulose cell walls will be saturated with water and it would be easier for a cell below its maximum turgor to absorb water by osmosis from the cell wall than from its neighbouring cell.

QUESTIONS

4 A 10 per cent solution of copper sulphate is separated by a partially permeable membrane from a 5 per cent solution of copper sulphate. Will water diffuse from the 10 per cent to the 5 per cent solution, or from the 5 per cent to the 10 per cent solution?

5 If a fresh beetroot is cut up, the pieces washed in water and then left for an hour in a beaker of water, little or no red pigment escapes from the cells into the water. If the beetroot is boiled first, the pigment does escape into the water. Bearing in mind the properties of a living cell membrane, offer an explanation for this difference.

6 When doing experiments with animal tissues (e.g. the experiment with muscle on page 30) they are usually bathed in Ringer's solution which has a concentration similar to that of blood or tissue fluid. Why do you think this is necessary?

7 Why does a dissolved substance reduce the number of 'free' water molecules in a solution?

8 When a plant leaf is in daylight, its cells make sugar from carbon dioxide and water (see p. 44). The sugar is at once turned into starch and deposited in plastids. What is the osmotic advantage of doing this? (Sugar is soluble in water, starch is not. See p. 16.)

PRACTICAL WORK

EXPERIMENTS ON DIFFUSION AND OSMOSIS

1. Diffusion of gases

Moisten eight small squares of red litmus paper with water and push them into a wide glass tube by means of a glass rod or piece of wire so that they are equally spaced out. The squares will stick to the inside of the glass because they are wet.

Close both ends of the tube with corks, one of which contains a pad of cotton wool soaked in ammonia solution (Fig. 16). Ammonia is an alkali which turns red litmus blue.

CAUTION Ammonia solution and ammonia vapour can be very harmful to the nose and eyes. The solutions must be made up and dispensed by the teacher.

Result The squares of litmus paper will gradually go blue, starting with the ones nearest the source of ammonia.

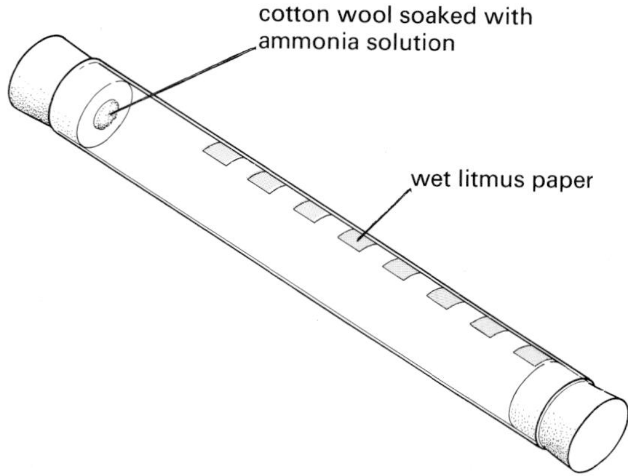

Fig. 16 Diffusion of ammonia in air

Interpretation Molecules of ammonia must have passed down the tube to reach the litmus squares. Since air currents have been prevented by closing the ends of the tube, the ammonia molecules must have travelled by diffusion.

2. Diffusion in a liquid

Diffusion in a liquid is slow and liable to be affected by convection and other movements in the liquid. In this experiment the water is 'kept still', so to speak, by dissolving gelatin in it.

Pour a warm 10 per cent solution of gelatin into a test-tube to half fill it. In a separate test-tube, colour a little of the liquid gelatin with methylene blue. When the first layer of gelatin has set, pour a little of the blue gelatin to make a thin layer on top of it. Once the blue layer is cold and firm, fill the rest of the test-tube with gelatin solution so that the blue layer is sandwiched between two clear layers (Fig. 17), and leave for a week.

Result After a week, the blue dye will be seen to have spread into the clear gelatin above and below the original blue layer.

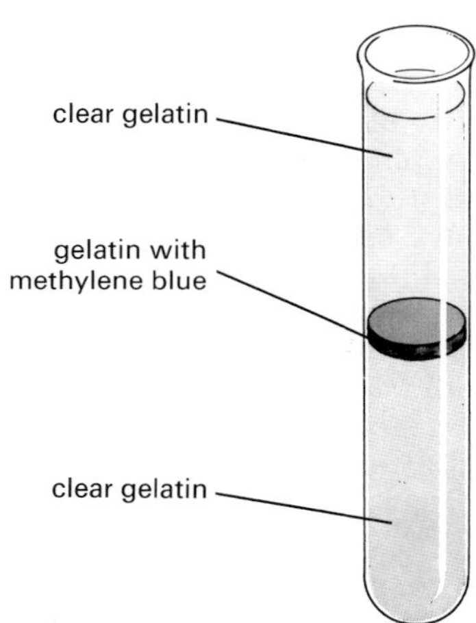

Fig. 17 Diffusion in a liquid

Interpretation The methylene blue molecules have passed, by diffusion, into the clear gelatin. They are not significantly affected by gravity since they have passed equally upwards and downwards.

3. Partial permeability (dialysis)

The next three experiments use 'Visking' dialysis tubing. It is made from cellulose and is partially permeable, allowing water molecules to diffuse through freely, but restricting the passage of dissolved substances to varying extents. It is used in kidney machines because it lets the small molecules of harmful waste products (e.g. urea, p. 160) out of the blood but retains the blood cells and large protein molecules.

Take a 15 cm length of dialysis tubing which has been soaked in water and tie a knot tightly at one end. Use a dropping pipette to partly fill the tubing with 1 per cent starch solution. Put the tubing in a test-tube and hold it in place with an elastic band as shown in Fig. 18. Rinse the tubing and test-tube under the tap to remove all traces of starch solution from the outside of the dialysis tube.

Fill the test-tube with water and add a few drops of iodine solution to colour the water yellow. Leave for 10–15 minutes.

Result The starch inside the dialysis tubing goes blue but the iodine outside stays yellow.

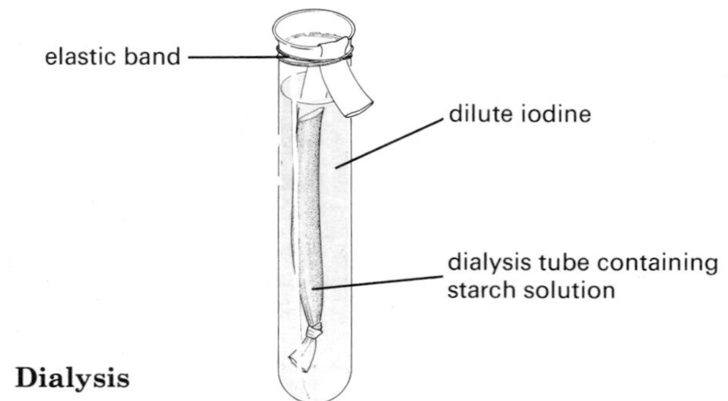
elastic band
dilute iodine
dialysis tube containing starch solution

Fig. 18 Dialysis

Interpretation The blue colour is characteristic of the reaction which takes place between starch and iodine, and is used as a test for starch (see p. 122). The results show that iodine molecules have passed through the dialysis tubing into the starch but the starch molecules have not moved out into the iodine. This is what we would expect if the dialysis tubing is partially permeable on the basis of its pore size. Starch molecules are very large (see p. 16) and probably cannot get through the pores. Iodine molecules are much smaller and can, therefore, get through.

4. Osmosis and turgor

Take a 20 cm length of dialysis tubing which has been soaked in water and tie a knot tightly at one end. Place 3 cm^3 of a strong sugar solution in the tubing using a graduated pipette (Fig. 19*a*) and then knot the open end of the tube (Fig. 19*b*). The partly filled tube should be quite floppy (Fig. 19*c*). Place the tubing in a test-tube of water for 30–45 minutes. After this time, remove the dialysis tubing from the water and look for any changes in how it looks or feels.

Result The tubing will now be firm, distended by the solution inside.

Interpretation The dialysis tubing is partially permeable and the solution inside has fewer free water molecules than outside. Water has, therefore, diffused in and increased the volume and the pressure of the solution inside.

This is a crude model of what is thought to happen to a plant cell when it becomes turgid. The sugar solution represents the cell sap and the dialysis tubing represents the cell membrane and cell wall combined.

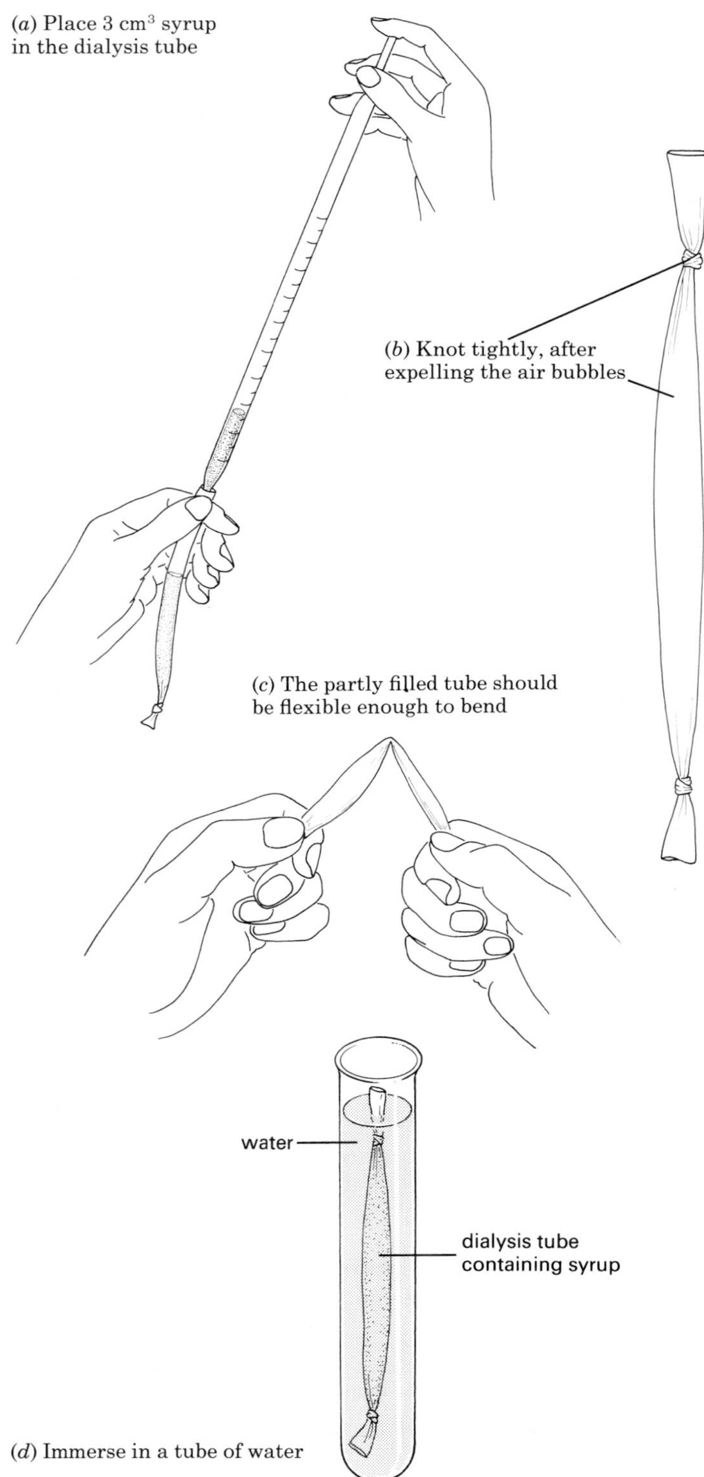
(*a*) Place 3 cm^3 syrup in the dialysis tube
(*b*) Knot tightly, after expelling the air bubbles
(*c*) The partly filled tube should be flexible enough to bend
water
dialysis tube containing syrup
(*d*) Immerse in a tube of water

Fig. 19 Experiment to illustrate turgor in a plant cell

5. Osmosis and water flow

Tie a knot in one end of a piece of soaked dialysis tubing and fill it with sugar solution as in the previous experiment. Then fit it over the end of a length of capillary tubing and hold it in place with an elastic band. Push the capillary tubing into the dialysis tubing until the sugar solution enters the capillary. Now clamp the capillary tubing so that the dialysis tubing is totally immersed in a beaker of water as shown in Fig. 20. Watch the level of liquid in the capillary tubing over the next 10 or 15 minutes.

Result The level of liquid in the capillary tube will be seen to rise.

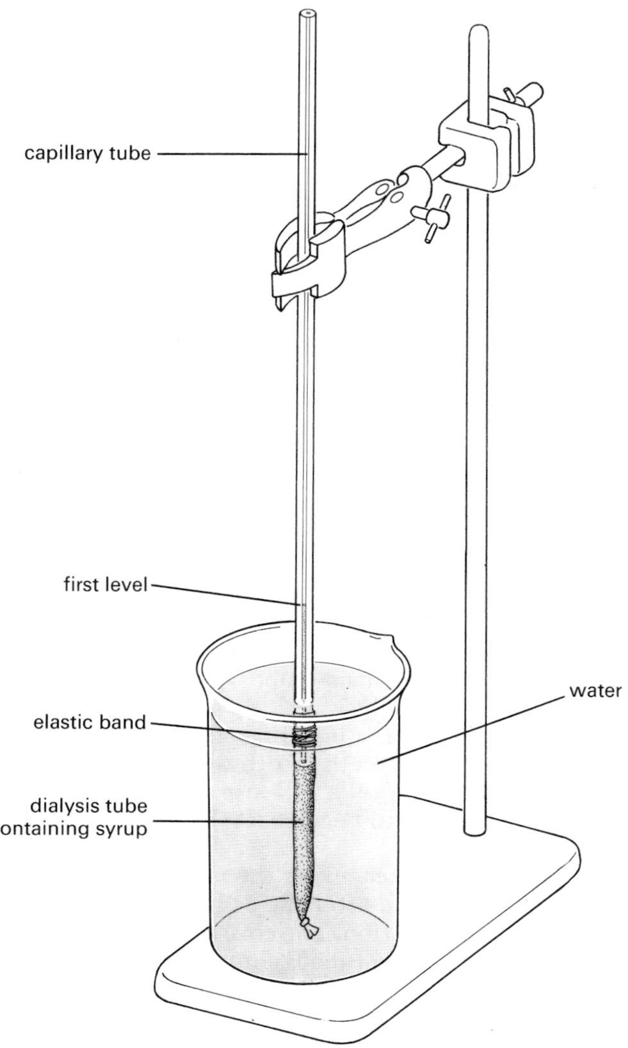

Fig. 20 Demonstration of osmosis

Interpretation Water must be passing into the sugar solution from the beaker. This is what you would expect when a concentrated solution is separated from water by a partially permeable membrane. The results are similar to those in Experiment 4 but instead of the expanding solution distending the dialysis tube, it escapes up the capillary.

A process similar to this might be partially responsible for moving water from the roots to the stem of a plant. (See Experiment 8, p. 79.)

6. Plasmolysis

Peel a small piece of epidermis (outer layer of cells) from a red area of a rhubarb stalk (Fig. 22*a*, p. 12). Place the epidermis on a slide with a drop of water and cover with a cover slip (Fig. 22*c*, p. 12). Place a 30 per cent solution of sugar at one edge of the cover slip with a pipette and then draw the solution under the cover slip by placing a piece of blotting-paper on the opposite side (Fig. 21). While you are doing this, study a small group of cells under the microscope and watch for any changes in their appearance.

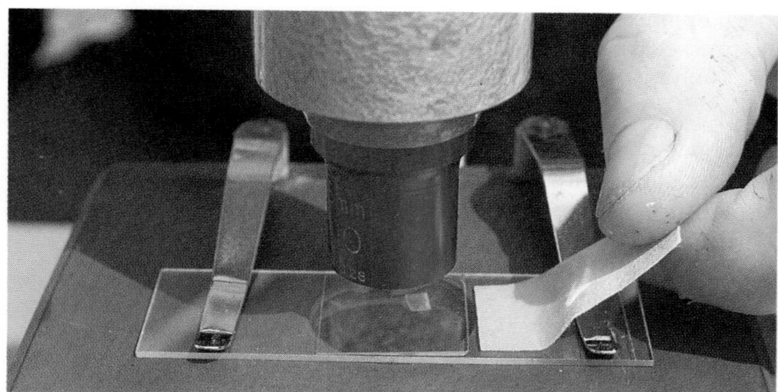

Fig. 21 Changing the water for sugar solution

Result The red cell sap will appear to shrink and get darker and pull the cytoplasm away from the cell wall leaving clear spaces. (It will not be possible to see the cytoplasm but its presence can be inferred from the fact that the red cell sap seems to have a distinct outer boundary in those places where it has separated from the cell wall.)

Interpretation The interpretation in terms of osmosis is outlined in Fig. 22. The cells are said to be **plasmolysed**.

The plasmolysis can be reversed by drawing water under the cover slip in the same way that you drew the sugar solution under. It may need two or three lots of water to flush out all the sugar. If you watch a group of cells, you should see their vacuoles expanding to fill the cells once again.

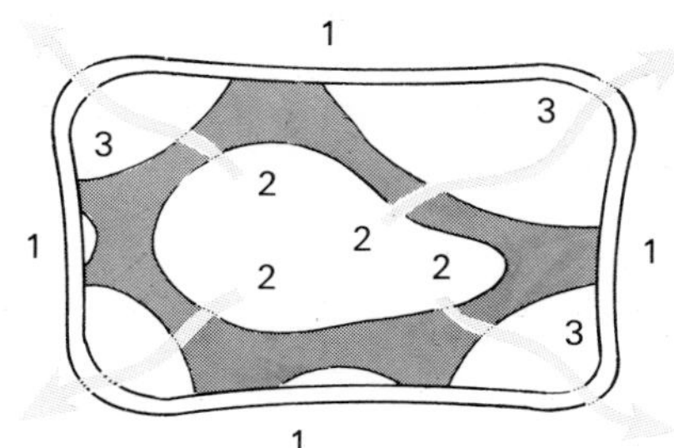

1 The solution outside the cell is more concentrated than the cell sap

2 Water diffuses out of the vacuole

3 The vacuole shrinks, pulling the cytoplasm away from the cell wall, leaving the cell flaccid

Fig. 22 Plasmolysis

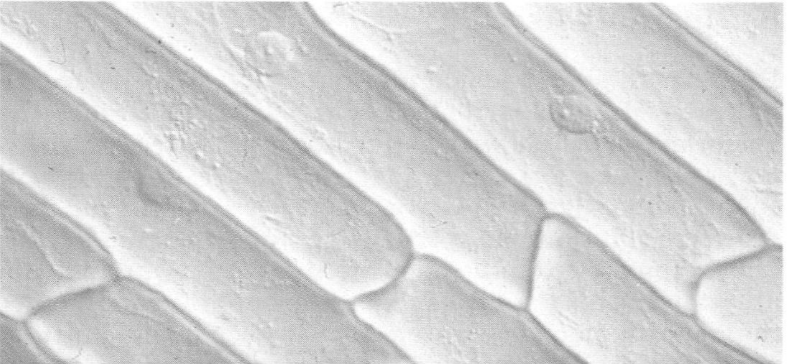

Fig. 23 (*a*) **Turgid cells** (×100). The cells are in a strip of epidermis from an onion scale. The cytoplasm is pressed against the inside of the cell wall by the vacuole. Nuclei can be seen in three of the cells.

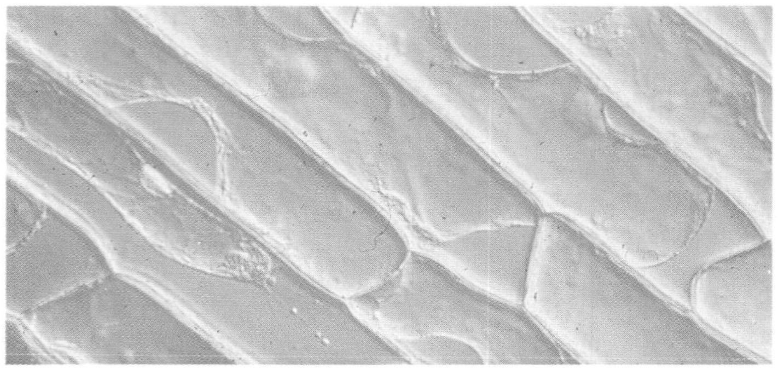

(*b*) **Plasmolysed cells** (×100). The same cells as they appear after treatment with salt solution. The vacuole has lost water by osmosis, shrunk and pulled the cytoplasm away from the cell wall.

Rhubarb is used for this experiment because the coloured cell sap shows up, but if rhubarb is not available the epidermis from an onion scale can be used with results similar to those in Fig. 23.

7. Turgor in potato tissue

Push a No. 4 or No. 5 cork borer into a large potato.

CAUTION Do not hold the potato in your hand but use a board as in Fig. 24*a*.

Push the potato tissue out of the cork borer using a pencil as in Fig. 24*b*. Prepare a number of potato cylinders in this way and choose the two longest (at least 50 mm). Cut these two accurately to the same length, e.g. 50, 60 or 70 mm. Measure carefully.

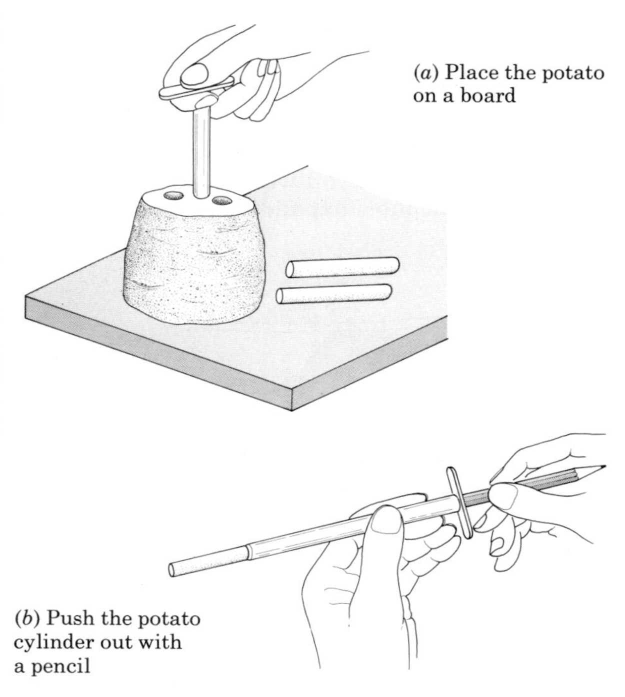

(*a*) Place the potato on a board
(*b*) Push the potato cylinder out with a pencil

Fig. 24 Obtaining cylinders of potato tissue

Label two test-tubes A and B and place a potato cylinder in each. Cover the potato tissue in tube A with water; cover the tissue in B with a 20 per cent sugar solution. Leave the tubes for a day.

After this time, remove the cylinder from tube A and measure its length. Notice also whether it is firm or flabby. Repeat this for the potato in tube B, but rinse it in water before measuring it.

Result The cylinder from tube A should have gained a millimetre or two and feel firm. The cylinder from tube B should be a millimetre or two shorter and feel flabby.

Interpretation If the potato cells were not fully turgid at the beginning of the experiment, they would take up water by osmosis (tube A), and cause an increase in length.

In tube B, the sugar solution is stronger than the cell sap of the potato cells, so these cells will lose water by osmosis. The cells will lose their turgor and the potato cylinder will become flabby and shorter.

An alternative to measuring the potato cores is to weigh them before and after the 24 hours immersion in water or sugar solution. The core in tube A should gain weight and that in tube B should lose weight. It is important to blot the cores dry with a paper towel before each weighing.

Whichever method is used, it is a good idea to pool the results of the whole class since the changes may be quite small. A gain in length of 1 or 2 mm might be due to an error in measurement, but if most of the class record an increase in length, then experimental error is unlikely to be the cause.

These experiments and eight others are. described in greater detail in *Experimental Work in Biology* (see p. 364).

QUESTIONS

9 In Experiment 1 (Fig. 16) it may take 20 minutes for the last piece of litmus paper on the right to go blue. What changes could you make in the design of the experiment in order to reduce this time interval?

10 In Experiment 5 (Fig. 20), what do you think would happen (*a*) if a much stronger sugar solution was placed in the cellulose tube, (*b*) if the beaker contained a weak sugar solution instead of water, (*c*) if the sugar solution was in the beaker and the water was in the cellulose tube?

11 In Experiment 4 (Fig. 19), what might happen if the cellulose tube filled with sugar solution was left in the water for several hours?

12 Figure 22 explains why the vacuole shrinks in Experiment 6. Give a brief explanation of why it swells up again when the cell is surrounded by water.

13 An alternative interpretation of the results of Experiment 3 on page 39 might be that the dialysis tubing allowed molecules (of any size) to pass in but not out. Describe an experiment to test this possibility and say what results you would expect (*a*) if it were correct and (*b*) if it were false.

14 In Experiment 5 on page 40, the column of liquid accumulating in the capillary tube exerts an ever-increasing pressure on the solution in the dialysis tube. Bearing this in mind and assuming a very long capillary, at what stage would you expect the net flow of water from the beaker into the dialysis tubing to cease?

Models

One way in which scientists explain their theories is by making 'models'. Experiment 4 on page 40 was a 'model' of a plant cell to help explain the phenomenon of turgor. The model omitted many features of a plant cell and concentrated solely on the partially permeable membrane and the solution inside it. The model probably helped you to visualize what might happen in a cell, but in many respects it is a poor model. For example, if you heated the dialysis tubing to 50 °C the experiment would still work. A cell membrane heated to this temperature would lose its selective properties. The dialysis tubing, therefore, fails to model the cell membrane very accurately.

Very often a model is no more than a kind of diagram to represent the way a hypothesis might work in practice. Figure 8 on page 35 is a model to explain how the hypothesis of active transport might work. It does not mean that we know for sure that it works in this way. The change of shape in the drawing of the carrier molecule is an over-simplification and may not be representative of what really happens.

Since we cannot get inside cells to see what is happening, most of the explanations given at this level are in the form of models which become more detailed and more refined as more information emerges. The danger is that a model may be taken to represent the 'truth' rather than an informed guess. For example, in recent years, the model for the structure of the cell membrane has changed from Fig. 25*a* to Fig. 25*b*. Those of us who thought that 25*a* *was* the structure of the cell membrane were quite surprised to learn that it was merely a model.

In a book of this kind it is tedious to have to keep reading 'It is thought that . . .', 'There is evidence to show . . .', 'Some biologists believe . . .', etc. So, many of the explanations of function are offered simply as if they were 'Gospel Truth'. Try to bear in mind that what is generally believed at the moment is based on our current level of knowledge.

Some explanations, such as the description of photosynthesis on page 44, have stood the test of time and, although they have been elaborated, have not been seriously challenged. Others, such as the account of active transport on page 35, are quite likely to undergo fundamental changes in the next few years.

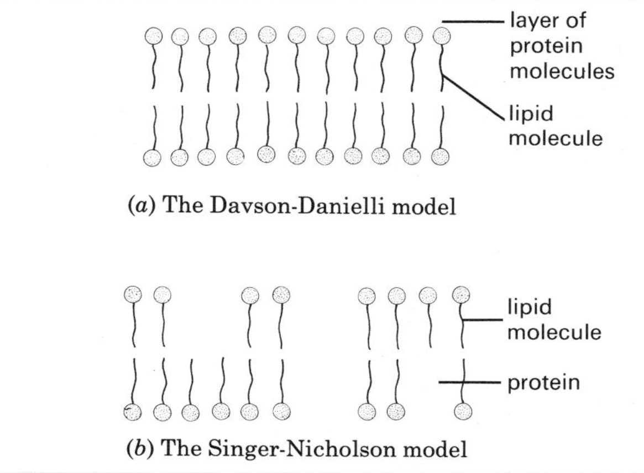

Fig. 25 Models of the cell membrane's structure

CHECK LIST

- **Diffusion is the result of molecules of liquid, gas or dissolved solid moving about.**
- **The molecules of a substance diffuse from a region where they are very concentrated to a region where they are less concentrated.**
- **Substances may enter cells by simple diffusion, controlled diffusion, active transport or endocytosis.**
- **Osmosis is the diffusion of water through a partially permeable membrane.**
- **Water diffuses from a dilute solution of salt or sugar to a concentrated solution because the concentrated solution contains fewer free water molecules.**
- **Cell membranes are partially permeable and cytoplasm and cell sap contain many substances in solution.**
- **Cells take up water from dilute solutions but lose water to concentrated solutions because of osmosis.**
- **Osmosis maintains turgor in plant cells.**

5 Photosynthesis and Nutrition in Plants

PHOTOSYNTHESIS

Introductory explanation; definition and equation.

PRACTICAL WORK

Experiments to test photosynthesis: principles of experimental design; destarching and testing a leaf for starch; need for chlorophyll, light and carbon dioxide; production of oxygen.

PROCESS OF PHOTOSYNTHESIS

The process in leaf cells.

GASEOUS EXCHANGE

Uptake of carbon dioxide and output of oxygen in photosynthesis. Uptake of oxygen and output of carbon dioxide in respiration. Compensation point.

ADAPTATION OF LEAVES FOR PHOTOSYNTHESIS

Shape and internal structure.

EFFECTS OF EXTERNAL FACTORS ON RATE OF PHOTOSYNTHESIS

Light intensity; carbon dioxide concentration; temperature. Limiting factors.

USE OF PHOTOSYNTHETIC PRODUCTS IN PLANTS

Storage, energy and growth.

SOURCES OF MINERAL ELEMENTS

Nitrates, sulphates and fertilizers.

WATER CULTURES

Experiment to demonstrate the importance of certain minerals.

All living organisms need food. They need it as a source of raw materials to build new cells and tissues as they grow. They also need food as a source of energy. Food is a kind of 'fuel' which drives essential living processes and brings about chemical changes (see pp. 24 and 114). Animals take in food, digest it, and use the digested products to build their tissues or to produce energy.

Plants also need energy and raw materials but, apart from a few insect-eating species, plants do not appear to take in food. The most likely source of their raw materials would appear to be the soil. However, experiments show that the weight gained by a growing plant is far greater than the weight lost by the soil it is growing in. So there must be additional sources of raw materials.

These additional sources can only be water and air. A hypothesis to explain the source of food in a plant is that it **makes it** from air, water and soil salts. Carbohydrates (p. 15) contain the elements carbon, hydrogen and oxygen, as in glucose ($C_6H_{12}O_6$). The carbon and oxygen could be supplied by carbon dioxide (CO_2) from the air, and the hydrogen could come from the water (H_2O) in the soil. The nitrogen and sulphur needed for making proteins (p. 14) could come from nitrates and sulphates in the soil.

This building up of complex food molecules from simpler substances is called a **synthesis** and it needs enzymes and energy to make it happen. The enzymes are present in the plant's cells and the energy for the first stages in the synthesis comes from sunlight. The process is, therefore, called **photosynthesis** ('photos' means 'light'). There is evidence to suggest that the green substance **chlorophyll**, in the chloroplasts of plant cells, plays a part in photosynthesis. Chlorophyll absorbs sunlight and makes the energy from sunlight available for chemical reactions. Thus, in effect, the function of chlorophyll is to convert light energy to chemical energy.

Our working hypothesis for photosynthesis is, therefore, the building-up of food compounds from carbon dioxide and water by green plants using energy from sunlight which is absorbed by chlorophyll.

A chemical equation for photosynthesis would be

$$\underset{\text{carbon dioxide}}{6CO_2} + \underset{\text{water}}{6H_2O} \xrightarrow{\text{light energy}} \underset{\text{glucose}}{C_6H_{12}O_6} + \underset{\text{oxygen}}{6O_2}$$

In order to keep the equation simple, glucose is shown as the food compound produced. This does not imply that it is the only substance synthesized by photosynthesis.

PRACTICAL WORK

EXPERIMENTS TO TEST PHOTOSYNTHESIS

The design of biological experiments was discussed on page 30 and this should be revised before studying the next section. It would also be helpful to read the section on 'Hypothesis testing' (p. 31) if you have not already done so.

A **hypothesis** is an attempt to explain certain observations. In this case the hypothesis is that plants make their food by photosynthesis. The equation given below is one way of stating the hypothesis and is used here to show how it might be tested.

$$6CO_2 + 6H_2O \xrightarrow[\text{chlorophyll}]{\text{sunlight}} C_6H_{12}O_6 + 6O_2$$

$6CO_2$	$6H_2O$	$C_6H_{12}O_6$	$6O_2$
↑ uptake of carbon dioxide	↑ uptake of water	↓ production of sugar (or starch)	↓ release of oxygen

If photosynthesis is going on in a plant, then the leaves should be producing sugars. In many leaves, as fast as sugar is produced, it is turned into starch. Since it is easier to test for starch than for sugar, we regard the production of starch in a leaf as evidence that photosynthesis has taken place.

The first three experiments described below are designed to see if the leaf can make starch without carbon dioxide, or sunlight, or chlorophyll, in turn. If the photosynthesis story is sound, then the lack of any one of these three conditions should stop photosynthesis, and so stop the production of starch. If a leaf without a supply of carbon dioxide can still produce starch, then the hypothesis is no good and must be altered or rejected.

In designing the experiments, it is very important to make sure that only *one* condition is altered. If, for example, the method of keeping light from a leaf also cuts off its carbon dioxide supply, it would be impossible to decide whether it was the lack of light or lack of carbon dioxide which stopped the production of starch. To make sure that the experimental design has not altered more than one condition, a **control** is set up in each case. This is an identical situation, except that the condition missing from the experiment, e.g. light, carbon dioxide or chlorophyll, is present in the control (see p. 30).

Destarching a plant If the production of starch is your evidence that photosynthesis is taking place, then you must make sure that the leaf does not contain any starch at the beginning of the experiment. This is done by **destarching** the leaves. It is not possible to remove the starch chemically, without damaging the leaves, so a plant is destarched simply by leaving it in darkness for 2 or 3 days. Potted plants are destarched by leaving them in a dark cupboard for a few days. In the darkness, any starch in the leaves will be changed to sugar and carried away from the leaves to other parts of the plant. For plants in the open, the experiment is set up on the day before the test. During the night, most of the starch will be removed from the leaves. Better still, wrap the leaves in aluminium foil for 2 days while they are still on the plant. Then test one of the leaves to see that no starch is present.

Testing a leaf for starch Iodine solution (yellow) and starch (white) form a deep blue colour when they mix. The test for starch, therefore, is to add iodine solution to a leaf to see if it goes blue. First, however, the leaf has to be treated as follows:

1. The leaf is detached and dipped in boiling water for half a minute. This kills the cytoplasm and destroys the enzymes in it, so preventing any further chemical changes. It also makes the cell more permeable to iodine solution.
2. The leaf is boiled in alcohol (ethanol), using a water bath (Fig. 1), until all the chlorophyll is dissolved out. This turns the leaf whitish and makes any colour changes caused by iodine easier to see.

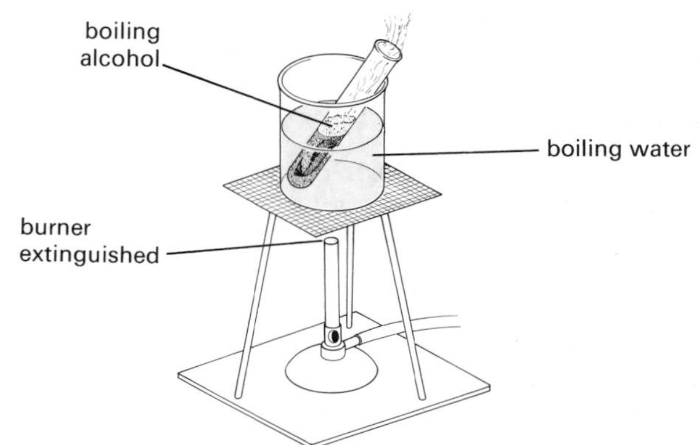

Fig. 1 To remove chlorophyll from a leaf

3. Alcohol makes the leaf brittle and hard, but it can be softened by dipping it once more into the hot water. Then it is spread flat on a white surface such as a glazed tile.
4. Iodine solution is placed on the leaf. Any parts which turn blue have starch in them. If no starch is present, the leaf is merely stained yellow or brown by the iodine.

Experiment 1. Is chlorophyll necessary for photosynthesis?

It is not possible to remove chlorophyll from a leaf without killing it, and so a **variegated** leaf, which has chlorophyll only on patches, is used. A leaf of this kind is shown in Fig. 2*a*. The white part of the leaf serves as the experiment, because it lacks chlorophyll, while the green part with chlorophyll is the control. After being destarched, the leaf—still on the plant—is exposed to daylight for a few hours. It is then removed from the plant, drawn carefully to show where the chlorophyll is (i.e. the green parts), and tested for starch as described above.

Result Only the parts that were previously green turn blue with iodine. The parts that were white stain brown (Fig. 2*b*).

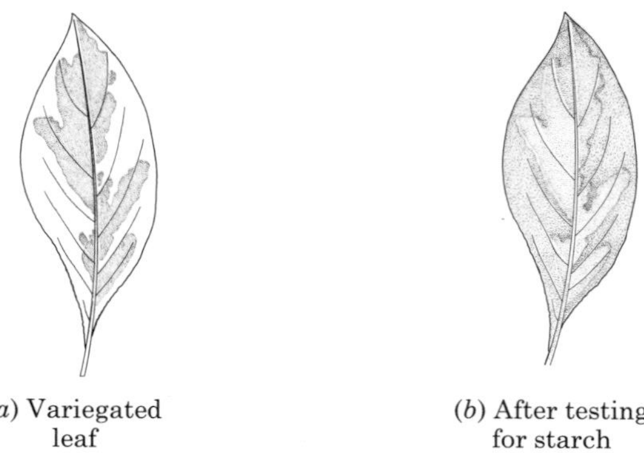

Fig. 2 To show that chlorophyll is necessary

Interpretation Since starch is present only in the parts which originally contained chlorophyll, it seems reasonable to suppose that chlorophyll is needed for photosynthesis.

It must be remembered, however, that there are other possible interpretations which this experiment has not ruled out; for example, starch could be made in the green parts and sugar in the white parts. Such alternative explanations could be tested by further experiments.

Experiment 2. Is light necessary for photosynthesis?

A simple shape is cut out from a piece of aluminium foil to make a stencil which is attached to a destarched leaf (Fig. 3*a*). After 4 to 6 hours of daylight, the leaf is removed from the plant and tested for starch.

Result Only the areas which had received light go blue with iodine.

Interpretation As starch has not formed in the areas which received no light, it seems that light is needed for starch formation and thus for photosynthesis.

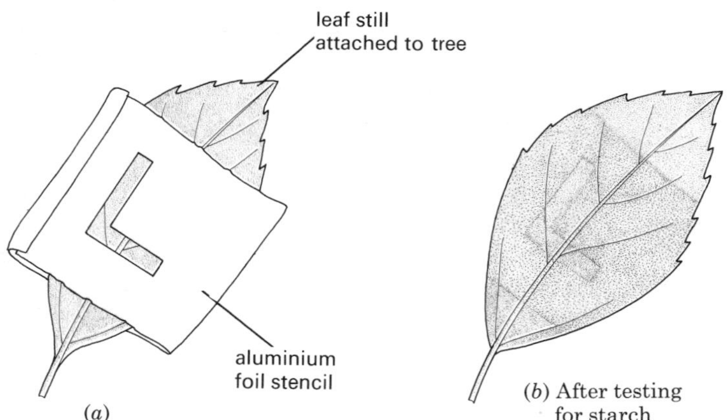

Fig. 3 To show that light is necessary

You could argue that the aluminium foil had stopped carbon dioxide from entering the leaf and that it was shortage of carbon dioxide rather than absence of light which prevented photosynthesis taking place. A further control could be designed, using transparent material instead of aluminium foil for the stencil.

Experiment 3. Is carbon dioxide needed for photosynthesis?

Two destarched potted plants are watered and the shoots enclosed in polythene bags. One of the bags contains soda-lime to absorb carbon dioxide from the air (the experiment). The other has sodium hydrogencarbonate solution to produce carbon dioxide (the control), as shown in Fig. 4. Both plants are placed in light for several hours and a leaf from each is then removed and tested for starch.

Result The leaf which had no carbon dioxide does not turn blue. The one from the polythene bag containing carbon dioxide does turn blue.

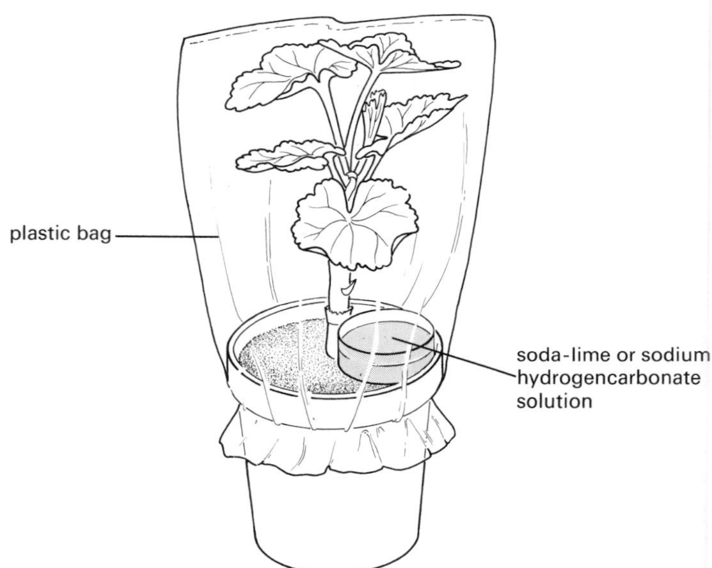

Fig. 4 To show that carbon dioxide is necessary

Interpretation The fact that starch was made in the leaves which had carbon dioxide, but not in the leaves which had no carbon dioxide, suggests that this gas must be necessary for photosynthesis. The control rules out the possibility that high humidity or high temperature in the plastic bag prevents normal photosynthesis.

Experiment 4. Is oxygen produced during photosynthesis?

A short-stemmed funnel is placed over some Canadian pondweed in a beaker of water. A test-tube filled with water is placed upside-down over the funnel stem (Fig. 5). The funnel is raised above the bottom of the beaker to allow the water to circulate. The apparatus is placed in sunlight, and bubbles of gas soon appear from the cut stems and collect in the test-tube. A control experiment should be set up in a similar way but placed in a dark cupboard. Little or no gas will be collected. When sufficient gas has collected from the plant in the light, the test-tube is removed and a glowing splint inserted.

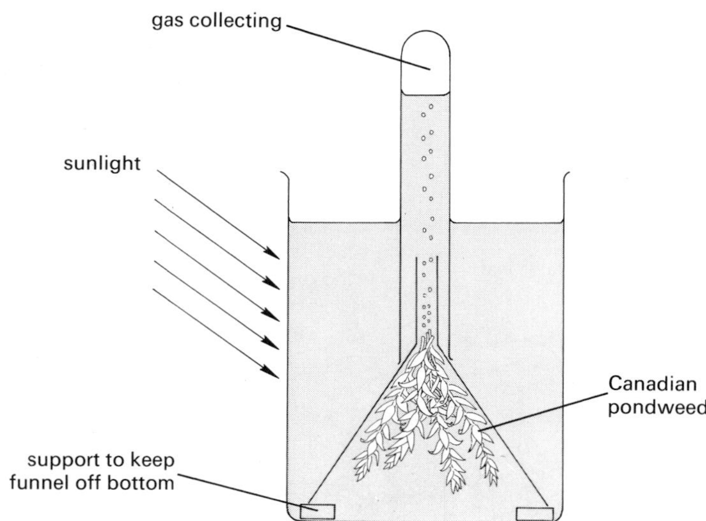

Fig. 5 To show that oxygen is produced

Result The glowing splint bursts into flames.

Interpretation The relighting of a glowing splint does not prove that the gas collected in the test-tube is *pure* oxygen, but it does show that it contains extra oxygen and this must have come from the plant. The oxygen is given off only in the light.

You should bear in mind that water contains dissolved oxygen, carbon dioxide and nitrogen. These gases may diffuse in or out of the bubbles as they pass through the water and collect in the test-tube. The composition of the gas in the test-tube may not be the same as that in the bubbles leaving the plant.

The results of these four experiments support the hypothesis of photosynthesis as stated on page 44 and as represented by the equation. Starch formation (our evidence for photosynthesis) does not take place in the absence of light, chlorophyll or carbon dioxide, and oxygen production occurs only in the light.

If starch or oxygen production had occurred in the absence of any one of these conditions, we should have to change our hypothesis about the way plants obtain their food. Bear in mind, however, that although our results support the photosynthesis theory, they do not prove it. For example, it is now known that many stages in the production of sugar and starch from carbon dioxide do not need light (the 'dark' reaction).

QUESTIONS

1 Which of the following are needed for starch production in a leaf: (*a*) carbon dioxide, (*b*) oxygen, (*c*) nitrates, (*d*) water, (*e*) chlorophyll, (*f*) soil, (*g*) light?
2 In Experiment 1 (concerning the need for chlorophyll), why was it not necessary to set up a separate control experiment?
3 What is meant by 'destarching' a leaf? Why is it necessary to destarch leaves before setting up some of the photosynthesis experiments?
4 In Experiment 3 (concerning the need for carbon dioxide), what were the functions of (*a*) the soda-lime, (*b*) the sodium hydrogencarbonate, (*c*) the polythene bag?
5 Why do you think a pondweed, rather than a land plant, is used for Experiment 4 (concerning production of oxygen)? In what way might this choice make the results less useful?
6 A green plant makes sugar from carbon dioxide and water. Why do we not try the experiment of depriving a plant of water to see if that stops photosynthesis?
7 Does the method of destarching a plant take for granted the results of Experiment 2? Explain your answer.
8 In Experiment 2, an extra control was suggested to see whether the aluminium foil stencil had prevented carbon dioxide as well as light from getting into a leaf. If the stencil was made of clear plastic, (*a*) should its effect differ from that of the aluminium foil stencil, and (*b*) what result would you expect (i) if the stencil *had* interfered with the supply of carbon dioxide and (ii) if it *had not*?

THE PROCESS OF PHOTOSYNTHESIS

Although the details of photosynthesis vary in different plants, the hypothesis as stated in this chapter has stood up to many years of experimental testing and is universally accepted. The next section describes how photosynthesis takes place in a plant.

The process takes place mainly in the cells of the leaves and is summarized in Fig. 6. In land plants water is absorbed from the soil by the roots and carried in the water vessels of the veins, up the stem to the leaf. Carbon dioxide is absorbed from the air through the stomata (pores in the leaf, see p. 60). In the leaf cells, the carbon dioxide and water are combined to make sugar; the energy for this reaction comes from sunlight which has been absorbed by the green pigment **chlorophyll**. The chlorophyll is present in the **chloroplasts** of the leaf cells and it is inside the chloroplasts that the reaction takes place. Chloroplasts (Fig. 6*d*) are small, green structures present in the cytoplasm of the leaf cells. Chlorophyll is the substance which gives leaves and stems their green colour. It is able to absorb energy from light and use it to split water molecules into hydrogen and oxygen (the 'light' reaction). The oxygen escapes from the leaf and the hydrogen molecules are added to carbon dioxide molecules to form sugar.

There are four types of chlorophyll which may be present in various proportions in different species.

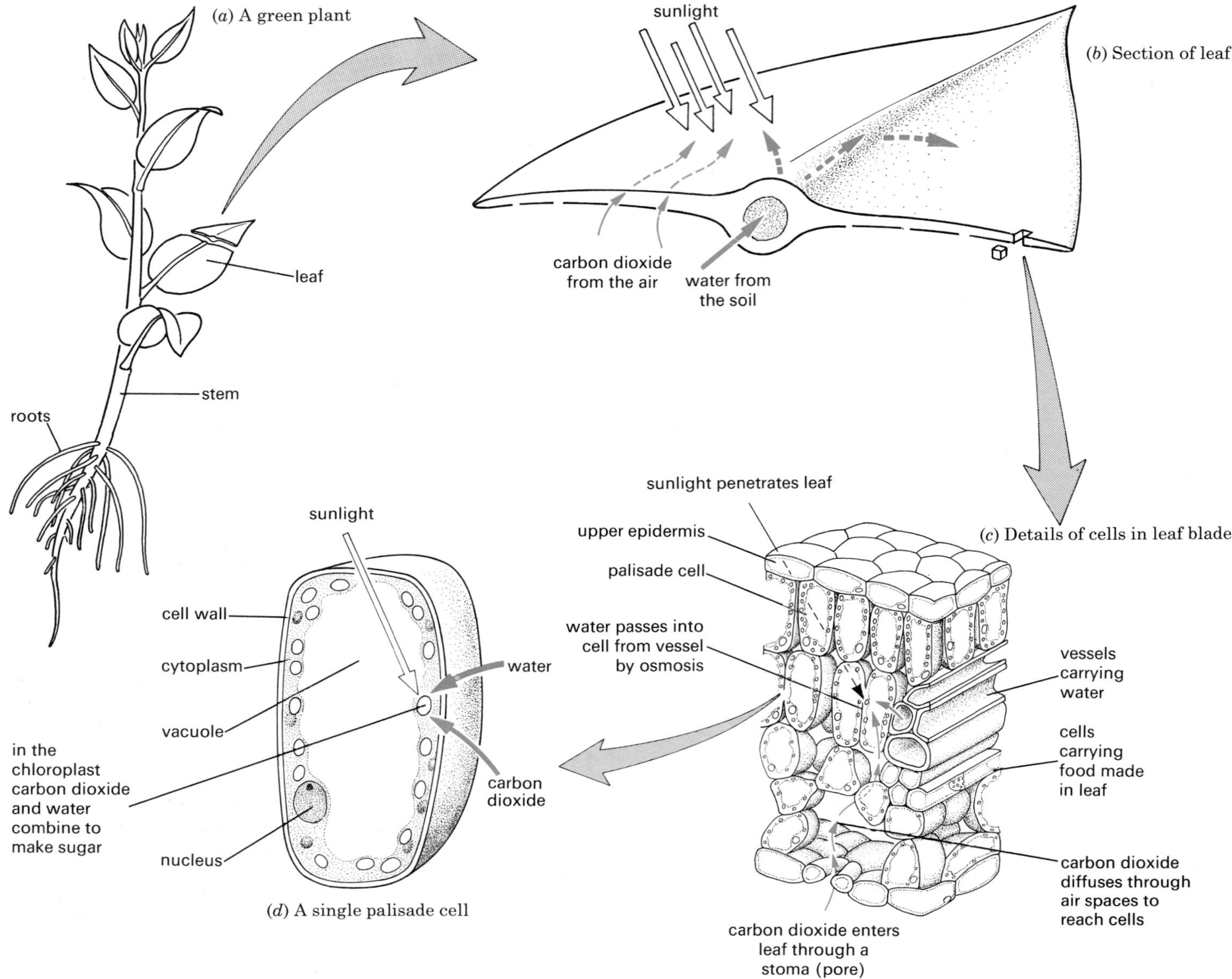

Fig. 6 Photosynthesis in a leaf

There are also a number of photosynthetic pigments, other than chlorophyll, which may mask the colour of chlorophyll even when it is present, e.g. the brown and red pigments which occur in certain seaweeds.

QUESTIONS

9 What substances must a plant take in, in order to carry on photosynthesis? Where does it get each of these substances from?
10 Look at Fig. 7*a* on page 61. Identify the palisade cells, the spongy mesophyll cells and the cells of the epidermis. In which of these would you expect photosynthesis to occur (*a*) most rapidly, (*b*) least rapidly, (*c*) not at all? Explain your answer.
11 (*a*) What provides a plant with energy for photosynthesis?
(*b*) What chemical process provides a plant with energy to carry on all other living activities?

GASEOUS EXCHANGE IN PLANTS

Air contains the gases, nitrogen, oxygen, carbon dioxide and water vapour. Plants and animals take in or give out these last three gases and this process is called **gaseous exchange**.

You can see from the equation for photosynthesis (p. 44) that one of its products is oxygen. Therefore, in daylight, when photosynthesis is going on in green plants, they will be taking in carbon dioxide and giving out oxygen. This exchange of gases is the opposite of that resulting from respiration (p. 24) but it must not be thought that green plants do not respire. The energy they need for all their living processes—apart from photosynthesis—comes from respiration and this is going on all the time, using up oxygen and producing carbon dioxide.

During the daylight hours, plants are photosynthesizing as well as respiring, so that all the carbon dioxide produced by respiration is used up by photosynthesis. At the same time, all the oxygen needed by respiration is provided by photosynthesis. Only when the rate of photosynthesis is faster than the rate of respiration will carbon dioxide be taken in and the excess oxygen be given out (Fig. 7).

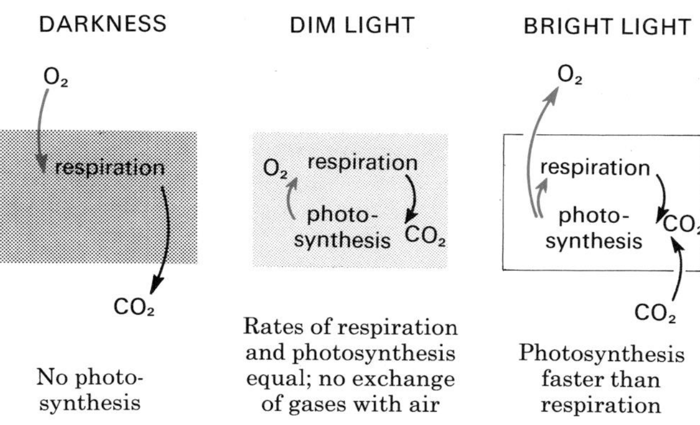

Fig. 7 Respiration and photosynthesis

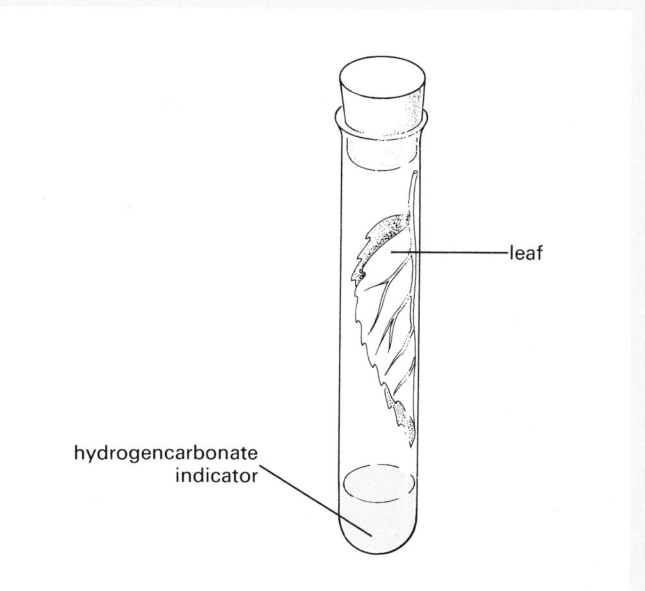

Fig. 8 Gaseous exchange during photosynthesis and respiration

Compensation point As the light intensity increases during the morning and fades during the evening, there will be a time when the rate of photosynthesis exactly matches the rate of respiration. At this point, there will be no net intake or output of carbon dioxide or oxygen. This is the compensation point. The sugar produced by photosynthesis exactly compensates for the sugar broken down by respiration.

Experiment 5. Gaseous exchange during photosynthesis

Wash three test-tubes first with tap-water, then with distilled water and finally with hydrogencarbonate indicator. Then place 2 cm^3 hydrogencarbonate indicator (see p. 363) in each tube.

Place a green leaf in tubes 1 and 2 so that it is held against the walls of the tube and does not touch the indicator (Fig. 8). Close the three tubes with bungs. Cover tube 1 with aluminium foil and place all three in a rack in direct sunlight, or a few centimetres from a bench lamp, for about 40 minutes.

Result The indicator (which was originally orange) should not change colour in tube 3, the control; that in tube 1, with the leaf in darkness, should turn yellow; and in tube 2 with the illuminated leaf, the indicator should be scarlet or purple.

Interpretation Hydrogencarbonate indicator is a mixture of dilute sodium hydrogencarbonate solution with the dyes cresol red and thymol blue. It is a pH indicator in equilibrium with the atmospheric carbon dioxide, i.e. its original colour represents the acidity produced by the carbon dioxide in the air. Increase in atmospheric carbon dioxide makes it more acid and it changes colour from orange to yellow. Decrease in atmospheric carbon dioxide makes it less acid and causes a colour change to red or purple.

The results, therefore, provide evidence that in darkness (tube 1) leaves produce carbon dioxide (from respiration), while in light (tube 2) they use up more carbon dioxide in photosynthesis than they produce in respiration. Tube 3 is the control, showing that it is the presence of the leaf which causes a change in the atmosphere in the test-tube.

The experiment can be criticized on the grounds that the hydrogencarbonate indicator is not a specific test for carbon dioxide but will respond to any change in acidity or alkalinity. In tube 2 there would be the same change in colour if the leaf produced an alkaline gas such as ammonia, and in tube 1, any acid gas produced by the leaf would turn the indicator yellow. However, a knowledge of the metabolism of the leaf suggests that these are less likely events than changes in the carbon dioxide concentration.

QUESTIONS

12 What gases would you expect a leaf to be (i) taking in, (ii) giving out, (*a*) in bright sunlight, (*b*) in darkness?

13 Measurements on a leaf show that it is giving out carbon dioxide and taking in oxygen. Does this prove that photosynthesis is *not* going on in the leaf? Explain your answer.

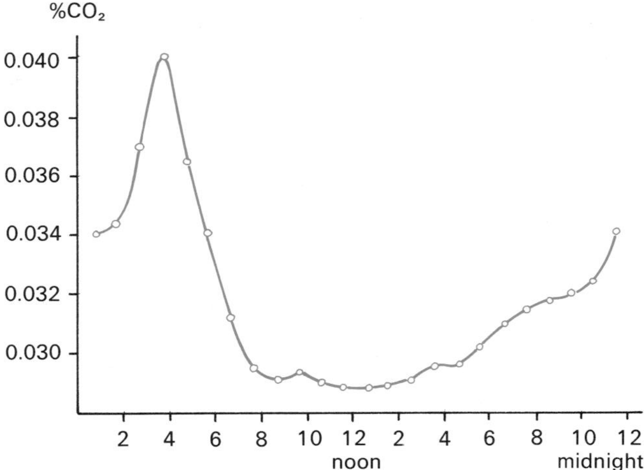

Fig. 9 Daily changes in concentration of carbon dioxide one metre above a plant crop. (From Verma and Rosenberg, *Span*, 1979.)

14 Figure 9 is a graph showing the average daily change in the carbon dioxide concentration, 1 metre above an agricultural crop in July. From what you have learned about photosynthesis and respiration, try to explain the changes in the carbon dioxide concentration.

15 How could you adapt the experiment with hydrogen-carbonate indicator to find the light intensity which corresponded to the compensation point?

How would you expect the compensation points to differ between plants growing in a wood and those growing in a field?

Adaptation of leaves for photosynthesis

When biologists say that something is **adapted**, they mean that its structure is well suited to its function. The detailed structure of the leaf is described on pages 59 to 61, and although there are wide variations in leaf shape the following general statements apply to a great many leaves, and are illustrated in Figs 6*b* and *c*.

1. Their broad, flat shape offers a large surface area for absorption of sunlight and carbon dioxide.
2. Most leaves are thin and the carbon dioxide has to diffuse across only short distances to reach the inner cells.
3. The large spaces between cells inside the leaf provide an easy passage through which carbon dioxide can diffuse.
4. There are many stomata (pores) in the lower surface of the leaf. These allow the exchange of carbon dioxide and oxygen with the air outside.
5. There are more chloroplasts in the upper (palisade) cells than in the lower (spongy mesophyll) cells. The palisade cells, being on the upper surface, will receive most sunlight and this will reach the chloroplasts without being absorbed by too many cell walls.
6. The branching network of veins provides a good water supply to the photosynthesizing cells. No cell is very far from a water-conducting vessel in one of these veins.

Although photosynthesis takes place mainly in the leaves, any part of the plant which contains chlorophyll will photosynthesize. Many plants have green stems in which photosynthesis takes place.

The rate of photosynthesis

The rate of the light reaction will depend on the light intensity. The brighter the light, the faster will water molecules be split in the chloroplasts. The 'dark' reaction will be affected by temperature. A rise in temperature will increase the rate at which carbon dioxide is combined with hydrogen to make carbohydrate.

Limiting factors If you look at Fig. 10*a*, you will see that an increase in light intensity does indeed speed up photosynthesis, but only up to a point. Beyond that point, any further increase in light intensity has only a small effect. This limit on the rate of increase could be because all available chloroplasts are fully occupied in light absorption. So, no matter how much the light intensity increases, no more light can be absorbed and used. Alternatively, the limit could be imposed by the fact that there is not enough carbon dioxide in the air to cope with the increased supply of hydrogen atoms produced by the light reaction. Or, it may be that low temperature is restricting the rate of the 'dark' reaction.

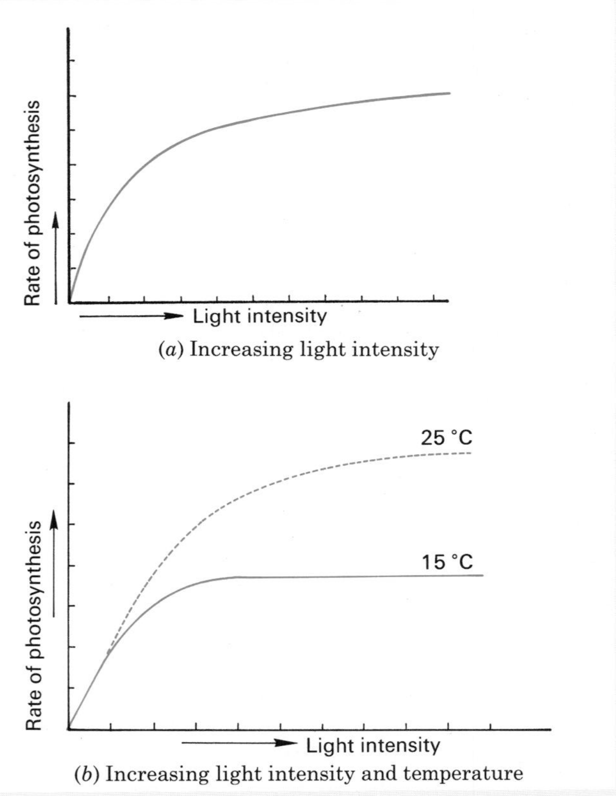

Fig. 10 Limiting factors in photosynthesis

Figure 10*b* shows that, if the temperature of a plant is raised, then the effect of increased illumination is not limited so much. Thus, in Fig. 10*a*, it seems likely that the increase in the rate of photosynthesis could have been limited by the temperature. Any one of the external factors, temperature, light intensity or carbon dioxide concentration, may limit the effects of the other two. A temperature rise may cause photosynthesis to speed up, but only to the point where the light intensity limits further increase. In such conditions, the external factor which restricts the effect of the others is called the **limiting factor**.

Since there is only 0.03 per cent of carbon dioxide in the air, it might seem that shortage of carbon dioxide could be an important limiting factor. Indeed, experiments do show that an increase in carbon dioxide concentration does allow a faster rate of photosynthesis. However, recent work in plant physiology has shown that the extra carbon dioxide affects reactions other than photosynthesis.

The main effect of extra carbon dioxide is to slow down the rate of oxidation of sugar by a process called **photorespiration** and this produces the same effect as an increase in photosynthesis.

Although carbon dioxide concentration limits photosynthesis only indirectly, artificially high levels of carbon dioxide in greenhouses do effectively increase yields of crops (Fig. 11).

Greenhouses also maintain a higher temperature and so reduce the effect of low temperature as a limiting factor.

Fig. 11 Effect of extra carbon dioxide on growth. Only 12 lettuces from the CO_2 enriched greenhouse will fit in a standard box compared with 15 from the control greenhouse.

The concept of limiting factors does not apply only to photosynthesis. Adding fertilizer to the soil, for example, may increase crop yields, but only up to the point where the roots can take up all the nutrients and the plant can build them into proteins, etc. The uptake of mineral ions is limited by the absorbing area of the roots, rates of respiration, aeration of the soil and availability of carbohydrates from photosynthesis.

Currently there is debate about whether athletic performance is limited by the ability of the heart and lungs to supply oxygenated blood to muscles, or by the ability of the muscles to take up and use the oxygen.

The role of the stomata The stomata (p. 60) in a leaf may affect the rate of photosynthesis according to whether they are open or closed. When photosynthesis is taking place, carbon dioxide in the leaf is being used up and its concentration falls. At low concentrations of carbon dioxide, the stomata will open. Thus, when photosynthesis is most rapid, the stomata are likely to be open, allowing carbon dioxide to diffuse into the leaf. When the light intensity falls, photosynthesis will slow down and the build-up of carbon dioxide from respiration will make the stomata close. In this way, the stomata are normally regulated by the rate of photosynthesis rather than photosynthesis being limited by the stomata. However, if the stomata close during the daytime as a result of excessive water loss from the leaf, their closure will restrict photosynthesis by preventing the inward diffusion of atmospheric carbon dioxide.

Normally the stomata are open in the daytime and closed at night. Their closure at night, when intake of carbon dioxide is not necessary, reduces the loss of water vapour from the leaf (p. 73).

Experiment 6. The effect of light intensity on the rate of photosynthesis

Fill a beaker or jar with tap-water and add about 5 cm^3 saturated sodium hydrogencarbonate solution. (This is to maintain a good supply of carbon dioxide.)

Select a pondweed shoot about 5–10 cm long. Partly prize open a small paper-clip and slide it over the tip of the shoot and drop the shoot into the jar of water. The paper-clip should hold the shoot under water as shown in Fig. 12.

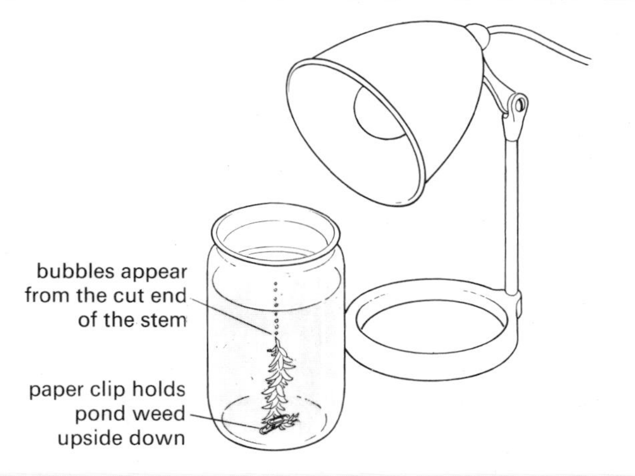

Fig. 12 Light intensity and oxygen production

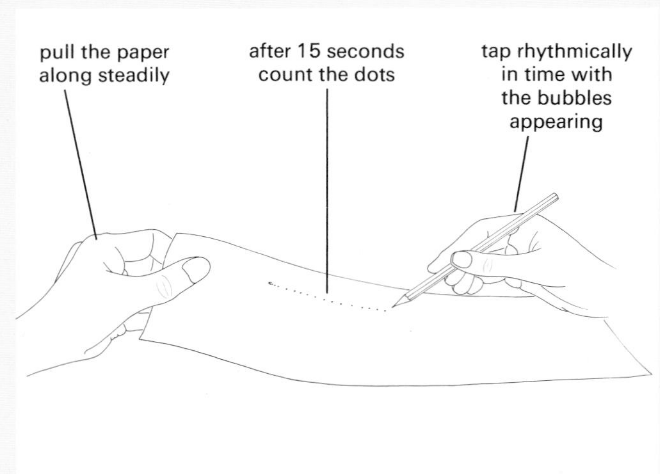
pull the paper along steadily
after 15 seconds count the dots
tap rhythmically in time with the bubbles appearing

Fig. 13 Estimating the rate of bubble production

Switch on a bench lamp and bring it close to the jar. After a minute or two, bubbles should appear from the cut end of the stem. If they do not, try a different piece of pondweed.

When the bubbles are appearing steadily, switch off the bench lamp and observe any change in the production of bubbles.

Now place the lamp about 25 cm from the jar. Switch on the lamp and try to count the number of bubbles coming off in a minute. Move the lamp to about 10 cm from the jar and count the bubbles again.

If the bubbles appear too rapidly to count, try tapping a pen or pencil on a sheet of paper at the same rate as the bubbles appear and get your partner to slide the paper slowly along for 15 seconds. Then count the dots. (Fig. 13.)

Result When the light is switched off, the bubbling should stop. The rate of bubbling should be faster when the lamp is closer to the plant.

Interpretation Assuming that the bubbles contain oxygen produced by photosynthesis, it seems that an increase in light intensity produces an increase in the rate of photosynthesis. We are assuming also that the bubbles do not change in size during the experiment. A fast stream of small bubbles might represent the same volume of gas as a slow stream of large bubbles.

Land plants Experiment 5 can be modified to investigate the effect of light intensity on the rate of photosynthesis. Two tubes with a leaf and hydrogen-carbonate indicator are set up as described in the experiment (Fig. 8) and placed at 10 cm and 20 cm respectively from the light source. The shorter the time the indicator takes to turn red or purple, the faster is the rate of carbon dioxide uptake. This is a measure of the rate of photosynthesis.

THE PLANT'S USE OF PHOTOSYNTHETIC PRODUCTS

The glucose molecules produced by photosynthesis are quickly built up into starch molecules and added to the growing starch granules in the chloroplast. If the glucose concentration was allowed to increase in the mesophyll cells of the leaf, it could disturb the osmotic balance between the cells (p. 38). Starch is a relatively insoluble compound and so does not alter the osmotic potential of the cell contents.

The starch, however, is steadily broken down to sucrose (p. 16) and this soluble sugar is transported out of the cell into the food-carrying cells (see p. 64) of the leaf veins. These veins will distribute the sucrose to all parts of the plant which do not photosynthesize, e.g. the growing buds, the ripening fruits, the roots and the underground storage organs.

The cells in these regions will use the sucrose in a variety of ways (Fig. 14).

Respiration The sugar can be used to provide energy. It is oxidized by respiration (p. 24) to carbon dioxide and water, and the energy released is used to drive other chemical reactions such as the building up of proteins described below.

Storage Sugar which is not needed for respiration is turned into starch and stored. Some plants store it as starch grains in the cells of their stems or roots. Other plants such as the potato or parsnip have special storage organs (tubers) for holding the reserves of starch (p. 102). Sugar may be stored in the fruits of some plants; grapes, for example, contain a large amount of glucose.

Synthesis of other substances As well as sugars for energy and starch for storage, the plant needs cellulose for its cell walls, lipids for its cell membranes, proteins for its cytoplasm and pigments for its flower petals, etc. All these substances are built up (synthesized) from the sugar molecules and other molecules produced in photosynthesis.

By joining hundreds of glucose molecules together, the long chain molecules of cellulose (Fig. 3, p. 16) are built up and added to the cell walls.

Amino acids (see p. 14) are made by combining **nitrogen** with sugar molecules or smaller carbohydrate molecules. These amino acids are then joined together to make the proteins which form the enzymes and the cytoplasm of the cell. The nitrogen for this synthesis comes from **nitrates** which are absorbed from the soil by the roots.

Proteins also need **sulphur** molecules and these are absorbed from the soil in the form of **sulphates** (SO_4). **Phosphorus** is needed for nucleic acids (p. 17) and for energy transfer reactions (p. 25). It is taken up as **phosphates** (PO_4).

The chlorophyll molecule needs **magnesium** (Mg). This metallic element is also obtained in salts from the soil (see the salts listed under 'Water cultures'.

Many other elements, e.g. iron, manganese, boron, in very small quantities, are also needed for healthy growth. These are often referred to as **trace elements**.

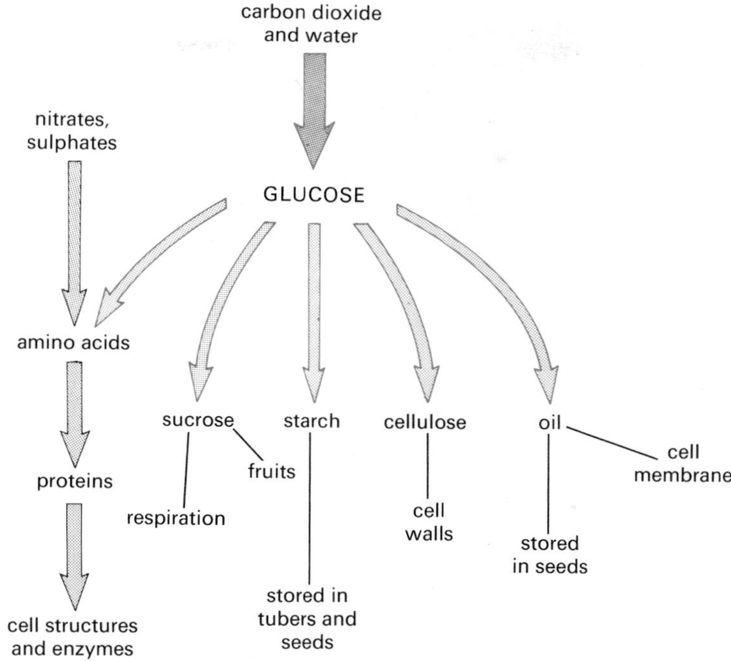

Fig. 14 Green plants can make all the materials they need from carbon dioxide, water and salts.

The metallic and non-metallic elements are all taken up in the form of their ions (p. 17).

All these chemical processes, such as the uptake of salts and the building up of proteins, need energy from respiration to make them happen.

QUESTIONS

16 What substances does a green plant need to take in, to make (*a*) sugar, (*b*) proteins? What must be present in the cells to make reactions (*a*) and (*b*) work?
17 A molecule of carbon dioxide enters a leaf cell at 4 p.m. and leaves the same cell at 6 p.m. What is likely to have happened to the carbon dioxide molecule during the 2 hours it was in the leaf cell?
18 In a partially controlled environment such as a greenhouse, (*a*) how could you alter the external factors to obtain maximum photosynthesis, (*b*) which of these alterations might not be cost-effective?

THE SOURCES OF MINERAL ELEMENTS

The mineral elements needed by plants are absorbed from the soil in the form of salts. For example, a plant's needs for potassium (K) and nitrogen (N) might be met by absorbing the salt **potassium nitrate** (KNO_3). Salts like this come originally from rocks which have been broken down to form the soil (p. 254). They are continually being taken up from the soil by plants or washed out of the soil by rain. They are replaced partly from the dead remains of plants and animals. When these organisms die and their bodies decay, the salts they contain are released back into the soil. This process is explained in some detail, for nitrates, on page 247.

In arable farming, the ground is ploughed and whatever is grown is removed. There are no dead plants left to decay and replace the mineral salts. The farmer must replace them by spreading animal manure, sewage sludge or artificial fertilizers in measured quantities over the land.

Three manufactured fertilizers in common use are ammonium nitrate, superphosphate and compound NPK.

Ammonium nitrate (NH_4NO_3) The formula shows that ammonium nitrate is a rich source of nitrogen but no other plant nutrients. It is sometimes mixed with calcium carbonate to form a compound fertilizer such as 'Nitro-chalk'.

Superphosphates These fertilizers are mixtures of minerals. They all contain calcium and phosphate and some have sulphate as well.

Compound NPK fertilizer 'N' is the chemical symbol for nitrogen, 'P' for phosphorus and 'K' for potassium. NPK fertilizers are made by mixing ammonium sulphate, ammonium phosphate and potassium chloride in varying proportions. They provide the ions of nitrate, phosphate and potassium which are the ones most likely to be below the optimum level in an agricultural soil.

Water cultures

It is possible to demonstrate the importance of the various mineral elements by growing plants in water cultures. A full water culture is a solution containing the salts which

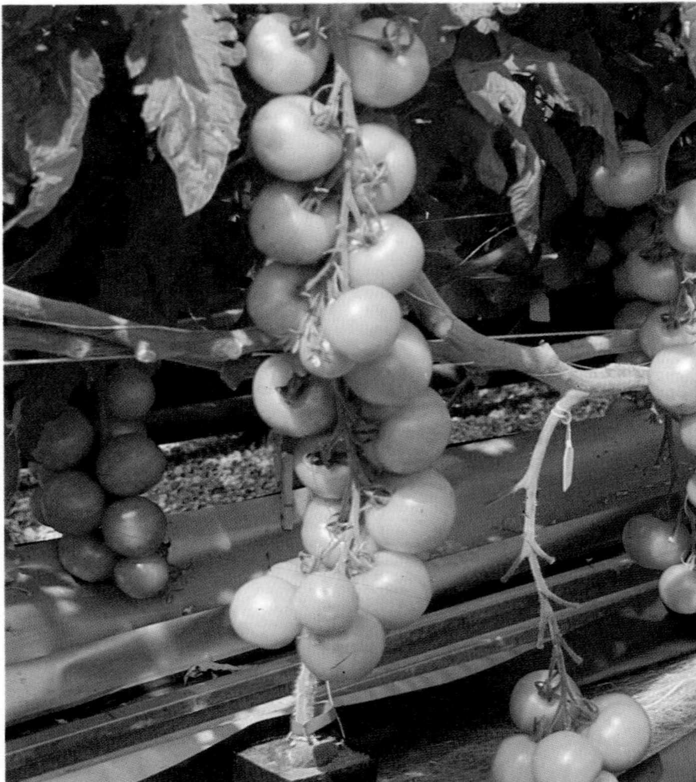

Fig. 15 Soil-less culture. The tomato plants are growing in a nutrient solution circulated through troughs of polythene. The network of roots can be seen in the polythene.

provide all the necessary elements for healthy growth, e.g.

potassium nitrate for potassium and nitrogen
magnesium sulphate for magnesium and sulphur
potassium phosphate for potassium and phosphorus
calcium nitrate for calcium and nitrogen

From these elements, plus the carbon dioxide, water and sunlight needed for photosynthesis, a green plant can make all the substances it needs for a healthy existence.

Some branches of horticulture, e.g. growing of glasshouse crops, make use of water cultures on a large scale. Tomatoes may be grown with their roots in flat polythene tubes. The appropriate water culture solution is pumped along these tubes (Fig. 15). This method has the advantage that the yield is increased and the need to sterilize the soil each year, to destroy pests, is eliminated. This kind of technique is sometimes described as hydroponics or soil-less culture.

Experiment 7. The importance of different mineral elements

Place wheat seedlings in test-tubes containing water cultures as shown in Fig. 16. Cover the tubes with aluminium foil to keep out light and so stop green algae from growing in the solution (p. 329). Some of the solutions

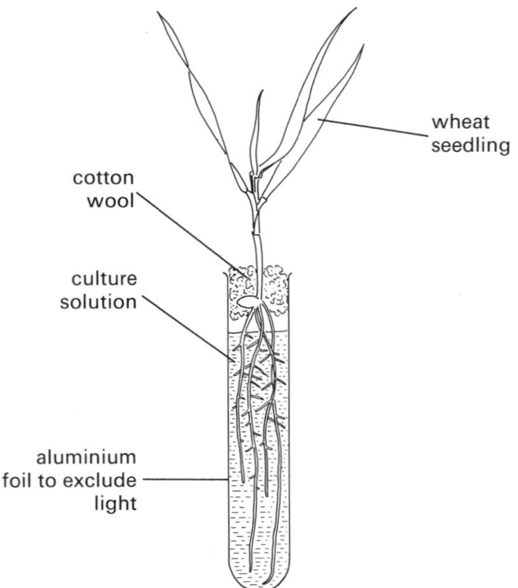

Fig. 16 To set up a water culture

have one of the elements missing. For example, in the list of chemicals above, magnesium chloride is used instead of magnesium sulphate and so the solution will lack sulphur. In a similar way, solutions lacking nitrogen, potassium and phosphorus can be prepared.

Leave the seedlings to grow in these solutions for a few weeks, keeping the tubes topped up with distilled water.

Result Figure 17 shows the kind of results which might be expected from wheat seedlings. Generally, the plants in a complete culture will be tall and sturdy, with large, dark green leaves. The plants lacking nitrogen will usually be stunted and have small, pale leaves. In the absence of magnesium, chlorophyll cannot be made, and these plants will be small with yellow leaves.

Figure 18 shows the results of a larger scale experiment.

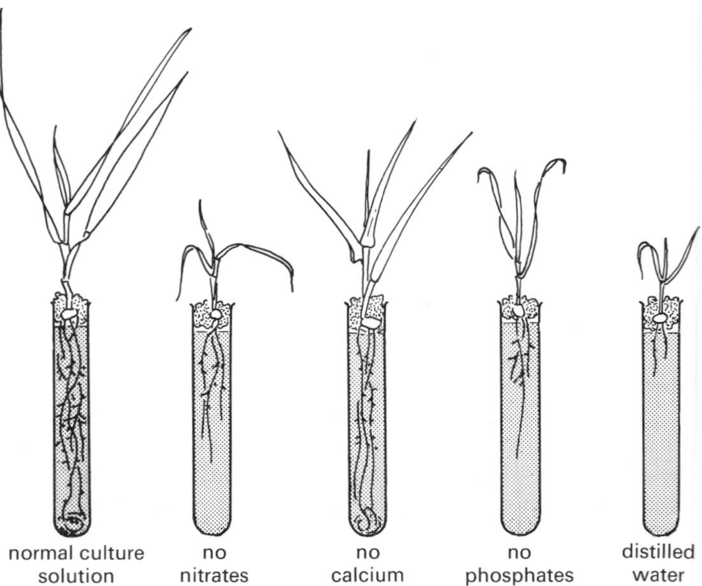

Fig. 17 Result of water culture experiment

Fig. 18 Effect of mineral salts on growth. The chrysanthemum plants on the left were planted at the same time as those on the right but have been deprived of phosphate.

Interpretation The healthy plant in the full culture is the control and shows that this method of raising plants does not affect them. The other, less healthy plants show that a full range of mineral elements is necessary for normal growth.

Quantitative results Although the effects of mineral deficiency can usually be seen simply by looking at the wheat seedlings, it is better if actual measurements are made.

The height of the shoot, or the total length of all the leaves on one plant can be measured. The total root length can also be measured, though this is difficult if root growth is profuse.

Alternatively, the dry weight (see p. 26) of the shoots and roots can be measured. In this case, it is best to pool the results of several experiments. All the shoots from the complete culture are placed in a labelled container; all those from the 'no nitrate' culture solution are placed in another container, and so on for all the plants from the different solutions. The shoots are then dried at 110 °C for 24 hours and weighed. The same procedure can be carried out for the roots.

You would expect the roots and shoots from the complete culture to weigh more than those from the nutrient-deficient cultures.

QUESTIONS

19 What salts would you put in a water culture which is to contain *no* nitrogen?
20 What mineral elements do you think are provided by (*a*) bone meal (p. 185), (*b*) dried blood (p. 137)?
21 How can a floating pond plant, such as duckweed, survive without having its roots in soil?
22 In the water culture experiment, why should (*a*) lack of nitrate, (*b*) lack of phosphate cause reduced growth?

CHECK LIST

- **Photosynthesis is the way plants make their food.**
- **They combine carbon dioxide and water to make sugar.**
- **To do this, they need energy from sunlight, which is absorbed by chlorophyll.**
- **Chlorophyll converts light energy to chemical energy.**
- **The equation to represent photosynthesis is**

$$6CO_2 + 6H_2O \xrightarrow[\text{absorbed by chlorophyll}]{\text{energy from sunlight}} C_6H_{12}O_6 + 6O_2$$

- **Plant leaves are adapted for the process of photosynthesis by being broad and thin, with many chloroplasts in their cells.**
- **From the sugar made by photosynthesis, a plant can make all the other substances it needs, provided it has a supply of mineral salts like nitrate, phosphate and potassium.**
- **In daylight, respiration and photosynthesis will be taking place in a leaf; in darkness, only respiration will be taking place.**
- **In daylight a plant will be taking in carbon dioxide and giving out oxygen.**
- **In darkness a plant will be taking in oxygen and giving out carbon dioxide.**
- **Experiments to test photosynthesis are designed to exclude light, or carbon dioxide, or chlorophyll, to see if the plant can still produce starch.**
- **The rate of photosynthesis may be limited by light intensity and temperature.**

Examination Questions

Section 1: Some Principles of Biology

Do not write on this page. Where necessary copy drawings, tables or sentences.

1 Which one of the following is found only in plant cells?
A a nucleus
B a cell membrane
C a cellulose wall
D cytoplasm (N)

2 The drawings show some cells drawn to scale.

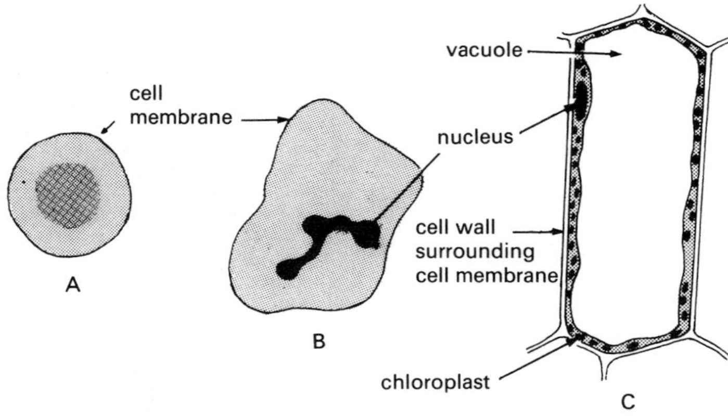

(*a*) State **three** differences you can see between Cell A and Cell B.
(*b*) State **three** differences you can see between Cell B and Cell C. (W)

3 (*a*) Complete the following word equation for photosynthesis.

$$\underline{\qquad\qquad} + \text{water} \xrightarrow[\text{chlorophyll}]{\text{light and}} \text{sugar} + \underline{\qquad\qquad}$$

(*b*) When you test a leaf for starch you have to remove chlorophyll. Name the chemical you would use to remove the chlorophyll and state why you have to be careful when you do this.
(*c*) The figure below shows a section through a leaf.

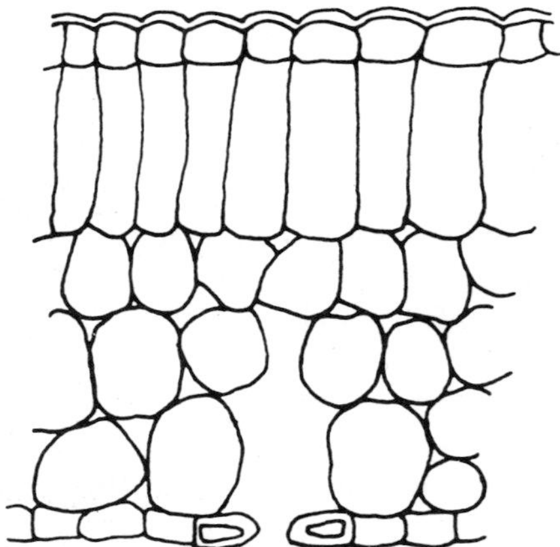

(i) Label a guard cell and a spongy mesophyll cell.
(ii) Describe **two** things about the spongy mesophyll tissue which make it easy for gases to move in and out of cells.
(*d*) What does the term diffusion mean? (M)

4 The apparatus below is used to measure the rate of a process taking place during the germination of peas.

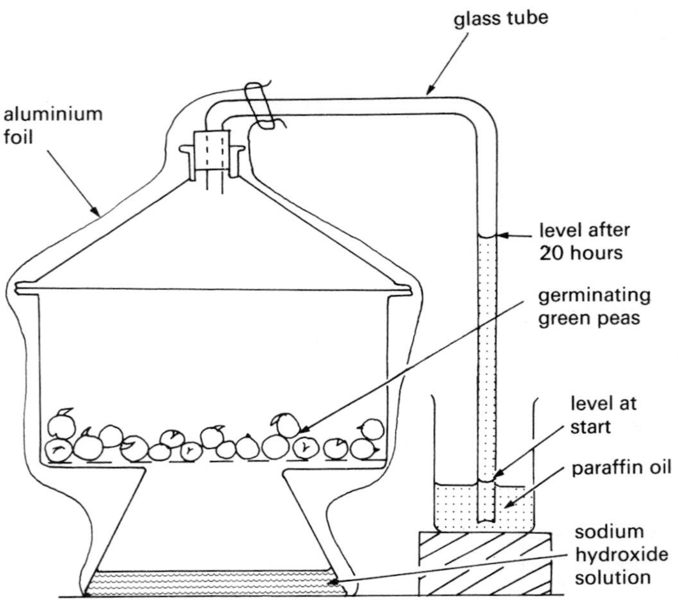

(*a*) State
(i) the name of the process being measured,
(ii) the purpose of the sodium hydroxide,
(iii) the reason for covering the apparatus with aluminium foil,
(iv) the name of the gas you would expect to be released,
(v) the name of the solution you would use to identify this gas,
(vi) what caused the oil to move up the tube.
(*b*) What else would you expect to be produced by the peas during this process?
(*c*) Describe one change you would make to carry out a controlled experiment. (W)

5 A biological washing powder is one which contains an enzyme. The enzyme removes stains, such as blood, from clothes which are soaked in water with the powder.
(*a*) Explain what an enzyme is.
(*b*) (i) Why would stains be removed faster with the powder in water at 30 °C rather than at 15 °C?
(i) Why is boiling the clothes with washing powder less likely to remove the stains? (L)

6 A manufacturer of the dialysis tubing used in artificial kidney machines has produced a new type of tubing. You are asked by the manufacturer to find out if the new tubing is permeable to urea at body temperature (37 °C).
You are given: ordinary laboratory apparatus,
some of the new tubing,
urea solution,
a colourless dye which turns blue when mixed with urea.
(*a*) Make a large labelled diagram of the apparatus with solutions, set up ready for your experiment.
(*b*) What result would show that urea can pass through the tubing? (N)

SECTION 2
FLOWERING PLANTS

6 Plant Structure and Function

LEAF

Epidermis, stomata, mesophyll, veins.

STEM

Epidermis, vascular bundles and their function.

ROOT

Outer layer and root hairs.

GROWTH IN ROOTS AND STEMS

Cell division and expansion at root and shoot tip. Buds.

PRACTICAL WORK

Tension in stems, cells and vessels. Stomata. Region of growth in roots and stems.

Figure 1 shows a young sycamore plant. It is typical of many flowering plants in having a **root system** below the ground and a **shoot** above ground. The shoot consists of an upright stem, with leaves and buds. The buds on the side of the stem are called **lateral buds**. When they grow, they will produce branches. The bud at the tip of the shoot is the **terminal bud** and when it grows, it will continue the upward growth of the stem. The lateral buds and the terminal buds may also produce flowers.

The region of stem from which leaves and buds arise is called a **node**. The region of stem between two nodes is the **internode**.

The leaves make food by photosynthesis (p. 44) and pass it back to the stem.

The stem carries this food to all parts of the plant which need it and also carries water and dissolved salts from the roots to the leaves and flowers.

In addition, the stem supports and spaces out the leaves so that they can receive sunlight and absorb carbon dioxide which they need for photosynthesis.

An upright stem also holds the flowers above the ground, helping the pollination by insects or the wind (p. 83). A tall stem may help in seed dispersal later on (p. 89).

The roots anchor the plant in the soil and prevent it falling over or being blown over by the wind. They also absorb the water and salts which the plant needs for making food in the leaves.

The structure and functions of the plant organs will be considered in more detail in this chapter.

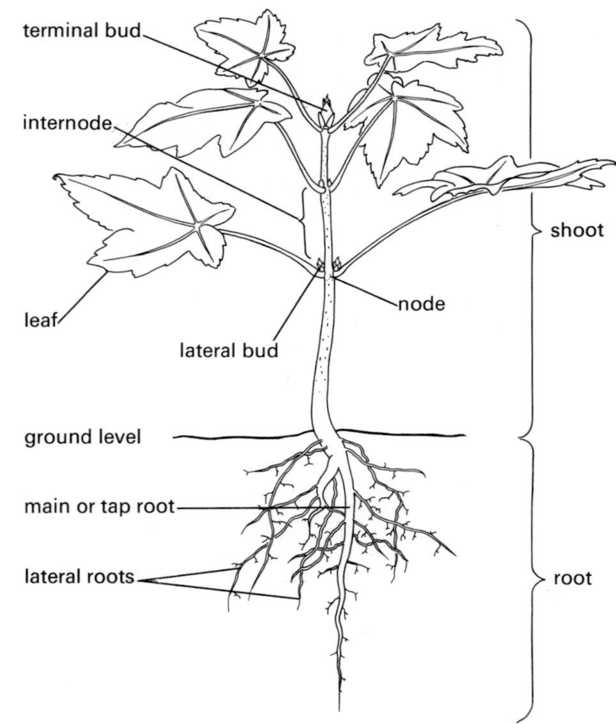

Fig. 1 Structure of a typical flowering plant

LEAF

Figure 2*a* shows a typical leaf of a broad-leaved plant. It is attached to the stem by a **leaf stalk** which continues into the leaf as a **midrib**. Branching from the midrib is a network of veins which deliver water and salts to the leaf cells and carry away the food made by them.

As well as carrying food and water, the network of veins forms a kind of skeleton which supports the softer tissues of the leaf blade.

The **leaf blade** (or **lamina**) is broad and thin. Figure 2*c* shows a vertical section through a small part of a leaf blade, and Fig. 3 is a photograph of a leaf section under the microscope.

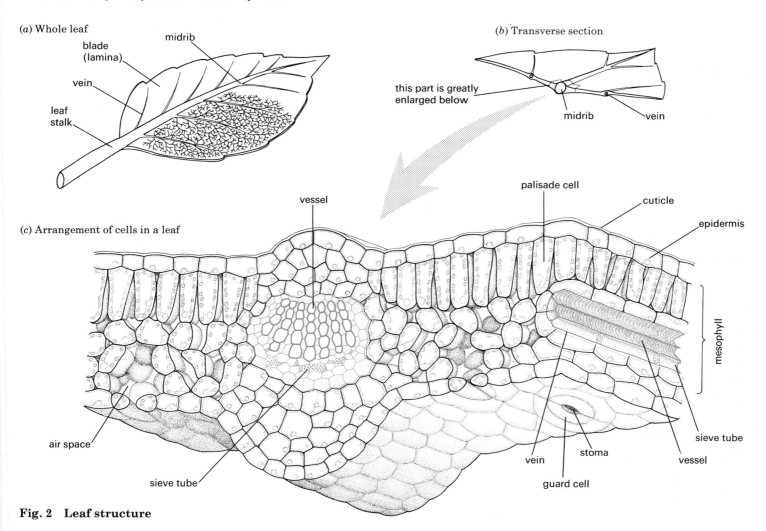

(*a*) Whole leaf

blade (lamina)
midrib
vein
leaf stalk

(*b*) Transverse section

this part is greatly enlarged below
midrib
vein

(*c*) Arrangement of cells in a leaf

vessel
palisade cell
cuticle
epidermis
mesophyll
air space
sieve tube
vein
stoma
guard cell
sieve tube
vessel

Fig. 2 Leaf structure

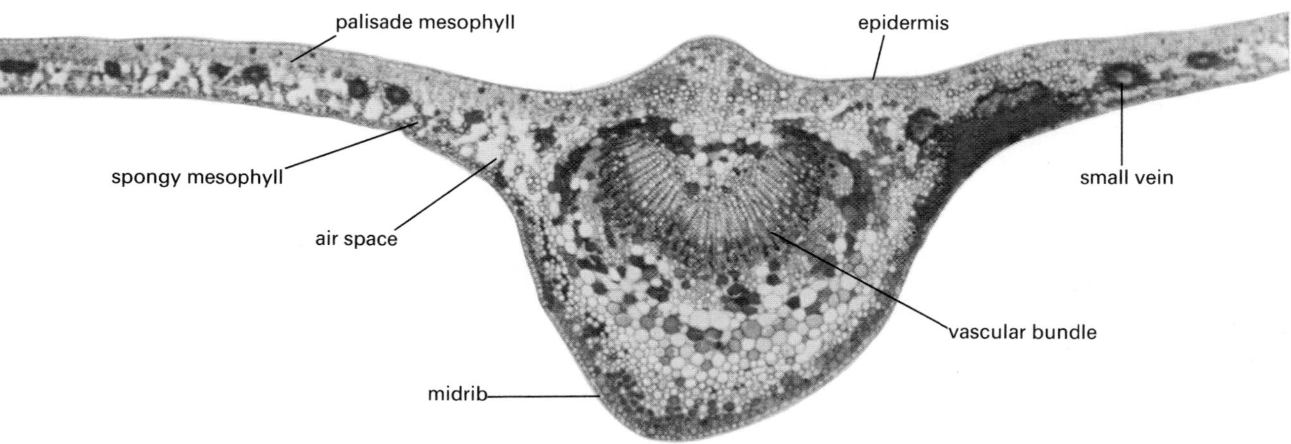

Fig. 3 Transverse section through a leaf (×30)

Epidermis

The **epidermis** is a single layer of cells on the upper and lower surface of the leaf. The epidermis helps to keep the leaf's shape. The closely fitting cells (Fig. 2*c*) reduce evaporation from the leaf and prevent bacteria and fungi from getting in. There is a thin waxy layer called the **cuticle** over the epidermis which helps to reduce water loss.

Stomata

In the leaf epidermis there are structures called **stomata** (singular = stoma). A stoma consists of a pair of **guard cells** (Fig. 4) surrounding an opening, or stomatal pore. Changes in the turgor (p. 38) and shape of the guard cells can open or close the stomatal pore. In most dicotyledons (i.e. the broad-leaved plants; see p. 306), the stomata occur only in the lower epidermis. In monocotyledons (i.e. narrow-leaved plants such as grasses) the stomata are equally distributed on both sides of the leaf.

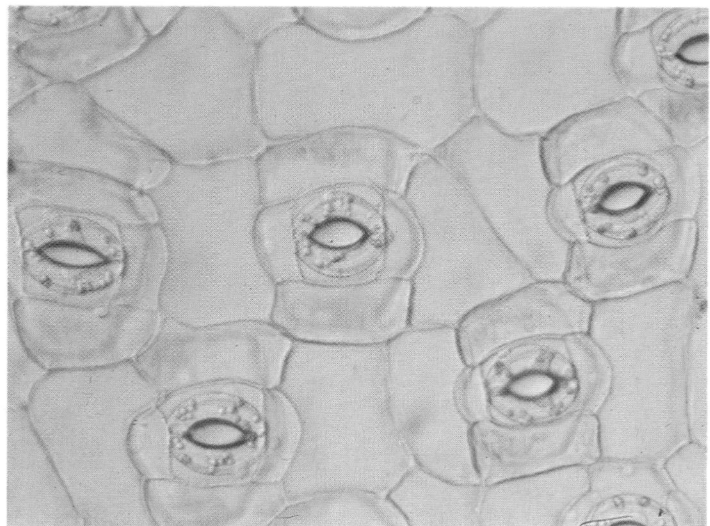

Fig. 4 Stomata in the lower epidermis of a leaf (× 350)

In very general terms, stomata are open during the hours of daylight but closed during the evening and most of the night (Fig. 5). This pattern, however, varies greatly with the plant species. A satisfactory explanation of stomatal rhythm has not been worked out, but when the stomata are open (i.e. mostly during daylight), they allow carbon dioxide to diffuse into the leaf where it is used for photosynthesis.

If the stomata close, the carbon dioxide supply to the leaf cells is virtually cut off and photosynthesis stops. However, in many species, the stomata are closed during the hours of darkness, when photosynthesis is not taking place anyway.

It seems, therefore, that stomata allow carbon dioxide into the leaf when photosynthesis is taking place and prevent excessive loss of water vapour (see pp. 72 and 73) when photosynthesis stops, but the story is likely to be more complicated than this.

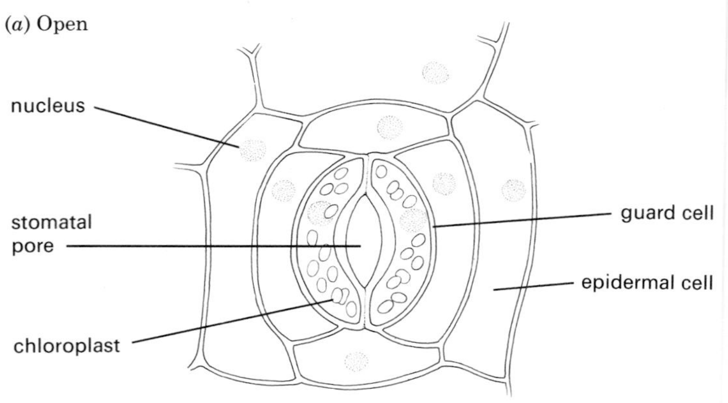

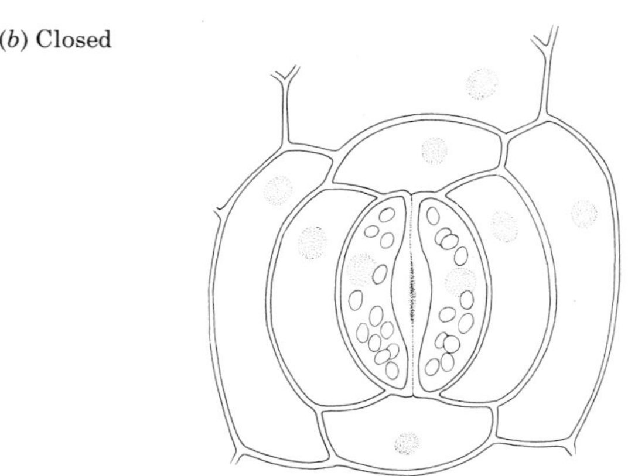

Fig. 5 Stoma

The detailed mechanism by which stomata open and close is not fully understood, but it is known that, in the light, the potassium concentration in the guard cell vacuoles increases. This lowers the water potential of the cell sap and water enters the guard cells by osmosis from their neighbouring epidermal cells. This inflow of water raises the turgor pressure inside the guard cells.

The cell wall next to the stomatal pore is thicker than elsewhere in the cell and is less able to stretch (Fig. 6). So, although the increased turgor tends to expand the whole guard cell, the thick inner wall cannot expand. This causes the guard cells to curve in such a way that the stomatal pore between them is opened.

When potassium ions leave the guard cell, the water potential rises, water passes out of the cells by osmosis, the turgor pressure falls and the guard cells straighten up and close the stoma.

Where the potassium ions come from and what triggers their movement into or out of the guard cells is still under active investigation.

You will notice from Figs 5 and 6 that the guard cells are the only epidermal cells containing chloroplasts. At one time it was thought that the chloroplasts built up sugar by photosynthesis during daylight, that the sugars made the cell sap more concentrated and so caused the increase in turgor. In

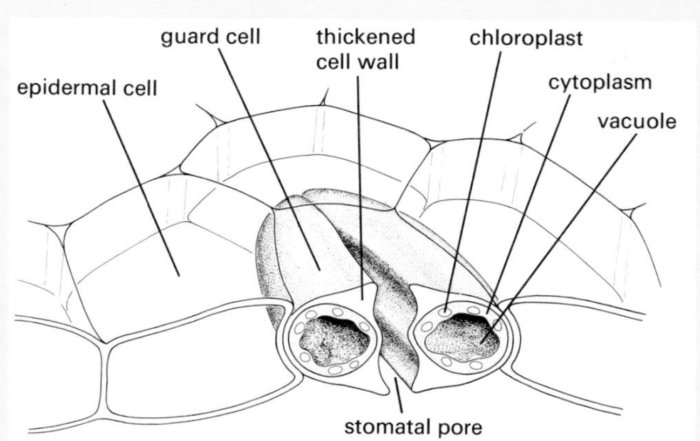

Fig. 6 Structure of guard cells

fact, little or no photosynthesis takes place in these chloroplasts and their function has not been explained, though it is known that starch accumulates in them during the hours of darkness. In some species of plants, the guard cells have no chloroplasts.

Mesophyll

The tissue between the upper and lower epidermis is called **mesophyll** (Fig. 2*c*). It consists of two zones: the upper, **palisade mesophyll** and the lower, **spongy mesophyll** (Fig. 7). The palisade cells are usually long and contain many chloroplasts. The spongy mesophyll cells vary in shape and fit loosely together, leaving many air spaces between them.

The function of the palisade cells and—to a lesser extent—of the spongy mesophyll cells is to make food by photosynthesis. Their chloroplasts absorb sunlight and use its energy to join carbon dioxide and water molecules to make sugar molecules as described on page 44.

In daylight, when photosynthesis is rapid, the mesophyll cells are using up carbon dioxide. As a result, the concentration of carbon dioxide in the air spaces falls to a low level and more carbon dioxide diffuses in (p. 33) from the outside air, through the stomata (Fig. 7). This diffusion continues through the air spaces, up to the cells which are using carbon dioxide. These cells are also producing oxygen as a by-product of photosynthesis. When the concentration of oxygen in the air spaces rises, it diffuses out through the stomata.

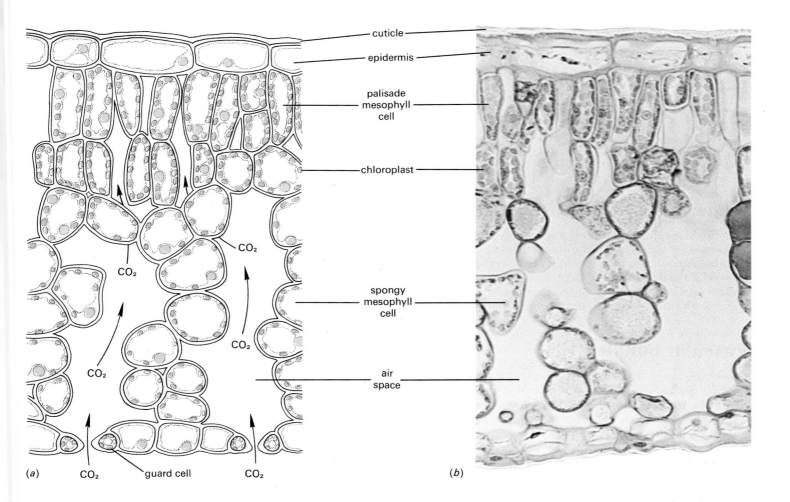

Fig. 7 Vertical section through a leaf blade (× 300)

Veins (vascular bundles)

The water needed for making sugar by photosynthesis is brought to the mesophyll cells by the **veins**. The cells take in the water by osmosis (p. 36) because the concentration of free water molecules in a leaf cell, which contains sugars, will be less than the concentration of water in the water vessels of a vein, which does not contain sugars. The branching network of leaf veins means that no cell is very far from a water supply.

The sugars made in the mesophyll cells are passed to the phloem cells (see below) of the veins, and these cells carry the sugars away from the leaf into the stem.

The ways in which a leaf is thought to be well adapted to its function of photosynthesis are listed on page 50.

QUESTIONS

1 What are the functions of (*a*) the epidermis, (*b*) the mesophyll of a leaf?
2 Look at Fig. 7. Why do you think that photosynthesis does not take place in the cells of the epidermis?
3 During bright sunlight, what gases are (*a*) passing out of the leaf through the stomata, (*b*) entering the leaf through the stomata?
4 What types of leaves do you know which do not have any midrib?
5 In some plants, the stomata close for a period at about midday. Suggest some possible advantages and disadvantages of this to the plant.

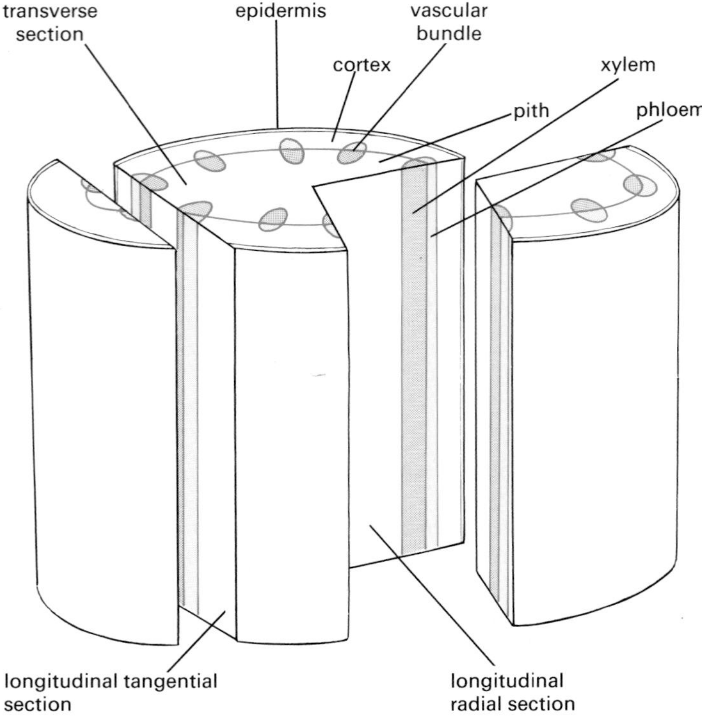

Fig. 8 Structure of plant stem

STEM

Figure 8 is a diagram of a stem cut across (transversely) and down its length (longitudinally) to show its internal structure.

Epidermis

Like the leaf epidermis, this is a single layer of cells which helps to keep the shape of the stem and cuts down the loss of water vapour. There are stomata in the epidermis which allow the tissues inside to take up oxygen and get rid of carbon dioxide. In woody stems, the epidermis is replaced by bark which consists of many layers of dead cells.

Vascular bundles

These are made up of groups of specialized cells which conduct water, dissolved salts and food up or down the stem (Fig. 10). The vascular bundles in the roots, stem, leaf stalks and leaf veins all connect up to form a transport system throughout the entire plant (Fig. 9). The two main tissues in the vascular bundles are called **xylem** and **phloem**. Food substances travel in the phloem; water and salts travel mainly in the xylem. The cells in each tissue form elongated tubes called **vessels** (in the xylem) or **sieve tubes** (in the phloem) and they are surrounded and supported by other cells.

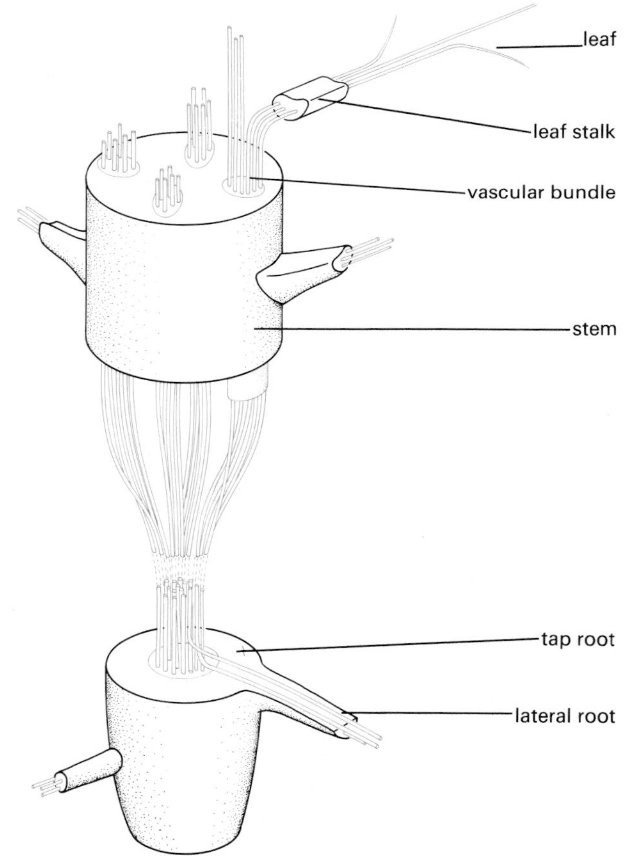

Fig. 9 Distribution of veins from root to leaf

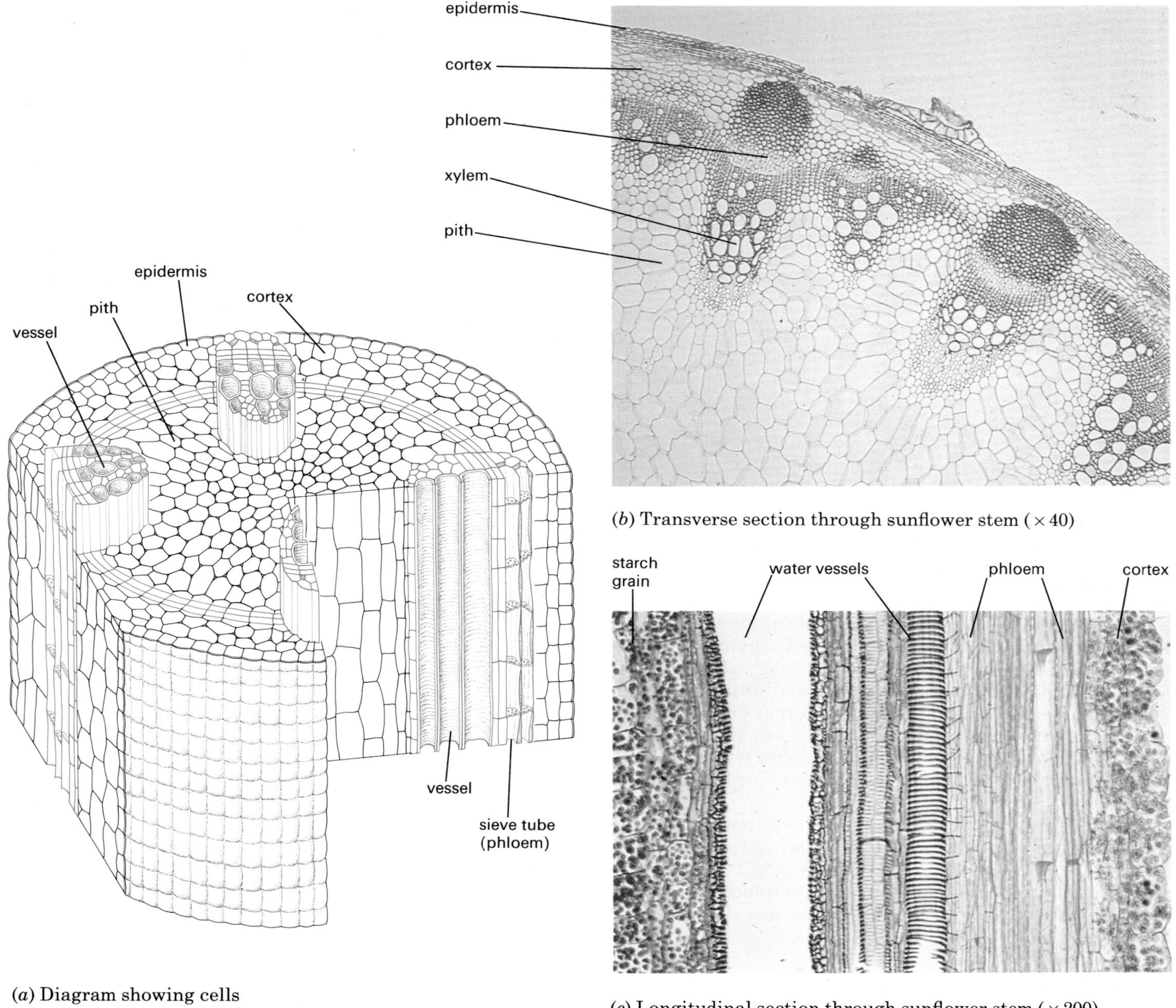

(*b*) Transverse section through sunflower stem (×40)

(*a*) Diagram showing cells

(*c*) Longitudinal section through sunflower stem (×200)

Fig. 10 Structure of plant stem

Vessels The cells in the xylem which carry water form vessels. A vessel is made up of a series of long cells joined end to end (Fig. 11*a*). Once a region of the plant has ceased growing, the end walls of these cells are digested away to form a continuous, fine tube (Fig. 10*c*). At the same time, the cell walls are thickened and impregnated with a substance called **lignin**, which makes the cell wall very strong and impermeable. Since these lignified cell walls prevent the free passage of water and nutrients, the cytoplasm dies. This does not affect the passage of water in the vessels. Xylem also contains many elongated, lignified supporting cells called **fibres**.

Sieve tubes The conducting cells in the phloem remain alive and form sieve tubes. Like vessels, they are formed by vertical columns of cells (Fig. 11*b*). Perforations appear in the end walls, allowing substance to pass from cell to cell, but the cell walls are not lignified and the cell contents do not die, although they do lose their nuclei. The perforated end walls are called **sieve plates**.

Phloem contains supporting cells as well as sieve tubes.

Functions of vascular bundles In general, water travels up the stem in the xylem from the roots to the leaves. Food may travel either up or down the stem in the phloem, from the leaves where it is made, to any part of the plant which is using or storing it.

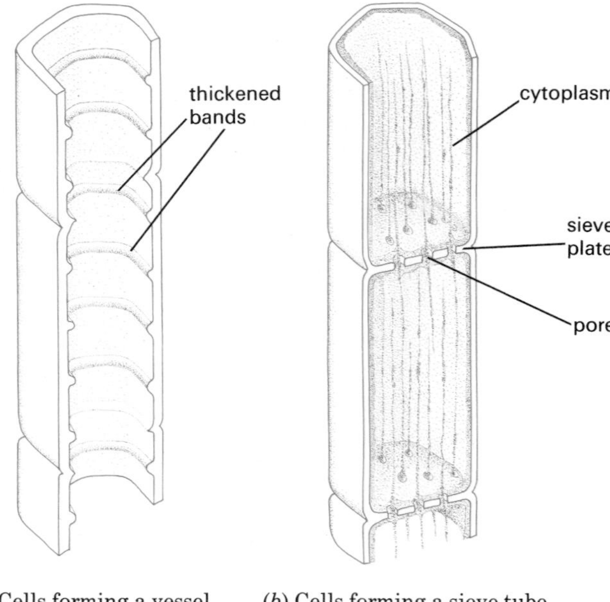

(a) Cells forming a vessel (b) Cells forming a sieve tube

Fig. 11 Conducting structures in a plant

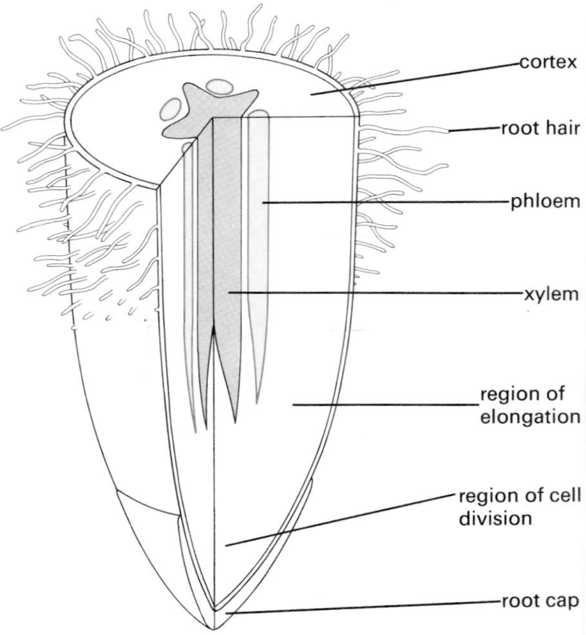

Fig. 12 Root structure

Vascular bundles have a supporting function as well as a transport function, because they contain vessels, fibres and other thick-walled elongated cells. In many stems, the vascular bundles are arranged in a cylinder, a little way in from the epidermis. This pattern of distribution helps the stem to resist the sideways bending forces caused by the wind. In a root, the vascular bundles are in the centre (Fig. 13) where they resist the pulling forces which the root is likely to experience when the shoot is being blown about by the wind.

The network of veins in many leaves supports the soft mesophyll tissues and resists stresses which could lead to tearing.

The methods by which water, salts and food are moved through the vessels and sieve tubes are discussed in Chapter 7 (p. 71).

Cortex and pith

The tissue between the vascular bundles and the epidermis is called the **cortex**. Its cells often store starch. In green stems, the outer cortex cells contain chloroplasts and make food by photosynthesis. The central tissue of the stem is called **pith**. The cells of the pith and cortex act as packing tissues and help to support the stem in the same way as a lot of blown-up balloons packed tightly into a plastic bag would form quite a rigid structure.

QUESTIONS

6 Make a list of the types of cells or tissues you would expect to find in a vascular bundle.
7 What structures help to keep the stem's shape and upright position?
8 What are the differences between xylem and phloem (a) in structure (b) in function?

ROOT

Figure 12 shows the internal structure of a typical root. The vascular bundle is in the centre of the root (Fig. 13), unlike the stem where the bundles form a cylinder in the cortex.

The xylem carries water and salts from the root to the stem. The phloem will bring food from the stem to the root, to provide the root cells with substances for their energy and growth.

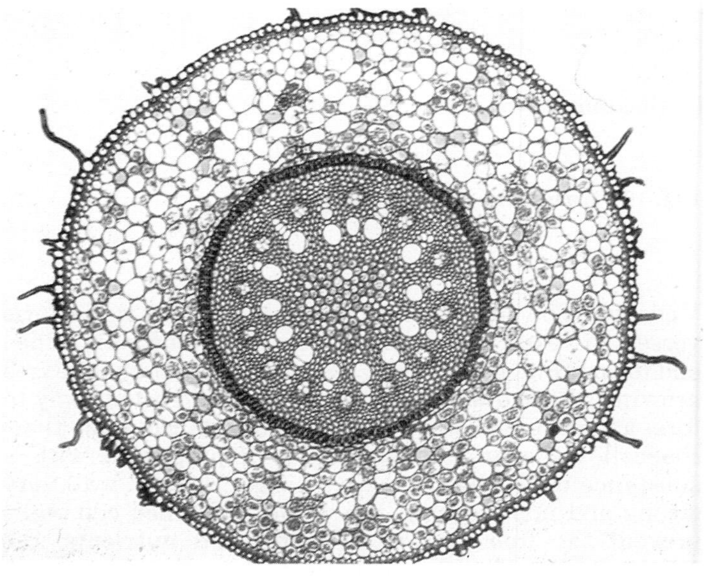

Fig. 13 Transverse section through a root (× 40). Notice that the vascular tissue is in the centre. Some root hairs can be seen in the outer layer of cells.

Outer layer and root hairs

There is no distinct epidermis in a root. At the root tip are several layers of cells forming the **root cap**. These cells are continually replaced as fast as they are worn away when the root tip is pushed through the soil.

In a region above the root tip, where the root has just stopped growing, the cells of the outer layer produce tiny, tube-like outgrowths called **root hairs** (Fig. 7, p. 75). These can just be seen as a downy layer on the roots of seedlings grown in moist air (Fig. 14). In the soil, the root hairs grow between the soil particles and stick closely to them. The root hairs take up water from the soil by osmosis, and absorb mineral salts (as ions) by diffusion or active transport (p. 35).

The large number of tiny root hairs greatly increases the absorbing surface of a root system. The surface area of the root system of a mature rye plant has been estimated at about 200 m^2. The additional surface provided by the root hairs was calculated to be 400 m^2.

Root hairs remain alive for only a short time. The region of root just below a root hair zone is producing new root hairs, while the root hairs at the top of the zone are shrivelling (Fig. 15). Above the root hair zone, the cell walls of the outer layer become less permeable. This means that water cannot get in so easily.

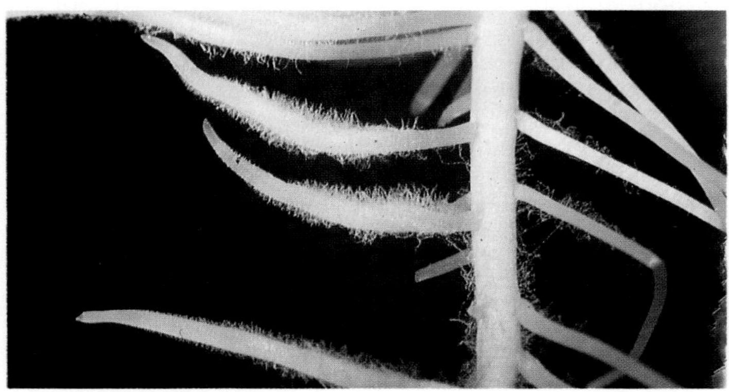

Fig. 14 Root hairs (×5) as they appear on a root grown in moist air.

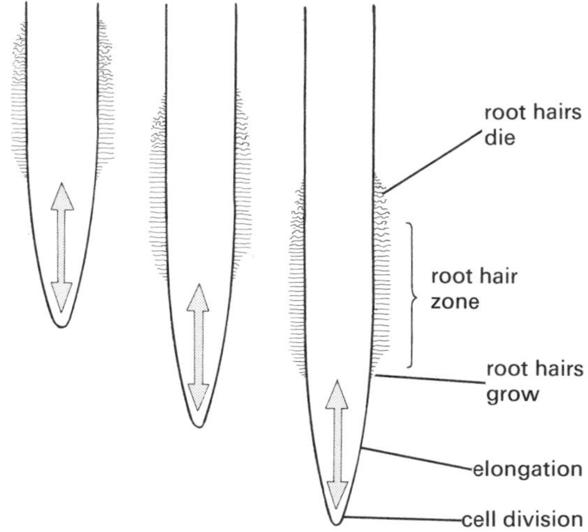

Fig. 15 The root hair zone changes as the root grows

QUESTIONS

9 State briefly the functions of the following: xylem, palisade cell, root hair, root cap, stoma, epidermis.
10 If you were given a cylindrical structure cut from part of a plant, how could you tell whether it was a piece of stem or a piece of root (*a*) with the naked eye, (*b*) with the aid of a microscope or hand lens?
11 Describe the path taken by (*a*) a carbon dioxide molecule from the air and (*b*) a water molecule from the soil, until they reach a mesophyll cell of a leaf to be made into sugar.
12 Why do you think that root hairs are produced only on the parts of the root system that have stopped growing?
13 Discuss whether you would expect to find a vascular bundle in a flower petal.

GROWTH IN ROOTS AND STEMS

Root

At the tip of the root, the cells are dividing rapidly and producing a large number of new cells (Fig. 16). Just above this region, the cells start to absorb water by osmosis. At this stage the cell walls are quite soft and as the vacuole expands it will make the cell longer (see Fig. 17, p. 8). Hundreds of cells getting longer at the same time will push the root tip down through the soil. Growth is thus the result of cell division and cell expansion.

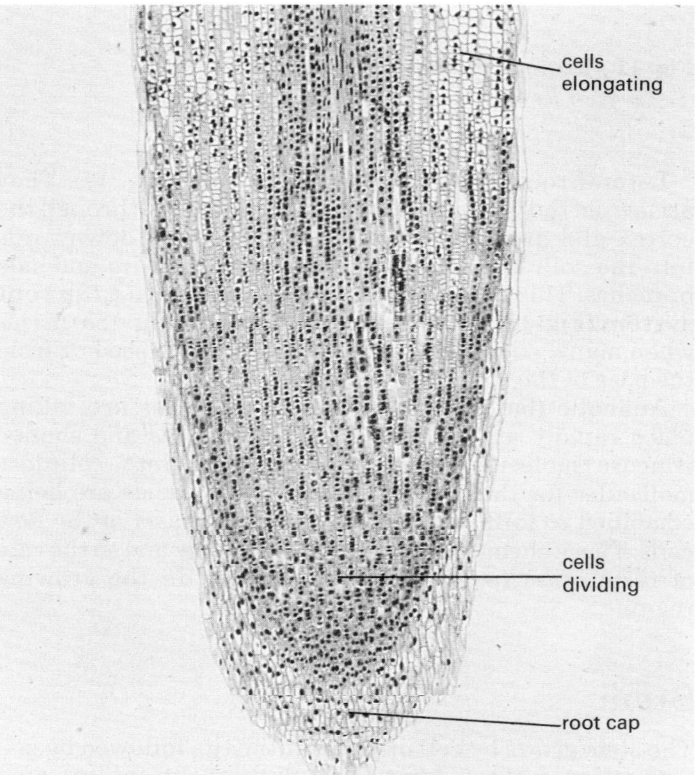

Fig. 16 Longitudinal section through the root tip of an onion (×60)

In addition, **cell differentiation** is taking place. This means that certain cells undergo changes in structure and function which suit them to perform a particular function, e.g. a conducting function. Cells in the centre of a growing root become vessels, and groups of cells just outside the centre differentiate into sieve tubes as described on page 63.

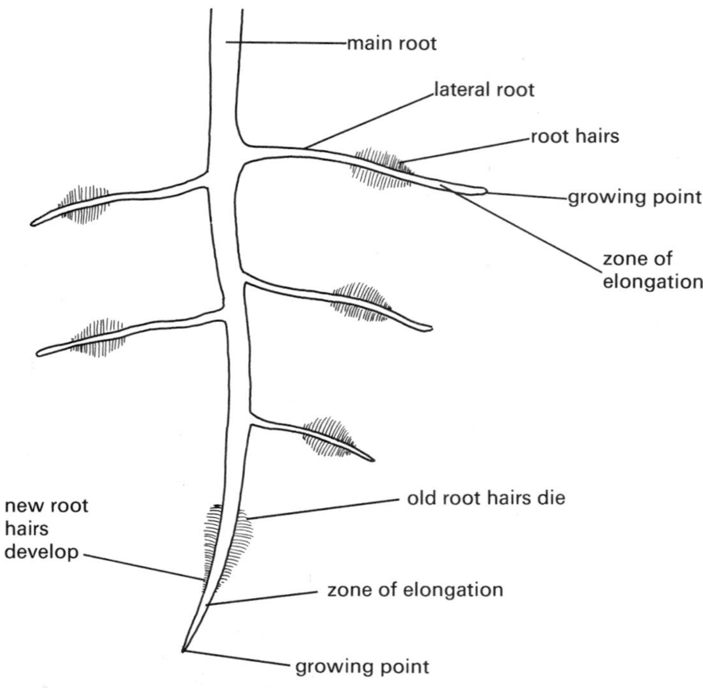

Fig. 17 Regions of a root system

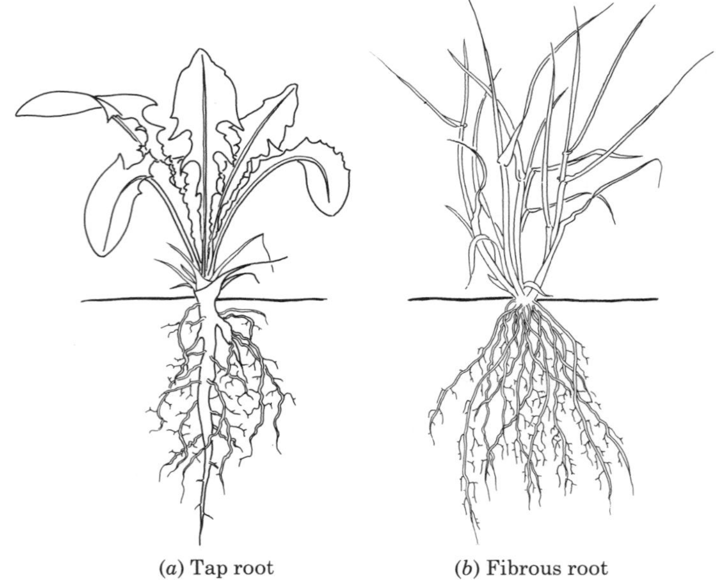

Fig. 18 Types of root system

Lateral roots grow from the main root (Fig. 17). They arise from the vascular bundle, push their way through the cortex and grow horizontally or diagonally downwards into the soil. They, too, will develop root hairs and side branches. This type of root growth gives rise to a **tap root system** (Fig. 18*a*). A **fibrous root system** (Fig. 18*b*) arises when many, equal-sized roots grow from the seed or from the base of the stem.

Anabolic (building-up) chemical reactions are taking place rapidly at the growing point of roots and shoots. Glucose molecules are being built up into cellulose molecules for the new cell walls. Amino acids are being combined to form proteins for the protoplasm of the new cells. These chemical reactions need energy and so the rate of respiration (p. 24) will also be high in the growing region.

Stem

The stem grows by cell division at the tip, followed by cell elongation further down. Cell division at the growing point is more complex than in the root because it produces new leaves as well as new stem.

Buds In woody plants, cell division at the tip of the shoot produces tiny, closely packed leaves on a short stem, so forming a **bud** (Fig. 19). The outermost leaves of the bud may become thick and tough, forming **bud scales**, which protect the more delicate inner leaves from frost and from attack by insects and fungi. The new leaves and stem pass the winter in the bud. In the spring, the bud's stem elongates very rapidly by cell expansion, pushing the bud scales apart and spacing out the new leaves (Fig. 20). The bud scales fall off but the new leaves expand and grow to full size.

All this growth takes place in a few weeks, after which the stem and leaves do not grow any more that year. New **terminal buds** are formed at the shoot tips; **lateral buds** develop at the junction of the leaf stalk and stem. In the following season, the growth of the terminal bud will increase the length of the shoot. The lateral buds, when they grow, will form branches. Terminal or lateral buds may produce flowers as well as leaves.

QUESTIONS

14 In which part of a growing root are the cells (*a*) dividing, (*b*) extending?
15 Which type of bud might (*a*) form a branch, (*b*) produce a flower, (*c*) continue growth in length of the stem?
16 If you removed 3 mm from a root tip, what further growth, if any, would you expect to occur?
17 Assuming that all the buds in Fig. 1 develop next season, make a sketch to show the appearance of the shoot. Indicate by colour which regions are last year's and which are this year's growth.
18 Cutting a hedge removes all the terminal buds. Suggest why it is that the hedge will (*a*) continue to grow in height, (*b*) become more bushy.
19 A farmer hammers a long nail into a tree-trunk 2 metres above ground-level. Have a guess at where the nail will be in 20 years' time.

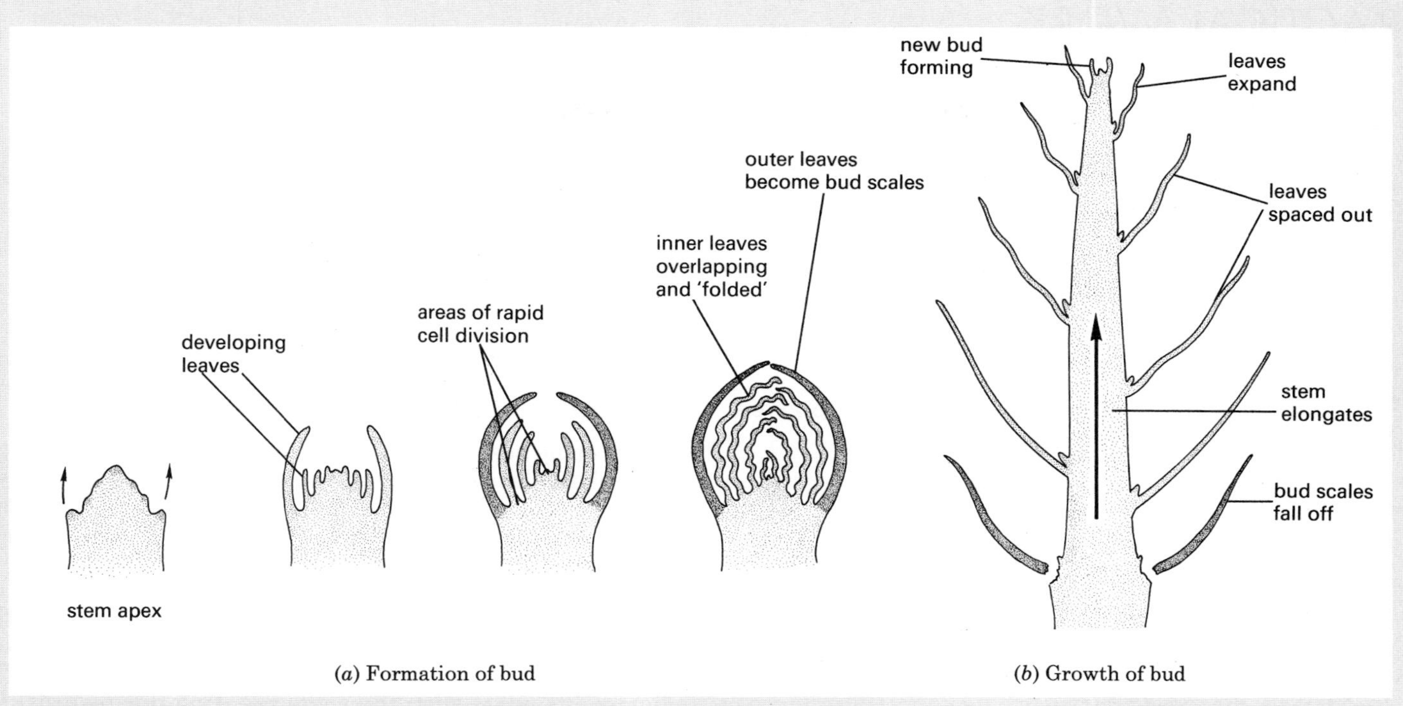

(*a*) Formation of bud

(*b*) Growth of bud

Fig. 19 Bud formation and growth

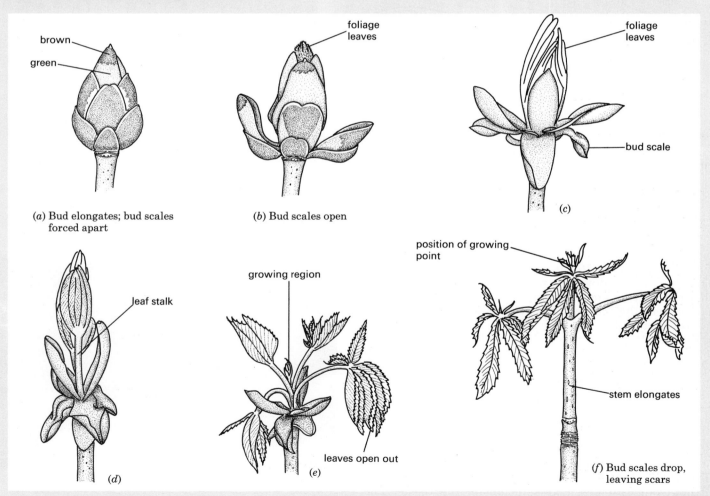

(*a*) Bud elongates; bud scales forced apart

(*b*) Bud scales open

(*c*)

(*d*)

(*e*)

(*f*) Bud scales drop, leaving scars

Fig. 20 Stages in the growth of a horse-chestnut terminal bud

PRACTICAL WORK

1. To show the tension in stems

Partly remove a strip of epidermis from a rhubarb stalk as shown in Fig. 21*a*. When you replace the epidermis in position it will have become too short, showing the shrinking stress that exists in the epidermis.

Push a cork borer into the pith and withdraw it without removing any tissue. The cylinder of pith so formed, freed from the constraint of the epidermis, expands and protrudes slightly. This shows the elongating tendency of the inner tissues (Fig. 21*b*).

These two opposing stresses in the stem help to give it rigidity when the cells are turgid (p. 38).

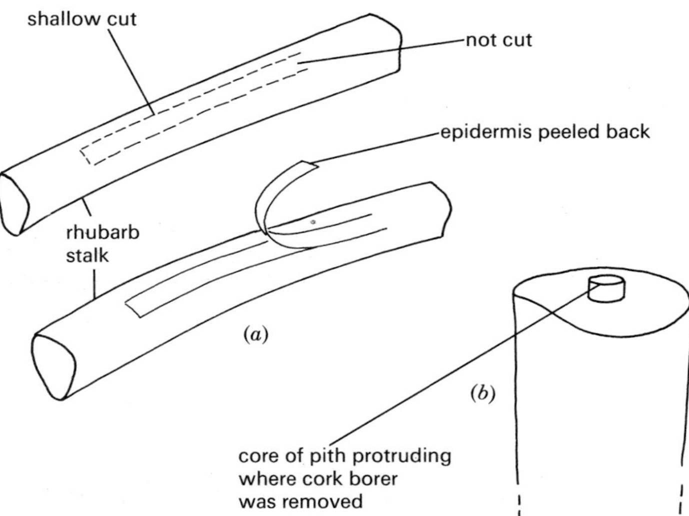

Fig. 21 To show opposite tensions in pith and epidermis

2. Cells and vessels

Some celery or rhubarb stalks which have been cut up into pieces about 2 cm long, are left for a few days in a macerating fluid of 10 per cent chromic and nitric acids. This fluid breaks down the material between the cells, so that the cells can be teased apart with mounted needles. The macerating acids will have been washed off before you use the material.

Place a little of the material on a slide with a little water and pull it to pieces with a pair of mounted needles. When you study it under the microscope, you should see individual cells and vessels.

3. Stomata

Strip off a piece of the lower epidermis of a rhubarb leaf and place it on a slide, in a little water, under the microscope (Fig. 22). The stomata can be made to close by putting a little strong salt or sugar solution on the tissue. The solution withdraws water from the guard cells by osmosis and they lose their turgor.

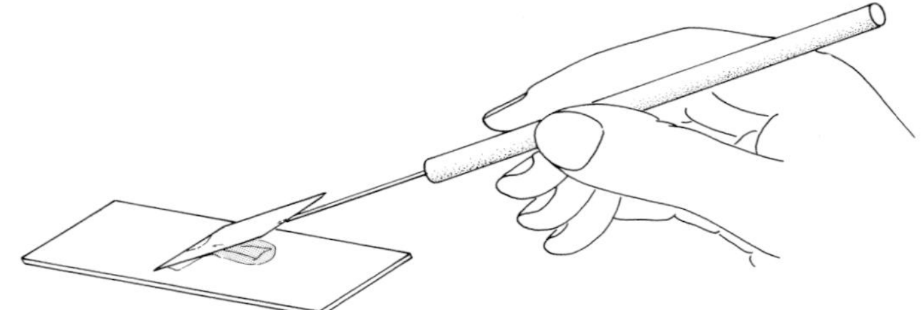

Fig. 22 Lower the cover slip carefully to exclude air bubbles

4. Region of growth in a root

Leave some peas in water for 24 hours and then wrap them in a roll of blotting-paper as shown in Fig. 23. After 3 days, the roots will have grown about 10 mm. Choose seedlings with straight roots and mark the roots with ink lines about 2 mm apart. Figure 24 shows one way of doing this. Place three marked seedlings between two strips of moist cotton wool in a Petri dish so that the seeds are held firmly but the roots are exposed and easily seen (Fig. 25*a*). Keep the lid of the Petri dish in place with an elastic band and leave the dish on its edge, with the roots pointing downwards, for 2 days.

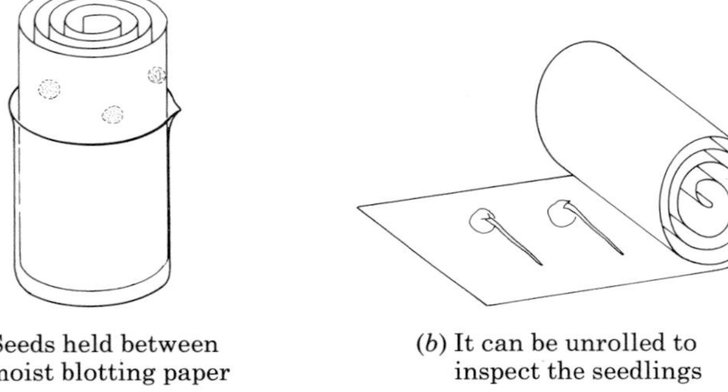

(*a*) Seeds held between moist blotting paper

(*b*) It can be unrolled to inspect the seedlings

Fig. 23 Growing seedlings with straight roots

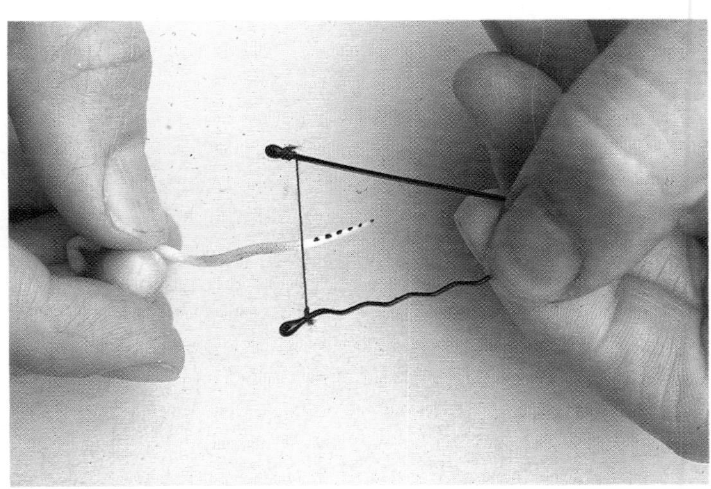

Fig. 24 Marking a root. A piece of cotton is held by the hairpin and dipped into black ink.

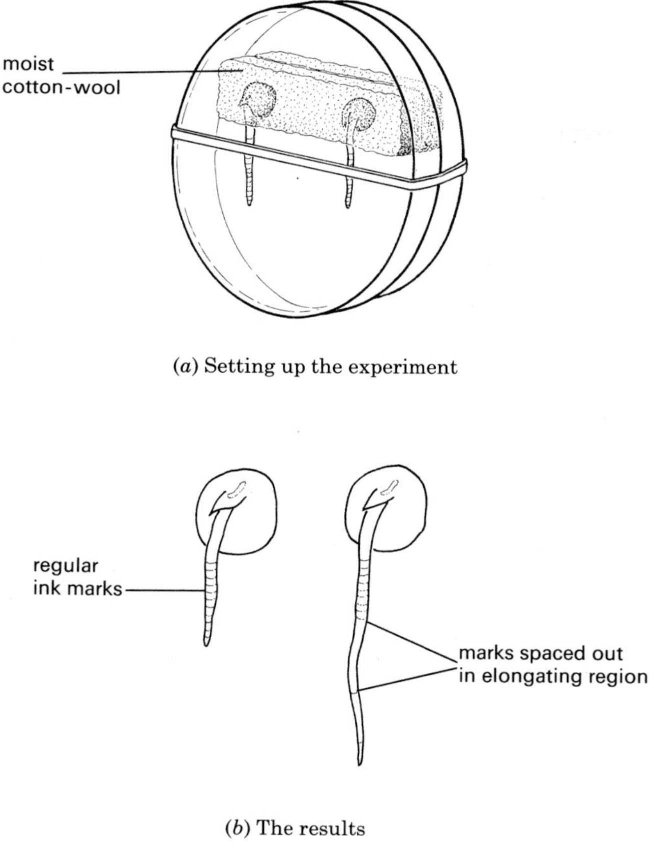

(*a*) Setting up the experiment

(*b*) The results

Fig. 25 Region of growth in a root

Result The ink marks will have become most widely spaced in the region just behind the root tip (Fig. 25*b*).

Interpretation The wide spacing of the marks shows that the region of most rapid growth is just behind the tip. In fact it is known that cell division takes place in the tip of the root but cell elongation occurs a short distance behind the tip (see Fig. 16, p. 65).

5. Growth in stems

Grow some bean seedlings as described for peas on page 68. Select one of the seedlings which has a straight stem and place it alongside a ruler. Use a fine marker pen to draw lines at 2 or 3 mm intervals all the way up the stem (Fig. 26), above and below the cotyledons (if these are still present).

Wrap a little cotton wool round the base of the stem and put the seedling in a test-tube of water. Leave it in a position where it will not receive strong light, for 2 days or more.

The spacing of the marks after this period will give an indication of the region of maximum elongation. The results will depend on the species of plant used and its age, but in general the region of elongation is usually just behind the tip of the shoot.

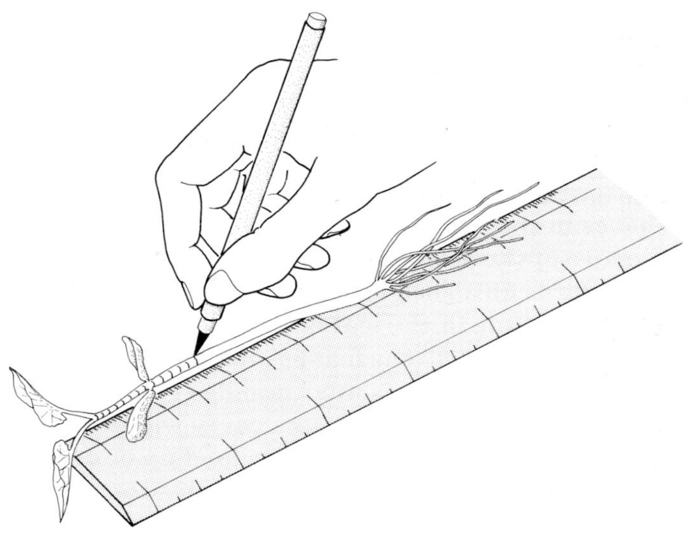

Fig. 26 Marking a French bean stem

QUESTION

20 Figure 25*b* shows the result of an experiment to find the region of maximum growth in a root. Draw a diagram to show how the result would have appeared if (*a*) the root grew simply by adding new cells at the tip, and (*b*) if the root grew mainly at the point just below its attachment to the cotyledons.

STRUCTURE AND FUNCTION

It is always tempting when studying an organism, to ascribe some function to the structures which are being observed. In some cases this is easy. It is obvious, for example, that the function of a mammal's hind limb is locomotion because the limb can be seen in action.

The functions of internal organs are not so obvious and guesses about their function may be quite inaccurate. At one time it was thought that the arteries carried air because the arteries seen in the dissection of dead animals often contained no blood. The chloroplasts in the guard cells (p. 60) were assumed for a long time to be the site of photosynthesis and it was only after conducting experiments that this assumption was shown to be false.

The functions of tissues described in this chapter have been stated as if they were certainly known and without offering any evidence for these functions. In Chapter 8 experiments will be described which do provide some evidence for the statements.

In general, although function can be guessed at from studying anatomy of dead organisms, it cannot be confirmed without experiments to test the guesses (hypotheses). It is usually most unwise to assume a particular function from simply studying anatomy.

CHECK LIST

- **The shoot of a plant consists of the stem, leaves, buds and flowers.**
- **The leaf makes food by photosynthesis in its mesophyll cells.**
- **The water for photosynthesis is carried in the leaf's veins.**
- **The carbon dioxide for photosynthesis enters the leaf through the stomata and diffuses through the air spaces in the leaf.**
- **Closure of the stomata stops the entry of carbon dioxide into a leaf but also reduces water loss.**
- **Sunlight is absorbed by the chloroplasts in the mesophyll cells.**
- **The food made in the leaf is carried away in the phloem cells.**
- **The stem supports the leaves and flowers.**
- **The stem contains vascular bundles (veins).**
- **The water vessels in the veins carry water up the stem to the leaves.**
- **The phloem in the veins carries food up or down the stem to wherever it is needed.**
- **The position of vascular bundles helps the stem to withstand sideways bending and the root to resist pulling forces.**
- **The roots hold the plant in the soil and absorb the water and mineral salts needed by the plant for making sugars and proteins.**
- **The root hairs make very close contact with soil particles and are the main route by which water and mineral salts enter the plant.**
- **Growth in the length of stems and roots takes place by cell division at the tip followed by cell elongation behind the tip.**
- **Buds are condensed shoots; their immature foliage leaves are protected inside thick bud scales.**
- **Buds grow by rapid elongation of the stem and expansion of the leaves.**

7 Transport in Plants

TRANSPORT OF WATER

Transpiration: its function and control. Rates of transpiration. Root pressure.

TRANSPORT OF SALTS

TRANSPORT OF FOOD

Movement of solutes in the phloem: evidence for this.

UPTAKE OF WATER AND SALTS

Osmosis in the roots. Possible mechanisms for salt uptake.

TRANSPORT OF GASES

Diffusion through stomata and intercellular spaces.

PRACTICAL WORK

Experiments on transport in plants: the potometer; rates of transpiration; role of vascular bundles; evaporation from a leaf; pathway for gases; root pressure.

TRANSPORT OF WATER

Transpiration

The main force which draws water from the soil and through the plant is caused by a process called **transpiration**. Water evaporates from the leaves and causes a kind of 'suction' which pulls water up the stem (Fig. 1). The water travels up the vessels in the vascular bundles (see Fig. 9, p. 62) and this flow of water is called the **transpiration stream**. Figure 2 shows the cells in part of a leaf blade. As explained on page 38, the cell sap in each cell is exerting a turgor pressure outwards on the cell wall. This pressure forces some water out of the cell wall and into the air space between the cells. Here the water evaporates and

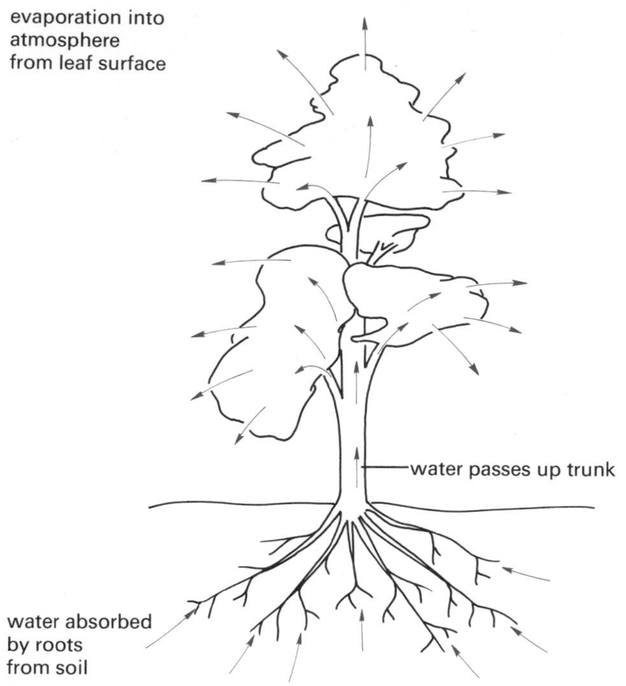

Fig. 1 The transpiration stream

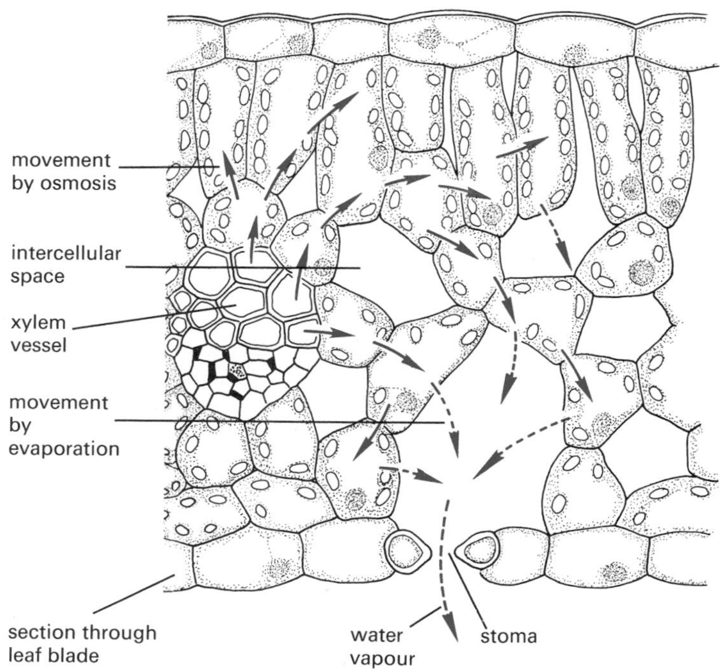

Fig. 2 Movement of water through a leaf

the water vapour passes by diffusion through the air spaces in the mesophyll and out of the stomata. It is this loss of water vapour from the leaves which is called 'transpiration'.

The cell walls which are losing water in this way replace it by drawing water from the nearest vein. Most of this water travels along the cell walls without actually going inside the cells (Fig. 3). Thousands of leaf cells are evaporating water like this and drawing water to replace it from the xylem vessels in the veins. As a result, water is pulled through the xylem vessels and up the stem from the roots. This transpiration pull is strong enough to draw up water 50 metres or more in trees (Fig. 4).

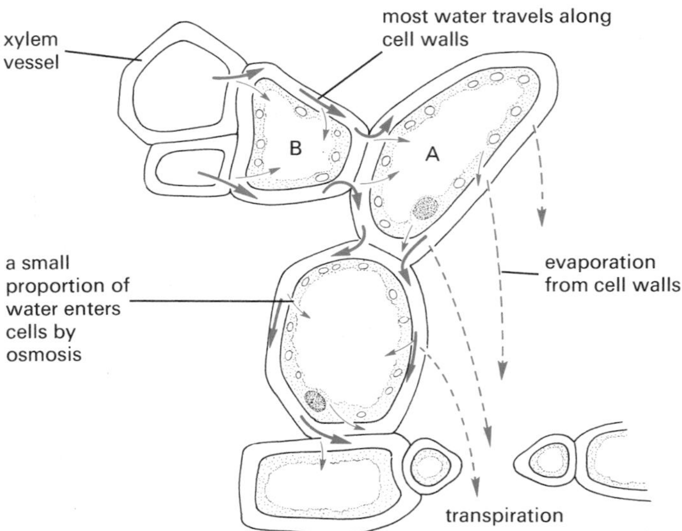

Fig. 3 Probable pathway of water through leaf cells

In addition to the water passing along the cell walls, a small amount will pass right through the cells. When leaf cell A (Fig. 3) loses water, its turgor pressure will fall. This fall in pressure allows the water in the cell wall to enter the vacuole and so restore the turgor pressure. In conditions of water shortage, cell A may be able to get water by osmosis from cell B more easily than B can get it from the xylem vessels. In this case, all the mesophyll cells will be losing water faster than they can absorb it from the vessels, and the leaf will wilt (see p. 38).

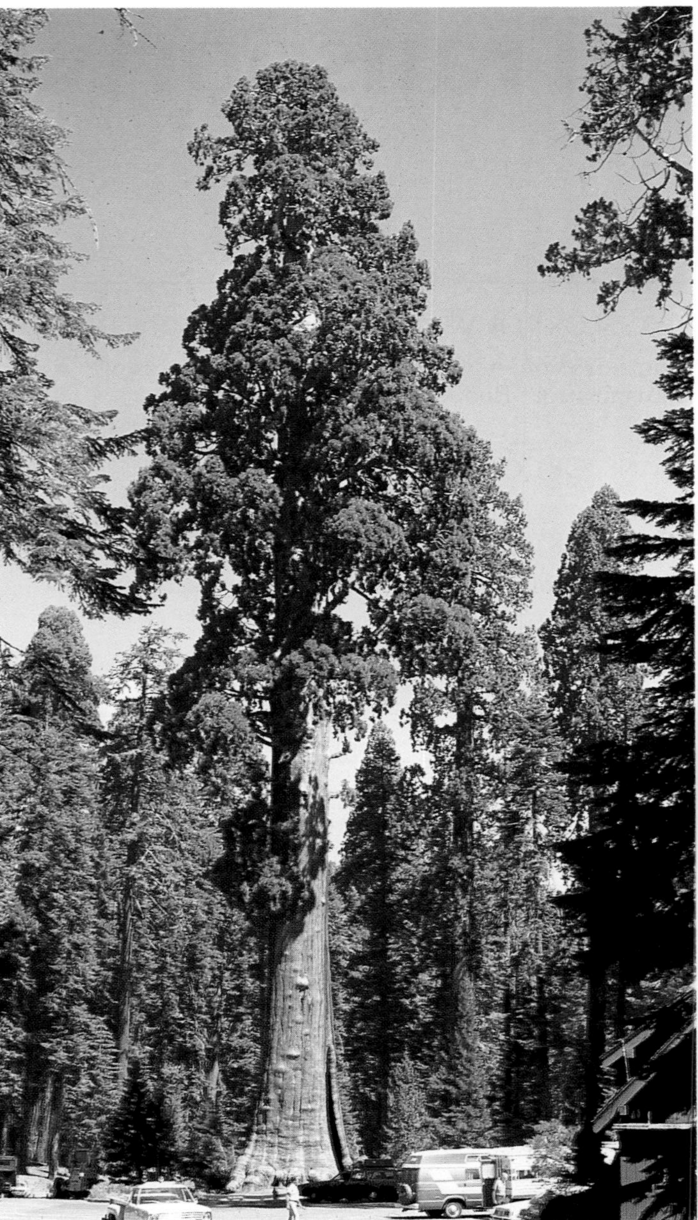

Fig. 4 Californian redwoods. Some of these trees are over 100 metres tall. Transpiration from their leaves pulls water up the trunk.

Importance of transpiration A tree, on a hot day, may draw up hundreds of litres of water from the soil. Most of this water evaporates from the leaves; only a tiny fraction is retained for photosynthesis and to maintain the turgor of the cells. The advantage to the plant of this excessive evaporation is not clear. A rapid water flow may be needed to obtain sufficient mineral salts, which are in very dilute solution in the soil. Evaporation may also help to cool the leaf when it is exposed to intense sunlight.

Against the first possibility, it has to be pointed out that, in some cases, an increased transpiration rate does not increase the uptake of minerals.

The second possibility, the cooling effect, might be very important. A leaf exposed to direct sunlight will absorb heat and its temperature may rise to a level which could kill the cytoplasm. Water evaporating from a leaf absorbs its latent heat and cools the leaf down. This is probably one value of transpiration. However, there are plants whose stomata close at around midday, greatly reducing transpiration. How do these plants avoid overheating?

Many biologists regard transpiration as an inevitable consequence of photosynthesis. In order to photosynthesize, a leaf has to take in carbon dioxide from the air. The pathway which allows carbon dioxide in will also let water vapour out whether the plant needs to lose water or not. In all probability, plants have to maintain a careful balance between the optimum intake of carbon dioxide and a damaging loss of water. Plants achieve this balance in different ways, some of which are described below.

Control of transpiration

Stomata Since most of the water vapour is lost through the stomata, the closure of these will greatly reduce transpiration. However, there is little or no evidence to suggest that stomata close in response to water loss except in extreme conditions. When loss of water greatly exceeds uptake and the plant wilts, the leaf cells, including the guard cells, become flaccid (flabby) and the stomata close, preventing further evaporation. Usually the movements of the stomata depend on the light intensity (but see p. 60), so that they are generally open during the day and closed at night. Less water vapour is lost during darkness, therefore, when photosynthesis is impossible and carbon dioxide is not needed.

Leaf fall Deciduous trees shed their leaves in winter: if they retained them, transpiration would still tend to go on although the supply of water from a cold or frozen soil, or through a frozen trunk or stem, would be very limited.

Leaf shape and cuticle Leaves with a small surface area, such as pine needles, will transpire less rapidly than the broad, flat deciduous leaves. Waxy cuticles and stomata sunk below the epidermis level are also modifications thought to be associated with reduced transpiration. They are often found in plants which grow in dry or cold conditions or in situations where water is difficult to obtain. Most evergreen plants have one or more of these leaf characteristics and this probably enables them to retain their leaves during the winter months.

Rate of transpiration

Transpiration is the evaporation of water from the leaves, so any change which increases or reduces evaporation will have the same effect on transpiration.

Light intensity Light itself does not affect evaporation, but in daylight the stomata (p. 60) of the leaves are open. This allows the water vapour in the leaves to diffuse out into the atmosphere. At night, when the stomata close, transpiration is greatly reduced.

Generally speaking, then, transpiration speeds up when light intensity increases because the stomata respond to changes in light intensity.

Sunlight may also warm up the leaves and increase evaporation (see below).

Humidity If the air is very humid, i.e. contains a great deal of water vapour, it can accept very little more from the plants and so transpiration slows down. In dry air, the diffusion of water vapour from the leaf to the atmosphere will be rapid.

Air movements In still air, the region round a transpiring leaf will become saturated with water vapour so that no more can escape from the leaf. In these conditions, transpiration would slow down. In moving air, the water vapour will be swept away from the leaf as fast as it diffuses out. This will speed up transpiration.

Temperature Warm air can hold more water vapour than cold air. Thus evaporation or transpiration will take place more rapidly into warm air.

Furthermore, when the sun shines on the leaves, they will absorb heat as well as light. This warms them up and increases the rate of evaporation of water.

Experiments which investigate the effect of some of these conditions on the rate of transpiration are described on page 77 (Experiment 2).

QUESTIONS

1 What kind of climate and weather conditions do you think will cause a high rate of transpiration?
2 What would happen to the leaves of a plant which was losing water by transpiration faster that it was taking it up from the roots?
3 In what two ways does sunlight increase the rate of transpiration?
4 Apart from drawing water through the plant, what else may be drawn up by the transpiration stream?
5 Transpiration has been described in this chapter as if it took place only in leaves. What other parts of a plant might transpire?

Water movement in the xylem

You may have learned in Physics that you cannot draw water up by 'suction' to a height of more than about 10 metres. Many trees are taller than this yet they can draw up water effectively. The explanation offered is that, in long vertical columns of water in very thin tubes, the attractive forces between the water molecules are greater than the forces trying to separate them. So, in effect, the transpiration stream is pulling up thin threads of water which resist the tendency to break.

There are still problems, however. It is likely that the water columns in some of the vessels do have air breaks in them and yet the total water flow is not affected.

Evidence for the pathway of water

Experiment 3 on page 77 uses a dye to show that, in a cut stem, the dye, and, therefore, presumably the water, travels in the vascular bundles. Closer examination with a microscope would show that it travels in the xylem vessels.

Removal of a ring of bark (which includes the phloem) does not affect the passage of water along a branch (Experiment 4). Killing parts of a branch by heat or poisons does not interrupt the flow of water, but anything which blocks the vessels does stop the flow.

The evidence all points to the non-living xylem vessels as the main route by which water passes from the soil to the leaves.

Root pressure

In Experiment 8 on page 79 it is demonstrated that liquid may be forced up a stem by pressure from the root system. The usual explanation for this is that the cell sap in the root hairs is more concentrated than the soil water and so water enters the root by osmosis (see p. 36). The water passes from cell to cell by osmosis and is finally forced into the xylem vessels in the centre of the root and up the stem.

This is rather an elaborate model from very little evidence. For example, a gradient of falling water potentials from the outside to the inside of a root has not been demonstrated. However, there is some supporting evidence for the movement of water as a result of root pressure.

Killed root systems do not exhibit any root pressure and so it is clearly dependent on intact cytoplasm and living processes in the roots. Reducing the water potential of soil water also reduces root pressure as would be expected if the osmosis story is correct.

Root pressures of 1–2 atmospheres have been recorded, and these would support columns of water 10 or 20 metres high. Some workers claim pressures of up to 8 atmospheres (i.e. 80 metres of water).

However, root pressure seems to occur mainly in young herbaceous (i.e. non-woody) plants or in woody plants early in the growing season and though in many species it must contribute to water movements in the stem, the observed rates of flow are too fast to be explained by root pressure alone.

TRANSPORT OF SALTS

The liquid which travels in the xylem is not, in fact, pure water. It is a very dilute solution, containing from 0.1 to 1.0 per cent dissolved solids, mostly amino acids, other organic acids and mineral salts. The organic acids are made in the roots; the mineral salts come from the soil. The faster the flow in the transpiration stream, the more dilute is the xylem sap.

Experimental evidence suggests that salts are carried from the soil to the leaves mainly in the xylem vessels.

TRANSPORT OF FOOD

The xylem sap is always a very dilute solution, but the phloem sap may contain up to 25 per cent of dissolved solids, the bulk of which consists of sucrose and amino acids.

There is a good deal of evidence to support the view that sucrose, amino acids and many other substances are transported in the phloem.

The movement of water and salts in the xylem is always upwards, from soil to leaf, but in the phloem, the solutes may be travelling up or down the stem. The carbohydrates made in the leaf during photosynthesis are converted to sucrose and carried out of the leaf to the stem. From here, the sucrose may pass upwards to growing buds and fruits or downwards to the roots and storage organs. All parts of a plant which cannot photosynthesize will need a supply of nutrients brought by the phloem. It is quite possible for substances to be travelling upwards and downwards at the same time in the phloem.

There is no doubt that substances travel in the sieve tubes (p. 63) of the phloem but the mechanism by which they are moved is not fully understood.

The transport of mineral salts and food in plants is sometimes called **translocation**.

Evidence of translocation in the phloem

If a leaf is supplied with radioactive carbon dioxide, the radioactive carbon (p. 31) soon appears in sucrose in the phloem. Furthermore, if the phloem of the stem below the leaf is killed by a jet of steam or removed by cutting away a ring of bark, the substances containing radioactive carbon are found to move only up the stem (Fig. 5). When the phloem above the leaf is killed or removed by ringing, conduction is only down the stem. If the phloem above and below the leaf is killed or removed, the radioactive substances do not appear anywhere in the stem. Similarly if the oxygen supply to the phloem is cut off, translocation of sugars ceases.

This is evidence which clearly shows that (*a*) sugars travel in the phloem, (*b*) they can travel in either direction and (*c*) some living process is involved.

There are several theories which attempt to explain how sucrose and other solutes are transported in the phloem but none of them is entirely satisfactory.

QUESTIONS

6 How do sieve tubes and vessels differ (*a*) in the substances they transport, (*b*) in the directions these substances are carried?

7 A complete ring of bark cut from round the circumference of a tree-trunk causes the tree to die. The xylem continues to carry water and salts to the leaves which can make all the substances needed by the tree. So why does the tree die?

8 Make a list of all the non-photosynthetic parts of a plant which need a supply of sucrose and amino acids.

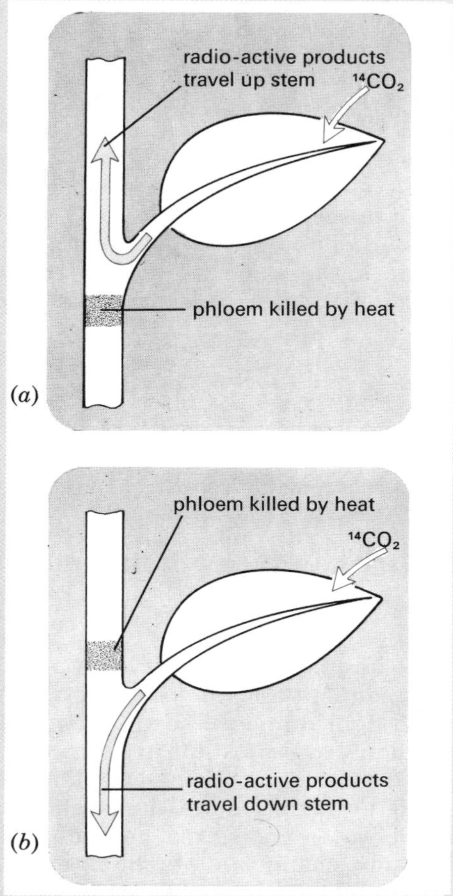

radio-active products travel up stem

$^{14}CO_2$

phloem killed by heat

(*a*)

phloem killed by heat

$^{14}CO_2$

radio-active products travel down stem

(*b*)

Fig. 5 Experimental evidence for transport in phloem (after Rabideau & Burr, *Amer. J. Bot.*, **32**, 1945)

UPTAKE OF WATER AND SALTS

Uptake of water

The water tension developed in the vessels by a rapidly transpiring plant is thought to be sufficient to draw water through the root from the soil. The precise pathway taken by the water is the subject of some debate, but the path of least resistance seems to be in or between the cell walls rather than through the cells.

When transpiration is slow, e.g. at night time or just before bud burst in a deciduous tree, then osmosis may play a more important part in the uptake of water. Figure 6 shows a root hair in the soil. The cytoplasm of the root hair is partially permeable to water. The soil water is more dilute than the cell sap and so water passes by osmosis (p. 36) from the soil into the cell sap of the root hair cell. This flow of water into the root hair cell raises the cell's turgor pressure (p. 38). So water is forced out through the cell wall into the next cell and so on, right through the cortex of the root to the xylem vessels (Fig. 7).

One problem for this explanation is that it has not been possible to demonstrate that there is an osmotic gradient across the root cortex which could produce this flow of water from cell to cell. Nevertheless, root pressure developed probably by osmosis can be shown to force water up the root system and into the stem (see pp. 38 and 79).

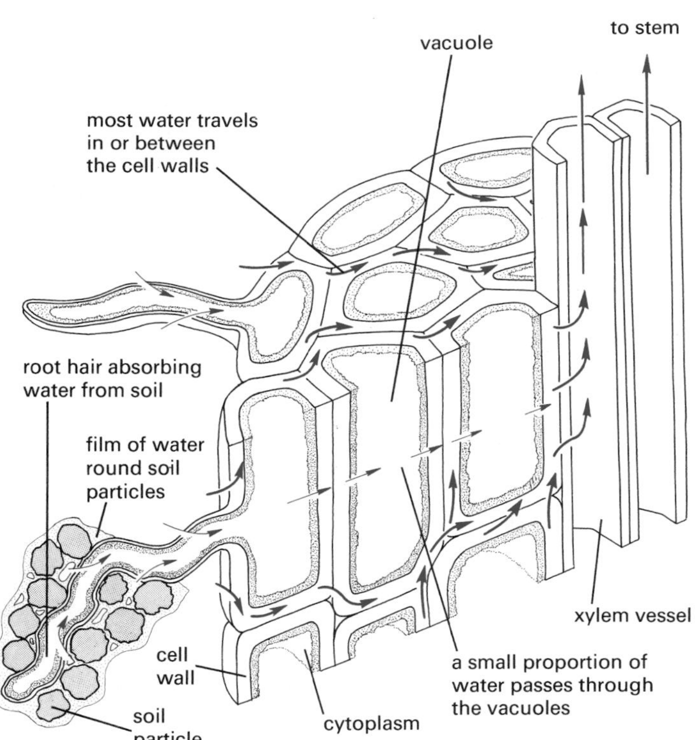

Fig. 6 The probable pathway of water through a root

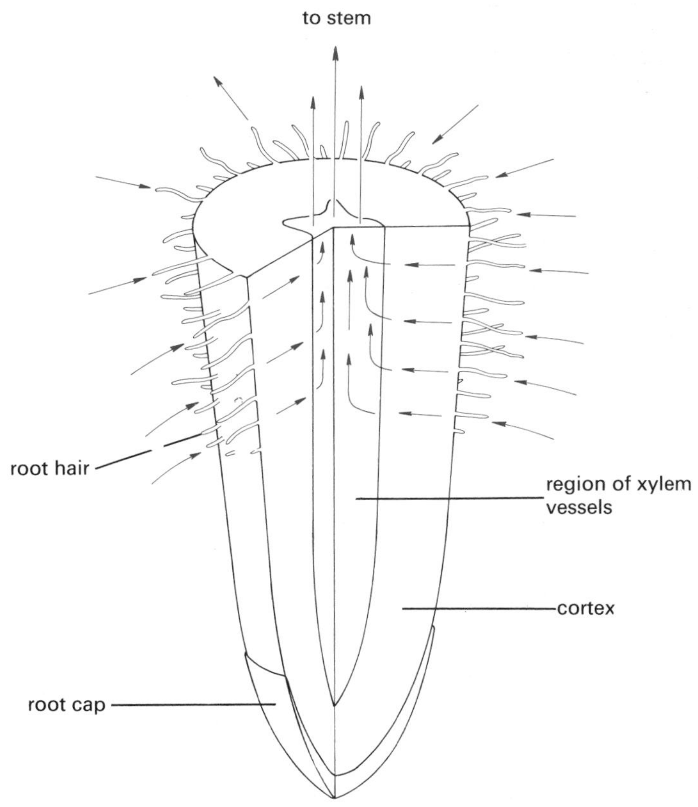

Fig. 7 Diagrammatic section of root to show passage of water from the soil

Uptake of salts

The methods by which roots take up salts from the soil are not fully understood. Some salts may be carried in with the water drawn up by transpiration and pass mainly along the cell walls in the root cortex and into the xylem.

It may be that diffusion from a relatively high concentration in the soil to a lower concentration in the root cells accounts for uptake of some individual salts, but it has been shown (*a*) that salts can be taken from the soil even when their concentration is below that in the roots and (*b*) that anything which interferes with respiration impairs the uptake of salts. This suggests that 'active transport' (p. 35) plays an important part in the uptake of salts.

The growing region of the root and the root hair zone seem to be most active in taking up salts. Most of the salts appear to be carried at first in the xylem vessels, though they soon appear in the phloem as well.

The salts are used by the plant's cells to build up essential molecules. Nitrates, for example, are combined with carbohydrates to make amino acids in the roots. These amino acids are used later to make proteins.

QUESTIONS

9 If root hairs take up water from the soil by osmosis, what would you expect to happen if so much nitrate fertilizer was put on the soil that the soil water became a stronger solution than the cell sap of the root hairs?
10 A plant's roots may take up water and salts less efficiently from a waterlogged soil than from a fairly dry soil. Revise 'Active transport' (p. 35) and 'Pore spaces' in the soil (p. 255) and suggest reasons for this.
11 Why do you think that, in a deciduous tree in spring, transpiration is negligible before bud burst?

TRANSPORT OF GASES

The process of diffusion described on page 33 accounts for the movement of gases in and out of a plant. During respiration, oxygen is taken in and carbon dioxide given out. When photosynthesis is faster than respiration, carbon dioxide diffuses in and oxygen diffuses out (Fig. 7, p. 49). In leaves and green stems, the gases enter and leave through the stomata (p. 60). Then they diffuse through the air spaces between the cells to reach all parts of the plant shoot. In woody plants, the stems have no stomata and the gases have to pass through small openings in the bark called **lenticels.**

Roots obtain their oxygen from the air spaces in the soil (p. 255). Much of this oxygen will be dissolved in the soil water which enters the root through the growing region and the root hairs.

ABSORPTION BY LEAVES

Leaves are able to absorb certain substances if these are sprayed on to them. Mineral ions in solution can be absorbed through the cuticle or stomata and, for some crops, this is a method of applying 'fertilizer'. (The process is called **foliar feeding**.)

Some insecticide and fungicide sprays are absorbed through the leaves and translocated through the plant. Such pesticides are called **systemic** because they enter the plant's system. A caterpillar which ate part of a leaf treated with a systemic insecticide would be poisoned by the chemical in the cells of the leaf.

It is important, of course, that systemic pesticides are broken down to harmless compounds by the plant, long before its leaves or fruits are used for human consumption.

PRACTICAL WORK

EXPERIMENTS ON TRANSPORT IN PLANTS

1. To measure the rate of transpiration

Since transpiration is the *loss* of water vapour, the rate of transpiration can be measured by weighing a plant at intervals to see how much weight it loses.

Water a potted plant and enclose the pot and soil in a plastic bag and tie it round the plant's stem (Fig. 8). This makes sure that any losses are due to evaporation from the shoot and not from the soil. Weigh the plant at intervals. If it loses, say, 56 grams in 4 hours, the rate of transpiration is 14 grams of water transpired per hour. This result assumes that any change in weight is entirely due to transpiration. In fact, there may be small gains in weight due to absorption of carbon dioxide in photosynthesis, or small losses due to the escape of carbon dioxide from respiration. In practice, over a few hours, these changes are very small compared with the losses due to transpiration.

A control experiment should be carried out. A pot of soil is watered and wrapped in a plastic bag exactly as before but without having a plant. If this is weighed over the same period as the experimental plant, there should be little or no change in weight. This shows that, in the experiment, the plant was responsible for the loss in weight.

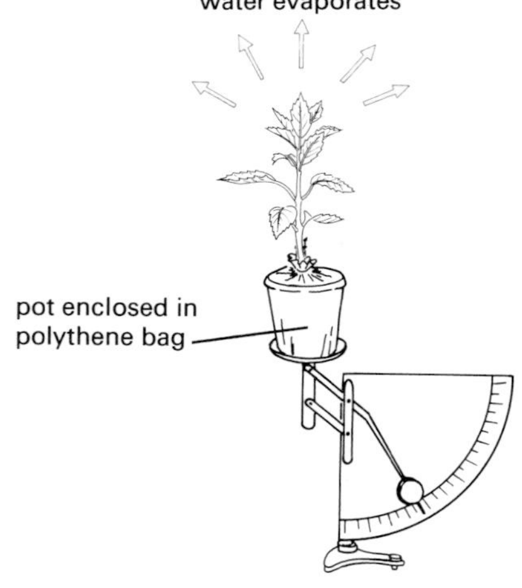

Fig. 8 Measuring the rate of transpiration in a potted plant

2. Rates of water uptake in different conditions

The apparatus shown in Fig. 10 is called a **potometer**. It is designed to measure the rate of uptake of water in a cut shoot.

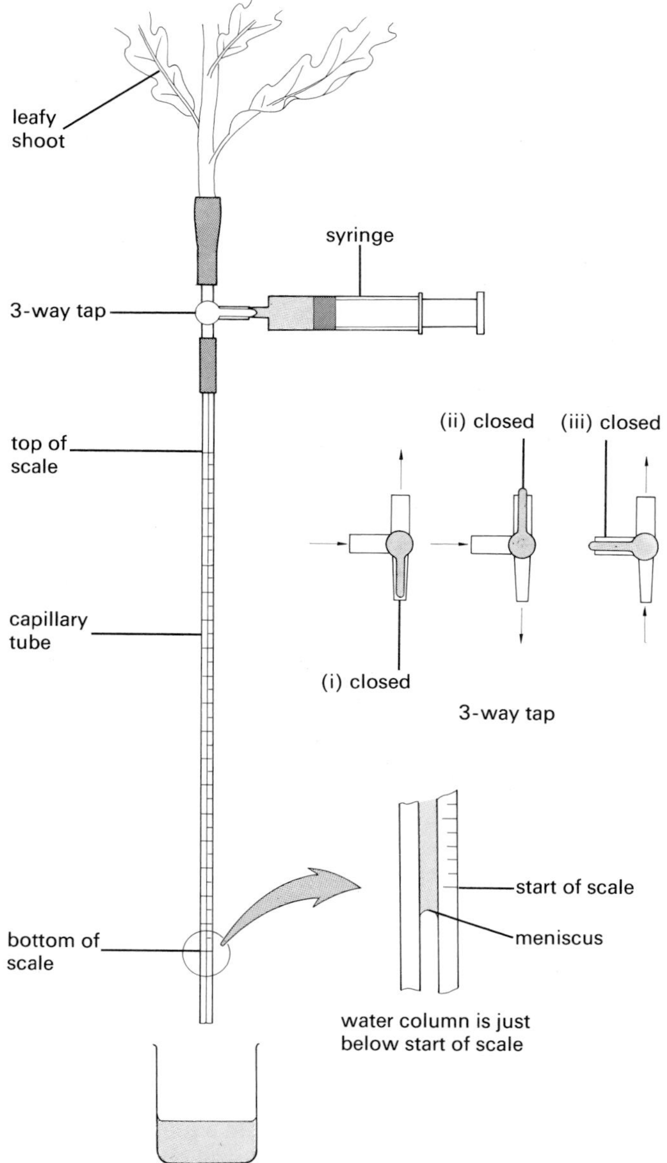

Fig. 9 A potometer

Fill the syringe with water and attach it to the side arm of the 3-way tap. Turn the tap downwards (i) and press the syringe until water comes out of the rubber tubing at the top.

Collect a leafy shoot and push its stem into the rubber tubing as far as possible. Set up the apparatus in a part of the laboratory that is not receiving direct sunlight.

Turn the tap up (ii) and press the syringe till water comes out of the bottom of the capillary tube. Turn the tap horizontally (iii).

As the shoot transpires, it will draw water from the capillary tube and the level can be seen to rise. Record the distance moved by the water column in 30 seconds or a minute.

Turn the tap up and send the water column back to the bottom of the capillary. Turn the tap horizontally and make another measurement of the rate of uptake. In this way obtain the average of three readings.

The conditions can now be changed in one of the following ways:

1. Move the apparatus into sunlight or under a fluorescent lamp.
2. Blow air past the shoot with an electric fan or merely fan it with an exercise book.
3. Cover the shoot with a plastic bag.

After the change of conditions, take three more readings of the rate of uptake and notice whether they represent an increase or a decrease in the rate of transpiration.

You might expect results as follows:

1. An increase in light intensity should make the stomata open and allow more rapid transpiration.
2. Moving air should increase the rate of evaporation and, therefore, the rate of uptake.
3. The plastic bag will cause a rise in humidity round the leaves and suppress transpiration.

Ideally, you should change only one condition at a time. If you took the experiment outside, you would be changing the light intensity, the temperature and the air movement. When the rate of uptake increased, you would not know which of these three changes was mainly responsible.

To obtain reliable results, you should really keep taking readings until three of them are nearly the same. A change in conditions may take 10 or 15 minutes before it produces a new, steady rate of uptake. In practice, you may not have time to do this, but even your first three readings should indicate a trend towards increased or decreased uptake.

Limitations of the potometer Although we use the potometer to compare rates of transpiration, it is really the rates of uptake that we are observing. Not all the water taken up will be transpired; some will be used in photosynthesis; some may be absorbed by cells to increase their turgor. However, these quantities are very small compared with the volume of water transpired and they can be disregarded.

The rate of uptake of a cut shoot may not reflect the rate in the intact plant. If the root system were present, it might offer resistance to the flow of water or it could be helping the flow by means of its root pressure.

3. Transport in the vascular bundles

Place the shoots of several leafy plants in a solution of 1 per cent methylene blue. 'Busy Lizzie' (*Impatiens*) or celery stalks with leaves are usually effective. Leave the shoots in the light for 30 minutes or more.

Result In some cases, after this time, the blue dye will appear in the leaf veins. If some of the stems are cut across, the dye will be seen in the vascular bundles (see Fig. 3, p. 3).

Interpretation These results show that the dye and, therefore, probably also the water, travels up the stem in the vascular bundles. Closer study would show that they travel in the xylem vessels.

4. Transport of water in the xylem

Cut three leafy shoots from a deciduous tree or shrub. Each shoot should have about the same number of leaves.

On one twig remove a ring of bark about 5 mm wide, about 100 mm up from the cut base. With the second shoot, smear a layer of Vaseline over the cut base so that it blocks the vessels. The third twig is a control.

Place all three twigs in a jar with a little water. The water-level must be below the region from which you removed the ring of bark. Leave the twigs where they can receive direct sunlight.

Result After an hour or two, you will probably find that the twig with blocked vessels shows signs of wilting. The other two twigs should still have turgid leaves.

Interpretation Removal of the bark (including the phloem) has not prevented water from reaching the leaves, but blocking the xylem vessels has. The vessels of the xylem, therefore, offer the most likely route for water passing up the stem.

5. Production of water vapour by transpiration

Enclose the shoot of a recently watered potted plant, or a plant in the garden, in a transparent polythene bag which is then tied round the base of the stem (Fig. 10). Leave the plant for an hour or two in direct sunlight.

The water vapour transpired by the plant will soon saturate the atmosphere inside the bag and drops of water will condense on the inside.

Remove the bag and shake all the condensed water into a corner and then pour a few drops of it on to some anhydrous copper sulphate or a piece of blue cobalt chloride paper.

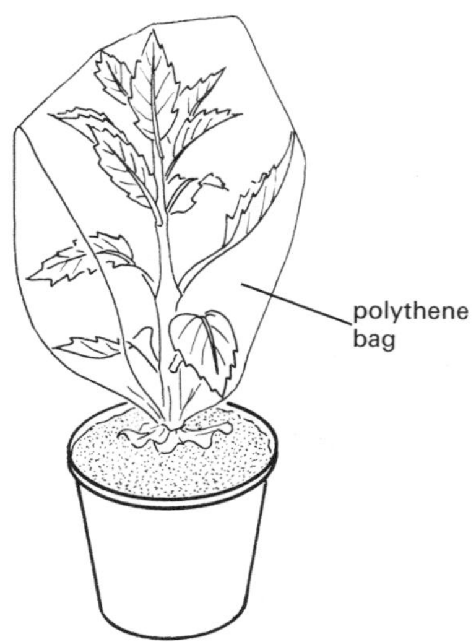

Fig. 10 Water produced in transpiration

If the copper sulphate goes blue, or the cobalt chloride turns pink, this is evidence of the presence of water in the condensed liquid (though it does not prove it is pure water).

A control can be set up using a shoot in a similar situation but from which all the leaves and flowers have been detached. Little or no water should condense in the plastic bag.

6. To find which surface of a leaf loses more water vapour

Filter paper is soaked in a 5 per cent solution of cobalt chloride, dried and cut into 5 mm squares. When dry the paper is blue, but it changes to pink when damp.

Take two squares of this cobalt chloride paper and, holding them with forceps, dry each in turn over a bench lamp or a small Bunsen flame until they are blue. Then stick them to a leaf using transparent sticky tape (Fig. 11). Stick one square on the upper surface and the other on the lower surface and note the time it takes for the cobalt chloride paper to turn pink.

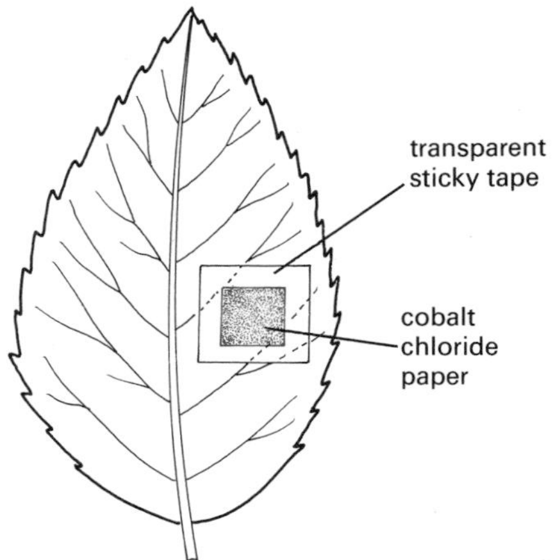

Fig. 11 To find which surface of a leaf loses more water vapour

Paint a small area on the upper and lower surface of the leaf with clear nail varnish and leave it to dry for 10 minutes. Cut a small piece (about 10 mm square) from part of the leaf which has the dried varnish and press the lower surface on to a piece of transparent adhesive tape. Use fine forceps to peel the leaf away from the tape, leaving the varnish peel behind. Stick the tape to a glass slide and examine the varnish peel under the microscope. The varnish film will bear an impression of the leaf epidermis and show the stomata. Count the number of stomata visible in the field or a fraction of the field of the microscope.

Repeat this operation with the varnish peel from the upper surface.

You would expect the side with more stomata to release more water vapour. This side, therefore, should be the first to turn the cobalt chloride paper pink.

In the leaves of most trees and shrubs, the stomata are on the underneath surface only. The leaves of grasses and related plants have stomata on both sides.

7. The pathway for gases

Heat a beaker of water to about 70 °C. Use forceps to put a leaf under the water (Fig. 12) and watch both surfaces of the leaf for air bubbles. The heat from the water expands the air inside the leaf and forces it out of the stomata. By observing the number of air bubbles on each side, it should be possible to see which side of the leaf has more stomata. Try comparing a rose leaf with a grass leaf.

You can also do this experiment with a piece of woody stem to see the air forced out of the lenticels, but you must first seal the cut ends of the twig with Plasticine to stop air escaping this way.

If you take an onion leaf (which is tubular) and blow down the cut end while holding the leaf under water, you will see air escaping from the stomata.

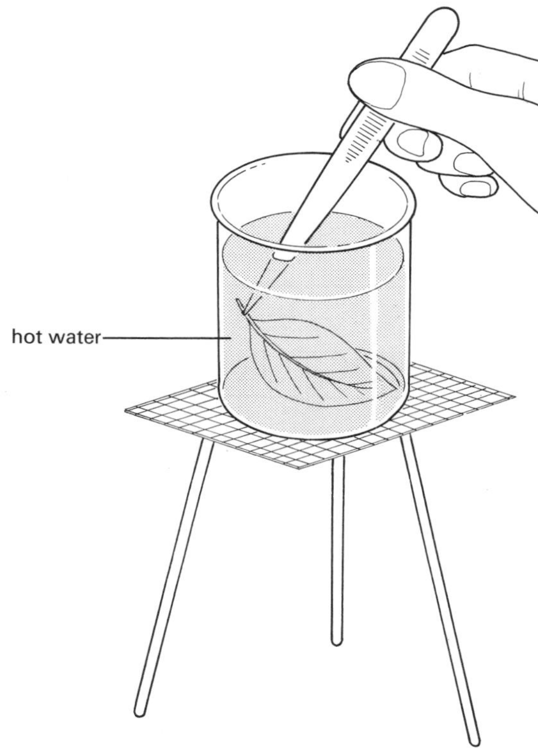

Fig. 12 The route for gaseous exchange in a leaf

8. To demonstrate root pressure

Connect a piece of glass tubing by means of rubber tubing to the freshly cut stem of a potted plant (Fig. 13). Place a little coloured water in the tube and mark its level. If the roots are kept well watered, the coloured water will rise a few centimetres in the tube. This demonstrates root pressure.

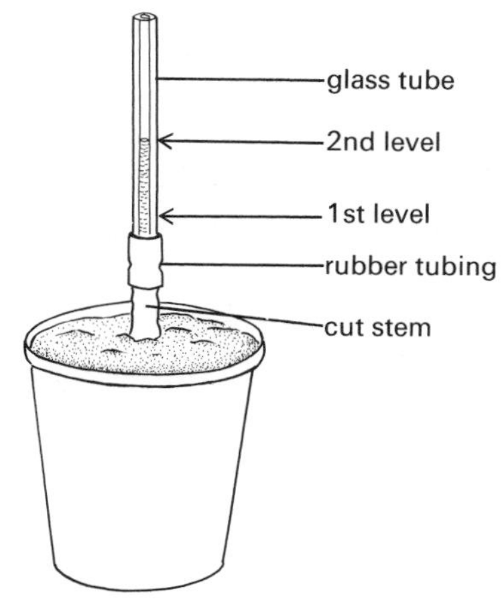

Fig. 13 Root pressure

QUESTIONS

12 A leafy shoot, plus the beaker of water in which it is placed, weighs 275 grams. Two hours later, it weighs 260 grams. An identical beaker of water, with no plant, loses 3 grams over the same period of time. What is the rate of transpiration of the shoot?

13 In Experiment 6, suggest (*a*) why forceps are used to handle the cobalt chloride paper squares, (*b*) why it was water vapour from the leaf and not water vapour from the air which made the cobalt chloride paper change from blue to pink.

14 An interesting experiment is to use a syringe, attached to the cut end of a woody shoot, to force air through the xylem vessels. If the shoot is held under water while you press the syringe, would you expect to see air escaping from the stomata? If you cut away the top half of each leaf, what would you expect to see when the syringe is pressed? Explain your reasoning.

CHECK LIST

- **Transpiration is the evaporation of water from the leaves of a plant.**
- **Transpiration produces the force which draws water up the stem.**
- **The water travelling in the transpiration stream will contain dissolved salts.**
- **Closure of stomata and shedding of leaves may help to regulate the transpiration rate.**
- **The rate of transpiration is increased by sunlight, high temperature, low humidity and air movements.**
- **Root pressure forces water up the stem as a result of osmosis in the roots.**
- **Salts are taken up from the soil by roots, and are carried in the xylem vessels.**
- **Water and salts move up the stem from the roots to the leaves.**
- **Food made in the leaves moves up or down the stem in the phloem.**
- **Oxygen and carbon dioxide move in or out of the leaf by diffusion through the stomata.**

8 Sexual Reproduction in Flowering Plants

FLOWER STRUCTURE

Sepals, petals, stamens, carpel, receptacle. Wallflower and lupin. Composite flowers. Grasses.

POLLINATION

Cross- and self-pollination. Wallflower, lupin and grass. Adaptation to wind and insect pollination.

FERTILIZATION AND FRUIT FORMATION

Fertilization: the pollen tube. Fruit and seed formation; the ovary forms the fruit.

DISPERSAL OF FRUIT AND SEEDS

Wind, animal and explosive methods.

PRACTICAL WORK

Growth of pollen tubes.

Flowers are reproductive structures; they contain the reproductive organs of the plant. The male organs are the **stamens** which produce pollen. The female organs are the **carpels**. After fertilization, part of the carpel becomes the fruit of the plant and contains the seeds. In the flowers of most plants there are both stamens and carpels. These flowers are, therefore, male and female at the same time, a condition known as **bisexual** or **hermaphrodite**.

Some species of plants have unisexual flowers, i.e. any one flower will contain either stamens or carpels but not both. Sometimes both male and female flowers are present on the same plant, e.g. the hazel, which has male and female catkins on the same tree. In the willow tree, on the other hand, the male and female catkins are on different trees.

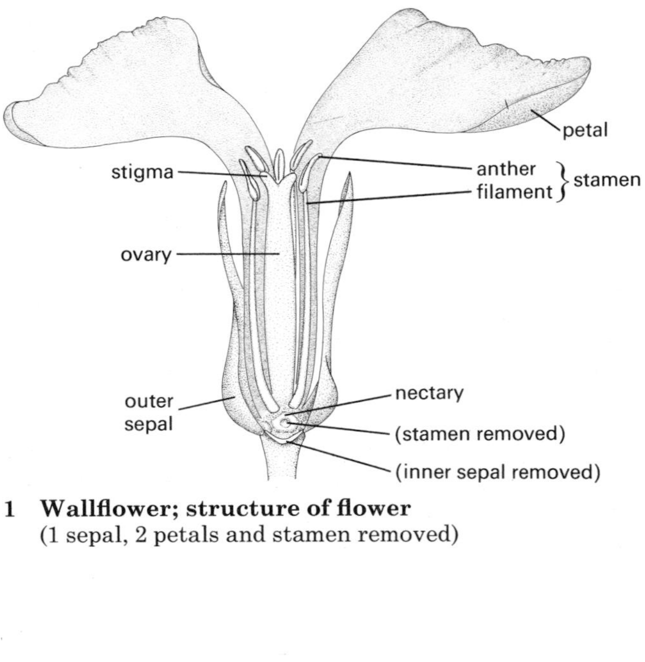

Fig. 1 Wallflower; structure of flower
(1 sepal, 2 petals and stamen removed)

FLOWER STRUCTURE

The basic structure of a flower is shown in Figs 1 and 2.

Petals These are usually brightly coloured and sometimes scented. They are arranged in a circle (Fig. 1) or a cylinder. Most flowers have from four to ten petals. Sometimes they are joined together to form a tube (Fig. 3) and the individual petals can no longer be distinguished. The colour and scent of the petals attract insects to the flower; the insects may bring about pollination (p. 83).

The flowers of grasses and many trees do not have petals but small, leaf-like structures which enclose the reproductive organs.

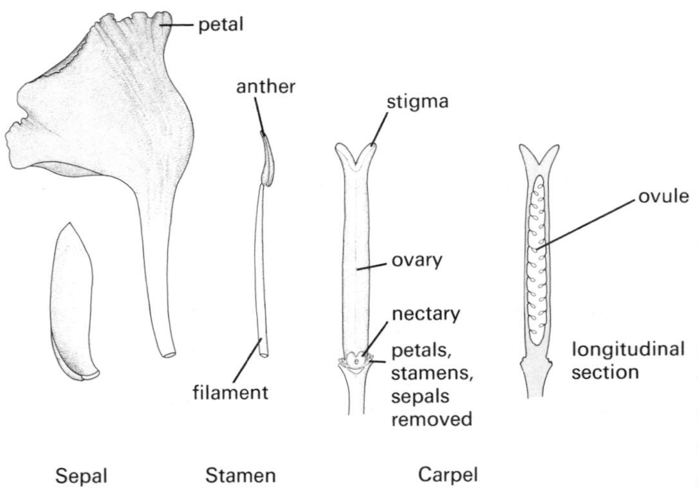

Fig. 2 Floral parts of wallflower

Fig. 3 Daffodil flower cut in half. The inner petals form a tube. Three stamens are visible round the long style and the ovary contains many ovules.

Sepals Outside the petals is a ring of sepals. They are often green and much smaller than the petals. They may protect the flower when it is in the bud.

Stamens The stamens are the male reproductive organs of a flower. Each stamen has a stalk called the **filament**, with an **anther** on the end. Flowers such as the buttercup and blackberry have many stamens; others such as the tulip have a small number, often the same as, or double, the number of petals or sepals. Each anther consists of four **pollen sacs** in which the pollen grains are produced by cell division. When the anthers are ripe, the pollen sacs split open and release their pollen (Fig. 11).

Carpels These are the female reproductive organs. Flowers such as the buttercup and blackberry have a large number of carpels while others, such as the lupin, have a single carpel. Each carpel consists of an **ovary**, bearing a **style** and a **stigma**.

Inside the ovary there are one or more **ovules**. Each blackberry ovary contains one ovule but the lupin ovary contains several. The ovule will become a **seed**, and the whole ovary will become a **fruit**. (In biology, a fruit is the fertilized ovary of a flower, not necessarily something to eat.)

The style and the stigma project from the top of the ovary. The stigma has a sticky surface and pollen grains will stick to it during pollination. The style may be quite short (wallflower, Fig. 1) or very long (lupin, Fig. 4).

Receptacle The flower structures just described are all attached to the expanded end of a flower stalk. This is called the **receptacle** and, in a few cases after fertilization, it becomes fleshy and edible, e.g. strawberry (Fig. 24), apple and pear.

Wallflower

The wallflower (Fig. 1) has four sepals (two outer and two inner), and four petals. The sepals and the bases of the petals are not joined up but form a kind of tube which encloses the stamens and ovary.

You might expect to find four or eight stamens but, in fact, there are six; two outer and an inner ring of four.

At the base of each outer stamen there is a swelling which forms a **nectary**. This produces a sugary solution called **nectar** which collects in the cup-like base of the outer sepals. Insects visit the flower to pick up this nectar.

The ovary consists of two carpels joined to make a fairly long pod-like ovary with a short style and a two-lobed stigma at the top. There are two rows of about ten ovules in each carpel.

In order to show the structure of a flower, it is usual to draw a 'half flower'. This shows the flower as it would appear if cut vertically down the middle. In the wallflower, the cut would have to pass between the petals or the sepals. So, Fig. 1 is not a true vertical section or half flower but a flower from which one sepal, two petals and one outer stamen have been removed to show the structure.

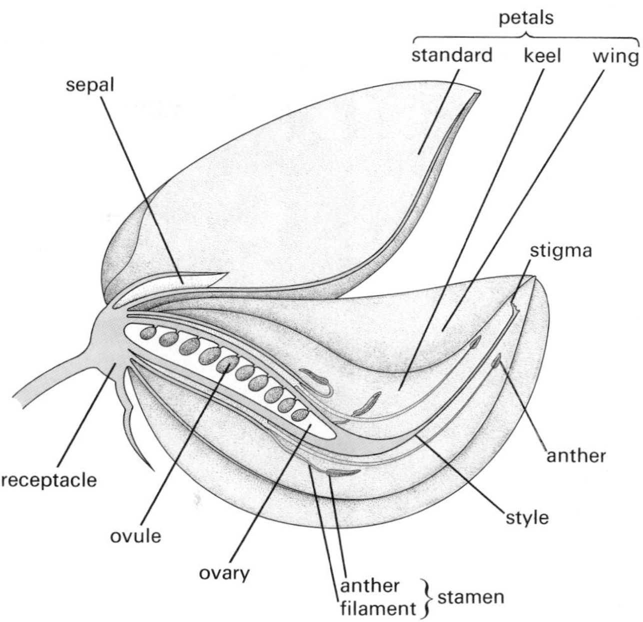

Fig. 4 Half-flower of lupin

Lupin

The lupin flower is shown in Figs 4–6. There are five sepals but these are joined together forming a short tube. The five petals are of different shapes and sizes. The uppermost, called the **standard**, is held vertically. Two petals at the sides are called **wings** and are partly joined together. Inside the wings are two more petals joined together to form a boat-shaped **keel**.

The single carpel is long, narrow and pod-shaped, with about ten ovules in the ovary. The long style ends in a stigma just inside the pointed end of the keel. There are ten stamens, five long ones and five short ones. Their filaments are joined together at the base to form a sheath round the ovary.

The flowers of peas and beans are very similar to those of lupins.

The shoots or branches of a plant carrying groups of flowers are called **inflorescences**. The flowering shoots of the lupin in Fig. 6 are inflorescences, each one carrying about a hundred individual flowers.

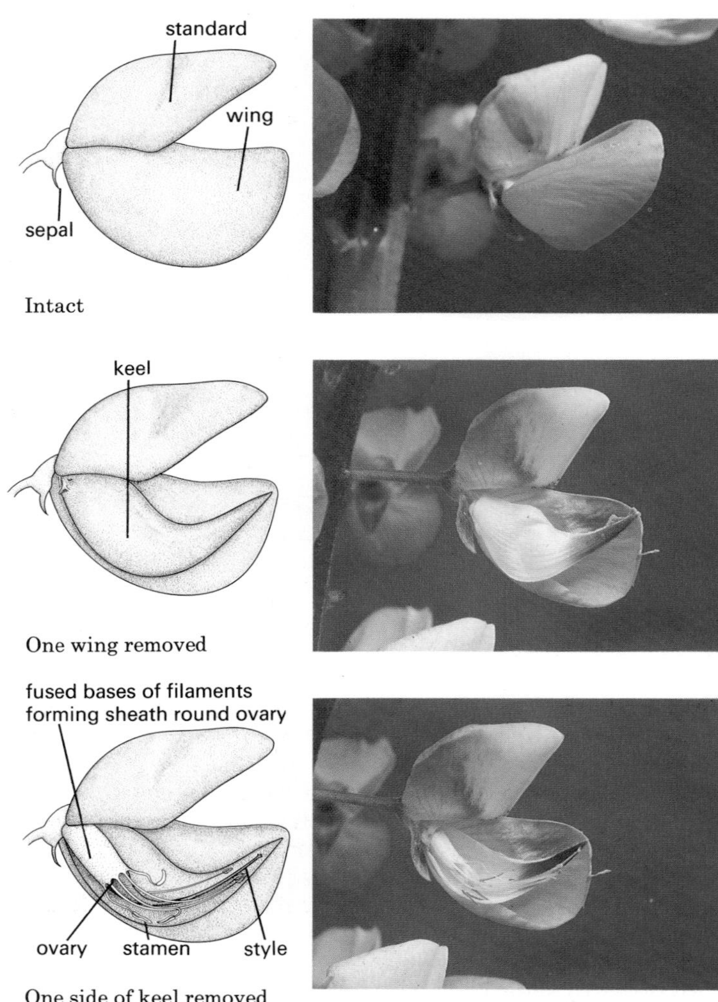

Fig. 5 Lupin flower dissected

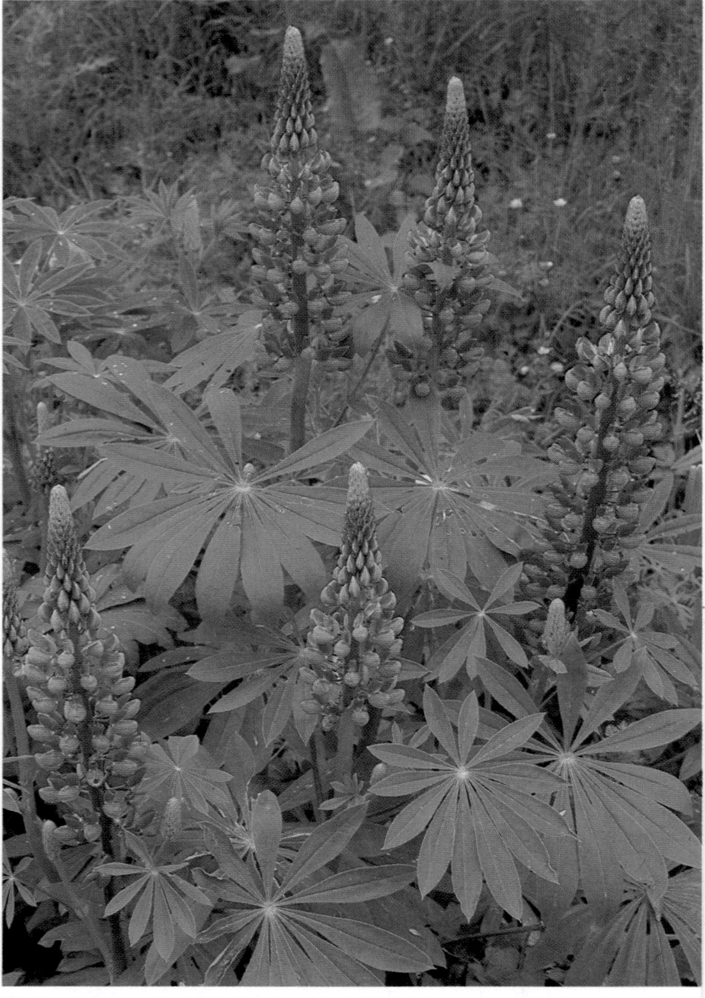

Fig. 6 Lupin inflorescence. There are a hundred or more flowers in each inflorescence. The youngest flowers, at the top, have not yet opened. The oldest flowers are at the bottom and have already been pollinated.

Composite flowers

The flowers of the **Compositae** family (daisies, dandelions, hawkweeds, etc.) are arranged in dense inflorescences (Fig. 7). What at first appears to be a petal in the flower head is actually a complete flower, often called a **floret** (Fig. 8).

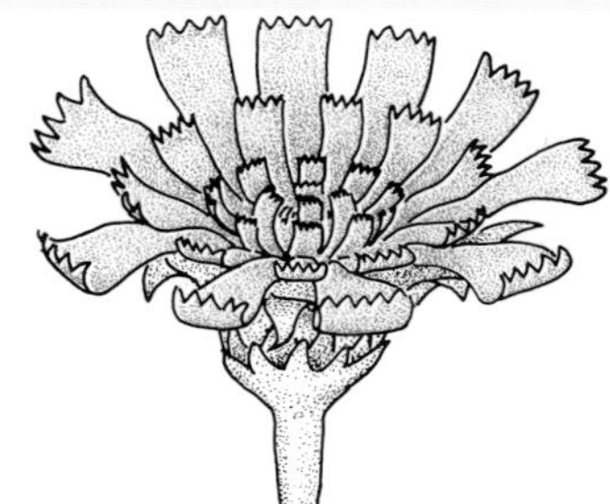

Fig. 7 A hawkweed. These composite flowers are inflorescences, that is, groups of many flowers.

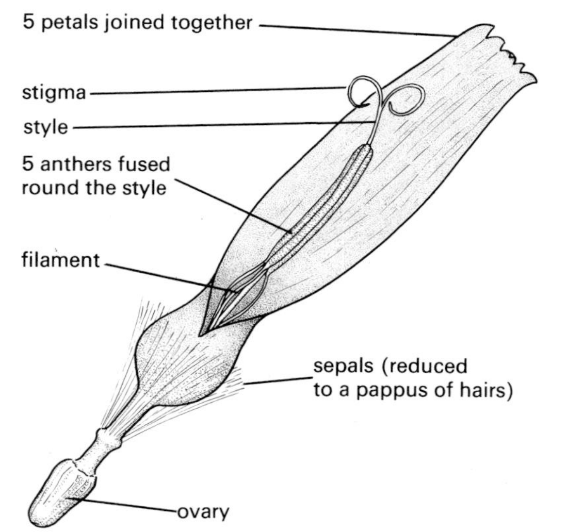

Fig. 8 Single floret of dandelion

Dandelion In the outer florets of the dandelion inflorescence there are five petals joined together. At the base, the fused petals form a tube but open out to a ribbon-like structure at the top. The sepals are reduced to a ring of fine hairs; the five fused anthers are grouped round the style, which has a forked stigma. There is a single ovule in the ovary.

In some Compositae, the outer florets with conspicuous petals are unisexual or have no reproductive organs at all. The inner florets, with tiny, joined petals, carry the male and female reproductive organs. The various daisies are examples of this type of inflorescence.

Grasses

The flowers of the grasses are tiny, inconspicuous and in dense inflorescences. There are no petals or sepals in the usual sense, but the reproductive organs are enclosed in two green, leaf-like structures called **bracts**. The ovary contains one ovule and bears two styles with feathery stigmas. There are three stamens, the anthers of which, when ripe, hang outside the bracts.

The cereals wheat, oats, barley, maize, etc. are grasses especially bred and cultivated by humans for the sake of the food stored in the fruits or seeds of their flowers.

Rye grass The inflorescence and flower of this grass are illustrated in Fig. 9.

QUESTIONS

1 Working from outside to inside, list the parts of a bisexual flower.
2 What features of flowers attract insects?
3 When you next get the chance, study the large white inflorescences of hedge parsnip or hogweed (Umbelliferae family). How do these inflorescences differ from those of the Compositae?

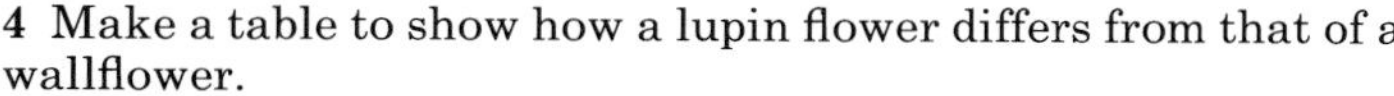
4 Make a table to show how a lupin flower differs from that of a wallflower.

	Lupin flower	*Wallflower*
Sepals Petals Stamens Ovary		

POLLINATION

The transfer of pollen from the anthers to the stigma is called 'pollination'. The anthers split open, exposing the microscopic pollen grains (Figs 10 and 11). The pollen grains are then carried away on the bodies of insects, or simply blown by the wind, and may land on the stigma of another flower. In **self-pollinating** plants, the pollen which reaches the stigma comes from the same flower or another flower on the same plant. In **cross-pollination**, the pollen is carried from the anthers of one flower to the stigma in a flower on another plant of the same species.

If a bee carried pollen from one of the younger flowers near the middle of the lupin inflorescence (Fig. 6) to an older flower near the bottom, this would be self-pollination. If, however, the bee flew off to visit a separate lupin plant and pollinated one of its flowers, this would be cross-pollination.

The term 'cross-pollination', strictly speaking, should be applied only if there are genetic differences between the two plants involved. The flowers on a single plant all have the same genetic constitution. The flowers on plants growing from the same rhizome or rootstock (p. 100) will also have the same genetic constitution. Pollination between such flowers is little different from self-pollination in the same flower.

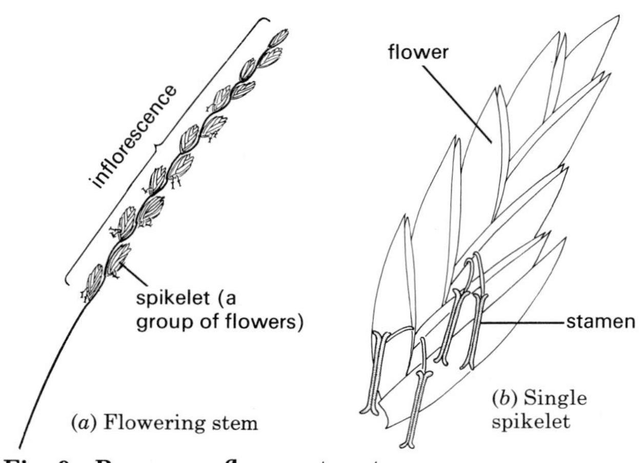

(*a*) Flowering stem

(*b*) Single spikelet

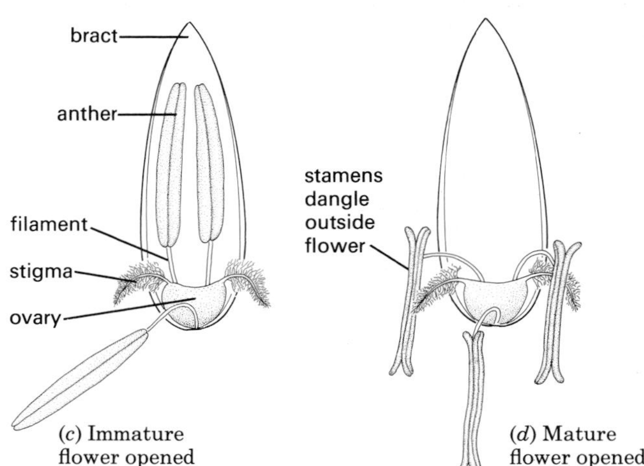

(*c*) Immature flower opened

(*d*) Mature flower opened

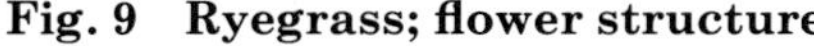
Fig. 9 Ryegrass; flower structure

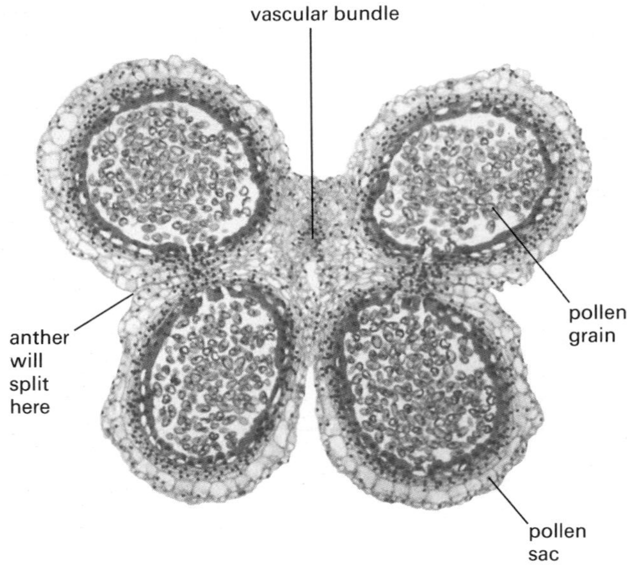

Fig. 10 Transverse section through an anther (×75)

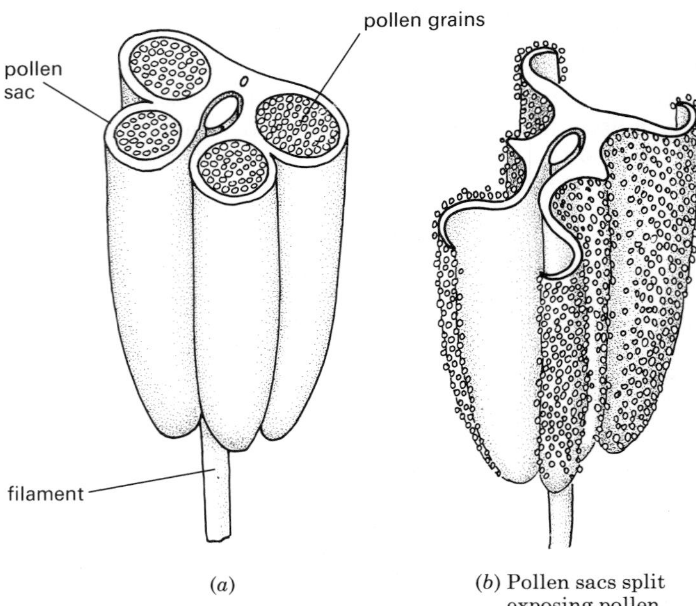

Fig. 11 Structure of anther (top cut off)

Pollination of the wallflower

Nectar is produced by the nectaries at the base of the outer stamens and collects in the cup-like base of the two outer sepals. The sepals and bases of the petals form a narrow tube which is partially blocked by the stamens and ovary. Thus only insects with long probosces ('tongues') can reach the nectar.

Fig. 12 Pollination of the wallflower. The bee pushes past the petals to reach the nectary and pollen is deposited on its body.

Long-tongued bees still have to push some way into the petal tube (Fig. 12) and, in so doing, become dusted with pollen from the ripe anthers. When the same bee visits an older flower, some of the pollen adhering to its body will be picked up by the sticky surface of the stigma.

Butterflies and moths have longer probosces than bees do. Even so, when a butterfly collects nectar from the flower, pollen is likely to adhere to its proboscis and be transferred to another flower.

Pollination of the lupin

Lupin flowers have no nectar. The bees which visit them come to collect pollen which they take back to the hive for food. Other members of the lupin family (Leguminosae), e.g. clover, do produce nectar.

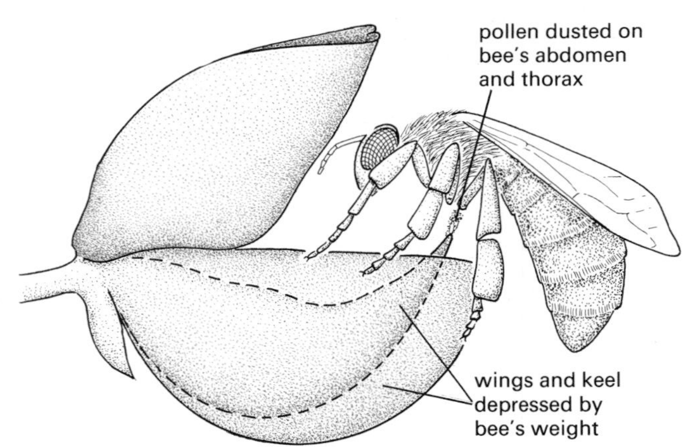

Fig. 13 Pollination of the lupin

The weight of the bee, when it lands on the flower's wings, pushes down these two petals and the petals of the keel. The pollen from the anthers has collected in the tip of the keel and as the petals are pressed down, the stigma and long stamens push the pollen out from the keel on to the underside of the bee (Fig. 13). The bee, with pollen grains sticking to its body, then flies to another flower. If this flower is older than the first one, it will already have lost its pollen. When the bee's weight pushes the keel down, only the stigma comes out and touches the insect's body, picking up pollen grains on its sticky surface.

Lupin and wallflower are examples of **insect-pollinated** flowers.

Pollination of the grasses

Grasses are pollinated, not by insects, but by air currents. This is called **wind pollination**.

At first, the feathery stigmas protrude from the flower, and pollen grains floating in the air are trapped by them. Later, the anthers hang outside the flower (Figs 9*b* and 14), the pollen sacs split, and the wind blows the pollen away. This sequence varies with species.

If the branches of a birch or hazel tree with ripe male catkins, or the flowers of the ornamental pampas grass, are shaken, a shower of pollen can easily be seen (Fig. 15).

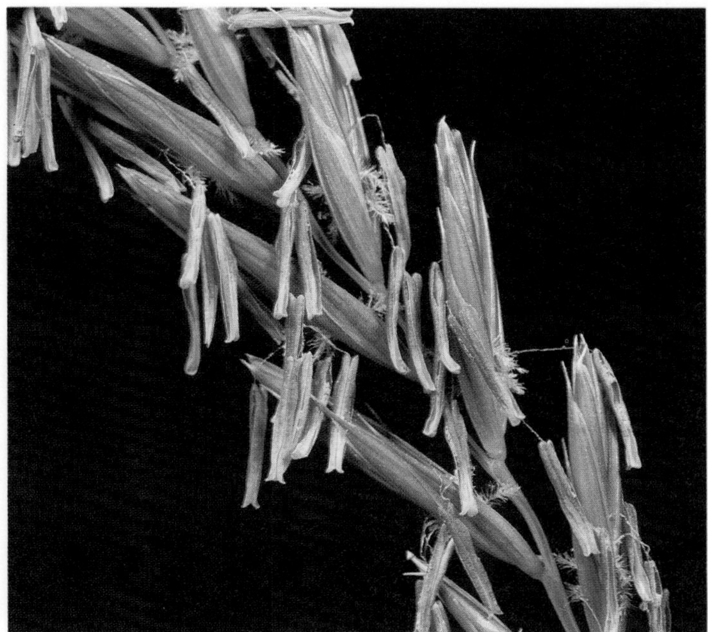

Fig. 14 Grass flowers. The stigmas are the fluffy white structures. Notice that the anthers and stigmas protrude from the flowers.

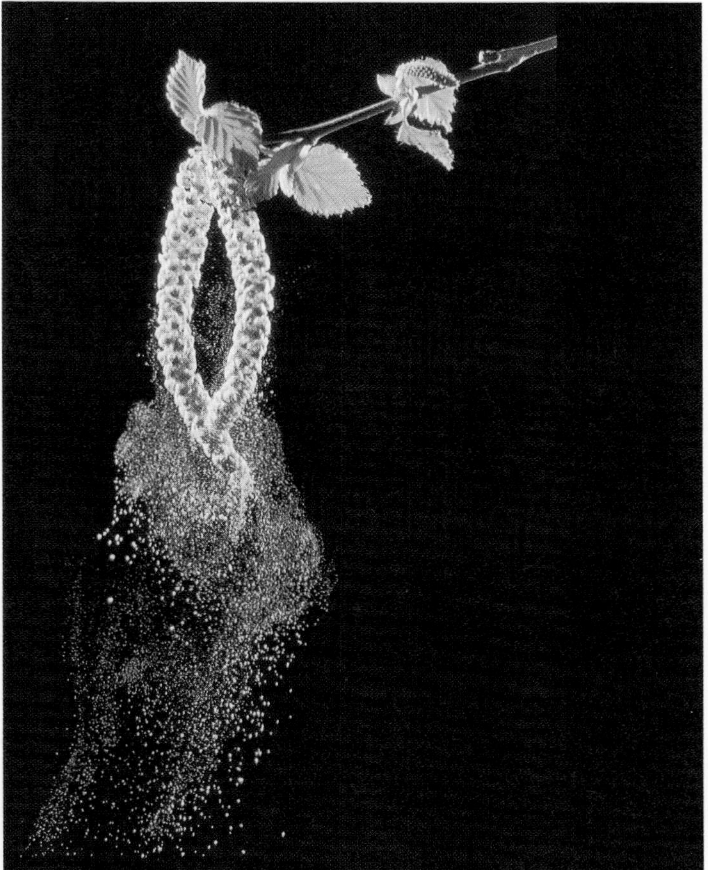

Fig. 15 Birch catkins shedding pollen

Adaptation

Insect-pollinated flowers are considered to be adapted in various ways to their method of pollination. The term '**adaptation**' implies that, in the course of evolution, the structure and physiology of a flower have been modified in ways which improve the chances of successful pollination by insects.

Most insect-pollinated flowers have brightly coloured petals and scent which attract a variety of insects. Some flowers produce nectar which is also attractive to many insects. The dark lines ('honey guides') on petals are believed to help direct the insects to the nectar source and thus bring it into contact with the stamens and stigma.

These features are adaptations to insect pollination in general, but are not necessarily associated with any particular insect species. The various petal colours and the nectaries of the wallflower attract a variety of insects. Many flowers, however, have modifications which adapt them to pollination by only one type or species of insect. The lupin can be pollinated only by insects heavy enough to depress the 'wing' petals and so cause the pollen or stigma to be extruded. Flowers such as the honeysuckle, with narrow, deep petal tubes are likely to be pollinated only by moths or butterflies whose long 'tongues' can reach down the tube to the nectar.

Tube-like flowers such as foxgloves need to be visited by fairly large insects to effect pollination. The petal tube is often lined with dense hairs which impede small insects that would take the nectar without pollinating the flower. A large bumble-bee, however, pushing into the petal tube, is forced to rub against the anthers and stigma.

Some orchid flowers resemble certain species of bee so closely in appearance and odour that male bees try to copulate with the flower and, in doing so, encounter the anthers and stigma.

Many tropical and sub-tropical flowers are adapted to pollination by birds or even by mammals, e.g. bats and mice.

Wind-pollinated flowers are adapted to their method of pollination by producing large quantities of light pollen, and having anthers and stigmas which project outside the flower. Many grasses have anthers which are not rigidly attached to the filaments and can be shaken by the wind. The stigmas of grasses are feathery and act as a net which traps passing pollen grains.

QUESTIONS

5 Put the following events in the correct order for pollination in a lupin plant: (*a*) bee gets dusted with pollen, (*b*) pollen is deposited on stigma, (*c*) bee visits older flower, (*d*) bee visits young flower, (*e*) anthers split open.

6 Why do you think a large insect such as a bee can pollinate a lupin flower, while a small insect like a fly cannot?
7 Which of the following trees would you expect to be pollinated by insects: apple, hazel, oak, cherry, horse-chestnut, sycamore?
8 In what ways do you think (*a*) an antirrhinum flower, (*b*) a nasturtium flower are adapted to insect pollination?
9 In the course of evolution, some flowers may have become adapted to pollination by certain insect species. Discuss whether the insects are likely to have become adapted to the flowers. What sort of adaptation might you expect?

(*a*) From hollyhock (insect-pollinated)

(*b*) From pine (wind-pollinated)

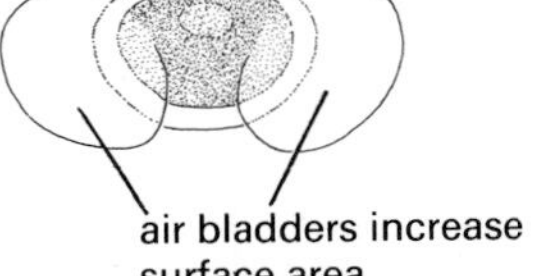

Fig. 16 Pollen grains

Wind- and Insect-pollinated flowers compared

Wind-pollinated	**Insect-pollinated**
1. Small, inconspicuous flowers; petals often green. No scent or nectar.	1. Relatively large flowers or conspicuous inflorescences. Petals brightly coloured and scented; nectaries often present.

Insects respond to the stimulus of colour and scent and are 'attracted' to the flowers. When in the flower, they collect or eat the nectar from the nectaries, or pollen from the anthers.

Wind-pollinated	**Insect-pollinated**
2. Anthers large and often loosely attached to filament so that the slightest air movement shakes them. The whole inflorescence often dangles loosely (hazel male catkins), and the stamens hang out of the flower exposed to the wind (Figs 14 and 15).	2. Anthers not so large and are firmly attached to the filament. They are often in a position within the petals where insects are likely to brush against them.

The wind is more likely to dislodge pollen from exposed, dangling anthers than from those enclosed in petals.

Wind-pollinated	**Insect-pollinated**
3. Large quantities of smooth, light pollen grains produced by the anthers.	3. Smaller quantities of pollen produced. The grains are often spiky or sticky and so stick to the insect's body (Fig. 16).

With wind pollination, only a very small proportion of pollen grains is likely to land on a ripe stigma. If large quantities of pollen are not shed, the chances of successful pollination become very poor. Smooth, light grains are readily carried in air currents and do not stick together.
In pollination by insects, fewer of the pollen grains will be wasted. The patterned or sticky pollen grains are more likely to adhere to the body of the insect.

Wind-pollinated	**Insect-pollinated**
4. Stigmas projecting outside the flower (Fig. 9*d*).	4. Flat or lobed, sticky stigmas inside the flower.

The feathery stigmas of grasses form a 'net' of relatively large area in which flying pollen grains may be trapped.

NOTE. These points of comparison apply to plants which are strongly adapted to wind pollination (e.g. grasses) or insect pollination (e.g. sweet pea). There are many exceptions to these generalizations.

FERTILIZATION AND FRUIT FORMATION

Pollination is complete when pollen from an anther has landed on a stigma. If the flower is to produce seeds, pollination has to be followed by a process called **fertilization**. In all living organisms, fertilization happens when a male sex cell and a female sex cell meet and join together (they are said to **fuse** together). The cell which is formed by this fusion develops into an embryo of an animal or a plant (Fig. 17). The sex cells of all living organisms are called **gametes**.

In animals, the male gamete is the sperm and the female gamete is the egg or ovum (p. 170).

In flowering plants, the male gamete is in the pollen grain; the female gamete, called the **egg cell**, is in the ovule. For fertilization to occur, the nucleus of the male cell from the pollen grain has to reach the female nucleus of the egg cell in the ovule, and fuse with it. The following account explains how this happens.

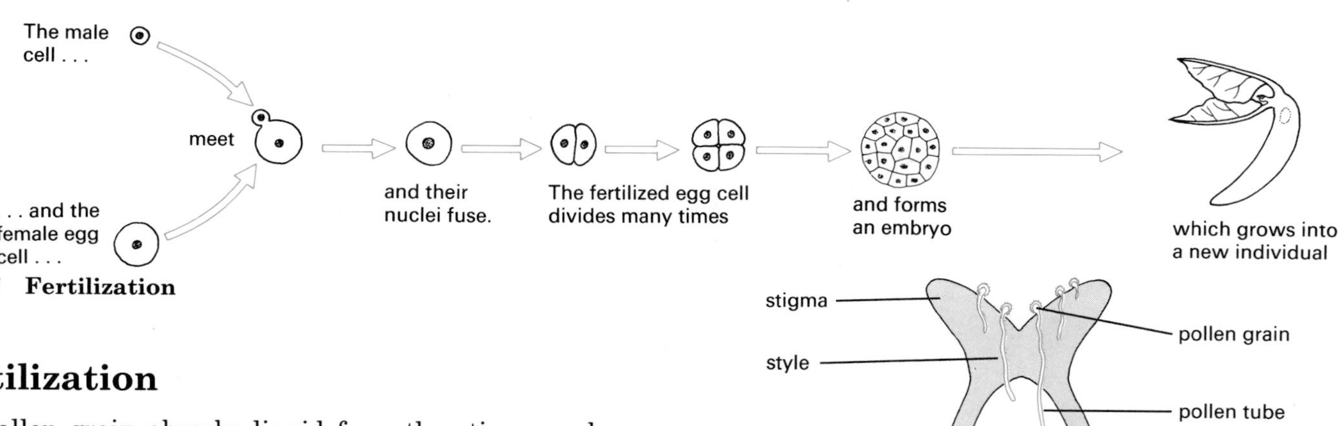
The male
cell . . .
meet
. . . and the
female egg
cell . . .
and their
nuclei fuse.
The fertilized egg cell
divides many times
and forms
an embryo
which grows into
a new individual

Fig. 17 Fertilization

Fertilization

The pollen grain absorbs liquid from the stigma and a microscopic **pollen tube** grows out of the grain (Fig. 18). This tube grows down the style and into the ovary where it enters a small hole, the **micropyle**, in an ovule (Fig. 19). The nucleus of the pollen grain travels down the pollen tube and enters the ovule. Here it combines with the nucleus of the egg cell. Each ovule in an ovary needs to be fertilized by a separate pollen grain.

Although pollination must occur before the ovule can be fertilized, pollination does not necessarily result in fertilization. A bee may visit many flowers on a Bramley apple tree, transferring pollen from one flower to the other. The Bramley, however, is 'self-sterile'; pollination with its own pollen will not result in fertilization. Pollination with pollen from a Worcester or a Cox's apple tree, however, can result in successful fertilization and fruit formation.

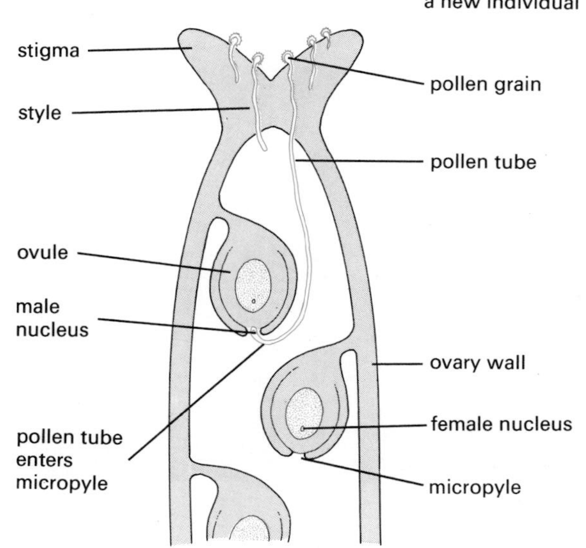
stigma
style
ovule
male
nucleus
pollen tube
enters
micropyle
pollen grain
pollen tube
ovary wall
female nucleus
micropyle

Fig. 19 Diagram of fertilization, e.g. wallflower

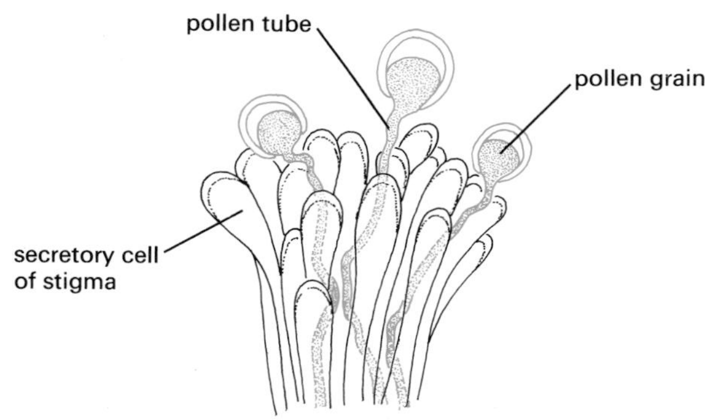
pollen tube
pollen grain
secretory cell
of stigma

Fig. 18 Pollen grains growing on stigma of crocus

Fruit and seed formation

After the pollen nucleus has fused with the egg nucleus, the egg cell divides many times and produces a miniature plant called an **embryo**. The embryo consists of a tiny root and shoot with two special leaves called **cotyledons**. In dicot plants (p. 306) food made in the leaves of the parent plant is carried in the phloem to the cotyledons. The cotyledons grow so large with this stored food that they completely enclose the embryo (Fig. 2*b*, p. 93). In monocot plants (p. 306) the food store is laid down in a special tissue called endosperm (p. 95) which is outside the cotyledons. In both cases the outer wall of the ovule becomes thicker and harder, and forms the seed coat or **testa** (see p. 92).

As the seeds grow, the ovary also becomes much larger and the petals and stamens shrivel and fall off (Fig. 20). The ovary is now called a **fruit**. The biological definition of a fruit is a fertilized ovary; it is not necessarily edible. In the lupin, the fertilized ovary forms a dry, hard pod, but in a related plant—the runner bean—the ovary wall becomes fleshy and edible before drying out. The wallflower ovary forms a dry capsule.

Fig. 20 Lupin flower after fertilization. The ovary (still with the style and stigma attached) has grown much larger than the flower and the petals have shrivelled.

A plum is a good example of a fleshy, edible fruit. Tomatoes (Fig. 21) and cucumbers are also fruits although they are classed as vegetables in the shops. Blackberries (Figs 22 and 23) and raspberries are formed by many small fruits clustered together. In strawberries (Fig. 24), the fruits are the pips and the edible part is formed by the receptacle of the flower. In the apple (Fig. 25) and pear the edible part consists of the swollen receptacle surrounding and fused to the ovary wall.

(*a*) Tomato flowers

(*b*) After fertilization

(*c*) Ripe fruit

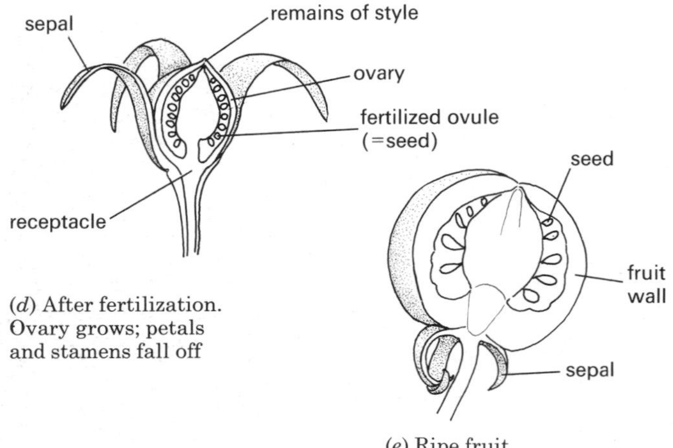

(*d*) After fertilization. Ovary grows; petals and stamens fall off

(*e*) Ripe fruit

Fig. 21 Tomato; fruit formation

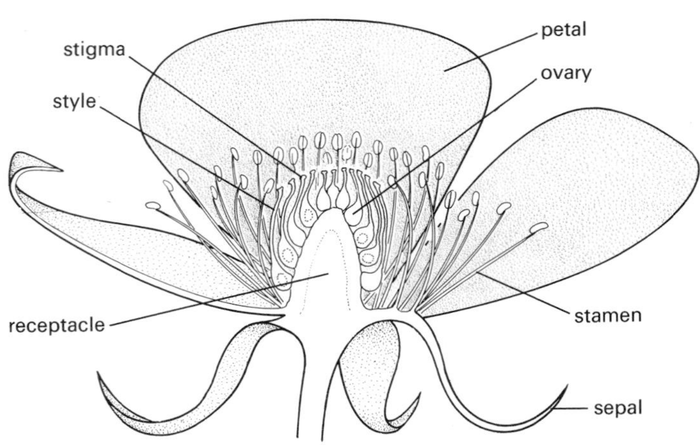

(*a*) Half flower

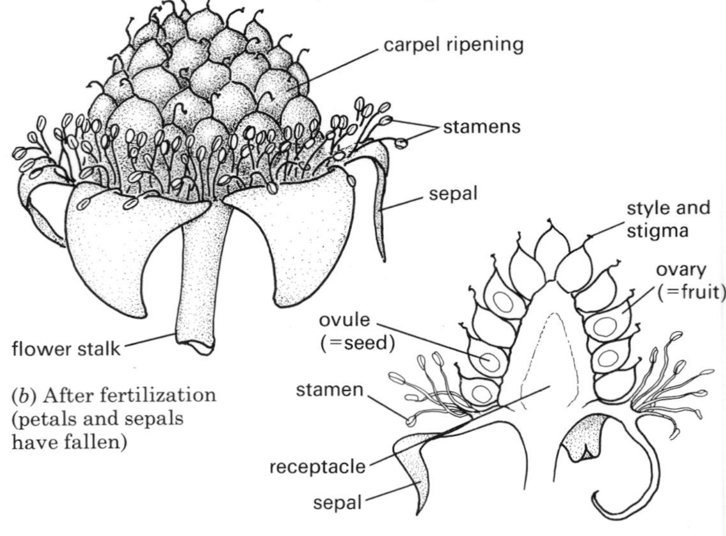

(*b*) After fertilization (petals and sepals have fallen)

(*c*) Longitudinal section of young fruit

Fig. 22 Blackberry; fruit formation

Fig. 23 Blackberry flowers and fruits. Most of the flowers have been pollinated and fertilized. The petals have dropped and the carpels are growing to form fruits.

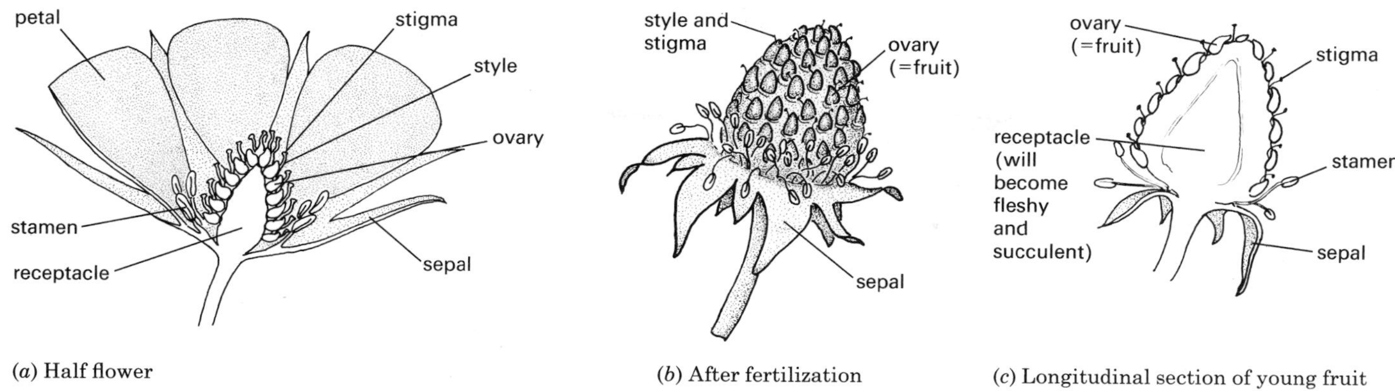

(*a*) Half flower

(*b*) After fertilization

(*c*) Longitudinal section of young fruit

Fig. 24 Strawberry; fruit formation

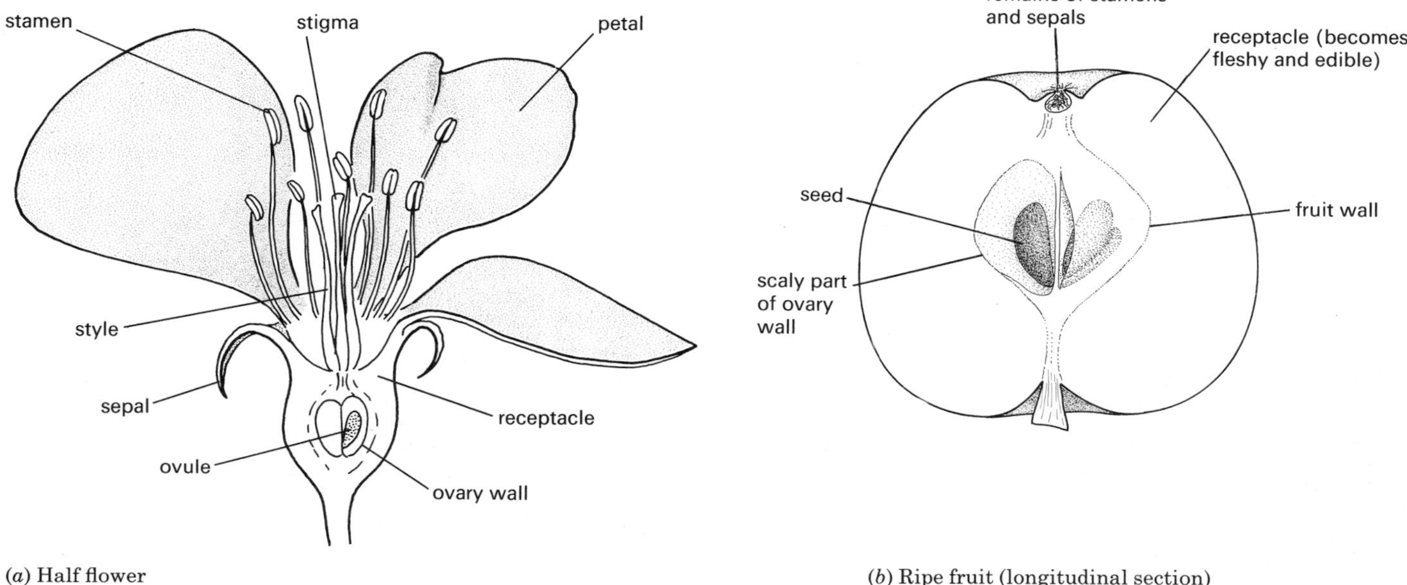

(*a*) Half flower

(*b*) Ripe fruit (longitudinal section)

Fig. 25 Apple; fruit formation

QUESTIONS

10 Which structures in a flower produce (*a*) the male gametes, (*b*) the female gametes?
11 In not more than two sentences, show that you understand the difference between pollination and fertilization.
12 In flowering plants, (*a*) can pollination occur without fertilization, (*b*) can fertilization occur without pollination?
13 Which parts of a tomato flower (*a*) grow to form the fruit, (*b*) fall off after fertilization, (*c*) remain attached to the fruit?
14 Which of the following edible plant products do you think are, biologically, (*a*) fruits, (*b*) seeds, (*c*) neither: runner beans, peas, grapes, baked beans, marrow, rhubarb, tomatoes?
15 Make a drawing to show what you think a ripe pea pod might look like if only four out of ten ovules in the ovary had been fertilized after pollination. (A pea flower is very similar to a lupin flower.)
16 Describe an experiment you could carry out with a lupin plant to demonstrate that pollination is necessary before fruit formation can occur.
17 Why do you think that some fruit-growers pay bee-keepers to set up their hives in their orchards during the spring? Why would an arable farmer not do this?

DISPERSAL OF FRUITS AND SEEDS

When the seeds are mature, the whole fruit or the individual seeds fall from the parent plant to the ground and the seeds may then germinate (p. 94). In many plants, the fruits or seeds are adapted in such a way that they are carried a long distance from the parent plant. This reduces competition for light and water between members of the same species. It may also result in plants growing in new places. The main adaptations are for dispersal by the wind and by animals but some plants have 'explosive' pods that scatter the seeds.

Wind dispersal

'Pepper-pot' effect Examples are the white campion, poppy (Fig. 26*a*) and antirrhinum. The flower .stalk is

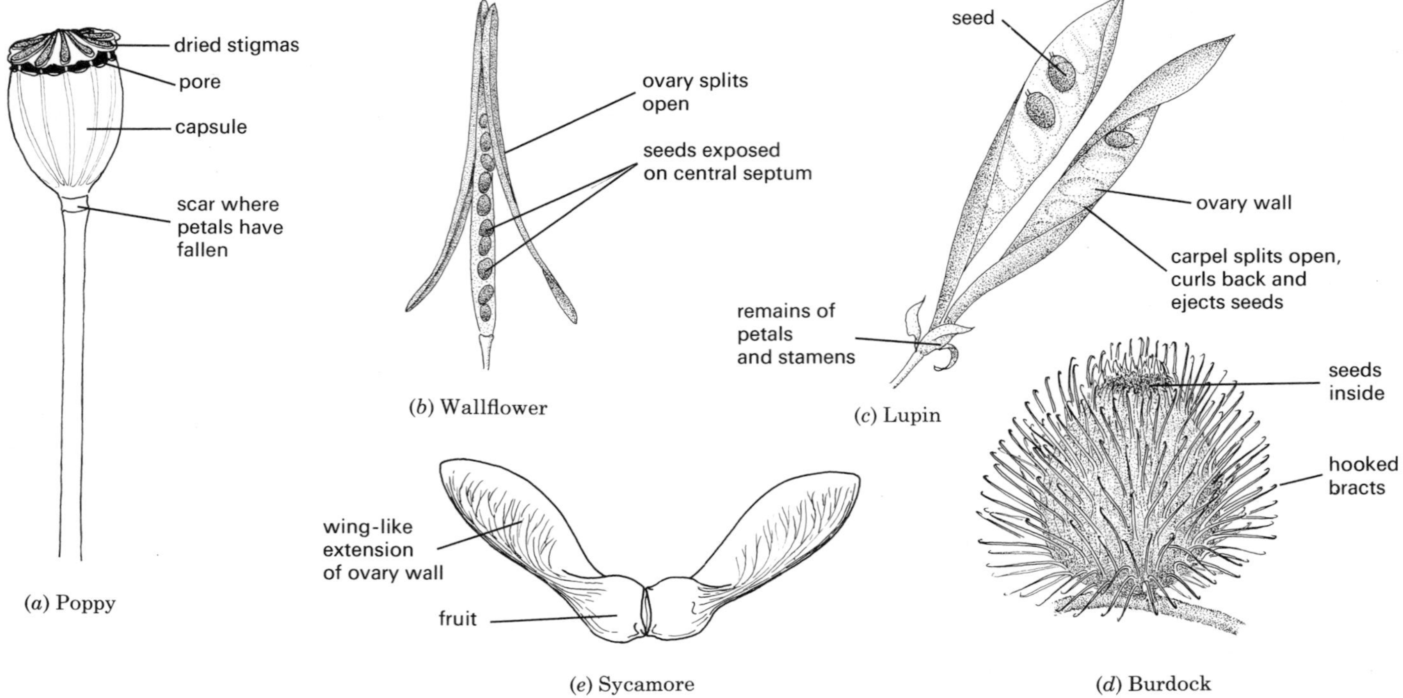

Fig. 26 Fruit and seed dispersal

usually long and the ovary becomes a dry, hollow capsule with one or more openings. The wind shakes the flower stalk and the seeds are scattered on all sides through the openings in the capsule.

In the wallflower, the fruit splits open from the bottom (Fig. 26*b*), exposing the seeds, which remain attached to a central septum. When the wind shakes the plant, the seeds become detached from the septum and are scattered, though not very far, away from the parent plant.

'Parachute' fruits and seeds Clematis, thistles, willow herb and dandelion (Fig. 27) have seeds or fruits of this kind. Feathery hairs project from the fruit or seed and so increase its surface area. As a result, the seed 'floats' over long distances before sinking to the ground. It is, therefore, likely to be carried a long way from the parent plant by slight air currents.

'Winged' fruits Fruits of the lime, sycamore (Fig. 26*e*) and ash trees have wing-like outgrowths from the ovary wall, or leaf-like structures on the flower stalk. These 'wings' cause the fruit to spin as it falls from the tree and so slow down its fall. This delay increases the chances of the fruit being carried away in air currents.

Fig. 27 Wind dispersal. Dandelion fruits being blown away.

Animal dispersal

Mammals and hooked fruits The inflorescence of the burdock is surrounded by bracts which form hooks (Fig. 26*d*). These hooks catch in the fur of passing mammals and the seeds fall out as the mammal moves about.

In other hooked fruits, e.g. agrimony, herb bennet and goosegrass, hooks develop from different parts of the flower.

Birds, mammals and succulent fruits Fruits such as the blackberry and elderberry are eaten by birds and mammals (Fig. 28). The hard pips containing the seed are not digested and so pass out with the droppings of the bird away from the parent plant. The soft texture and, in some cases, the bright colour of these fruits may be regarded as an adaptation to this method of dispersal.

Fig. 28 Animal dispersal. Blackbird eating a hawthorn berry.

'Explosive' fruits

The pods of flowers in the pea family, e.g. gorse, broom, lupin and vetches, dry in the sun and shrivel. The tough fibres in the fruit wall shrink and set up a tension. When the fruit splits in half down two lines of weakness, the two halves curl back suddenly and flick out the seeds (Fig. 26*c*).

QUESTIONS

18 What are the advantages of a plant dispersing its seeds over a wide area? What disadvantages might there be?
19 Which methods of dispersal are likely to result in seeds travelling the greatest distances? Explain your answer.
20 Students sometimes confuse 'wind pollination' and 'wind dispersal'. Write a sentence or two about each to make the difference clear.

PRACTICAL WORK

The growth of pollen tubes

(1) Make a solution of 15 g sucrose and 0.1 g sodium borate in 100 cm^3 water. Put a drop of this solution on a cavity slide and scatter some pollen grains on the drop. This can be done by scraping an anther (which must already have opened to expose the pollen) with a mounted needle, or simply by touching the anther on the liquid drop.

Cover the drop with a cover slip and examine the slide under the microscope at intervals of about 15 minutes. In some cases, pollen tubes may be seen growing from the grains.

Suitable plants include lily, narcissus, tulip, bluebell, lupin, wallflower, sweet pea or deadnettle, but a 15 per cent sucrose solution may not be equally suitable for all of them. It may be necessary to experiment with solutions ranging from 5 to 20 per cent.

(2) Cut the stigma from a mature flower, e.g. honeysuckle, crocus, evening primrose or chickweed and place it on a slide in a drop of 0.5 per cent methylene blue. Squash the stigma under a cover slip (if the stigma is large, it may be safer to squash it between two slides), and leave it for 5 minutes.

Put a drop of water on one side of the slide, just touching the edge of the cover slip and draw it under the cover slip by holding a piece of filter paper against the opposite edge (see Fig. 21, p. 41). This will remove excess stain.

If the squash preparation is now examined under the microscope, pollen tubes may be seen growing between the spread-out cells of the stigma (Fig. 18).

CHECK LIST

- **Flowers contain the reproductive organs of plants.**
- **The stamens are the male organs. They produce pollen grains which contain the male gamete.**
- **The carpels are the female organs. They produce ovules which contain the female gamete and will form the seeds.**
- **The flowers of most plant species contain male and female organs. A few species have unisexual flowers.**
- **Brightly-coloured petals attract insects, which pollinate the flower.**
- **Pollination is the transfer of pollen from the anthers of one flower to the stigma of another.**
- **Pollination may be done by insects or by the wind.**
- **Flowers which are pollinated by insects are usually brightly coloured and have nectar. The shape of the petals may adapt the flower to pollination by one type of insect.**
- **Flowers which are pollinated by the wind are usually small and green. Their stigmas and anthers hang outside the flower where they are exposed to air movements.**
- **Fertilization occurs when a pollen tube grows from a pollen grain into the ovary and up to an ovule. The pollen nucleus passes down the tube and fuses with the ovule nucleus.**
- **After fertilization, the ovary grows rapidly to become a fruit and the ovules become seeds.**
- **Seeds and fruits may be dispersed by the wind, by animals, or by an 'explosive' method.**
- **Dispersal scatters the seeds so that the plants growing from them are less likely to compete with each other and with their parent plant.**

9 Seed Germination

SEED STRUCTURE AND GERMINATION

Seed structure: French bean; broad bean.
Germination: stages in the French bean. Other forms of germination: broad bean and wheat.

PRACTICAL WORK 1

Experiments on germination: the role of the cotyledons; food reserves. Food reserves and their mobilization; conversion of starch to sugar. Decrease in dry weight; using up reserves. Region of growth in radicles; marking the roots.

CONDITIONS FOR GERMINATION
PRACTICAL WORK 2

Experiments on the need for water, oxygen and suitable temperature. Controlling the variables; experimental design.

THE IMPORTANCE OF WATER, OXYGEN AND TEMPERATURE

Uses of water. Uses of oxygen. Importance of temperature. Germination and light.

DORMANCY

The previous chapter described how a seed is formed from the ovule of a flower as a result of fertilization, and is then dispersed from the parent plant. If the seed lands in a suitable place it will **germinate**, i.e. grow into a mature plant.

Flowering plants can be divided into two major groups, the **monocotyledons** and the **dicotyledons**. A **cotyledon** is a modified leaf in a seed. The cotyledon plays a part in supplying food to the growing plant embryo. Monocotyledons (usually abbreviated to 'monocots') have only one of these seed leaves in the seed. Dicotyledons ('dicots') have two cotyledons in the seed and these two cotyledons store food.

Other characteristics and examples of monocots and dicots are discussed on page 306.

To follow the process of germination, the structure of a dicot seed, in this case the French bean, will first be described. The structure of the broad bean is basically the same though its shape is different (Fig. 3).

SEED STRUCTURE AND GERMINATION

Seed structure (French bean)

The seed (Figs 1 and 2) contains a miniature plant, the **embryo**, which consists of a root or **radicle**, and a shoot or **plumule**. The embryo is attached to two leaves called the **cotyledons**, which are swollen with stored food. This stored food, mainly starch, is used by the embryo when it starts to grow. The embryo and cotyledons are enclosed in a tough seed coat or **testa**. The **micropyle**, through which the pollen tube entered (p. 87), remains as a small hole in the testa and is an important route for the entry of water in some seeds. The **hilum** is the scar left where the seed was attached to the pod.

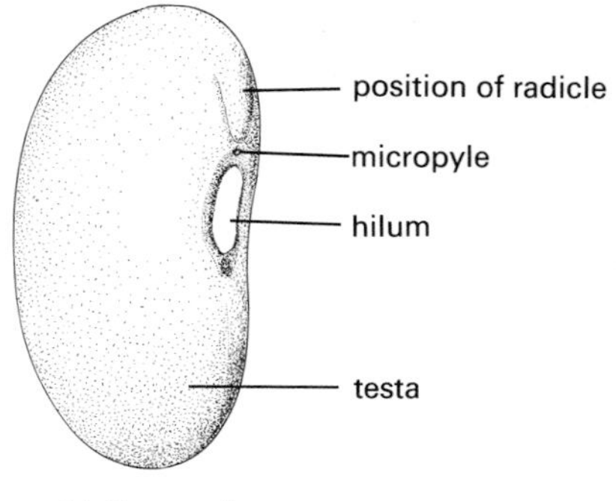

(*a*) External appearance

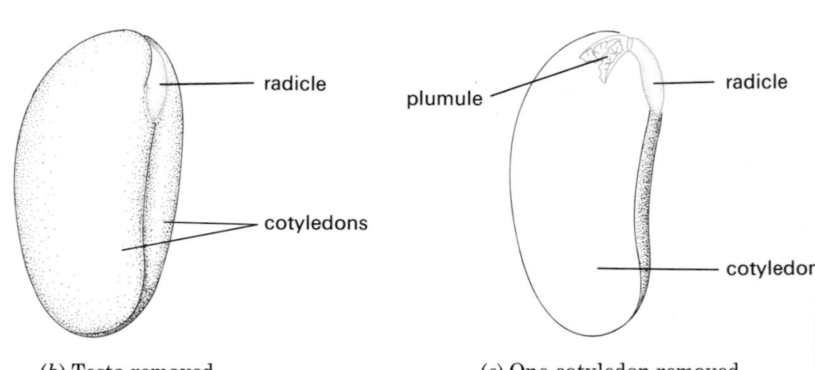

(*b*) Testa removed

(*c*) One cotyledon removed

Fig. 1 The French bean seed

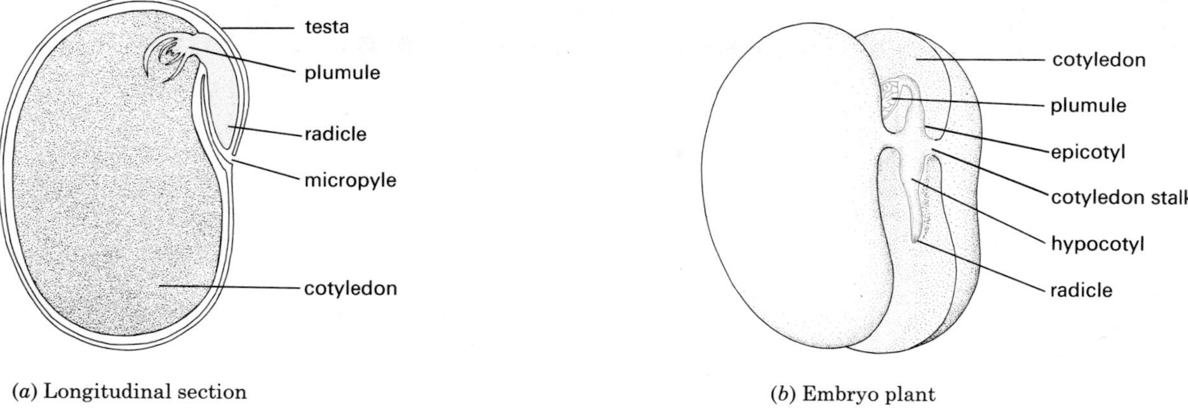

(*a*) Longitudinal section

(*b*) Embryo plant

Fig. 2 Seed structure of a dicotyledon

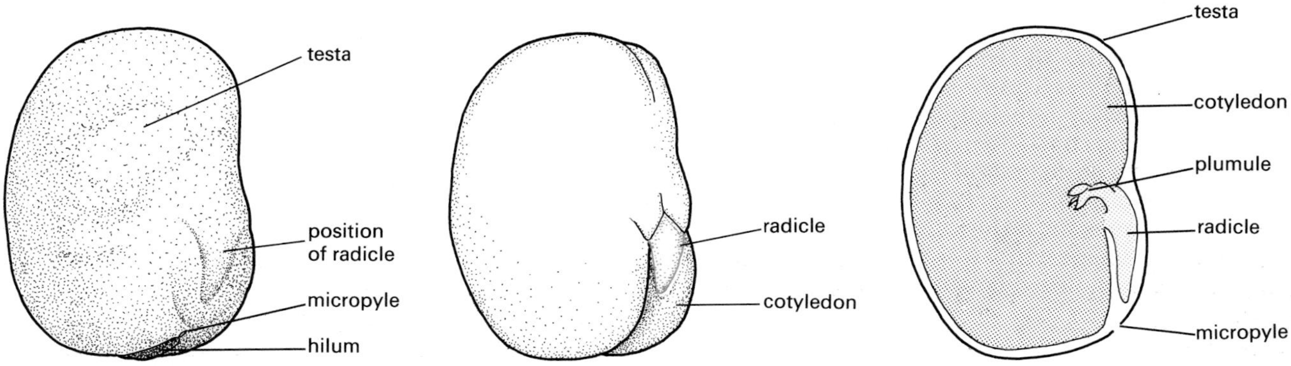

(*a*) External appearance

(*b*) Testa removed

(*c*) Longitudinal section

Fig. 3 Broad bean seed

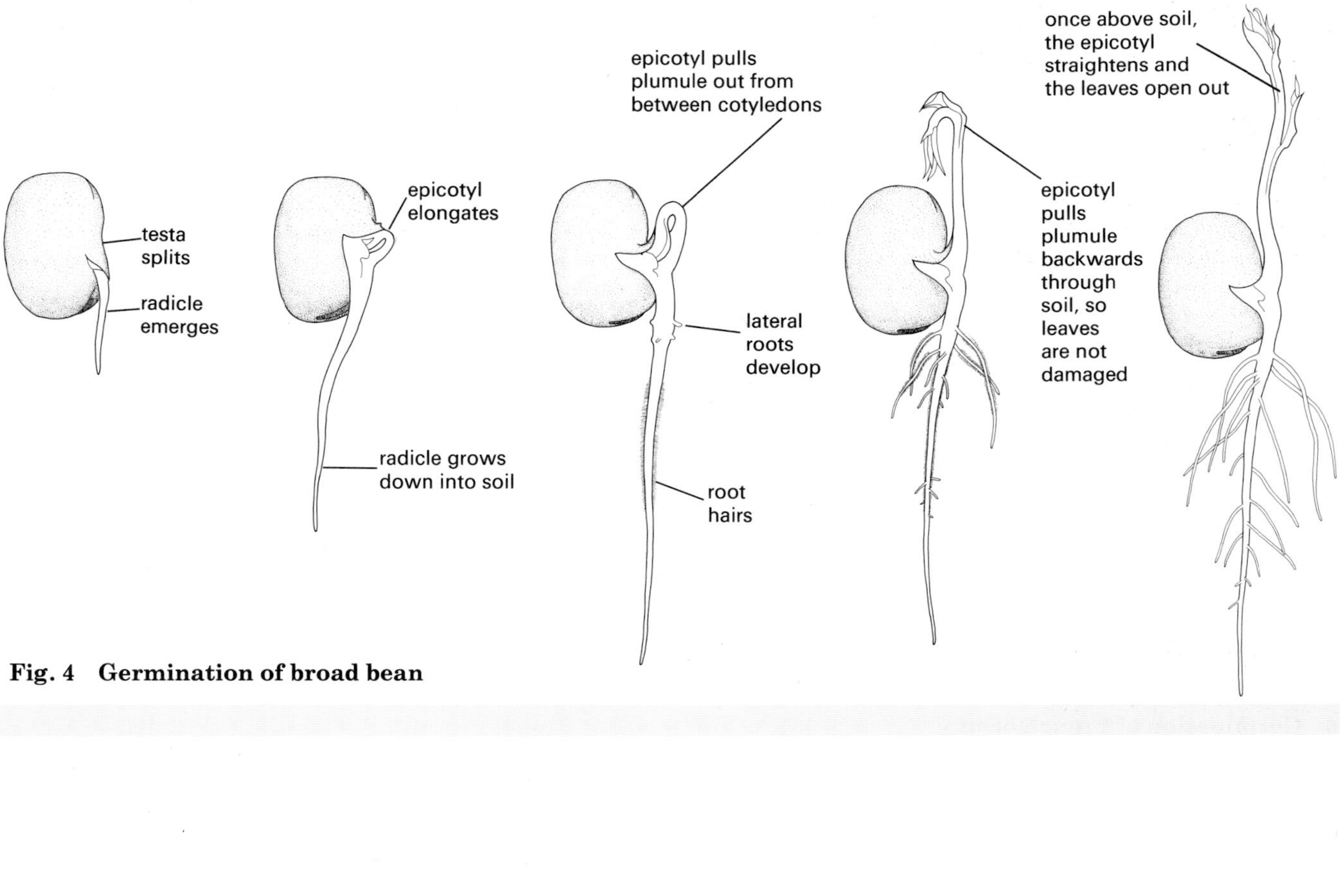

Fig. 4 Germination of broad bean

Germination

The stages of germination of a French bean are shown in Fig. 6. The corresponding stages in the broad bean are shown in Fig. 4.

A seed just shed from its parent plant contains only 5–20 per cent water, compared with 80–90 per cent in mature plant tissues. Once in the soil, some seeds will absorb water and swell up, but will not necessarily start to germinate until other conditions are suitable (see pp. 98–9).

The radicle grows first and bursts through the testa. The radicle continues to grow down into the soil, pushing its way between soil particles and small stones. Its tip is protected by the root cap (see p. 64). Branches, called **lateral roots**, grow out from the side of the main root and help to anchor it firmly in the soil. On the main root and the lateral roots, microscopic **root hairs** grow out. These are fine outgrowths from some of the outer cells. They make close contact with the soil particles and absorb water from the spaces between them (see p. 75).

In the French bean a region of the embryo's stem, the **hypocotyl**, just above the radicle (Fig. 2*b*), now starts to elongate. The radicle is by now firmly anchored in the soil, so the rapidly growing hypocotyl arches upwards through the soil, pulling the cotyledons with it (Fig. 6). Sometimes the cotyledons are pulled out of the testa, leaving it below the soil, and sometimes the cotyledons remain enclosed in the testa for a time. In either case, the plumule is well protected from damage while it is being pulled through the soil, because it is enclosed between the cotyledons.

Fig. 5 Germinating seeds. The brown testa is shed, the cotyledons turn green and the plumule expands.

Once the cotyledons are above the soil, the hypocotyl straightens up and the leaves of the plumule open out. Up to this point, all the food needed for making new cells and producing energy has come from the cotyledons.

The main type of food stored in the cotyledons is starch. Before this can be used by the growing shoot and root, the starch has to be turned into a soluble sugar. In this form, it can be transported by the phloem cells. The change from starch to sugar in the cotyledons is brought about by enzymes which become active as soon as the seed starts to germinate. The cotyledons shrivel as their food reserve is used up, and they fall off altogether soon after they have been brought above the soil.

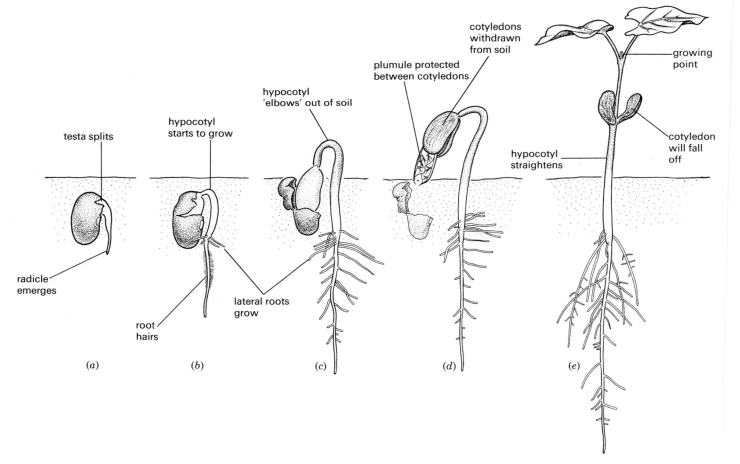

Fig. 6 Germination of French bean

By now the plumule leaves have grown much larger, turned green and started to absorb sunlight and make their own food by photosynthesis (p. 44). Between the plumule leaves is a growing point which continues the upward growth of the stem and the production of new leaves. The embryo has now become an independent plant, absorbing water and mineral salts from the soil, carbon dioxide from the air and making food in its leaves.

Other forms of germination

Not all seeds germinate in the same way as the French bean. In peas, broad beans and runner beans, for example, it is the **epicotyl** (Fig. 2*b*) which elongates, leaving the cotyledons in the soil (Fig. 4).

In grasses and cereals, the food reserve is not in the cotyledon but in a region called the **endosperm**. There is only one cotyledon (hence the classification as monocots), and this remains in the seed, digesting, absorbing and transferring the stored food in the endosperm (Fig. 7). During germination, the shoot grows straight up through the soil, protected at first by a sheath-like **coleoptile**. Once above ground, the coleoptile stops growing and the first leaf bursts through (Fig. 8).

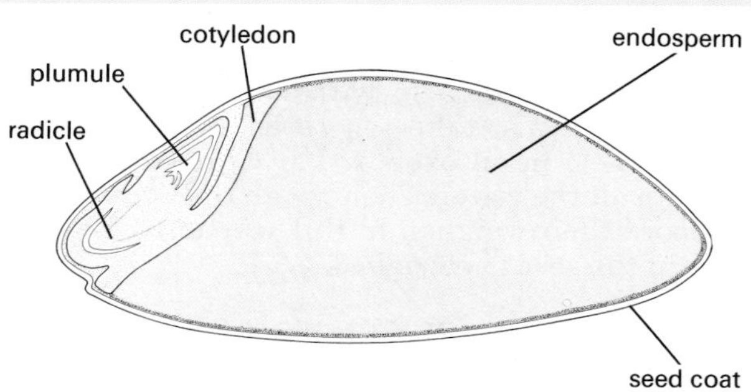

Fig. 7 Wheat seed; vertical section

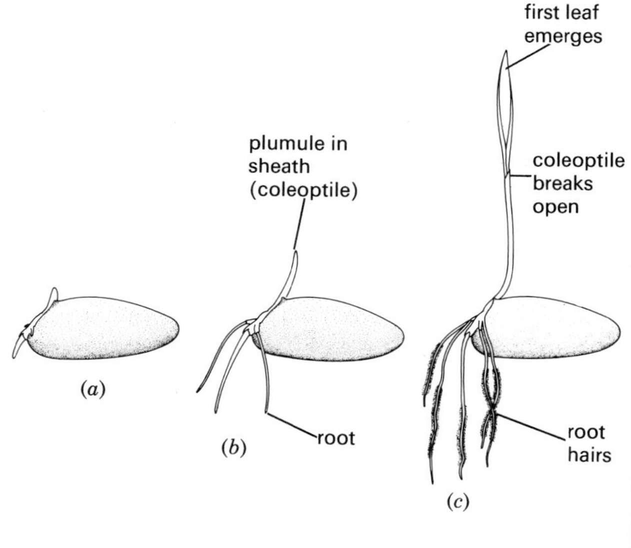

Fig. 8 Germination of wheat

QUESTIONS

1 What are the functions of (*a*) the radicle, (*b*) the plumule and (*c*) the cotyledons of a seed?

2 Rewrite the following in the correct order for germination in a French bean seed: (*a*) cotyledons photosynthesize, (*b*) radicle grows down into the soil, (*c*) hypocotyl straightens, (*d*) hypocotyl starts to elongate, (*e*) root hairs develop, (*f*) radicle emerges from testa, (*g*) cotyledons pulled out of soil, (*h*) testa splits.

3 During germination of the French bean, how are (*a*) the plumule, (*b*) the radicle protected from damage as they are forced through the soil?

4 List all the possible purposes for which a growing seedling might use the food stored in its cotyledons.

5 At what stage of development is a seedling able to stop depending on the cotyledons for its food?

6 What do you think are the advantages to a germinating seed of having its radicle growing some time before the shoot starts to grow?

PRACTICAL WORK

EXPERIMENTS ON GERMINATION

1. The role of the cotyledons in the bean seed

Soak four runner bean seeds in water for 24 hours. Peel off the testas and separate the cotyledons. Discard the cotyledons which do not have embryos attached to them. Pin one of the cotyledons to a piece of expanded polystyrene ceiling tile covered with blotting-paper (Fig. 9*a*). Cut away and discard three quarters of the second cotyledon and pin the remaining piece, with the embryo, to the polystyrene. Cut away as much as possible of the third cotyledon, leaving a small piece attached to the embryo. Pin this to the polystyrene. Finally, dissect off the embryo from the fourth cotyledon and pin it to the polystyrene.

Place the polystyrene strip in a tall, screw-top jar with a little water, as shown in Fig. 9*a*, and leave it for a week.

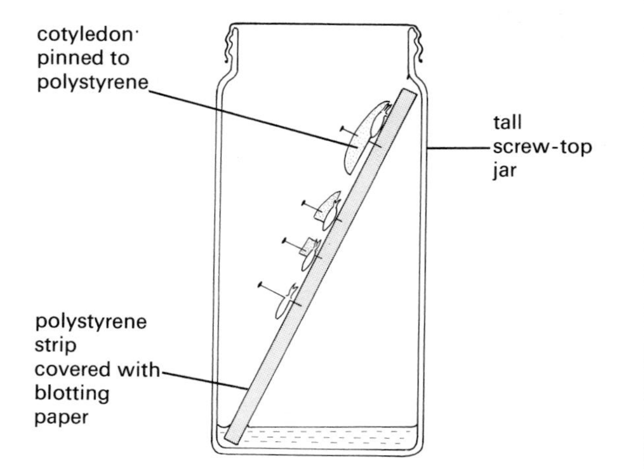

Fig. 9(a) The importance of cotyledons

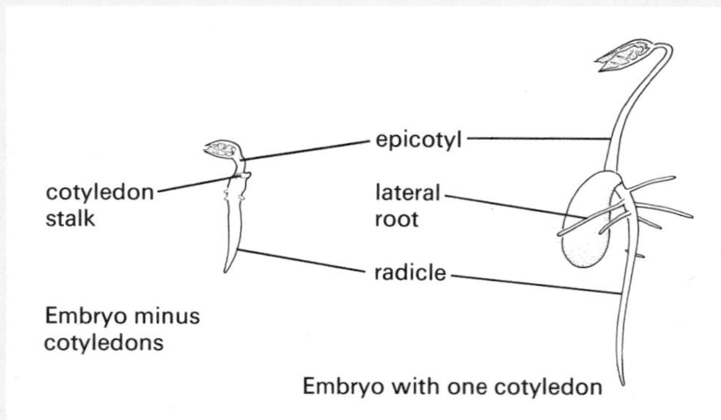

Fig. 9(b) Seedlings after 1 week

Result The embryo without its cotyledons will grow a little (Fig. 9*b*), but not as much as the others which will probably grow in proportion to the amount of cotyledon attached to them. It may take another week, however, to see a difference between the whole and the quarter cotyledon.

Interpretation The rate of growth of the embryo seems to depend on the amount of cotyledon. This could be because the cotyledon supplies the embryo with food or because it contains a growth-promoting substance.

2. Food reserves and their mobilization

You will need about 20 wheat grains which have been soaked overnight and the same number of wheat seedlings which have been germinating for about a week.

Grind the ungerminated grains with a little sand and water in a mortar and place the watery mixture in a test-tube labelled A.

Cut the shoots from the seedlings, grind them with a little sand and water and put the mixture in test-tube B.

Cut and discard the roots from the seedlings and grind the seeds with water and sand as before. Put this mixture into test-tube C.

Leave all three tubes in a boiling water bath (p. 45) for 10 minutes and then cool them under the cold tap.

Tip half the contents of tube A into a clean test-tube and add a few drops of iodine. Add Benedict's solution to the rest of the liquid in tube A and heat it in the water bath for 5 minutes.

Repeat these steps with tubes B and C, keeping a note of any colour changes and what these signify. (See 'Food tests' on page 122.)

Results (A). The extract from the ungerminated seeds should give a positive result for starch but probably not for sugar, i.e. iodine turns the mixture blue but heating with Benedict's solution does not give a yellow, orange or red precipitate.

(B) The shoot extract will probably give a positive result with Benedict's solution but not with iodine.

(C) The extract from the germinating seeds will probably give a blue colour with iodine and also produce an orange precipitate with Benedict's solution.

Interpretation The results suggest that ungerminated seeds contain starch, germinating seeds contain starch and sugar and the shoots contain only sugar. (Note: the tests reveal nothing about proteins, or fats.)

This is consistent with the suggestion that the food reserve in the seed is starch and that, during germination, the starch is changed to sugar and transported to the growing shoot.

3. Decrease in dry weight (using up carbohydrate)

If living material is converting food, e.g. carbohydrate, to carbon dioxide and water, its weight (mass) should decrease. However, it is the dry weight which must be measured, since any material, living or non-living, may lose weight by evaporation of water into the atmosphere.

Soak one hundred wheat seeds in water for 24 hours. Kill half of the seeds by boiling them. These are the **controls** (p. 30). Place the living seeds in a dish containing moist cotton wool and the dead seeds in an identical dish also with moist cotton wool. At 2-day intervals, take ten seeds from each dish and heat them in an oven at 120 °C for 12 hours to evaporate all the water. Then weigh the two samples and record their weights. In this way only the solid matter in the seeds is weighed.

Result The dry weight of the living seeds decreases over the 10 days. The dry weight of killed seeds does not change (Fig. 10). (There will, of course, be variations because the individual grains are not all exactly the same size, but there will be no downward trend.)

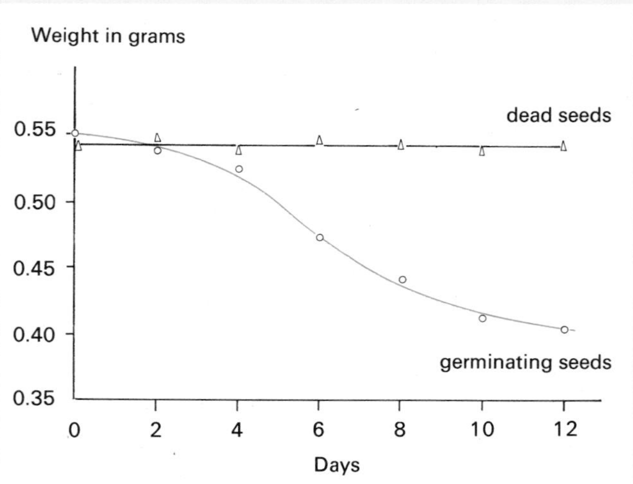

Fig. 10 Decrease in dry weight

Interpretation The living seeds are respiring and oxidizing their starch stores to produce the energy needed in germination. The loss of carbon dioxide to the atmosphere results in a decrease in weight.

The weight of the dead seeds did not change and, therefore, you have shown that the loss in weight must be the result of the living processes involved in germination.

4. Region of growth in radicles

Radicles of germinating peas are marked with evenly spaced lines. In a day or two, the region of most rapid growth is revealed by the increased spacing between the lines. This experiment is described on p. 68.

QUESTIONS

7 In Experiment 1, two alternative interpretations of the results are given. Suggest an experiment which might help to distinguish between these two interpretations.

8 Explain why sugar can be found in germinating wheat grains but not in ungerminated grains.

EXPERIMENTS ON THE CONDITIONS FOR GERMINATION

The environmental conditions which might be expected to affect germination are temperature, light intensity and the availability of water and air. The relative importance of some of these conditions can be tested by the experiments which follow. The effects of these conditions are discussed on pp. 98–9.

5. The need for water

Label three containers A, B and C and put dry cotton wool in the bottom of each. Place a number of soaked seeds in B and C, and an equal number of dry seeds in A. Leave A quite dry; add water to B to make the cotton wool moist; add water to C until all the seeds are completely covered (Fig. 11). Put lids on the containers and leave them all at room temperature for a week.

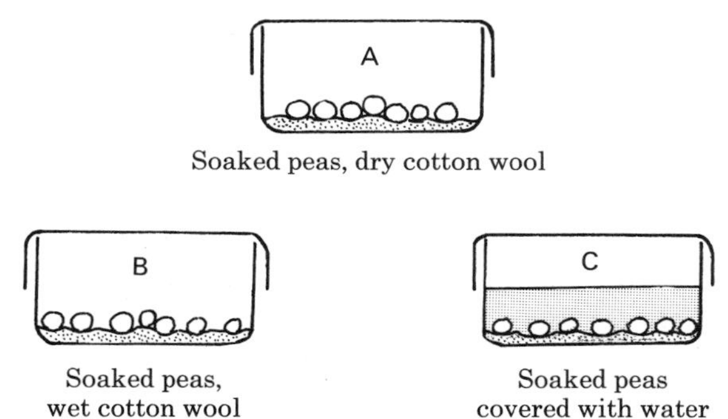

Fig. 11 The need for water in germination

Result The seeds in B will germinate normally. Those in A will not germinate. The seeds in C may have started to germinate but will probably not be as advanced as those in B and may have died and started to decay.

Interpretation Although water is necessary for germination, too much of it may prevent germination by cutting down the oxygen supply to the seed.

6. The need for oxygen

Set up the experiment as shown in Fig. 12. (**CARE:** Pyrogallic acid and sodium hydroxide is a caustic mixture. Use eye shields, handle the liquids with care and report any spillage at once.)

If the moist cotton wool is rolled in some cress seeds, they will stick to it. The bungs must make an airtight seal in the flask and the cotton wool must not touch the solution. Pyrogallic acid and sodium hydroxide absorb oxygen from the air, so the cress seeds in flask A are deprived of oxygen. Flask B is the control (see p. 30). This is to show that germination can take place in these experimental conditions provided oxygen is present. Leave the flasks for a week at room temperature.

Result The seeds in flask B will germinate but there will be little or no germination in flask A.

Interpretation The main difference between flasks A and B is that A lacks oxygen. Since the seeds in this flask have not germinated, it looks as if oxygen is needed for germination.

To show that the chemicals in flask A had not killed the seeds, the cotton wool can be swapped from A to B. The seeds from A will now germinate.

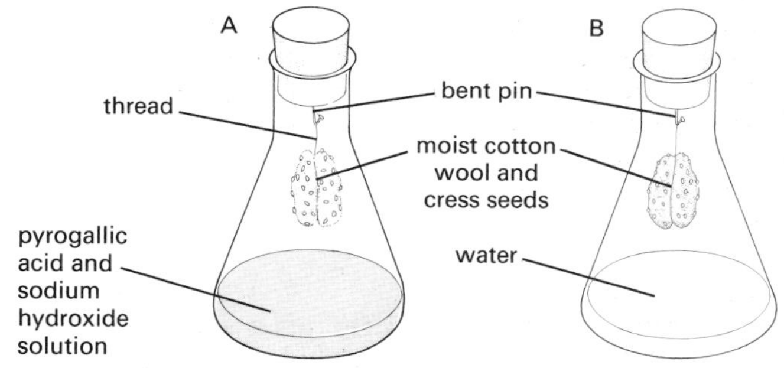

Fig. 12 The need for oxygen

Note Sodium hydroxide absorbs carbon dioxide from the air. The mixture (sodium hydroxide + pyrogallic acid) in flask A, therefore, absorbs both carbon dioxide and oxygen from the air in this flask. The control flask B, therefore, should really contain sodium hydroxide rather than water. The sodium hydroxide would absorb carbon dioxide but not oxygen. When the seeds in B germinate, it can then be argued that lack of carbon dioxide did not affect them, whereas lack of oxygen in A did.

7. Temperature and germination

Soak some maize grains for a day and then roll them up in three strips of moist blotting-paper as shown in Fig. 13. Put the rolls into plastic bags. Place one in a refrigerator (about 4 °C), leave one upright in the room (about 20 °C) and put the third one in a warm place such as over a radiator or—better—in an incubator set to 30 °C.

Because the seeds in the refrigerator will be in darkness, the other seeds must also be enclosed in a box or a cupboard, to exclude light. Otherwise it could be objected that it was lack of light rather than low temperature which affected germination.

After a week, examine the seedlings and measure the length of the roots and shoots.

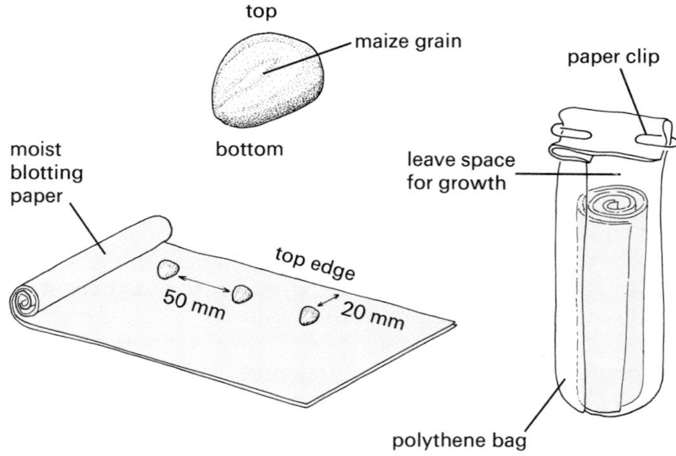

Fig. 13 Temperature and germination. Roll the seeds in moist blotting paper, and stand the rolls upright in plastic bags.

Result The seedlings kept at 30 °C will be more advanced than those at room temperature. The grains in the refrigerator may not have started to germinate at all.

Interpretation Seeds will not germinate below a certain temperature. The higher the temperature, the faster the germination, at least up to 35–40 °C.

Note These experiments and others are fully described in *Experimental Work in Biology*, Combined Edition. (See p. 364.)

QUESTIONS

9 List the external conditions necessary for germination.
10 Do any of the results of Experiments 5–7 suggest whether or not light is necessary for germination? Explain.
11 (*a*) In Experiment 7 it could be argued that the low temperature of the refrigerator had killed the seeds and that this explains why they did not germinate. How could you check on this possibility? (*b*) How could you modify Experiment 7, at least in theory, to find the minimum temperature at which maize seeds would germinate?

Controlling the variables

These experiments on germination illustrate one of the problems of designing biological experiments. You have to decide what conditions (the 'variables') could influence the results and then try to change only one condition at a time. The dangers are that (1) some of the variables might not be controllable, (2) controlling some of the variables might also affect the condition you want to investigate and (3) there might be a number of important variables you have not thought of.

1. In your germination experiments, you were unable to control the quality of the seeds, but had to assume that the differences between them would be small. If some of the seeds were dead or diseased, they would not germinate in any conditions and this could distort the results. This is one reason for using as large a sample as possible in the experiments.
2. You had to ensure that, when temperature was the variable, the exclusion of light from the seeds in the refrigerator was not an additional variable. This was done by putting all the seeds in darkness.
3. A variable you might not have considered could be the way the seeds were handled. Some seeds can be induced to germinate more successfully by scratching or chipping the testa.

THE IMPORTANCE OF WATER, OXYGEN AND TEMPERATURE

Use of water in the seedling

Most seeds, when first dispersed, contain very little water. In this dehydrated state, their metabolism is very slow and their food reserves are not used up. The dry seeds are also resistant to extremes of temperature and to desiccation. Before the metabolic changes necessary for germination can take place, the seeds must absorb water.

Water is absorbed firstly through the micropyle, in some species, and then through the testa as a whole. Once the radicle has emerged, it will absorb water from the soil, particularly through the root hairs. The water which reaches the embryo and cotyledons is used to:

1. activate the enzymes in the seed;
2. help the conversion of stored starch to sugar, and proteins to amino acids (p. 14);
3. transport the sugar in solution from the cotyledons to the growing regions;
4. expand the vacuoles of newly formed cells and so cause the root and shoot to grow and the leaves to expand (p. 65);
5. maintain the turgor (p. 38) of the cells and thus keep the shoot upright and the leaves expanded;
6. provide the water needed for photosynthesis once the plumule and young leaves are above ground;
7. transport salts from the soil to the shoot.

Uses of oxygen

In some seeds the testa is not very permeable to oxygen, and the early stages of germination are probably anaerobic (p. 24). The testa when soaked or split open allows oxygen to enter. The oxygen is used in aerobic respiration, which provides the energy for the many chemical changes involved in mobilizing the food reserves and making the new cytoplasm and cell walls of the growing seedling.

Importance of temperature

On page 19 it was explained that a rise in temperature speeds up most chemical reactions, including those taking place in living organisms. Germination, therefore, occurs more rapidly at high temperatures, up to about 40 °C. Above 45 °C, the enzymes in the cells are denatured and the seedlings would be killed. Below certain temperatures (e.g. 0–4 °C) germination may not start at all in some seeds. However, there is a considerable variation in the range of temperatures at which seeds of different species will germinate.

Germination and light

Since a great many cultivated plants are grown from seeds which are planted just below soil level, it seems obvious that light is not necessary for germination. Similarly, experiments in the laboratory, with seeds grown in glass dishes, show that light does not prevent germination. There are some species, however, in which the seeds need some exposure to light before they will germinate, e.g. foxgloves and some varieties of lettuce. In all seedlings, once the shoot is above ground, light is necessary for photosynthesis.

QUESTIONS

12 Describe the natural conditions in the soil that would be most favourable for germination. How could a gardener try to create these conditions? (See also pp. 256–7.)

13 Suppose you planted some bean seeds in the garden but no plants appeared after 3 weeks. List all the possible causes for this failure. How could you investigate some of these possibilities?

DORMANCY

When plants shed their seeds in summer and autumn, there is usually no shortage of water, oxygen and warmth. Yet, in a great many species, the seeds do not germinate until the following spring. These seeds are said to be **dormant**, i.e. there is some internal control mechanism which prevents immediate germination even though the external conditions are suitable.

If the seeds did germinate in the autumn, the seedlings might be killed by exposure to frost, snow and freezing conditions. The advantage of a dormant period is that germination is delayed until these adverse conditions are past.

The controlling mechanisms are very varied and are still the subject of investigation and discussion. The factors known to influence dormancy are plant growth substances (p. 109), the testa, low temperature and light, or a combination of these.

CHECK LIST

- **A dicot seed consists of an embryo with two cotyledons enclosed in a seed coat (testa).**
- **The embryo consists of a small root (radicle) and shoot (plumule).**
- **The cotyledons contain the food store that the embryo will use when it starts to grow.**
- **The food stored in the cotyledons has to be made soluble by enzymes and transported to the growing regions.**
- **When germination takes place, (*a*) the radicle of the embryo grows out of the testa and down into the soil and (*b*) the plumule is pulled backwards out of the soil.**
- **In many seeds, the cotyledons are also brought above the soil. They photosynthesize for a while before falling off.**
- **Germination is influenced by temperature and the amount of water and oxygen available.**
- **Some seeds have a dormant period after being shed. During this period they will not germinate.**

10 Vegetative Reproduction

VEGETATIVE PROPAGATION

Stolons; rhizomes; strawberry runners; bulbs and corms. Food storage. Advantages of vegetative reproduction; clones.

ARTIFICIAL PROPAGATION

Cuttings; grafting; preservation of useful characteristics.

In addition to reproducing by seeds, many plants are able to produce new individuals asexually, i.e. without gametes, pollination and fertilization. This form of asexual reproduction is also called **vegetative propagation**.

VEGETATIVE PROPAGATION

When vegetative propagation takes place naturally, it usually results from the growth of a bud on a stem which is close to, or under the soil. Instead of just making a branch, the bud produces a complete plant with roots, stem and leaves. When the old stem dies, the new plant is independent of the parent which produced it.

Stolons and rhizomes

The flowering shoots of plants such as the strawberry and the creeping buttercup are very short and, for the most part, below ground. The stems of shoots such as these are called **rootstocks**. The rootstocks bear leaves and flowers. After the main shoot has flowered, the lateral buds produce long shoots which grow horizontally over the ground (Fig. 1). These shoots are called **stolons** (or 'runners'), and have only small scale-leaves at their nodes and very long internodes. At each node there is a bud which can produce not only a shoot, but roots as well (Fig. 2). Thus a complete plant may develop and take root at the node, nourished for a time by food sent from the parent plant through the stolon. Eventually, the stolon dries up and withers, leaving an independent daughter plant growing a short distance away from the parent. In this way a strawberry plant can produce many daughter plants by vegetative propagation in addition to producing seeds (p. 89).

In many plants, horizontal shoots arise from lateral buds near the stem base, and grow under the ground. Such underground horizontal stems are called **rhizomes**. At the nodes of the rhizome are buds which may develop to produce shoots above the ground. The shoots become independent plants when the connecting rhizome dies.

Many grasses propagate by rhizomes; the couch grass (Fig. 4) is a good example. Even a small piece of

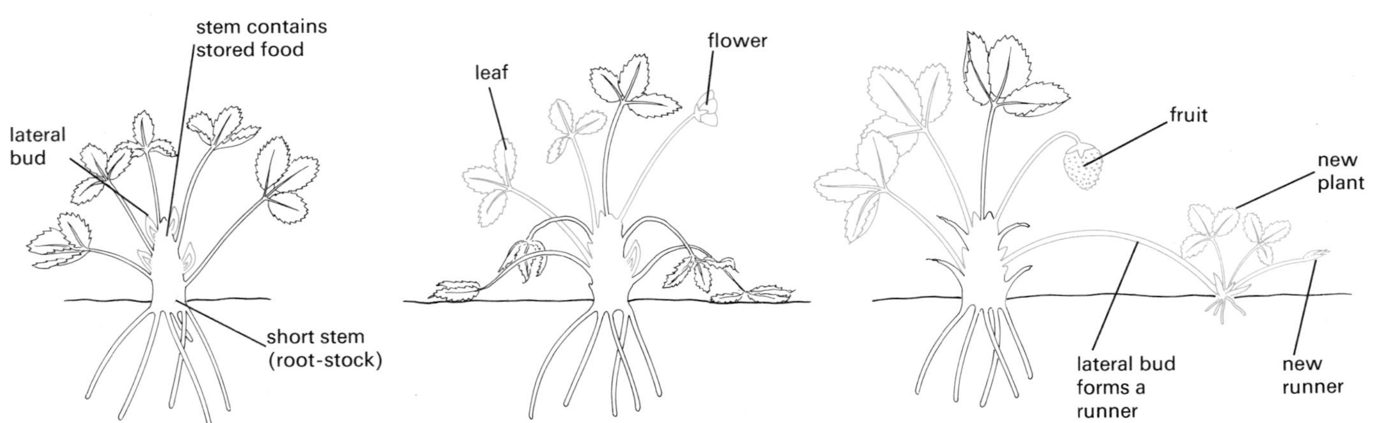

Fig. 1 Strawberry runner developing from root-stock

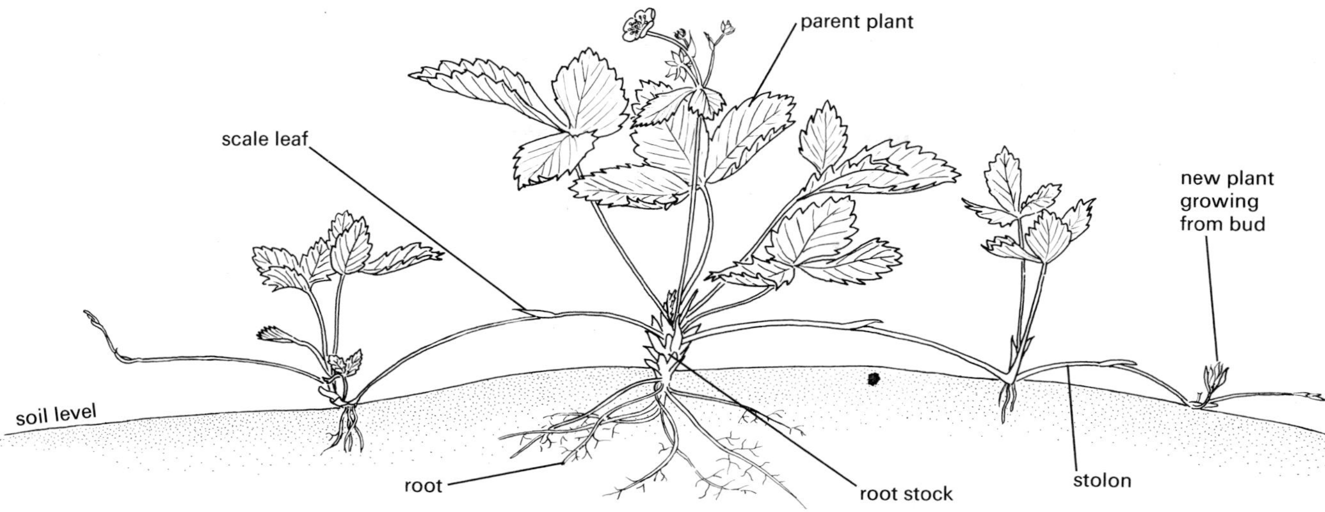

Fig. 2 Wild strawberry runner (Early summer)

rhizome, provided it has a bud, can produce a new plant.

In the bracken, the entire stem is horizontal and below ground (p. 331). The bracken fronds you see in summer are produced from lateral buds on a rhizome many centimetres below the soil.

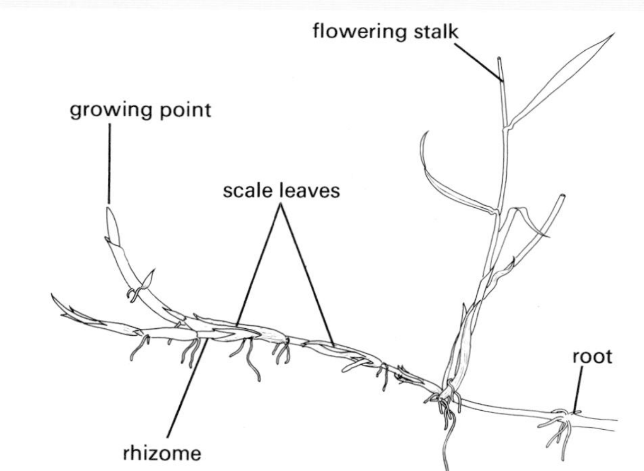

Fig. 3 Couch grass rhizome

Bulbs and corms

Bulbs such as those of the daffodil and snowdrop are very short shoots. The stem is only a few millimetres long and the leaves, which encircle the stem, are thick and fleshy with stored food.

In spring, the stored food is used by a rapidly growing terminal bud which produces a flowering stalk and a small number of leaves. During the growing season, food made in the leaves is sent to the leaf bases and stored. The leaf bases swell and form a new bulb ready for growth in the following year.

Vegetative reproduction occurs when some of the food is sent to a lateral bud as well as to the leaf bases. The lateral bud grows inside the parent bulb and, next year, will produce an independent plant (Figs 4 and 6).

The corms of crocuses and anemones have life cycles similar to those of bulbs but it is the stem, rather than the leaf bases, which swells with stored food. Vegetative reproduction takes place when a lateral bud on the short, fat stem grows into an independent plant.

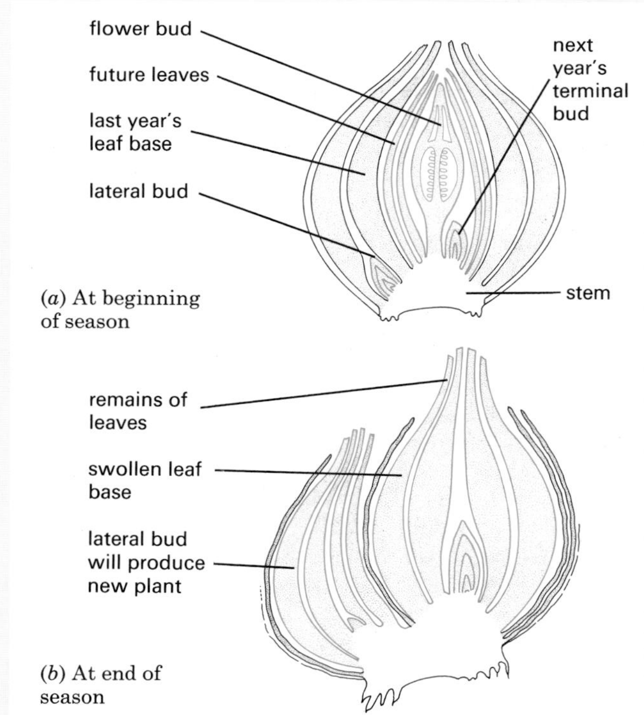

Fig. 4 Daffodil bulb; vegetative reproduction

Food storage

In many cases the organs associated with asexual reproduction also serve as food stores.

Food in the storage organs enables very rapid growth in the spring. A great many of the spring and early summer plants have bulbs, corms, rhizomes or tubers: e.g. daffodil, snowdrop and bluebell, crocus and cuckoo pint, iris and lily-of-the-valley and lesser celandine.

Potatoes are **stem tubers**. Lateral buds at the base of the potato shoot produce underground shoots (rhizomes). These rhizomes swell up with stored starch and form tubers (Fig. 5*a*). Because the tubers are stems, they carry buds. If the tubers are left in the ground or transplanted, the buds will produce shoots, using food stored in the tuber (Fig. 5*b*). In this way, the potato plant can propagate vegetatively.

The food reserve in plant storage organs is often starch but in some cases it is glucose (as in the onion) or sucrose (as in sugar-beet).

Early growth enables the plant to flower and produce seeds before competition with other plants (for water, mineral salts and light) reaches its maximum. This must be particularly important in woods where, in summer, the leaf canopy prevents much light from reaching the ground and the tree roots tend to drain the soil of moisture over a wide area.

In agriculture, humans have exploited many of these types of plant and bred them for bigger and more nutritious storage organs for their own consumption, e.g. onion, leek, potato, beetroot.

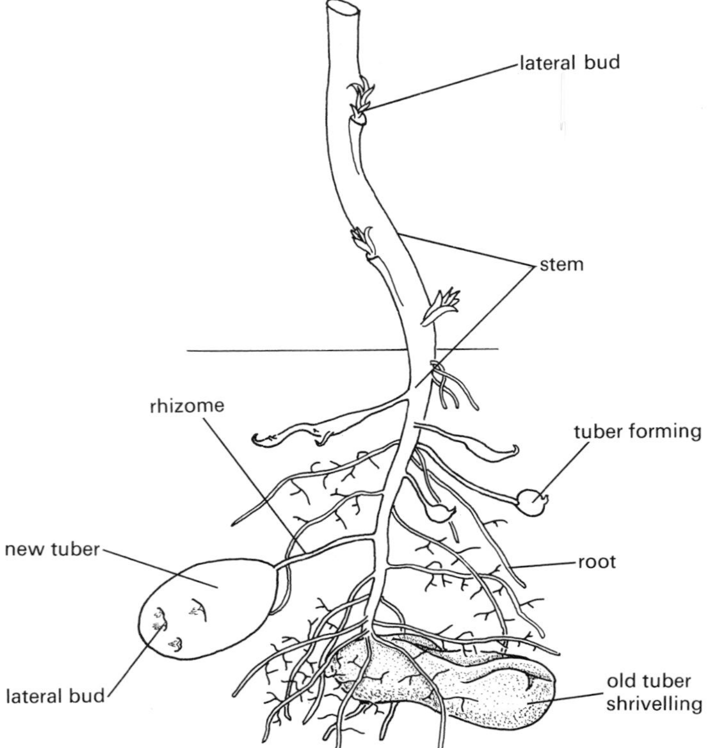

Fig. 5(a) Stem tubers growing on potato plant

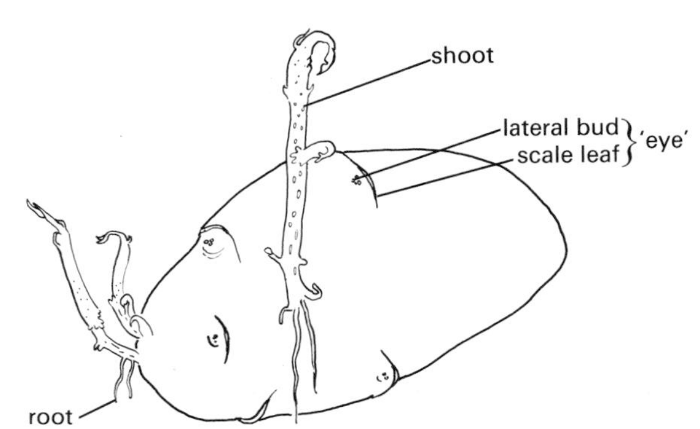

Fig. 5(b) Potato tuber sprouting

Advantages of vegetative propagation

The outcome of sexual reproduction in plants is the formation and dispersal of seeds. When seeds are dispersed from a parent plant, many of them will fall in areas quite unsuitable for germination and growth. The soil may be hard and dry or there may be inadequate light.

A plant which reproduces vegetatively must, itself, already be in a situation favourable for growth. The situation must, therefore, be equally suitable for its offspring. Thus some of the hazards of seed dispersal and germination are avoided.

However, plants which reproduce vegetatively also reproduce sexually and bear fruits and seeds.

Dispersal methods such as those described on page 89 spread seeds over a wide area and some of the seeds may eventually colonize new regions, away from competition with the parent plant. Vegetative reproduction produces daughter plants very close to the parent (Fig. 6), so the spread to new regions is very slow and there is probably some competition between individuals. Nevertheless, the dense colony of vegetatively produced plants usually excludes competitors of other species.

Fig. 6 Daffodils growing in clumps. Each clump is probably derived from one or two parent bulbs.

Clones Daughter plants, produced vegetatively, are all genetically identical with each other and their parent, i.e. they form **clones** (see p. 350). There will be no variation between individuals. In agriculture and horticulture, this has the advantage that all the good characteristics of a crop plant, such as potato, can be preserved from generation to generation. From the point of view of evolution (p. 237), however, the absence of different varieties may be a disadvantage if there is a change in climate or other environmental condition. For example, among the varieties produced by sexual reproduction, there may be one which is particularly resistant to drought. If the environment becomes very dry for a long period, some of these individuals will survive and reproduce. Vegetatively produced individuals, having produced no varieties of any kind, might not survive a prolonged drought.

However, although these distinctions are important from an agricultural viewpoint, they are of less significance in wild plants, many of which employ both methods of reproduction.

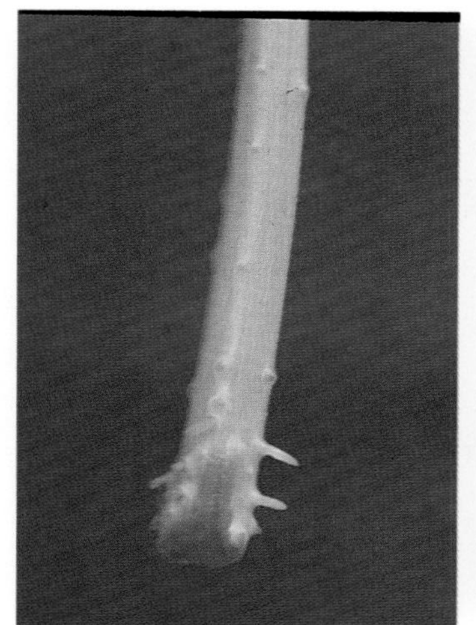

(*a*) Roots developing from Busy Lizzie stem.

(*b*) Roots growing from Coleus cutting.

Fig. 7 Rooted cuttings

QUESTIONS

1 Plants can often be propagated from stems but rarely from roots. What features of shoots account for this difference?
2 The plants which survive a heath fire are often those which have a rhizome (e.g. bracken). Suggest a reason why this is so.
3 Suggest why pulling up couch grass is unlikely to be effective in eliminating this weed from the garden.

ARTIFICIAL PROPAGATION

Agriculture and horticulture exploit vegetative reproduction in order to produce fresh stocks of plants. This can be done naturally, e.g. by planting potatoes, dividing up rootstocks or pegging down stolons at their nodes to make them take root. There are also methods which would not occur naturally in the plant's life cycle. Two important methods of artificial propagation are by taking cuttings and by grafting.

Cuttings

It is possible to produce new individuals from certain plants by putting the cut end of a shoot into water or moist earth. Roots (Fig. 7) grow from the base of the stem into the soil while the shoot continues to grow and produce leaves.

In practice, the cut end of the stem may be treated with a rooting 'hormone' (see p. 110) to promote root growth, and evaporation from the shoot is reduced by covering it with polythene or a glass jar. Carnations, geraniums and chrysanthemums are commonly propagated from cuttings.

Grafting

A bud or shoot from one plant is inserted under the bark on the stem of another, closely related, variety. The rooted portion is called the **stock**; the bud or shoot being grafted is the **scion** (Figs 8 and 9).

The stock is obtained by growing a plant from seed and then cutting away the shoot. The scion is a branch or a bud cut from a cultivated variety with the required characteristics of flower colour, fruit quality, etc.

Rose plants grown from seed would produce a wide variety of plants, only a few of which would retain all the desirable features of the parent plant. Most of them would be like wild roses. Similarly, most of the apple trees grown from seed would bear only small, sour 'crab-apples'. By taking cuttings and making grafts, the inbred characteristics of the plant are preserved and you can guarantee that all the new individuals produced by this kind of artificial propagation will be the same.

QUESTIONS

4 Draw up a table like the one partially shown here.

	Vegetative propagation	*Sexual reproduction (seeds)*
(*a*) Variations in offspring		

Make notes in columns 2 and 3 to compare the features mentioned in column 1 for the two methods of reproduction. Do the same for the following three features (*b*) colonizing new areas, (*c*) food supplies to offspring, (*d*) chance of reaching suitable habitat.
5 After making a graft, why is it desirable to remove any branches which grow from the stock?

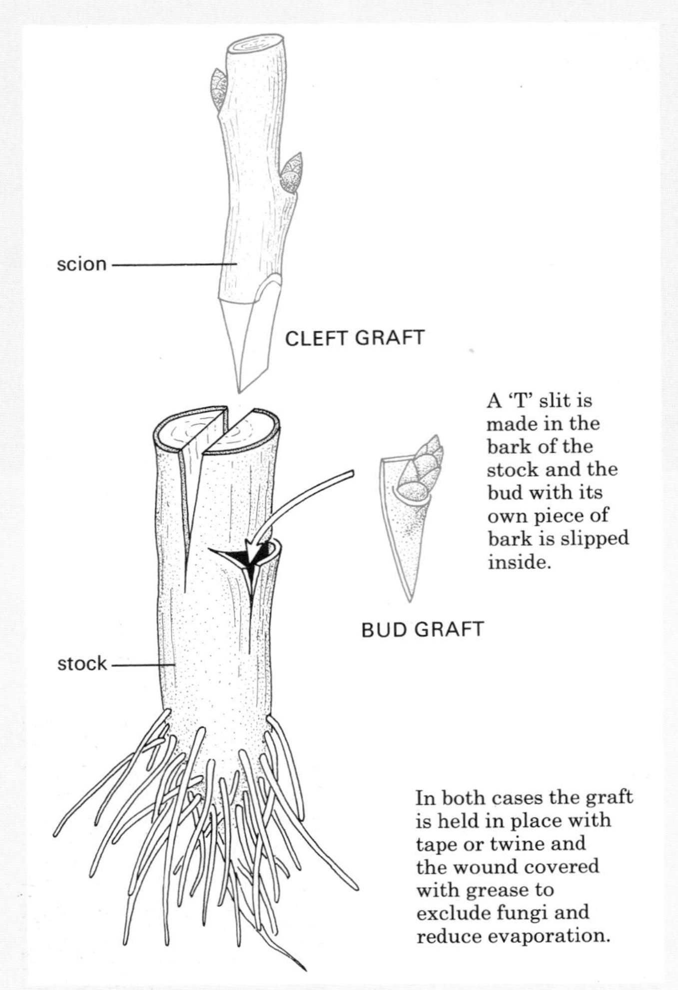

scion
CLEFT GRAFT
A 'T' slit is made in the bark of the stock and the bud with its own piece of bark is slipped inside.
BUD GRAFT
stock
In both cases the graft is held in place with tape or twine and the wound covered with grease to exclude fungi and reduce evaporation.

Fig. 8 **Two types of grafting**

Fig. 9 **Two cleft grafts on a pear tree stump**

EXPERIMENTS ON PROPAGATION FROM CUTTINGS

1. Effect of rooting hormone

An experiment on the effect of rooting hormone is described on p. 110 (Experiment 5).

2. Formation of adventitious roots

If small branches cut from a Busy Lizzie plant or a *Coleus* are placed in a jar of water, they will develop roots in a week or two. Roots which grow directly from a stem are called **adventitious roots**.

CHECK LIST

- **Asexual reproduction occurs without gametes or fertilization.**
- **Plants reproduce asexually when some of their buds grow into new plants.**
- **Asexual reproduction in plants is called vegetative propagation.**
- **The stolon of the strawberry plant is a horizontal stem which grows above the ground, takes root at the nodes and produces new plants.**
- **The couch grass rhizome is a horizontal stem which grows below the ground and sends up shoots from its nodes.**
- **Bulbs are condensed shoots with circular fleshy leaves. Bulb-forming plants reproduce asexually from lateral buds.**
- **Rhizomes, corms, bulbs and tap-roots may store food which is used to accelerate early growth.**
- **Vegetative propagation produces (genetically) identical individuals.**
- **A clone is a population of organisms produced asexually from a single parent.**
- **Artificial propagation from cuttings or grafts preserves the desirable characteristics of a crop plant.**

11 Plant Sensitivity

RESPONSES TO STIMULI

Tropic and nastic responses.

PHOTOPERIODISM

TROPISMS

Geotropism and phototropism. Tropism experiments. Advantages of tropic responses. The mechanism of tropisms; the auxin theory.

CHEMICAL CO-ORDINATION IN PLANTS

Plant 'growth' substances; 'hormones'.

PRACTICAL WORK

Regions of response to one-sided stimuli. Response of wheat coleoptiles to one-sided light. Use of hormone rooting powder.

One of the features of living organisms which distinguishes them from non-living things is their **sensitivity** (sometimes called 'irritability'), i.e. they respond to **stimuli** (singular = **stimulus**). A stimulus may be a touch, a change in light intensity or temperature, a sound or vibration, or various combinations of these. If you push a stone, it will move, but if you touch it lightly, nothing will happen. If you touch an earthworm, however, it will quickly retreat into its burrow; if you try to swat a fly, it will take off when the shadow of your hand falls on it. The fly and the earthworm have responded to stimuli and this demonstrates their sensitivity.

Plants are living organisms and they, too, respond to stimuli, though not so obviously or so rapidly as animals do. In plants, the response is made by certain organs, e.g. flowers, leaves, roots, stems or tendrils, whereas an animal usually responds as a whole. Important stimuli for plants are light, day length, gravity, temperature and touch. They respond by changing their direction of growth and by opening or closing their flowers and leaves.

Plant responses which are related to the **direction** of the stimulus are called **tropic responses** or **tropisms**; responses which are not directional are **nastic responses**; responses made to changes in day length are called **photoperiodism**.

Examples of nastic responses are the 'sleep movements' made by some plants at night or when light intensity falls, e.g.

1. the closing up of the petals of crocuses and tulips
2. the closing of the outer florets of daisies (Fig. 1)
3. the folding of the leaflets of wood sorrel.

In some cases, the movement is brought about by changes in growth rate. In other cases it is changes in the turgor (p. 38) of groups of cells which bring about the movement.

It is not entirely clear what the advantages of 'sleep movements' are.

(*a*) Daytime; inflorescences open

(*b*) Evening; inflorescences closed

Fig. 1 'Sleep movements' of daisies

PHOTOPERIODISM

There are many species of plant which will not produce flowers unless the day length is above or below a certain minimum. If spinach plants are kept in an artificial situation with 12 hours of darkness and 12 hours of light, they will grow well but will not produce flowers. Only if the period of light is increased to about 14 hours will they flower. Spinach is an example of a 'long-day' plant; it 'bolts' (i.e. flowering shoots suddenly grow tall and produce flowers) in the late spring when daylight extends from, say, 5 a.m. to 8 p.m.

Chrysanthemums, on the other hand, are 'short-day' plants and will not flower unless they receive more than 9 hours of continuous darkness. Horticulturists prevent their chrysanthemums from flowering by simply switching on an electric light for a few minutes in the night. As Christmas approaches, the plants are allowed 10 or more hours of unbroken darkness and they flower. Thus, blossoms are available when both the demand and the prices are high.

The response to day or night length is assumed to be an adaptation to the latitude and climatic conditions of the plant's habitat. Plants which will not flower until the day length exceeds 14 hours are usually summer-flowering species growing in northern latitudes. For example, delay of flowering till late in the year may give a plant a longer period of growth before producing seeds. Late flowering might also coincide with the maximum abundance of pollinating insects.

Note that long-day and short-day plants are only the extremes of a wide variety of plant responses to variations of day length.

QUESTIONS

1 Insectivorous plants often make movements which trap insects. What might be the stimuli which set off these movements?

2 Why do you think that short-day plants might not flourish in the Arctic?

TROPISMS

Tropisms are growth movements related to directional stimuli, e.g. a shoot will grow towards a source of light but away from the direction of gravity. Growth movements of this kind are usually in response to the direction of light, or gravity. Responses to light are called **phototropisms**; responses to gravity are **geotropisms** (or **gravitropisms**).

If the plant organ responds by growing towards the stimulus, the response is said to be 'positive'. If the response is growth away from the stimulus, it is said to be 'negative'.

For example, if a plant is placed horizontally, its stem will change its direction and grow upwards, away from gravity (Fig. 2). The shoot is **negatively geotropic**. The roots, however, will change their direction of growth to grow vertically downwards towards the pull of gravity (Experiment 1). Roots, therefore, are **positively geotropic**.

Phototropism and geotropism are best illustrated by some simple controlled experiments.

Fig. 2 Negative geotropism. The tomato plant has been left on its side for 24 hours.

EXPERIMENTS ON TROPISMS

1. Geotropism in pea radicles

Soak about 20 peas in water for a day and then let them germinate in a roll of moist blotting-paper (Fig. 23, p. 68). After 3 days, choose 12 seedlings with straight radicles and pin six of these to the turntable of a clinostat so that the radicles are horizontal. Pin another six seedlings to a cork that will fit in a wide-mouthed jar (Fig. 3). Leave the jar on its side. A **clinostat** is a clockwork or electric turntable which rotates the seedlings slowly about four times an hour. Although gravity is pulling sideways on their roots, it will pull equally on all sides as they rotate.

Place the jar and the clinostat in the same conditions of lighting or leave them in darkness for 2 days.

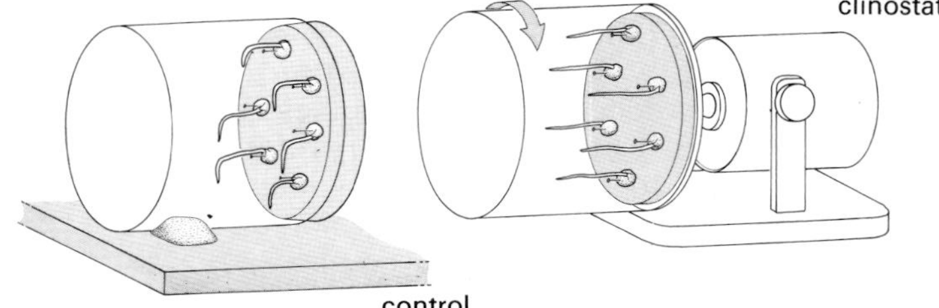

Fig. 3 Geotropism in roots

Result The radicles in the clinostat will continue to grow horizontally but those in the jar will have changed their direction of growth, to grow vertically downwards.

Interpretation The stationary radicles have responded to the stimulus of one-sided gravity by growing towards it. The radicles are positively geotropic.

The radicles in the clinostat are the controls. Rotation of the clinostat has allowed gravity to act on all sides equally and there is no one-sided stimulus, even though the radicles were horizontal.

2. Phototropism in shoots

Select two potted seedlings, e.g. sunflower or French bean, of similar size and water them both. Place one of them under a cardboard box with a window cut in one side so that light reaches the shoot from one direction only (Fig. 4). Place the other plant in an identical situation but on a clinostat. This will rotate the plant about 4 times per hour and expose each side of the shoot equally to the source of light. This is the control.

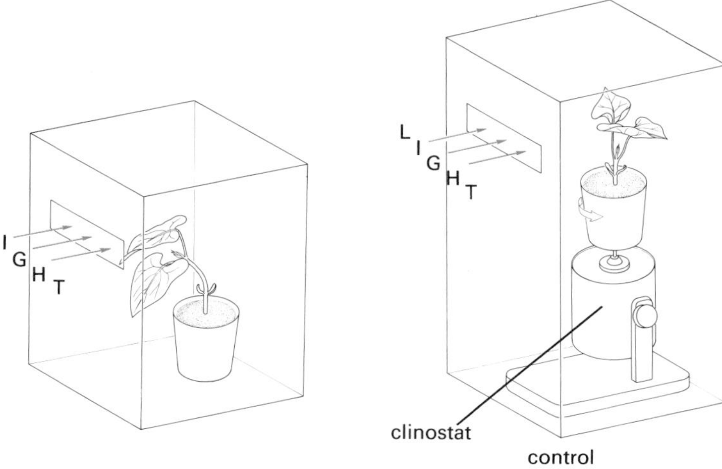

Fig. 4 Phototropism in a shoot

Result After 1 or 2 days, the two plants are removed from the boxes and compared. It will be found that the stem of the plant with one-sided illumination has changed its direction of growth and is growing towards the light (Fig. 5). The control shoot has continued to grow vertically.

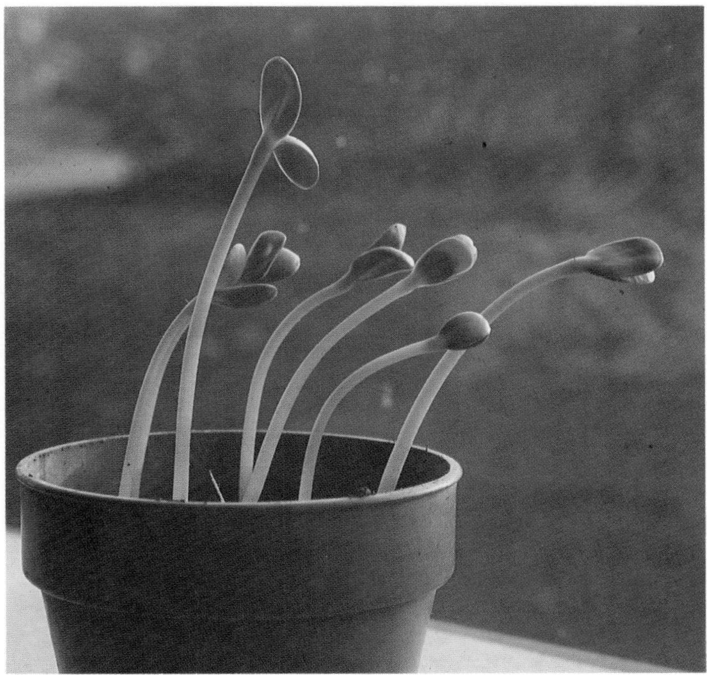

Fig. 5 Positive phototropism. The sunflower seedlings have received one-sided lighting for a day. (See Fig. 6 for a theoretical explanation.)

Interpretation The results suggest that the young shoot has responded to one-sided lighting by growing towards the light. The shoot is said to be positively phototropic because it grows towards the direction of the stimulus.

However, the results of an experiment with a single plant cannot be used to draw conclusions which apply to green plants as a whole. The experiment described here is more of an illustration than a critical investigation. To investigate phototropisms thoroughly, a large number of plants from a wide variety of species would have to be used.

Advantages of tropic responses

Positive phototropism of shoots By growing towards the source of light, a shoot brings its leaves into the best situation for photosynthesis. Similarly, the flowers are brought into an exposed position where they are most likely to be seen and pollinated by flying insects.

Negative geotropism in shoots Shoots which are negatively geotropic grow vertically. This lifts the leaves and flowers above the ground and helps the plant to compete for light and carbon dioxide. The flowers are brought into an advantageous position for insect or wind pollination. Seed dispersal may be more effective from fruits on a long, vertical stem. However, these advantages are a product of a tall shoot rather than negative geotropism. There are a great many very successful plants with short, vertical root-stocks, e.g. dandelion.

Stems which form rhizomes (p. 100) are not negatively geotropic; they grow horizontally below the ground, though the shoots which grow up from them are negatively geotropic.

Branches from upright stems are not negatively geotropic; they grow at 90 degrees or, usually, at a more acute angle to the directional pull of gravity. The lower branches of a potato plant must be partially **positively** geotropic when they grow down into the soil and produce potato tubers (p. 102).

Positive geotropism in roots By growing towards gravity, roots penetrate the soil which is their means of anchorage and their source of water and mineral salts. Lateral roots are not positively geotropic; they grow at right angles or slightly downwards from the main root. This **diageotropism** of lateral roots enables a large volume of soil to be exploited and helps to anchor the plants securely.

QUESTIONS

3 (*a*) To what directional stimuli do (i) roots, (ii) shoots respond?
(*b*) Name the plant organs which are (i) positively phototropic, (ii) positively geotropic, (iii) negatively geotropic.
4 Why is it incorrect to say (*a*) 'Plants grow towards the light', (*b*) 'If a root is placed horizontally, it will bend towards gravity'?
5 Explain why a clinostat is used for the controls in tropism experiments.
6 Look at Fig. 2. What will the shoot look like in 24 hours after the pot has been stood upright again? (Just draw the outline of the stem.)

7 (*a*) Why does a seed still germinate successfully even if you plant it 'upside-down'?

(*b*) Refer to Fig. 8 on page 95 and make a simple drawing to correspond to Fig. 8*c*, assuming that the wheat grain had been planted 'upside-down'.

8 What do you think might happen if a potted plant were placed on its side and the shoot illuminated from below (i.e. light and gravity are acting from the same direction)?

The mechanism of tropisms

There are a number of classical experiments, starting with Darwin in 1880, which have shaped the theory of plant responses to one-sided stimuli. Plant physiologists experimented with the rapidly growing coleoptiles (p. 95) of oat seedlings. From their results and from more modern studies a general model of tropic movements has grown up. This model is far from satisfactory but does serve as a working hypothesis for the time being.

Plants have been shown to produce growth-regulating substances in their tissues. One group of these, including **indole-acetic acid (IAA)**, are called **auxins**. They are thought to be produced at the tips of shoots and roots. When they pass back from the shoot tip, the auxins cause more rapid extension of cells in the region of the shoot behind the tip.

Phototropism If the shoot is illuminated from above, auxin passes from its tip to the region of cell extension, which grows uniformly and vertically. One-sided illumination affects the distribution of auxin. How the light affects the distribution of auxin is not clear, but the outcome is that less auxin reaches the illuminated side and more reaches the shaded side. The result is that cells on the shaded side extend more rapidly than those on the illuminated side, causing the shoot to grow in a curve towards the light source (Fig. 6).

The evidence in support of this explanation of phototropism is not entirely satisfactory. Experiments have shown that there certainly is a higher concentration of auxin on the dark side than on the illuminated side of a shoot, but its concentration and rate of accumulation do not seem high enough to account for the observed degree and rate of growth curvature.

In addition, the evidence from oat coleoptiles does not apply very well to dicotyledonous shoots. The coleoptile is a hollow structure, which grows at its base and not at its tip. It is very short-lived and serves only as a tube conducting the first leaf through the soil. This makes it a poor model for a dicot shoot which is solid, grows at the tip and produces structures which last throughout the life of the plant.

Geotropism of shoots If a plant is placed with its shoot horizontal, more auxin reaches the lower side of the shoot. (Again the reason for this is not clear.) The lower side of the shoot, therefore, elongates more rapidly than the upper side and the shoot curves upwards as it grows.

Geotropism in roots The root tip (in some cases the root cap) produces auxin. In a horizontal root, more auxin reaches the lower side. One theory associates this with the movement of starch grains to the lower side of the root cells, though how this affects the distribution of auxin is not known. The root is thought to be more sensitive to auxin than the shoot and cell extension is inhibited rather than enhanced by auxin. (There is supporting evidence for this.) Consequently, the lower side of the root grows more slowly than the upper side and makes the root curve downwards as it grows (Fig. 7).

There is evidence to suggest that auxin from the shoot may be translocated to the root and interact with other growth substances produced by the root tip. The simple models presented here are going to need considerable modification.

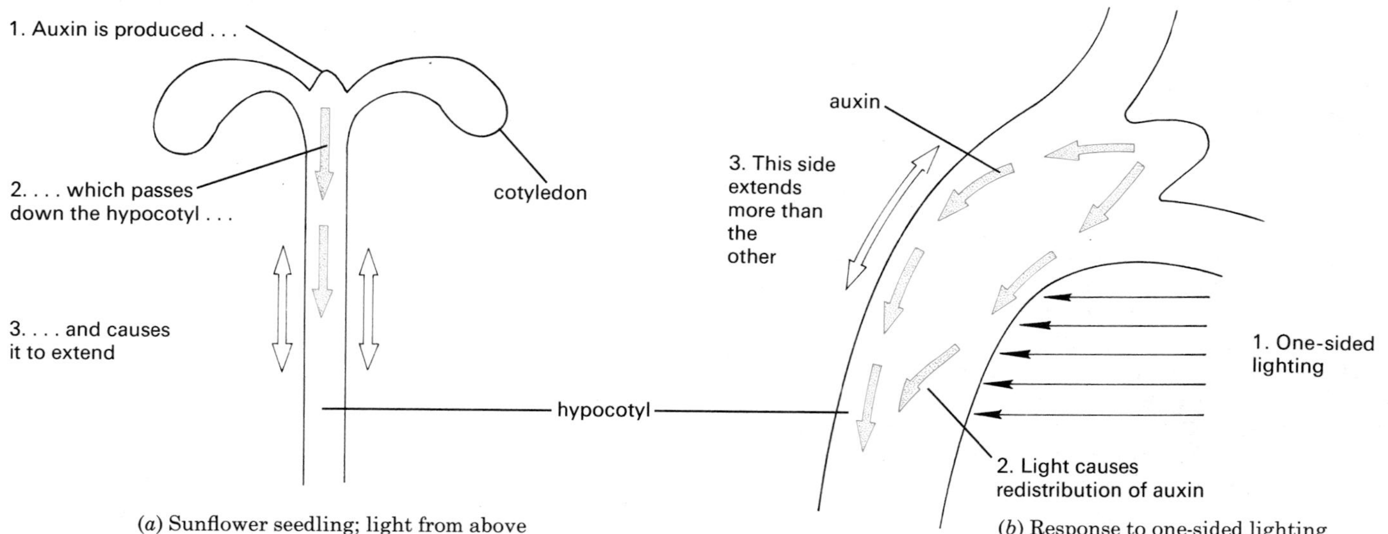
1. Auxin is produced . . .
2. . . . which passes down the hypocotyl . . .
3. . . . and causes it to extend
cotyledon
hypocotyl
auxin
3. This side extends more than the other
1. One-sided lighting
2. Light causes redistribution of auxin
(*a*) Sunflower seedling; light from above
(*b*) Response to one-sided lighting

Fig. 6 One way the auxin theory could explain phototropism

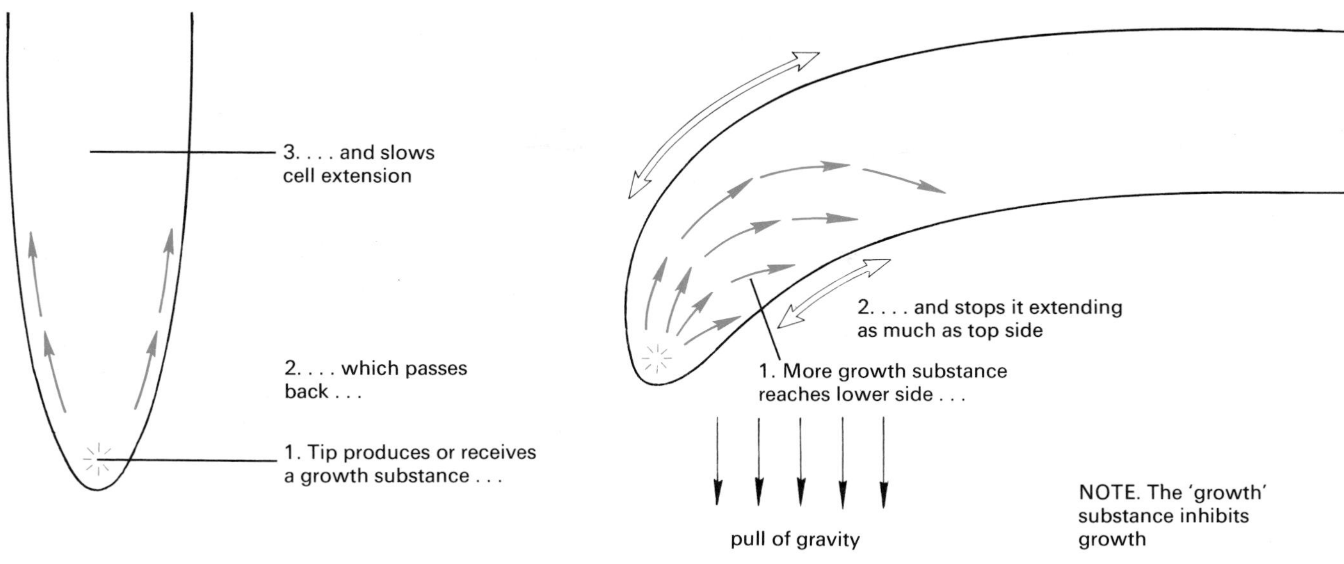
3. . . . and slows
cell extension
2. . . . which passes
back . . .
1. Tip produces or receives
a growth substance . . .
2. . . . and stops it extending
as much as top side
1. More growth substance
reaches lower side . . .
pull of gravity
NOTE. The 'growth'
substance inhibits
growth

(*a*) Normal growth; root vertical

(*b*) Response to one-sided gravity

Fig. 7 Possible explanation of positive geotropism in roots

QUESTIONS

9 Look at Fig. 2. Which side of the stem has (*a*) probably received more auxin, (*b*) grown faster?
10 Root tips and shoot tips are both presumed to produce auxin, yet a horizontal shoot grows upwards while a horizontal root grows downwards. Suggest an explanation for this difference in response to the same stimulus.
11 If cutting off the tip of a root removes the region of cell division, (*a*) why does this operation not immediately stop growth and (*b*) would you expect the root to respond to unidirectional gravity?

CHEMICAL CO-ORDINATION IN PLANTS

Although plants do not have nervous systems, they still need to co-ordinate their activities. For example, the stomata need to be open in the daylight and closed at night; food reserves in seeds need to be digested and transported to the rapidly growing roots and shoots at the right time.

Some of these activities are co-ordinated by chemicals such as the auxins which are thought to play a part in tropisms. In some cases, these chemicals are made in one part of the plant (e.g. the root cap), but produce their effect in another part (e.g. the growing region of the root). Only tiny quantities of the chemicals are needed to produce an effect. In these two respects, the plant chemicals are similar to the hormones (p. 212) of animals and they are sometimes called 'plant hormones'. However, they are not chemically related to animal hormones and they are not always produced at a distance from their point of action so the term 'hormone' is not really appropriate. They are usually called **plant growth substances**, though they influence many factors other than growth.

The auxins, for example, affect not only growth but also fruit formation, bud burst, leaf fall and root formation. Other growth substances regulate cell division and differentiation, root and shoot growth, flower formation, seed dormancy and the formation of winter buds. An auxin produced by the fertilized ovules in a flower plays a part in causing the fruit to develop.

Use of plant growth substances Chemicals can be manufactured which closely resemble natural growth substances and may be employed to control various aspects of growth and development of crop plants.

An artificial auxin sprayed on to tomato flowers will induce all of them to produce fruit, whether or not they have been pollinated. Another growth substance, sprayed on to fruit trees, prevents early fruit fall and enables all the fruit to be harvested at the same time.

The weed-killer, 2,4-D, is very similar to one of the auxins. When sprayed on a lawn, it affects the broad-leaved weeds (e.g. daisies and dandelions) but not the grasses. (It is called a 'selective weed-killer'.) Among other effects, it distorts the weeds' growth and speeds up their rate of respiration to the extent that they exhaust their food reserves and die.

PRACTICAL WORK

FURTHER EXPERIMENTS ON TROPIC RESPONSES

3. Region of response

Grow pea seedlings as described on page 68 and select six with straight radicles about 25 mm long. Mark all the radicles with lines about 1 mm apart as described on page 68. Use four strips of moist cotton wool to wedge three seedlings in each of two Petri dishes. Leave the dishes on their sides for two days, one (A) with the radicles horizontal and the other (B) with the radicles vertical.

Result The ink marks will be more widely spaced in the region of greatest extension (Fig. 8). By comparing the seedlings in the two dishes, it can be seen that the region of curvature in the B seedlings corresponds to the region of extension in the A seedlings.

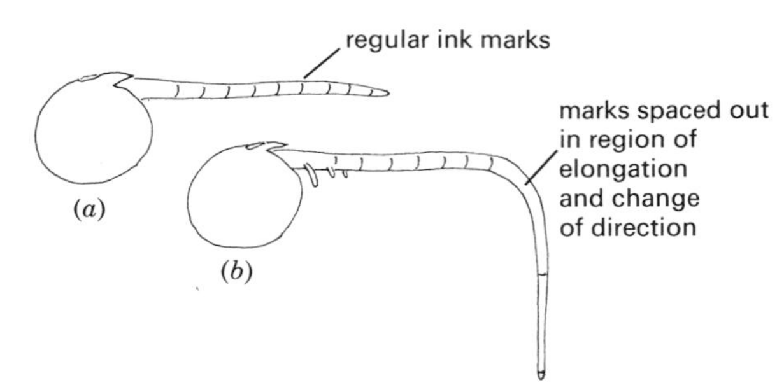

Fig. 8 The region of response to one-sided gravity

Interpretation The response to the stimulus of one-sided gravity takes place in the region of extension. It may be that this is also the region which detects the stimulus.

4. Response of coleoptiles to one-sided illumination

Soak about 20 wheat seeds in water for 24 hours and then place them on moist cotton wool in a Petri dish. Allow the wheat to germinate in darkness for about 6 days, checking each day that the cotton wool is moist and the coleoptiles do not exceed about 30 mm.

When most of the coleoptiles are between 20 and 30 mm, cut 5 mm from the tip of three or four seedlings. Put caps of aluminium foil on the tips of another three or four and leave the rest intact. (The foil caps can be made by rolling small pieces of aluminium foil round the tip of a cocktail stick. The caps should not be longer than 5 mm.)

Place the Petri dish in a light-proof box, with a slot cut in one side at the level of the coleoptiles, or illuminate the seedlings with a bench lamp 15–30 cm away and screen them from other light sources. Examine the seedlings after 3, 5 and 24 hours.

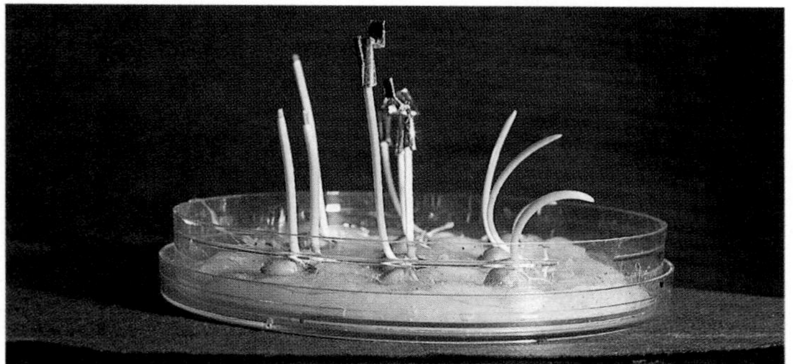

Fig. 9 Phototropism in coleoptiles. The shoots with tips covered or cut off have not responded to one-sided lighting.

Result After 3 hours, the intact coleoptiles should show a distinct growth curvature towards the light source. The foil-covered and decapitated coleoptiles should still be growing more or less vertically and straight (Fig. 9).

After 24 hours, the intact coleoptiles will exhibit a 'lean' towards the light source rather than a curvature and the other coleoptiles may show signs of directional growth, particularly if the first leaf has emerged and pushed off the foil cap.

Interpretation The intact coleoptiles have exhibited positive phototropism and the curvature appears 10 mm or more below the tip. Covering or removing the tip has impaired the phototropic response.

Since the response occurs below the tip, this result supports the hypothesis that the stimulus of one-sided lighting is perceived by the tip, but the response occurs at a distance from the tip. (More sophisticated experiments suggest that other parts of the coleoptile can detect the stimulus of one-sided light.)

Note These experiments and others are fully described in *Experimental Work in Biology*, Combined Edition. (See p. 364.)

5. Use of hormone rooting powder

Cut ten shoots, about 15 cm long, from the new growth of an untrimmed *Lonicera* hedge in June–August. (*Lonicera nitida* is bush honeysuckle; cuttings from flowering currant and laurel also work well but take longer than two weeks.) Remove the leaves from the lower 5 cm of the stems and, with five of the cuttings, dip the exposed stem first into water and then into hormone rooting powder. Tap off the excess powder and push the stems into a compost of sand and peat in a flower-pot labelled 'A'. Dip the other five stems into water only and push them into compost in a flower-pot labelled 'B'. Wash your hands to remove any traces of the powder.

Leave both pots in the same conditions of light and temperature and keep the compost moist for 2 weeks. After this time, remove the cuttings and wash the compost off the stem bases.

Result Most of the cuttings will have developed roots from the part of the stem in the compost, but the cuttings treated with hormone rooting powder will have many more roots. Some of the untreated cuttings may have failed to develop any roots at all.

Interpretation The hormone rooting powder contains a chemical similar to the auxin IAA (probably naphthyl acetic acid NAA). This chemical has promoted the formation of roots in the stem cuttings.

Note The rooting powder may also contain a fungicide, which makes interpretation of the results less certain.

CHECK LIST

- The roots or shoots of plants may respond to the stimuli of light or gravity.
- A response related to the direction of the stimulus is a tropism.
- Nastic responses are non-directional responses, e.g. the opening and closing of flower petals.
- The time of flowering of many plants is controlled by day length.
- Phototropism is a growth response to the direction of light.
- Geotropism is a growth response to the direction of gravity.
- Growth towards the direction of the stimulus is called 'positive'; growth away from the stimulus is called 'negative'.
- Tropic responses bring shoots and roots into the most favourable positions for their life-supporting functions.
- Auxin is a growth-promoting substance produced in plants.
- Tropic responses may be explained, in some cases, by the uneven distribution of auxin on opposite sides of a stem or root.
- Plants have a number of different 'growth' substances which help to co-ordinate their activities.
- These activities include growth, fruit formation, leaf fall, dormancy and bud burst.

Examination Questions

Section 2: Flowering Plants

Do not write on these pages. Where necessary copy drawings, tables or sentences.

1 The following diagram shows a section cut through part of a plant.

(*a*) Name parts A–D.

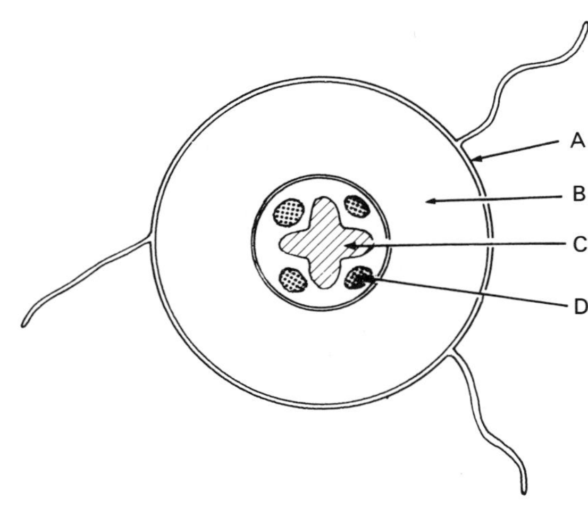

(*b*) (i) Name the part of the plant through which the section was cut.

(ii) State **one** reason for your answer to (*b*) (i).

(*c*) In what direction has the section been cut?

(*d*) If this plant had been grown in water coloured with red ink, which part on the diagram would be stained red?

(*e*) State the function of tissue D. (W)

2 The diagram below shows a section through a flower at two stages in its development.

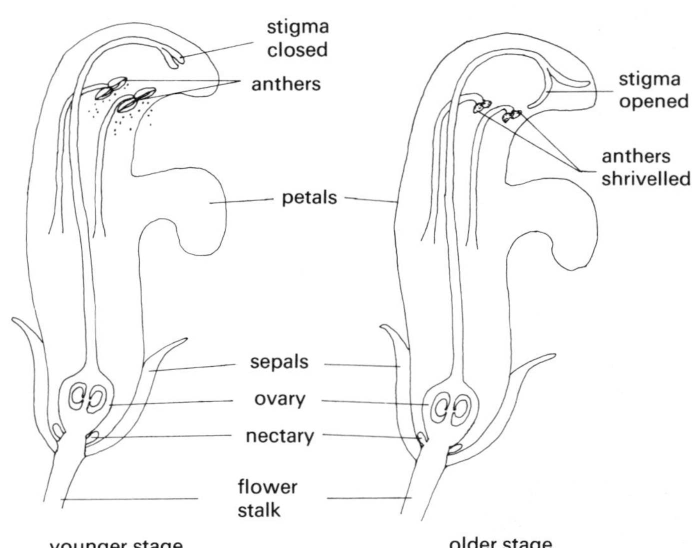

(*a*) Write down **two** features shown in the diagram which suggest that the flower is insect pollinated.

(*b*) How might your nose help to confirm that the flower is insect pollinated?

(*c*) Cell division by meiosis takes place in the flower. Name the TWO structures, labelled in the diagram, in which meiosis takes place.

(*d*) (i) What is cross-pollination?

(ii) What information shown in the diagrams would suggest that cross-pollination takes place in this flower? (L)

3 The diagram shows a fruit.

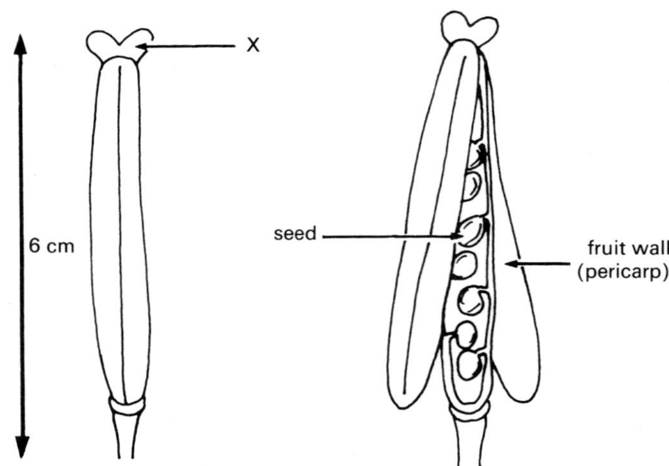

(*a*) Name the plant from which the fruit has come.
(*b*) From which parts of the flower have the following developed?
(i) the fruit wall (pericarp)
(ii) the seed
(iii) the part labelled X
(*c*) Explain how the seeds of this fruit are dispersed. (W)

4 The drawing shows a bean seed which is starting to grow.

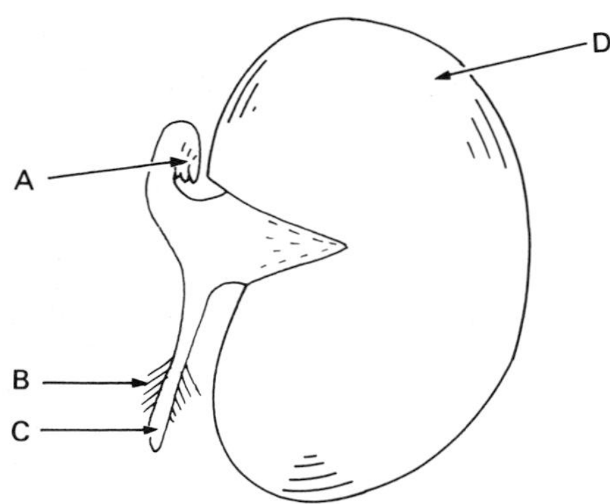

(*a*) Label parts A–D.
(*b*) State how parts A and C are protected as they grow through the soil.
(*c*) Given the correct conditions, how long would it take for the bean seed to grow to the stage shown?
8 days 8 hours 8 weeks 8 minutes
(*d*) Name the gas found in the air which is needed for germination.
(*e*) State **two** other conditions needed for germination. (W)

5 Seeds germinate faster in warm soil than in cold. This is because, in the seed, warmth speeds up

A photosynthesis **C** uptake of salts
B enzyme activity **D** transpiration (N)

6 The diagram shows a plant of couch grass (*Agropyron repens*).

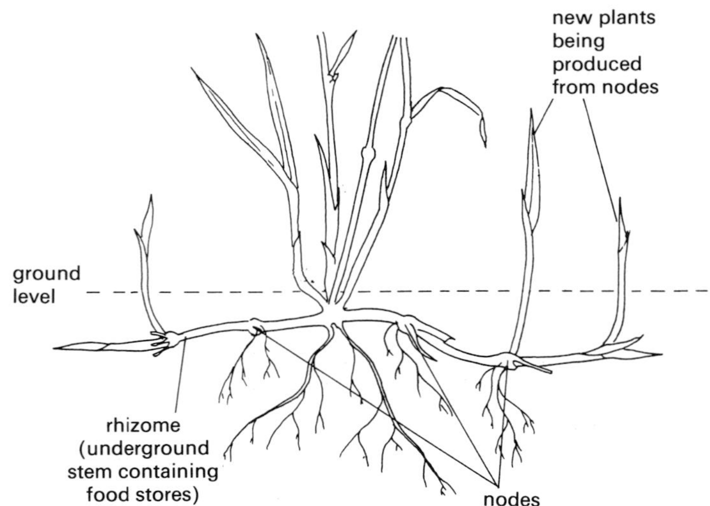

(*a*) The new plants being produced are the result of asexual reproduction. How does the diagram show this?
(*b*) Couch grass is a weed (unwanted plant) in farmland. Some farmers try to control its spread by cultivating the soil. In this process the plant is broken into small pieces and buried. Other farmers use herbicides (chemical weedkillers) to control the plant.
(i) Look carefully at the diagram and suggest why cultivating the soil is unlikely to get rid of couch grass.
(ii) Why may it be dangerous to use herbicides (weedkillers) to control couch grass? (S)

7 The graph below shows the rate of transpiration for four leaves A, B, C and D. These leaves are of the same size and from the same plant, but they have been treated differently.

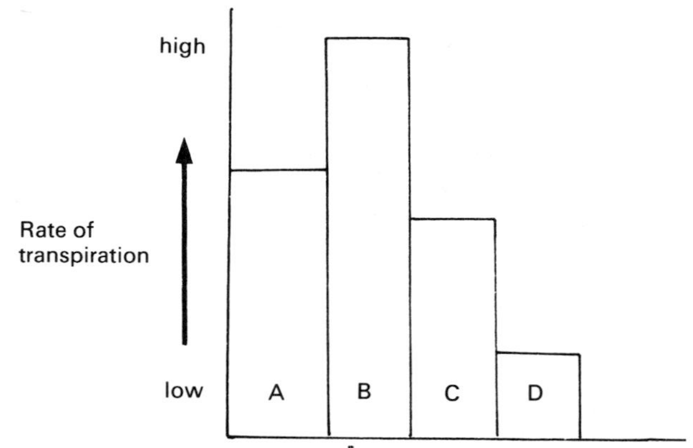

(*a*) Match the letters on the graph with the descriptions below.
(i) Lower surface coated with vaseline.
(ii) Upper surface of leaf coated with vaseline.
(iii) Leaf left untreated.
(iv) Leaf left near hair dryer blowing hot air.
(*b*) If you were given the following apparatus, describe how you could determine the distribution of stomata on the surfaces of a leaf:
microscope, slides, cover slips, nail varnish, leaves. (W)

SECTION 3
HUMAN PHYSIOLOGY

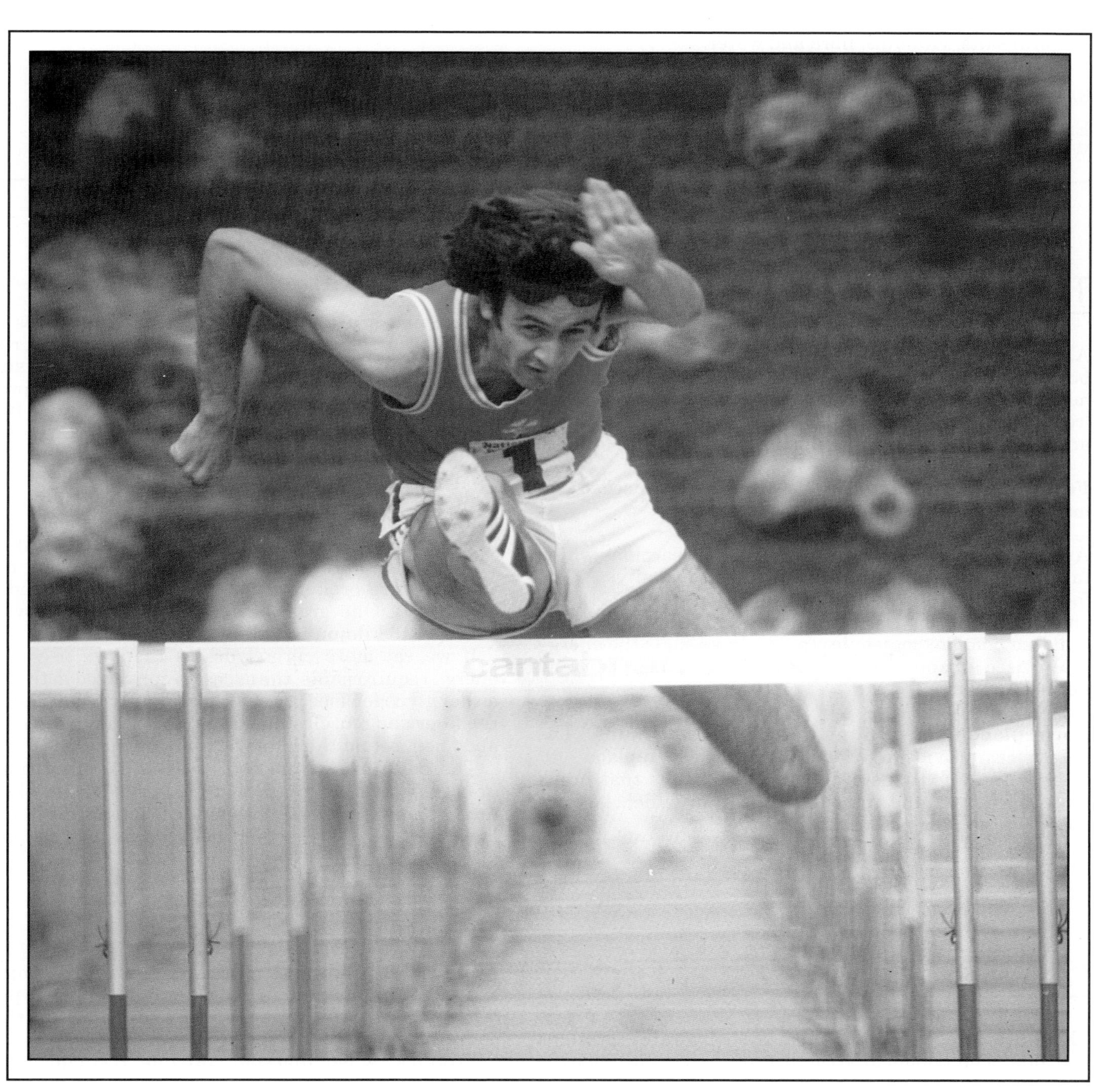

12 Food and Diet

CLASSES OF FOOD
Carbohydrates, proteins and fats.

DIET
Mineral salts, vitamins, fibre and water.

BALANCED DIETS
Energy needs and proteins. Special needs. Western diets.

PRESERVING AND PROCESSING
Additives.

WORLD FOOD

PRACTICAL WORK
Food tests. Energy from food.

THE NEED FOR FOOD

All living organisms need food. An important difference between plants and animals is that the green plants can make food in their leaves but animals have to take it in 'ready-made' by eating plants or the bodies of other animals. In all plants and animals, food is used as follows:

For growth It provides the substances needed for making new cells and tissues.

As a source of energy for the chemical reactions which take place in living organisms to keep them alive. When food is broken down during respiration (see p. 24), the energy from the food is used for chemical reactions such as building complex molecules (p. 14). In animals the energy is also used for activities such as movement, heart beat and nerve impulses.

For replacement of worn and damaged tissues The substances provided by food are needed to replace—for example—the millions of our red blood cells that break down each day, and to replace the skin which is worn away, and to repair wounds.

CLASSES OF FOOD

There are three classes of food: carbohydrates, proteins and fats. The chemical structure of these substances is described on pages 14–17.

Carbohydrates

Sugar and **starch** are important carbohydrates in our diet. Starch is abundant in potatoes, bread, maize, rice and other cereals. Sugar appears in our diet mainly as **sucrose** (table sugar) which is added to drinks and many prepared foods such as jam, biscuits and cakes. Glucose and fructose are sugars which occur naturally in many fruits and some vegetables.

Although all foods provide us with energy, carbohydrates are the cheapest and most readily available source of energy. They contain the elements carbon, hydrogen and oxygen (e.g. glucose is $C_6H_{12}O_6$). When carbohydrates are oxidized to provide energy by respiration they are broken down to carbon dioxide and water (see p. 24). One gram of carbohydrate can provide, on average, 16 kilojoules (kJ) of energy.

If we eat more carbohydrates than we need for our energy requirements, the excess is converted in the liver to either glycogen (see p. 131) or fat. The glycogen is stored in the liver and muscles; the fat is stored in fat depots in the abdomen, round the kidneys or under the skin (Fig. 1).

The **cellulose** in the cell walls of all plant tissues is a carbohydrate. We probably derive relatively little nourishment from cellulose but it is important in the diet as **fibre** (see p. 117) which helps to maintain a healthy digestive system.

Proteins

Lean meat, fish, eggs, milk and cheese are important sources of animal protein. All plants contain some protein, but beans or cereals like wheat and maize are the best sources.

Proteins, when digested, provide the chemical substances needed to build cells and tissues, e.g. skin, muscle,

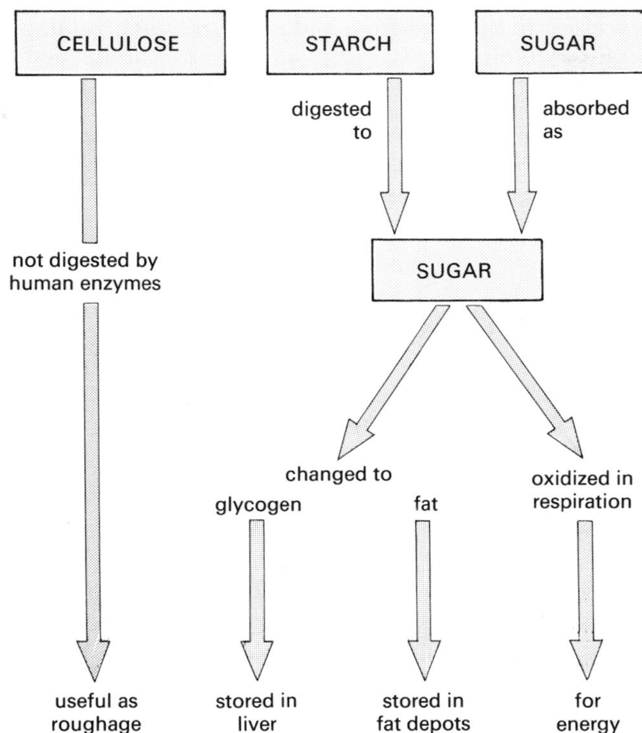

Fig. 1 Digestion and use of carbohydrate

ala—gly—gly—leu—val—cys—gly
S
S
glu—val—lys—cys—ala
leu
leu

Part of a plant protein of 14 amino acids

ala gly val cys val gly ala gly cys val leu glu lys leu

Digestion breaks up protein into its amino acids and . . .

glu—val—cys—gly
S
S
ala—leu—cys—val—gly
leu
lys
ala—leu—gly

. . . our body builds up the same 14 amino acids but into a protein that it needs

ala=alanine gly=glycine leu=leucine
cys=cysteine glu=glutamine lys=lysine val=valine

Fig. 2 Digestion and use of a protein molecule

blood and bones. Neither carbohydrates nor fats can do this and so it is essential to include some proteins in the diet.

Protein molecules consist of long chains of **amino acids** (see p. 14). When proteins are digested, the molecules are broken up into the constituent amino acids. The amino acids are absorbed into the bloodstream and used to build up different proteins. These proteins form part of the cytoplasm and enzymes of cells and tissues. Figure 2 shows such a rearrangement of amino acids.

The amino acids which are not used for making new tissues cannot be stored, but the liver removes their amino ($-NH_2$) groups and changes the residue to glycogen. The glycogen can be stored or oxidized to provide energy (p. 132).

Chemically, proteins differ from both carbohydrates and fats because they contain nitrogen and sulphur as well as carbon, hydrogen and oxygen.

Essential amino acids Although humans and other animals cannot make amino acids, they can change one amino acid into another. There are, however, at least eight amino acids that the body cannot produce in this way. These are called **essential amino acids** and they must be included in the diet. Lysine, valine and leucine are examples of essential amino acids.

Some plant proteins do not contain enough essential amino acids to supply the needs of the body. Protein from cereals, for example, has too little lysine. A diet that was almost entirely made up of rice or maize could lead to a protein-deficiency disease. Beans, however, are a rich source of lysine and by including these in the diet, the lysine deficiency is made up.

Most animal proteins contain all the essential amino acids (see Table 1 overleaf).

Fats

Animal fats are found in meat, milk, cheese, butter and egg-yolk. Plant fats occur as oils in fruits (e.g. palm oil) and seeds (e.g. sunflower seed oil), and are used for cooking and making margarine. Fats and oils are sometimes collectively called **lipids**.

Lipids are used in the cells of the body to form part of the cell membrane and other membrane systems (p. 4). Lipids can also be oxidized in respiration, to carbon dioxide and water. When used to provide energy in this way, 1 g fat gives 39 kJ of energy. This is more than twice as much energy as can be obtained from the same weight of carbohydrate or protein.

Fats can be stored in the body, so providing a means of long-term storage of energy in fat depots. The fatty tissue, **adipose tissue**, under the skin forms an insulating layer which reduces heat losses from the body.

QUESTIONS

1 What sources of protein-rich foods are available to a vegetarian who (*a*) will eat animal products but not meat itself, (*b*) will eat only plants and their products?
2 Why must all diets contain some protein?
3 In what sense can the fats in your diet be said to contribute to 'keeping you warm'?

4 The table below shows the number of milligrams of four essential amino acids provided by a fixed amount of protein. It also shows how much of these the body needs to get from that amount of protein. Of the seven proteins shown
(*a*) which one is the best source of essential amino acids;
(*b*) which is the best plant source; and
(*c*) which combinations of plant products will provide an adequate supply of essential amino acids?

Table 1 Four essential amino acids

	Amino acid (mg)			
	Try	Lys	Leu	Val
Body needs	90	270	300	270
Protein in these foods provides:				
Meat	80	510	490	330
Milk	90	490	630	440
Eggs	110	420	560	450
Wheat	80	170	400	270
Rice	90	220	510	370
Beans	60	450	530	370
Potatoes	65	230	290	270

(Try = tryptophan; Lys = lysine; Leu = leucine; Val = valine.)

DIET

In addition to proteins, carbohydrates and fats, the diet must include salts, vitamins, water and vegetable fibre (roughage). These substances are present in a balanced diet and do not normally have to be taken in separately.

Salts

These are sometimes called 'mineral salts' or just 'minerals'. Proteins, carbohydrates and fats provide the body with carbon, hydrogen, oxygen, nitrogen, sulphur and phosphorus but there are several more elements which the body needs and which occur as salts in the food we eat.

Iron The red blood cells contain the pigment haemoglobin (p. 137). Part of the haemoglobin molecule contains iron and this plays an important part in carrying oxygen round the body. Millions of red cells break down each day and their iron is stored by the liver and used to make more haemoglobin. However, some iron is lost and adults need to take in about 15 mg each day. Iron is needed also in the muscles and for enzyme systems in all the body cells.

Red meat, especially liver and kidney, is the richest source of iron in the diet, but eggs, groundnuts, bread, spinach and other green vegetables are also important sources.

If the diet is deficient in iron, a person may suffer from some form of anaemia. Insufficient haemoglobin is made and the oxygen-carrying capacity of the blood is reduced.

Calcium Calcium, in the form of calcium phosphate, is deposited in the bones and the teeth and makes them hard. It is present in blood plasma and plays an essential part in normal blood clotting (p. 146). Calcium is also needed for the chemical changes which make muscles contract and for the transmission of nerve impulses.

The richest sources of calcium are milk (liquid, skimmed or dried), and cheese, but calcium is present in most foods in small quantities and also in 'hard' water.

Many calcium salts are not soluble in water and may pass through the alimentary canal without being absorbed. Simply increasing the calcium in the diet may not have much effect unless the calcium is in the right form, the diet is balanced and the intestine is healthy. Vitamin D and bile salts (p. 129) are needed for efficient absorption.

The level of calcium in the blood is controlled by a gland in the neck, called the **parathyroid**.

Iodine This is needed in only small quantities, but it forms an essential part of the molecule of **thyroxine**. Thyroxine is a hormone (p. 212) produced by the thyroid gland in the neck.

Specially rich sources of iodine are sea-fish and shellfish but it is present in most vegetables, provided that the soil in which they grow is not deficient in the mineral. In some parts of the world, where soils have little iodine, potassium iodide may be added to table salt to bring the iodine in the diet to a satisfactory level.

Sodium and potassium These elements occur in all the cells and fluids of the body. Sodium is mainly in the blood and tissue fluid, while most of the potassium is inside the cells. Salts of sodium and potassium are needed to keep the blood and tissue fluids at the correct osmotic concentration (see p. 37) so that the tissues do not become waterlogged or dehydrated. Both sodium and potassium salts are excreted in the urine, and sodium is lost in the sweat. We need about 2–4 g of each of these salts every day.

Most foods contain sodium and potassium in sufficient quantities. The extra salt added during cooking, or shaken over the food, is probably unnecessary and may even be harmful for some people. Only people who lose a lot of sodium chloride in sweat may need a high intake.

Phosphorus Phosphorus is needed for the calcium phosphate of bone, and also for the nucleotides such as ADP and ATP, and the nucleic acids DNA and RNA (p. 17). It is present in nearly all food but is particularly abundant in cheese, meat and fish.

Fluoride Fluoride occurs naturally, in small quantities, in drinking-water. It plays a part in the formation of teeth. Although tooth decay results mainly from the continuous presence of sugar or other carbohydrate in the mouth (p. 191), fluoride helps to resist decay, particularly in children. For this reason, some toothpastes have fluoride added to them and some people dissolve fluoride tablets in their drinking-water.

Vitamins

All proteins are similar to each other in their chemical structure and so are all carbohydrates. Vitamins, on the other hand, are a group of organic substances quite unrelated to each other in their chemical structure. The features shared by them all are:

1. They are not digested or broken down for energy.
2. Mostly, they are not built into the body structures.
3. They are essential in very small quantities for normal health.
4. They are needed for chemical reactions in the cells, working in association with enzymes.

Plants can make these vitamins in their leaves, but animals have to take them in ready-made either from plants or from other animals.

If any one of the vitamins is missing from our diet, or at a low level, we will develop a vitamin-deficiency disease. These diseases can be cured, at least in the early stages, simply by adding the vitamin to the diet.

Fifteen or more vitamins have been identified and they are sometimes grouped into two classes: water-soluble or fat-soluble. The fat-soluble vitamins are found mostly in animal fats or vegetable oils, which is one reason why our diet should include some of these fats. The water-soluble vitamins are present in green leaves, fruits and cereal grains.

Table 2 describes vitamins, gives examples of foods which contain them and mentions the vitamin-deficiency diseases which result from their absence.

Dietary fibre (roughage)

When we eat vegetables and other fresh plant material, we take in a large quantity of plant cells. The cell walls of plants consist mainly of cellulose, but we do not have enzymes for digesting this substance. The result is that the plant cell walls reach the large intestine (colon) without being digested. This undigested part of the diet is called **fibre** or roughage. The colon contains many bacteria which can digest some of the substances in the plant cell walls to form fatty acids (see p. 15). Vegetable fibre, therefore, may supply some useful food material, but it has other important functions.

The fibre itself and the bacteria which multiply from feeding on it, add bulk to the contents of the colon and help it to retain water. This softens the faeces (p. 131) and reduces the time needed for the undigested residues to pass out of the body. Both of these effects help to prevent constipation and keep the colon in a healthy condition.

Most vegetables and whole cereal grains contain fibre, but white flour and white bread do not contain much. Good sources of dietary fibre are wholemeal bread and bran.

Water

About 70 per cent of most tissue consists of water; it is an essential part of cytoplasm. The body fluids, blood, lymph and tissue fluid (Chapter 14) are composed mainly of water.

Digested food, salts and vitamins are carried round the body as a watery solution in the blood (p. 145) and

Table 2 Vitamins

Name and source of vitamin	*Diseases and symptoms caused by lack of vitamin*	*Notes*
Retinol (vitamin A; fat-soluble): Liver, cheese, butter, margarine, milk, eggs. **Carotene** (vitamin A precursor; water-soluble): Fresh green leaves and carrots.	Reduced resistance to disease, particularly those which enter through the epithelium. Poor night vision. Cornea of eyes becomes dry and opaque leading to **keratomalacia** and blindness.	The yellow pigment, carotene, present in green leaves and carrots is turned into retinol by the body. Retinol forms part of the light-sensitive pigment in the retina (p. 196). Retinol is stored in the liver.
Folic acid (water-soluble): Liver, spinach, fish, beans, peas.	**Vitamin deficiency anaemia:** Not enough red blood cells are made.	Likely to affect pregnant women on poor diets.
Ascorbic acid (vitamin C; water-soluble): Oranges, lemons, grapefruit, tomatoes, fresh green vegetables, potatoes.	Fibres in connective tissue of skin and blood vessels do not form properly, leading to bleeding under the skin, particularly at the joints, swollen, bleeding gums and poor healing of wounds. These are all symptoms of **scurvy**.	Possibly acts as a catalyst in cell respiration. Scurvy is only likely to occur when fresh food is not available. Cows' milk and milk powders contain little ascorbic acid so babies may need additional sources. Cannot be stored in the body; daily intake needed.
Calciferol (vitamin D; fat-soluble): Butter, milk, cheese, egg-yolk, liver, fish-liver oil.	Calcium is not deposited properly in the bones, causing **rickets** in young children because the bones remain soft and are deformed by the child's weight. Deficiency in adults causes **osteomalacia**; fractures are likely.	Calciferol helps the absorption of calcium from the intestine and the deposition of calcium salts in the bones. Natural fats in the skin are converted to a form of calciferol by sunlight.
The B vitamins There are ten or more water-soluble vitamins which occur together, particularly in whole cereals, peas and beans. A deficiency of any one of these vitamins is likely to occur only in communities living on restricted diets such as maize or milled rice.		
There are several other substances classed as vitamins, e.g. **riboflavin** (B_2), **tocopherol** (E), **phylloquinone**, but these are either (1) unlikely to be missing from the diet, or (2) not known to be important in the human diet.		

excretory products such as excess salt and urea are removed from the body in solution by the kidneys (p. 159). Water thus acts as a solvent and as a transport medium for these substances.

Digestion is a process which uses water in a chemical reaction to break down insoluble substances to soluble ones (p. 126). These products then pass, in solution, into the bloodstream. In all cells there are many reactions in which water plays an essential part as a reactant and a solvent.

Since we lose water by evaporation, sweating, urinating and breathing, we have to make good this loss by taking in water with the diet.

QUESTIONS

5 Which tissues of the body need (*a*) iron, (*b*) glucose, (*c*) calcium, (*d*) protein, (*e*) iodine?
6 Figure 3 shows some examples of the food that would give a balanced diet. Consider each sample in turn and say what class of food or item of diet is mainly present. For example, the meat is mainly protein but will also contain some iron.
7 What is the value of leafy vegetables, such as cabbage and lettuce, in the diet?
8 Why is a diet consisting mainly of one type of food, e.g. rice or potatoes, likely to be unsatisfactory even if it is sufficient to meet our energy needs?

Fig. 3 Examples of types of food in a balanced diet (see question 6)

BALANCED DIETS

A balanced diet must contain enough carbohydrates and fats to meet our energy needs. It must also contain enough protein of the right kind to provide the essential amino acids to make new cells and tissues for growth or repair. The diet must also contain vitamins and mineral salts, plant fibre and water. Figure 4 shows the composition of four food samples.

Energy requirements

Energy can be obtained from carbohydrates, fats and proteins. The cheapest energy-giving food is usually carbohydrate; the greatest amount of energy is available in fats; proteins give as much energy as carbohydrates but are expensive. Whatever mixture of carbohydrate, fat and protein makes up the diet, the total energy must be sufficient (1) to keep our internal body processes working (e.g. heart beating, breathing action), (2) to keep up our body temperature and (3) to meet the needs of work and other activities.

The amount of energy that can be obtained from food is measured in calories or joules. One gram of carbohydrate or protein can provide us with up to 17 kJ (kilojoules). A gram of fat can give 39 kJ. We need to obtain about 12 000 kJ of energy each day from our food. Table 3 shows how this figure is obtained. However, the figure will vary greatly according to our age, occupation and activity (Fig. 5). It is fairly obvious that a person who does hard manual work, such as digging, will use more energy than someone who sits in an office.

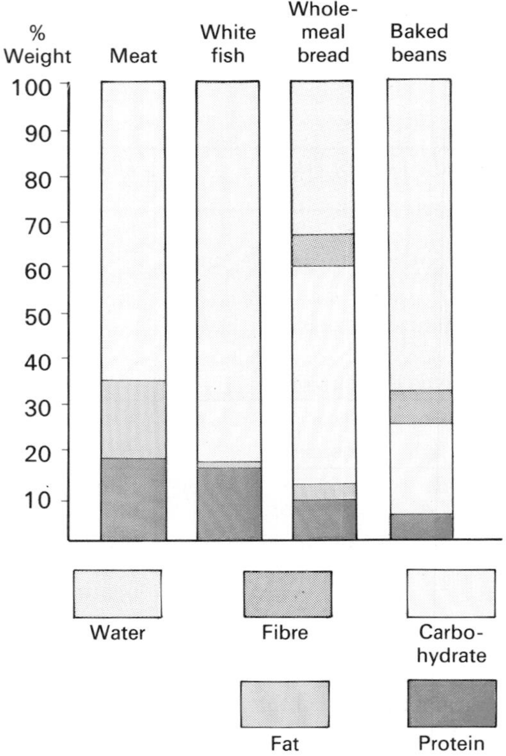

Note The % water includes any salts and vitamins. There are wide variations in the composition of any given food sample according to its source and the method of preservation and cooking. 'White fish' (e.g. cod, haddock, plaice) contains only 0.5% fat whereas herring and mackerel contain up to 14%. White bread contains only 2–3% fibre. Frying the food adds greatly to its fat content.

Fig. 4 An analysis of four food samples

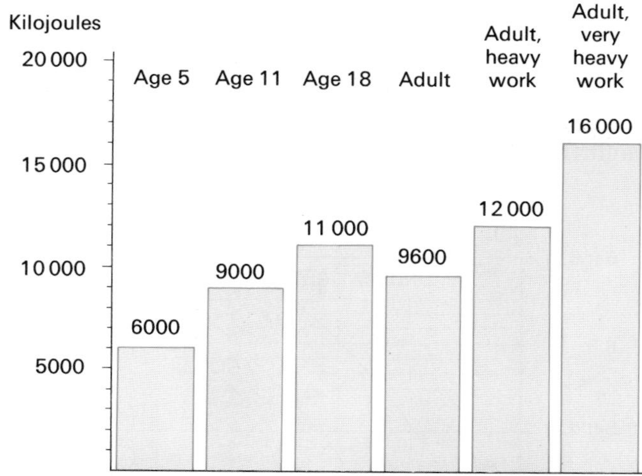

Fig. 5 The changing energy requirements with age and occupation

Table 3 Energy requirements in kJ

8 hours asleep	2400
8 hours awake; relatively inactive physically	3000
8 hours physically active	6600
Total	12 000

The 2400 kJ used during 8 hours sleep represents the energy needed for **basal metabolism** which maintains the circulation, breathing, body temperature, brain function and essential chemical processes in the liver and other organs.

If the diet includes more food than is needed to supply the energy demands of the body, the surplus food is stored either as glycogen in the liver (see p. 132) or as fat below the skin and in the abdomen.

Protein requirements

As explained on page 14, proteins are an essential part of the diet because they supply the amino acids needed to build up our own body structures. Estimates of how much protein we need have changed over the last few years. A recent FAO/WHO report recommended that an average person needs 0.57 g protein for every kilogram of body weight. That is, a 70 kg person would need $70 \times 0.57 = 39.9$, i.e. about 40 g protein per day. This could be supplied by about 200 g (7 ounces) lean meat or 500 g bread but 2 kg potatoes would be needed to supply this much protein and even this will not contain all the essential amino acids.

Special needs

Pregnancy A pregnant woman who is already receiving an adequate diet needs to take in no extra food. Her body's metabolism will adapt to the demands of the growing baby. If, however, her diet is deficient in protein, calcium, iron or vitamin D, she will need to increase her intake of these substances to meet the needs of the baby.

The baby needs protein for making its tissues; calcium and vitamin D are needed for bone development, and iron is used to make the haemoglobin in its blood.

Lactation 'Lactation' means the production of breast milk for feeding the baby. The production of milk, rich in proteins and minerals, makes a large demand on the mother's resources. If her diet is already adequate, her metabolism will adjust to these demands. Otherwise, she may need to increase her intake of proteins, vitamins and calcium to produce milk of adequate quality and quantity.

Growing children Most children up to the age of about 12 years need less food than adults, but they need more in proportion to their body weight. For example, an adult may need 0.57 g protein per kg body weight, but a 6–11 month baby needs 1.53 g per kg, and a 10-year-old needs 0.8 g per kg. The extra protein is needed for making new tissues as the child grows.

In addition to protein, children need extra calcium for their growing bones, iron for their red blood cells, vitamin D to help calcification of their bones and vitamin A for disease resistance.

Western diets

In the affluent societies of the USA, USSR and Europe, there is no general shortage of food and most people can afford a diet with an adequate energy and protein content. Consequently there are few people who suffer from malnutrition, but it seems very likely that many people eat too much food of the wrong kind. This causes illnesses in middle age and old age.

Too much sugar White sugar (sucrose) is made by refining (purifying) the sugar from sugar-cane or sugar-beet. Refined sugar is used in sweets, jam, biscuits and soft drinks. It is also added to many other processed foods.

There is plenty of evidence to show that sugar is an important cause of tooth decay (p. 191), but refined sugar also affects us in many other ways. It is a very concentrated source of energy. You can absorb a lot of sugar from biscuits, ice-cream, sweets, soft drinks, tinned fruits and sweet tea without ever feeling 'full up', so you tend to take in more sugar than your body needs. A high intake of refined sugar, therefore, causes some people to become overweight, and this leads to other forms of illness as described below.

Too much fat The fatty layer which forms in the lining of arteries and leads to coronary heart disease contains fats and a substance called cholesterol. The more fat and cholesterol you have in your blood, the more likely you are to suffer a coronary heart attack. Many doctors and scientists think that if you eat too much fat, you raise the level of fats and cholesterol in the blood and so put yourself at risk. There is still some argument about this, but until more is known, it seems to be a good idea to keep a low level of fats in your diet.

There are reasons to believe that animal fats such as butter, cream, some kinds of cheese, egg-yolk and the fat present in meat might be more harmful than some vegetable oils. Animal fats are digested to give what are called **saturated** fatty acids (because of the structure of their molecules). Many of the fats and oils from plants, such as the oil from sunflower seeds, contain **unsaturated**

Fig. 6 Margarine from sunflower oil. Notice that the cholesterol level is low and the unsaturated fatty acids (polyunsaturates) are high.

fatty acids (Fig. 6). These are thought to be less likely to cause fatty deposits in the arteries. For this reason, it seems to be better to fry food in vegetable oil and to use margarine from certain vegetable oils rather than butter, though this is still a matter for controversy.

Too little fibre We eat too many processed foods, such as sugar, which have been purified from their vegetable sources, and too much white bread, from which the bran has been removed. Unprocessed foods such as potatoes, vegetables and fruit contain a large amount of fibre. Fibre prevents constipation and probably other disorders and diseases of the intestine, including cancer. Eating a diet with a lot of fibre makes you feel 'full up' and so stops you from over-eating. A 100-gram portion of boiled potato provides only 340 kJ. (A potato about the size of an egg weighs 50–70 grams.) You could feel quite full after eating 300 grams of potatoes but would take in only 1000 kJ. A 100-gram portion of milk chocolate will give you 2400 kJ but it is not filling. So a high fibre diet helps to keep your weight down without leaving you feeling hungry all the time.

Too much of everything If you eat more food than your body requires for its energy needs or for building tissues, you are likely to store the surplus as fat and so become overweight (Fig. 7). An overweight person is much more likely to suffer from high blood pressure, coronary heart disease and diabetes than a person whose weight is about right. Being fat also makes you less willing to take exercise because you have to carry the extra weight around.

Whether you put on weight or not depends to some extent on genetics. You may inherit the tendency to get fat. Some people seem able to 'burn off' their excess food as heat and never get fat, no matter how much they eat. You can't change your genetics but you can avoid putting on too much weight by controlling your diet. This does not necessarily mean eating less but simply eating differently. Avoid sugar and all processed food with a high sugar level, such as sweets, cakes and biscuits, and include more vegetables, fruit and wholemeal bread in your diet. Your teeth, waistline, intestines and health in general will benefit from such a change in diet.

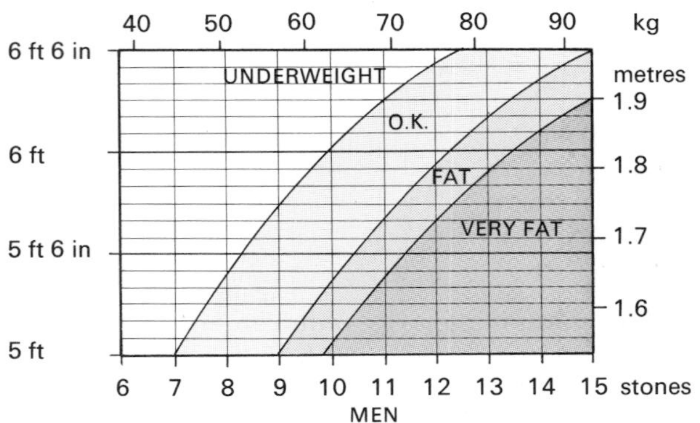

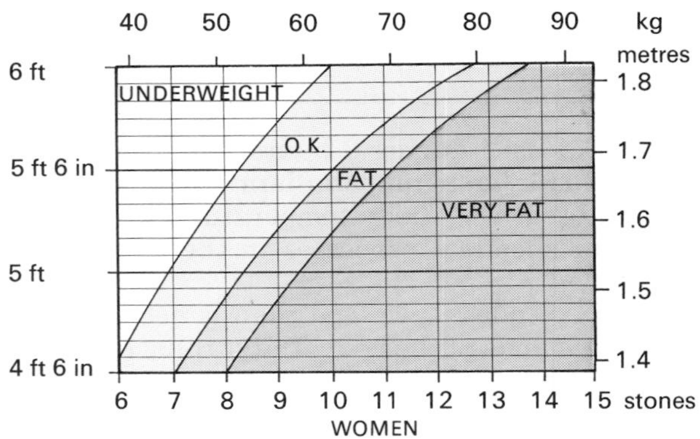

Fig. 7 Height and weight. These tables are intended for adults who have reached their full height. They would find their height on the vertical scales and look along the line till they reached their weight on the horizontal scales.
(By permission of The Health Education Council, London)

Alcohol

Alcohol in drinks such as beer, wine and spirits can be used as a source of energy (29 kJ per g) or converted to fats. However, it is not a desirable food source because it lacks other nutrients and has adverse effects on the body.

The short-term effects of alcohol are:

(*a*) Depression of brain function. This tends to reduce inhibitions and helps people to overcome shyness and nervous tension but it also leads to slower thought processes and less rational behaviour. It blurs the judgement and increases reaction time, which is why alcohol in the blood makes it unsafe to drive. In 1974, 35 per cent of drivers killed in accidents had been drinking alcohol.

(*b*) It causes vaso-dilation (p. 168). The blood vessels in the skin dilate and allow more blood to flow near the surface. This makes you *feel* warm but in fact leads to a more rapid loss of heat from the body.

Alcohol taken in moderation seems to have little

harmful effect (except in pregnant women) but taken in excess can lead to irrational and anti-social behaviour. There is evidence to show that pregnant women who take as little as one alcoholic drink a day have a greater risk of spontaneous abortion (p. 177) and are more likely to produce babies of below average birth weight.

Heavy drinking over a long period can lead to liver damage, irritation of the stomach lining, ulcers, weakened heart muscle and alcoholism (addiction, see p. 216).

QUESTIONS

9 Make a list of the food and other substances that are needed to make up a balanced diet.

10 Why is it better to take in regular, small amounts of protein rather than to eat a large amount of protein at one meal? (Revise page 115.)

11 Select one food class and one mineral salt which are particularly important in the diet of *all three* of the following: pregnant woman, woman breast-feeding a baby, growing child.

12 (*a*) If you feel 'peckish' between meals, why is it better to eat an apple than a bar of chocolate?

(*b*) If you are going to do a long-distance walk, why is it better to take chocolate bars than apples?

13 One hundred grams of boiled potato will give you 340 kJ, but 100 grams of chips give you 900 kJ. Why do you think there is such a big difference?

14 Why should a 'high fibre' diet help to stop you putting on weight?

15 It is sometimes believed that a person who does hard, physical work needs to eat a lot of protein. Try to explain why this is not true.

16 How much protein would a 5 kg baby need each day?

17 From Table 3 on page 119 work out the approximate minimum amount of energy needed each day to maintain your basal metabolism.

PRESERVING AND PROCESSING

Food preservation

If food is kept for any length of time before it is eaten it may start to 'go off'. This may be because it is attacked by its own enzymes, oxidized by the air or, more important, decomposed by bacteria and fungi. All these processes make the food taste and smell unpleasant, but the greatest harm is likely to result from the fungi and bacteria.

Both these organisms may produce poisonous compounds (toxins) which make us ill if we eat them, e.g. salmonella poisoning by bacteria. Cooking the food may kill the organisms but will not necessarily destroy the toxins they have already produced.

Methods of food preservation try to prevent the food's own enzymes from working and to stop the growth of fungi and bacteria.

Drying Removal of water stops the enzymes from working and also prevents the growth of fungi and bacteria.

Salting Impregnating food with salt, e.g. sodium chloride or sodium nitrate, lowers its water potential. Any bacteria that land on the food will be killed because the salt extracts water from them by osmosis (p. 37).

Syrup Immersing food in concentrated sugar solutions has the same osmotic effect on bacteria as does salt.

Refrigeration and freezing Most refrigerators are kept at 4 °C. At this temperature bacteria reproduce only very slowly but they are not killed. The activity of any enzymes in the food is also slowed down. The temperature of −20 °C in the freezer stops the growth and reproduction of bacteria but it still does not kill them all.

Pickling The ethanoic (acetic) acid in vinegar lowers the pH to the extent that bacteria cannot grow and enzymes are inhibited.

Blanching and pasteurization Heating food to 90 °C denatures its enzymes and kills most bacteria. This process is sometimes called 'blanching'.

Milk is heated to 72 °C for 15 seconds to destroy most of the bacteria and improve its keeping qualities. This is called 'pasteurization'.

Food additives

About 3500 different chemicals may be used by the processed food industry. These chemicals have no food value but are added to food (1) to stop it going bad, (2) to 'improve' its colour or (3) to alter or enhance its flavour.

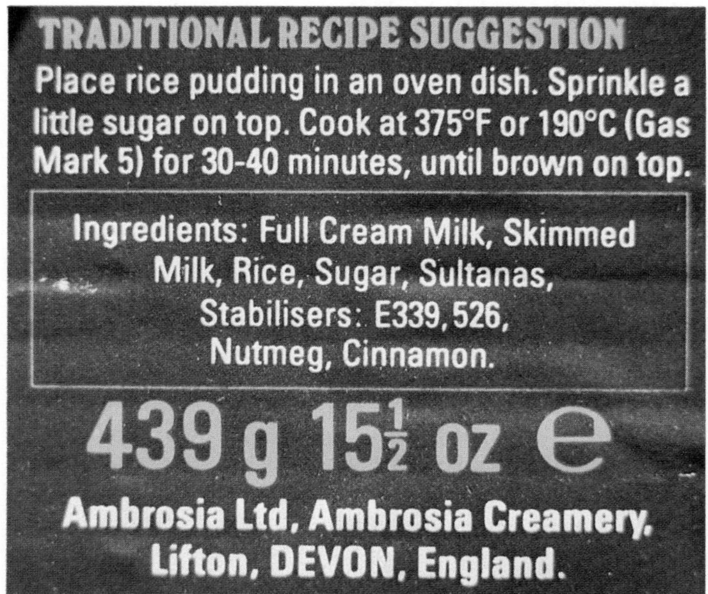

Fig. 8 Food additives. E339 is sodium dihydrogen phosphate; E526 is calcium hydroxide. These chemicals maintain the pH, keep the liquid components in a stable emulsion and stop them separating out.

There is an unavoidable time lag between harvesting a perishable food, processing it, packaging and despatching it to the food shops. The food also spends time on the shelf in the shop and pantry. Thus, for foods other than dried or frozen food, some chemical method of preservation is needed to stop bacterial growth. The need for artificial colouring and flavours is less obvious, though some soft drinks consist of nothing else, apart from water and carbon dioxide (Fig. 8).

Possible harmful effects Only about 300 of the additives are regulated by law, but the food industry tests all the additives, usually by feeding them in large doses to animals. The results, however, are not necessarily applicable to humans. Some workers, for example, estimate that tests for cancer-causing properties may be only 37 per cent successful in animal trials.

There is little widely accepted evidence to show that food additives are harmful. Most allergic reactions to food are caused by naturally occurring substances in the food, though the yellow dye, tartrazine (E102), does cause an allergic reaction in a small number of people. Eczema in children, in some cases, has been relieved by eliminating artificial colour and flavour from their diets. The nitrates and nitrites used to 'cure' ham and bacon are sometimes suspected, without much evidence, as possible carcinogens (cancer-causers).

Most of us eat from 3 to 7 kg of additives each year with no obvious harm, but it is very difficult to know whether the additives have a long-term effect. The risks of cancer from eating cured meats are far lower than the risk of serious bacterial poisoning from eating unprocessed meat. Nevertheless, some people are uneasy that there are so many food additives, whose long-term effects are not known for sure, and which are not really essential for a safe and healthy diet.

WORLD FOOD

The world population is increasing at the rate of about 50 million people per year. Whether this rate of increase will continue is impossible to say but at the moment, the rate of increase is itself accelerating. United Nations experts suggest that by AD 2000 the world population may have increased from its present 4700 million to 7000 million. Such a rate of increase introduces many problems, but one of the most urgent is the supply of food. Whether increase in food production can match this increase in numbers is another controversial matter, both in theory and in practice. Theoretically a ten- or twenty-fold increase in world food production seems possible, and yet in the 10 years from 1960 to 1970 the Food and Agriculture Organization of the United Nations (FAO) estimated that although food production increased by 2.7 per cent per year, the demand for food by an increasing population rose by 3.9 per cent per year. In Africa and the Near East, food production had not kept up with population growth and in Latin America it had barely kept pace. Clearly, a large proportion of the world population receives inadequate food, a fact made obvious by the droughts and famines in Africa in recent years.

Theoretically, the parts of the world which produce surplus food could make up the shortfall in the Third World, but this is not a satisfactory long-term solution.

Famine is caused by a number of interrelated events. Long-term climatic changes may gradually make agriculture impossible. For example, meteorological evidence suggests that the rainfall in Ethiopia, Sudan and Chad has been declining for the last 200 years. Attempts to continue cultivation in these arid conditions leads to the spread of the desert margins.

Overpopulation puts more pressure on poor quality land, leading to soil erosion and the spread of deserts (see p. 269). The land in these over-grazed, over-cultivated regions is also much more susceptible to the effects of drought. In some cases, the adverse changes in climate can be attributed to removal of forests and other plant cover to make way for agriculture. Certainly, many disastrous floods, which wash away topsoil, are the result of deforestation of mountainous areas.

Economic pressures also cause food shortages. In the simplest case, a country may have an adequate food supply but the people are just too poor to buy it. Some Third World countries use their best quality land for raising cash crops, such as coffee and tea. These have no food value but are exported to earn foreign currency.

In many Third World countries, even if the people do get enough to eat in terms of energy content, their diets are unbalanced and lack necessary proteins and vitamins. These people will suffer from malnutrition even though they are not starving.

PRACTICAL WORK

1. Food tests

See page 363 for the preparation of iodine and Benedict's solutions.

(*a*) **Test for starch** Shake a little starch powder in a test-tube with some cold water and then boil it to make a clear solution. When the solution is cold, add 3 or 4 drops of **iodine solution**. A dark blue colour should be produced.

(*b*) **Test for glucose** Heat a little glucose with some **Benedict's solution** in a test-tube. The heating is done by placing the test-tube in a beaker of boiling water (see Fig. 9). The solution will change from clear blue to cloudy green, then yellow and finally to a red precipitate (deposit) of copper(I) oxide.

(*c*) **Test for protein** (Biuret test) To a 1 per cent solution of albumen (the protein of egg-white) add 5 cm^3 dilute sodium hydroxide (*CARE:* this solution is caustic), followed by 5 cm^3 1 per cent copper sulphate solution. A purple colour indicates protein.

(*d*) **Test for fat** Shake 2 drops of cooking oil with about 5 cm^3 ethanol in a dry test-tube until the fat dissolves. Pour the alcoholic solution into a test-tube containing a few cm^3 water. A cloudy white emulsion will form. This shows that the solution contained some fat or oil.

(*e*) **Test for vitamin C** Draw up 2 cm^3 fresh lemon juice into a plastic syringe. Add this juice drop by drop to 2 cm^3 of a 0.1 per cent solution of PIDCP (a blue dye) in a test-tube. The PIDCP will become colourless quite suddenly as the juice is added. The amount of juice added from the syringe should be noted down. Repeat the experiment but with orange juice in the syringe. If it takes more orange juice than lemon juice to decolourize the PIDCP, the orange juice must contain less vitamin C.

2. Application of the food tests

The tests can be used on samples of food such as milk, potato, raisins, onion, beans, egg-yolk, peanuts, to find what food materials are present (Fig. 9). The solid samples are crushed in a mortar and shaken with warm water to extract the soluble products. Separate samples of the watery mixture of crushed food are tested for starch, glucose or protein as described above. To test for fats, the food must first be crushed in ethanol, not water, and then filtered. The clear filtrate is poured into water to see if it goes cloudy, indicating the presence of fats.

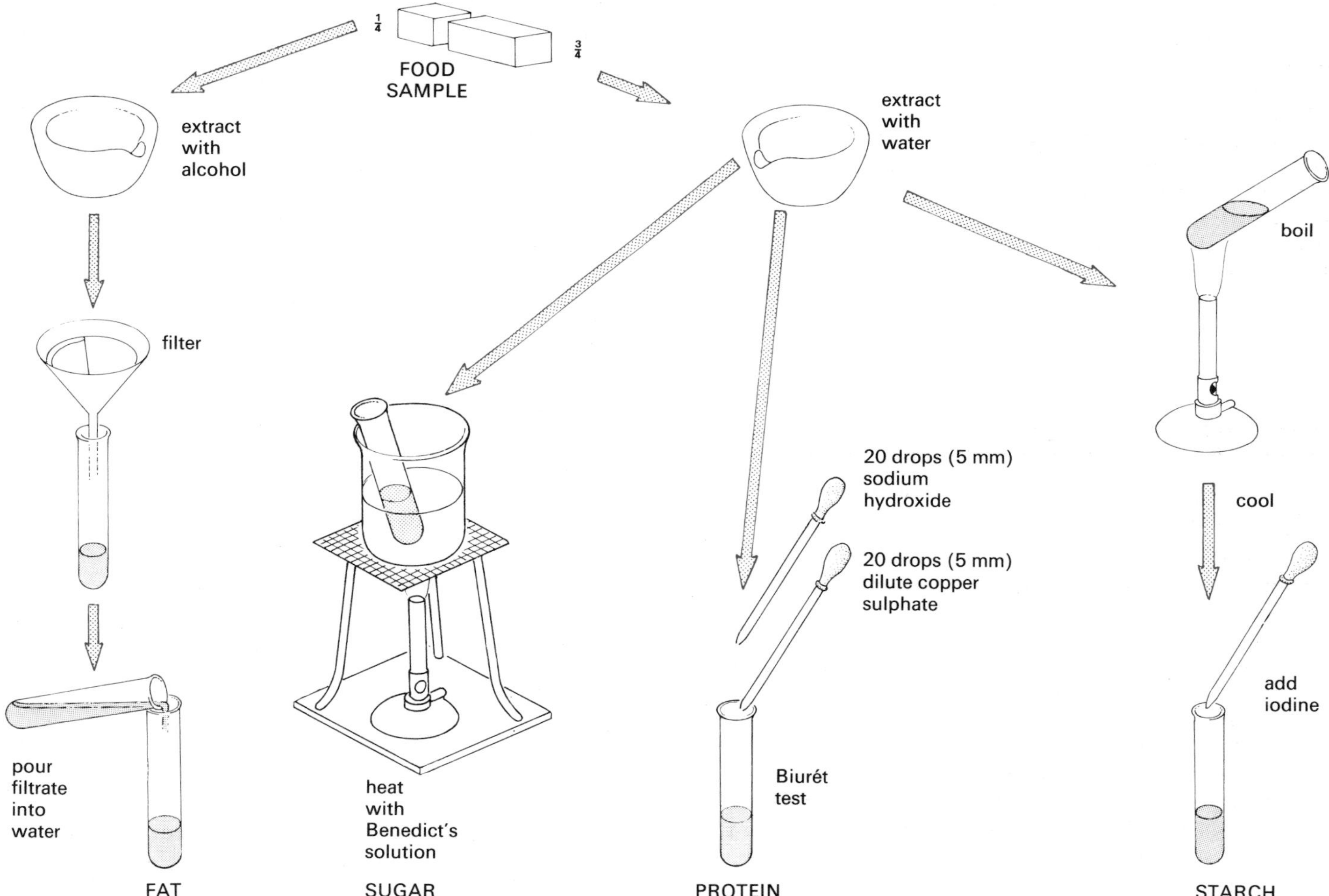

Fig. 9 Food tests

3. Energy from food

Arrange the apparatus as shown in Fig. 10. Use a measuring cylinder to place 100 cm^3 cold water in the can. With a thermometer, find the temperature of the water and make a note of it. In the nickel crucible or tin lid place 1 g sugar and heat it with the Bunsen flame until it begins to burn. As soon as it starts burning, slide the crucible under the can so that the flames heat the water. If the flame goes out, do not apply the Bunsen burner to the crucible while it is under the can, but return the crucible to the Bunsen flame to start the sugar burning again and replace the crucible beneath the can as soon as the sugar catches light. When the sugar has finished burning and cannot be ignited again, gently stir the water in the can with the thermometer and record its new temperature. Calculate the rise in temperature by subtracting the first from the second temperature. Work out the quantity of energy transferred to the water from the burning sugar as follows.

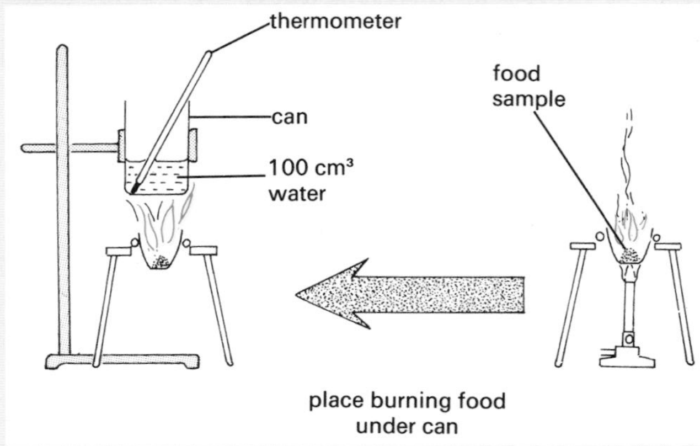

Fig. 10 Energy from food

4.2 joules raise 1 g water 1 °C
100 cm^3 cold water weighs 100 g
Let the rise in temperature be T °C

To raise 1 g water 1 °C needs 4.2 joules
$\therefore$ To raise 100 g water 1 °C needs 100×4.2 joules
$\therefore$ To raise 100 g water T °C needs $T \times 100 \times 4.2$ joules
$\therefore$ 1 g burning sugar produced $420 \times T$ joules

The experiment may now be repeated using 1 g vegetable oil instead of sugar and replacing the warm water in the can with 100 cm^3 cold water.

(**Note** The experiment is very inaccurate because much of the heat from the burning food escapes into the air without reaching the water, but since the errors are about the same for both samples, the results can at least be used to compare the energy released from sugar and oil.)

CHECK LIST

- **Our diets must contain proteins, carbohydrates, fats, minerals, vitamins, fibre and water.**
- **Fats, carbohydrates and proteins provide energy.**
- **Proteins provide amino acids for the growth and replacement of the tissues.**
- **Mineral salts like calcium and iron are needed in tissues such as bone and blood.**
- **Vegetable fibre helps to maintain a healthy intestine.**
- **Adolescents and adults need about 10–12 thousand kilojoules of energy each day from their food.**
- **Vitamins are essential in small quantities for chemical reactions in cells.**
- **The fat-soluble vitamins A and D occur mainly in animal products.**
- **Most cereals contain vitamins of the B group.**
- **Vitamin C occurs in certain fruits and in green leaves.**
- **Lack of vitamin A can lead to blindness; shortage of C causes scurvy; inadequate D causes rickets.**
- **Growing children have special dietary needs.**
- **Western diets often contain too much sugar and fat and too little fibre.**
- **Methods of food preservation aim to stop enzymes working and to suppress growth of fungi and bacteria.**

13 Digestion, Absorption and Use of Food

THE ALIMENTARY CANAL
Structure, function, peristalsis.

DIGESTION
Definition. Mouth, stomach, small intestine.

ABSORPTION
Absorption in the small and large intestine.

USE AND STORAGE OF DIGESTED FOOD
Glucose, fats, amino acids.

THE LIVER
Functions.

PRACTICAL WORK
Experiments on digestion. Salivary amylase and pepsin.

Feeding involves taking food into the mouth, chewing it and swallowing it down into the stomach. This satisfies our hunger, but for food to be of any use to the whole body it has first to be **digested** and **absorbed**. This means that the food is dissolved, passed into the bloodstream and carried by the blood all round the body. In this way, the blood delivers dissolved food to the living cells in all parts of the body such as the muscles, brain, heart and kidneys. This chapter describes how the food is digested and absorbed. Chapter 14 describes how the blood carries it round the body.

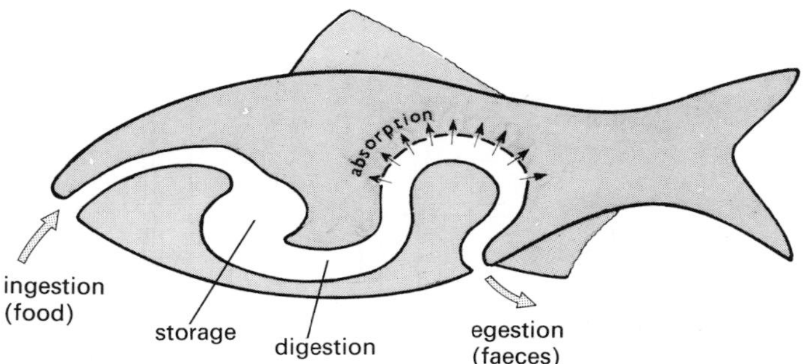

Fig. 1 The alimentary canal (generalized)

THE ALIMENTARY CANAL

The **alimentary canal** is a tube, running through the body. Food is digested in the alimentary canal. The soluble products are absorbed and the indigestible residues expelled (egested). Figure 1 shows a simplified diagram of an alimentary canal.

The inside of the alimentary canal is lined with layers of cells forming what is called an **epithelium**. New cells in the epithelium are being produced all the time to replace the cells worn away by the movement of the food. There are also cells in the lining which produce **mucus**. Mucus is a slimy liquid that lubricates the lining of the canal and protects it from wear and tear. Mucus may also protect the lining from attack by the **digestive enzymes** which are released into the alimentary canal.

Some of the digestive enzymes are produced by cells in the lining of the alimentary canal, as in the stomach lining. Others are produced by **glands** which are outside the alimentary canal but pour their enzymes through tubes (called **ducts**) into the alimentary canal. The **salivary glands** (Fig. 6) and the **pancreas** (Fig. 9) are examples of such digestive glands.

The alimentary canal has a great many blood vessels in its walls, close to the lining. These bring oxygen needed by the cells and take away the carbon dioxide they produce. They also absorb the digested food from the alimentary canal.

Peristalsis

The alimentary canal has layers of muscle in its walls (Fig. 2). The fibres of one layer of muscles run round the canal (**circular muscle**) and the others run along its length (**longitudinal muscle**). When the circular muscles in one region contract, they make the alimentary canal narrow in that region.

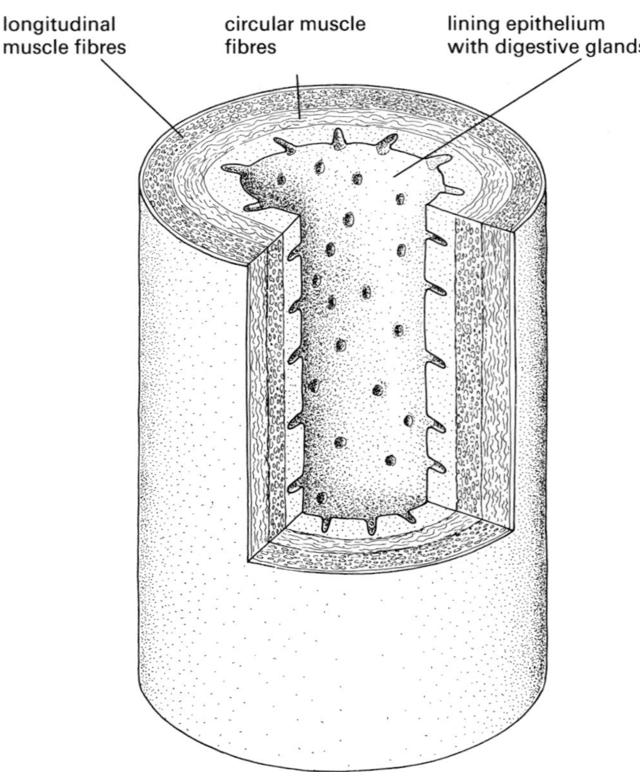
longitudinal muscle fibres
circular muscle fibres
lining epithelium with digestive glands

Fig. 2 The general structure of the alimentary canal

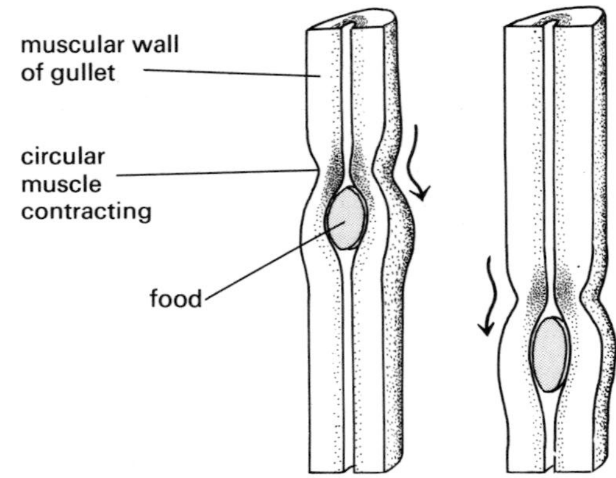
muscular wall of gullet
circular muscle contracting
food

Fig. 3 Diagram to illustrate peristalsis

A contraction in one region of the alimentary canal is followed by another contraction just below it so that a wave of contraction passes along the canal pushing food in front of it.

The wave of contraction, called **peristalsis**, is illustrated in Fig. 3.

QUESTIONS

1 What three functions of the alimentary canal are shown in Fig. 1?
2 Into what parts of the alimentary canal do (*a*) the pancreas, (*b*) the salivary glands, pour their digestive juices?
3 Starting from the inside, name the layers of tissue that make up the alimentary canal.

DIGESTION

Digestion is a chemical process and consists of breaking down large molecules to small molecules. The large molecules are usually not soluble in water, while the smaller ones are. The small molecules can pass through the epithelium of the alimentary canal, through the walls of the blood vessels and into the blood.

Some food can be absorbed without digestion. The glucose in fruit juice, for example, could pass through the walls of the alimentary canal and enter the blood vessels. Most food, however, is solid and cannot get into blood vessels. Digestion is the process by which solid food is dissolved to make a solution.

The chemicals which dissolve the food are **enzymes**, described on page 17. A protein might take 50 years to dissolve if just placed in water but is completely digested by enzymes in a few hours. All the solid starch in foods such as bread and potatoes is digested to **glucose** which is soluble in water. The solid proteins in meat, egg and beans are digested to soluble substances called **amino acids**. Fats are digested to two soluble products called **glycerol** and **fatty acids** (see p. 15).

The chemical breakdown usually takes place in stages. For example, the starch molecule is made up of hundreds of carbon, hydrogen and oxygen atoms. The first stage of digestion breaks it down to a 12-carbon sugar molecule called **maltose**. The last stage of digestion breaks the

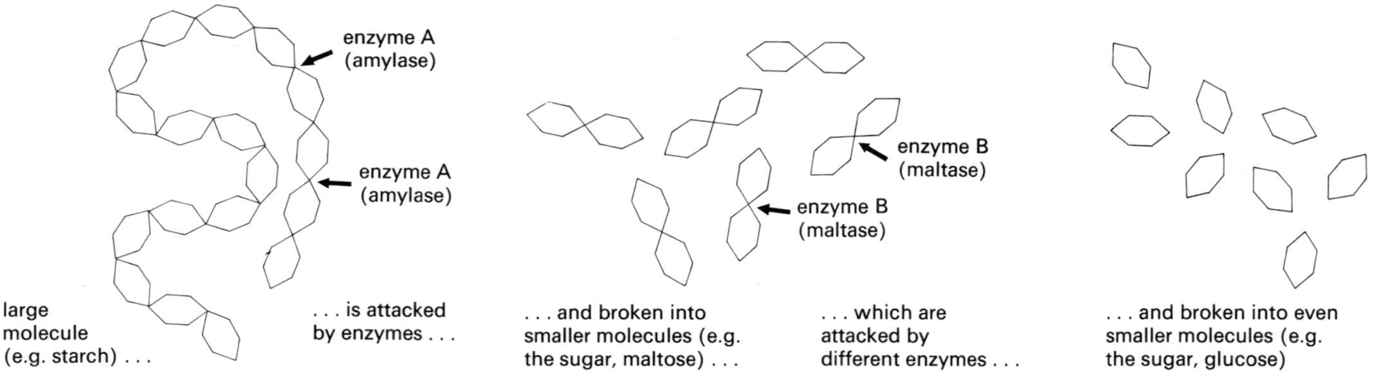
enzyme A (amylase)
enzyme A (amylase)
enzyme B (maltase)
enzyme B (maltase)
large molecule (e.g. starch) . . .
. . . is attacked by enzymes . . .
. . . and broken into smaller molecules (e.g. the sugar, maltose) . . .
. . . which are attacked by different enzymes . . .
. . . and broken into even smaller molecules (e.g. the sugar, glucose)

Fig. 4 Enzymes acting on starch

maltose molecule into two 6-carbon sugar molecules called glucose (Fig. 4). Protein molecules are digested first to smaller molecules called **peptides** and finally into completely soluble molecules called amino acids.

starch → maltose → glucose

protein → peptide → amino acid

These stages take place in different parts of the alimentary canal. The progress of food through the canal and the stages of digestion will now be described (Figs 5 and 6).

The mouth

The act of taking food into the mouth is called **ingestion**. In the mouth, the food is chewed and mixed with **saliva**. The chewing breaks the food into pieces which can be swallowed and it also increases the surface area for the enzymes to work on later. Saliva is a digestive juice produced by three pairs of glands whose ducts lead into the mouth (Fig. 6). It helps to lubricate the food and make the small pieces stick together. Saliva contains one enzyme, **salivary amylase** (sometimes called **ptyalin**), which acts on cooked starch and begins to break it down into maltose.

Strictly speaking, the 'mouth' is the aperture between the lips. The space inside, containing the tongue and teeth, is called the **buccal cavity**. Beyond the buccal cavity is the 'throat' or **pharynx**.

Swallowing

By studying Fig. 6*a*, it can be seen that for food to enter the gullet (oesophagus), it has to pass over the windpipe. All the complicated actions which occur during swallowing ensure that food does not enter the windpipe and cause choking.

1. The tongue presses upwards and back against the roof of the mouth, forcing a pellet of food, called a **bolus**, to the back of the mouth.

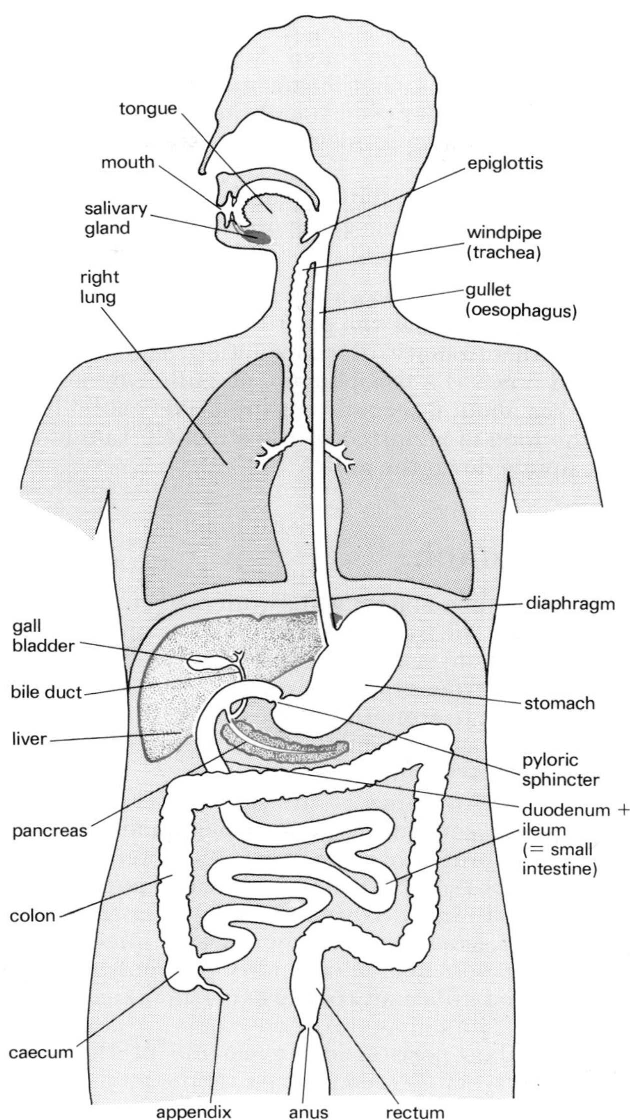

Fig. 5 The alimentary canal

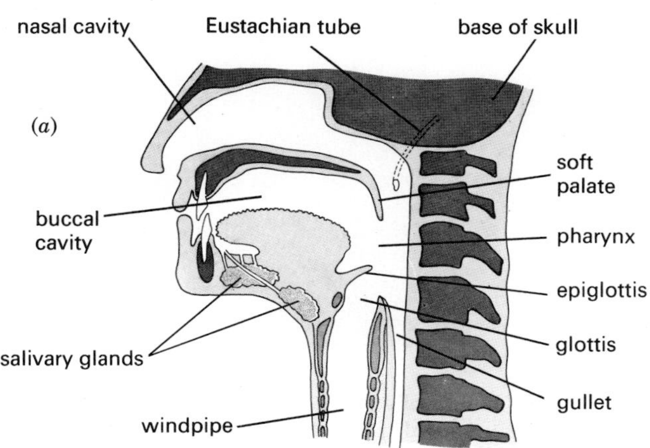

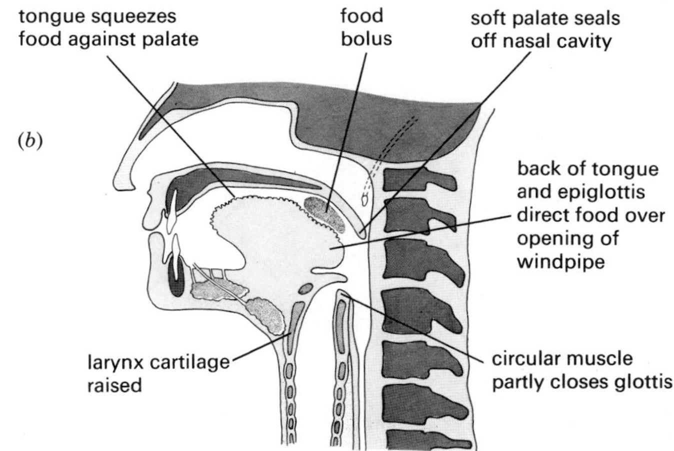

Fig. 6 Section through head to show swallowing action

2. The soft palate closes the nasal cavity at the back.
3. The larynx cartilage round the top of the windpipe is pulled upwards so that the opening of the windpipe (the **glottis**) lies under the back of the tongue.
4. The glottis is also partly closed by the contraction of a ring of muscle.
5. The **epiglottis**, a flap of cartilage (gristle) helps to prevent the food from going down the windpipe instead of the gullet.

The beginning of the swallowing action is voluntary, but once the food reaches the back of the mouth, swallowing becomes an automatic or reflex action. The food is forced into and down the oesophagus, or gullet, by peristalsis. This takes about 6 seconds with relatively solid food and then the food is admitted to the stomach. Liquid travels more rapidly down the gullet.

The stomach

The stomach has elastic walls which stretch as the food collects in it. The **pyloric sphincter** is a circular band of muscle at the lower end of the stomach which stops solid pieces of food from passing through. The main function of the stomach is to store the food from a meal, turn it into a liquid and release it in small quantities at a time to the rest of the alimentary canal.

Glands in the lining of the stomach (Fig. 7) produce **gastric juice** containing the enzyme **pepsin**. Pepsin is a **protease** (or proteinase), i.e. it acts on proteins and breaks them down into soluble compounds called **peptides** (p. 15). The stomach lining also produces hydrochloric acid which makes a weak solution in the gastric juice. This acid provides the best degree of acidity for pepsin to work in (see p. 19) and kills many of the bacteria taken in with the food.

The regular, peristaltic movements of the stomach, about once every 20 seconds, mix up the food and gastric juice into a creamy liquid. How long food remains in the stomach depends on its nature. Water may pass through in a few minutes; a meal of carbohydrate such as porridge may be held in the stomach for less than an hour, but a mixed meal containing protein and fat may be in the stomach for 1 or 2 hours.

The pyloric sphincter lets the liquid products of digestion pass, a little at a time, into the first part of the small intestine called the **duodenum**.

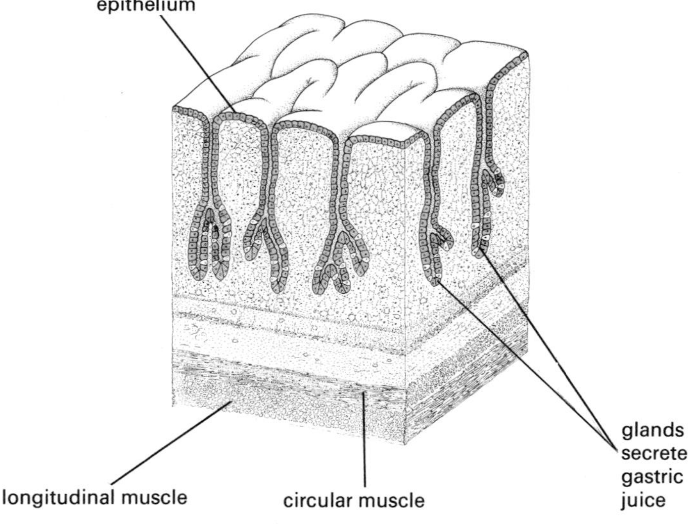

Fig. 7 Diagram of section through stomach wall

The pyloric sphincter closes as each wave of peristalsis passes down the stomach and also when the acid contents of the stomach enter the duodenum. It then remains closed until the duodenal contents are partially neutralized.

The stomach contractions can be monitored by swallowing a small balloon attached to a long tube. The balloon is inflated and the pressure changes are measured by connecting the tube to a manometer.

The peristaltic movements of the alimentary canal in general can be followed by observing the passage of a barium-containing meal by using X-rays (Fig. 8). Barium sulphate is opaque to X-rays and can be seen travelling along the alimentary canal by using an X-ray screen.

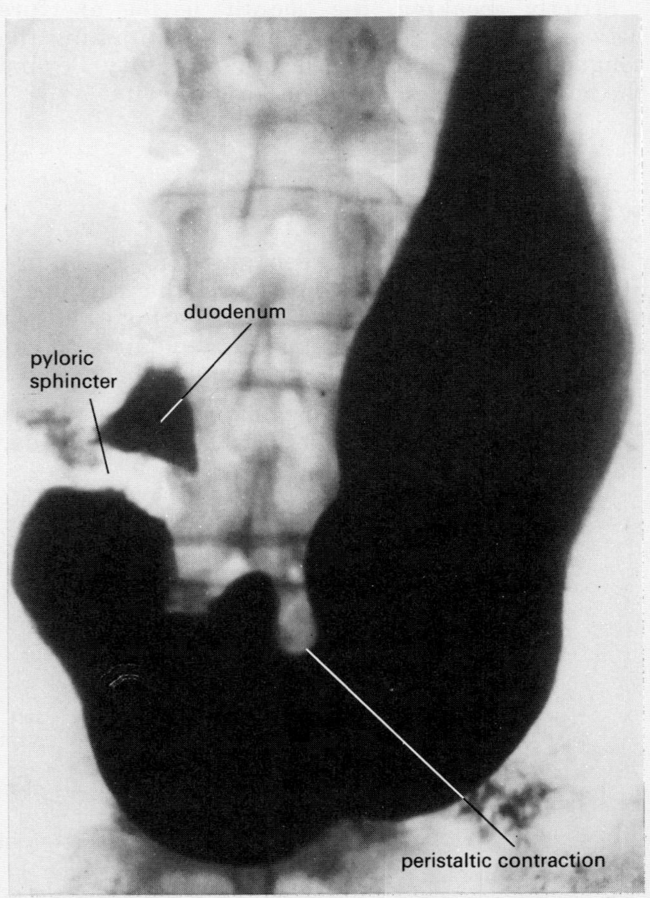

Fig. 8 X-ray of stomach containing a barium meal. The barium meal has started to enter the duodenum. The clear gap in this region represents the pyloric sphincter.

The small intestine

A digestive juice from the pancreas (**pancreatic juice**) and bile from the liver are poured into the duodenum to act on food there. The pancreas is a digestive gland lying below the stomach (Fig. 9). It makes a number of enzymes, which act on all classes of food. There are several proteases which break down proteins to peptides and **amino acids**. **Pancreatic amylase** attacks starch and converts it to maltose. **Lipase** digests fats (lipids) to fatty acids and glycerol.

Pancreatic juice contains sodium hydrogencarbonate which partly neutralizes the acid liquid from the stomach. This is necessary because the enzymes of the pancreas do not work well in acid conditions.

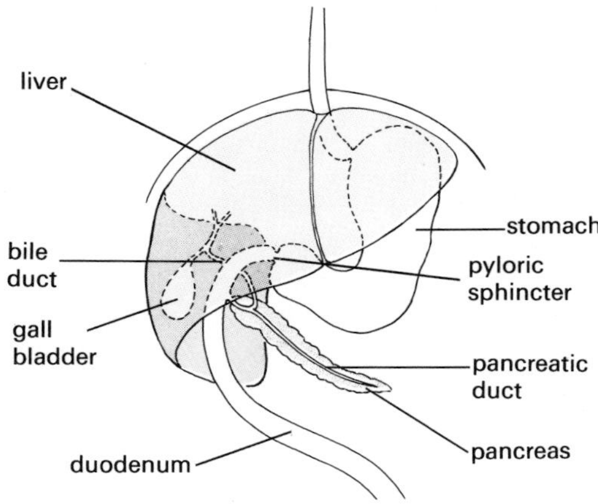

Fig. 9 Relationship between stomach, liver and pancreas

Bile is a green, watery fluid made in the liver, stored in the gall-bladder and delivered to the duodenum by the bile duct (Fig. 9). It contains no enzymes, but its green colour is caused by bile pigments which are formed from the breakdown of haemoglobin in the liver. Bile also contains bile salts which act on fats rather like a detergent. The bile salts **emulsify** the fats. That is, they break them up into small drops which are more easily digested by lipase.

All the digestible material is thus changed to soluble compounds which can pass through the lining of the intestine and into the bloodstream. The final products of digestion are:

food		**final products**
starch	⟶	glucose
proteins	⟶	amino acids
fats (lipids)	⟶	fatty acids and glycerol

The small intestine itself does not appear to liberate digestive enzymes. The structure labelled 'crypt' in Fig. 12 is not a digestive gland, though some of its cells do produce mucus and other secretions. The main function of the crypts is to produce new epithelial cells (see 'Absorption' below) to replace those lost from the tips of the villi.

The epithelial cells of the villi contain, in their cell membranes, enzymes which complete the breakdown of sugars and peptides, before they pass through the cells on their way to the bloodstream. For example, the enzyme **maltase** converts the disaccharide maltose into the monosaccharide, glucose.

Prevention of self-digestion The gland cells of the stomach and pancreas make protein-digesting enzymes (proteases) and yet the proteins of the cells which make these enzymes are not digested. One reason for this is that the proteases are secreted in an inactive form. Pepsin is produced as **pepsinogen** and does not become the active enzyme until it encounters the hydrochloric acid in the stomach. The lining of the stomach is protected from the action of pepsin probably by the layer of mucus.

Similarly, trypsin, one of the proteases from the pancreas, is secreted as the inactive **trypsinogen** and is activated by **enterokinase**, an enzyme secreted by the lining of the duodenum.

Control of secretion

The sight, smell and taste of food set off nerve impulses from the sense organs to the brain. The brain relays these impulses to the stomach and initiates gastric secretion. When the food reaches the stomach, it stimulates the stomach lining to produce a hormone (p. 212) called **gastrin**. This hormone circulates in the blood and when it returns to the stomach in the bloodstream, it stimulates the gastric glands to continue secretion. Thus, gastric secretion is maintained all the time food is present.

In a similar way, the pancreas is affected first by nervous impulses and then by the hormone **secretin**. Secretin is released into the blood from cells in the duodenum when they are stimulated by the acid contents of the stomach. When secretin reaches the pancreas, it stimulates it to produce pancreatic juice.

Caecum and appendix

In humans, the caecum and appendix are small structures, possibly without digestive functions. In grass-eating animals (herbivores) like the cow and the rabbit, however, the caecum and appendix are much larger and it is here that digestion of the cellulose in plant cell walls takes place, largely as a result of bacterial activity.

QUESTIONS

4 Why can you not breathe while you are swallowing?
5 Why is it necessary for our food to be digested? Why do plants not need a digestive system? (See p. 44.)
6 In which parts of the alimentary canal is (*a*) starch (*b*) protein digested?
7 Study the characteristics of enzymes on pages 17–19.
(*a*) Suggest a more logical name for pepsin.
(*b*) In what ways does pepsin show the characteristics of an enzyme?

ABSORPTION

The small intestine consists of the duodenum and the **ileum**. Nearly all the absorption of digested food takes place in the ileum which is efficient at this for the following reasons:

1. It is fairly long and presents a large absorbing surface to the digested food.
2. Its internal surface is greatly increased by circular folds (Fig. 10) bearing thousands of tiny projections called **villi** (singular = villus) (Figs 11 and 12). These villi are about 0.5 millimetre long and may be finger-like or flattened in shape.
3. The lining epithelium is very thin and the fluids can pass rapidly through it. The outer membrane of each epithelial cell has microvilli (p. 34) which increase by 20 times the exposed surface of the cell.
4. There is a dense network of blood capillaries (tiny blood vessels, see p. 142) in each villus (Fig. 12).

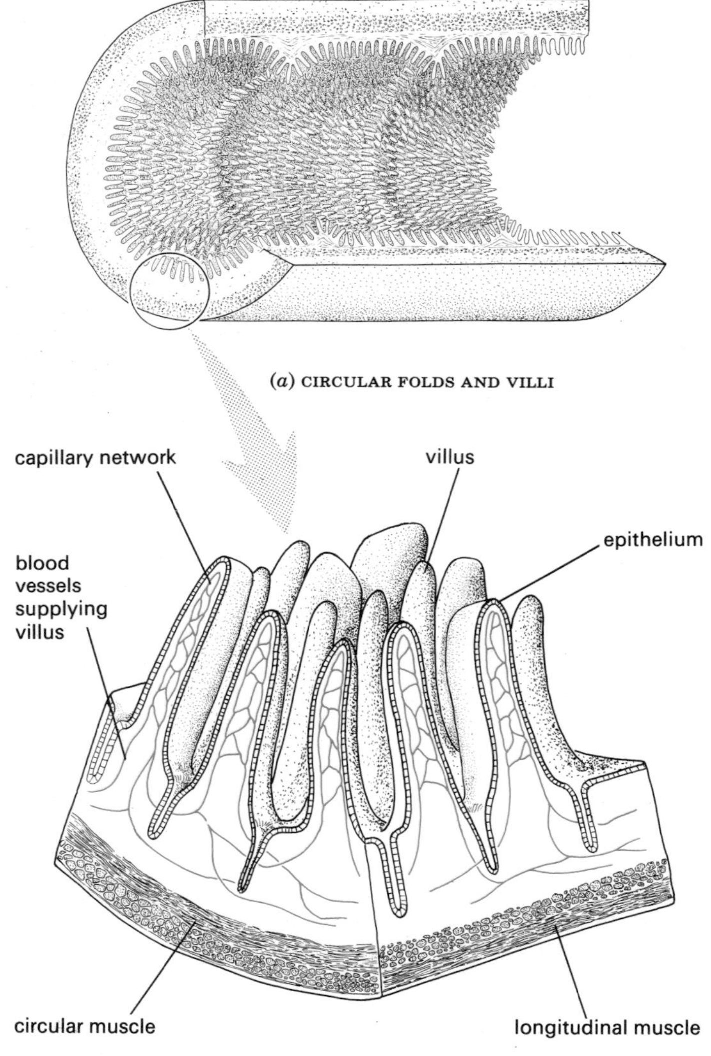

Fig. 10 The absorbing surface of the ileum

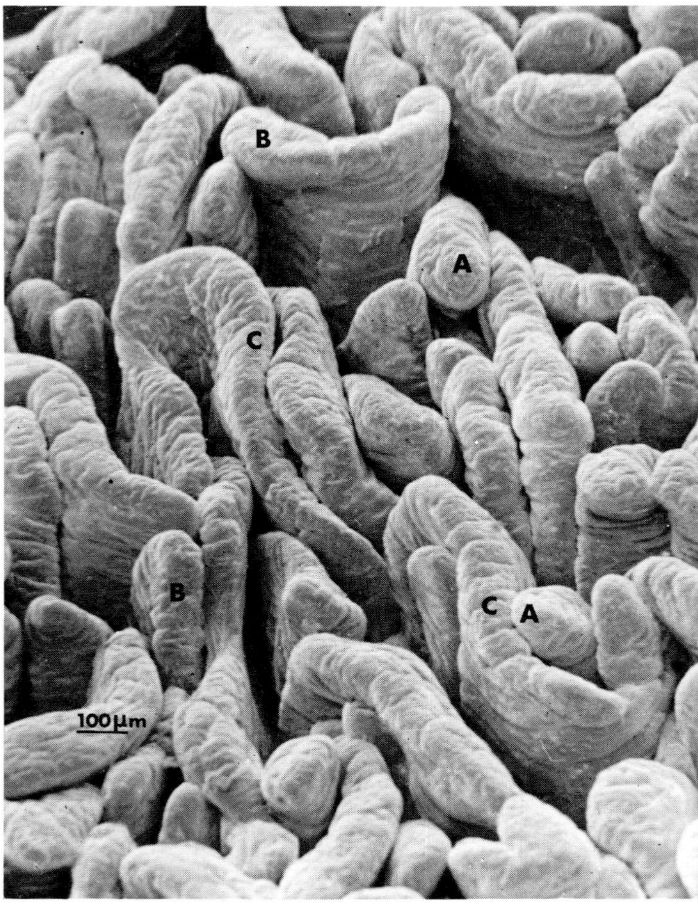

Fig. 11 Scanning electron micrograph of the human intestinal lining (×60). The villi are about 0.5 mm long. In the duodenum they are mostly leaf-like (C), but further towards the ileum they become narrower (B), and in the ileum they are mostly finger-like (A). This micrograph is of a region in the duodenum.

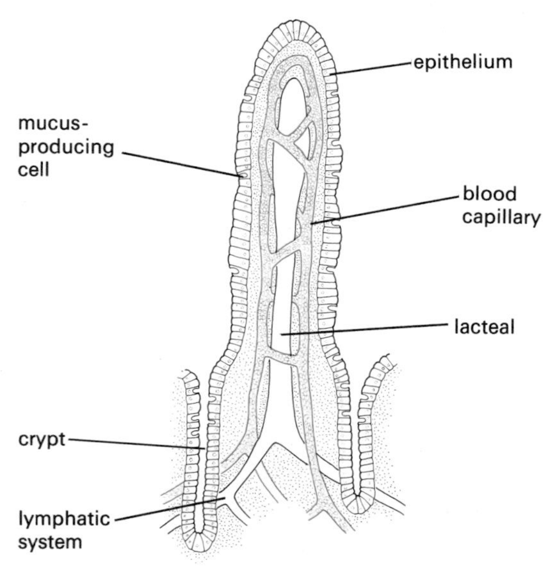

Fig. 12 Structure of a single villus

The small molecules of the digested food, for example, glucose and amino acids, pass into the epithelial cells and then through the wall of the capillaries in the villus and into the bloodstream. They are then carried away in the capillaries which join up to form veins. These veins unite to form one large vein, the **hepatic portal vein** (see Fig. 10 on page 141). This vein carries all the blood from the intestine to the liver, which may store or alter any of the digestion products. When these products are released from the liver, they enter the general blood circulation.

Some of the fatty acids and glycerol from the digestion of fats enter the blood capillaries of the villi. However, a large proportion of the fatty acids and glycerol may be combined to form fats again in the intestinal epithelium. These fats then pass into the **lacteals** (Fig. 12). The fluid in the lacteals flows into the **lymphatic system** which forms a network all over the body and eventually empties its contents into the bloodstream (see p. 144).

Absorption of the products of digestion and other dietary items is not just a matter of simple diffusion, except perhaps for alcohol and, sometimes, water. Although the mechanisms for transport across the intestinal epithelium have not been fully worked out, it seems likely that various forms of active transport (p. 35) are involved. Even water can cross the epithelium against an osmotic gradient (p. 37). Amino acids, sugars and salts are, almost certainly, taken up by active transport. Glucose, for example, crosses the epithelium faster than fructose although their rates of diffusion would be about the same.

Water-soluble vitamins may diffuse into the epithelium but fat-soluble vitamins are carried in the microscopic fat droplets that enter the cells. The ions of mineral salts are probably absorbed by active transport. Calcium ions need vitamin D for their effective absorption.

The epithelial cells of the villi are constantly being shed into the intestine. Rapid cell division in the epithelium of the crypts (Fig. 12), replaces these lost cells. In effect there is a steady procession of epithelial cells moving up from the crypts to the villi.

The large intestine (colon and rectum)

The material passing into the large intestine consists of water with undigested matter, largely cellulose and vegetable fibres (roughage), mucus and dead cells from the lining of the alimentary canal. The large intestine secretes no enzymes but the bacteria in the colon digest part of the fibre to form fatty acids which the colon can absorb. Bile salts are absorbed and returned to the liver by the blood circulation. The colon also absorbs much of the water from the undigested residues. About 7 litres of digestive juices are poured into the alimentary canal each day. If the water from these was not absorbed by the ileum and colon, the body would soon be dehydrated.

The semi-solid waste, the **faeces** or 'stool', is passed into the rectum by peristalsis and is expelled at intervals through the anus. The residues may spend from 12 to 24 hours in the intestine. The act of expelling the faeces is called **egestion** or **defecation**.

QUESTIONS

8 What are the products of digestion of (*a*) starch, (*b*) protein, (*c*) fats, which are absorbed by the ileum?
9 What characteristics of the small intestine enable it to absorb digested food efficiently?

USE OF DIGESTED FOOD

The products of digestion are carried round the body in the blood. From the blood, cells absorb and use glucose, fats and amino acids. This uptake and use of food is called **assimilation**.

Glucose During respiration in the cells, glucose is oxidized to carbon dioxide and water (see p. 24). This reaction provides energy to drive the many chemical processes in the cells which result in, for example, the building up of proteins, contraction of muscles or electrical changes in nerves.

Fats These are built into cell membranes and other cell structures. Fats also form an important source of energy for cell metabolism. Fatty acids produced from stored fats or taken in with the food, are oxidized in the cells to carbon dioxide and water. This releases energy for processes such as muscle contraction. Fats can provide twice as much energy as sugars.

Amino acids These are absorbed by the cells and built up, with the aid of enzymes, into proteins. Some of the proteins will become plasma proteins in the blood (p. 139). Others may form structures such as the cell membrane or they may become enzymes which control the chemical activity within the cell. Amino acids not needed for making cell proteins are converted by the liver into glycogen which can then be used for energy.

QUESTIONS

10 State briefly what happens to a protein molecule in food, from the time it is swallowed, to the time its products are built up into the cytoplasm of a muscle cell.
11 List the chemical changes which a starch molecule undergoes from the time it reaches the duodenum to the time its carbon atoms become part of carbon dioxide molecules. Say where in the body these changes occur.

STORAGE OF DIGESTED FOOD

If more food is taken in than the body needs for energy or for building tissues, such as bone or muscle, it is stored in one of the following ways:

Glucose The sugar not required immediately for energy in the cells is changed in the liver to **glycogen**. The glycogen molecule is built up by combining many glucose molecules into a long chain molecule similar to that of starch (see p. 16). Some of this insoluble glycogen is stored in the liver and the rest in the muscles. When the blood sugar falls below a certain level, the liver changes its glycogen back to glucose and releases it into the circulation. The muscle glycogen is not returned to the circulation but is used by muscle cells as a source of energy during muscular activity.

The glycogen in the liver is a 'short-term' store, sufficient for only about 6 hours. Excess glucose not stored as glycogen is converted to fat and stored in the fat depots.

Fat Unlike glycogen, there is no limit to the amount of fat stored and because of its high energy value (p. 115) it is an effective 'long-term' store. The fat is stored in adipose tissue in the abdomen, round the kidneys and under the skin. These are the **fat depots**. Figures 1–3 on page 166 show the adipose tissue of the skin.

Certain cells can accumulate drops of fat in their cytoplasm. As these drops increase in size and number, they join together to form one large globule of fat in the middle of the cell, pushing the cytoplasm into a thin layer and the nucleus to one side (Fig. 13). Groups of fat cells form adipose tissue.

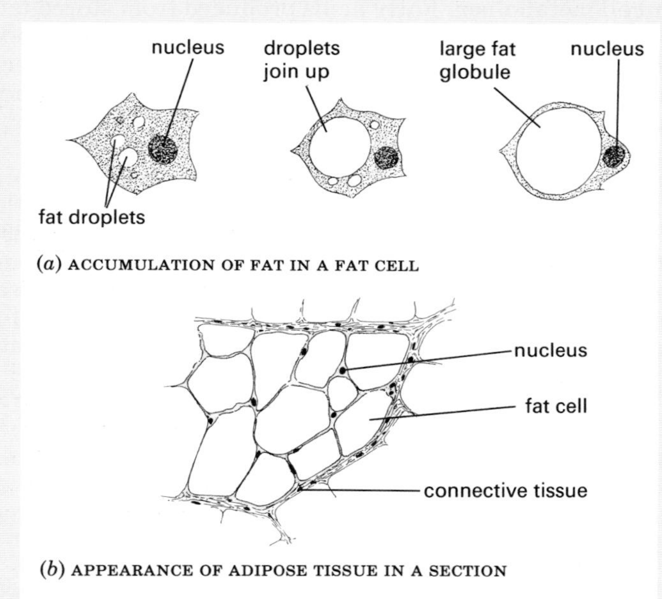

(*a*) ACCUMULATION OF FAT IN A FAT CELL

(*b*) APPEARANCE OF ADIPOSE TISSUE IN A SECTION

Fig. 13 Adipose tissue

Amino acids Amino acids are not stored in the body. Those not used in protein formation are **de-aminated** (see below). The protein of the liver and other tissues can act as a kind of protein store to maintain the protein level in the blood, but absence of protein in the diet soon leads to serious disorders.

All these conversions of one substance to another need specific enzymes to make them happen.

Body weight

The rate at which glucose is oxidized or changed into glycogen and fat is controlled by hormones (p. 212). When intake of carbohydrate and fat is more than enough to meet the energy requirements of the body, the surplus will be stored mainly as fat. Some people never seem to get fat no matter how much they eat, while others start to lay down fat when their intake only just exceeds their needs. Putting on weight is certainly the result of eating more food than the body needs, but it is not clear why people should differ so much in this respect. The explanation probably lies in the balance of hormones which, to some extent, is determined by heredity.

QUESTIONS

12 List the ways in which the body can store an excess of carbohydrates taken in with the diet.
13 If you were deprived of food for several days, how would your body meet the demands for energy by your heart and other organs?

THE LIVER

The liver has been mentioned several times in connection with the digestion, use and storage of food. This is only one aspect of its many important functions, some of which are listed below. It is a large, reddish-brown organ which lies just beneath the diaphragm and partly overlaps the stomach. All the blood from the blood vessels of the alimentary canal passes through the liver, which adjusts the composition of the blood before releasing it into the general circulation (Fig. 14).

Regulation of blood sugar After a meal, the liver removes excess glucose from the blood and stores it as glycogen. In the periods between meals, when the glucose concentration in the blood starts to fall, the liver converts some of its stored glycogen into glucose and releases it into the bloodstream. In this way, the concentration of sugar in the blood is kept at a fairly steady level.

The concentration of glucose in the blood of a person who has not eaten for 8 hours is usually between 90 and 100 mg/100 cm^3 blood. After a meal containing carbohydrate, the blood sugar level may rise to 140 mg/100 cm^3 but 2 hours later, the level returns to about 95 mg as the liver has converted the excess glucose to glycogen.

About 100 g glycogen is stored in the liver of a healthy man. If the concentration of glucose in the blood falls below about 80 mg/100 cm^3 blood, some of the glycogen stored in the liver is converted by enzyme action into glucose and it enters the circulation. If the blood sugar level rises above 160 mg/100 cm^3, glucose is excreted by the kidneys. A blood glucose level below 40 mg/100 cm^3 affects the brain cells adversely, leading to convulsions and coma. By helping to keep the glucose concentration between 80 and 150 mg, the liver prevents these undesirable effects and so contributes to the homeostasis (see below) of the body. (See Fig. 10 on page 141 for the circulatory supply to liver.)

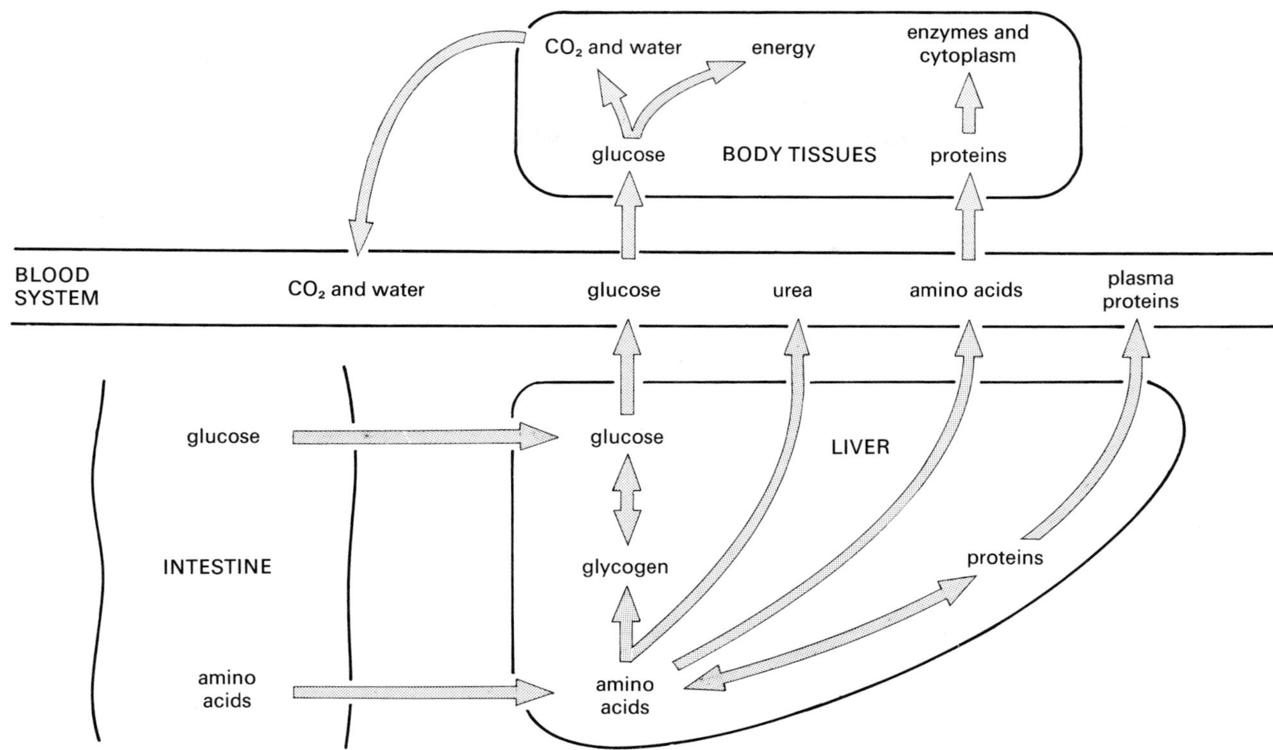

Fig. 14 Some functions of the liver

Production of bile Cells in the liver make bile continuously and this is stored in the gall-bladder until it is discharged through the bile duct into the duodenum. The green colour of the bile results from a pigment, **bilirubin**, which comes from the breakdown of haemoglobin from worn-out red blood cells.

The bile also contains bile salts which assist the digestion of fats as described on page 129.

A large proportion of the bile salts is reabsorbed in the ileum along with the fats they have helped to emulsify. Bile salts are also absorbed in the colon.

De-amination The amino acids not needed for making proteins are converted to glycogen in the liver. During this process, the nitrogen-containing, amino part (NH_2) of the amino acid is removed and changed to **urea**, which is later excreted by the kidneys (see p. 160).

When the $-NH_2$ group is removed from certain amino acids it forms ammonia, NH_3 (or, more strictly, the ammonium ion $-NH_4^+$). Ammonia is very poisonous to the body cells, and the liver converts it at once to urea, $(NH_2)_2CO$, which is a comparatively harmless substance.

Storage of iron Millions of red blood cells break down every day. The iron from their haemoglobin (p. 137) is stored in the liver.

Manufacture of plasma proteins The liver makes most of the proteins found in blood plasma, including fibrinogen which plays an important part in the clotting action of the blood (p. 146).

Detoxication Poisonous compounds, produced in the large intestine by the action of bacteria on amino acids, enter the blood, but on reaching the liver are converted to harmless substances, later excreted in the urine. Many other chemical substances normally present in the body or introduced as drugs are modified by the liver before being excreted by the kidneys. Hormones, for example, are converted to inactive compounds in the liver, so limiting their period of activity in the body.

Storage of vitamins The fat-soluble vitamins A and D are stored in the liver. This is the reason why animal liver is a valuable source of these vitamins in the diet.

Note Contrary to statements made in previous impressions (prior to 1992), the liver is not a net exporter of heat. Although some of its chemical reactions release heat energy, many require an input of energy.

QUESTION

14 Explain how the liver exercises control over the substances coming from the intestine and entering the general blood circulation.

Homeostasis

A complete account of the functions of the liver would involve a very long list. It is most important, however, to realize that the one vital function of the liver, embodying all the details outlined above, is that it helps to maintain the concentration and composition of the body fluids, particularly the blood.

Within reason, a variation in the kind of food eaten will not produce changes in the composition of the blood.

If this **internal environment**, as it is called, were not so constant, the chemical changes that maintain life would become erratic and unpredictable so that with quite slight changes of diet or activity the whole organization might break down. The maintenance of the internal environment is called **homeostasis** and is discussed again on pages 145, 163 and 214.

PRACTICAL WORK

1. The action of salivary amylase on starch

Rinse the mouth with water to remove traces of food. Collect saliva in two test-tubes, labelled A and B, to a depth of about 15 mm (see Fig. 15). Heat the saliva in tube B over a small flame until it boils for about 30 seconds and then cool the tube under the tap. Add about 2 cm^3 of a 2 per cent starch solution to each tube; shake each tube and leave them for 5 minutes.

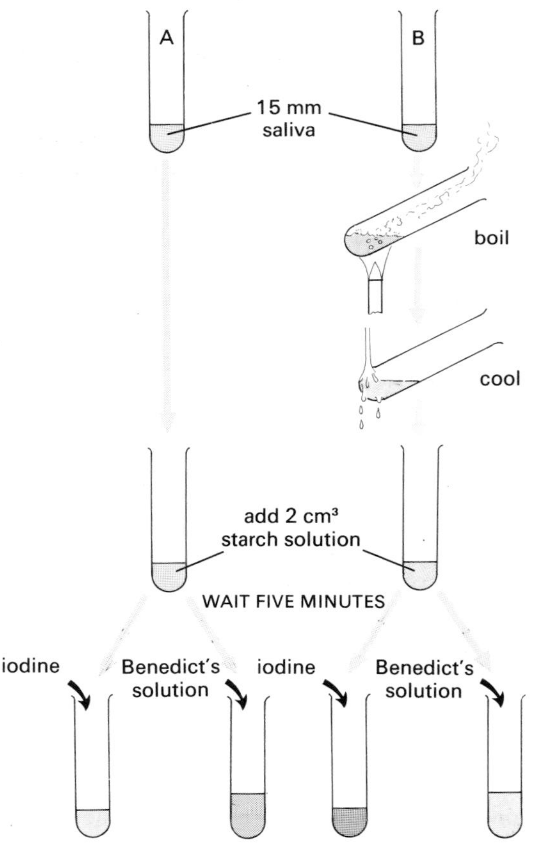

Fig. 15 Salivary amylase acting on starch

Share the contents of tube A between two clean test-tubes. To one of these add some iodine solution. To the other add some Benedict's solution and heat in a water bath as described on page 45. Test the contents of tube B in exactly the same way.

Results The contents of tube A fail to give a blue colour with iodine, showing that the starch has gone. The other half of the contents, however, gives a red or orange precipitate with Benedict's solution, showing that sugar is present.

The contents of tube B still give a blue colour with iodine but do not form a red precipitate on heating with Benedict's solution.

Interpretation The results with tube A suggest that something in saliva has converted starch into sugar. The fact that the boiled saliva in tube B fails to do this, suggests that it was an enzyme in saliva which brought about the change (see page 19), because enzymes are proteins and are destroyed by boiling. If the boiled saliva had changed starch to sugar, it would have ruled out the possibility of an enzyme being responsible.

This interpretation assumes that it is something in saliva which changes starch into sugar. However, the results could equally well support the claim that starch can turn unboiled saliva into sugar. Our knowledge of (1) the chemical composition of starch and saliva and (2) the effect of heat on enzymes, makes the first interpretation more plausible.

2. The action of pepsin on egg-white protein

A cloudy suspension of egg-white is prepared by stirring the white of one egg into 500 cm^3 tap water, heating it to boiling point and filtering it through glass wool to remove the larger particles.

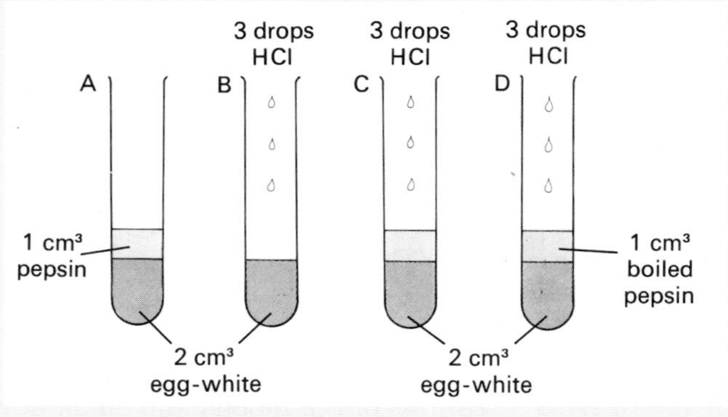

Fig. 16 Pepsin acting on egg-white

Label four test-tubes A, B, C and D and place 2 cm^3 egg-white suspension in each of them. Then add pepsin solution and/or dilute hydrochloric acid to the tubes as follows (Fig. 16):

A Egg-white suspension + 1 cm^3 pepsin solution (1%)
B Egg-white suspension + 3 drops dilute hydrochloric acid (HCl)
C Egg-white suspension + 1 cm^3 pepsin + 3 drops HCl
D Egg-white suspension + 1 cm^3 boiled pepsin + 3 drops HCl

Place all four tubes in a beaker of warm water at 35 °C for 10–15 minutes.

Result The contents of tube C go clear. The rest remain cloudy.

Interpretation The change from a cloudy suspension to a clear solution shows that the solid particles of egg protein have been digested to soluble products. The failure of the other three tubes to give clear solutions shows that:

A Pepsin will only work in acid solutions.
B It is the pepsin and not the hydrochloric acid which does the digestion.
D Pepsin is an enzyme, because its activity is destroyed by boiling.

3. A 'model' for digestion and absorption

Put about 30 mm 3 per cent starch solution in a test-tube. Take a 15 cm length of dialysis tubing which has been soaked in water and tie a knot securely in one end. Use a dropping pipette to fill the dialysis tubing with the starch solution and place the tubing in a test-tube labelled A. Secure the tubing to the rim of the test-tube with an elastic band as shown in Fig. 17. Wash away all traces of starch solution from the outside of the tube by filling the test-tube with water and emptying it several times. Finally, fill the test-tube with water and leave it in the rack.

Now add about 2 mm saliva to the remaining starch solution. Shake the mixture and fill a second dialysis tube exactly as before. Wash it thoroughly and secure it in a test-tube of water, labelled B.

After 24 hours pour about 30 mm of the water from tube A into each of two clean test-tubes. Add 5 drops of iodine to one of these tubes and test the other for sugar by heating it in a water bath with an equal volume of Benedict's solution.

Repeat this procedure for the water taken from tube B.

Result The water from tube A should not give a blue colour with iodine, nor should it give a red precipitate with Benedict's solution.

The water from tube B also should not give a blue colour with iodine but it should give a red precipitate with Benedict's solution.

Interpretation In tube A, the starch has been confined to the dialysis tubing and has not diffused out into the surrounding water. In tube B, starch has not diffused through the dialysis tubing. However, in tube B, salivary amylase has turned some of the starch into maltose which diffused through the dialysis tubing into the water.

This is an acceptable model for the digestion of starch, but a poor model for absorption in the ileum. It is doubtful whether much, if any, absorption in the

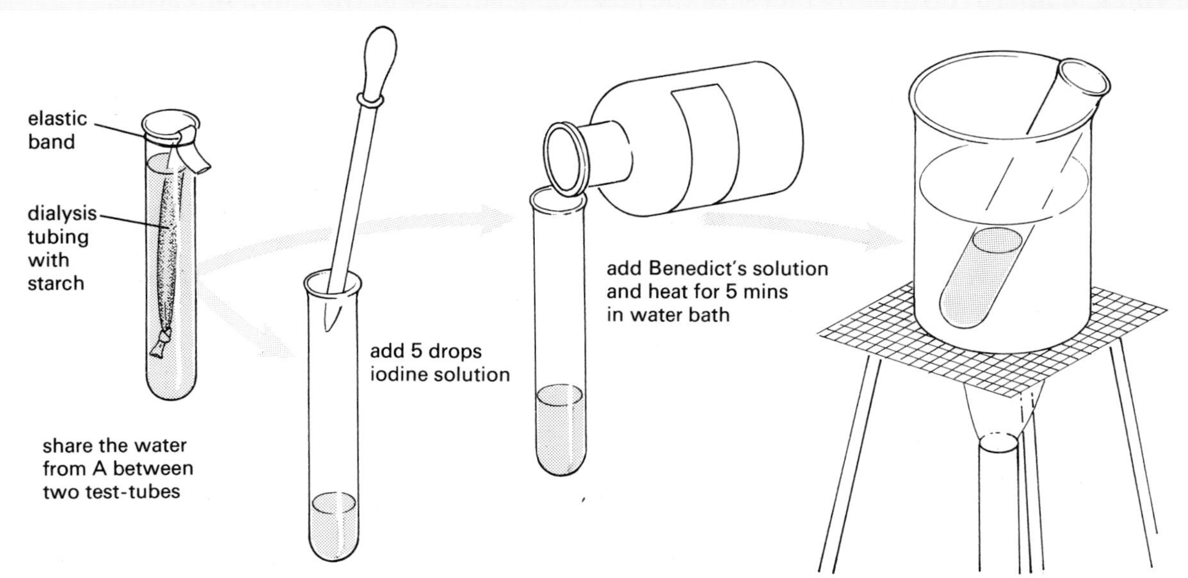
elastic band
dialysis tubing with starch
share the water from A between two test-tubes
add 5 drops iodine solution
add Benedict's solution and heat for 5 mins in water bath

Fig. 17 A model for digestion and absorption

ileum takes place by passive diffusion. The epithelial cell membranes play an active part in both the digestion and absorption of food in the ileum.

More rapid results can be obtained by placing a mixture of 3 per cent starch and 30 per cent glucose solutions in the dialysis tubing. The glucose can be detected in the dialysate (the liquid in the test-tube) after about 15 minutes, but the model is even less satisfactory than before because no digestion is involved. It could also be argued that the glucose, being 10 times more concentrated than the starch, would be expected to diffuse faster.

QUESTIONS

15 In experiments with enzymes, the control often involves the boiled enzyme. Suggest why this type of control is used?
16 In Experiment 2, why does the change from cloudy to clear suggest that digestion has occurred?
17 How would you modify Experiment 2 if you wanted to find the optimum temperature for the action of pepsin on egg-white?
18 It was suggested that an alternative interpretation of the result in Experiment 1 might be that starch has turned saliva into sugar. From what you know about starch, saliva and the design of the experiment, explain why this is a less acceptable interpretation.
19 Write down the menu for your breakfast and lunch (or supper). State the main food substances present in each item of the meal. State the final digestion product of each.

CHECK LIST

- **Digestion is the process which changes insoluble food into soluble substances.**
- **Digestion takes place in the alimentary canal.**
- **The changes are brought about by chemicals called digestive enzymes.**

Region of alimentary canal	*Digestive gland*	*Digestive juice produced*	*Enzymes in the juice*	*Class of food acted upon*	*Substances produced*
Mouth	salivary glands	saliva	salivary amylase	starch	maltose
Stomach	glands in stomach lining	gastric juice	pepsin	proteins	peptides
Duodenum	pancreas	pancreatic juice	proteases	proteins and peptides	peptides and amino acids
			amylase	starch	maltose
			lipase	fats	fatty acids and glycerol

- **Maltose and sucrose are changed to glucose by enzymes in the epithelium of the villi.**
- **The ileum absorbs amino acids, glucose and fats.**
- **These are carried in the bloodstream first to the liver and then to all parts of the body.**
- **Internal folds, villi and microvilli greatly increase the absorbing surface of the small intestine.**
- **The digested food is used or stored in the following ways:**
  Glucose is (1) oxidized for energy or (2) changed to glycogen or fat and stored.
  Amino acids are (1) built up into proteins or (2) de-aminated to urea and glycogen and used for energy.
  Fats are (1) oxidized for energy or (2) stored.
- **Glycogen in the liver and muscles acts as a short-term energy store; fat in the fat depots acts as a long-term energy store.**
- **The liver stores glycogen and changes it to glucose and releases it into the bloodstream to keep a steady level of blood sugar.**
- **The liver exercises control over many other aspects of blood composition and so helps maintain chemical stability in the body.**

14 The Blood Circulatory System

BLOOD COMPOSITION
Blood cells and plasma.

THE HEART
Structure and function.

CIRCULATION
Arteries, veins and capillaries.

LYMPHATIC SYSTEM
Spleen and thymus.

FUNCTIONS OF THE BLOOD
Homeostasis, transport and defence.

ANTIBODIES AND IMMUNITY
Vaccines.

BLOOD GROUPS AND TRANSFUSIONS
ABO blood groups.

CORONARY HEART DISEASE
Possible causes.

CORRELATION AND CAUSE

The previous chapter explained how food is digested to amino acids, glucose, etc., which are absorbed in the small intestine. These substances are needed in all living cells in the body such as the brain cells, leg muscle cells and kidney cells. The substances are carried from the intestine to other parts of the body by the blood system. In a similar way, the oxygen taken in by the lungs is needed by all the cells and is carried round the body in the blood.

COMPOSITION OF BLOOD

Blood consists of red cells, white cells and platelets floating in a liquid called **plasma**. There are between 5 and 6 litres of blood in the body of an adult.

Red cells

These are tiny, disc-like cells (Figs 1*a* and 2) which do not have nuclei. They are made of spongy cytoplasm enclosed in an elastic cell membrane. In their cytoplasm is the red pigment, **haemoglobin**, a protein combined with iron. Haemoglobin combines with oxygen in places where

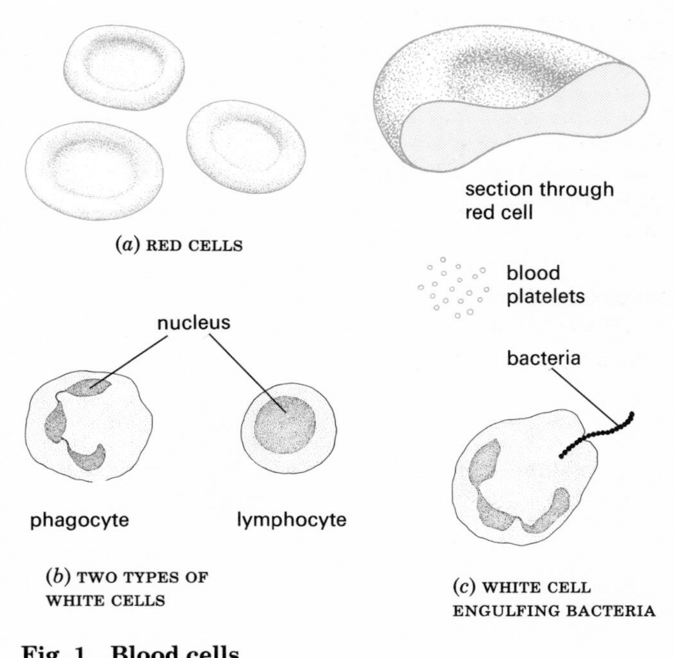

Fig. 1 Blood cells

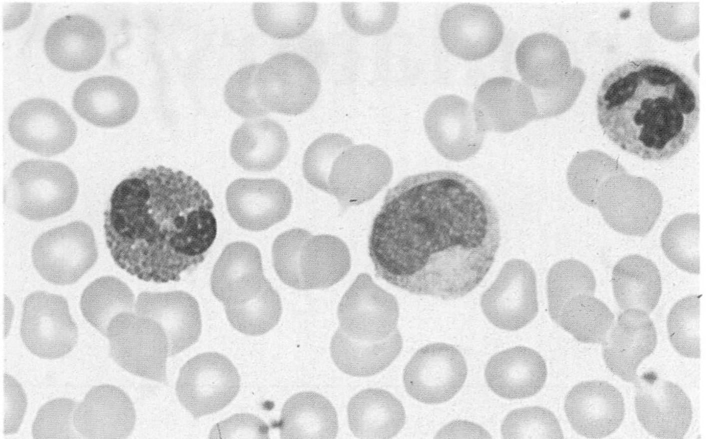

Fig. 2 Red and white cells from human blood (×2000). The large nucleus can be clearly seen in the white cells.

there is a high concentration of oxygen, to form **oxy-haemoglobin**. Oxy-haemoglobin is an unstable compound. It breaks down and releases its oxygen in places where the oxygen concentration is low (Fig. 3). This makes haemoglobin very useful in carrying oxygen from the lungs to the tissues.

Blood which contains mainly oxy-haemoglobin is said to be **oxygenated**. Blood with little oxy-haemoglobin is called **deoxygenated**.

Each red cell lives for about 4 months, after which it breaks down. The red haemoglobin changes to a yellow pigment, bilirubin, which is excreted in the bile. The iron from the haemoglobin is stored in the liver. About 200 000 million red cells wear out and are replaced each day. This is about 1 per cent of the total. Red cells are made by the red bone marrow of certain bones in the skeleton; in the ribs, vertebrae and breastbone for example.

White cells

There are several different kinds of white cell (Figs 1*b* and 2). Most are larger than the red cells, and they all have a nucleus. There is 1 white cell to every 600 red cells and they are made in the same bone marrow that makes red cells. Many of them undergo a process of maturation and development in the thymus gland, lymph nodes or spleen (p. 144). The two most numerous types of white cells are **phagocytes** and **lymphocytes**.

The phagocytes can move about by a flowing action of their cytoplasm and can escape from the blood capillaries into the tissues by squeezing between the cells of the capillary walls. They collect at the site of an infection, engulfing (**ingesting**) and digesting harmful bacteria and cell debris (Fig. 1*c*). In this way they prevent the spread of infection through the body.

One of the functions of lymphocytes is to produce antibodies (p. 147).

Platelets

These are pieces of special blood cells budded off in the red bone marrow. They help to clot the blood at wounds and so stop the bleeding (p. 146).

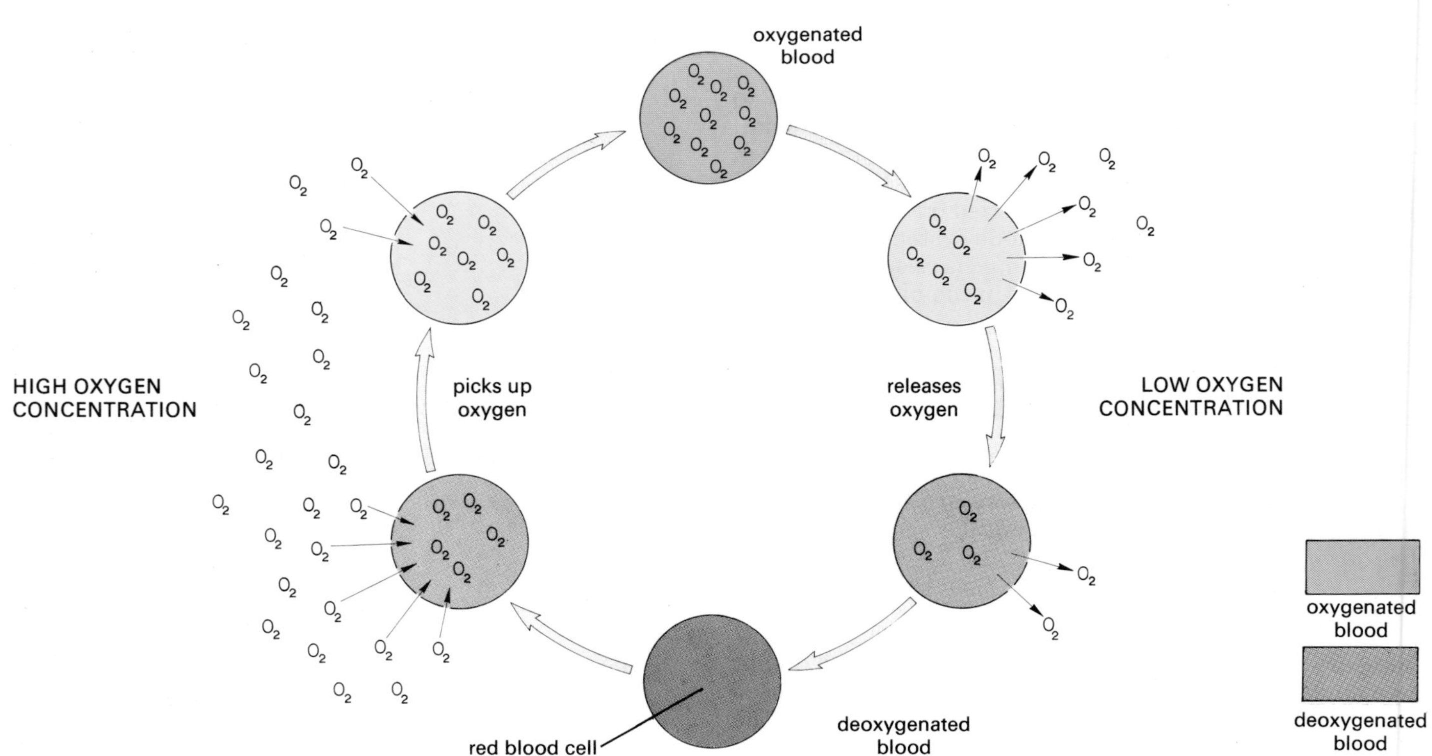

Fig. 3 The function of the red cells

Plasma

The liquid part of the blood is called plasma. It is water with a large number of substances dissolved in it. The ions of sodium, potassium, calcium, chloride and hydrogen-carbonate, for example, are present. Proteins such as fibrinogen, albumin and globulins constitute an important part of the plasma. Fibrinogen is needed for clotting (p. 146), and the globulin proteins include the antibodies which combat bacteria and other foreign matter (p. 147). The plasma will also contain varying amounts of food substances such as amino acids, glucose and lipids (fats). There may also be hormones (p. 212) present, depending on the activities taking place in the body. The excretory product, urea, is dissolved in the plasma.

The liver and kidneys keep the composition of the plasma more or less constant but the amount of digested food, salts and water will vary within narrow limits according to food intake and body activities.

QUESTIONS

1 In what ways are white cells different from red cells in (*a*) their structure, (*b*) their function?
2 (*a*) Where, in the body, would you expect haemoglobin to be combining with oxygen to form oxy-haemoglobin?
3 (*b*) In what parts of the body would you expect oxy-haemoglobin to be breaking down to oxygen and haemoglobin?
4 Why is it important for oxy-haemoglobin to be an unstable compound, i.e. easily changed to oxygen and haemoglobin?
What might be the effect on a person whose diet contained too little iron?

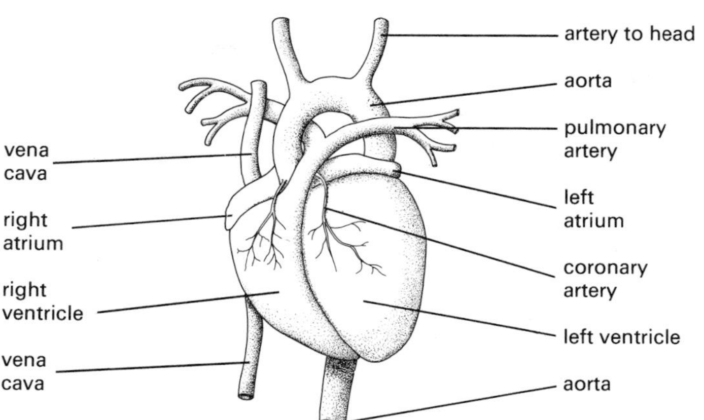

Fig. 4 External view of the heart

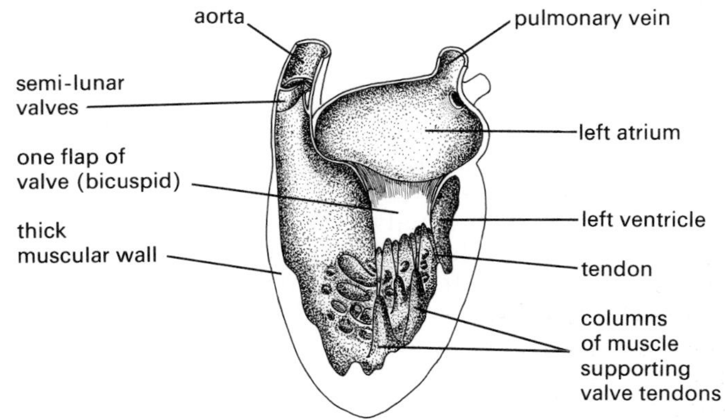

Fig. 5 Diagram of heart cut open (left side)

THE HEART

The heart pumps blood through the circulatory system all round the body. Figure 4 shows its appearance from the outside, Fig. 5 shows the left side cut open, while Fig. 6 is a diagram of a vertical section to show its internal structure. Since the heart is seen as if in a dissection of a person facing you, the left side is drawn on the right.

If you study Fig. 6 you will see that there are four chambers. The upper, thin-walled chambers are the **atria** (singular = atrium) and each of these opens into a thick-walled chamber, the ventricle, below.

Blood enters the atria from large veins. The **pulmonary vein** brings oxygenated blood from the lungs into the left atrium. The **vena cava** brings deoxygenated blood from the body tissues into the right atrium. The blood passes from each atrium to its corresponding ventricle, and the ventricle pumps it out into the arteries.

The artery carrying oxygenated blood to the body from the left ventricle is the **aorta**. The **pulmonary artery** carries deoxygenated blood from the right ventricle to the lungs.

In pumping the blood, the muscle in the walls of the atria and ventricles contracts and relaxes (Fig. 7). The walls of the atria contract first and force blood into the two ventricles. Then the ventricles contract and send blood into the arteries. The blood is stopped from flowing

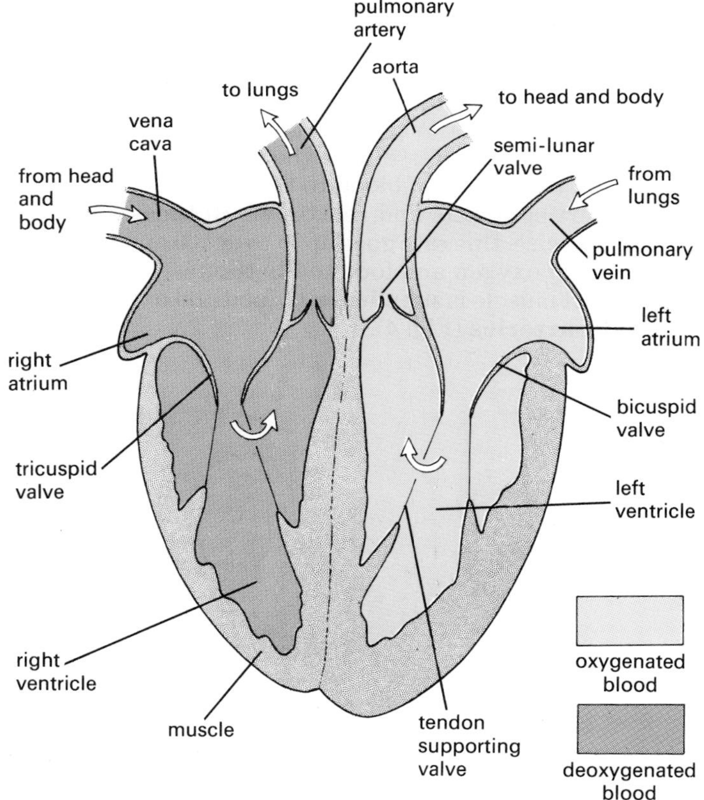

Fig. 6 Diagram of heart, longitudinal section

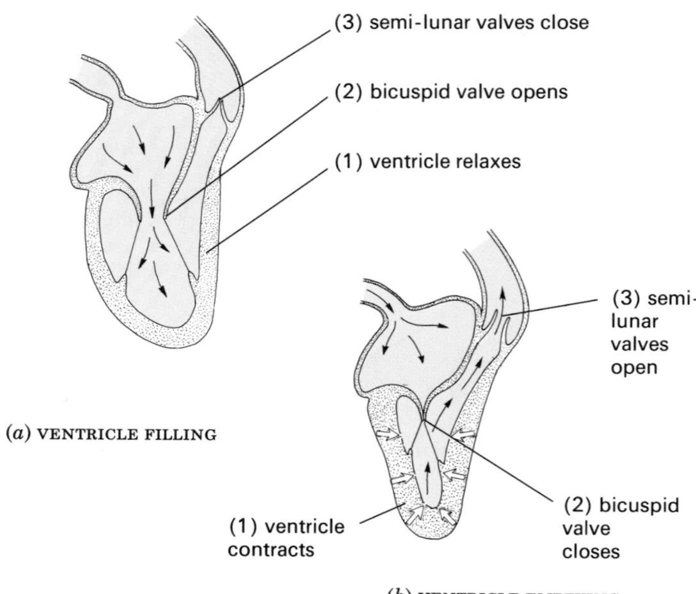

Fig. 7 Diagram of heart beat (only the left side is shown)

backwards by four sets of valves. Between the right atrium and the right ventricle is the **tricuspid** (= three flaps) valve. Between the left atrium and left ventricle is the **bicuspid** (= two flaps) valve. The flaps of these valves are shaped rather like parachutes, with 'strings' called **tendons** or **cords** to prevent their being turned inside out.

In the pulmonary artery and aorta are the **semi-lunar** (= half-moon) valves. These each consist of three pockets which are pushed flat against the artery walls when blood flows one way. If blood tries to flow the other way, the 'pockets' fill up and meet in the middle to stop the flow of blood (Fig. 8).

When the ventricles contract, blood pressure closes the bicuspid and tricuspid valves and these prevent blood returning to the atria. When the ventricles relax, the blood pressure in the arteries closes the semi-lunar valves so preventing the return of blood to the ventricles.

The heart contracts and relaxes 60–80 times a minute. During exercise, this rate goes up to over 100 and increases the supply of oxygen and food to the tissues.

The heart muscle is supplied with food and oxygen by the **coronary arteries** (Fig. 4).

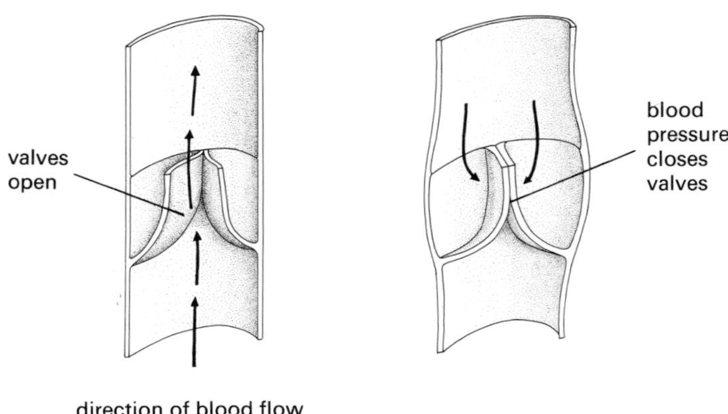

Fig. 8 Action of the semi-lunar valves

From the description above, it may seem that the ventricles are filled with blood as a result of the contraction of the atria. In fact, when the ventricles relax, their internal volume increases and they draw in blood from the pulmonary vein or vena cava through the relaxed atria. Atrial contraction then forces the final amount of blood into the ventricles just before ventricular contraction.

Control of heart beat

At rest, the normal heart rate may lie between 50 and 100 beats per minute, according to age, sex and other factors. During exercise, the rate may increase to 200 per minute.

The heart beat is initiated by the **'pace-maker'**, a small group of specialized muscle cells at the top of the right atrium. The pace-maker receives two sets of nerves from the brain. One group of nerves speeds up the heart rate and the other group slows it down. By this means, the heart rate is adjusted to meet the needs of the body at times of rest, exertion and excitement.

QUESTIONS

5 Which parts of the heart (*a*) pump blood into the arteries, (*b*) stop blood flowing the wrong way?

6 Put the following in the correct order: (*a*) blood enters arteries, (*b*) ventricles contract, (*c*) atria contract, (*d*) ventricles relax, (*e*) blood enters ventricles, (*f*) semi-lunar valves close, (*g*) tri- and bicuspid valves close.

7 Why do you think that (*a*) the walls of the ventricles are more muscular than the walls of the atria and (*b*) the muscle of the left ventricle is thicker than that of the right ventricle? (Consult Fig. 10.)

8 Which important veins are not shown in Fig. 4?

9 Why is a person whose heart valves are damaged by disease unable to take part in active sport?

THE CIRCULATION

The blood, pumped by the heart, travels all round the body in blood vessels. It leaves the heart in arteries and returns in veins. Figure 9 shows the route of the circulation as a diagram. The blood passes twice through the heart during one complete circuit; once on its way to the body and again on its way to the lungs. The circulation through the lungs is called the **pulmonary** circulation; the circulation round the rest of the body is called the **systemic** circulation. On average, a red cell would go round the whole circulation in 45 seconds. Figure 10 is a more detailed diagram of the circulation.

Arteries

These are fairly wide vessels (Figs 11*a* and 12) which carry blood from the heart to the limbs and organs of the body (Fig. 13*a*). The blood in the arteries, except for the pulmonary arteries, is oxygenated.

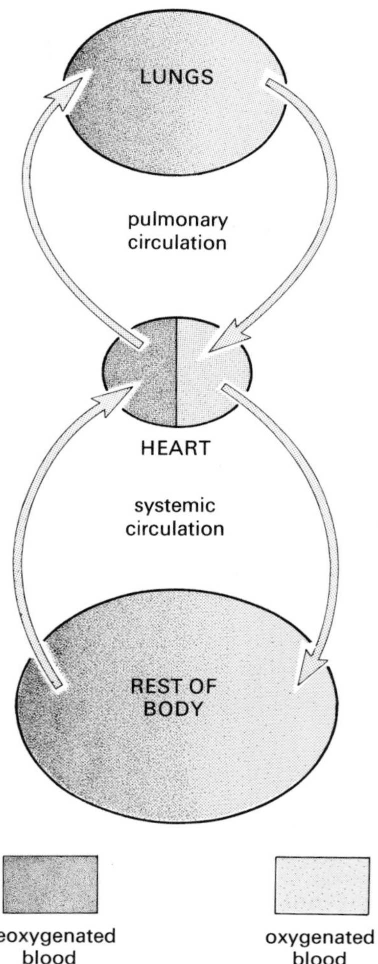

Fig. 9 Blood circulation

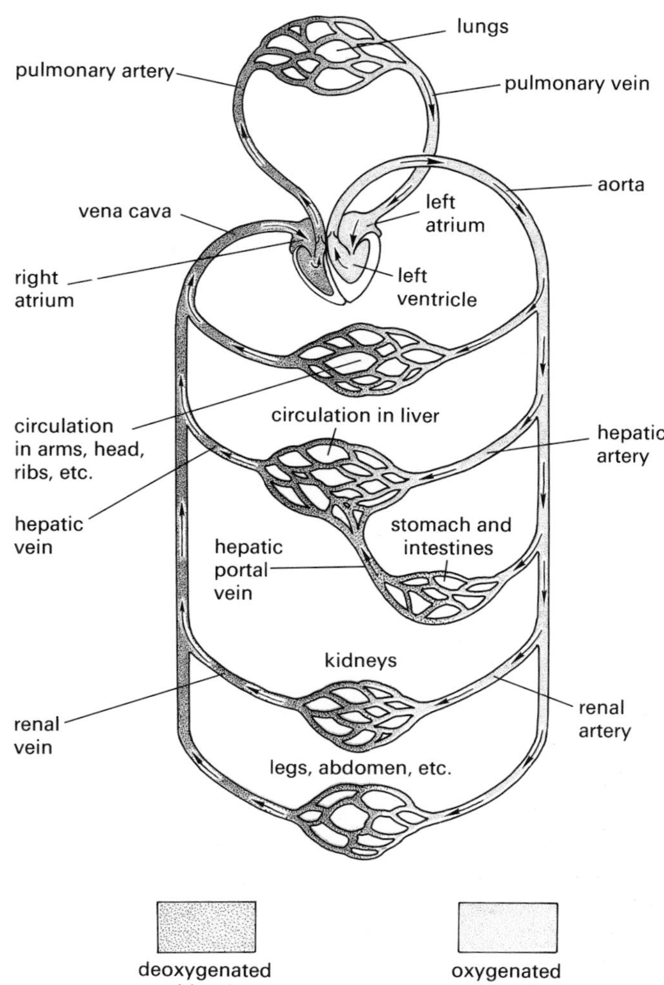

Fig. 10 Diagram of human circulation

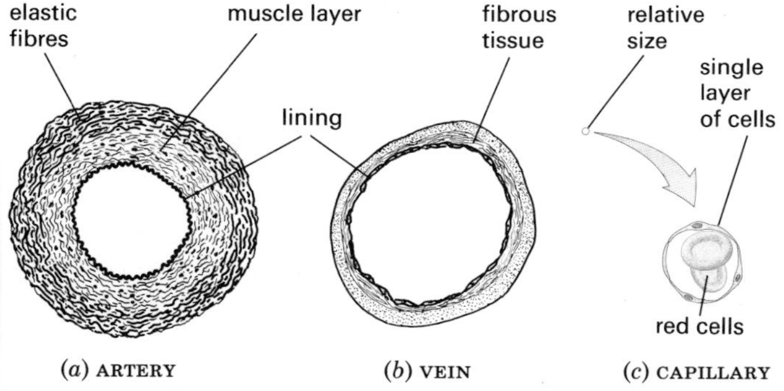

Fig. 11 Blood vessels, transverse sections

Arteries have elastic tissue and muscle fibres in their thick walls.

The large arteries, near the heart, have a greater proportion of elastic tissue which allows these vessels to stand up to the surges of high pressure caused by the heart beat. The ripple of pressure which passes down an artery as a result of the heart beat can be felt as a 'pulse' when the artery is near the surface of the body. You can feel the pulse in your radial artery by pressing the finger-tips of one hand on the wrist of the other (Fig. 14).

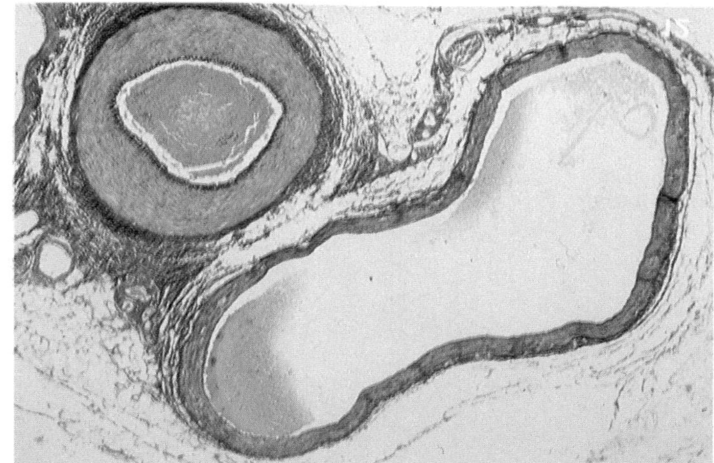

Fig. 12 Transverse section through a vein and artery. The vein is on the right, the artery on the left. Notice that the wall of the artery is much thicker than that of the vein. The material filling the artery is formed from coagulated red blood cells. These are also visible in two regions of the vein.

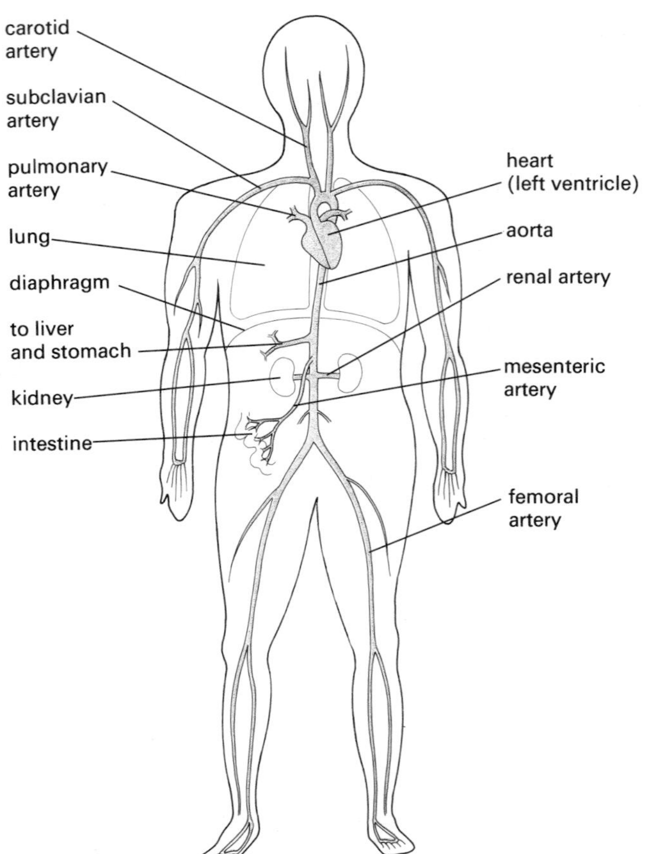

Fig. 13(*a*) Diagram of the arterial system

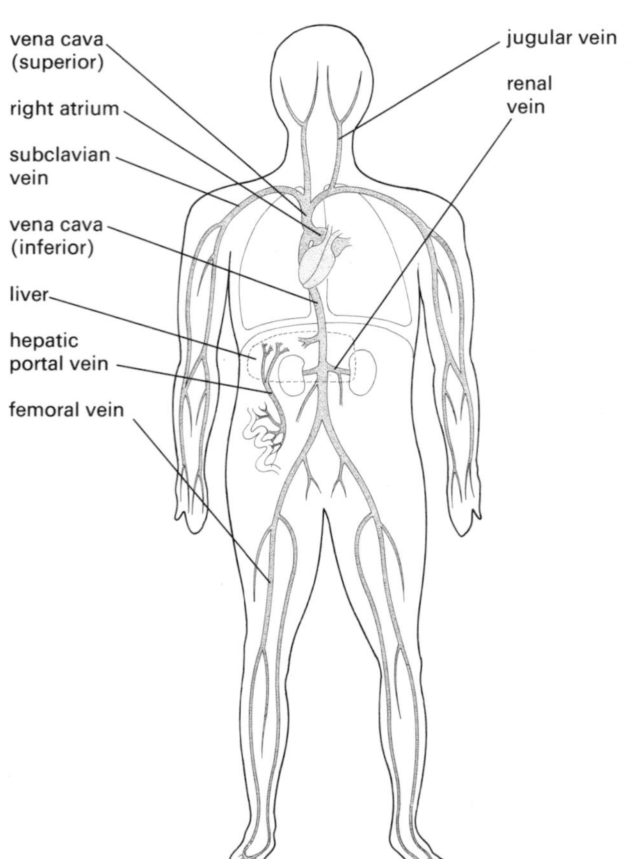

Fig. 13(*b*) Diagram of the venous system

The arteries divide into smaller vessels called **arterioles**. The small arteries and the arterioles have proportionately less elastic tissue and more muscle fibres than the great arteries. When the muscle fibres of the arterioles contract, they make the vessels narrower and restrict the blood flow. In this way, the distribution of blood to different parts of the body can be regulated. (See p. 168 for an example of this.)

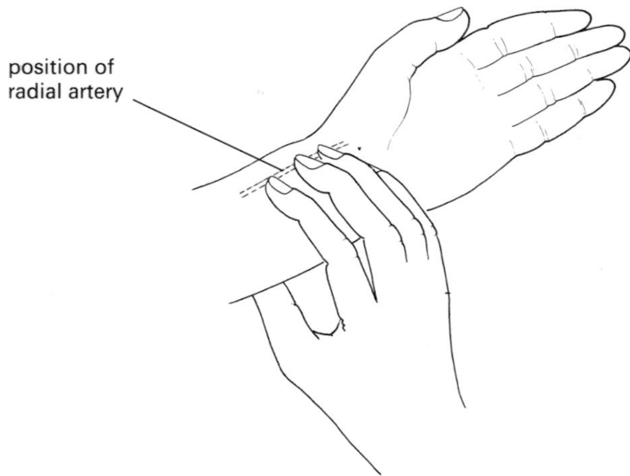

Fig. 14 Taking the pulse

The arterioles divide repeatedly to form a branching network of microscopic vessels passing between the cells of every living tissue. These final branches are called **capillaries**.

Capillaries

These are tiny vessels, often as little as 0.001 mm in diameter and with walls only 1-cell thick (Figs 11*c* and 15). Although the blood as a whole cannot escape from the capillary, the thin capillary walls allow some liquid to pass through, i.e. they are permeable. Blood pressure in the capillaries forces part of the plasma out through the walls. The fluid which escapes is not blood, nor plasma, but **tissue fluid**. Tissue fluid is similar to plasma but contains less protein. This fluid bathes all the living cells of the body and since it contains dissolved food and oxygen from the blood, it supplies the cells with their needs (Figs 16 and 17). The tissue fluid eventually seeps back into the capillaries, having given up its oxygen and dissolved food to the cells, but it has now received the waste products of the cells, such as carbon dioxide, which are carried away by the bloodstream.

The capillary network is so dense that no living cell is far from a supply of oxygen and food. The capillaries join up into larger vessels, called **venules**, which then combine to form **veins**.

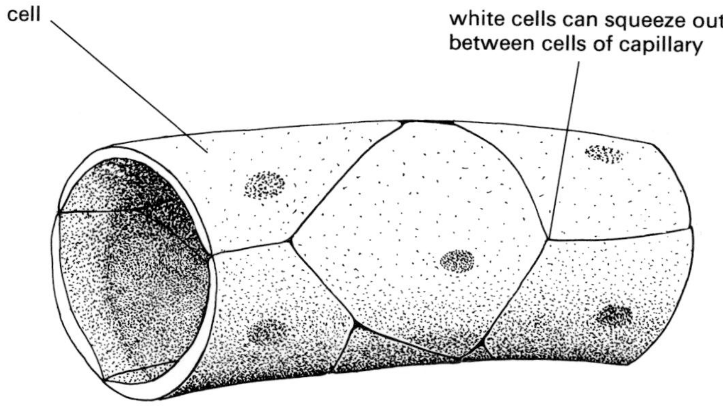

Fig. 15 Diagram of blood capillary

Veins

Veins return blood from the tissues to the heart (Fig. 13*b*). The blood pressure in them is steady and is less than that in the arteries. They are wider and their walls are thinner, less elastic and less muscular than those of the arteries (Figs 11*b* and 12). They also have valves in them similar to the semi-lunar valves (Fig. 8).

Contraction of body muscles, particularly in the limbs, compresses the thin-walled veins. The valves in the veins prevent the blood flowing backwards when the vessels are compressed in this way.

The blood in most veins is deoxygenated and contains less food but more carbon dioxide than the blood in most arteries. This is because respiring cells have used the oxygen and food and produced carbon dioxide. The pulmonary veins, which return blood from the lungs to the heart, contain oxygenated blood.

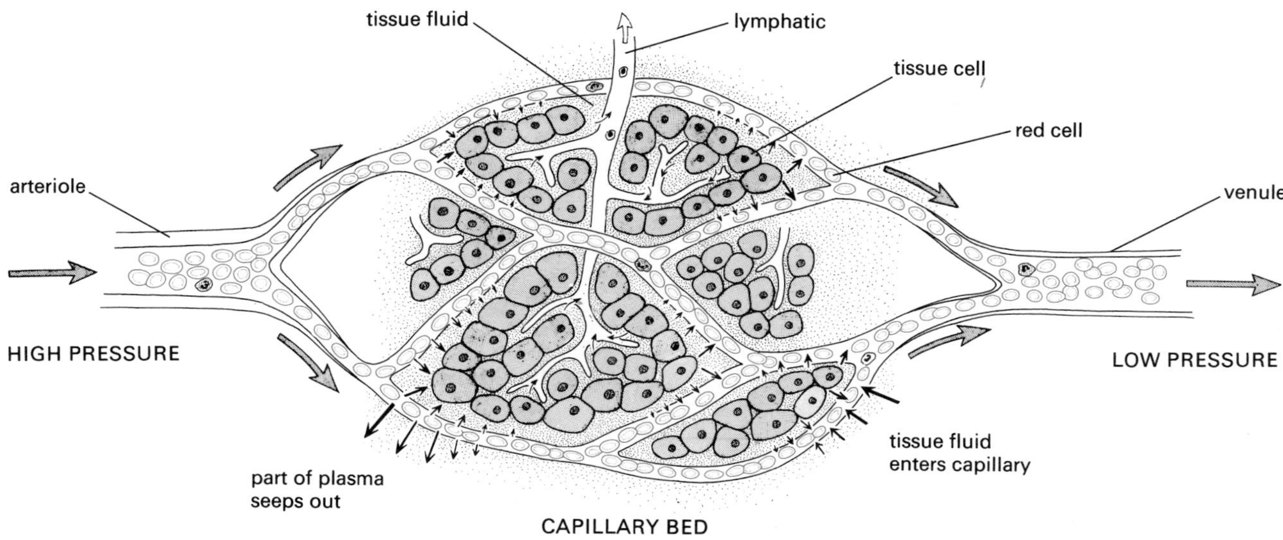

Fig. 16 Relationship between capillaries, cells and lymphatics

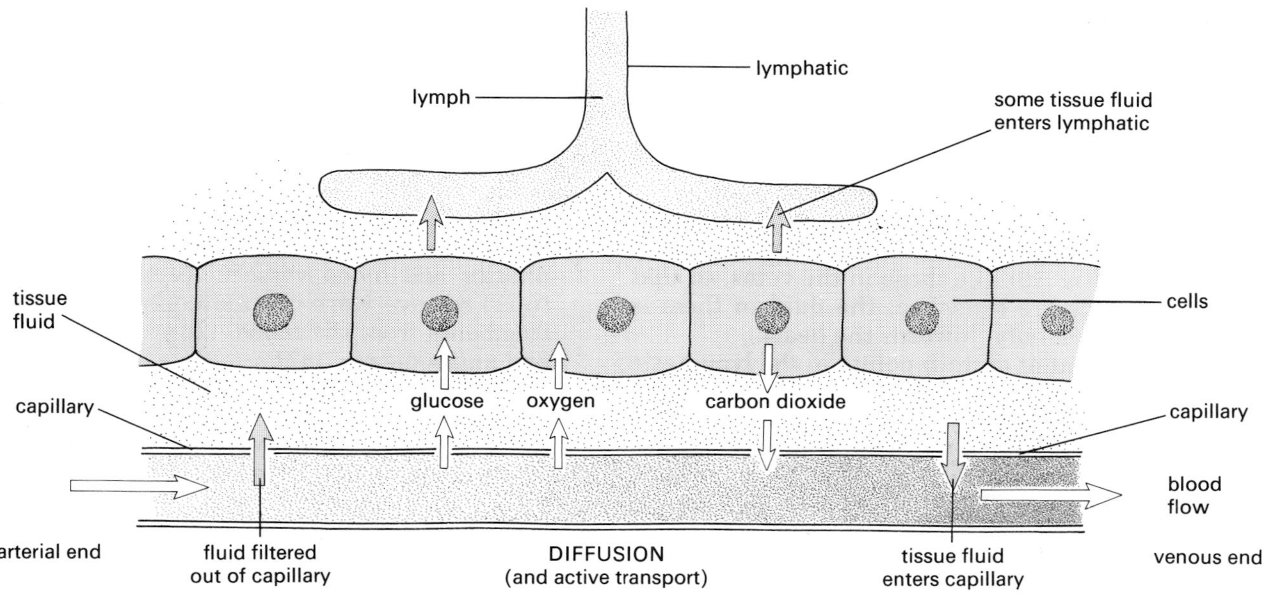

Fig. 17 Blood, tissue fluid and lymph

Blood pressure

The pumping action of the heart produces a pressure which drives blood round the circulatory system. In the arteries, the pressure fluctuates with the heart beat, and the pressure wave can be felt as a pulse. The millions of tiny capillaries offer resistance to the blood flow and, by the time the blood enters the veins, the surges due to the heart beat are lost and the blood pressure is greatly reduced.

In medical terms, 'blood pressure' means the two pressures, as measured in the arteries, when the heart is contracting and relaxing.

QUESTIONS

10 Starting from the left atrium, put the following in the correct order for circulation of the blood: (*a*) left atrium, (*b*) vena cava, (*c*) aorta, (*d*) lungs, (*e*) pulmonary artery, (*f*) right atrium, (*g*) pulmonary vein, (*h*) right ventricle, (*i*) left ventricle.
11 Why is it not correct to say that all arteries carry oxygenated blood and all veins carry deoxygenated blood?
12 How do veins differ from arteries in (*a*) their function, (*b*) their structure?
13 How do capillaries differ from other blood vessels in (*a*) their structure, (*b*) their function?
14 Describe the path taken by a molecule of glucose, from the time it is absorbed in the small intestine, and the path taken by a molecule of oxygen absorbed in the lungs, to the time when they both meet in a muscle cell of the leg (use Fig. 10).

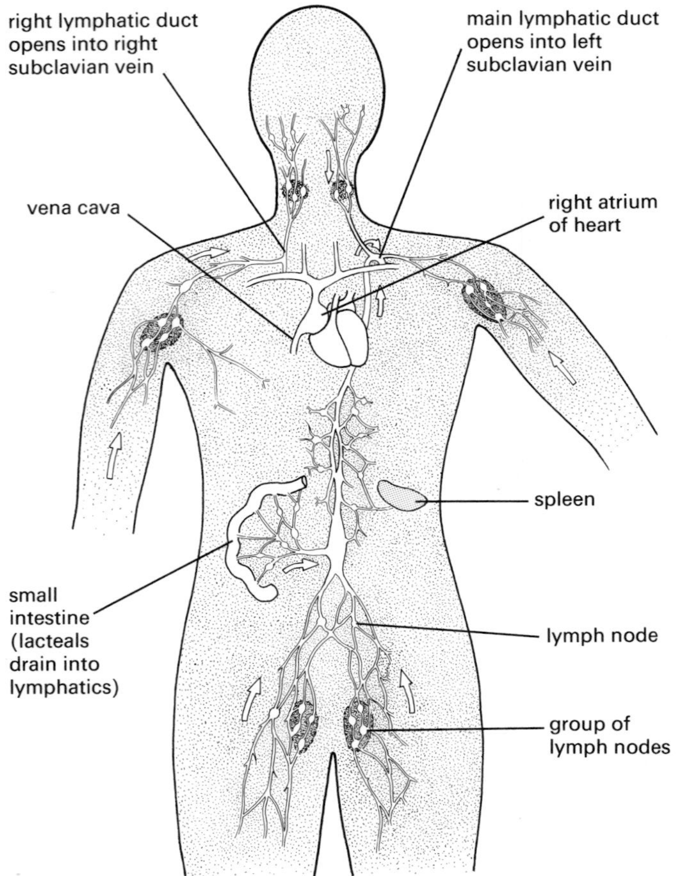

Fig. 18 Main drainage routes of the lymphatic system

THE LYMPHATIC SYSTEM

Not all the tissue fluid returns to the capillaries. Some of it enters blindly-ending, thin-walled vessels called **lymphatics** (Fig. 16). The lymphatics from all parts of the body join up to make two large vessels which empty their contents into the blood system as shown in Fig. 18.

The lacteals from the villi in the small intestine (p. 130) join up with the lymphatic system, so most of the fats absorbed in the intestine reach the circulation by this route. The fluid in the lymphatic vessels is called **lymph** and is similar in composition to tissue fluid.

Some of the larger lymphatics can contract, but most of the lymph flow results from the vessels being compressed from time to time when the body muscles contract in movements such as walking or breathing. There are valves in the lymphatics (Fig. 19) like those in the veins, so that when the lymphatics are squashed, the fluid in them is forced in one direction only: towards the heart.

Figure 18 shows that at certain points in the lymphatic vessels, there are swellings called **lymph nodes**. Lymphocytes are stored in the lymph nodes and released into the lymph to reach, eventually, the blood system. There are also phagocytes in the lymph nodes. If bacteria enter a wound and are not ingested by the white cells of the blood or lymph, they will be carried in the lymph to a lymph node and white cells there will ingest them. The lymph nodes thus form part of the body's defence system against infection.

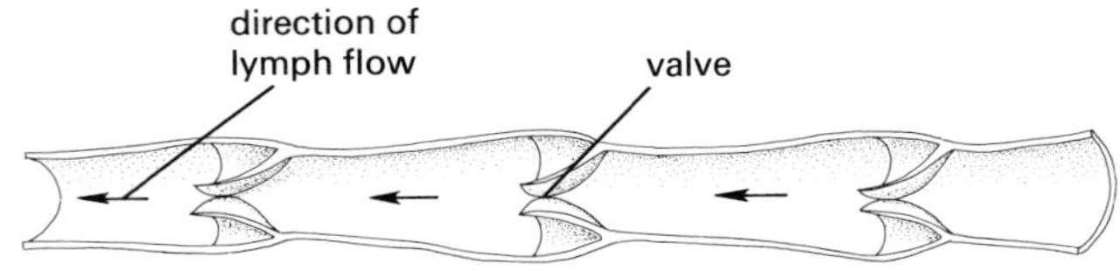

Fig. 19 Lymphatic vessel cut open to show valves

Spleen

The spleen is the largest organ in the adult lymphatic system. It is a solid, deep red body about 12 cm long and lies in the left side of the upper abdomen, between the lower ribs and the stomach. It contains lymphatics and blood vessels. Its main functions are to (1) remove worn-out red cells, bacteria and cell fragments from the blood, (2) produce lymphocytes and antibodies (p. 147).

The spleen contains many phagocytes which ingest the spent red cells and turn their haemoglobin into the compounds **bilirubin** and **ferritin** which are released into the blood circulation. The yellow pigment, bilirubin, is excreted in the bile (p. 129) but ferritin is a protein which contains the iron from the haemoglobin and is used by the red bone marrow to make more haemoglobin.

If bacteria or their antigens (p. 147) reach the spleen, the lymphocytes there start to make antibodies against them.

Thymus

The thymus gland lies at the top of the thorax, partly over the heart and lungs. It is an important lymphoid organ particularly in the newborn where it controls the development of the spleen and lymph nodes. The thymus produces lymphocytes and is the main centre for providing immunity against harmful micro-organisms.

After puberty, the thymus becomes smaller but is still an important immunological organ. White cells from the bone marrow are stored in the thymus. Here they undergo cell division to produce a large population of lymphocytes which can be 'programmed' to make antibodies against specific micro-organisms.

QUESTIONS

15 List the things you would expect to find if you analysed a sample of lymph.

16 Describe the course taken by a molecule of fat from the time it is absorbed in the small intestine to the time it reaches the liver to be oxidized for energy. (Use Fig. 12 on p. 130, Fig. 10 on p. 141 and Fig. 18 on p. 144.)

FUNCTIONS OF THE BLOOD

It is convenient, at this point, to distinguish between the functions of the blood (1) as the agent replenishing the tissue fluid surrounding the cells, i.e. its role in homeostasis (p. 134), (2) as a circulatory transport system and (3) as a defence mechanism against harmful bacteria, viruses and foreign proteins.

Homeostatic functions

All the cells of the body are bathed by tissue fluid which is derived from plasma. Tissue fluid supplies the cells with the food and oxygen necessary for their living chemistry. It also removes unwanted substances produced by the cell's metabolism.

The composition of the blood plasma is regulated by the liver and kidneys so that, within narrow limits, the living cells are soaked in a liquid of almost unvarying composition. This provides them with the environment they need and enables them to live and grow in the most favourable conditions. By delivering oxygen and nutrients to the tissue fluid and removing the excretory products, the blood fulfils a homeostatic function (p. 134), maintaining the constancy of the internal environment. (See p. 163 for further details.)

Transport

Transport of oxygen from the lungs to the tissues In the lungs, the concentration of oxygen is high and so the oxygen combines with the haemoglobin in the red cells, forming oxy-haemoglobin. The blood is now said to be **oxygenated**. When this oxygenated blood reaches tissues where oxygen is being used up, the oxy-haemoglobin breaks down and releases its oxygen to the tissues. Oxygenated blood is a bright red colour; **deoxygenated** blood is dark red.

Transport of carbon dioxide from the tissues to the lungs The blood picks up carbon dioxide from actively respiring cells and carries it to the lungs. In the lungs, the carbon dioxide escapes from the blood and is breathed out (see p. 154).

The carbon dioxide is carried in the form of hydrogen-carbonate ions ($-HCO_3^-$). Some of the hydrogencarbonate is carried in the red cells, but most of it is dissolved in the plasma.

Transport of digested food from the intestine to the tissues The soluble products of digestion pass into the capillaries of the villi lining the small intestine (p. 130). They are carried in solution by the plasma and, after passing through the liver, enter the main blood system. Glucose, salts, vitamins and some proteins pass out of the capillaries and into the tissue fluid. The cells bathed by this fluid take up the substances they need for their living processes.

Transport of nitrogenous waste from the liver to the kidneys When the liver changes amino acids into glycogen (p. 131), the amino part of the molecules ($-NH_2$) is changed into the nitrogenous waste product, urea. This substance is carried away in the blood circulation. When the blood passes through the kidneys, much of the urea is removed and excreted (p. 160).

Transport of hormones Hormones are chemicals made by certain glands in the body (see p. 212). The blood carries these chemicals from the glands which make them, to the organs (target organs) where they affect the rate of activity. For example, a hormone called **insulin**, made in the pancreas, is carried by the blood to the liver and controls how much glucose is stored as glycogen (p. 132).

Table 1 Transport by the blood system

Substance	*From*	*To*
Oxygen	lungs	whole body
Carbon dioxide	whole body	lungs
Urea	liver	kidneys
Hormones	glands	target organs
Digested food	intestine	whole body
Heat	liver and muscles	whole body

Note that the blood is not directed to a particular organ. A molecule of urea may go round the circulation many times before it enters the renal artery, by chance, and is removed by the kidneys.

Transport of heat The limbs and head lose heat to the surrounding air. Chemical changes elsewhere in the body produce heat. The blood carries the heat from the warm places to the cold places and so helps to keep an even temperature in all regions. Also by opening or closing blood vessels in the skin, the blood system helps to control the body temperature (see p. 168).

QUESTION

17 What substance would the blood (*a*) gain, (*b*) lose, on passing through (i) the kidneys, (ii) the lungs, (iii) an active muscle? Remember that respiration (p. 24) is taking place in all these organs.

Defence against infection

Clotting When tissues are damaged and blood vessels cut, platelets clump together and block the smaller capillaries. The platelets and damaged cells at the wound also produce a substance which acts, through a series of enzymes, on the plasma protein called **fibrinogen**. As a result of this action, the fibrinogen is changed into **fibrin**, which forms a network of fibres across the wound. Red cells become trapped in this network and so form a blood clot. The clot not only stops further loss of blood, but also prevents the entry of harmful bacteria into the wound (Figs 20 and 21).

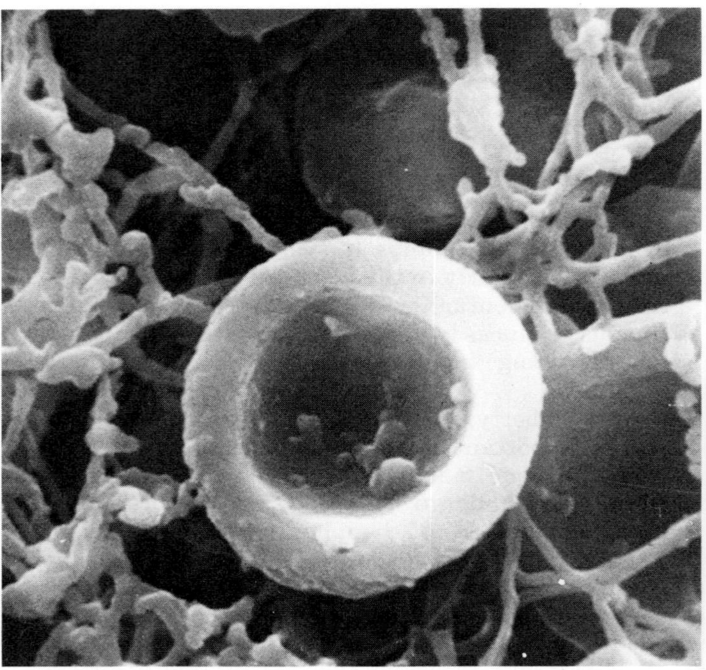

Fig. 20 A red cell trapped in a fibrin network (×8800)

White cells White cells at the site of the wound, in the blood capillaries or in lymph nodes (p. 144) may ingest harmful bacteria and so stop them entering the general circulation. White cells can squeeze through the walls of capillary vessels and so attack bacteria which get into the tissues, even though the capillaries themselves are not damaged.

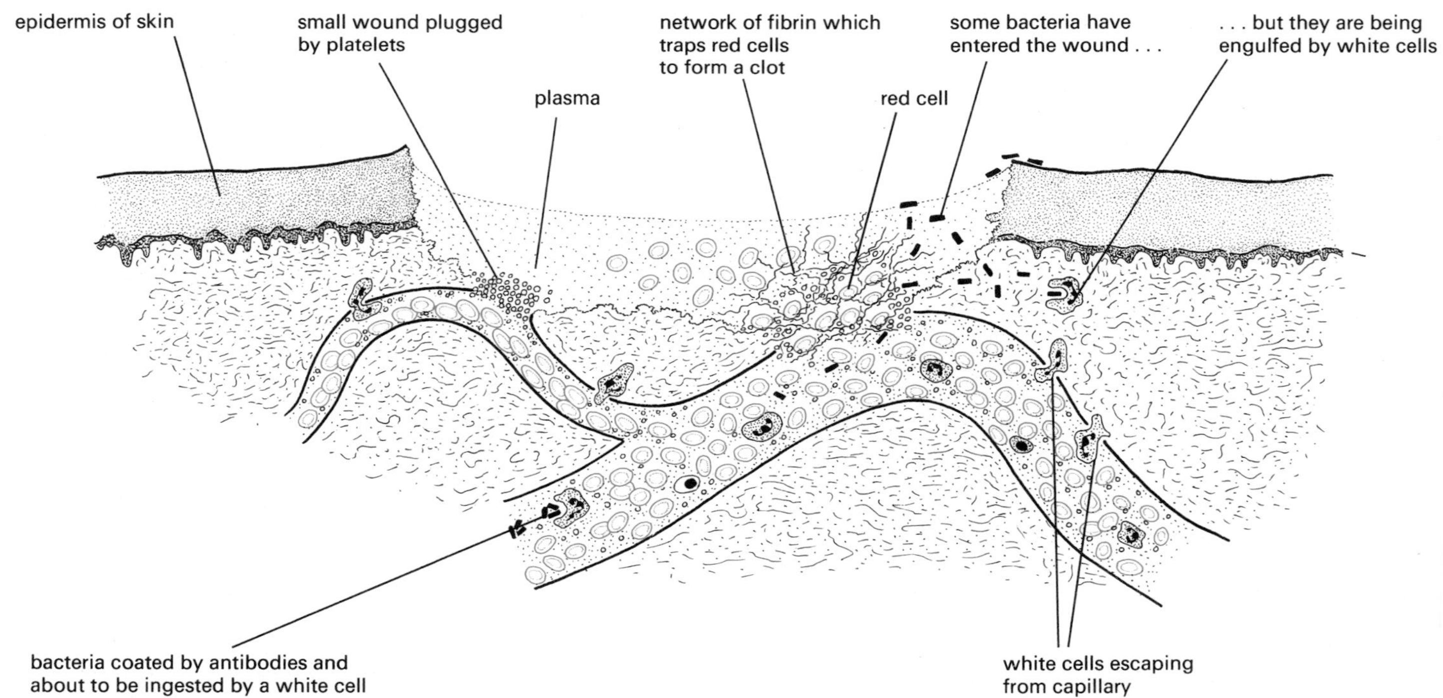

Fig. 21 The defence against infection. An area of skin has been damaged and two capillaries broken open

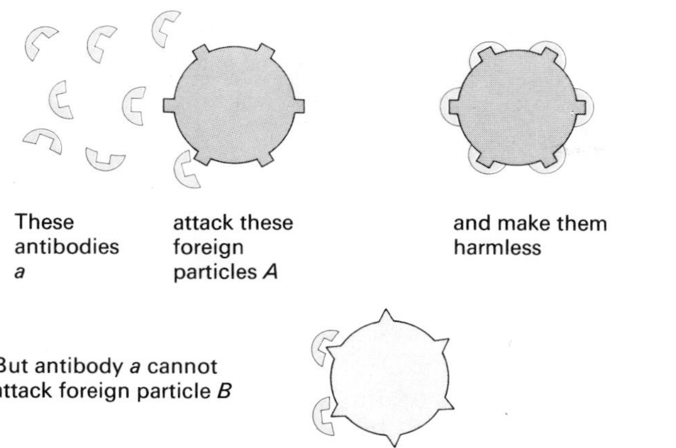
These antibodies a
attack these foreign particles A
and make them harmless
But antibody a cannot attack foreign particle B

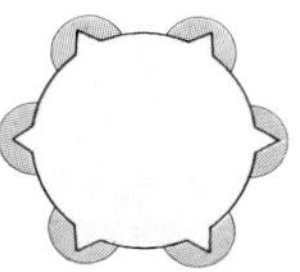
These antibodies b
attack these foreign particles B
and make them harmless
and antibody b is not effective against foreign particle A

Fig. 22 Antibodies are specific

Antibodies On the surface of bacteria there are chemical substances called **antigens**. Certain types of white cells, lymphocytes, produce chemicals called **antibodies** which attack the antigens of bacteria and other foreign proteins which get into the body. Antibodies are themselves proteins, released into the plasma by the lymphocytes. They may attach to the surface of the bacteria and make them easier for the phagocytes to ingest, or simply neutralize the poisonous proteins (toxins) produced by the bacteria.

Each antibody is very **specific**. This means that an antibody which attacks a typhoid bacterium will not affect a pneumonia bacterium. Figure 22 illustrates this in the form of a diagram.

Once an antibody has been made by the blood, it may remain in the circulation for some time. This means that the body has become **immune** to the disease, because the antibody will attack the bacteria or viruses as soon as they get into the body. Even if the antibodies do not remain for long in the circulation, the lymphocytes can usually make them again very quickly, so giving the person some degree of immunity. This explains why, once you have had measles or chicken pox, for example, you are very unlikely to catch the same disease again. This is called **natural immunity**.

When you are **inoculated** (vaccinated) against a disease, a harmless form of the bacteria or viruses is introduced into your body. The white cells make the correct antibodies, so that if the real micro-organisms get into the blood, the antibody is already present or very quickly made by the blood.

The material which is injected or swallowed is called a **vaccine** and is either

(1) a harmless form of the micro-organism, e.g. the BCG inoculation against tuberculosis and the Sabin oral vaccine against polio. (Oral, in this context, means 'taken by mouth'.)

(2) the killed micro-organisms, e.g. the Salk anti-polio vaccine and the whooping cough vaccine,

(3) a **toxoid**, i.e. the inactivated toxin from the bacteria, e.g. the diphtheria and tetanus vaccines. (A toxin is the poisonous substance produced by certain bacteria and which causes the disease symptoms.)

The immunity produced by a vaccine is called **artificial immunity**.

Antibodies are also present in the sera used to treat certain diseases. Serum is plasma with the fibrinogen removed. Sera are prepared from the plasma given by blood donors. People who have recently received an anti-tetanus inoculation will have made anti-tetanus antibodies in their blood. Some of these people volunteer to donate more blood than usual, but their plasma is separated at once and the red cells returned to their circulation. The anti-tetanus antibodies are then extracted from the plasma and used to treat patients who are at risk of contracting tetanus. In a similar way, antibodies against chicken-pox and rabies can be produced.

QUESTIONS

18 What part do white cells play in the defence of the body against infection?

19 Why is it necessary to inoculate a person against a disease before he catches it rather than wait until he catches it?

BLOOD GROUPS AND TRANSFUSIONS

If somebody loses a lot of blood as a result of an injury or surgical operation, he can be given a blood transfusion. Blood taken from a healthy person, the **donor**, is fed into one of the patient's veins. For a transfusion to be successful the blood type of the donor has to match the blood of the patient. If the two blood types do not match, the donor's red cells are clumped in the patient's blood vessels and cause serious harm. The red cells are clumped because they carry antigens on their cell membranes and if the blood types do not match, the antibodies in the patient's blood will act on the donor's red cells and clump them together.

For the purposes of transfusion, people can be put into one of four groups called group A, group B, group AB and group O. The red cells of group A people have antigen A on their cell membranes and the plasma contains anti-B antibodies. Since antibodies are

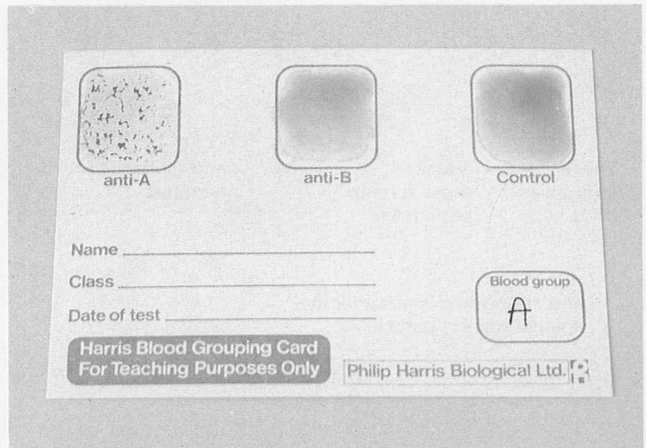

Fig. 23*a* ABO blood grouping. A blood sample is mixed with the dried serum on the card. Clumping has occurred in anti-A but not in anti-B serum, so the sample is from a group A person. The control ensures that clumping was not caused by some other factor.

specific, as explained on page 147, the anti-B antibodies do not attack the A antigens on the red cells.

The red cells of group B people carry the B antigen and their plasma has the anti-A antibody. So if group A cells are introduced into a group B person, they will be clumped by the anti-A antibody. Group AB people have both A and B antigens on their cells but no antibodies in their plasma. The red cells of group O people have neither A nor B antigens but their plasma contains both anti-A and anti-B antibodies (Table 2).

Table 2 Antigens and antibodies

Group	*Antigen on cells*	*Antibody in plasma*
A	A	anti-B
B	B	anti-A
AB	A and B	neither
O	neither	anti-A and anti-B

The red cells from group O people can be given to any other group because they have neither the A nor B antigens and so cannot be clumped. Group O people, on the other hand, can receive blood only from their own group because their plasma contains both anti-A and anti-B antibodies.

Group AB people, having neither anti-A nor anti-B antibodies in their plasma, can receive blood from any group. Table 3 shows the acceptable pattern of giving and receiving for the four groups.

Table 3 Blood transfusion

Group	*Can donate blood to*	*Can receive blood from*
A	A and AB	A and O
B	B and AB	B and O
AB	AB	all groups
O	all groups	O

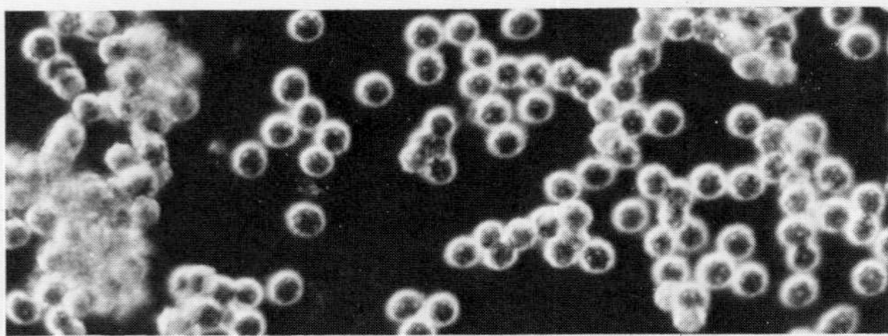

Fig. 23*b* The red cells are being clumped ($\times$800).

It is possible to find out a person's blood group by mixing a drop of his or her blood with anti-A serum and anti-B serum (Fig. 23). Group AB cells will clump in both anti-sera; group O cells will clump in neither; A cells will clump only in anti-A, and B cells only in anti-B.

A donor gives 420 cm^3 of blood from a vein in the arm (Fig. 24). The blood is led into a sterilized bottle containing sodium citrate which prevents clotting.

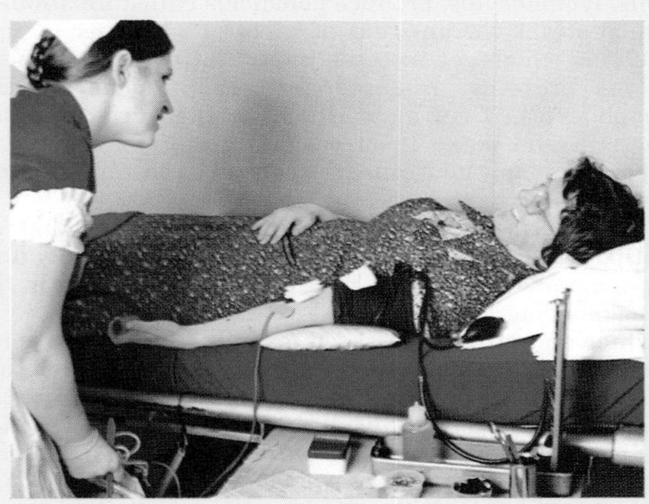

Fig. 24 Blood donor. The veins in the upper arm are compressed by using an inflatable 'cuff'. The donor's blood is then tapped from a vein near the inside of the elbow. It takes 5–10 minutes to fill the bottle.

The blood is then stored at 5 °C for 10 days, or longer if glucose is added. Before blood is transfused, even though both groups are known, it is carefully tested against the patient's blood to make sure of a good match. Then it is fed into one of the patient's arm veins at the correct rate and temperature. In a few hours, the donor will have made up his or her blood to the normal volume and in a week or two the red cells will have been replaced.

QUESTIONS

20 One of the ABO blood groups is sometimes called the 'universal donor'. Which group do you think this is and why?
21 A drop of a person's blood shows clumping in anti-B serum but not in anti-A. What is his blood group?
22 A drop of blood from a donor is clumped in serum taken from a group B person. What blood groups might the donor be?

CORONARY HEART DISEASE

In the lining of the large and medium arteries, deposits of a fatty substance, called **atheroma**, are laid down in patches. This happens to everyone and the patches get more numerous and extensive with age but until one of them actually blocks an important artery, the effects are not noticed. It is not known how or why the deposits form. Some doctors think that fatty substances in the blood pass into the lining. Others believe that small blood clots form on damaged areas of the lining and are covered over by the atheroma patches. The patches may join up to form a continuous layer which reduces the internal diameter of the vessel (Fig. 25).

The surface of a patch of atheroma sometimes becomes rough and causes fibrinogen in the plasma to deposit fibrin on it, so causing a blood clot (a **thrombus**) to form. If the blood clot blocks the **coronary artery** (Fig. 4) which supplies the muscles of the ventricles with blood, it starves the muscles of oxygenated blood and the heart may stop beating. This is a severe heart attack from **coronary thrombosis**. A thrombus might form anywhere in the arterial system, but its effects in the coronary artery and in parts of the brain are the most drastic.

In the early stages of coronary heart disease, the atheroma may partially block the coronary artery and reduce the blood supply to the heart. This can lead to **angina**, i.e. a pain in the chest which occurs during exercise or exertion. This is a warning to the person that he is at risk and should take precautions to avoid a coronary heart attack.

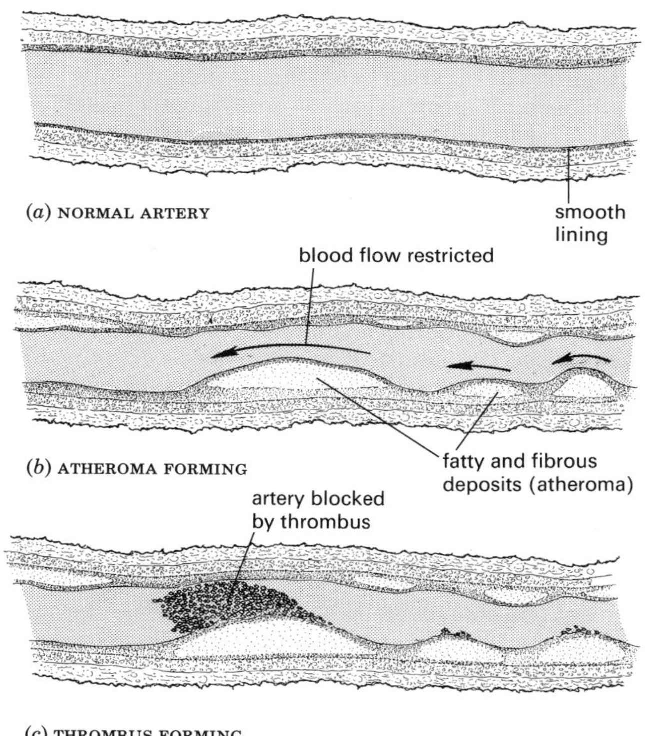

Fig. 25 Atheroma and thrombus formation

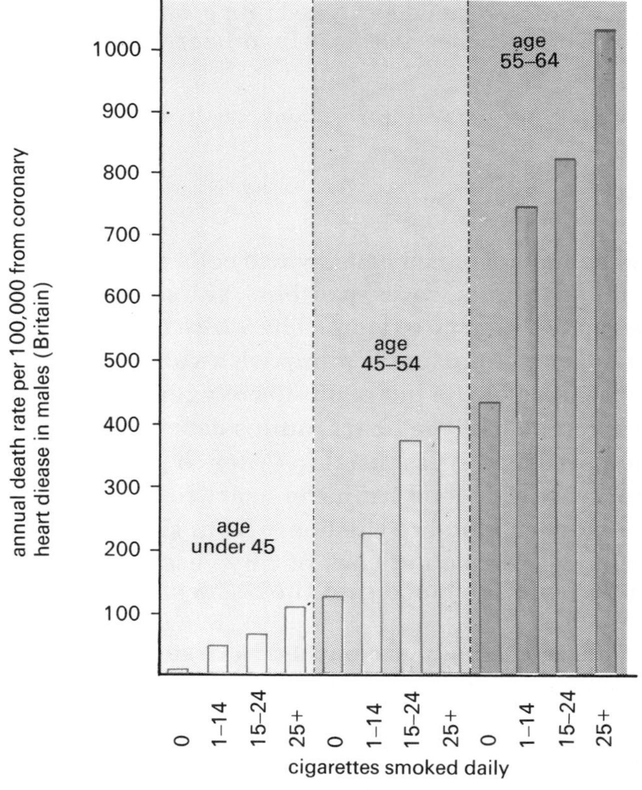

Fig. 26 Smoking and heart disease. Obviously, as you get older you are more likely to die from a heart attack, but notice that in any age group, the more you smoke, the higher your chances of dying from heart disease. (From *Smoking or Health: a report of the Royal College of Physicians*, Pitman Medical Publishing Co. Ltd)

Causes of heart disease

Atheroma and thrombus formation are the immediate causes of a heart attack but the long-term causes, which give rise to these conditions, are not well understood.

There is an inherited tendency towards the disease but the disease has increased very significantly in affluent countries in recent years. This makes us think that some features of 'Western' diets or life-styles might be causing it. Although there is very little direct evidence, the main factors are thought to be smoking, fatty diet, stress and lack of exercise.

Smoking Statistical studies suggest that smokers are 2–3 times more likely to die from a heart attack than are non-smokers of a similar age (Fig. 26). The carbon monoxide and other chemicals in cigarette smoke may damage the lining of the arteries, allowing atheroma to form, but there is not much direct evidence for this.

Fatty diet The atheroma deposits contain cholesterol, which is present, combined with proteins, in the blood. Cholesterol plays an essential part in our physiology, but it is known that people with high levels of blood cholesterol are more likely to suffer from heart attacks than people with low cholesterol levels.

Blood cholesterol can be influenced, to some extent, by the amount and type of fat in the diet. Many doctors and dieticians believe that animal fats (milk, cream, butter, cheese, egg-yolk, fatty meat) are more likely to raise the blood cholesterol than are the vegetable oils which contain a high proportion of unsaturated fatty acids (p. 119). This is still a matter of some controversy.

Stress Emotional stress often leads to a raised blood pressure. High blood pressure may increase the rate at which atheroma is formed in the arteries.

Lack of exercise There is some evidence that regular, vigorous exercise reduces the chances of a heart attack. This could be the result of an improved coronary blood flow. A sluggish blood flow, resulting from lack of exercise, may allow atheroma to form in the arterial lining but, once again, the direct evidence for this is slim.

Correlation and cause

It is not possible or desirable to conduct experiments on humans to find out, more precisely, the causes of heart attack. The evidence has to be collected from long-term studies on populations of individuals, e.g. smokers and non-smokers. Statistical analysis of these studies will often show a **correlation**, e.g. more smokers, within a given age band, suffer heart attacks than do non-smokers of the same age. This correlation does *not* prove that smoking **causes** heart attacks. It could be argued that people who are already prone to heart attacks for other reasons (e.g. high blood pressure) are more likely to take up smoking. This may strike you as implausible, but until it can be shown that substances in tobacco smoke do cause an increase in atheroma, the correlation cannot be used on its own to claim a cause and effect.

Nevertheless, there are so many other correlations between smoking and ill-health (bronchitis, emphysema, lung cancer) that the circumstantial evidence against smoking is very strong.

Another example of a positive correlation is between the possession of a television set and heart disease. Nobody would seriously claim that television sets cause heart attacks. The correlation probably reflects an affluent way of life, associated with over-eating, fatty diets, lack of exercise and other factors which may contribute to coronary heart disease.

QUESTIONS

23 (*a*) What positive steps could you take and (*b*) what things should you avoid to reduce your risk of coronary heart disease in later life?

24 About 95 per cent of patients with disease of the leg arteries are cigarette-smokers. Arterial disease of the leg is the most frequent cause of leg amputation.

(*a*) Is there a correlation between smoking and leg amputation?

(*b*) Does smoking cause leg amputation?

(*c*) In what way could smoking be a possible cause of leg amputation?

CHECK LIST

- **Blood consists of red cells, white cells and platelets suspended in plasma.**
- **Plasma contains water, proteins, salts, glucose and lipids.**
- **The red cells carry oxygen. The white cells attack bacteria.**
- **The heart is a muscular pump with valves, which sends blood round the circulatory system.**
- **The left side of the heart pumps oxygenated blood round the body.**
- **The right side of the heart pumps deoxygenated blood to the lungs.**
- **Blood pressure is essential in order to pump blood round the body.**
- **Arteries carry blood from the heart to the tissues.**
- **Veins return blood to the heart from the tissues.**
- **Capillaries form a network of tiny vessels in all tissues. Their thin walls allow dissolved food and oxygen to pass from the blood into the tissues, and carbon dioxide and other waste substances to pass back into the blood.**
- **All cells in the body are bathed in tissue fluid which is derived from plasma.**
- **Lymph vessels return tissue fluid to the lymphatic system and finally into the blood system.**
- **One function of the blood is to carry substances round the body, e.g. oxygen from lungs to body, food from intestine to body and urea from the liver to the kidneys.**
- **Lymph nodes, the spleen and the thymus are important immunological organs.**
- **Antibodies are chemicals made by white cells in the blood. They attack any micro-organisms or foreign proteins which get into the body.**
- **In blood transfusions, it is essential to match the A, B, O blood group of donor and recipient.**
- **Blockage of the coronary arteries in the heart leads to a heart attack.**
- **Smoking, fatty diets, stress and lack of exercise may contribute to heart disease.**

15 Breathing

LUNG STRUCTURE
Air passages and alveoli.

VENTILATION OF THE LUNGS
Inhaling, exhaling, lung capacity.

GASEOUS EXCHANGE
Uptake of oxygen; removal of carbon dioxide.

RESPIRATORY SURFACES
Characteristics.

SMOKING
Effect of smoking on the lungs and circulatory system.

PRACTICAL WORK
The composition of exhaled air. Lung volume.

All the processes carried out by the body, such as movement, growth and reproduction, require energy. In animals, this energy can be obtained only from the food they eat. Before the energy can be used by the cells of the body, it must be set free from the chemicals of the food by a process called 'respiration' (see p. 24). Respiration needs a supply of oxygen and produces carbon dioxide as a waste product. All cells, therefore, must be supplied with oxygen and must be able to get rid of carbon dioxide.

In man and other mammals, the oxygen is obtained from the air by means of the lungs. In the lungs, the oxygen dissolves in the blood and is carried to the tissues by the circulatory system (p. 141).

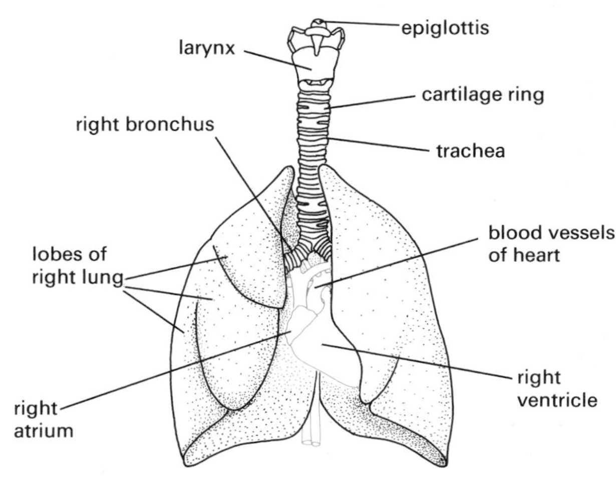

Fig. 1 Diagram of lungs, showing position of heart

LUNG STRUCTURE

The lungs are enclosed in the thorax (chest region) (see Fig. 5, p. 127). They have a spongy texture and can be expanded and compressed by movements of the thorax in such a way that air is sucked in and blown out. The lungs are joined to the back of the mouth by the windpipe or **trachea** (Fig. 1). The trachea divides into two smaller tubes, **bronchi** (singular = bronchus), which enter the lungs and divide into even smaller branches. When these branches are only about 0.2 mm in diameter, they are called **bronchioles** (Fig. 2*a*). These fine branches end up in a mass of little, thin-walled, pouch-like air sacs called **alveoli** (Figs 2*b*, 2*c* and 3).

Rings of gristle (cartilage) stop the trachea and bronchi collapsing when we breathe in. The **epiglottis** and other structures at the top of the trachea stop food and drink from entering the air passages when we swallow (see p. 127).

The epithelium which lines the inside of the trachea, bronchi and bronchioles consists of ciliated cells (p. 9). There are also cells which secrete mucus. The mucus forms a thin film over the internal lining. Dust particles and bacteria become trapped in the sticky mucus film and the mucus is carried upwards, away from the lungs, by the flicking movements of the cilia. In this way, harmful particles are prevented from reaching the alveoli. When the mucus reaches the top of the trachea, it passes down the gullet during normal swallowing.

The alveoli have thin elastic walls, formed from a single cell layer or **epithelium**. Beneath the epithelium is a dense network of capillaries (Fig. 2*c*) supplied with deoxygenated blood (p. 145). This blood, from which the body has taken oxygen, is pumped from the right ventricle, through the pulmonary artery (see Fig. 10, page 141). In humans, there are about 350 million alveoli, with a total absorbing surface of about 90 m^2. This large absorbing surface makes it possible to take in oxygen and give out carbon dioxide at a rate to meet the body's needs.

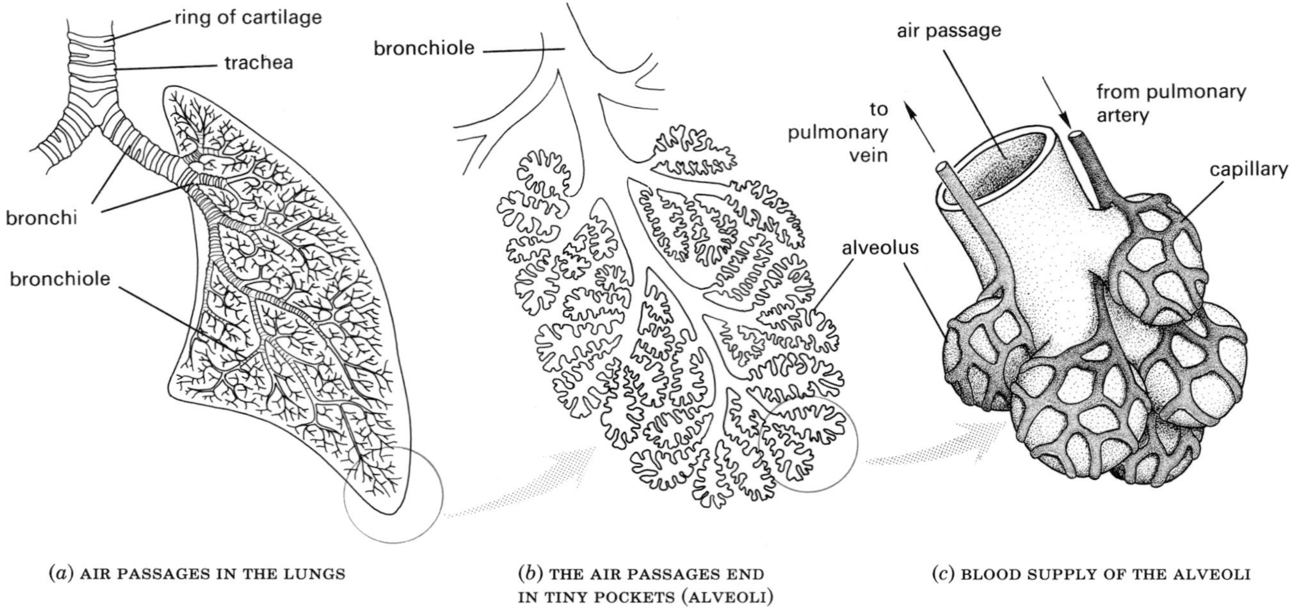

Fig. 2 Lung structure

QUESTIONS

1 Place the following structures in the order in which air will reach them when breathing in: bronchus, trachea, nasal cavity, alveolus.
2 One function of the small intestine is to absorb food (p. 130). One function of the lungs is to absorb oxygen. Point out the basic similarities in these two structures which help to speed up the process of absorption.

VENTILATION OF THE LUNGS

The movement of air into and out of the lungs, called **ventilation**, renews the oxygen supply in the lungs and removes the surplus carbon dioxide from them. The lungs contain no muscle fibres and are made to expand and contract by movements of the ribs and diaphragm.

The **diaphragm** is a sheet of tissue which separates the thorax from the abdomen (see Fig. 5, p. 127). When relaxed, it is domed slightly upwards. The ribs are moved by the **intercostal muscles** which run from one rib to the next (Fig. 4). Figure 5 shows how the contraction of the intercostal muscles makes the ribs move upwards.

Inhaling

1. The diaphragm muscles contract and pull it down (Fig. 7*a*).
2. The intercostal muscles contract and pull the rib cage upwards and outwards (Fig. 6*a*).

These two movements make the space in the thorax bigger, so forcing the lungs to expand and draw air in through the nose and trachea.

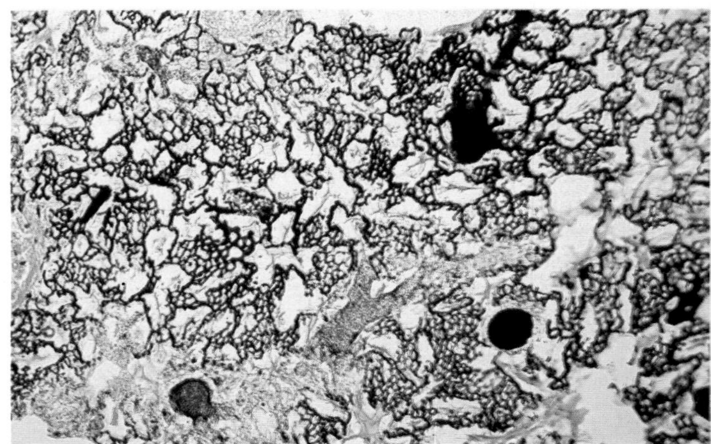

Fig. 3 Small piece of lung tissue (×40). The capillaries have been injected with red and blue dye. The networks surrounding the alveoli can be seen.

Exhaling

1. The diaphragm muscles relax, allowing the diaphragm to return to its domed shape (Fig. 7*b*).
2. The intercostal muscles relax, allowing the ribs to move downwards under their own weight (Fig. 6*b*).

The lungs are elastic and shrink back to their relaxed size, forcing air out again.

The outside of the lungs and the inside of the thorax are lined with a smooth membrane called the **pleural membrane**. This produces a thin layer of liquid called **pleural fluid** which reduces the friction between the lungs and the inside of the thorax.

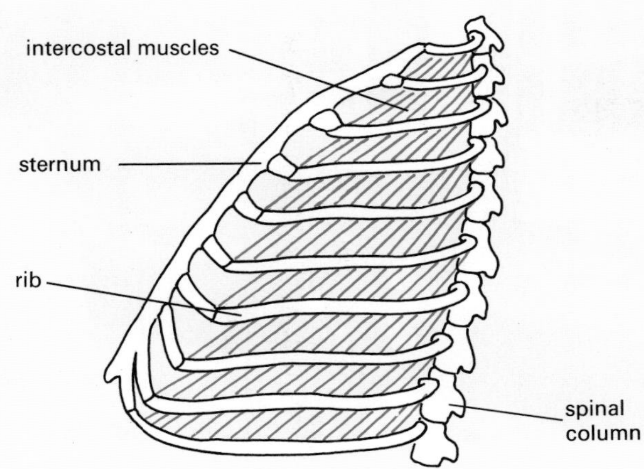

Fig. 4 Rib cage seen from left side, showing intercostal muscles

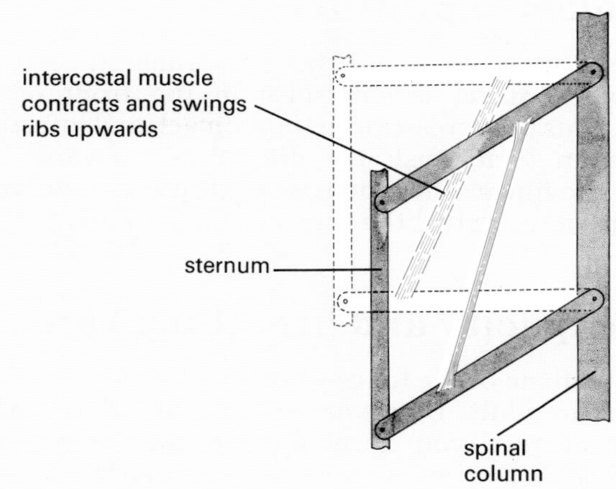

Fig. 5 Model to show action of intercostal muscles

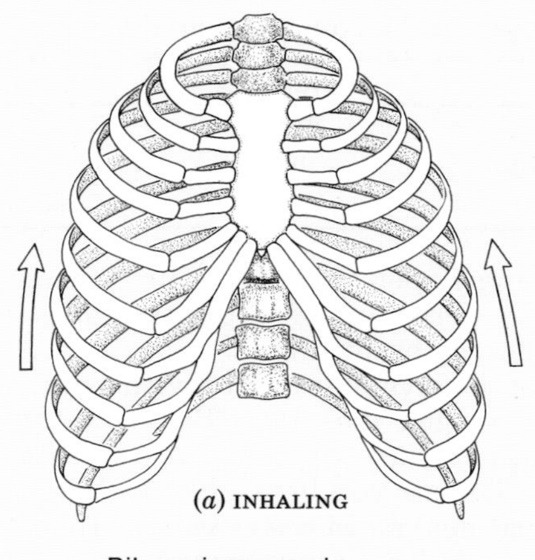

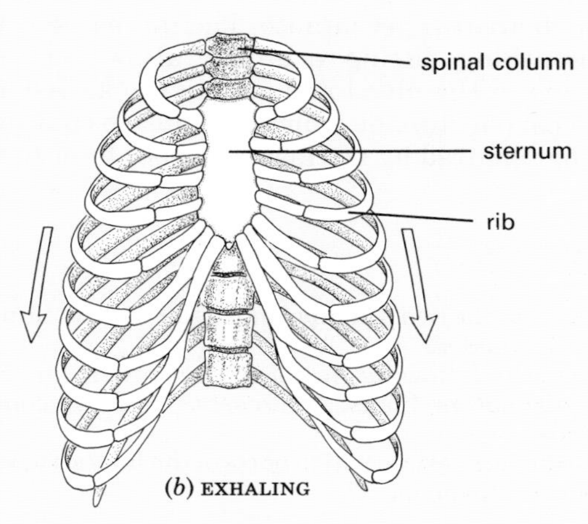

Fig. 6 Movement of rib cage during breathing

3 air drawn in

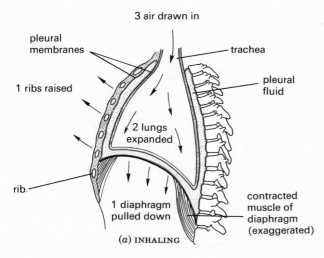

pleural membranes

trachea

1 ribs raised

pleural fluid

2 lungs expanded

rib

1 diaphragm pulled down

contracted muscle of diaphragm (exaggerated)

(*a*) INHALING

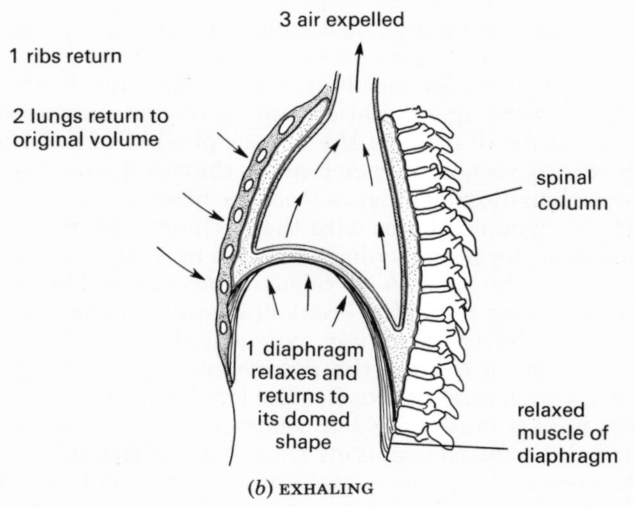

Fig. 7 Diagrams of thorax to show mechanism of breathing

Controlled expiration

When we speak, sing, play a wind instrument, cough or sneeze, we contract the muscles in the front of the abdomen. This puts pressure on the stomach and intestines and they, in turn, push the diaphragm upwards and compress the lungs. The glottis at the top of the windpipe can be closed to control the rate at which air escapes.

Lung capacity and breathing rate

The total volume of the lungs when fully inflated is about 5 litres in an adult. However, in quiet breathing, when asleep or at rest, you normally exchange only about 500 cm^3 (Fig. 8). During exercise you can take in and expel an extra 3 litres. There is a residual volume of 1.5 litres which cannot be expelled no matter how hard you breathe out.

At rest, you normally inhale and exhale about 16 times per minute. During exercise, the breathing rate may rise to 20 or 30 breaths per minute. The increased rate and depth of breathing during exercise, allows more oxygen to dissolve in the blood and supply the active muscles. The extra carbon dioxide which the muscles put into the blood will be removed by the faster, deeper breathing.

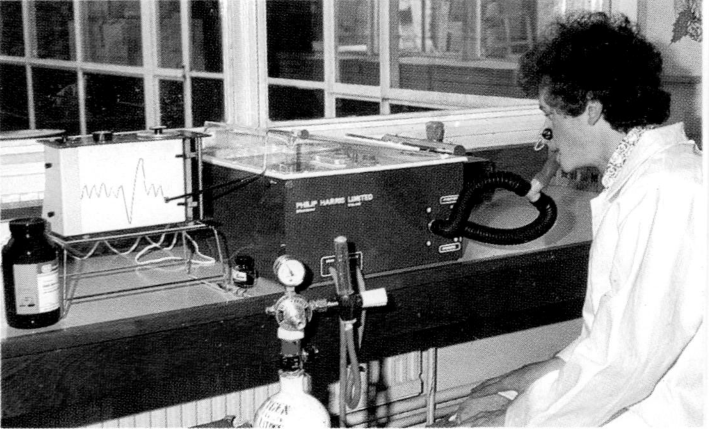

Fig. 8 A spirometer. This instrument measures the volume of air breathed in and out of the lungs. The first part of the chart shows quiet breathing. In the middle, the chart shows a deep breath in and a deep breath out.

Table 1 Changes in the composition of breathed air

	Inhaled %	*Exhaled %*
Oxygen	21	16
Carbon dioxide	0.04	4
Water vapour	variable	saturated

QUESTIONS

3 What are the two principal muscular contractions which cause air to be inhaled?
4 Place the following in the correct order: lungs expand, ribs rise, air enters lungs, intercostal muscles contract, thorax expands.
5 During inhalation, which parts of the lung structure would you expect to expand most?

GASEOUS EXCHANGE

Ventilation refers to the movement of air into and out of the lungs. Gaseous exchange refers to the exchange of oxygen and carbon dioxide which takes place between the air and the blood vessels in the lungs.

The 1.5 litres of residual air in the alveoli is not exchanged during ventilation and the oxygen has to reach the blood capillaries by the slower process of diffusion. Figure 9 shows how oxygen reaches the red blood cells and how carbon dioxide escapes from the blood.

The oxygen combines with the haemoglobin in the red blood cells, forming oxy-haemoglobin (p. 138). The carbon dioxide in the plasma is released when the hydrogen-carbonate ions ($-HCO_3$) break down to CO_2 and H_2O.

The capillaries carrying oxygenated blood from the alveoli join up to form the pulmonary vein (see Fig. 10, p. 141), which returns blood to the left atrium of the heart. From here it enters the left ventricle and is pumped all round the body, so supplying the tissues with oxygen.

The process of gaseous exchange in the alveoli does not remove all the oxygen from the air. The air breathed in contains about 21 per cent of oxygen; the air breathed out still contains 16 per cent of oxygen (see Table 1).

The remaining 79 per cent of the air consists mainly of nitrogen, whose percentage composition does not change significantly during breathing.

The lining of the alveoli is coated with a film of moisture in which the oxygen dissolves. Some of this moisture evaporates into the alveoli and saturates the air with water vapour. The air you breathe out, therefore, always contains a great deal more water vapour than the air you breathe in. The exhaled air is warmer as well, so in cold and temperate climates, you lose heat to the atmosphere by breathing.

Sometimes the word **respiration** or **respiratory** is used in connection with breathing. The lungs, trachea and

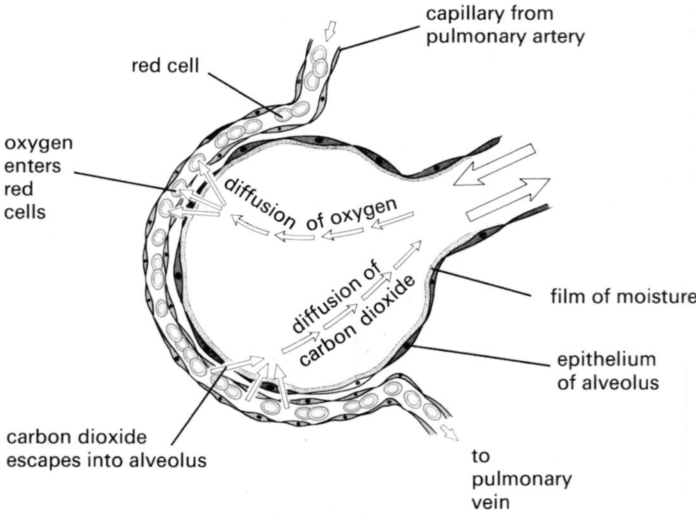

Fig. 9 Gaseous exchange in the alveolus

bronchi are called the **respiratory system**; a person's rate of breathing may be called his **respiration rate**. This use of the word should not be confused with the biological meaning of respiration, namely the release of energy in cells (p. 24). This chemical process is sometimes called **tissue respiration** or **internal respiration** to distinguish it from breathing. Gaseous exchange and ventilation are sometimes called **external respiration**.

Characteristics of respiratory surfaces

The exchange of oxygen and carbon dioxide across a respiratory surface, as in the lungs or over the gills of a fish, depends on the diffusion of these two gases. Diffusion occurs more rapidly if (1) there is a large surface exposed to the gas, (2) the distance across which diffusion has to take place is small, (3) there is a big difference in the concentrations of the gas at two points brought about by ventilation and (4) there is a rich supply of blood capillaries.

Large surface The presence of millions of alveoli in the lungs provides a very large surface for gaseous exchange. The many branching filaments in a fish's gills have the same effect.

Thin epithelium There is only a 2-cell layer, at the most, separating the air in the alveoli from the blood in the capillaries (Fig. 9). Thus, the distance for diffusion is very short.

Ventilation Ventilation of the lungs helps to maintain a steep diffusion gradient (p. 34), between the air at the end of the air passages and the alveolar air. The concentration of the oxygen in the air at the end of the air passages is high, because the air is constantly replaced by the breathing actions.

Capillary network The continual removal of oxygen by the blood in the capillaries lining the alveoli keeps its concentration low. In this way, a steep diffusion gradient is maintained which favours the rapid diffusion of oxygen from the air passages to the alveolar lining.

The continual delivery of carbon dioxide from the blood, into the alveoli and its removal from the air passages by ventilation, similarly maintains a diffusion gradient which promotes the diffusion of carbon dioxide from the alveolar lining into the bronchioles.

The respiratory surfaces of land-dwelling mammals are invariably moist. Oxygen has to dissolve in the thin film of moisture before passing across the epithelium.

QUESTIONS

6 Try to make a clear distinction between 'respiration' (p. 24), 'gaseous exchange' and 'ventilation'. Say how one depends on the other.

7 Describe the path taken by a molecule of oxygen from the time it is breathed in through the nose, to the time it enters the heart in some oxygenated blood.

8 Figure 9 shows oxygen and carbon dioxide diffusing across an alveolus. What causes them to diffuse in opposite directions? (See p. 34.)

9 In 'mouth to mouth' resuscitation, air is breathed from the rescuer's lungs into the lungs of the person who has stopped breathing. How can this 'used' air help to revive the person?

SMOKING

The short-term effects of smoking cause the bronchioles to constrict and the cilia lining the air passages to stop beating. The smoke also makes the lining produce more mucus. The long-term effects may take many years to develop but they are severe, disabling and often lethal.

Lung cancer

Although all forms of air pollution are likely to increase the chances of lung cancer, many scientific studies show, beyond all reasonable doubt, that the vast increase in lung cancer (4000 per cent in the last century) is almost entirely due to cigarette-smoking (Fig. 10).

There are at least 17 substances in tobacco smoke known to cause cancer in experimental animals, and it is now thought that 90 per cent of lung cancer is caused by smoking. Table 2 shows the relationship between smoking cigarettes and the risk of developing lung cancer.

Table 2 Cigarette-smoking and lung cancer

Number of cigarettes per day	*Increased risk of lung cancer*
1–14	×8
15–24	×13
25+	×25

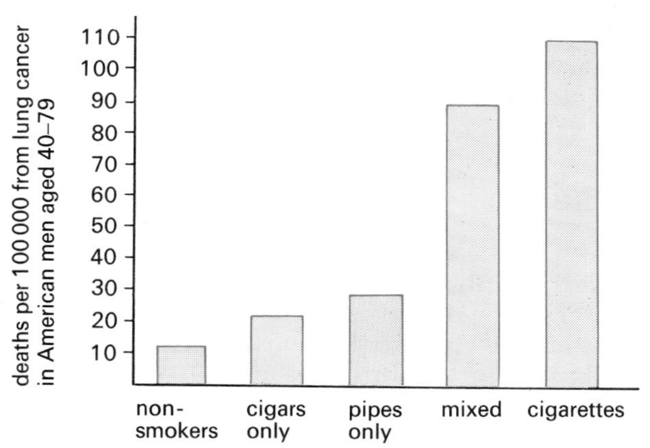

Fig. 10 Smoking and lung cancer. Cigar and pipe smokers are probably at less risk because they often do not inhale. But notice that their death rate from lung cancer is still twice that of non-smokers. (From *Smoking and Health Now: a report of the Royal College of Physicians,* Pitman Medical Publishing Co. Ltd.)

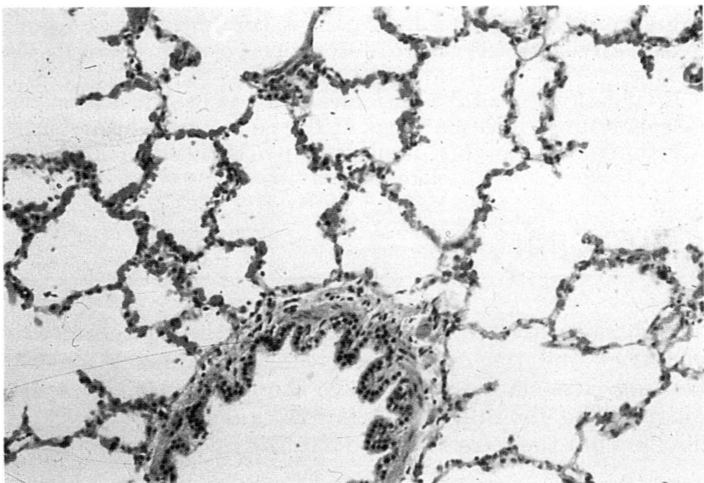

(*a*) Normal lung tissue showing a bronchiole and about 25 alveoli (×200).

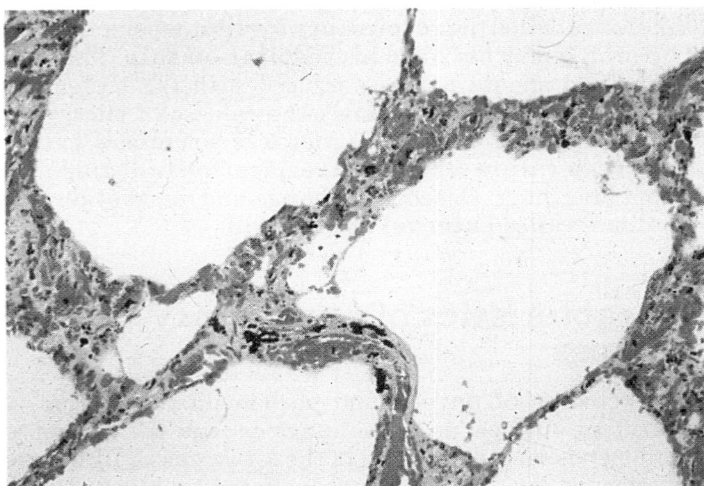

(*b*) Lung tissue from a person with emphysema. This is the same magnification as (*a*). The alveoli have broken down leaving only about five air sacs which provide a much reduced absorbing surface.

Fig. 11 Emphysema

Emphysema

Emphysema is a breakdown of the alveoli. The action of one or more of the substances in tobacco smoke weakens the walls of the alveoli. The irritant substances in the smoke cause a 'smokers' cough' and the coughing bursts some of the weakened alveoli. In time, the absorbing surface of the lungs is greatly reduced (Fig. 11). Then the smoker cannot oxygenate his blood properly and the least exertion makes him breathless and exhausted.

Chronic bronchitis

The smoke stops the cilia in the air passages from beating and so the irritant substances in the smoke and the excess mucus collect in the bronchi. This leads to the inflammation known as **bronchitis**. Over 95 per cent of people suffering from bronchitis are smokers and they have a 20 times greater chance of dying from bronchitis than non-smokers.

Heart disease

Coronary heart disease is the leading cause of death in most developed countries. It results from a blockage of coronary arteries by fatty deposits. This reduces the supply of oxygenated blood to the heart muscle and sooner or later leads to heart failure (see p. 149). High blood pressure, diets with too much animal fat, and lack of exercise are also thought to be causes of heart attack, but about a quarter of all deaths due to coronary heart disease are thought to be caused by smoking (Fig. 26, p. 149).

The nicotine and carbon monoxide from cigarette smoke increase the tendency for the blood to clot and so block the coronary arteries, already partly blocked by fatty deposits. The carbon monoxide increases the rate at which the fatty material is deposited in the arteries.

Other risks

About 95 per cent of patients with disease of the leg arteries are cigarette-smokers and this condition is the most frequent cause of leg amputations.

Strokes due to arterial disease in the brain are more frequent in smokers.

Cancer of the bladder, ulcers in the stomach and duodenum, tooth decay, gum disease and tuberculosis all occur more frequently in smokers.

Babies born to women who smoke during pregnancy are smaller than average, probably as a result of reduced oxygen supply caused by the carbon monoxide in the blood. In smokers, there is twice the frequency of miscarriages, a 50 per cent higher still-birth rate and a 26 per cent higher death rate of babies (Fig. 12).

In 1976 two famous doctors predicted that one in every three smokers will die as a result of their smoking habits. Those who do not die at an early age will probably be seriously disabled by one of the conditions described above.

Passive smoking It is not only the smokers themselves who are harmed by tobacco smoke. Non-smokers in the same room are also affected. One study has shown that children whose parents both smoke, breathe in as much nicotine as if they were themselves smoking 80 cigarettes a year.

Statistical studies also suggest that the non-smoking wives of smokers have an increased chance of lung cancer.

Reducing the risks

By giving up smoking, a person who smokes up to 20 cigarettes a day will, after 10 years, be at no greater risk than a non-smoker of the same age. The pipe- or cigar-smoker, provided he does not inhale, is at less risk than a cigarette-smoker but still at greater risk than a non-smoker. The risk of disease or death is also reduced by changing to low-tar cigarettes, leaving longer stubs, inhaling less and taking fewer puffs.

Fig. 12 Cigarette smoke can harm the unborn baby. Pregnant women who smoke may have smaller babies and a higher chance of miscarriage or stillbirth.

Correlations and causes

On page 150 it was explained that a correlation between two variables does not prove that one of the variables causes the other. The fact that a higher risk of dying from lung cancer is correlated with heavy smoking does not actually prove that smoking is the cause of lung cancer. The alternative explanation is that people who become heavy smokers are, in some way, exposed to other potential causes of lung cancer, e.g. they live in areas of high air pollution or they have an inherited tendency to cancer of the lung. These alternatives are not very convincing, particularly when there is such an extensive list of ailments associated with smoking.

This is not to say that smoking is the only cause of lung cancer or that everyone who smokes will eventually develop lung cancer. There are likely to be complex interactions between life-styles, environments and genetic backgrounds which could lead, in some cases, to lung cancer. Smoking may be only a part, but a very important part, of these interactions.

QUESTIONS

10 What are (*a*) the immediate effects and (*b*) the long-term effects of tobacco smoke on the trachea, bronchi and lungs?
11 Why does a regular smoker get out of breath sooner than a non-smoker of similar age and build?
12 If you smoke 20 cigarettes a day, by how much are your chances of getting lung cancer increased?
13 Apart from lung cancer, what other diseases are probably caused by smoking?

PRACTICAL WORK

1. Oxygen in exhaled air

Place a large screw-top jar on its side in a bowl of water (Fig. 13*a*). Put a rubber tube in the mouth of the jar and then turn the jar upside-down, still full of water and with the rubber tube still in it. Start breathing out and when you feel your lungs must be about half empty, breathe the last part of the air down the rubber tubing so that the air collects in the upturned jar and fills it (Fig. 13*b*). Put the screw top back on the jar under water, remove the jar from the bowl and place it upright on the bench.

Light the candle on the special wire holder (Fig. 13*c*), remove the lid of the jar, lower the burning candle into the jar and count the number of seconds the candle stays alight. Now take a fresh jar, with ordinary air, and see how long the candle stays alight in this.

Results The candle will burn for about 15–20 seconds in a large jar of ordinary air. In exhaled air it will go out in about 5 seconds.

Interpretation Burning needs oxygen. When the oxygen is used up, the flame goes out. It looks as if exhaled air contains much less oxygen than atmospheric air.

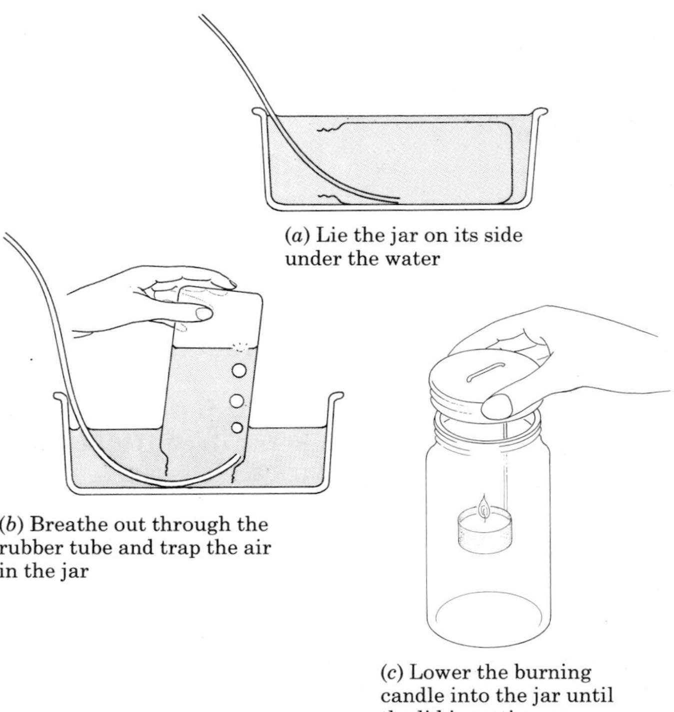

(*a*) Lie the jar on its side under the water
(*b*) Breathe out through the rubber tube and trap the air in the jar
(*c*) Lower the burning candle into the jar until the lid is resting on the rim

Fig. 13 Testing exhaled air for oxygen

2. Carbon dioxide in exhaled air

Prepare two large test-tubes as shown in Fig. 14, each containing a little clear lime water. Put the ends of both rubber tubes at the same time in your mouth and breathe in and out gently through the tubes for about 15 seconds. Notice which tube is bubbling when you breathe out and which one bubbles when you breathe in.

If after 15 seconds there is no difference in the appearance of the lime water in the two tubes, continue breathing through them for another 15 seconds.

Results The lime water in tube B goes milky. The lime water in tube A stays clear.

Interpretation Carbon dioxide turns lime water milky. Exhaled air passes through tube B. Inhaled air passes through tube A. Exhaled air must, therefore, contain more carbon dioxide than inhaled air.

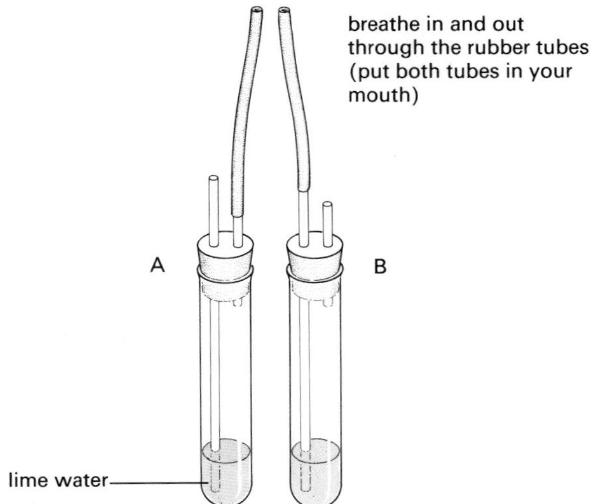

Fig. 14 Comparing the carbon dioxide content of inhaled and exhaled air

3. Volume of air in the lungs

Calibrate a large (5 litre) plastic bottle by filling it with water, half a litre at a time, and marking the water levels on the outside. Fill the bottle with water and put on the stopper. Put about 50 mm depth of water in a large plastic bowl. Hold the bottle upside-down with its neck under water and remove the screw top. Some of the water will run out but this does not matter. Push a rubber tube into the mouth of the bottle (Fig. 15) to position A, shown on the diagram. Take a deep breath and then exhale as much air as possible down the tubing into the bottle. The final water level inside the bottle will tell you how much air you can exchange in one deep breath.

Now push the rubber tubing further into the bottle, to position B (Fig. 15), and blow out any water left in the tube. Support the bottle with your hand and breathe quietly in and out through the tube, keeping the water level inside and outside the bottle the same. This will give you an idea of how much air you exchange when breathing normally.

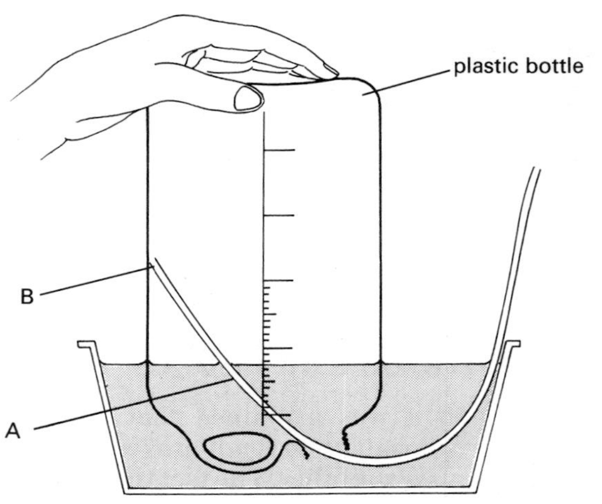

Fig. 15 Measuring the volume of air exhaled from the lungs. 'A' shows the position of the tube when measuring the maximum usable lung volume. 'B' is the position for measuring the volume exchanged in quiet breathing.

CHECK LIST

- **Ventilation is inhaling and exhaling air.**
- **The ribs, rib muscles and diaphragm make the lungs expand and contract. This causes inhaling and exhaling.**
- **Air is drawn into the lungs through the trachea, bronchi and bronchioles.**
- **The vast number of air pockets (alveoli) give the lungs an enormous internal surface area. This surface is moist and lined with capillaries.**
- **The blood in the capillaries picks up oxygen from the air in the alveoli and gives out carbon dioxide. This is called gaseous exchange.**
- **Ventilation exchanges the air in the air passages but not in the alveoli.**
- **Exchange of oxygen and carbon dioxide in the alveoli takes place by diffusion.**
- **The oxygen is carried round the body by the blood and used by the cells for their respiration.**
- **During exercise, the rate and depth of breathing increase. This supplies extra oxygen to the muscles and removes their excess carbon dioxide.**
- **Tobacco smoke causes the bronchioles to constrict, the cilia in their lining to stop beating and excessive mucus to be produced.**
- **Smoking is correlated with heart disease, bronchitis, emphysema and lung cancer.**

16 Excretion and the Kidneys

EXCRETION

Definition.

EXCRETORY ORGANS

Lungs, kidneys, liver.

KIDNEYS

Structure. Function. Selective reabsorption.

OSMO-REGULATION

Controlling the blood concentration.

THE DIALYSIS MACHINE

HOMEOSTASIS

The stability of the internal environment.

A great number of chemical reactions take place inside the cells of an organism in order to keep it alive. The products of some of these reactions are poisonous and must be removed from the body. For example, the breakdown of glucose during respiration (p. 24) produces carbon dioxide. This is carried away by the blood and removed in the lungs. Excess amino acids are de-aminated in the liver to form glycogen and **urea**, as explained on page 133. The urea is removed from the tissues by the blood, and expelled by the kidneys.

Urea and similar waste products, like **uric acid**, from the breakdown of proteins, contain the element nitrogen. For this reason they are often called **nitrogenous waste products**.

During feeding, more water and salts are taken in with the food than are needed by the body. So these excess substances need to be removed as fast as they build up.

The hormones produced by the endocrine glands (p. 212) affect the rate at which various body systems work. Adrenaline, for example, speeds up the heart beat. When hormones have done their job, they are modified in the liver and excreted by the kidneys.

The nitrogenous waste products, excess salts and spent hormones are excreted by the kidneys as a watery solution called **urine**.

Excretion is the name given to the removal from the body of

1. the waste products of its chemical reactions,
2. the excess water and salts taken in with the diet, and
3. spent hormones.

Excretion also includes the removal of drugs or other foreign substances taken into the alimentary canal and absorbed by the blood. The term 'excretion' should not usually be applied to the passing out of faeces (p. 131), because most of the contents of the faeces, apart from the bile pigments, have not taken part in reactions in the cells of the body.

QUESTIONS

1 Write a list of the substances that are likely to be excreted from the body during the day.
2 Why do you think that urine analysis is an important part of medical diagnosis?

EXCRETORY ORGANS

Lungs The lungs supply the body with oxygen, but they are also excretory organs because they get rid of carbon dioxide. They also lose a great deal of water vapour, but this loss is unavoidable and is not a method of controlling the water content of the body.

Kidneys The kidneys remove urea and other nitrogenous waste from the blood. They also expel excess water, salts, hormones (p. 212) and drugs.

Liver The yellow/green bile pigment, bilirubin, is a breakdown product of haemoglobin (see p. 133). Bilirubin is excreted with the bile into the small intestine and expelled with the faeces. The pigment undergoes changes in the intestine and is largely responsible for the brown colour of the faeces.

Skin Sweat consists of water, with sodium chloride and

traces of urea dissolved in it. When you sweat, you will expel these substances from your body and so, in one sense, they are being excreted. However, sweating is a response to a rise in temperature and not to a change in the blood composition. In this sense, therefore, skin is not an excretory organ like the lungs and kidneys.

THE KIDNEYS

Structure

The two kidneys are fairly solid, oval structures. They are red-brown, enclosed in a transparent membrane and attached to the back of the abdominal cavity (Fig. 1). The **renal artery** branches off from the aorta and brings oxygenated blood to them. The **renal vein** takes deoxygenated blood away from the kidneys to the vena cava (see Fig. 10, p. 141). A tube, called the **ureter**, runs from each kidney to the bladder in the lower part of the abdomen.

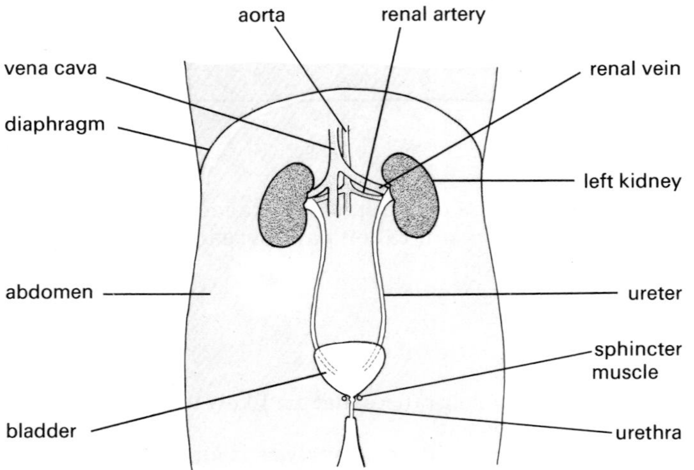

Fig. 1 Position of the kidneys in the body

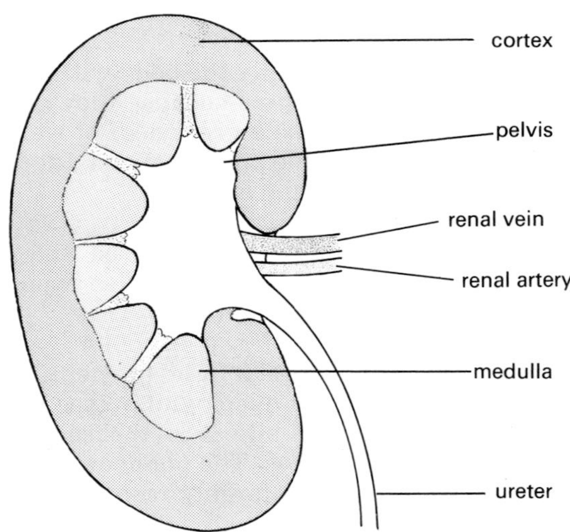

Fig. 2 Section through the kidney to show regions

The kidney tissue consists of many capillaries and tiny tubes, called **renal tubules**, held together with connective tissue. If the kidney is cut down its length (sectioned), it is seen to have a dark, outer region called the **cortex** and a lighter, inner zone, the **medulla**. Where the ureter joins the kidney there is a space called the **pelvis** (Fig. 2).

The renal artery divides up into a great many arterioles and capillaries, mostly in the cortex (Fig. 3). Each arteriole leads to a **glomerulus**. This is a capillary repeatedly divided and coiled, making a knot of vessels (Fig. 4). Each glomerulus is almost entirely surrounded by a cup-shaped organ called a **renal capsule**, which leads to a coiled renal tubule. This tubule, after a series of coils and loops, joins a **collecting duct** which passes through the medulla to open into the pelvis (Fig. 5). There are thousands of glomeruli in the kidney cortex and the total surface area of their capillaries is very great.

A **nephron** is a single glomerulus with its renal capsule, renal tubule and blood capillaries (see Fig. 6).

Function of the kidneys

The blood pressure in a glomerulus causes part of the blood plasma to leak through the capillary walls. The red blood cells and the plasma proteins are too big to pass out of the capillary, so the fluid that does filter through is plasma without the protein, i.e. similar to tissue fluid (see p. 142). The fluid thus consists mainly of water with dissolved salts, glucose, urea and uric acid. The process by which the fluid is filtered out of the blood by the glomerulus is called **ultra-filtration**.

The filtrate from the glomerulus collects in the renal capsule and trickles down the renal tubule (Fig. 6). As it does so, the capillaries which surround the tubule absorb back into the blood those substances which the body needs. First, all the glucose is reabsorbed, with much of the water. Then some of the salts are taken back to keep the correct concentration in the blood. The process of absorbing back the substances needed by the body is called **selective reabsorption**.

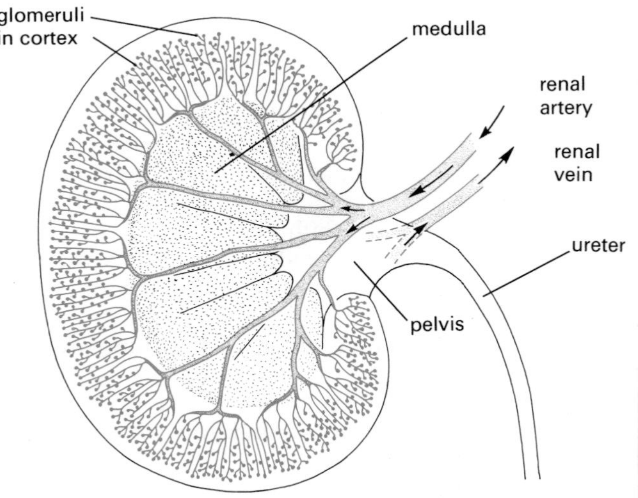

Fig. 3 Section through kidney to show distribution of glomeruli

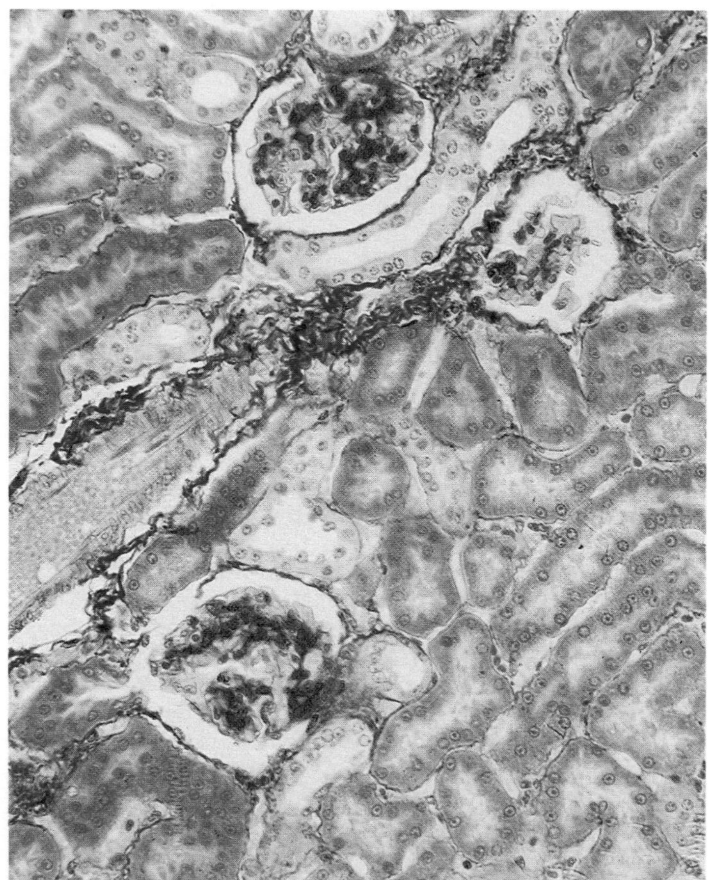
Fig 4 Glomeruli in the kidney cortex (× 300). The three glomeruli are surrounded by kidney tubules sectioned at different angles. The light space round each glomerulus represents the renal capsule.

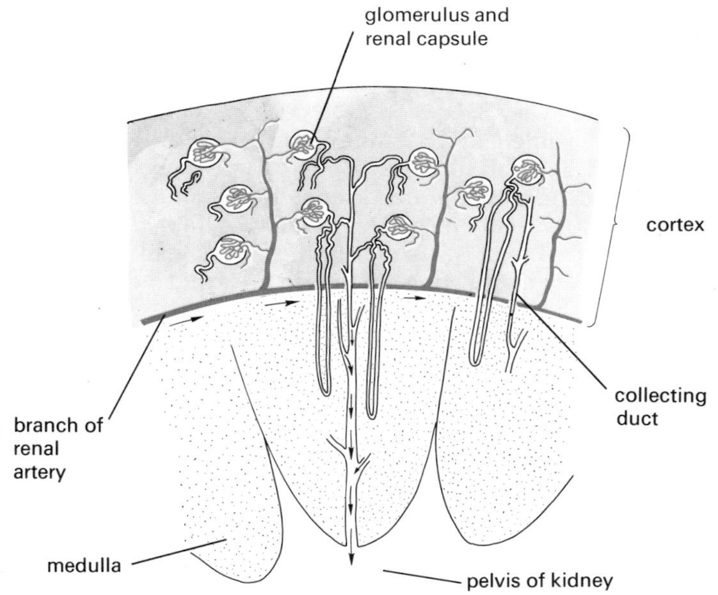

Fig. 5 Section through cortex and medulla

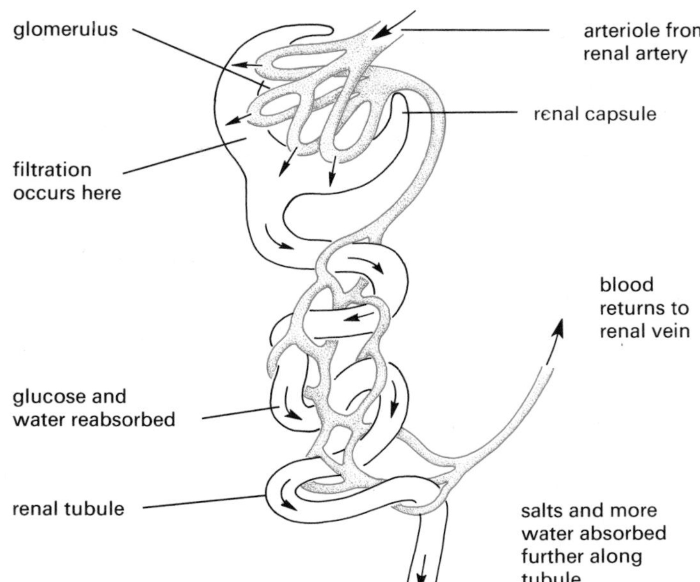

Fig. 6 Part of a nephron (glomerulus, renal capsule and renal tubule)

Salts not needed by the body are left to pass on down the kidney tubule together with the urea and uric acid. So, these nitrogenous waste products, excess salts and water continue down the renal tube into the pelvis of the kidney. From here the fluid, now called **urine**, passes down the ureter to the bladder.

The following table shows some of the differences in composition between the blood plasma and the urine. The figures represent average values because the composition of the urine varies a great deal according to the diet, activity, temperature and intake of liquid.

Table 1 Composition of blood plasma and urine

	Plasma %	*Urine %*
Water	90–93	95
Urea	0·03	2
Uric acid	0·003	0·05
Ammonia	0·0001	0·05
Sodium	0·3	0·6
Potassium	0·02	0·15
Chloride	0·37	0·6
Phosphate	0·003	0·12

The **bladder** can expand to hold about 400 cm^3 urine. The urine cannot escape from the bladder because a band of circular muscle, called a sphincter, is contracted, so shutting off the exit. When this sphincter muscle relaxes, the muscular walls of the bladder expel the urine through the **urethra**. Adults can control this sphincter muscle and relax it only when they want to urinate. In babies, the sphincter relaxes by a reflex action (p. 207), set off by pressure in the bladder. By 3 years old, most children can control the sphincter voluntarily.

Water balance and osmo-regulation

Your body gains water from food and drink. It loses water by evaporation, urination and defecation (p. 131). Evaporation from the skin takes place all the time but is particularly rapid when we sweat. Air from the lungs is saturated with water vapour which is lost to the atmosphere every time we exhale. Despite these gains and losses of water, the concentration of body fluids is kept within very narrow limits by the kidneys, which adjust the concentration of the blood flowing through them.

If the blood is too dilute (i.e. contains too much water), less water is reabsorbed from the renal tubules, leaving more to enter the bladder. Thus, after drinking a lot, a large volume of dilute urine is produced.

If the blood is too concentrated, more water is absorbed back into the blood from the kidney tubules. So, if the body is short of water, e.g. after sweating profusely, only a small quantity of concentrated urine is produced.

A rise in the blood concentration is thought to stimulate a 'thirst' centre in the brain. The drinking which follows this stimulation restores the blood to its correct concentration.

These regulatory processes keep the blood at a steady concentration and together are called **osmo-regulation** because they regulate the osmotic strength (see p. 36) of the blood. Osmo-regulation is one example of the process of **homeostasis** which is described on page 163.

Changes in the concentration of the blood are detected by an area in the brain called the **hypothalamus**. If the blood passing through the brain is too concentrated, the hypothalamus stimulates the pituitary gland beneath it to secrete into the blood a hormone (p. 212) called anti-diuretic hormone (ADH). When this hormone reaches the kidneys, it causes the kidney tubules to absorb more water from the glomerular filtrate back into the blood. Thus the urine becomes more concentrated and the further loss of water from the blood is reduced. If blood passing through the hypothalamus is too dilute, production of ADH from the pituitary is suppressed and less water is absorbed from the glomerular filtrate.

QUESTIONS

3 Why should a fall in blood pressure sometimes lead to kidney failure?

4 In what ways would you expect the composition of blood in the renal vein to differ from that in the renal artery? (Remember that the cells in the kidney will be respiring.)

5 Where, in the urinary system, do the following take place (answer as precisely as possible): filtration, reabsorption, storage of urine, transport of urine, osmo-regulation?

6 In hot weather, when you sweat a great deal, you urinate less often and the urine is a dark colour. In cold weather, when you sweat little, urination occurs more often and the urine is pale in colour. Use your knowledge of kidney function to explain these observations.

7 Trace the path taken by a molecule of urea from the time it is produced in the liver, to the time it leaves the body in the urine (see also p. 141).

The dialysis machine ('artificial kidney')

Kidney failure may result from an accident involving a drop in blood pressure, or from a disease of the kidneys. In the former case, recovery is usually spontaneous, but if it takes longer than 2 weeks, the patient may die as a result of a potassium imbalance in the blood, which causes heart failure. In the case of kidney disease, the patient can survive with only one kidney, but if both fail the patient's blood composition has to be regulated by a dialysis machine. Similarly, the accident victim can be kept alive on a dialysis machine until his or her blood pressure is restored.

In principle, a dialysis machine consists of a long cellulose tube coiled up in a water bath. The patient's blood is led from a tube in the radial artery and pumped through the cellulose (dialysis) tubing (Figs 7 and 8). The sub-microscopic pores in the dialysis tubing allow small molecules, such as those of salts, glucose and urea to leak out into the water bath. Blood cells and protein molecules are too large to get through the pores (see Experiment 3, p. 39). This stage is similar to the filtration process in the glomerulus.

To prevent a loss of glucose and essential salts from the blood, the liquid in the water bath consists of a solution of salts and sugar of the correct composition, so that only the substances above this concentration can diffuse out of the blood into the bathing solution. Thus, urea, uric acid and excess salts are removed.

The bathing solution is also kept at body temperature and is constantly changed as the unwanted blood solutes accumulate in it. The blood is then returned to the patient through a vein in the arm.

A patient with total kidney failure has to spend 2 or 3 nights each week connected to the machine (Fig. 8). With this treatment and a carefully controlled diet, the patient can lead a fairly normal life. A kidney transplant, however, is a better solution because the patient is not obliged to return to the dialysis machine every 3 days or so.

The problem with kidney transplants is to find enough suitable donors of healthy kidneys and to prevent the transplanted kidney from being rejected.

The donor may be a close relative who is prepared to donate one of his or her kidneys (you can survive adequately with one kidney). Alternatively, the donated kidney may be taken from a healthy person who dies, for example, as a result of a road accident. Unless the accident victim is carrying a kidney donor card which gives permission for his or her kidneys to be used, the relatives must give their permission.

The problem with rejection is that the body reacts to any transplanted cells or tissues as it does to all foreign proteins and produces lymphocytes which attack and destroy them (see p. 147). This rejection can be overcome by (1) choosing a donor whose tissues are as similar as possible to those of the patient, e.g. a close relative, and (2) using immuno-

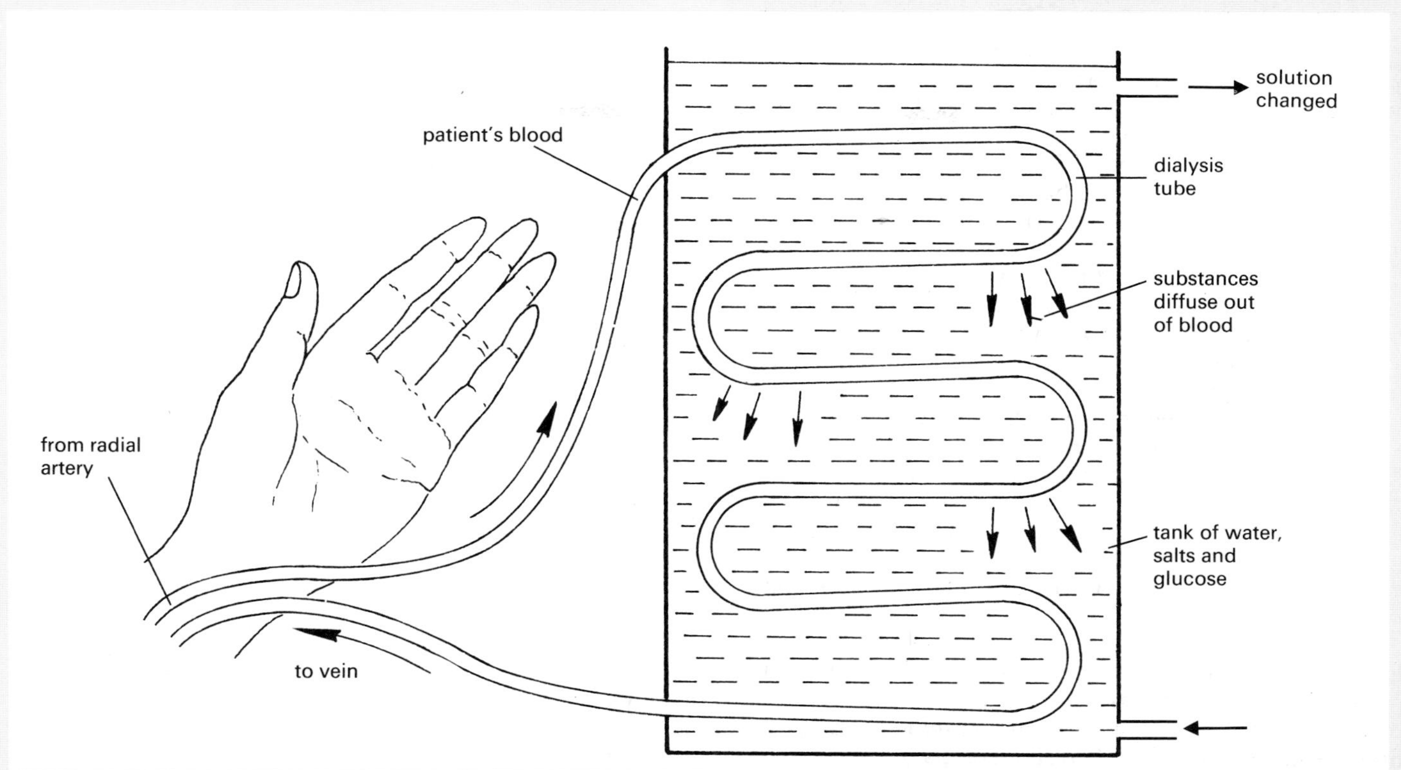

Fig. 7 The principle of the kidney dialysis machine

suppressive drugs which suppress the production of lymphocytes and their antibodies against the transplanted organ.

HOMEOSTASIS

Homeostasis means 'staying the same'. It refers to the fact that the composition of the tissue fluid (p. 142) in the body is kept very steady. Its concentration, acidity and temperature are being adjusted all the time to prevent any big changes.

On page 17 it was explained that in living cells, all the chemical reactions are controlled by enzymes. The enzymes are very sensitive to the conditions in which they work. A slight fall in temperature or a rise in acidity (p. 19) may slow down or stop an enzyme from working and thus prevent an important reaction from taking place in the cell.

The cell membrane controls the substances which enter and leave the cell, but it is the tissue fluid which supplies or removes these substances, and it is therefore important to keep the composition of the tissue fluid as steady as possible. If the tissue fluid were too concentrated, it would withdraw water from the cells by osmosis (p. 36) and the body would be dehydrated. If the tissue fluid were too dilute, the cells would take up too much water from it by osmosis and the tissues would become waterlogged and swollen.

Many systems in the body contribute to homeostasis (Fig. 9). The obvious example is the kidneys, which remove substances that might poison the enzymes. The kidneys also control the level of salts, water and acids in the blood. The composition of the blood affects the tissue fluid which, in turn, affects the cells.

Another example of a homeostatic organ is the liver, which regulates the level of glucose in the blood (p. 132).

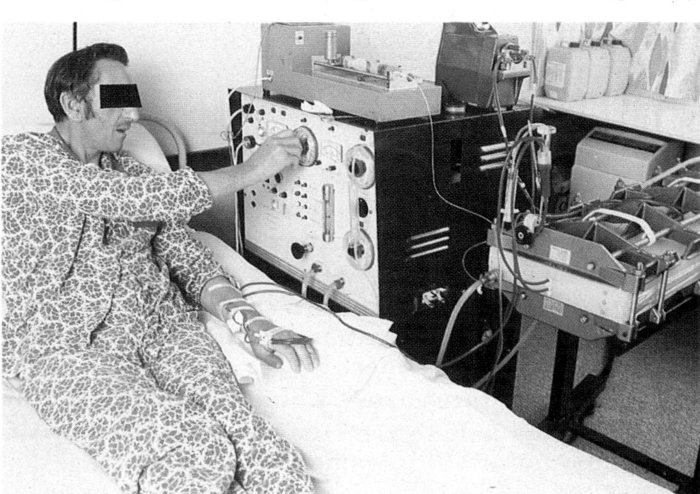

Fig. 8 Kidney dialysis machine. The patient's blood is sent by the pump (top right) to the dialyser (bottom right). The patient is adjusting the control box which regulates the temperature and concentration of the dialysing liquid.

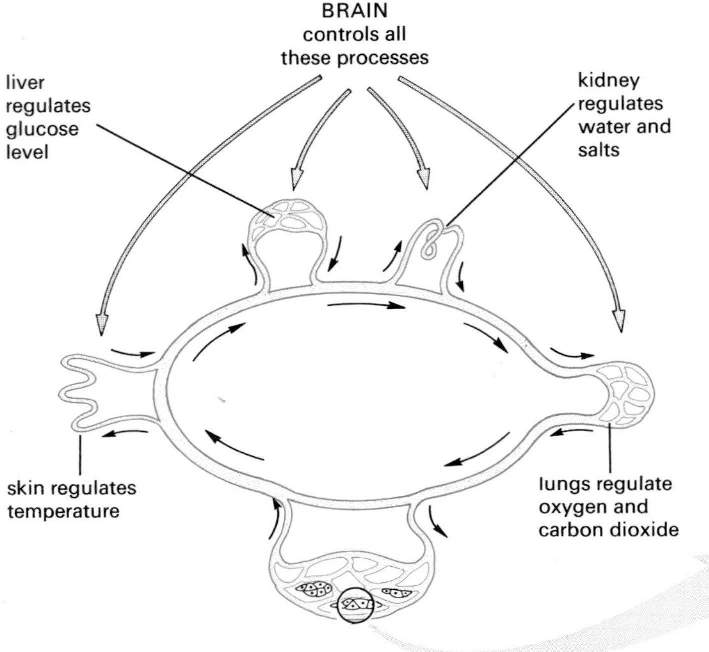

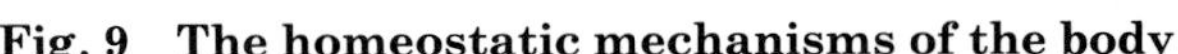

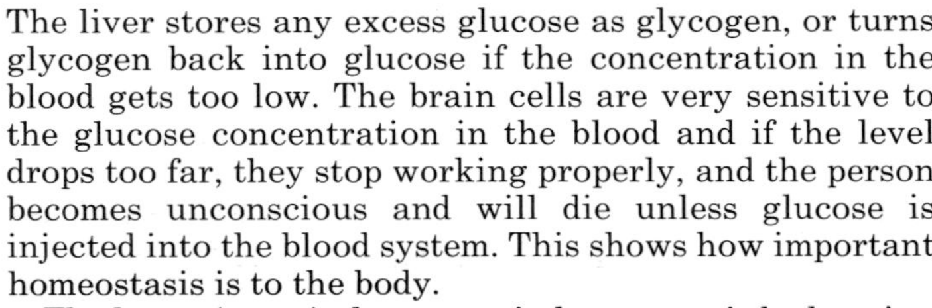

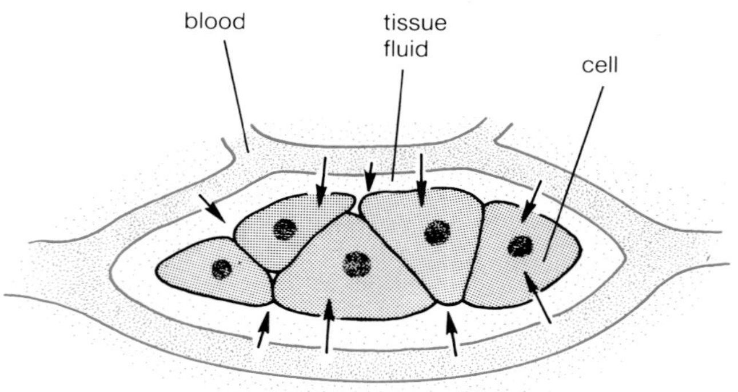

This tissue fluid, with its carefully controlled composition, provides the best conditions for the cell's enzymes to work in

Fig. 9 The homeostatic mechanisms of the body

The liver stores any excess glucose as glycogen, or turns glycogen back into glucose if the concentration in the blood gets too low. The brain cells are very sensitive to the glucose concentration in the blood and if the level drops too far, they stop working properly, and the person becomes unconscious and will die unless glucose is injected into the blood system. This shows how important homeostasis is to the body.

The lungs (p. 151) play a part in homeostasis by keeping the concentrations of oxygen and carbon dioxide in the blood at the best level for the cell's chemical reactions, especially respiration.

The next chapter describes the way in which the skin regulates the temperature of the blood. If cells were to get too cold, the chemical reactions would become too slow to maintain life. If they became too hot, the enzymes would be destroyed.

The brain has over-all control of the homeostatic processes in the body. It checks the composition of the blood flowing through it and if it is too warm, too cold, too concentrated or has too little glucose, nerve impulses or hormones are sent to the organs concerned, causing them to make the necessary adjustments.

QUESTION

8 Where will the brain send nerve impulses or hormones if the blood flowing through it (*a*) has too much water, (*b*) contains too little glucose, (*c*) is too warm, (*d*) has too much carbon dioxide?

CHECK LIST

- **Excretion is getting rid of unwanted substances from the body.**
- **The lungs excrete carbon dioxide.**
- **The kidneys excrete urea, unwanted salts and excess water.**
- **Part of the blood plasma entering the kidneys is filtered out by the capillaries. Substances which the body needs, like glucose, are absorbed back into the blood. The unwanted substances are left to pass down the ureters into the bladder.**
- **The bladder stores urine, which is discharged at intervals.**
- **The kidneys help to keep the blood at a steady concentration by excreting excess salts and by adjusting the amount of water (osmo-regulation).**
- **The kidneys, lungs, liver and skin all help to keep the blood composition the same (homeostasis).**

17 The Skin, and Temperature Control

FUNCTIONS OF THE SKIN
Protection, sensitivity, temperature control.

STRUCTURE OF THE SKIN
Epidermis, dermis, hair, sweat glands.

HEAT BALANCE
Gain and loss of heat.

TEMPERATURE CONTROL
Vaso-dilation, vaso-constriction, sweating.

HYPOTHERMIA

The skin forms a continuous layer over the entire body. It makes it difficult for harmful substances, bacteria and fungi to get into the body. It reduces the loss of water from the body and it helps to regulate the body temperature.

FUNCTIONS OF THE SKIN

Protection

A brown or black pigment in the skin absorbs the harmful ultra-violet rays from sunlight. Skin containing only a little pigment may be damaged by these rays. This happens when white-skinned people become sunburned. Many white-skinned people, however, produce extra pigment and so acquire a 'sun-tan' which helps to protect the skin from the effects of ultra-violet light.

The layer of dead cells at the surface of the skin (1) stops harmful bacteria getting into the living tissues beneath and (2) greatly reduces the evaporation of water from the body, so helping to maintain the composition of the body fluids.

Sensitivity

Scattered through the skin are a large number of tiny sense organs which give rise to sensations of touch, pressure, heat, cold and pain. These make us aware of changes in our surroundings and enable us to take action to avoid damage, to recognize objects by touch and to manipulate objects with our hands (see p. 193).

Temperature regulation

The way in which the skin helps to keep the body temperature constant is described on pages 168–9.

QUESTION

1 To what dangers is the body exposed if (*a*) a small area of skin is damaged, (*b*) a large area of skin is damaged?

STRUCTURE OF THE SKIN

There are two main layers in the skin, an outer **epidermis** and an inner **dermis**. The thickness of these two layers depends on which part of the body they are covering. The skin on the palms of the hands and soles of the feet has a very thick epidermis and no hairs (Fig. 2). Over the rest of the body, the epidermis is thinner and has hairs (Fig. 3). The diagram of a section through the skin (Fig. 1) is a generalized one. It shows all the structures that are in skin as a whole, even though some may be absent in a particular area.

Epidermis

Malpighian layer This is the innermost layer of cells in the epidermis. These cells contain the pigment that gives the skin its colour and helps to absorb ultra-violet light from the sun. The cells of the Malpighian layer keep

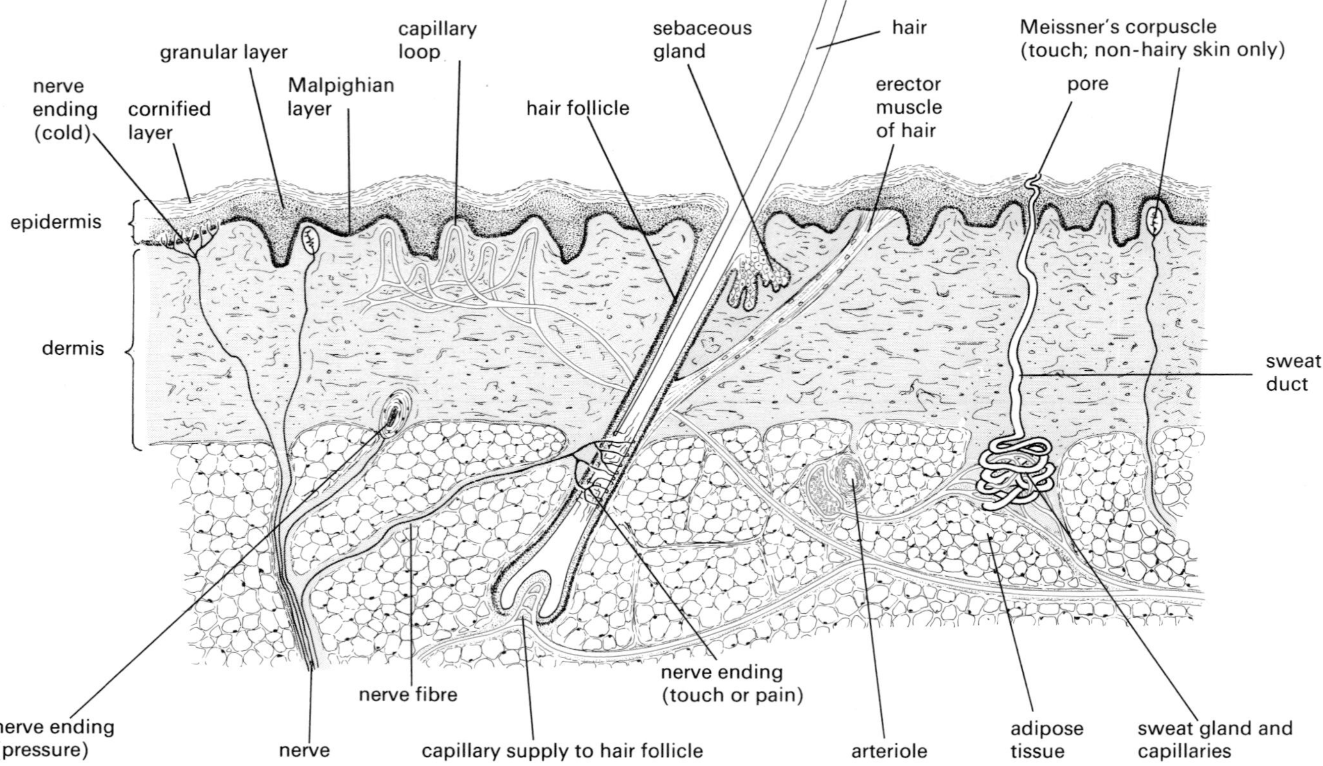

Fig. 1 Generalized section through the skin

dividing and producing new cells which are pushed towards the outside of the skin, forming the granular layer.

Granular layer The cells produced by the Malpighian layer move through the granular layer and then die, so forming the cornified layer.

Cornified layer This consists of dead cells. It is the outermost layer of the epidermis which helps to cut down evaporation and keep bacteria out. The cells are constantly worn away and replaced from below by the Malpighian and granular layers (Fig. 4).

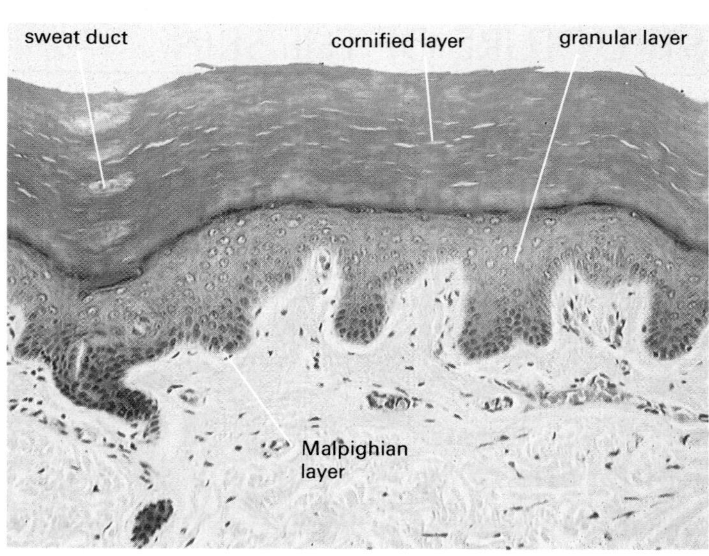

Fig. 2 Section through non-hairy skin (×300). The sweat ducts are contorted as they pass through the cornified layer.

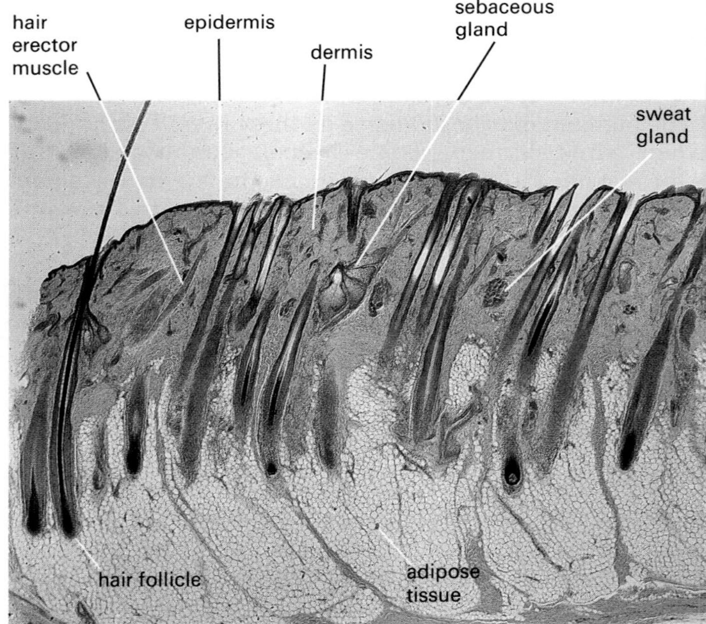

Fig. 3 Section through hairy skin (×20)

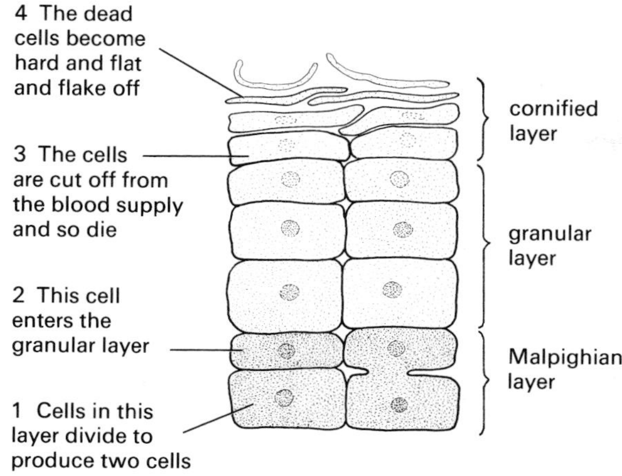
4 The dead cells become hard and flat and flake off
3 The cells are cut off from the blood supply and so die
2 This cell enters the granular layer
1 Cells in this layer divide to produce two cells
cornified layer
granular layer
Malpighian layer

Fig. 4 Growth of the epidermis

Dermis

The dermis is a layer of connective tissue containing capillaries, sensory nerve endings, lymphatics, sweat glands and hair follicles.

Capillaries These bring oxygen and food to the skin and remove its carbon dioxide and nitrogenous waste. They supply the hair follicles and sweat glands. The capillary loops close to the surface, i.e. just beneath the epidermis, play an important part in regulating heat loss from the body (see p. 168).

Sweat glands A sweat gland is a coiled tube deep in the dermis. When the body temperature is too high, the sweat gland takes up water from the capillaries around it. The water collects in the gland, travels up the **sweat duct**, comes out of a **pore** in the epidermis and on to the skin surface. When the sweat evaporates, it takes heat from the body and so cools it down. Unless the sweat evaporates, it has no cooling effect.

Sweat is mainly water but there are some dissolved salts and urea in it. Some people regard this as a form of excretion (p. 159), but sweating occurs in response to a rise in temperature and not because there is too much water or salt in the body.

Hair follicles A hair follicle is a deep pit lined with granular and Malpighian cells. The Malpighian cells keep dividing and adding cells to the base of the hair, making it grow. A hair is a lot of cornified cells formed into a tube. The hair follicle has nerve endings which respond when the hair is touched, or give a sensation of pain if the hair is pulled.

In furry mammals, the hairs trap a layer of air close to the body. Air is a bad conductor of heat and so this layer of air insulates the body against heat loss. When the **hair erector muscle** contracts, it pulls the hair more upright. In mammals, this makes the fur stand up more and so provides a thicker layer of insulating air in cold weather. Most of our body is covered with short hairs, but they trap very little air and when the hair erector muscles contract, they produce only 'goose pimples'.

Sebaceous glands The sebaceous glands open into the top of the hair follicles and produce an oily substance that keeps the epidermis waterproof and stops it drying out.

Sensory nerve endings These are described more fully on page 193.

Fat layer The fat stored in the adipose tissue beneath the skin not only provides a store of food, but also forms an insulating layer and reduces the heat lost from the body.

QUESTIONS

2 Is a hair made by the epidermis or the dermis?
3 Why do you think dead cells are better than living cells in reducing water loss from the skin?
4 Describe the path taken by a water molecule that enters the skin in the blood plasma of a capillary and ends up in sweat on the surface of the skin.

HEAT BALANCE

Body temperature

Different parts of the body are at different temperatures. The skin temperature is lower than the liver temperature and the feet and hands may be colder than the abdomen. However, 'body temperature' usually means the temperature deep inside the body. It is not possible to put a thermometer far inside the body so it is usually placed in the mouth, under the tongue, and held there for 2 minutes with the mouth closed.

The body temperature varies during the day and there is no single 'correct' body temperature. Variations in the range of 35.8–37.7 °C (96.4–99.8 °F) are normal. Body temperatures below 34 °C (93 °F) and above 40 °C (104 °F), if maintained for long, are considered to be dangerous (see p. 169).

The body may gain or lose heat as a result of internal changes or external influences as explained below.

Heat gain

Internal Many of the chemical reactions in cells release heat. The chief heat producers are the abdominal organs, the brain and contracting muscles. An increase in the muscular activity of the body will result in more heat being produced.

External Direct heat from the sun will be absorbed by the body. If the air temperature is above 37 °C, the body will absorb heat. Hot food and drink also add heat to the body.

Heat loss

Heat is lost to the air from the exposed surfaces of the body by conduction, convection and radiation. Evaporation from the skin and lungs takes place all the time and is a cause of heat loss. The cold air breathed into the lungs and cold food or drink taken into the stomach all absorb heat from the body.

To a large extent, the heat lost from the body is balanced by the heat absorbed or produced. However, changes in the temperature of the surroundings or in the rate of activity by the animal may upset this balance. In humans, any change in the temperature balance is regulated mainly by changes in the skin.

We also control our heat loss and gain by taking conscious action. We remove clothing or move into the shade to cool down, or put on more clothes or take exercise to keep warm.

QUESTIONS

5 What conscious actions do we take to reduce the heat lost from the body?

6 What sort of chemical reaction in the liver and in active muscle will produce heat? How does this heat get to other parts of the body? (See pp. 24, 133 and 141.)

7 Draw up a balance sheet to show all the possible ways the human body can gain or lose heat. Make two columns, with 'Gains' on the left and 'Losses' on the right.

TEMPERATURE CONTROL

Skin structure

The structures in the skin which play a part in temperature control are the small blood vessels, the sweat glands and, in furry mammals, the hairs and hair muscles.

The arterioles and capillaries near the surface of the skin can increase or decrease in width and so increase or reduce the amount of blood flowing through them. An increase in width of the vessel is called **vaso-dilation**; a reduction in width is **vaso-constriction**.

The sweat glands are not active when the loss and gain of heat are balanced, but as soon as the body temperature starts to rise, the sweat glands of the trunk, limbs and face produce sweat as described above. Sweating from the hands, feet and armpits is a response to emotional stress rather than overheating.

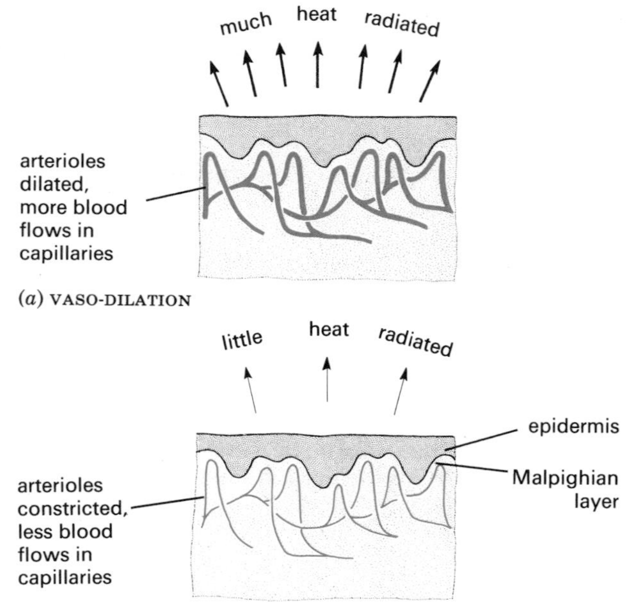

Fig. 5 Vaso-dilation and vaso-constriction

Overheating

If the body gains or produces heat faster than it is losing it, the following processes occur:

Vaso-dilation The widening of the blood vessels in the dermis allows more warm blood to flow near the surface of the skin and so lose more heat (Fig. 5*a*).

Sweating The sweat glands pour sweat on to the skin surface. When this layer of liquid evaporates it takes heat (latent heat) from the body and so cools it down (Fig. 6).

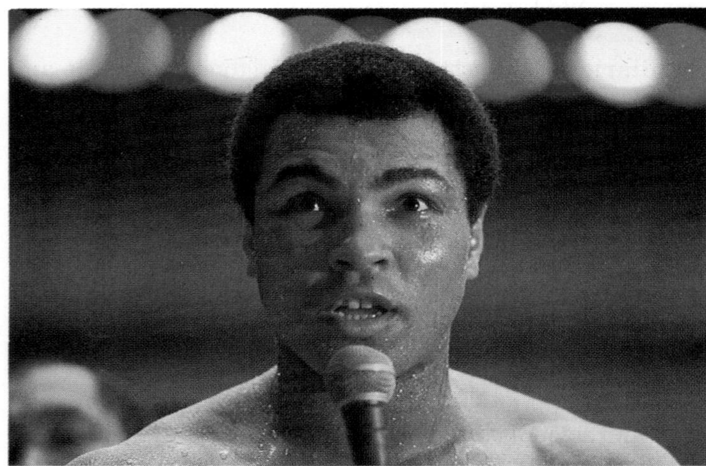

Fig. 6 Sweating. During vigorous activity the sweat evaporates from the skin and helps to cool the body. When the activity stops, continued evaporation of sweat may over-cool the body unless it is towelled off.

Overcooling

If the body begins to lose heat faster than it can produce it, the following changes occur:

Sweat production stops Thus the heat lost by evaporation is reduced.

Vaso-constriction Vaso-constriction of the blood vessels in the skin reduces the amount of warm blood flowing near the surface (Fig. 5*b*).

Shivering Uncontrollable bursts of rapid muscular contraction in the limbs release heat as a result of the chemical changes in the muscles.

As a result of these processes, the body temperature of an adult does not usually vary by more than 1 °C either side of 37 °C even when the external temperature changes. The average temperature recorded by a thermometer under the tongue is 36.7 °C and is called the 'core' temperature of the body. The temperature of the skin can change through very wide limits without affecting the core temperature. It is the sensory nerve endings in your dermis which make you feel hot or cold. You cannot consciously detect the small changes in your core temperature.

Hypothermia

If the body loses heat faster than the tissues can produce it, the core temperature may fall. Below 35 °C there is an impairment of normal function (e.g. slurred speech, unreasonable behaviour, defective vision) and below 32 °C there is loss of consciousness. This lowering of core temperature is called **hypothermia** and unless the body is rewarmed in a controlled manner, it can result in death.

In young people, hypothermia may occur in conditions such as immersion in water or exposure in wet clothing. In elderly people, hypothermia may result simply from inactivity in inadequately heated surroundings, and from insufficient intake of food.

Prevention of hypothermia The chance of suffering hypothermia can be reduced by eating a good meal before taking part in outdoor pursuits such as fell-walking, climbing or sailing. The food provides the source of energy for heat production.

Warm clothing helps to reduce heat losses and, in particular, should include an outer layer which is wind-proof and rain-proof.

Hypothermia in the elderly can be avoided if they are visited regularly to ensure that they are eating well and keeping their rooms at a suitable temperature (above 20 °C).

QUESTIONS

8 (*a*) Which structures in the skin of a furry mammal help to reduce heat loss?

(*b*) What changes take place in the skin of man to reduce heat loss?

9 If your body temperature hardly changes at all, why do you sometimes feel hot and sometimes cold?

10 Sweating cools you down only if the sweat can evaporate.

(*a*) In what conditions might the sweat be unable to evaporate from your skin? (See p. 73.)

(*b*) What conditions might speed up the evaporation of sweat and so make you feel very cold?

CHECK LIST

- **Skin consists of an outer layer of epidermis and an inner dermis.**
- **The epidermis is growing all the time and has an outer layer of dead cells.**
- **The dermis contains the sweat glands, hair follicles, sense organs and capillaries.**
- **Skin (1) protects the body from bacteria and drying out, (2) contains sense organs which give us the sense of touch, warmth, cold and pain, and (3) controls the body temperature.**
- **Chemical activity in the body, and muscular contractions produce heat.**
- **Heat is lost to the surroundings by conduction, convection, radiation and evaporation.**
- **If the body temperature rises too much, the skin cools it down by sweating and vaso-dilation.**
- **If the body loses too much heat, vaso-constriction and shivering help to keep it warm.**

18 Reproduction

REPRODUCTIVE SYSTEMS

Male and female organs. Gamete production.

FERTILIZATION AND DEVELOPMENT

Mating and fertilization. Development of embryo. Placenta. Twins.

BIRTH AND PARENTAL CARE

Normal labour and delivery. Surgical intervention. Breast-feeding.

PUBERTY AND THE MENSTRUAL CYCLE

Hormones and secondary sexual characteristics. Menstruation and menopause.

FAMILY PLANNING AND FERTILITY

Birth control. In-vitro fertilization.

WORLD POPULATION

Reproduction is the process of producing new individuals. Some single-celled creatures can reproduce by simply dividing into two. Some many-celled animals can produce offspring by a process of 'budding', in which part of their body breaks away and grows into a new individual. These are methods of asexual reproduction (see p. 350).

Most animals reproduce sexually. The two sexes, male and female, each produce special types of reproductive cells, called **gametes**. The male gametes are the **sperms** (or **spermatozoa**) and the female gametes are the **ova** (singular = ovum) or eggs (Fig. 1).

To produce a new individual, a sperm has to reach an ovum and join with it (**fuse** with it). The sperm nucleus then passes into the ovum and the two nuclei also fuse. This is called **fertilization**.

The cell formed after the fertilization of an ovum by a sperm is called a **zygote**. A zygote will grow by cell

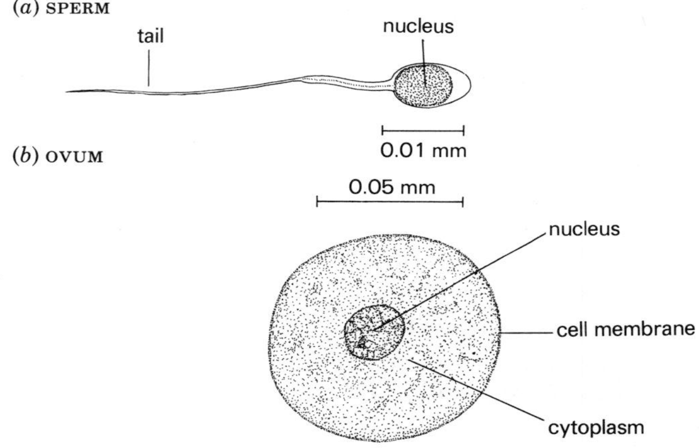

Fig. 1 Human gametes

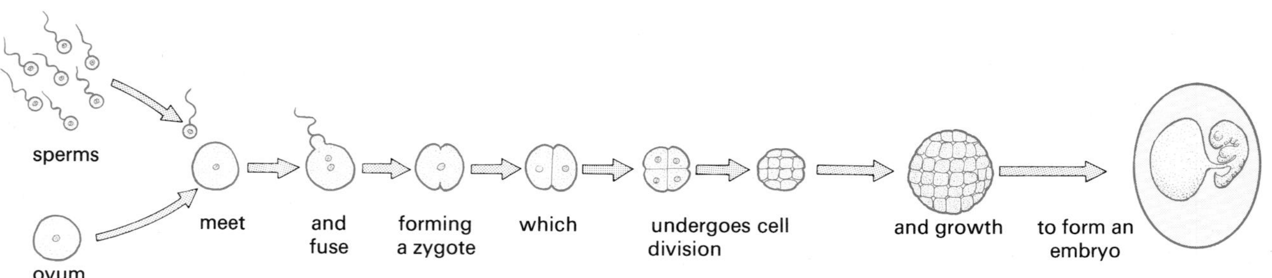

Fig. 2 Fertilization and development

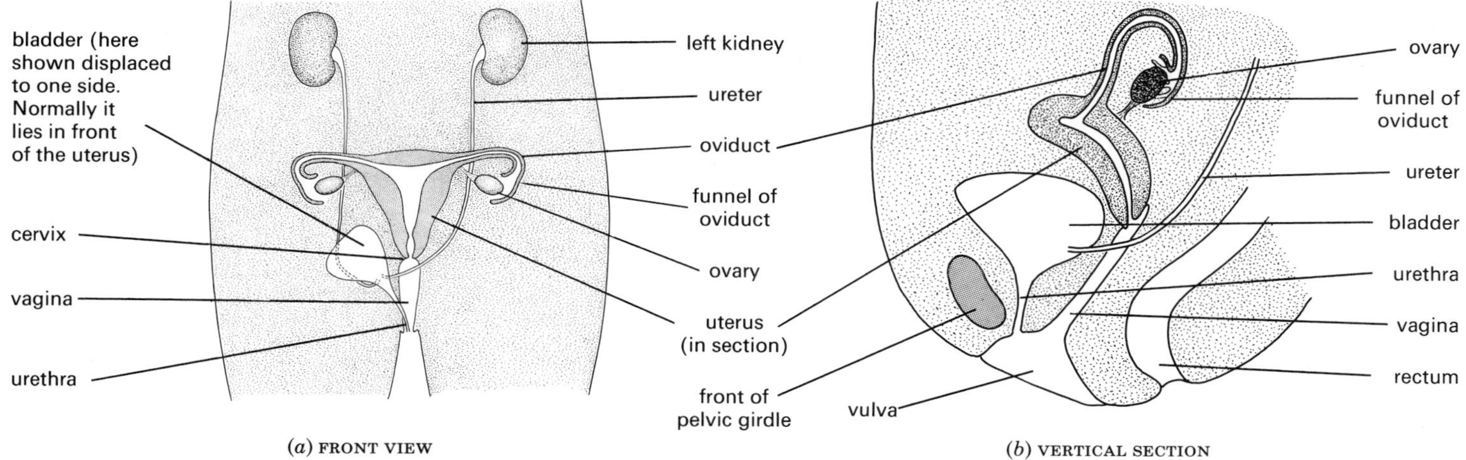

Fig. 3 The female reproductive organs

division to produce first an **embryo** and then a fully formed animal (Fig. 2).

The male animal always produces a large number (millions) of sperms, while the female produces a smaller number of eggs. In some animals, such as fish and frogs, many eggs are fertilized and there are a large number of offspring. In mammals, a small number of eggs is fertilized, from one to twenty. In humans, usually only one egg is fertilized at a time; two eggs being fertilized produces twins.

To bring the sperms close enough to the ova for fertilization to take place, there is an act of mating or **copulation**. In mammals this act results in sperms from the male animal being injected into the female. The sperms swim inside the female's reproductive system and fertilize any eggs which are present. The zygote then grows into an embryo inside the body of the female.

THE HUMAN REPRODUCTIVE SYSTEM

Female

The eggs are produced from the female reproductive organs called **ovaries**. These are two whitish oval bodies, 3–4 cm long. They lie in the lower half of the abdomen, one on each side of the **uterus** (Fig. 3*a* and *b*). Close to each ovary is the expanded, funnel-shaped opening of the **oviduct**, the tube down which the ova pass when released from the ovary. The oviduct is sometimes called the **Fallopian tube**.

The oviducts are narrow tubes that open into a wider tube, the uterus or womb, lower down in the abdomen. When there is no embryo developing in it, the uterus is only about 80 mm long. It leads to the outside through a muscular tube, the **vagina**. The **cervix** is a ring of muscle closing the lower end of the uterus where it joins the vagina. The urethra, from the bladder, opens into the **vulva** just in front of the vagina.

Male

Sperms are produced in the male reproductive organs (Figs 4 and 5), called the **testes** (singular = testis). These lie outside the abdominal cavity in a special sac called the **scrotum**. In this position they are kept at a temperature slightly below the rest of the body. This is the best temperature for sperm production.

The testes consist of a mass of sperm-producing tubes (Fig. 6). These tubes join to form ducts leading to the **epididymis**, a coiled tube about 6 metres long on the outside of each testis. The epididymis, in turn, leads into a muscular **sperm duct**. The two sperm ducts, one from each testis, open into the top of the urethra just after it leaves the bladder. A short, coiled tube called the **seminal vesicle** branches from each sperm duct just before it enters the **prostate gland**, which surrounds the urethra at this point.

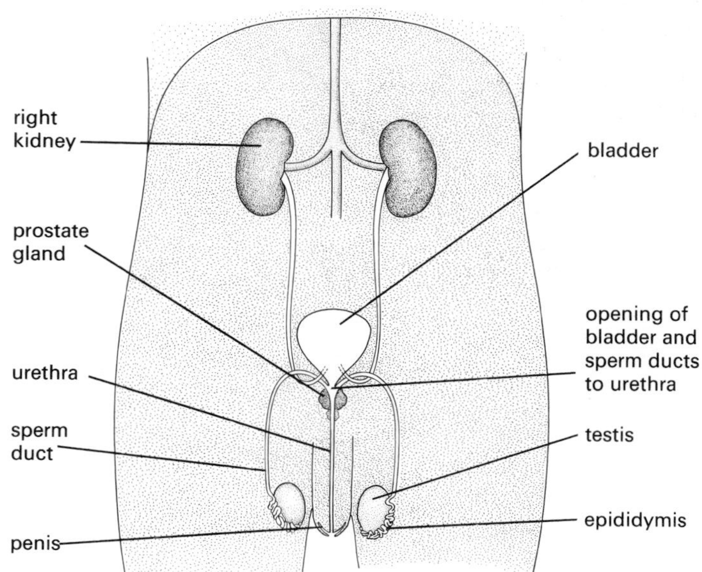

Fig. 4 The male reproductive organs; front view

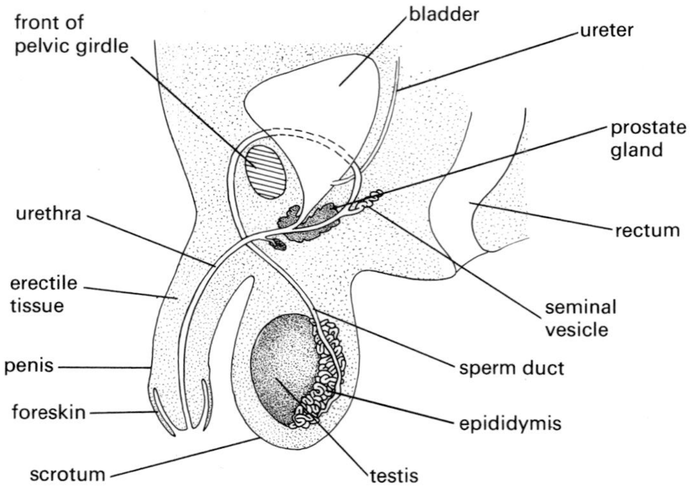

Fig. 5 Male reproductive organs; side view

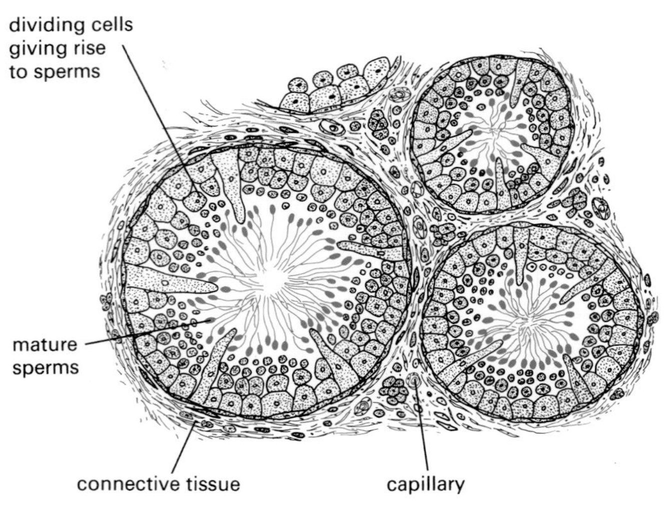

Fig. 6 Section through tubules

The urethra passes through the **penis** and may conduct either urine or sperms at different times. The penis consists of connective tissue with many blood spaces in it. This is called **erectile tissue**.

QUESTIONS

1 How do sperms differ from ova in their structure (see Fig. 1)?
2 List the structures, in the correct order, through which the sperms must pass from the time they are produced in the testis, to the time they leave the urethra.
3 What structures are shown in Fig. 5 which are not shown in Fig. 4?
4 In what ways does a zygote differ from any other cell in the body?

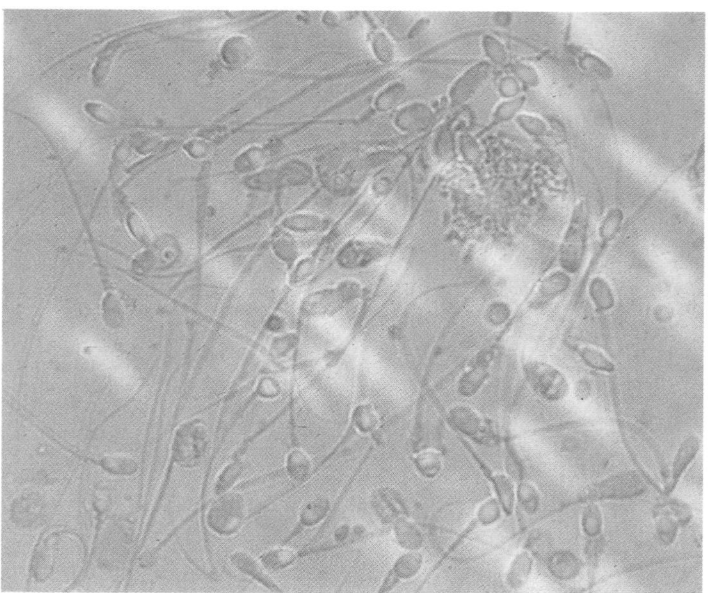

Fig. 7 Human sperms (×700). The head of the sperm has a slightly different appearance when seen in 'side' view or in 'top' view.

PRODUCTION OF GAMETES

Sperm production

The lining of the sperm-producing tubules in the testis consists of rapidly dividing cells (Fig. 6). After a series of cell divisions, the cells grow long tails and become sperms (Fig. 7) which pass into the epididymis.

During copulation, the epididymis and sperm ducts contract and force sperms out through the urethra. The prostate gland and seminal vesicle add fluid to the sperms. This fluid plus the sperms it contains is called **semen**, and the ejection of sperms through the penis is called **ejaculation**.

Ovulation

The egg cells (ova) are present in the ovary from the time of birth. No more are formed during the lifetime, but between the ages of 10 and 14 the egg cells start to ripen and are released, one at a time about every 4 weeks from alternate ovaries. As each ovum matures, the cells round it divide rapidly and produce a fluid-filled sac. This sac is called a **follicle** (Fig. 8) and when mature, it projects from the surface of the ovary like a small blister (Fig. 9). Finally, the follicle bursts and releases the ovum with its coating of cells into the funnel of the oviduct. This is called **ovulation**. From here, the ovum is wafted down the oviduct by the action of cilia (p. 9) in the lining of the tube. If the ovum meets sperm cells in the oviduct, it may be fertilized by one of them.

The released ovum is enclosed in a jelly-like coat called the **zona pellucida** and is still surrounded by a layer of follicle cells. Before fertilization can occur, sperms have to get through this layer of cells and the successful sperm has to penetrate the zona pellucida with the aid of enzymes secreted by the head of the sperm.

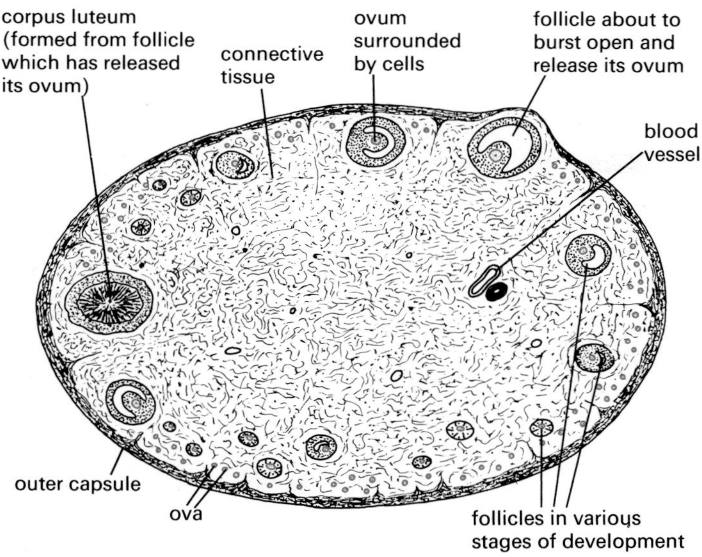

Fig. 8 Section through an ovary

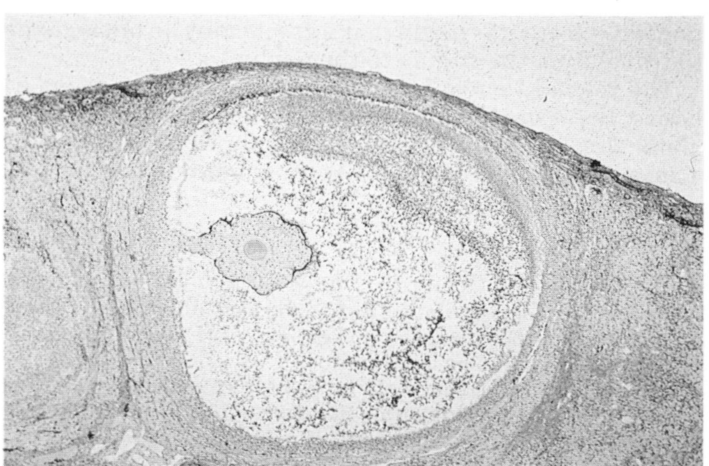

Fig. 9 Mature follicle as seen in a section through part of an ovary (×30). The ovum is surrounded by follicle cells. These produce the fluid which occupies much of the space in the follicle.

MATING AND FERTILIZATION

Mating

As a result of sexual stimulation, the male's penis becomes erect. This is due to blood flowing into the erectile tissue round the urethra. In the female, the lining of the vagina produces mucus which makes it possible for the penis to enter. The sensory stimulus (sensation) produced by copulation causes a reflex (p. 207) in the male which results in the ejaculation of semen into the top of the vagina.

Fertilization

The sperms swim through the cervix and into the uterus by wriggling movements of their tails. They pass through the uterus and enter the oviduct, but the method by which they do this is not known for certain. If there is an ovum in the oviduct, one of the sperms may bump into it and stick to its surface. The sperm then enters the cytoplasm of the ovum and the male nucleus of the sperm fuses with the female nucleus. This is the moment of fertilization and is shown in more detail in Fig. 10. Although a single ejaculation may contain about five hundred million sperms, only one will fertilize the ovum. The function of the others is not

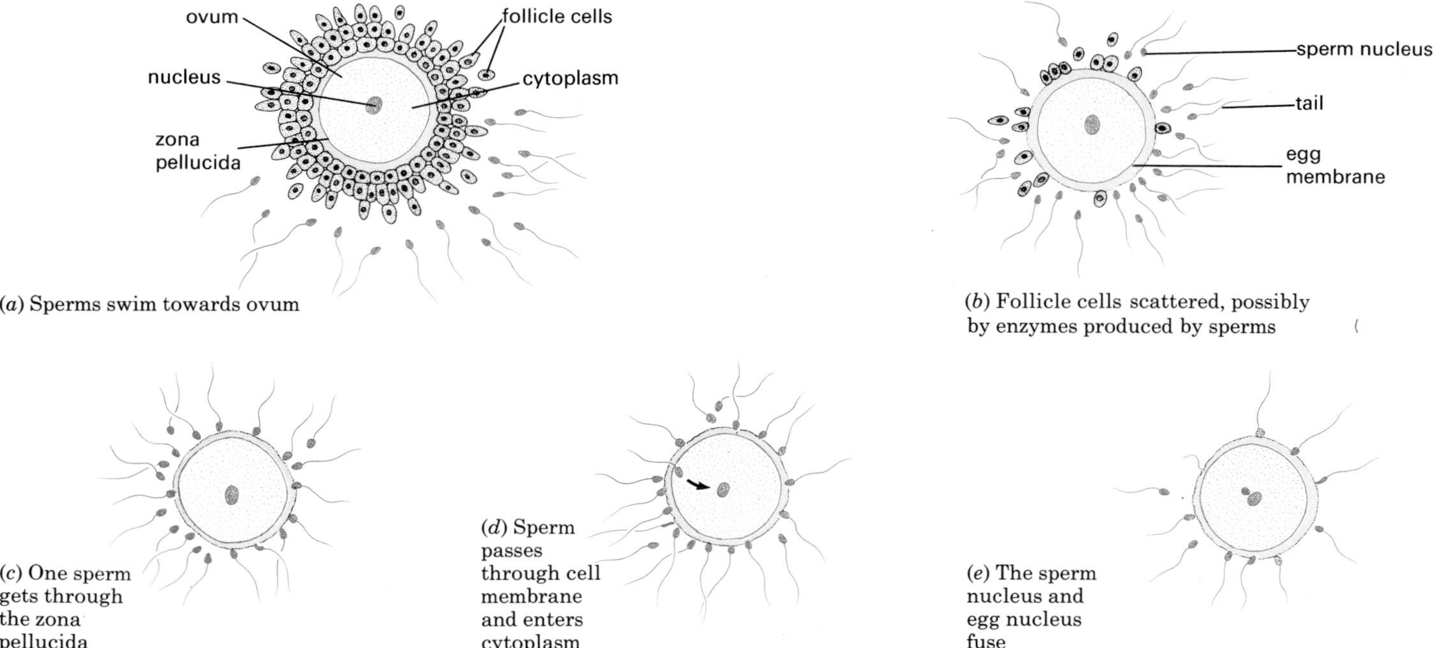

(*a*) Sperms swim towards ovum

(*b*) Follicle cells scattered, possibly by enzymes produced by sperms

(*c*) One sperm gets through the zona pellucida

(*d*) Sperm passes through cell membrane and enters cytoplasm

(*e*) The sperm nucleus and egg nucleus fuse

Fig. 10 Fertilization of an ovum. (The diagrams show what is thought to happen, but the details are not known for certain.)

understood but it is likely that a great many do not manage to travel from the vagina to the oviduct.

The released ovum is thought to survive for about 24 hours; the sperms might be able to fertilize an ovum for about 2 or 3 days. So there is only a short period of about 4 days each month when fertilization might occur.

QUESTIONS

5 If a woman starts ovulating at 13 and stops at 50, (*a*) how many ova are likely to be released from her ovaries, (*b*) about how many of these are likely to be fertilized?
6 List, in the correct order, the parts of the female reproductive system through which sperms must pass before reaching and fertilizing an ovum.
7 State exactly what happens at the moment of fertilization.
8 If mating takes place (*a*) 2 days before ovulation, (*b*) 2 days after ovulation, is fertilization likely to occur? Explain your answers.

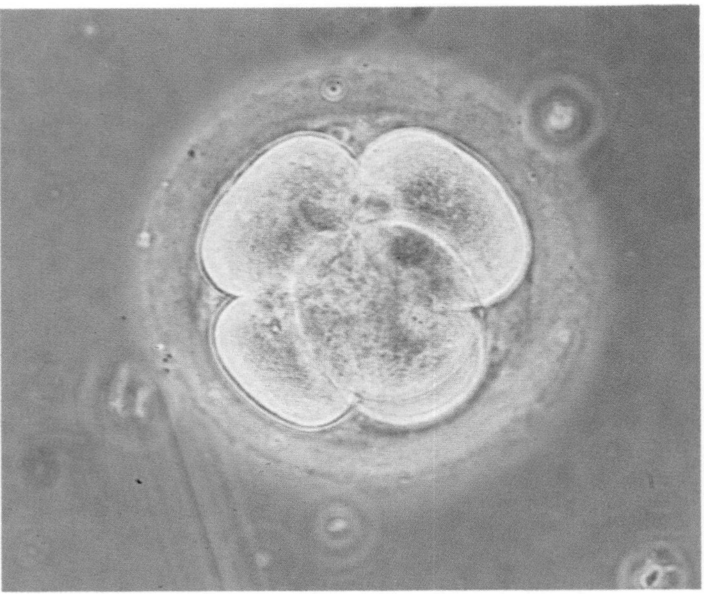

Fig. 11 Human embryo at the 5-cell stage (×230). The embryo is surrounded by the zona pellucida.

PREGNANCY AND DEVELOPMENT

The fertilized ovum first divides into two cells. Each of these divides again, so producing four cells. The cells continue to divide in this way to produce a solid ball of cells (Fig. 11), an early stage in the development of the **embryo**. This early embryo travels down the oviduct to the uterus. Here it sinks into the lining of the uterus, a process called **implantation** (Fig. 12*a*). The embryo continues to grow and produces new cells which form tissues and organs (Fig. 13). After eight weeks, when all the organs are formed, the embryo is called a **fetus**. One of the first organs to form is the heart, which pumps blood round the body of the embryo.

As the embryo grows, the uterus enlarges to contain it. Inside the uterus the embryo becomes enclosed in a fluid-filled sac called the **amnion** or water sac, which protects it from damage and prevents unequal pressures from acting on it (Figs 12*b* and *c*). The oxygen and food needed to keep the embryo alive and growing are obtained from the mother's blood by means of a structure called the placenta.

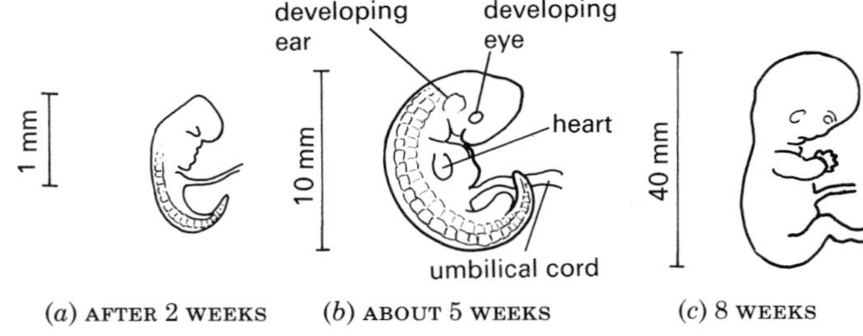

Fig. 13 Human embryo: the first 8 weeks

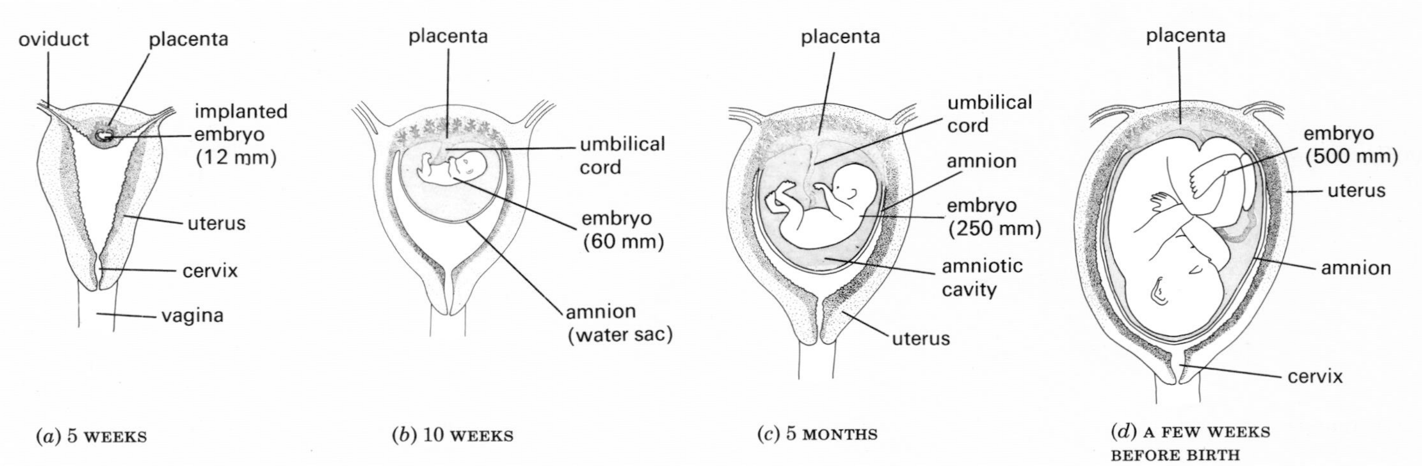

Fig. 12 Growth and development in the uterus (not to scale)

Placenta

Soon after the ball of cells reaches the uterus, some of the cells, instead of forming the organs of the embryo, grow into a disc-like structure, the **placenta** (Fig. 12*c*). The placenta becomes closely attached to the lining of the uterus and is attached to the embyro by a tube called the **umbilical cord** (Fig. 12*c*). After a few weeks, the embryo's heart has developed and is circulating blood through the umbilical cord and placenta as well as through its own tissues. The blood vessels in the placenta are very close to the blood vessels in the uterus so that oxygen, glucose, amino acids and salts can pass from the mother's blood to the embryo's blood (Fig. 14*a*). So the blood flowing in the umbilical vein from the placenta carries food and oxygen to be used by the living, growing tissues of the embryo. In a similar way, the carbon dioxide and urea in the embryo's blood escape from the vessels in the placenta and are carried away by the mother's blood in the uterus (Fig. 14*b*). In this way the embryo gets rid of its excretory products.

There is no direct communication between the mother's blood system and the embyro's. The exchange of substances takes place across the thin walls of the blood vessels. In this way, the mother's blood pressure cannot damage the delicate vessels of the embryo and it is possible for the placenta to select the substances allowed to pass into the embyro's blood. The placenta can prevent some harmful substances in the mother's blood from reaching the embryo. It cannot prevent all of them, however, as is shown by the effects of cigarette smoke and alcohol described below.

The placenta produces hormones, including oestrogens and progesterone. It is assumed that these hormones play an important part in maintaining the pregnancy and preparing for birth, but their precise function is not known. They may influence the development and activity of the muscle layers in the wall of the uterus and prepare the mammary glands in the breasts for milk production.

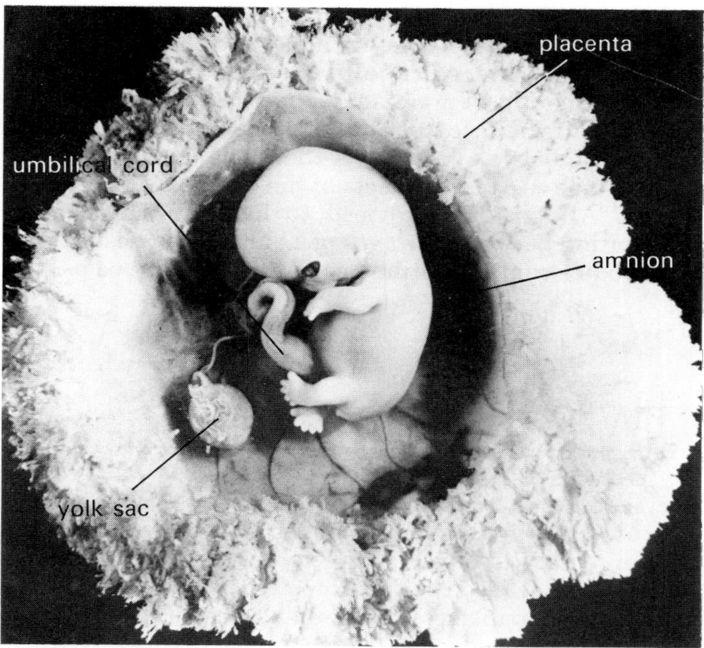

Fig. 15 Human embryo, 7 weeks (×1.5). The embryo is enclosed in the amnion. Its limbs, eye and ear-hole are clearly visible. The amnion is surrounded by the placenta; the fluffy-looking structures are the placental villi which penetrate into the lining of the uterus. The umbilical cord connects the embryo to the placenta.

Care during pregnancy During pregnancy, the mother-to-be should ensure that her diet is adequate (see p. 119) and avoid drinking alcohol or smoking. It has been shown that pregnant women who drink alcohol or smoke cigarettes run a higher risk of producing babies with low birth weights. These babies are more prone to disease than babies with normal birth weights.

There is also evidence that smoking or drinking during pregnancy can lead to a higher rate of miscarriage and there is strong suspicion that heavy drinking may damage the developing brain of the fetus.

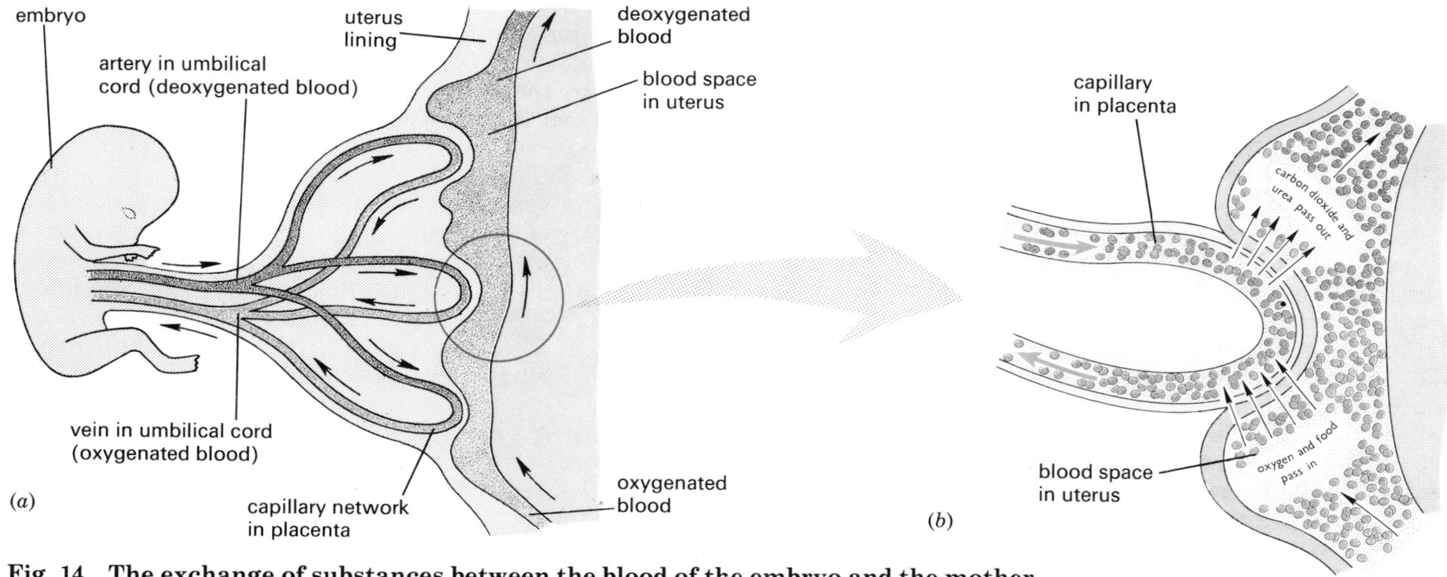

Fig. 14 The exchange of substances between the blood of the embryo and the mother

During pregnancy, it is also advisable not to take any drugs unless strictly necessary and prescribed by a doctor.

If a woman catches **rubella** (German measles) during the first 4 months of pregnancy, there is a danger that the virus will affect the foetus and cause abortion or still-birth. Even if the baby is born alive, there is a chance that the virus will have caused defects of the eyes, ears or nervous system. All girls are advised to have an anti-rubella vaccination between the ages of 11 and 13, to ensure that their bodies contain antibodies to the disease (see p. 147).

TWINS

Sometimes a woman releases two ova when she ovulates. If both ova are fertilized, they may form twin embryos, each with its own placenta and amnion. Because the twins come from two separate ova, each fertilized by a different sperm, it is possible to have a boy and a girl. Twins formed in this way are called **fraternal twins**. Although they are both born within a few minutes of each other, they are no more alike than other brothers or sisters.

Another cause of twinning is when a single fertilized egg, during an early stage of cell division, forms two separate embryos. Sometimes these may share a placenta and amnion. Twins formed from a single ovum and sperm must be the same sex, because only one sperm (X or Y, see p. 232) fertilized the ovum. These 'one-egg' twins are sometimes called **identical twins** because, unlike fraternal twins, they will closely resemble each other in every respect.

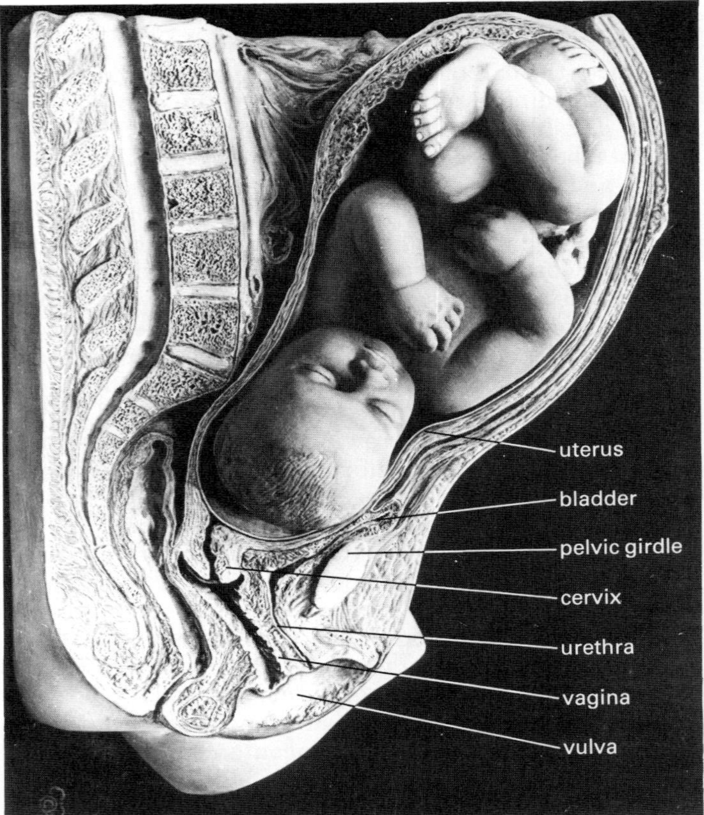

Fig. 16 Model of human fetus just before birth. The cervix and vagina seem to provide narrow channels for the baby to pass through but they widen quite naturally during labour and delivery.

QUESTIONS

9 In what ways will the composition of the blood in the umbilical vein differ from that in the umbilical artery?
10 An embryo is surrounded with fluid, its lungs are filled with fluid and it cannot breathe. Why doesn't it suffocate?
11 If a mother gives birth to twin boys, does this mean that they are identical twins? Explain.

BIRTH

From fertilization to birth takes about 38 weeks in humans. This is called the **gestation** period. A few weeks before the birth, the foetus has come to lie head downwards in the uterus, with its head just above the cervix (Figs 12*d* and 16). When the birth starts, the uterus begins to contract rhythmically. This is the beginning of what is called 'labour'. These regular rhythmic contractions become stronger and more frequent. The opening of the cervix gradually widens enough to let the baby's head pass through and the contractions of the uterus are assisted by muscular contractions of the abdomen. The water sac breaks at some stage in labour and the fluid escapes through the vagina. Finally, the muscular contractions of the uterus and abdomen push the baby head-first through the widened cervix and vagina (Fig. 17). The umbilical cord which still connects the child to the placenta is tied and cut. Later, the placenta breaks away from the uterus and is pushed out separately as the 'after-birth'.

The sudden fall in temperature felt by the newly born baby stimulates it to take its first breath and it usually cries. In a few days, the remains of the umbilical cord attached to the baby's abdomen shrivel and fall away, leaving a scar in the abdominal wall, called the navel.

Caesarian section If you study Fig. 5*a* on page 184, you will see the hole in the pelvic girdle through which the baby's head has to pass. During the course of pregnancy, the tissues holding the right and left sides of the girdle together soften and allow the girdle to widen. The bony plates in the baby's skull can overlap during birth and so reduce the diameter of its head.

If the pelvic girdle is too narrow to allow the baby's head to pass through, the baby can be delivered by Caesarian section. The mother is anaesthetized and her abdomen and uterus are opened surgically to extract the baby.

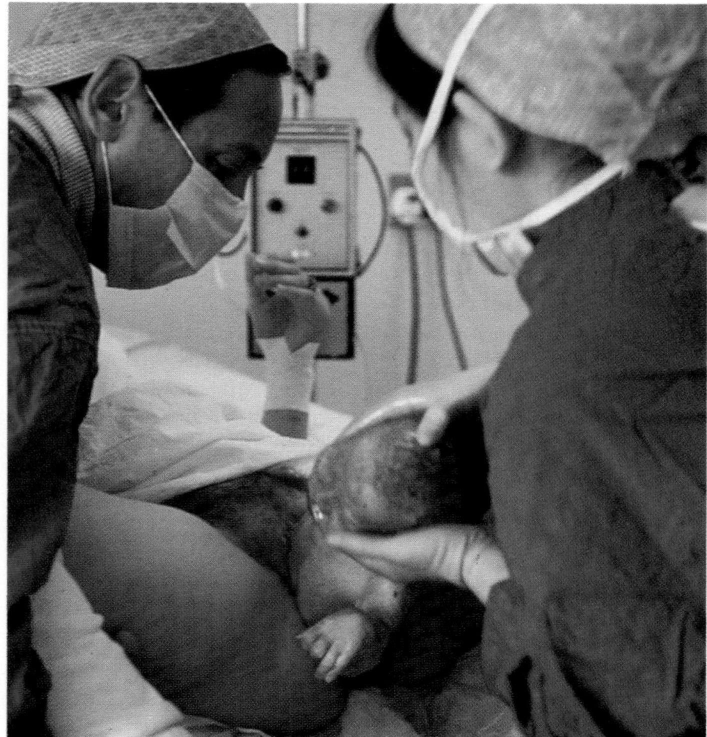

Fig. 17 Delivery of a baby. The nurse supports the baby's head while the mother's contractions push out the rest of the body.

Breech delivery A baby sometimes fails to turn head downwards in the uterus and is delivered feet first or bottom first. If, in the 'feet-first' position, its legs are straight, the birth is not impeded but there is a risk of suffocation if the head does not follow quickly.

A breech delivery may often be avoided by externally manipulating the baby so that it lies head down. A 'bottom-first' delivery can be avoided by straightening the legs in the early stages of birth. In either case, it may be decided to use a Caesarian section.

Abortion

The ending of a pregnancy before 24 weeks is called an **abortion** because the fetus cannot survive on its own at this early stage. A birth between 24 and 38 weeks might be called **premature** if the baby has not developed sufficiently to maintain its body temperature or oxygenate its blood without help.

Spontaneous abortion (miscarriage) This occurs usually because the embryo is defective in some way. The embryo dies and is expelled from the uterus. This may occur at such an early stage in the pregnancy that it is unnoticed or confused with menstruation.

Perhaps as many as 30 per cent of fertilized ova fail to implant or develop properly, though such figures are difficult to ascertain.

Induced abortion This is a deliberate termination of the pregnancy because it is considered medically desirable, e.g. there is a risk to the mother's health if the pregnancy continues or the fetus is defective, for example, after an attack of rubella in the first 3 months of the pregnancy.

A fetus of less than 12 weeks is sucked out of the uterus through a tube inserted via the cervix. After 12 weeks, abortion is more difficult and may involve surgical removal of the fetus as in a Caesarian section. Developments are now taking place in the use of chemicals (prostaglandins) which induce early labour to expel the fetus.

The use of abortion as a means of birth control is very controversial since it involves the destruction of a fetus which has reached a stage where it might be considered as an actual or a potential human being.

FEEDING AND PARENTAL CARE

About 24 hours after birth, the baby starts to suck at the breast. During pregnancy the mammary glands (breasts) enlarge as a result of an increase in the number of milk-secreting cells. No milk is secreted during pregnancy, but the hormones which start the birth process also act on the milk-secreting cells of the breasts. The breasts are stimulated to release milk by the first sucklings. The continued production of milk is under the control of hormones, but the amount of milk produced is related to the quantity taken by the child during suckling.

Fig. 18 Breast-feeding helps to establish an emotional bond between mother and baby

Fig. 19 Parental care. Human parental care includes a long period of education.

Milk contains nearly all the food, vitamins and salts that babies need for their energy requirements and tissue-building, but there is no iron present for the manufacture of haemoglobin. All the iron needed for the first weeks or months is stored in the body of the foetus during gestation.

The mother's milk supply increases with the demands of the baby, up to 1 litre per day. It is gradually supplemented and eventually replaced entirely by solid food, a process known as **weaning**.

Cows' milk is not wholly suitable for human babies. It has more protein, sodium and phosphorus, and less sugar, vitamin A and vitamin C than human milk. Manufacturers modify the components of dried cows' milk to resemble human milk more closely and this makes it more acceptable if the mother cannot breast-feed her baby.

Cows' milk and proprietary dried milk both lack human antibodies, whereas the mother's milk contains antibodies to any diseases from which she has recovered. It also carries white cells which produce antibodies or ingest bacteria. These antibodies are important in defending the baby against infection at a time when its own immune responses are not fully developed. Breast-feeding provides milk free from bacteria, whereas bottle-feeding carries the risk of introducing bacteria which cause intestinal diseases. Breast-feeding also offers emotional and psychological benefits to both mother and baby (Fig. 18).

Most young mammals are independent of their parents after a few weeks or months even though they may stay together as a family group. In man, however, the young are dependent on their parents for food, clothing and shelter for many years. During this long period of dependence, the young learn to talk, read and write and learn a great variety of other skills that help them to survive and be self-sufficient (Fig. 19).

QUESTION

12 Apart from learning to talk, what other skills might a young human develop that would help him when he is no longer dependent on his parents in (*a*) an agricultural society, (*b*) an industrial society?

PUBERTY AND THE MENSTRUAL CYCLE

Puberty

Although the ovaries of a young girl contain all the ova she will ever produce, they do not start to be released until she reaches an age of about 10–14 years. This stage in her life is known as **puberty**.

At about the same time as the first ovulation, the ovary also releases female sex hormones into the bloodstream. These hormones are called **oestrogens** and when they circulate round the body, they bring about the development of **secondary sexual characteristics**. In the girl these are the increased growth of the breasts, a widening of the hips and the growth of hair in the pubic region and in the armpits. There is also an increase in the size of the uterus and vagina. Once all these changes are complete, the girl is capable of having a baby.

Puberty in boys occurs at about the same age as in girls. The testes start to produce sperms for the first time and also release a hormone, called **testosterone**, into the bloodstream. The male secondary sexual characteristics which begin to appear at puberty are enlargement of the testes and penis, deepening of the voice, growth of hair in the pubic region, armpits, chest and, later on, the face. In both sexes there is a rapid increase in the rate of growth during puberty.

The menstrual cycle

The ovaries release an ovum about every 4 weeks. As each follicle develops, the amount of oestrogens produced by the ovary increases. The oestrogens act on the uterus and cause its lining to become thicker and develop more blood vessels. These are changes which help an early embryo to implant as described on page 174.

Once the ovum has been released, the follicle which produced it develops into a solid body called the **corpus luteum**. This produces a hormone called **progesterone**, which affects the uterus lining in the same way as the oestrogens, making it grow thicker and produce more blood vessels.

If the ovum is fertilized, the corpus luteum continues to release progesterone and so keeps the uterus in a state suitable for implantation. If the ovum is not fertilized, the corpus luteum stops producing progesterone. As a result, the thickened lining of the uterus breaks down and loses blood which escapes through the cervix and vagina. This is known as a **menstrual period**. The appearance of the first menstrual period is one of the signs of puberty in girls. The events in the menstrual cycle are shown in Fig. 20.

Menopause Between the ages of 40 and 55, the ovaries cease to release ova or produce hormones. As a consequence, menstrual periods cease, the woman can no longer have children, and sexual desire is gradually reduced.

QUESTIONS

13 From the list of changes at puberty in girls, select those which are related to child-bearing and say what part you think they play.
14 One of the first signs of pregnancy is that the menstrual periods stop. Explain why you would expect this.

FAMILY PLANNING AND FERTILITY

As little as 4 weeks after giving birth, it is possible, though unlikely, that a woman may conceive again. Frequent breast-feeding may reduce the chances of conception. Nevertheless, it would be possible to have children at about 1-year intervals. Most people do not want, or cannot afford, to have as many children as this. All human communities, therefore, practise some form of birth control to space out births and limit the size of the family.

Natural methods of family planning

If it were possible to know exactly when ovulation occurred, intercourse could be avoided for 3–4 days before, and 1 day after ovulation (see p. 172). At the moment, however, there is no simple, reliable way to recognize ovulation, though it is usually 12–16 days before the onset of the next menstrual period. By keeping careful records of the intervals between menstrual periods, it is possible to calculate a potentially fertile period of about 10 days in mid-cycle, when sexual intercourse should be avoided if children are not wanted.

On its own, this method is not very reliable but there are some physiological clues which help to make it more accurate. During or soon after ovulation, a woman's temperature rises about 0.5 °C. It is reasonable to assume that one day after the temperature returns to normal, a woman will be infertile. Another clue comes from the type of mucus secreted by the cervix and lining of the vagina. As the time for ovulation approaches, the mucus becomes more fluid. Women can learn to detect these changes and so calculate their fertile period.

By combining the 'calendar', 'temperature' and 'mucus' methods, it is possible to achieve about 80 per cent 'success', i.e. only 20 per cent unplanned pregnancies. Highly motivated couples may achieve better rates of success and, of course, it is a very helpful way of finding the fertile period for couples who do want to conceive.

Artificial methods of family planning

The sheath or condom A thin rubber sheath is placed on the erect penis before sexual intercourse. The sheath traps the sperms and prevents them from reaching the uterus.

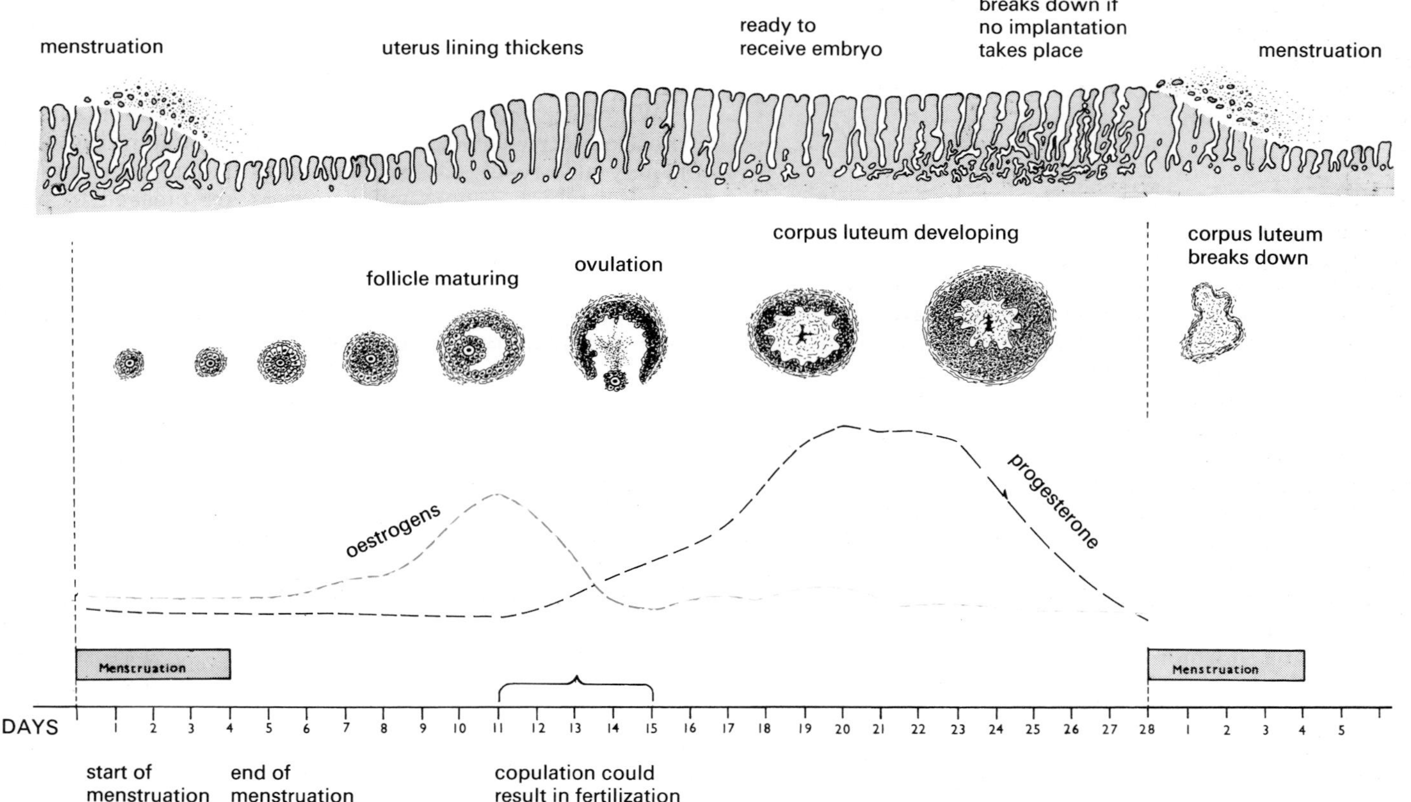

Fig. 20 The menstrual cycle. (After G. W. Corner, *The Hormones in Human Reproduction*, Princeton.)

The diaphragm A thin rubber disc, placed in the vagina before intercourse, covers the cervix and stops sperms entering the uterus. Condoms and diaphragms, used in conjunction with chemicals that immobilize sperms, are about 95 per cent effective.

Intra-uterine device (IUD) A small metal or plastic strip bent into a loop or coil is inserted and retained in the uterus, where it probably prevents implantation of a fertilized ovum. It is about 98 per cent effective but there is a small risk of developing uterine infections, particularly if sexual relations are promiscuous.

The contraceptive pill The pill contains chemicals which have the same effect on the body as the hormones oestrogen and progesterone. When mixed in suitable proportions these hormones suppress ovulation (see 'Feedback', p. 214) and so prevent conception. The pills need to be taken each day for the 21 days between menstrual periods.

There are many varieties of contraceptive pill in which the relative proportions of oestrogen- and progesterone-like chemicals vary. They are 99 per cent effective but long-term use of some types is thought to increase the risk of cancer of the breast and cervix.

Vasectomy This is a simple and safe surgical operation in which the man's sperm ducts are cut and the ends sealed. This means that his semen contains the secretions of the prostate gland and seminal vesicle but no sperms and so cannot fertilize an ovum. Sexual desire, erection, copulation and ejaculation are quite unaffected.

The testis continues to produce sperms and testosterone. The sperms are removed by white cells as fast as they form. The testosterone ensures that there is no loss of masculinity.

The sperm ducts can be rejoined by surgery but this is not always successful.

In-vitro fertilization ('test-tube babies')

'In-vitro' means literally 'in glass' or, in other words, the fertilization is allowed to take place in laboratory glassware rather than 'in-vivo', i.e. in the living body.

One cause of infertility in women is a blockage of the oviducts. Sometimes this can be cured by surgery. When surgery is ineffective or inappropriate, the woman may be offered the chance of in-vitro fertilization, a process which has only recently been developed and which has received considerable publicity since the first 'test-tube baby' was born in 1978.

The woman may be given injections of follicle-stimulating hormone (FSH), and luteinizing hormone (LH) (see p. 214) which cause her ovaries to release several mature ova simultaneously. These ova are then collected by laparoscopy, i.e. they are sucked up in a fine tube inserted through the abdominal wall. The ova are then mixed with the husband's seminal fluid and watched under the microscope to see if cell division takes place. (Figure 11 on p. 174 is a photograph of such an 'in-vitro' fertilized ovum.)

One or more of the dividing zygotes are then introduced to the woman's uterus by means of a tube inserted through the cervix. Usually, only one (or none) of the zygotes develops though, occasionally, there are multiple births. The success rate for in-vitro fertilization is probably little better than 20–30 per cent but the main controversy seems to be about the fate of the 'spare' embryos. Some people believe that since these embryos are potential human beings, they should not be destroyed or used for research. In some cases, the embryos have been frozen and used later, if the first transplants did not work. The 1990 Human Fertilization and Embryology Bill permits limited research on embryos up to 14 days after fertilization.

World population

Because of the great improvement in drugs and medical knowledge, fewer and fewer people die from infectious diseases such as typhoid and cholera. But the birth rate in many countries has not gone down, so their population is doubling every 50 years or less.

Another theory to account for the increase in population in the Third World involves breast-feeding. There is evidence that breast-feeding at frequent intervals, i.e. 6 or more times a day, suppresses ovulation. If breast-feeding continues for 3 years, as it does in some communities, births are spaced at about 4-year intervals. The tendency to abandon breast-feeding in favour of bottle-feeding, together with early weaning (i.e. changing to solid food), reduces birth intervals to just over 1 year in populations which do not have access to birth control methods.

Whatever the cause, the rapid growth of population in these countries leads to shortages of food and living space. It may also result in environmental damage, e.g. the soil erosion which occurs when forests are cleared to plant crops.

There must be a limit to the number of people who can live on the Earth and, therefore, it is necessary to try and make people understand the need to prevent the population increase from getting out of hand (Fig. 21).

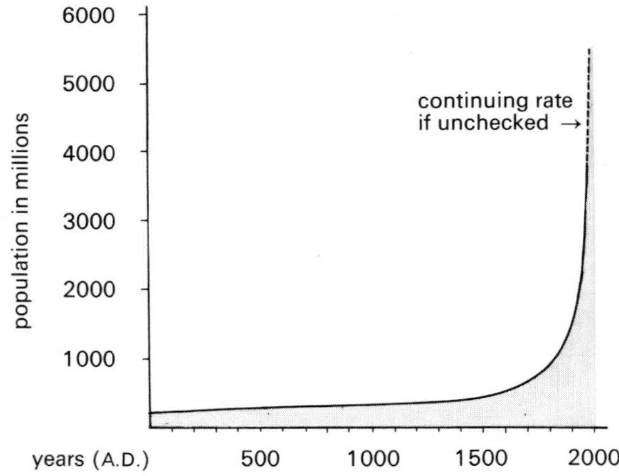

Fig. 21 World population growth in the last 2000 years

CHECK LIST

- The male reproductive cells (gametes) are sperms. They are produced in the testes and expelled through the urethra and penis during mating.
- The female reproductive cells (gametes) are ova (eggs). They are produced in the ovaries. One is released each month. If sperms are present, the ovum may be fertilized as it passes down the oviduct to the uterus.
- Fertilization happens when a sperm enters an ovum and the sperm and egg nuclei join up (fuse).
- The fertilized ovum (zygote) divides into many cells and becomes embedded in the lining of the uterus. Here it grows into an embryo.
- The embryo gets its food and oxygen from its mother.
- The embryo's blood is pumped through blood vessels in the umbilical cord to the placenta, which is attached to the uterus lining. The embryo's blood comes very close to the mother's blood so that food and oxygen can be picked up and carbon dioxide and nitrogenous waste can be got rid of.
- When the embryo is fully grown, it is pushed out of the uterus through the vagina by contractions of the uterus and abdomen.
- Each month, the uterus lining thickens up in readiness to receive the fertilized ovum. If an ovum is not fertilized, the lining and some blood is lost through the vagina. This is menstruation.
- The release of ova and the development of an embryo are under the control of hormones like oestrogen and progesterone.
- Twins may result from two ova being fertilized at the same time or from a zygote forming two embryos.
- At puberty, (1) the testes and ovaries start to produce mature gametes, (2) the secondary sexual characteristics develop.
- Human milk and breast-feeding are best for babies.
- Young humans are dependent on their parents for a long time. During this period much essential learning takes place.
- The world population is doubling every 50 years to the detriment of food supplies, the environment and quality of life.
- There are effective natural and artificial methods for spacing births and limiting the size of a family.

19 The Skeleton, Muscles and Movement

STRUCTURE OF THE SKELETON
Vertebral column, skull, limbs and joints.

FUNCTIONS OF THE SKELETON
Support, protection and movement.

BONE, CARTILAGE AND MUSCLE
Skeletal and smooth muscle. Muscle contraction.

MOVEMENT AND LOCOMOTION
Muscles produce movement. Lever action of limbs. Locomotion.

PRACTICAL WORK
Structure of bone.

STRUCTURE OF THE SKELETON

Figure 2 shows a human skeleton. It consists of a vertebral column (sometimes called the 'backbone', 'spine' or 'spinal column') which supports the skull. Twelve pairs of ribs are attached to the upper part of the vertebral column and the limbs are attached to it by means of **girdles**.

The hip girdle (pelvic girdle) is joined rigidly to the lower end of the vertebral column. The shoulder girdle (pectoral girdle) consists of a pair of collar-bones and shoulder-blades which are not rigidly fixed to the vertebral column but held in place by muscles.

The upper arm bone (humerus) fits into a socket in the shoulder-blade; the thigh bone (femur) fits into a socket in the hip girdle.

Vertebral column

This forms the central supporting structure of the skeleton. It consists of 33 individual bones called **vertebrae** (singular = vertebra), separated by discs of fibrous cartilage (Fig. 1*b*). These discs allow the vertebrae to move slightly and so enable the vertebral column to bend backwards and forwards or from side to side.

The spinal cord (p. 210) runs through an arch of bone (**neural arch**) formed by the vertebrae. In this position the cord is protected from damage (Fig. 1*a*).

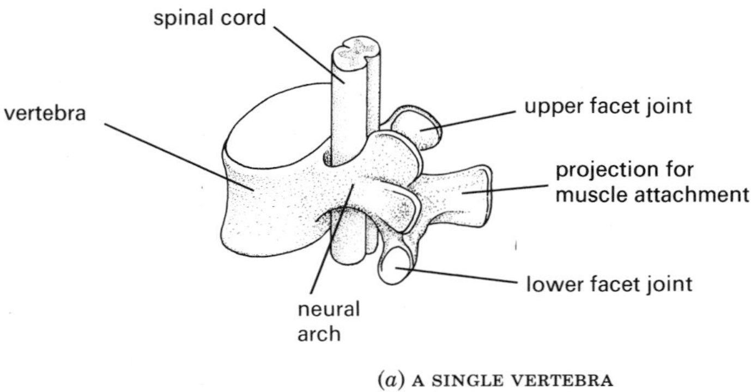

(*a*) A SINGLE VERTEBRA

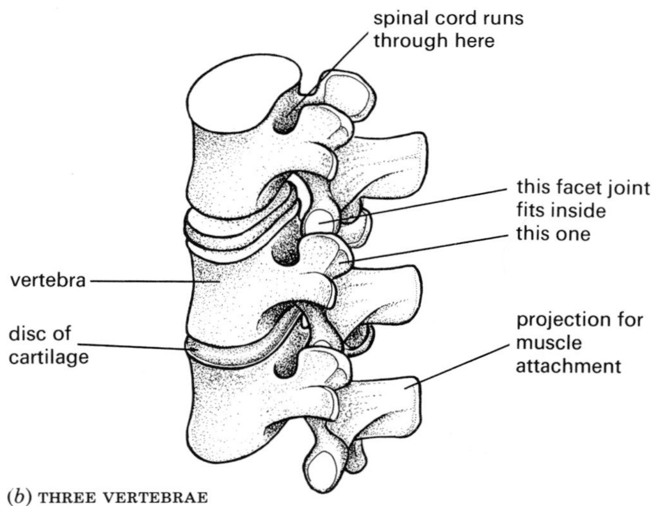

(*b*) THREE VERTEBRAE

Fig. 1 The vertebral column

The skull

This is made up of many bony plates joined together (Fig. 3). It encloses and protects the brain and also carries and protects the main sense organs, the eyes, ears and nose. The upper jaw is fixed to the skull but the lower jaw is hinged to it in a way which allows chewing.

The base of the skull makes a joint with the top vertebra of the vertebral column. This joint allows the head to make nodding and rotational movements.

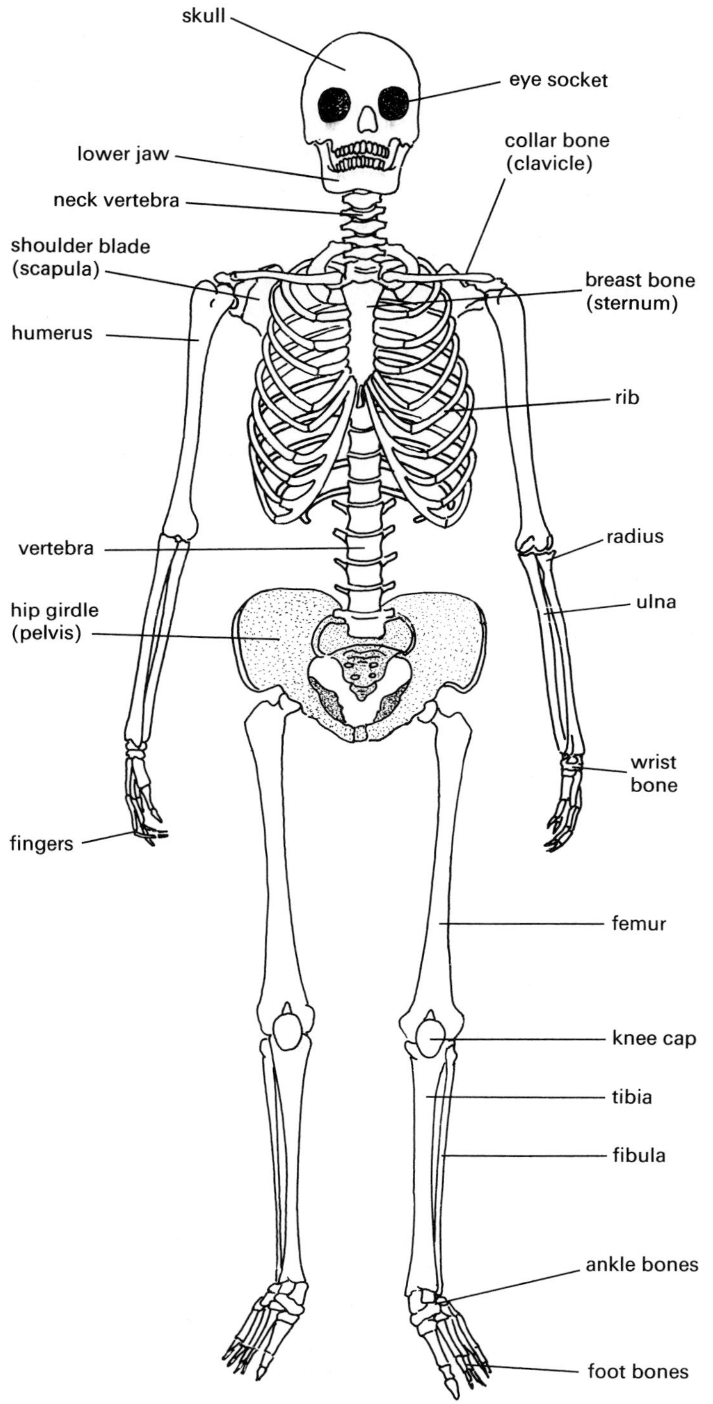

Fig. 2 The skeleton

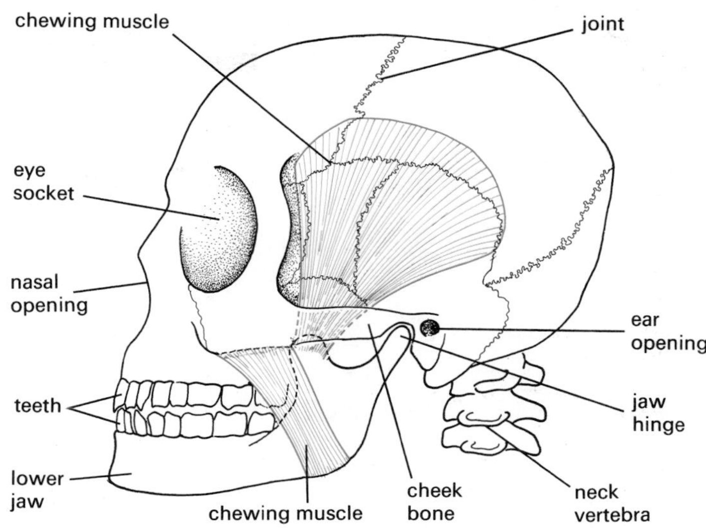

Fig. 3 The skull and chewing muscles

The limbs

Arm The upper arm bone is the **humerus**. It is attached by a hinge joint to the lower arm bones, the **radius** and **ulna** (Fig. 4). These two bones make a joint with a group of small wrist bones which in turn join to a series of five hand and finger bones. The ulna and radius can partly rotate round each other so that the hand can be held palm up or palm down.

Leg The thigh bone or **femur** is attached at the hip to the pelvic girdle by a ball joint and at the knee it makes a hinge joint with the **tibia**. The **fibula** runs parallel to the tibia but does not form part of the knee joint. The ankle, foot and toe bones are similar to those of the wrist, hand and fingers.

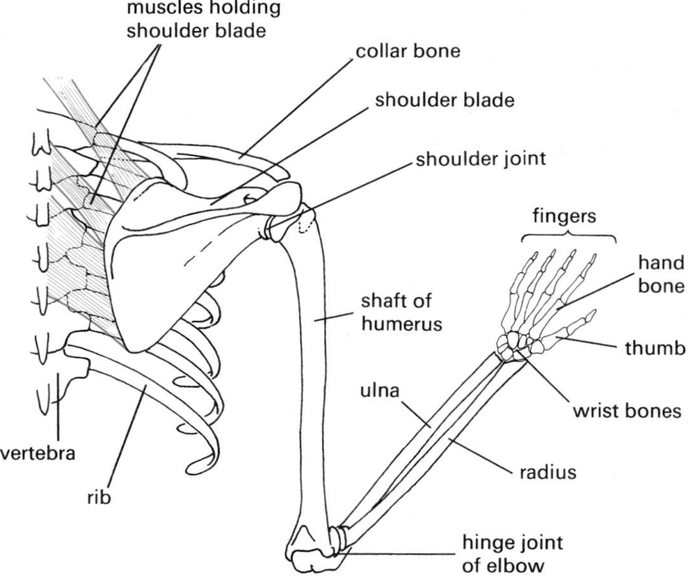

Fig. 4 Skeleton of arm and shoulder

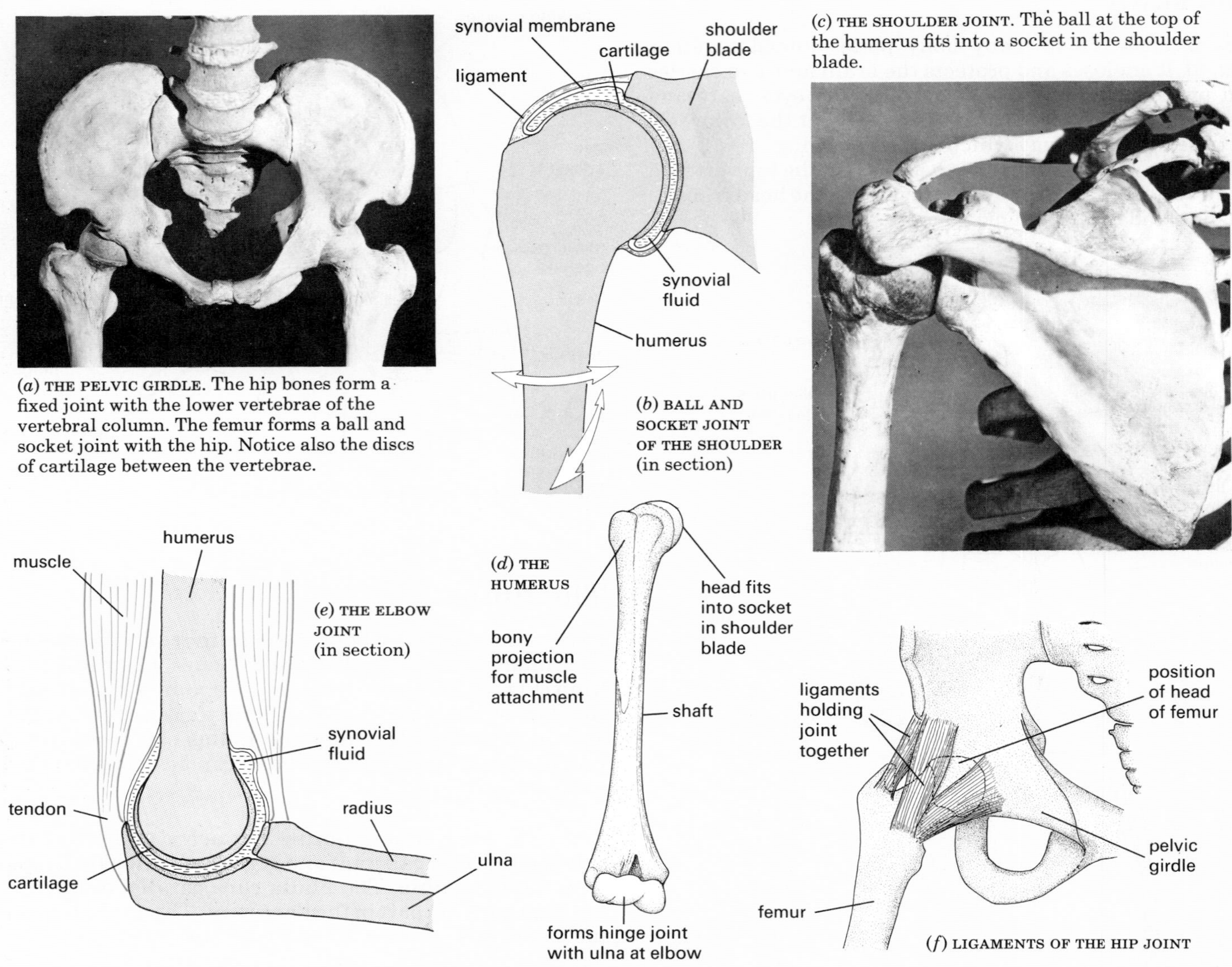

(*a*) THE PELVIC GIRDLE. The hip bones form a fixed joint with the lower vertebrae of the vertebral column. The femur forms a ball and socket joint with the hip. Notice also the discs of cartilage between the vertebrae.

(*b*) BALL AND SOCKET JOINT OF THE SHOULDER (in section)

(*c*) THE SHOULDER JOINT. The ball at the top of the humerus fits into a socket in the shoulder blade.

Fig. 5 The joints of the hip, shoulder and elbow

Joints

Where two bones meet they form a joint. It may be a **fixed joint** as in the junction of the hip girdle and the vertebral column (Fig. 5*a*) or a **movable joint** as in the knee. Two important types of movable joint already mentioned are the **ball and socket joints** of the hip and the shoulder (Fig. 5*b*), and the **hinge joints** of the elbow (Fig. 5*e*) and knee. The ball and socket joint allows movement forwards, backwards and sideways, while the hinge joint allows movement in only one direction.

Where the surfaces of the bones in a joint rub over each other, they are covered with smooth cartilage which reduces the friction between them. Friction is also reduced by a thin layer of lubricating fluid called **synovial fluid** (Figs. 5*b* and *e*). (Movable joints are sometimes called **synovial joints**.) The bones forming the joint are held in place by tough bands of fibrous tissues called **ligaments** (Fig. 5*f*). Ligaments keep the bones together but do not stop their various movements.

FUNCTIONS OF THE SKELETON

Support The skeleton holds the body off the ground and keeps its shape even when muscles are contracting to produce movement.

Protection The brain is protected from injury by being enclosed in the skull. The heart, lungs and liver are protected by the rib cage, and the spinal cord is enclosed inside the neural arches of the vertebrae.

Movement Many bones of the skeleton act as levers. When muscles pull on these bones, they produce movements such as the raising of the ribs during breathing (see p. 153), or the chewing action of the jaws. For a skeletal muscle to produce movement, both its ends need to have a firm attachment. The skeleton provides suitable points of attachment for the ends of muscles.

Production of blood cells The red marrow of some bones, e.g. the vertebrae, ribs, breastbone and the heads of the limb bones, produce both red and white blood cells (pp. 137 and 138).

QUESTIONS

1 Study Fig. 2 and then write the biological names of the following bones: upper arm bone, upper leg bone, hip bone, breastbone, 'backbone', lower arm bones.
2 Apart from the elbow and knee, what other joints in the body function like hinge joints?
3 What types of joints are visible in Fig. 5*a*? Name the bones which form each type of joint that you mention.
4 Which parts of the skeleton are concerned with both protection and movement?

CARTILAGE, BONE AND MUSCLE

All these tissues are formed by living cells, though bone and cartilage do contain some non-living components.

Cartilage

This occurs in several forms. One form is firm and semi-transparent. It makes up the rings which keep the trachea and bronchi open (p. 151), it covers the surfaces of movable joints, reducing friction and wear, and it supports that part of the nose which protrudes from the face.

Fibrous cartilage contains many fibres as well as living cells. The fibrous cartilage which forms the external ear pinna and the epiglottis is elastic and flexible. Less flexible but very strong fibres are found in the fibro-cartilage which contributes to some ligaments and attaches tendons to bones. Fibro-cartilage also forms part of the inter-vertebral discs.

When a foetus is developing in the first few weeks (p. 174), its skeleton is first formed in cartilage. The cartilage is then gradually replaced by bone in the course of development, before the baby is born.

Cartilage does not have its own blood supply but depends on oxygen and food diffusing from the capillaries of nearby tissues.

Bone

Bone is a harder tissue than cartilage and less flexible. It contains living cells and non-living fibres. The fibrous tissue between the cells becomes hardened by a deposit of calcium salts such as calcium phosphate.

Although bone contains a high proportion of non-living material it is penetrated by blood vessels which keep the cells alive and allow growth and repair to take place.

Muscle

There are three main types of muscle. One kind is called **skeletal muscle** (or striated, or voluntary muscle). Another kind is called **smooth muscle** (or unstriated, or involuntary muscle). A third kind occurs only in the heart.

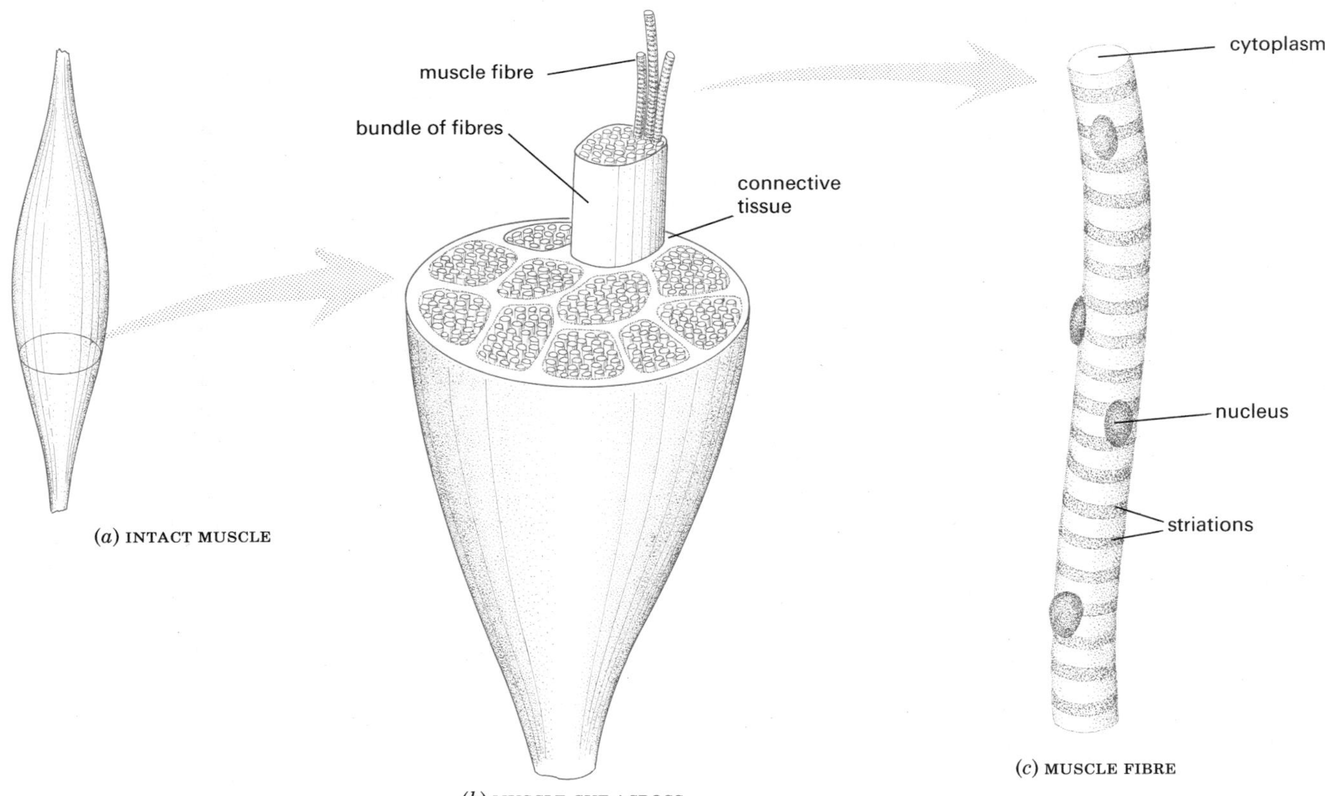

Fig. 6 Skeletal (striated) muscle

Skeletal muscle This is made up of long fibres. Each fibre is formed from many cells, but the cells have fused together. The cell boundaries cannot be seen but the individual nuclei are still present (Fig. 6*c*).

The muscle fibres are arranged in bundles which form distinct muscles (Fig. 6*b*). Most of these are attached to bones, and produce movement as described below. Each muscle has a nerve supply. When a nerve impulse is sent to a muscle, it makes the muscle contract (i.e. get shorter and fatter). We can usually control most of our skeletal muscles. For this reason they are called **voluntary muscles**.

Smooth muscle This is made up of elongated cells which are not fused together to form fibres (see Fig. 19*c*, p. 10). The cells make layers of muscle tissue rather than distinct muscles, e.g. in the walls of the alimentary canal (p. 126), the uterus (p. 171) and the arterioles (p. 142).

When the cells are arranged at right angles to the organ, they form a layer of circular muscle. When circular muscle contracts it makes the gut or the arteriole narrower. This is the basis for the process of peristalsis (p. 126), vaso-constriction (p. 168) or labour (p. 176). A localized region of circular muscle may form a **sphincter**, e.g. the pyloric sphincter at the exit of the stomach (p. 128), the sphincter at the base of the bladder (p. 161), the anal sphincter at the anus and the sphincter muscle in the iris of the eye (p. 198).

We do not have conscious control over most of the smooth muscle in our bodies (though we do for the anal and bladder sphincters). This is why smooth muscle is sometimes called **involuntary muscle**.

Muscle contraction The fibres of skeletal muscle and the cells of smooth muscle have the special property of being able to contract, i.e. shorten, when stimulated by nerve impulses. However, the fibres and cells cannot elongate; they can only contract and relax. So, they have to be pulled back into their elongated shape by other muscles which work in the opposite direction (see 'Muscles and movement' below).

QUESTIONS

5 What are the principal differences in structure and function between skeletal and smooth muscle?

6 If a tendon is damaged, it may take a long time to heal. A damaged bone heals relatively more quickly. Suggest a reason for this difference.

MOVEMENT AND LOCOMOTION

Muscles and movement

The ends of the limb muscles are drawn out into **tendons** which attach each end of the muscle to the skeleton (Fig. 5*e*).

Figure 8 shows how a muscle is attached to a limb to make it bend at the joint. The tendon at one end is attached to a non-moving part of the skeleton while the tendon at the other end is attached to the movable bone close to the joint.

Fig. 7 Leg muscles used in running. Compare with Fig. 10.

When the muscle contracts it pulls on the bones and makes one of them move. The position of the attachment means that a small contraction of the muscle will produce a large movement at the end of the limb. Figure 9 is a model which shows how the shortening of muscle can move a limb. Figure 8 shows how a contraction of the **biceps muscle** bends (or **flexes**) the arm at the elbow, while the **triceps** straightens (or **extends**) the arm.

The non-moving end of the biceps is attached to the shoulder-blade while the moving end is attached to the ulna, near the elbow joint.

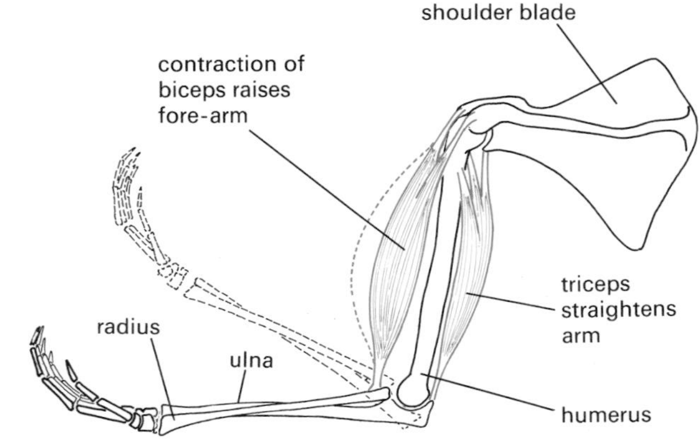

Fig. 8 Antagonistic muscles of fore-arm

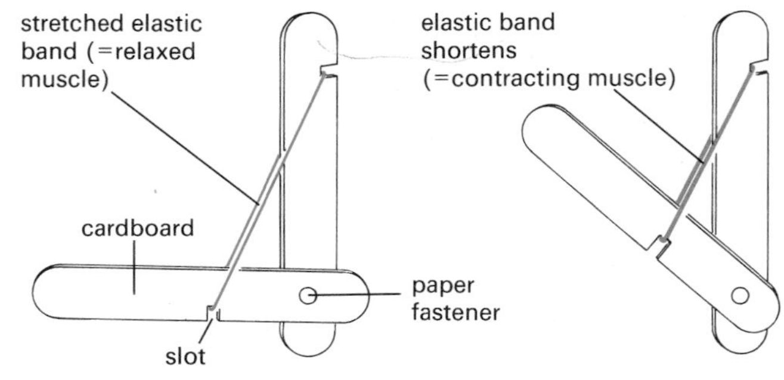

Fig. 9 Model to show how muscles pull on bones to produce movement

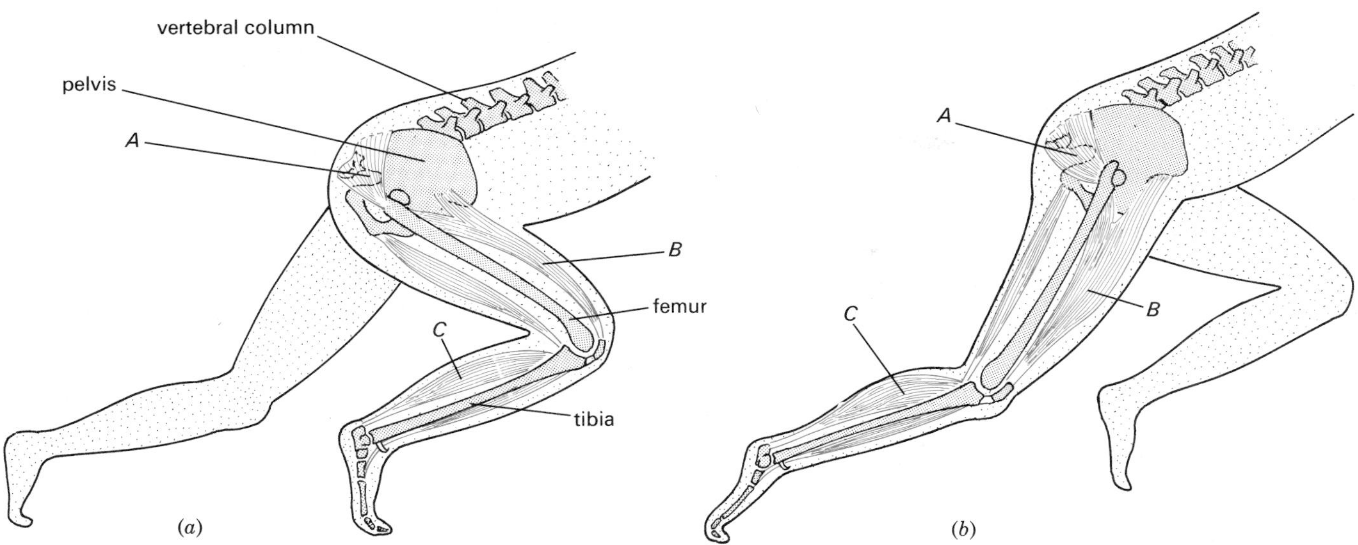

Fig. 10 Action of leg muscles. There are many other muscles which are not shown here. *A*, *B* and *C* are all contracting to straighten the leg and foot.

Limb muscles are usually arranged in pairs having opposite effects. This is because muscles can only shorten or relax, they cannot elongate, so the triceps is needed to pull the relaxed biceps back to its elongated shape after it has contracted Pairs of muscles like this are called **antagonistic** muscles. Antagonistic muscles are also important in holding the limbs steady, both muscles keeping the same state of tension or 'tone'.

The contraction of muscles is controlled by nerve impulses. The brain sends out impulses in the nerves so that the muscles are made to contract or relax in the right order to make a movement. For example, when a muscle contracts to bend the limb, its antagonistic muscle must be kept in a relaxed state.

There are many muscular activities which bring about movements but do not result in locomotion. Chewing, breathing, throwing, swallowing and blinking are examples of such movements.

All muscles need energy to make them contract. This energy comes from respiration (p. 24). In the muscle cells, glucose and oxygen, both brought by the bloodstream, are made to react together and provide the energy which causes the long muscle cells to get shorter.

On page 25 it was explained that ATP is the universal provider of energy. When ATP breaks down to ADP it provides the muscle fibres with the energy for repeated contraction. The glucose and oxygen are used to keep up the supply of ATP by converting ADP into ATP.

The chemical reactions that take place during respiration not only cause muscle contraction, they also produce heat. A contracting muscle will, therefore, become warm and the blood will carry the heat away and distribute it to other parts of the body. If this raises the general body temperature, it may lead to vaso-dilation and sweating (see p. 168).

Locomotion

Locomotion is brought about by the limb muscles contracting and relaxing in an orderly (i.e. co-ordinated) manner. Figure 7 shows a sprinter at the start of a race, and Fig. 10 shows how some of his leg muscles are acting on the bones to thrust him forward. When muscle A contracts, it pulls the femur backwards. Contraction of muscle B straightens the leg at the knee. Muscle C contracts and pulls the foot down at the ankle. When these three muscles contract at the same time, the leg is pulled back and straightened and the foot is extended, pushing the foot downwards and backwards against the ground. If the ground is firm, the straightening of the leg pushes upwards against the pelvic girdle which in turn pushes the vertebral column and so lifts the whole body upwards and forwards.

While muscles A, B and C are contracting to extend the leg, their antagonistic muscles are kept in a state of relaxation. At the end of the extension movement, muscles A, B and C relax, and their antagonistic partners contract to flex the leg.

There are at least two different types of muscle fibre in mammalian muscle, slow-contracting and fast-contracting, sometimes called type 1 (red muscle) and type 2 (white muscle). White muscle is powerful and contracts rapidly but it also fatigues quickly. Red muscle contracts more slowly and uses energy more economically.

In general terms, white muscle is important for vigorous bursts of activity, such as sprinting. Red muscle is associated with maintaining posture and with activities requiring endurance. There is much debate among physiologists about whether the ratio of red and white muscle in an individual is determined by heredity or whether the ratio can be changed by particular forms of training.

Limbs as levers

Most limb bones act as levers which greatly magnify the movement of the muscles which act on them (Fig. 11). Figure 11*c* shows how the lower arm works as a lever. The muscle supplies the **effort**; the **load** is an object held in the hand; and the elbow joint is the **fulcrum** or pivot.

(*a*) THE LEVER EFFECT

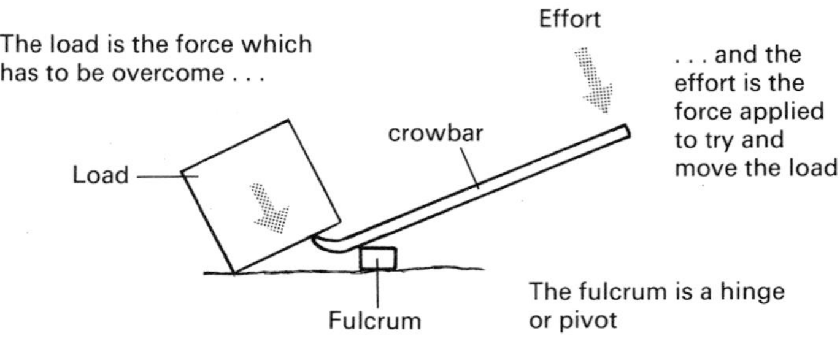

(*b*) LEVER EFFECT IN THE FOOT

(*c*) LEVER EFFECT IN THE ARM

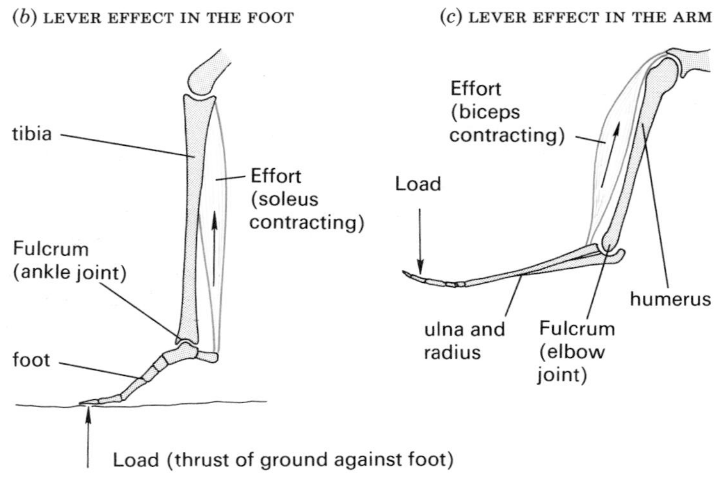

Fig. 11 The limb as a lever

PRACTICAL WORK

1. The structure of bone

Obtain two small bones, e.g. limb bones from a chicken. Place one of them in a test-tube or beaker, cover it with dilute hydrochloric acid and leave it for 24 hours, during which time the acid will dissolve most of the calcium salts in the bone. After 24 hours pour away the acid, wash the bone thoroughly with water and then try to bend it.

Take the second bone and hold one end of it in a hot Bunsen flame, heating it strongly for 2 minutes. At first the bone will char and then glow red as the fibrous organic material burns away, but it will retain its shape. Allow the bone to cool and then try crushing the heated end against the bench with the end of a pencil.

Result Both bones retain their shape after treatment, but the bone whose calcium salts had been dissolved in acid is rubbery and flexible because only the organic, fibrous connective tissue is left. The bone that had this fibrous tissue burned away is still hard but very brittle and easily shattered.

Interpretation The combination in bone of mineral salts and organic fibres produces a hard, strong and resilient structure.

QUESTIONS

7 What is the difference between the functions of a ligament and a tendon?
8 What is the main action of (*a*) your calf muscle, (*b*) the muscles in the front of your thigh and (*c*) the muscles in your forearm? If you don't already know the answer, try making the muscles contract and feel where the tendons are pulling.
9 In Fig. 10: (*a*) To what bone is the non-moving end of muscle B attached? (*b*) Which muscle is the antagonistic partner to C?

CHECK LIST

- **The vertebral column ('backbone') is made up of 33 vertebrae.**
- **The vertebral column forms the main support for the body and also protects the spinal cord.**
- **The legs are attached to the vertebral column by the hip girdle.**
- **The shoulder-blades and collar-bones form the shoulder girdle.**
- **The arms are attached to the shoulder-blades.**
- **The skull protects the brain, eyes and ears.**
- **The ribs protect the lungs, heart and liver, and also play a part in breathing.**
- **The limb joints are either ball and socket (e.g. hip and shoulder) or hinge (e.g. knee and elbow).**
- **The surfaces of the joints are covered with cartilage and lubricated with synovial fluid.**
- **Skeletal muscles are formed from fibres and can be consciously controlled.**
- **Smooth muscle is formed from layers of cells and cannot be consciously controlled.**
- **Limb muscles are attached to the bones by tendons.**
- **When the muscles contract, they pull on the bones and so bend and straighten the limb or move it forwards and backwards.**
- **Most limb muscles are arranged in antagonistic pairs, e.g. one bends and one straightens the limb.**
- **By straightening the leg and thrusting it against the ground, an animal can propel itself forwards.**

20 Teeth

FUNCTIONS OF TEETH
Incisors, canines, premolars and molars.

STRUCTURE OF TEETH
Enamel, dentine, pulp and cement.

MILK TEETH AND PERMANENT TEETH

DENTAL HEALTH
Tooth decay and gum disease: methods of prevention.

FLUORIDATION
Fluoride in drinking water.

FUNCTIONS OF TEETH

Before food can be swallowed, pieces have to be bitten off which are small enough to pass down the gullet into the stomach. Digestion is made easier if these pieces of food are crushed into even smaller particles by the action of chewing. Biting and chewing are actions carried out by our teeth, jaws and muscles (see Fig. 3, p. 183).

The teeth are given different names according to their position in the jaw (Fig. 1). In the front of each of the upper and lower jaws there are four **incisors**. Our top incisors pass in front of our bottom incisors and cut pieces off the food, as when biting into an apple or taking a bite out of a sandwich.

On each side of the incisors there is a **canine** tooth. In carnivorous mammals, like dogs, the canines are long and pointed but in humans they are similar to the incisors but a little more pointed. They function rather like extra incisors.

At the side of each jaw are the **premolars**, two on each side. They are larger than the canines and have one or two blunt points or **cusps.** At the back of each jaw are two or three **molars**. The molars are larger than the premolars and have four or more cusps.

The premolars and molars are similar in function. Their knobbly surfaces meet when the jaws are closed, and crush the food into small pieces.

As well as breaking the food into a suitable size for swallowing, the crushing action of the molars and premolars reduces food to quite small particles. Small particles of food are easier to digest than large pieces because they offer a greater surface area to the digestive enzymes.

QUESTIONS

1 How many incisors, canines, premolars and molars are there in a human upper jaw (see Fig. 1*b*)? How many teeth are there in a full set?

2 The words 'biting', 'crushing' and 'chewing' have been used several times in this section. Say exactly what you think each word means.

3 Look at the position of the jaw muscles in Fig. 3 on page 183. Which teeth do you think exert the greatest force on the food?

(*a*) TEETH AND JAWS, SIDE VIEW

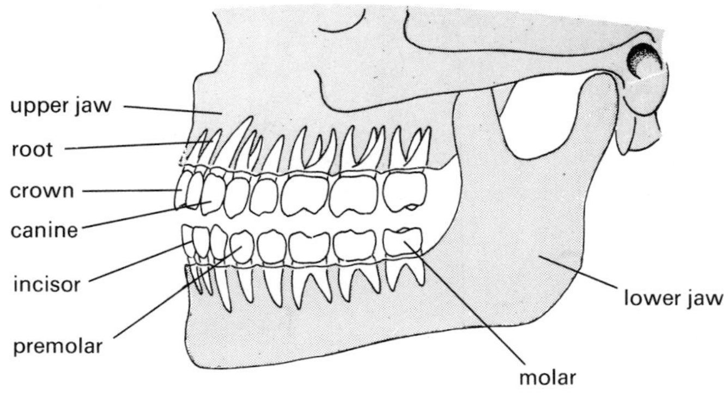

(*b*) TEETH IN UPPER JAW

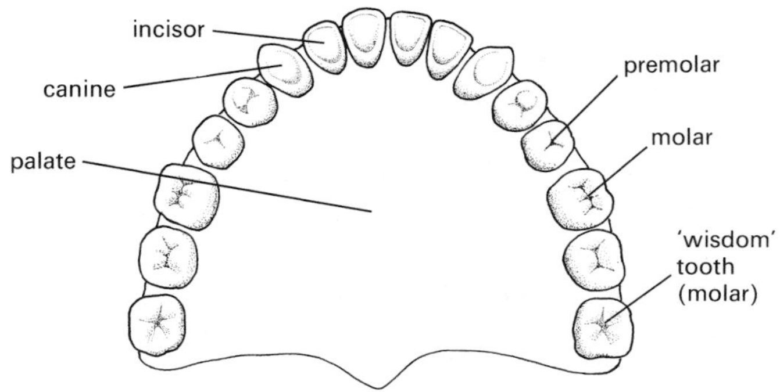

Fig. 1 Human dentition

STRUCTURE OF TEETH

Figure 2*a* shows the structure of an incisor or canine tooth as it would appear in vertical section (i.e. cut from top to bottom), and Fig. 2*b* shows a section through a molar tooth.

Enamel Enamel covers the exposed part, or **crown**, of the tooth and makes a hard biting surface. It is a non-living substance containing 97 per cent calcium salts and only 3 per cent organic matter. Although the enamel is laid down before the tooth emerges from the gum, it can be thickened and strengthened by salts deposited from the saliva or from food and drink. Fluoride ions taken up by the enamel increase its resistance to decay.

Dentine This is rather like bone and softer than enamel. It is a living tissue with threads of cytoplasm running through it. The hardness of both enamel and dentine depends on there being enough calcium in the diet and sufficient vitamin D to help the absorption of calcium in the intestine.

Pulp In the centre of the tooth is soft connective tissue. It contains cells which make the dentine and keep the tooth alive. In the pulp are blood vessels which bring food and oxygen, so that the tooth can grow at first and then remain alive when growth has stopped. There are also sensory nerve endings in the pulp, which are sensitive to heat and cold but give only the sensation of pain. If you plunge your teeth into an ice-cream, they do not feel cold but they do hurt.

Cement This is a bone-like substance which covers the root of the tooth. In the cement are embedded tough fibres which pass into the bone of the jaw and hold the tooth in place.

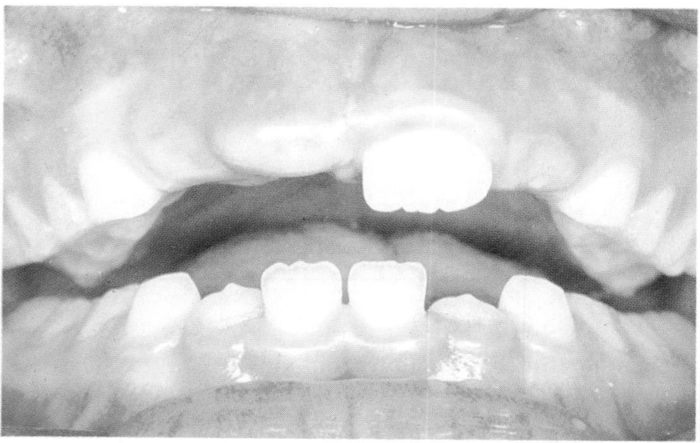

Fig. 3 Permanent teeth coming through. In the bottom jaw, the two outer incisors are emerging. Only one of the top incisors has come through.

Milk teeth and permanent teeth

Mammals have two sets of teeth in their lifetimes (Fig. 4). In humans, the first set, or **milk teeth**, grow through the gum during the first year of life and consist of four incisors, two canines and four premolars in each jaw. Between the ages of 6 and 12 years, these milk teeth gradually fall out (Fig. 3) and are replaced by the **permanent teeth**, including six molars in each jaw. The last of these molars, the 'wisdom teeth', may not grow until the age of 17 or later. In some cases they do not appear at all. If these permanent teeth are lost for any reason, they do not grow again.

QUESTIONS

4 Which of the permanent teeth are not represented in the set of milk teeth?

5 What are the differences between dentine and enamel?

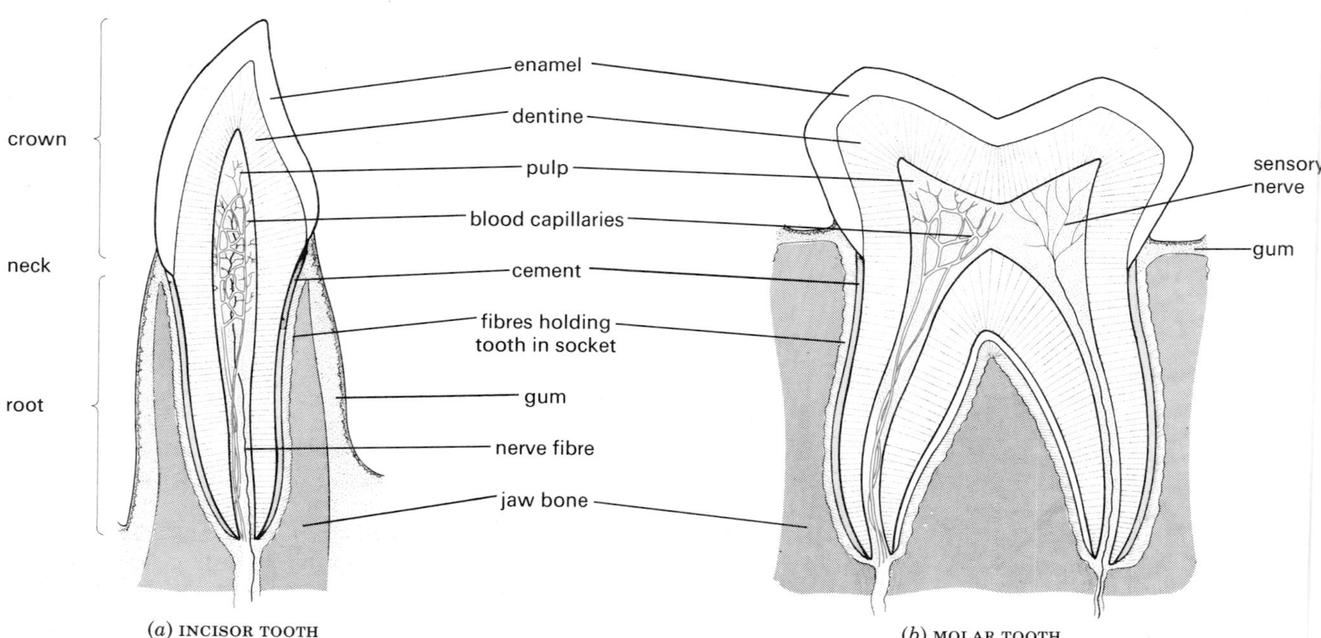

Fig. 2 Longitudinal sections through incisor and molar teeth

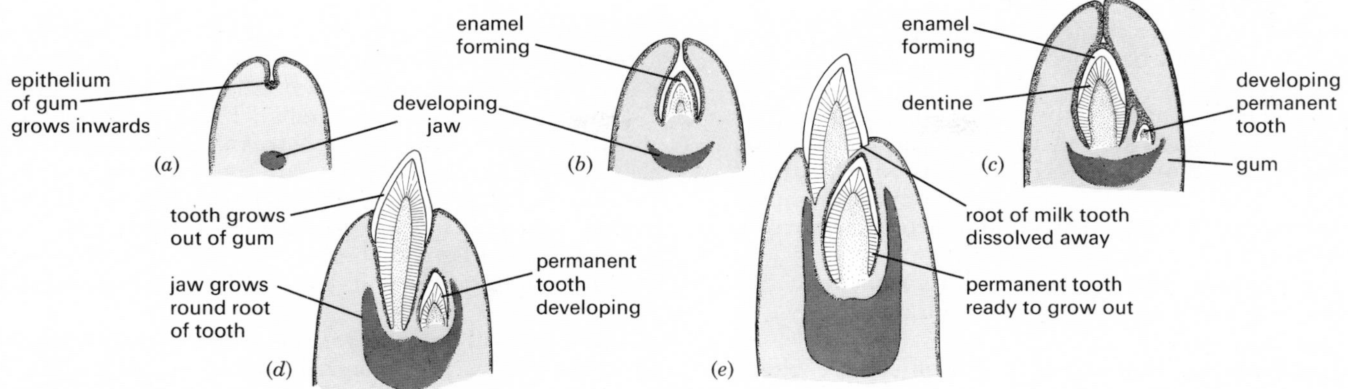

Fig. 4 Sections through lower jaw to show development and growth of milk tooth and permanent tooth

DENTAL HEALTH

The most likely reasons for the loss of teeth are dental decay and gum disease.

Dental decay (dental caries)

Decay begins when small holes (cavities) appear in the enamel (Fig. 5*b*). The cavities are caused by bacteria on the tooth surface. The bacteria produce acids which dissolve the calcium salts in the tooth enamel. The enamel and dentine are dissolved away in patches, forming cavities. The cavities reduce the distance between the outside of the tooth and the nerve endings. The acids produced by the bacteria irritate the nerve endings and cause toothache. If the cavity is not cleaned and filled by a dentist, the bacteria will get into the pulp cavity and cause a painful abscess at the root. Often, the only way to treat this is to have the tooth pulled out.

Although some people's teeth are more resistant to decay than others, it seems that it is the presence of refined sugar (sucrose) that contributes to decay.

Western diets contain a good deal of refined sugar and children suck sweets between one meal and the next. The high level of dental decay in Western society is thought to be caused mainly by the prolonged presence of sugar in the mouth.

The best way to prevent tooth decay, therefore, is to avoid eating sugar at frequent intervals either in the form of sweets or in sweet drinks such as orange squash or Coca Cola. It is often stated that eating hard fibrous food, such as raw vegetables, removes plaque and prevents decay but there is not much evidence to support this.

Brushing the teeth or rinsing the mouth does little to prevent dental decay. However, if a fluoride toothpaste is used it does help to increase the resistance of enamel to bacterial acids. Adding fluoride to drinking-water also reduces the incidence of dental decay but it is a controversial measure. Brushing the teeth is very important in the prevention of gum diseases, which causes more tooth loss than caries does.

Gum disease (periodontal disease)

There is usually a layer of saliva and mucus over the teeth. This layer contains bacteria which live on the food residues in the mouth, building up a coating on the teeth, called **plaque**. If the plaque is not removed, mineral salts of calcium and magnesium are deposited in it, forming a hard layer of 'tartar' or **calculus**. If the bacterial plaque which forms on teeth is not removed regularly, it spreads down the tooth into the narrow gap between the gum and enamel. Here it causes inflammation, called **gingivitis**, which leads to redness and bleeding of the gums and to bad breath. It also causes the gums to recede and expose the

(*a*) The teeth are free from cavities and the gums are healthy

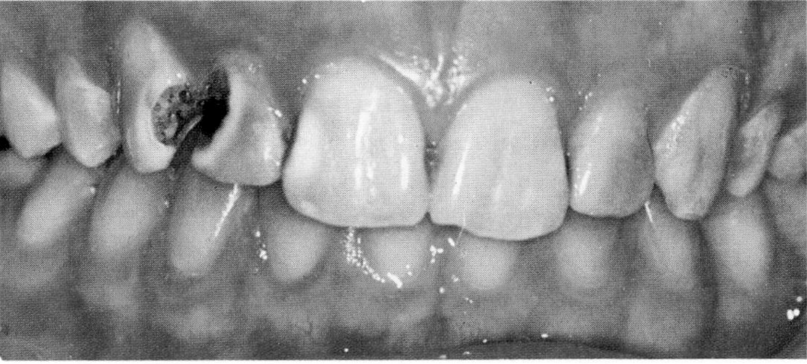

(*b*) Two of the upper teeth have cavities caused by decay

Fig. 5 Healthy and decayed teeth

cement. If gingivitis is not treated, it progresses to **periodontitis**; the fibres holding the tooth in the jaw are destroyed, so the tooth becomes loose and falls out or has to be pulled out.

There is evidence that cleaning the teeth does help to prevent gum disease. It is best to clean the teeth about twice a day using a toothbrush. No one method of cleaning has proved to be any better than any other but the cleaning should attempt to remove all the plaque from the narrow crevice between the gums and the teeth.

Drawing a waxed thread ('dental floss') between the teeth helps to remove plaque in these regions.

Disclosing tablets These contain a harmless dye which colours any plaque present on the teeth. You chew a disclosing tablet before cleaning your teeth and then rinse your mouth with water. Any plaque on your teeth will be stained red. If you now clean your teeth, you will be able to see how efficiently you remove the plaque.

Fluoridation

Fluoride ions are a fairly common constituent of drinking-water occurring naturally in concentrations of up to 5 parts per million (ppm) or more. It has been shown that in areas where water naturally contains fluoride ions, in concentrations of about 1 ppm, the incidence of decay is up to 60 per cent less than in areas containing little or no fluoride. Experiments were conducted in some American towns by adding fluoride to drinking-water in concentrations of 1 ppm, and the populations of these towns were compared with control areas with little fluoride in the water. A similar reduction in decay was found in the children, with no evidence of undesirable side effects. Concentrations of 2 ppm or more, however, tend to cause some degree of mottling. Experimental fluoridation has been carried out in the USA for over 25 years, and in Britain for over 20 years, with encouraging results.

Fluoride ions are taken up directly from the mouth by the surface layers of enamel. Fluoride which is swallowed gets into the bloodstream; 95 per cent of it is excreted by the kidneys but the fraction which remains is taken up by the teeth and bones. Fluoride makes enamel more resistant to decay and is particularly effective when the permanent teeth are still developing, but it has a beneficial effect at all ages.

Opposition to fluoridation has arisen mainly on the grounds that it is a measure forced on all people, giving them no choice in the matter. Biologically it may seem a rational adjustment of our environment to meet optimum demands. Teeth seem to need a supply of fluoride just as they need calcium and phosphate, and the best way, it is claimed, of obtaining this supply in continuous small doses is via the drinking-water.

Certainly, any adjustment of our environment which affects the health of millions of people needs very thorough consideration. Though the case for fluoridation seems to have received careful study, there are still wide differences of opinion (*a*) about the interpretations of the evidence and (*b*) about the desirability of interfering with the water supplies to achieve medical benefits as distinct from merely making it safe to drink.

If the fluoride level in your drinking-water is less than 0.3 ppm, it is possible to increase your intake by chewing fluoride tablets.

QUESTIONS

6 Does your diet include any refined sugar? If so make a list of the food substances or meals which contain it.
7 What are the most important things to do to avoid dental decay and gum disease?

CHECK LIST

- **The roots of our teeth are held in the jawbone.**
- **The crowns of the teeth are covered with enamel and project into the mouth.**
- **The pulp and the dentine are kept alive by blood vessels bringing food and oxygen.**
- **We have incisor, canine, premolar and molar teeth.**
- **The incisors and canines bite off pieces of food.**
- **The premolars and molars crush the food ready for swallowing.**
- **We have two sets of teeth in our lifetime, milk teeth (20) and permanent teeth (32).**
- **Bacteria in the mouth can cause cavities in teeth if sugar is present.**
- **Plaque is a layer which forms on the teeth; it consists of saliva, mucus, bacteria and bacterial products.**
- **If plaque is not removed, it may cause gum disease.**
- **Cleaning the teeth removes the plaque and helps prevent gum disease.**

21 The Senses

SKIN SENSES
Touch, pressure, cold.

TASTE AND SMELL
Taste buds. Chemoreceptors.

PROPRIOCEPTORS
Position of limbs. Tension of muscles.

SIGHT
Eye structure and function. Image formation. Accommodation. 3-D vision.

HEARING
Ear structure and function.

BALANCE
Semicircular canals and utriculus.

SENSORY IMPULSES
Sensations. Pain.

PRACTICAL WORK
Experiments on touch and taste.

Our senses make us aware of changes in our surroundings and in our own bodies. We have sense cells which respond to stimuli (singular = stimulus). A **stimulus** is a change in light, temperature, pressure, etc., which produces a reaction in a living organism. Structures which detect stimuli are called **receptors**. Some of these receptors are scattered through the skin while others are concentrated into special organs such as the eye and the ear.

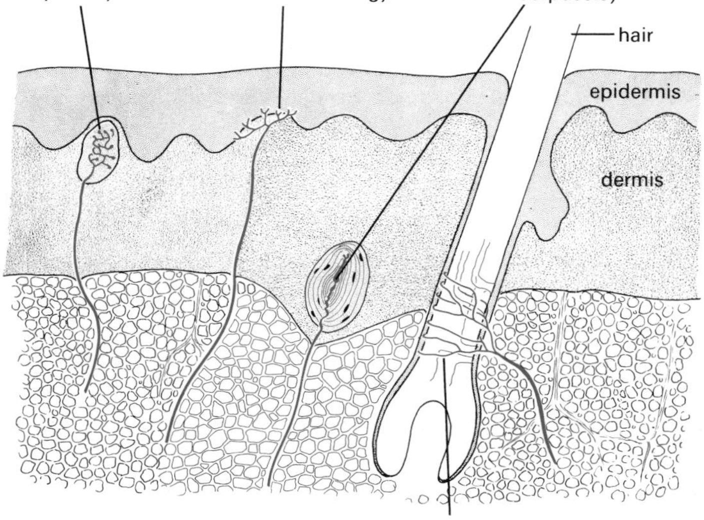

Fig. 1 The sense organs of the skin (generalized diagram)

The special property of sensory cells and sense organs is that they are able to convert one form of energy to another. Structures which can do this are called energy **transducers**. The eyes can convert light energy into the electrical energy of a nerve impulse. The ears convert the energy in sound vibrations into nerve impulses. The forms of energy which make up the stimuli may be very different, e.g. mechanical, chemical, light, but they are all transduced into pulses of electrical energy in the nerves.

SKIN SENSES

There are a great many sensory nerve endings in the skin which respond to the stimuli of touch, pressure, heat and cold, and some which cause a feeling of pain. These sensory nerve endings are very small; they can be seen only in sections of the skin when studied under the microscope (Fig. 1), and some have not yet been identified.

Some of the sensory nerve endings are **encapsulated**, i.e. they are enclosed in a capsule, e.g. the Meissner's corpuscles (touch) and the Pacinian corpuscles (pressure) (Fig. 2). Other sensory endings seem to consist only of fine branches from the nerve fibre, e.g. the 'cold' sensors or the hair plexuses. These are called **'free' nerve endings**.

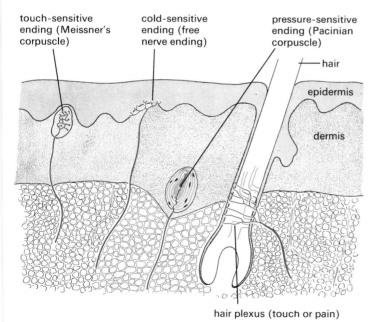

Certain regions of the skin have a greater concentration of sense organs than others. The finger-tips, for example, have a large number of touch organs, making them particularly sensitive to touch. The front of the upper arm is sensitive to heat and cold. Some areas of the skin have

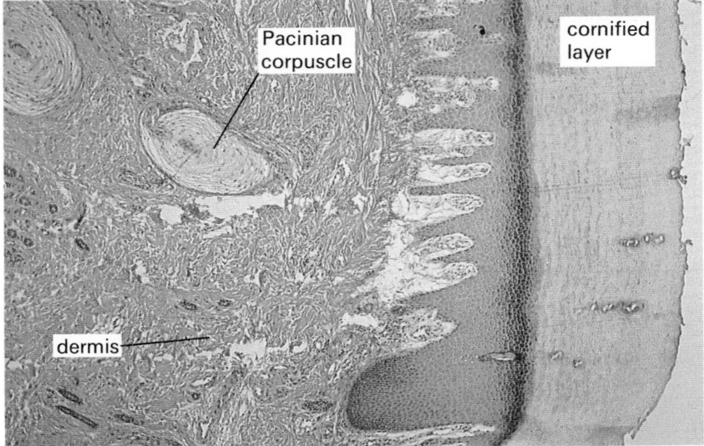

Fig. 2 Pacinian corpuscle in human skin (×60)

relatively few sense organs and can be pricked or burned in certain places without any sensation being felt.

When the nerve ending receives a stimulus, it sends a nerve impulse to the brain which makes us aware of the sensation. Generally, each type of nerve ending responds to only one kind of stimulus. For example, a heat receptor would send off a nerve impulse if its temperature were raised but not if it were touched.

QUESTIONS

1 What sensation would you expect to feel if a warm pin-head was pressed on to a touch receptor in your skin? Explain.
2 If a piece of ice is pressed on to the skin, which receptors are likely to send impulses to the brain?

TASTE AND SMELL

Taste In the lining of the nasal cavity and on the tongue are groups of sensory cells, called **chemo-receptors**, which respond to chemicals. On the tongue, these groups are called **taste-buds** and they lie mostly in the grooves of the tongue (Fig. 3). The receptor cells in the taste-buds can recognize only four classes of chemicals. These are the chemicals which give the taste sensations of sweet, sour, salt or bitter. Nearly all acids, for example, give the taste sensation we call 'sour', but a wide variety of quite different chemicals give the sensation of 'sweet'. Generally, the taste cells are sensitive to only one or two of these classes of chemical. For a substance to produce a sensation of taste, it must be able to dissolve in the film of water covering the tongue.

Smell The epithelium lining the top of the nasal cavity contains chemo-receptors. Fine processes from these chemo-receptors extend into the film of mucus which lines the nasal epithelium (Fig. 4). A wide range of airborne chemicals stimulates the nasal receptors and they send nerve impulses to the brain.

Our sense of smell is able to distinguish a great many more chemicals than our sense of taste. However, there is not yet a satisfactory classification of types of smell

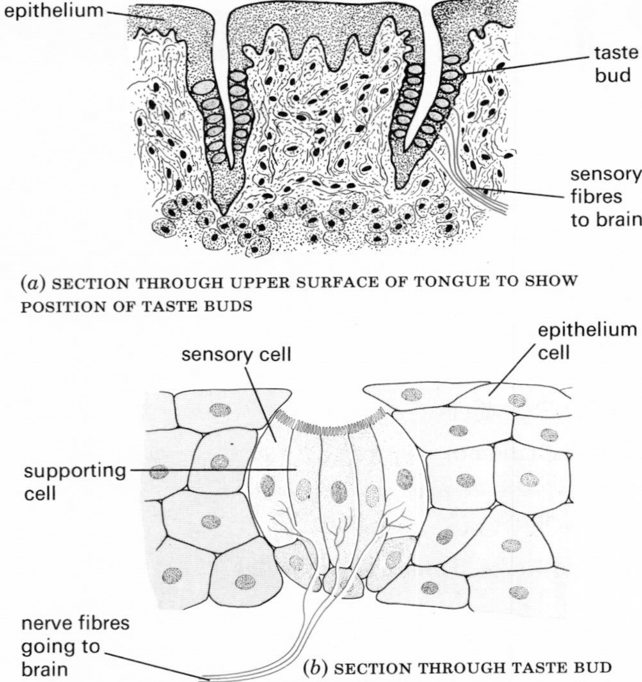

(*a*) SECTION THROUGH UPPER SURFACE OF TONGUE TO SHOW POSITION OF TASTE BUDS

(*b*) SECTION THROUGH TASTE BUD

Fig. 3 Sensory system of the tongue

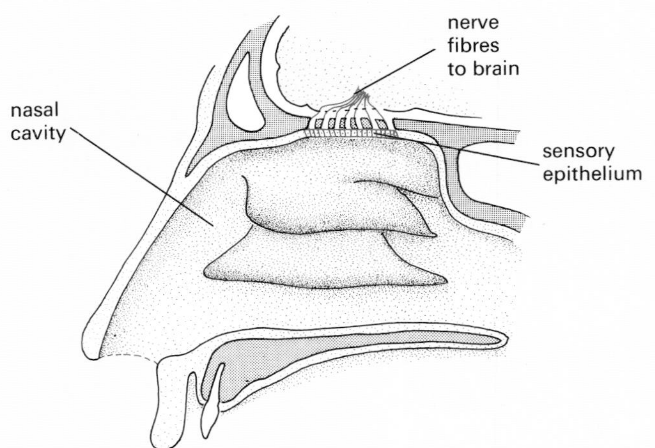

(*a*) POSITION OF CHEMO-RECEPTORS IN NASAL CAVITY

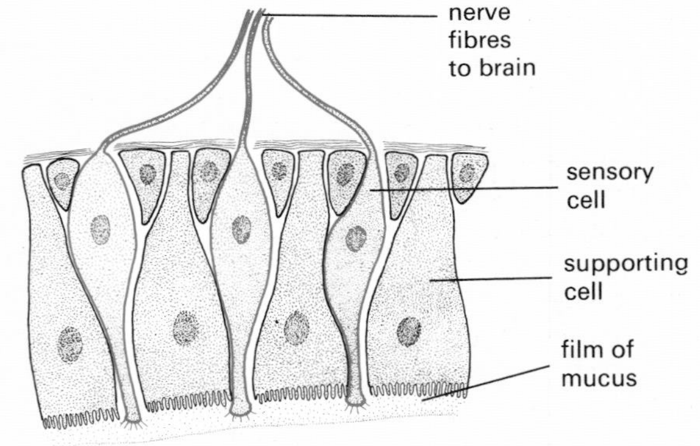

(*b*) SENSORY EPITHELIUM OF SMELL RECEPTOR

Fig. 4 Sense of smell

we can recognize, nor an explanation of how they are distinguished.

The sensation we call 'flavour' (as distinct from 'taste') is the result of the vapours, from the food in the mouth, reaching the chemo-receptors in the nose. Flavour is, therefore, largely 'smell' and we lose some of the sensation of flavour when the nose is blocked, as during a heavy cold.

The sense of smell is easily fatigued, that is, a smell experienced for a long period ceases to give any sensation and we become unaware of it, though a newcomer may detect it at once.

QUESTIONS

3 Which types of taste-buds are likely to be stimulated by lemonade?
4 Apart from the cells which detect chemicals, what other types of receptor must be present in the tongue?
5 What is the difference between taste, smell and flavour?

PROPRIOCEPTORS

Sometimes the term 'proprioceptors' refers to any kind of internal sense organ that responds to changes within the body, e.g. blood pressure. More commonly it is used to describe muscle receptors which respond to stretching. The **stretch-receptors** are specialized muscle fibres with a sensory nerve supply (Fig. 5).

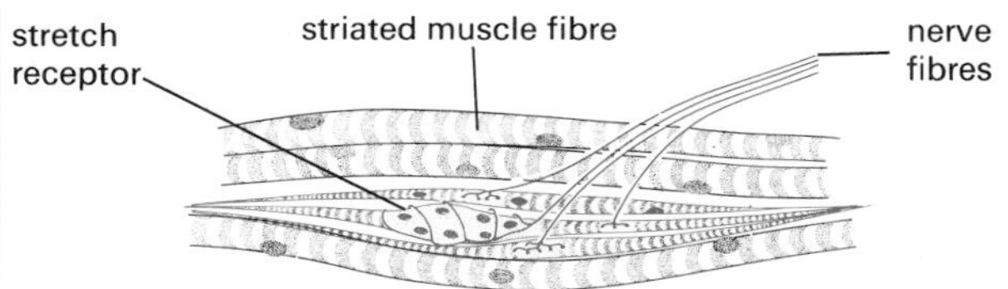

Fig. 5 Stretch receptor in muscle

They lie parallel to the other muscle fibres and are stretched when the relaxed muscle is extended by the contraction of its antagonistic partner (p. 187).

Stretch-receptors trigger off certain reflex actions (see p. 207) and control walking patterns in many land vertebrates. They enable us to 'know' about the position and movement of our limbs without having to watch them. In this way, the proprioceptors play an important part in co-ordinated movement.

Proprioceptors also represent an important means of **'feedback'** to the central nervous system (p. 210), indicating the degree of tension in the muscles and the position of the limbs. The human upright posture is maintained by keeping opposing sets of muscles in a state of tension (tone). However, it is impossible to stand perfectly still; we sway and tilt imperceptibly all the time. If we sway slightly forward, the stretch-receptors in the calf muscles will be stimulated and set off a reflex (p. 207) which tightens the calf muscles themselves and other muscles in the back of the legs. This will restore the upright position.

QUESTION

6 Refer back to Figs 8 and 10 on pages 186 and 187. In which muscles will the stretch-receptors be stimulated?

SIGHT

If you are not already familiar with the way in which lenses work, you are advised to study Fig. 6 before reading the next section.

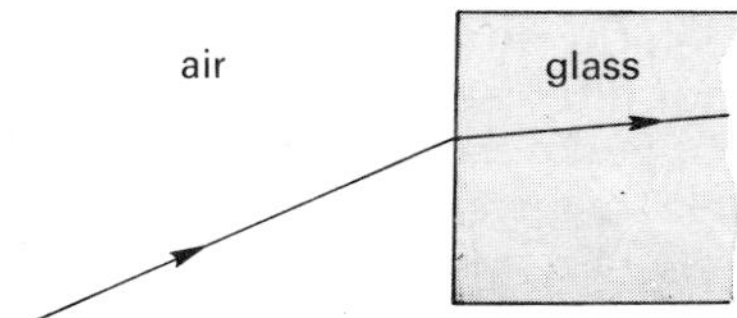

(*a*) When a ray of light passes, at an angle, from air to glass or from air to water, the ray is bent slightly, as shown. The bending is called 'refraction'.

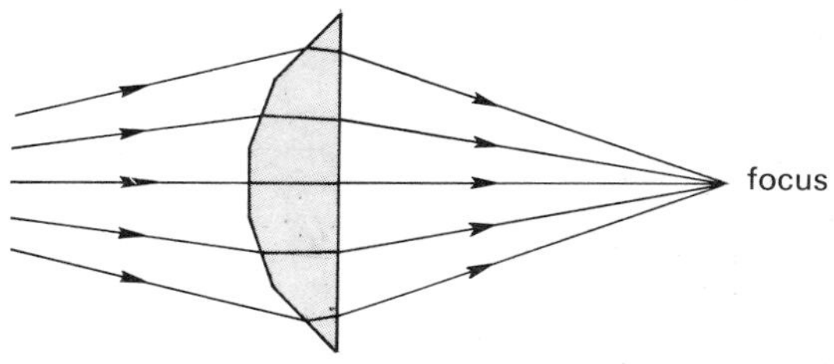

(*b*) If the rays pass through the curved glass surface shown here (a lens), they are bent towards each other and come to a focus.

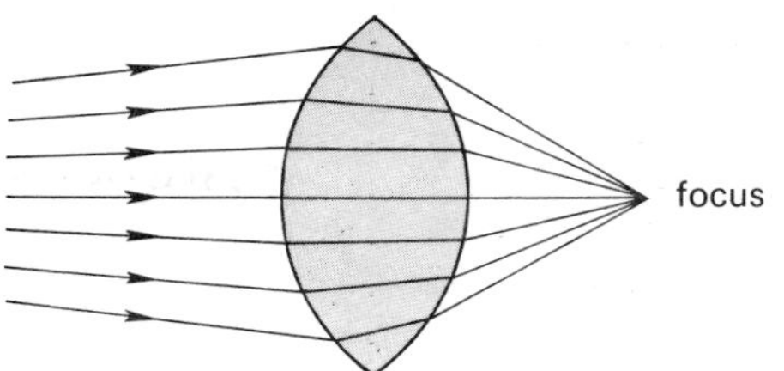

(*c*) A thick lens bends light more . . .

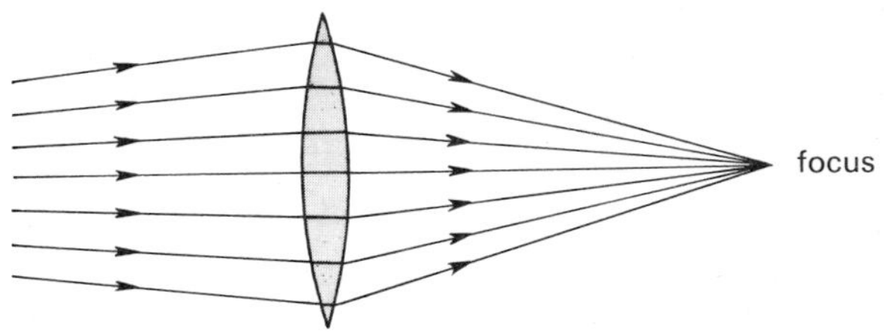

. . . than a thin lens.

Fig. 6 How a convex lens works

The eye

The structure of the eye is shown in Figs 7 and 8. The **sclera** is the tough, white outer coating. The front part of the sclera is clear and allows light to enter the eye. This part is called the **cornea**. The **conjunctiva** is a thin epithelium which lines the inside of the eyelids and the front of the sclera, and is continuous with the epithelium of the cornea.

The eye contains a clear liquid whose outward pressure on the sclera keeps the spherical shape of the eyeball. The liquid in the back part of the eye is jelly-like and called **vitreous humour**. The **aqueous humour** in the front part is watery.

The **lens** is a transparent structure, held in place by a ring of fibres called the **suspensory ligament**. Unlike the lens of a camera or a telescope, the eye lens is flexible and can change its shape. In front of the lens is a disc of tissue called the **iris**. It is the iris we refer to when we describe the colour of the eye as brown or blue. There is a hole in the centre of the iris called the **pupil**. This lets in light to the rest of the eye. The pupil looks black because all the light entering the eye is absorbed by the black pigment in the **choroid**. The choroid layer, which contains many blood vessels, lies between the retina and the sclera. In the front of the eyeball, it forms the iris and the **ciliary body** (see p. 198). The ciliary body produces aqueous humour.

The internal lining at the back of the eye is the **retina** and it consists of many thousands of cells which respond to light. When light falls on these cells, they send off nervous impulses which travel in nerve fibres, through the **optic nerve**, to the brain and so give rise to the sensation of sight.

Tear glands under the top eyelid produce tear fluid. This is a dilute solution of sodium chloride and sodium hydrogencarbonate. The fluid is spread over the eye surface by the blinking of the eyelids, keeping the surface moist and washing away any dust particles or foreign bodies. Tear fluid also contains an enzyme, **lysozyme**, which attacks bacteria.

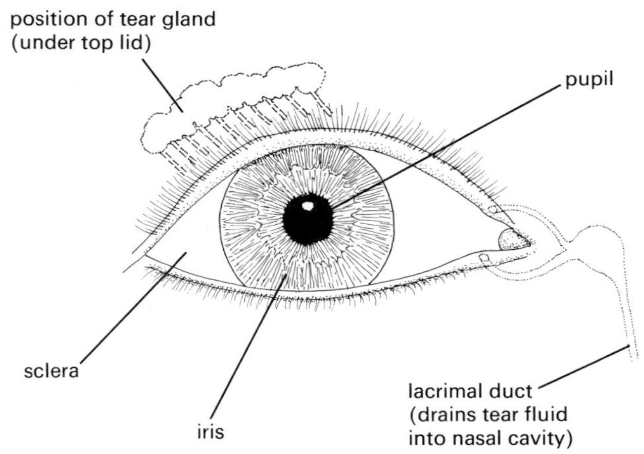

Fig. 7 Appearance of right eye from the front

Vision

Figures 9 and 10 explain how light from an object produces a focused **image** on the retina (like a 'picture' on a cinema screen). The curved surfaces of the cornea and lens both 'bend' the light rays which enter the eye, in such a way that each 'point of light' from the object forms a 'point of light' on the retina. These points of light will form an image, upside-down and smaller than the object.

The cornea and the aqueous and vitreous humours are mainly responsible for the 'bending' (refraction) of light. The lens makes the final adjustments to the focus (Fig. 9*b*).

The pattern of sensory cells stimulated by the image will produce a pattern of nerve impulses sent to the brain. The brain interprets this pattern, using its past experience and learning, and so forms an impression of the real size, distance and upright nature of the object.

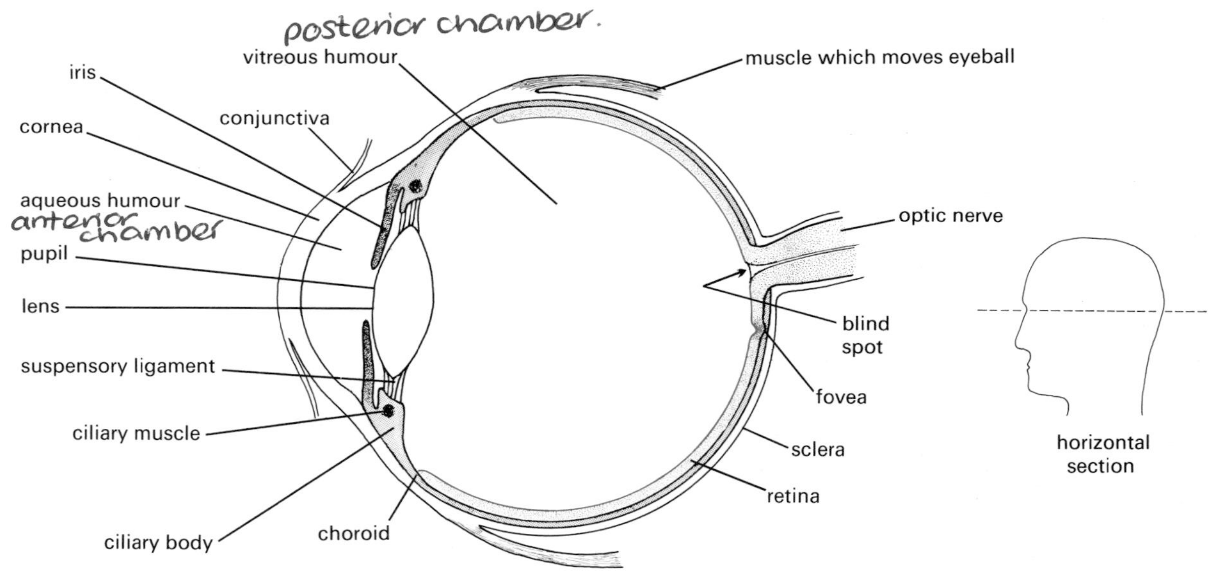

Fig. 8 Horizontal section through left eye

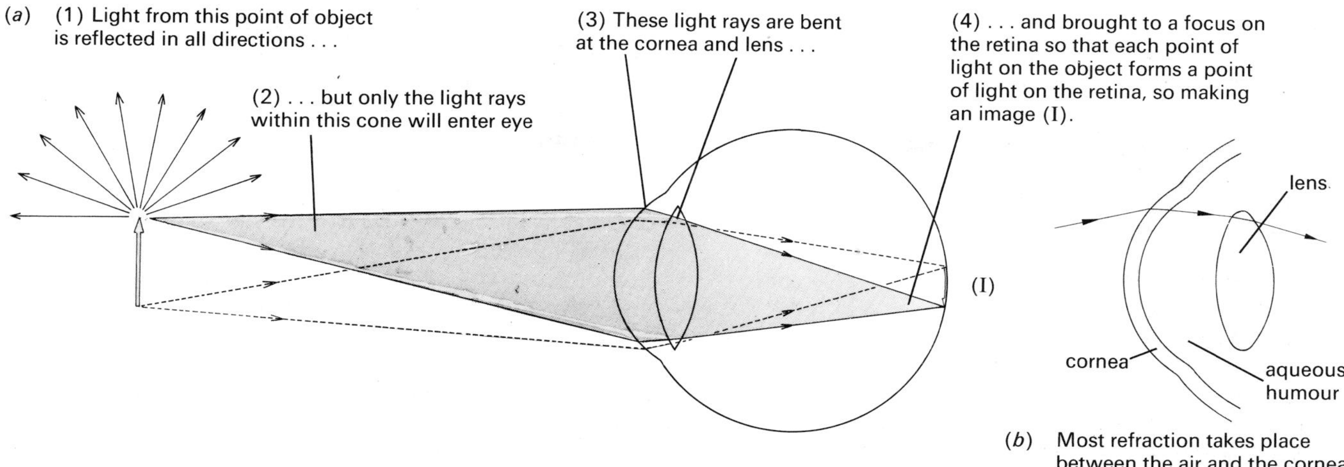
(a) (1) Light from this point of object is reflected in all directions . . .
(2) . . . but only the light rays within this cone will enter eye
(3) These light rays are bent at the cornea and lens . . .
(4) . . . and brought to a focus on the retina so that each point of light on the object forms a point of light on the retina, so making an image (I).
(I)
lens
cornea
aqueous humour
(b) Most refraction takes place between the air and the cornea

Fig. 9 Image formation on the retina

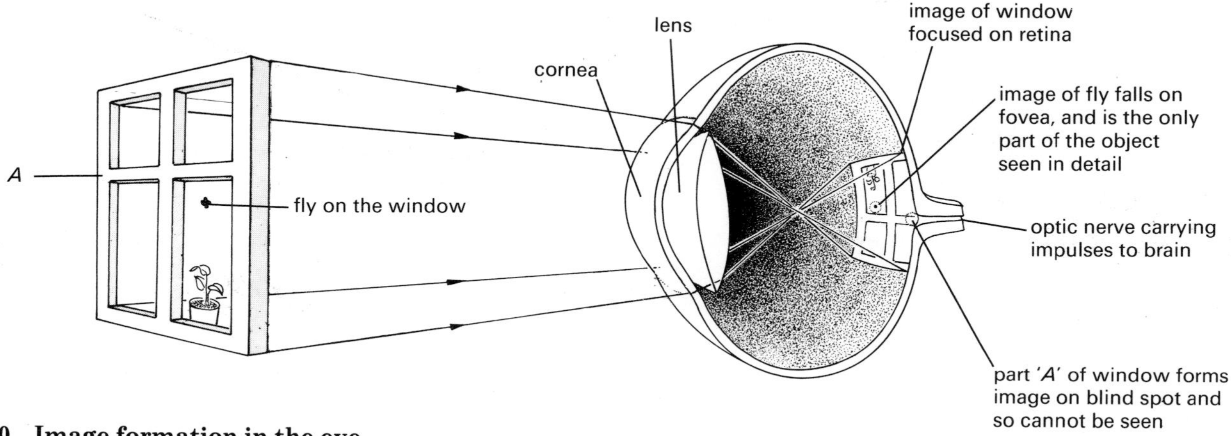
lens
cornea
image of window focused on retina
image of fly falls on fovea, and is the only part of the object seen in detail
A
fly on the window
optic nerve carrying impulses to brain
part 'A' of window forms image on blind spot and so cannot be seen

Fig. 10 Image formation in the eye

Blind spot At the point where the optic nerve leaves the retina, there are no sensory cells and so no information reaches the brain about that part of the image which falls on this **blind spot** (see Fig. 11).

Retina The millions of light-sensitive cells in the retina are of two kinds, the **rods** and the **cones** (according to shape). The cones enable us to distinguish colours. There are thought to be three types of cone cell. One type responds best to red light, one to green and one to blue. If all three types are equally stimulated we get the sensation of white. The cone cells are concentrated in a central part of the retina, called the **fovea** (Fig. 8), and when you study an object closely, you are making its image fall on the fovea.

Fovea Only in the fovea is it possible for the eye and brain to make a really detailed analysis of the colour and shape of an object. Although we can see objects included in a zone of about 100° from each eye, only those objects within a 2° zone can be seen in detail (Fig. 12). This is much less than people imagine and means, for example, that only about two letters in any word on this page can be studied in detail. It is the constant movement of the eyes which enables you to build up an accurate picture of a scene.

Fig. 11 The blind spot. Hold the book about 50 cm away. Close the left eye and concentrate on the cross with the right eye. Slowly bring the book closer to the face. When the image of the dot falls on the blind spot it will seem to disappear.

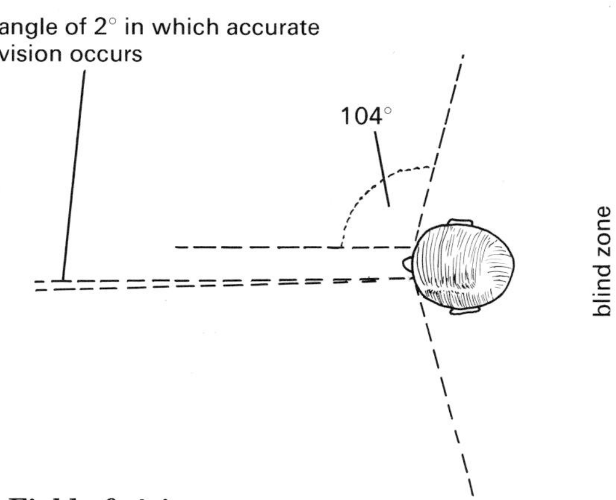
angle of 2° in which accurate vision occurs
104°
blind zone

Fig. 12 Field of vision

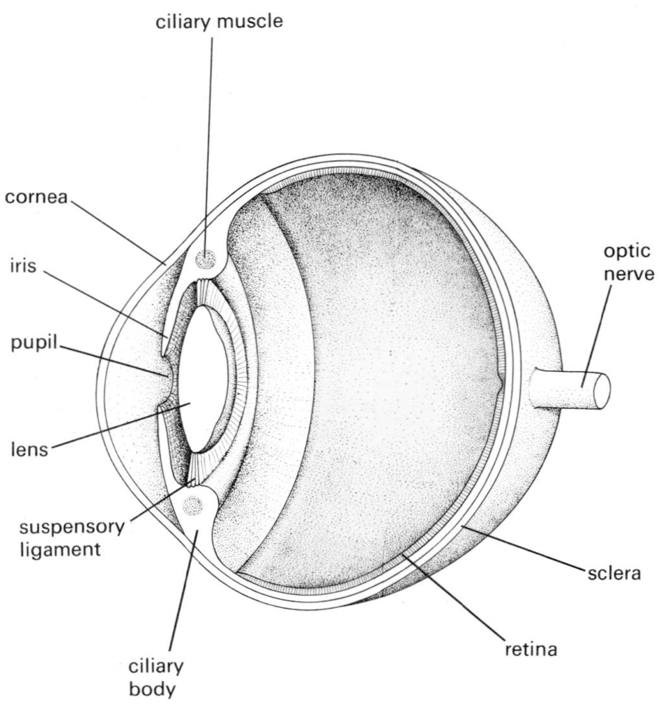

Fig. 13 Vertical section through the left eye

Accommodation (focusing)

The eye can produce a focused image of either a near object or a distant object. To do this the lens changes its shape, becoming thinner for distant objects and fatter for near objects. This change in shape is caused by contracting or relaxing the **ciliary muscle** which forms a circular band of muscle in the ciliary body (Figs 13 and 14). When the ciliary muscle is relaxed, the outward pressure of the humours on the sclera pulls on the suspensory ligament and stretches the lens to its thin shape. The eye is now accommodated (i.e. focused) for distant objects (Figs 14*a* and 15*a*). To focus a near object, the ciliary muscle contracts to a smaller circle and this takes the tension out of the suspensory ligament (Figs 14*b* and 15*b*). The lens is elastic and flexible and so is able to change to its fatter shape. This shape is better at bending the light rays from a close object.

Control of light intensity

The amount of light entering the eye is controlled by altering the size of the pupil. If the light intensity is high, it causes a contraction in a ring of muscle fibres (a sphincter) in the iris. This reduces the size of the pupil and cuts down the intensity of light entering the eye. High intensity light can damage the retina, so this reaction has a protective function.

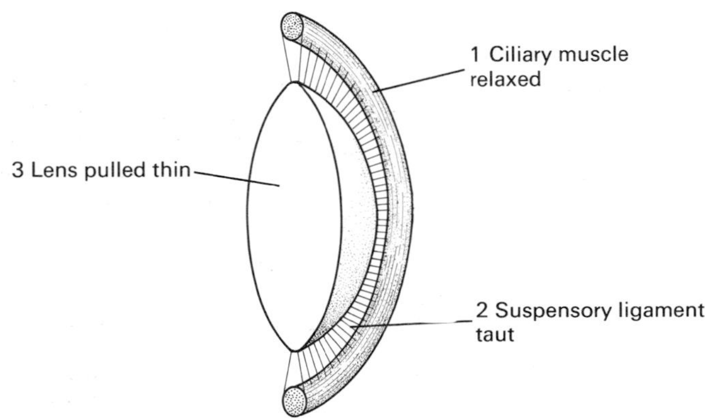

(*a*) ACCOMMODATED FOR DISTANT OBJECT

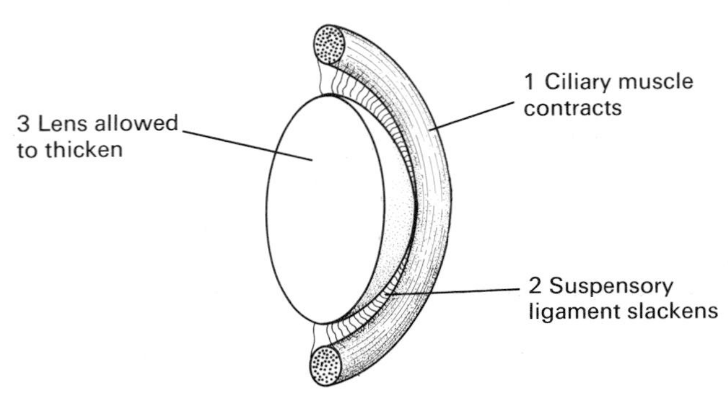

(*b*) ACCOMMODATED FOR NEAR OBJECT

Fig. 14 How accommodation is brought about

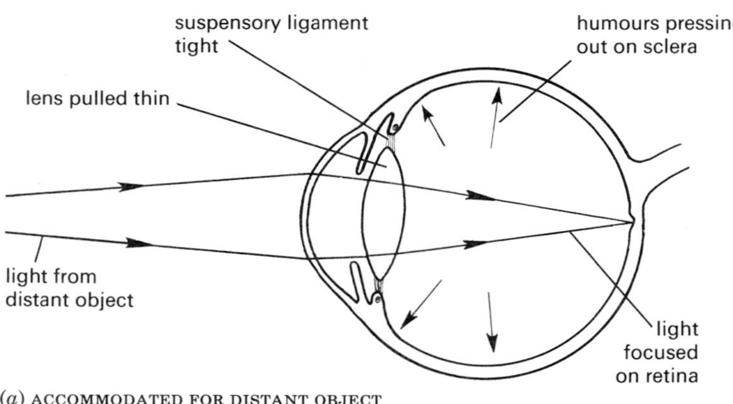

(*a*) ACCOMMODATED FOR DISTANT OBJECT

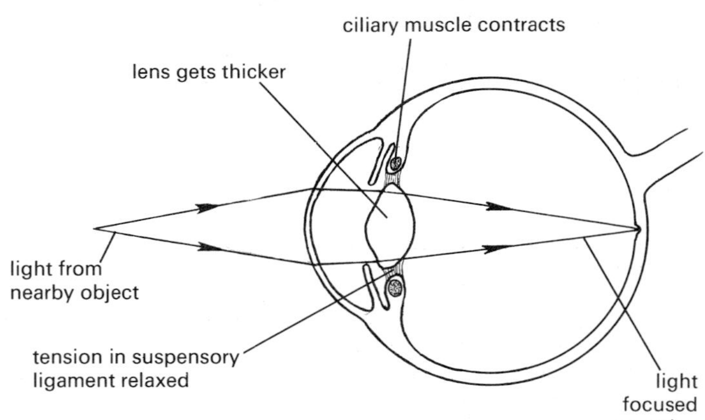

(*b*) ACCOMMODATED FOR NEAR OBJECT

Fig. 15 Accommodation

In low light intensities, the sphincter muscle of the iris relaxes and muscle fibres running radially (i.e. like wheel spokes) contract. This makes the pupil enlarge and admits more light (see Experiment 4, p. 203).

The change in size of the pupil is caused by an automatic reflex action (p. 207); you cannot control it consciously.

3-D vision and distance judgement

When we look at an object, each eye forms its own image of the object. So, two sets of impulses are sent to the brain. The brain somehow combines these impulses so that we see only one object and not two. However, because the eyes are spaced apart, they do not have the same view of the object. In Fig. 16, the left eye sees more of the left side of the box and the right eye sees more of the right side. When the brain combines the information from both eyes, it gives the impression that the object is three-dimensional (3-D) rather than flat.

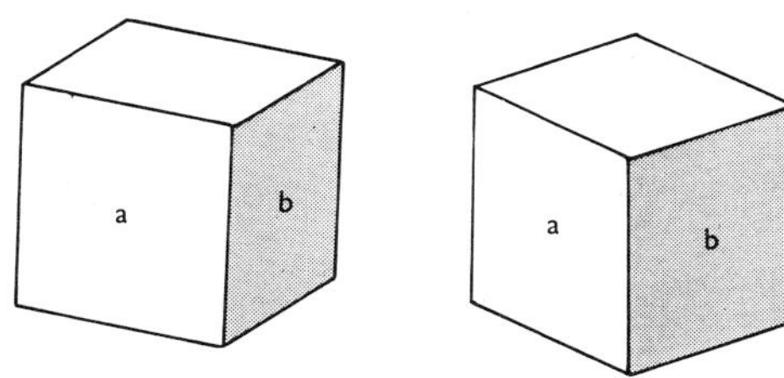

Fig. 16 Different views of the same cube seen by left and right eyes. The left eye sees more of side *a*. The right eye sees more of side *b*.

In order to look at a nearby object, the eyes have to turn inwards slightly. (Trying to focus on a very close object will make you look 'cross-eyed'.) Stretch-receptors in the eye muscles will send impulses to the brain and make it aware of how much the eyes are turning inwards. This is one way by which the brain may be able to judge how close an object is, but it is not likely to be much use for objects beyond 15 metres. We probably judge greater distances by comparing the sizes of familiar objects. The further away an object is, the smaller will be its image on the retina. We use many other clues to help us to judge distances.

We are able to experience 3-D vision and to judge distances effectively because our eyes are set in the front of our head. This gives us what is called **binocular vision**. Animals with eyes set at the side of the head, e.g. rabbits and many birds, do not have binocular vision.

QUESTIONS

7 In Fig. 10, what structures of the eye are not shown in the diagram?
8 In Fig. 9, explain what the broken lines are meant to represent.
9 (*a*) If your ciliary muscles are relaxed, are your eyes focused on a near or a distant object? Explain.
(*b*) If you dissected a cow's eye, would you expect the lens to be at its thinnest or its fattest shape? Explain.
10 How do you think animals without binocular vision are able to judge distances?

HEARING

The hearing apparatus is enclosed in bone on each side of the skull, just behind the jaw hinge. Sound vibrations travel through a short, wide tube (the outer ear) and are converted to nerve impulses by the apparatus in the middle ear and inner ear. Figure 18 on p. 200 shows the structure of the ear.

Outer ear

Sound is the name we give to the sensation we get as a result of vibrations in the air. These vibrations are pulses of compressed air. They enter the tube of the outer ear and hit the **ear-drum**, a thin membrane like a drum-skin across the inner end of the tube. The air vibrations cause the ear-drum to vibrate backwards and forwards. If there are 200 pulses of compressed air every second, the ear-drum will move backwards and forwards at the same rate.

The ear **pinnae** are the flaps of skin and elastic cartilage which project from the sides of the head and which we usually call our 'ears'. The ear pinnae of mammals, such as dogs and cats, help to direct sound vibrations into the outer ear and enable the animal to locate the source of a sound. It is possible that our ear pinnae have a similar function.

Middle ear

This is a cavity with air in it. It contains a chain of tiny bones or **ossicles**. The first of these ossicles, the **malleus**, is attached to the ear-drum, and the inner ossicle, the **stapes**, fits into a small hole in the skull called the **oval window**. The malleus is connected to the stapes by the

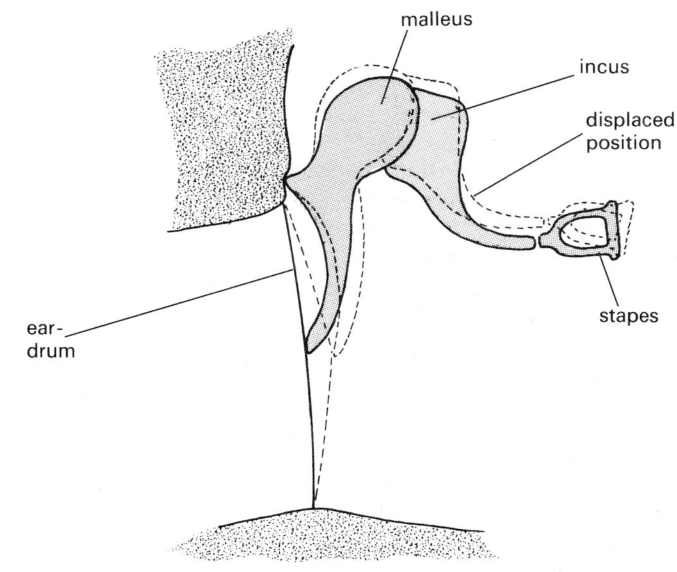

Fig. 17 Movement of the ear ossicles in transmitting sound

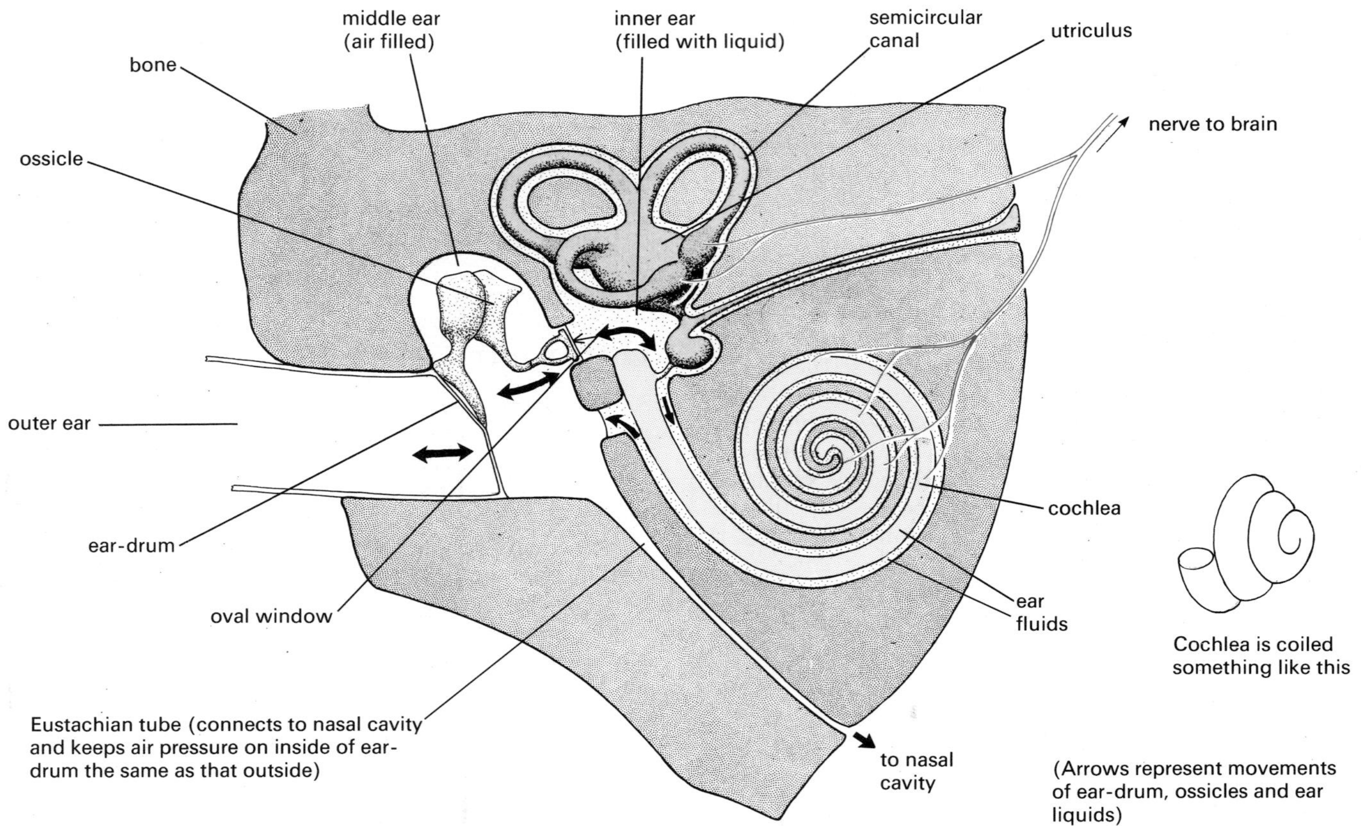

Fig. 18 Diagram of the ear

incus (Fig. 17). When the ear-drum vibrates back and forth, it forces the ossicles to vibrate in the same way, so that the stapes moves rapidly backwards and forwards like a tiny piston in the oval window. The way the ossicles are attached to each other increases the force of the vibrations and so amplifies the sound.

A narrow tube, called the **Eustachian tube**, connects the nasal cavity (see Fig. 6, p. 127) to the middle ear. Although the Eustachian tube is normally closed, it does open to admit or release air if the air pressure on the outside of the ear-drum changes. Air pressure falls when going up a mountain or in an aircraft. The action of swallowing often causes the Eustachian tube to open and gives a 'popping' sound if there has been a change of pressure.

Inner ear

This is where the vibrations are changed into nerve impulses. The inner ear contains liquid, and the vibrations of the ossicles are passed to this liquid. The sensitive part of the inner ear is the **cochlea**, a coiled tube with sensory nerve endings in it. When the liquid in the cochlea is made to vibrate, the nerve endings send off impulses to the brain. The nerve endings at the inner (top) end of the cochlea are sensitive to low-frequency vibration (low notes) and those at the outer end are sensitive to high-frequency vibrations (high notes). So, if the brain receives nerve impulses coming from the first part of the cochlea, it interprets this as a high-pitched noise or a high musical note. If the impulses come from the top end of the cochlea, the brain recognizes them as being caused by a low note.

In fact, the way in which vibrations of different frequencies are converted to nerve impulses by the cochlea has not been fully worked out. It is certainly more complicated than the simplified account given here.

QUESTIONS

11 What is the function of (*a*) the ear-drum, (*b*) the ossicles, (*c*) the liquid of the inner ear, (*d*) the cochlea?

12 (*a*) How does the function of a sensory cell in the retina differ from the function of a sensory cell in the cochlea?

(*b*) If nerve impulses from the cochlea were fed into the optic nerve, what sensations would you expect if somebody clapped their hands?

13 Sometimes, as a result of catching a cold, the middle ear becomes filled with a clear sticky fluid. Why do you think this causes deafness? (The fluid usually drains away through the Eustachian tube when the cold goes.)

BALANCE

Semicircular canals

These are in the inner ear (see Fig. 18), but do not play a part in hearing. There are two vertical (upright) canals at right angles to each other and one horizontal canal. Each canal has a swelling at one end called an **ampulla** (Fig. 19). In each ampulla there is a structure called a **cupula**. This is rather like a swing door which can be pushed either way by the liquid in the semicircular canal (Fig. 20). When the body turns round, the liquid in the canals lags slightly behind and pushes the cupula to one side. The cupula pulls on sensory hairs in the ampulla and these send nerve impulses to the brain.

The impulses to the brain inform it of the direction and speed of rotation. The horizontal canal responds best to rotations in the horizontal plane, e.g. twisting the body round while in an upright position. The vertical canals respond particularly to tilting movements of the body forwards, backwards or sideways.

Utriculus

This is a sac filled with liquid (Fig. 19). Resting on the floor of the sac is a small plate made of a dense, gelatinous (jelly-like) substance containing some chalky granules which increase its weight. Under the gelatinous plate there are sensory cells with hair-like projections, similar to cilia. These sensory 'hairs' are embedded in the jelly (Fig. 21). When the head tilts to one side, the gelatinous plate pulls on the sensory 'hairs' and their cells fire off nerve impulses to the brain. These impulses 'inform' the brain of the new position of the head. As a result, the brain may send impulses to the body to make certain muscles contract and bring the head into an upright position again.

The semicircular canals respond most strongly to rotational movements of the body. The utriculus responds mainly to changes of posture. Both these sense organs help us to hold our position when standing still and to keep our balance when moving about.

We also make use of information reaching the brain from our eyes, from pressure receptors in our feet and tension receptors in the muscles to build up an impression of our posture and movement.

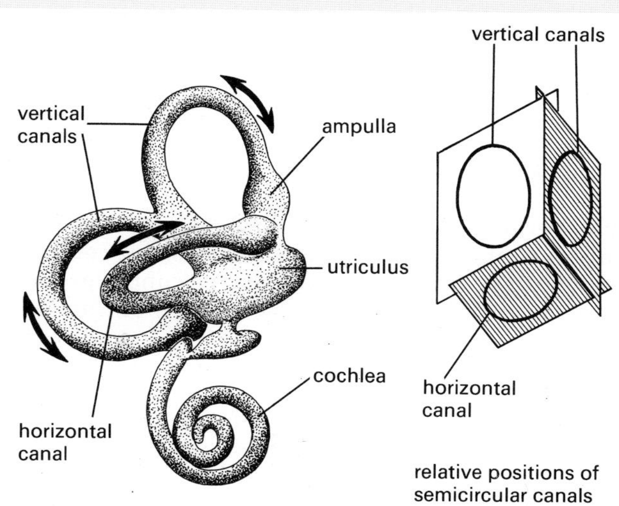

Fig. 19 Semicircular canals (arrows show direction of rotation which stimulates each canal).

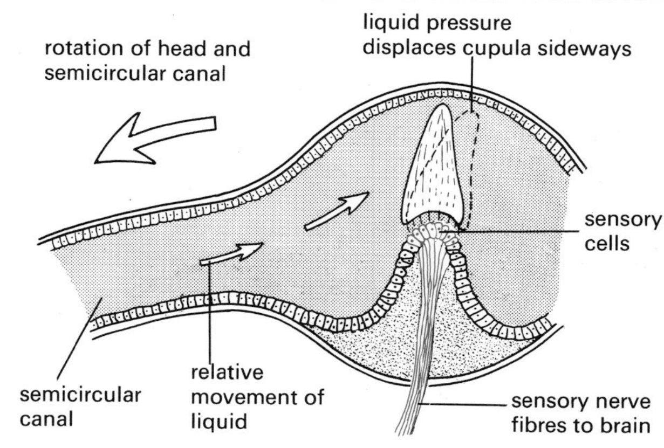

Fig. 20 Inside the ampulla

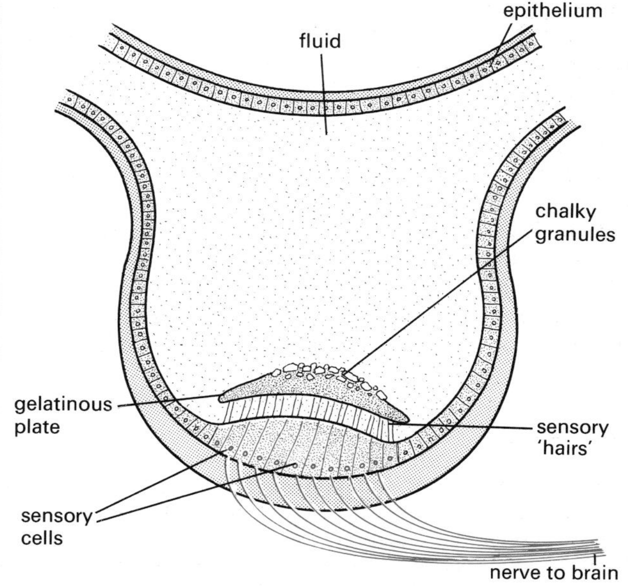

Fig. 21 Section through utriculus

QUESTIONS

14 (*a*) Which semicircular canals are likely to be stimulated the most while you are turning a somersault?

(*b*) What other sensory information will be reaching the brain while you are doing this?

15 Some people attribute 'sea-sickness' to the conflicting sensory information from the eyes and the organs of balance. Assuming you are below deck, in what way does the sensory information conflict?

SENSORY IMPULSES AND THEIR INTERPRETATION

Specificity of sense organs

In general, a particular sensory cell can respond to only one kind of stimulus. A light-sensitive cell in the eye does not respond to sound-waves and a pressure receptor in the skin is not affected by the stimulus of heat.

However, some sensory endings are less specific than this. The sensory nerve endings in the skin, which produce the sensation of pain, respond to a variety of stimuli, including pressure, heat and cold. The ear-lobe, which contains only free nerve endings and hair plexuses, can detect heat, cold and pressure. The function of a skin receptor cannot always be deduced from its structure.

Intensity of sensation

A strong stimulus usually produces a more pronounced sensation than a weak stimulus. This is probably due to (*a*) the increase in frequency of the nerve impulses generated, (*b*) the greater number of sense organs stimulated in the area, and (*c*) the stimulation of a number of sensory cells which do not respond at all unless the stimulus is intense. Many sensory organs are groups of cells, some of which are triggered off by the slightest stimulus while others need a powerful stimulus to affect them. When these latter are activated, the stimulus is recognized as being stronger than usual.

Vigorous stimulation does not affect the quality or intensity of the electrical impulse travelling in the nerve fibres but increases the total number of these impulses reaching the brain.

Stimulation and conduction of impulses

The sense organ or sense cell is connected to the brain or spinal cord by nerve fibres. When the sense organ receives an appropriate stimulus it sets off an electrical impulse which travels along the nerve fibre to the brain or spinal cord. When the impulse reaches one of these centres it may produce an automatic or reflex action, or record an impression by which we feel the nature of the stimulus and where it was applied.

The sense organs of one kind and in a definite area are connected with one particular region of the brain. It is the region of the brain to which the impulse comes that gives rise to the knowledge about the nature of the stimulus and where it was received. For example, if the regions of the brain receiving impulses from the right leg were suppressed by drugs, no amount of stimulation of the sensory endings would produce any sensation at all, although the sense organs would still be functioning normally. On the other hand, if a region of the brain dealing with impulses from sense organs in the leg is stimulated by any means, the sensations produced seem to be from the leg. Even when a limb has been amputated the region of the brain to which it originally sent impulses is still active and may produce sensations of pain from the 'phantom limb'.

Another important consideration is the fact that the impulses transmitted along the nerve fibres are fundamentally all exactly alike. It is not the sensations themselves that are carried but simply a surge of electricity, and this is so whether it is a heat organ or a touch organ that sets off the impulse. It is only in the brain that the stimulus is identified, according to the region of the brain which the impulse enters. For example, if the nerves from the arm and leg were changed over just before they entered the brain, stubbing one's toe would produce a sensation of pain in the arm or hand.

Pain

A sensation of pain usually accompanies a response to a potentially dangerous stimulus, e.g. touching something hot or sharp. Pain is not an essential part of the quick, reflex action (p. 207) which removes the hand from the source of the stimulus, but it does help us to avoid the same situation in the future. Pain, therefore, plays a part in learning and survival.

QUESTIONS

16 It is possible to connect a nerve fibre to an amplifier and hear a 'blip' every time a nerve impulse passes. Suppose the fibre comes from a touch receptor in the skin: (*a*) How would the sound heard from the amplifier differ for a light touch and a hard pinch? (*b*) How might the sound differ if the amplifier is connected to a pressure receptor?

17 Chemicals such as sugar and saccharine both taste sweet and yet they are chemically quite different. Middle C on the piano has a frequency of 264 vibrations per second whereas D has 297 vibrations per second. The difference is small and yet the two notes are easily distinguished.

What are the properties of the sense organs concerned which make for poor discrimination of chemicals and precise discrimination of sounds?

18 Most animals have a distinct head end and tail end. Why do you think the main sensory organs are concentrated at the head end?

PRACTICAL WORK

All the experiments suggested here are best done by three people working together: an **experimenter** who applies or offers the stimuli, the **subject** who is given the stimuli, and a **recorder** who watches and makes a note of the responses made by the subject.

1. Sensitivity to touch

The experimenter marks a regular pattern of dots on the back of the subject's hand. To do this he uses an ink-pad and a rubber stamp like the one in Fig. 22. He also stamps the same pattern on to a piece of paper for the recorder to use. The experimenter now tests the sensitivity of the subject's skin by pressing a fine bristle on to each dot in turn. The bristle, e.g. a horsehair, is glued to a wooden handle or held in forceps and pressed on to each mark until the bristle just starts to bend. The subject must not look, and simply says 'yes' if he feels the stimulus. The recorder marks on his pattern how many spots in each row were sensitive to touch.

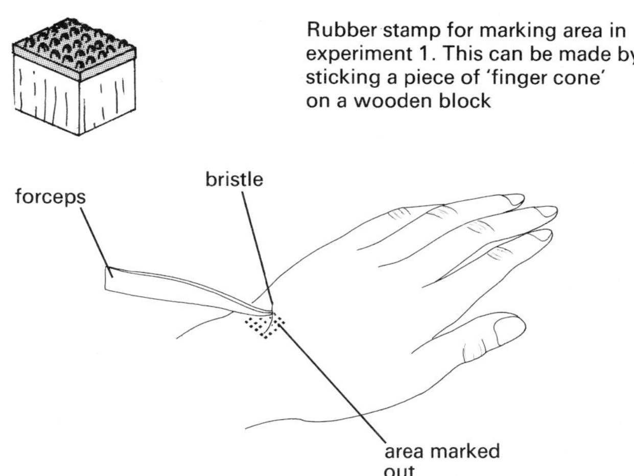

Fig. 22 Testing sensitivity to touch

The sensitivity of the back of the hand can now be compared with the sensitivity of the finger-tips or the back of the neck by repeating the experiment in these regions.

2. Ability to distinguish between two touch stimuli

A piece of wire is bent to a hairpin shape like the one in Fig. 23*a*. The experimenter adjusts the hairpin so that the points are exactly 5 mm apart and presses *one* of the points or *both at once* (Fig. 23*b*) on to the skin of the subject's hand, just enough to dent the skin. The subject must not watch the experiment and simply says 'one' or 'two' if he thinks he is being touched by one or two points. The recorder notes how many times the subject is correct. The experimenter must use one point or two points in a random way so that the subject does not recognize a pattern. The recorder should make up a programme of ten stimuli—for example, 1.1.2.1.1.2.2.2.1.2, with five single and five double stimuli. The experimenter carries out this programme and the recorder ticks the correctly recognized stimuli on his plan.

If the subject gets them all right, the experimenter should move the points closer together and try again. If the subject makes a lot of mistakes, the points should be opened out to 10 mm and the experiment repeated. The idea

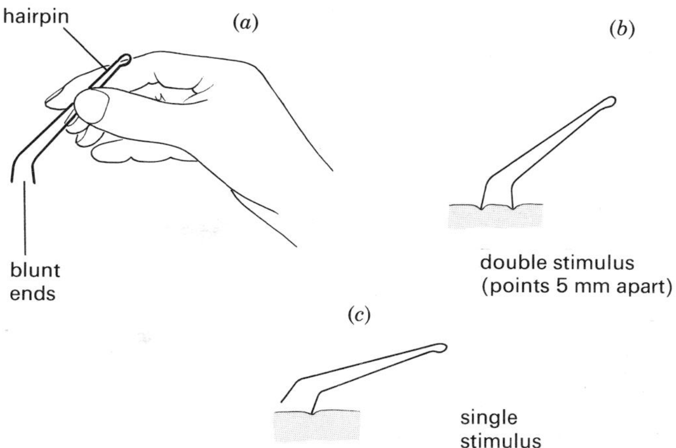

Fig. 23 Double and single stimulus

is to find the least distance the points have to be apart for the subject to recognize every time whether one or two points are touching him. Once this distance has been discovered for the back of the hand, the experiment is repeated for the finger-tips and the back of the neck.

3. To find which eye is used more

A pencil is held at arm's length in line with a distant object. First one eye is closed and opened and then the other. With one the pencil will seem to jump sideways. This shows which eye was used to line up the pencil in the first place.

4. The iris diaphragm

For this experiment the room needs to be darkened, but not blacked out.

If you are working in pairs, sit facing each other and take it in turns to hold a bench lamp or torch about 10 cm from the *side* of your partner's face. Switch on the torch or bench lamp and watch the pupil of the eye very carefully. Try switching the light on and off at about 3- or 4-second intervals.

If you are working on your own, you must look closely in a mirror while switching the light on and off.

5. Sensitivity of the tongue to different tastes

(*CAUTION:* It is normally forbidden to taste chemical substances in a laboratory. The substances in this experiment are harmless but should be made up exactly as described on p. 363 and the test-tubes and other apparatus used must be very clean.)

Sweet, sour, salt and bitter solutions are made up as described on p. 363. The subject sticks his tongue out and the experimenter picks up a drop of one of the solutions on the end of a drinking-straw (Fig. 24) and touches it on the subject's tongue. The subject must not know in advance which solution is being used, but must leave his tongue out while he decides whether he can recognize the taste or not. If he cannot recognize it, he shakes his head, still leaving

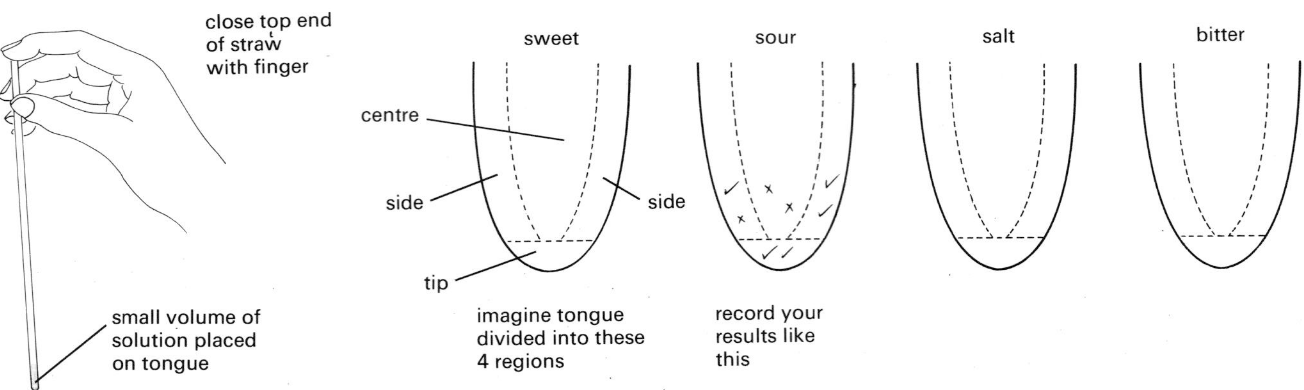

Fig. 24 Sensitivity to taste

his tongue out. The experimenter now tries a drop of the same solution on a different part of the tongue and keeps doing this until the subject recognizes the taste. At this point, the subject has to pull his tongue in to say what the taste is.

The recorder makes four charts of the tongue as shown in Fig. 24 and puts a tick where each taste was recognized and a cross where it was not. The experimenter now changes to a different solution (and a different drinking-straw) and repeats the experiment. He should try to test all four regions of the tongue equally with all four solutions and may have to go back several times to any one of the solutions during the course of the experiment if the subject recognizes the taste after only one or two trials.

It should be possible to build up a picture of which regions of the tongue are particularly sensitive or insensitive to the four tastes.

There are more experiments on human senses in *Experimental Work in Biology*, Combined Edition (see p. 364).

CHECK LIST

- **Our senses detect changes in ourselves and in our surroundings.**
- **The skin is a sense organ which detects heat, cold, touch and pressure.**
- **Sensory endings in the nose respond to chemical substances in the air and give us the sense of smell.**
- **The tongue responds to chemicals in food and drink and gives the sense of taste.**

SIGHT

- **The lens focuses light from the outside world to form a tiny image on the retina.**
- **The sensory cells of the retina are stimulated by the light and send nerve impulses to the brain.**
- **The brain interprets these nerve impulses and so gives us the sense of vision.**
- **The eye can focus on near or distant objects by changing the thickness of the lens.**

SOUND

- **The ear-drum is made to vibrate backwards and forwards by sound-waves in the air.**
- **The vibrations are passed on to the inner ear by the three tiny ear bones.**
- **Sensory nerve endings in the inner ear respond to the vibrations and send impulses to the brain.**
- **The brain interprets these impulses as sound.**

BALANCE

- **The semicircular canals are tubes filled with fluid, which detect turning movements of the head.**
- **When the head rotates, the fluid in the tubes lags behind slightly and pushes a cupula to one side.**
- **The movement of the cupula sends nerve impulses to the brain.**
- **A gelatinous plate in the utriculus pulls on sensory hairs when the head is tilted.**
- **The hair cells send nerve impulses to the brain.**
- **As a result of impulses from the semicircular canals and utriculus, the brain makes your body move in a way that helps to keep your balance.**

22 Co-ordination

NERVOUS SYSTEM
Nerve cells, synapses, nervous impulses.

REFLEX ACTION
Reflex arc.

CONDITIONED REFLEXES

VOLUNTARY ACTION

CENTRAL NERVOUS SYSTEM
Spinal cord and brain.

ENDOCRINE SYSTEM
Compared with nervous system. Thyroid, adrenal, pancreas, diabetes, reproductive organs.

HOMEOSTASIS AND FEEDBACK

MOOD-INFLUENCING DRUGS
Stimulants, depressants, analgesics.

TOLERANCE AND DEPENDENCE
Drug addiction.

Co-ordination is the way all the organs and systems of the body are made to work efficiently together (Fig. 1). If, for example, the leg muscles are being used for running, they will need extra supplies of glucose and oxygen. To meet this demand, the lungs breathe faster and deeper to obtain the extra oxygen and the heart pumps more rapidly to get the oxygen and glucose to the muscles more quickly.

The brain detects changes in the oxygen and carbon dioxide content of the blood and sends nervous impulses to the diaphragm, intercostal muscles and heart. In this example, the co-ordination of the systems is brought about by the **nervous system**.

The extra supplies of glucose needed for running come from the liver. Glycogen in the liver is changed to glucose which is released into the bloodstream (p. 132). The conversion of glycogen to glucose is stimulated by, among other things, a chemical called adrenaline (p. 213). Co-ordination by chemicals is brought about by the **endocrine system**.

The nervous system works by sending electrical impulses along nerves. The endocrine system depends on the release of chemicals, called **hormones**, from **endocrine glands**. Hormones are carried by the bloodstream. For example, insulin (p. 213), is carried from the pancreas to the liver by the circulatory system.

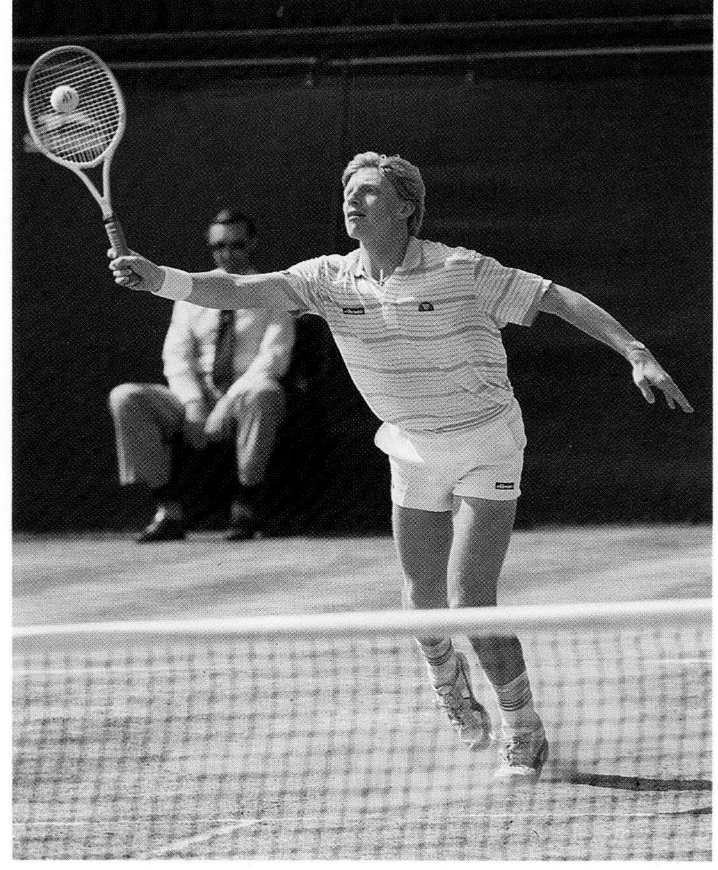

Fig. 1 Co-ordination. The tennis player's brain is receiving sensory impulses from his eyes, semicircular canals, utriculus and muscle stretch receptors. Using this information, the brain co-ordinates the muscles of his limbs so that even while running or leaping he can control his stroke.

THE NERVOUS SYSTEM

Figure 2 is diagram of the human nervous system. The brain and spinal cord together form the **central nervous system.** Nerves carry electrical impulses from the central nervous system to all parts of the body, making muscles contract or glands produce enzymes or hormones.

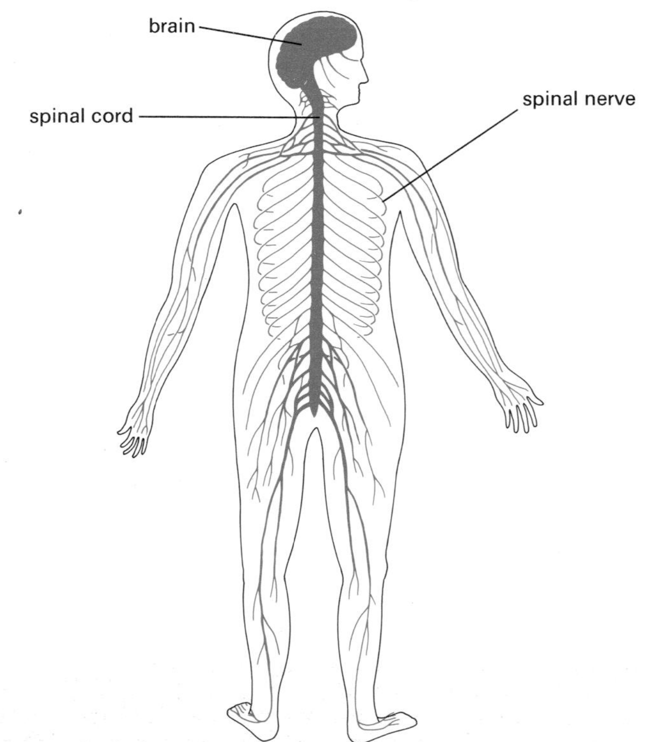

Fig. 2 The human nervous system

Glands and muscles are called **effectors** because they go into action when they receive nerve impulses or hormones. The biceps muscle (p. 186) is an effector which flexes the arm; the salivary gland is an effector which produces saliva when it receives a nerve impulse from the brain.

The nerves also carry impulses back to the central nervous system from the sense organs of the body. These impulses from the eyes, ears, skin, etc. make us aware of changes in our surroundings or in ourselves. Nerve impulses from the sense organs to the central nervous system are called **sensory impulses**; those from the central nervous system to the effectors, resulting in action of some kind, are called **motor impulses**.

The nerves which connect the body to the central nervous system make up the **peripheral nervous system**.

Nerve cells (neurones)

The central nervous system and the peripheral nerves are made up of nerve cells, called **neurones**. Figure 3 shows three types of neurone. The **motor neurones** carry impulses from the central nervous system to muscles and glands. The **sensory neurones** carry impulses from the sense organs to the central nervous system. The **multi-**

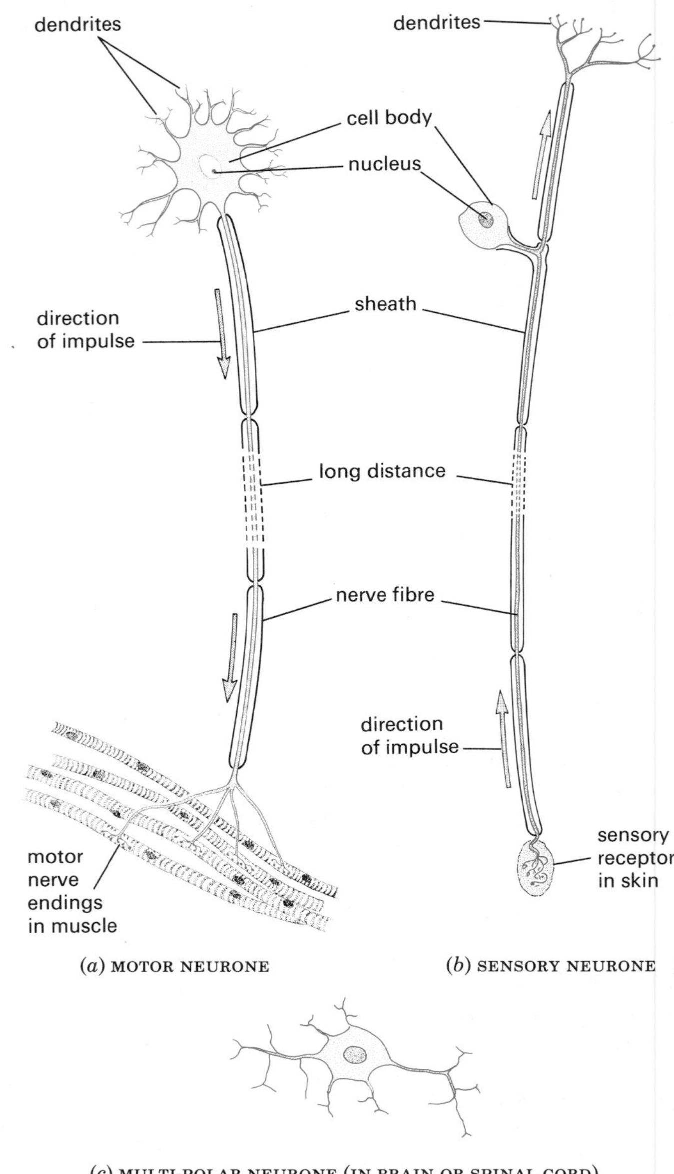

(*a*) MOTOR NEURONE (*b*) SENSORY NEURONE

(*c*) MULTI-POLAR NEURONE (IN BRAIN OR SPINAL CORD)

Fig. 3 Nerve cells (neurones)

polar neurones are neither sensory nor motor but make connections to other neurones inside the central nervous system.

Each neurone has a **cell body** consisting of a nucleus surrounded by a little cytoplasm. Branching fibres, called **dendrites**, from the cell body make contact with other neurones. A long filament of cytoplasm, surrounded by an insulating sheath, runs from the cell body of the neurone. This filament is called a **nerve fibre** (Fig. 3*a* and *b*). The cell bodies of the neurones are mostly located in the brain or in the spinal cord and it is the nerve fibres which run in the nerves. A **nerve** is easily visible, white, tough, and stringy and consists of hundreds of microscopic nerve fibres bundled together (Fig. 4). Most nerves will contain a mixture of sensory and motor fibres. So a nerve can carry many different impulses. These impulses will travel in one direction in sensory fibres and in the opposite direction in motor fibres.

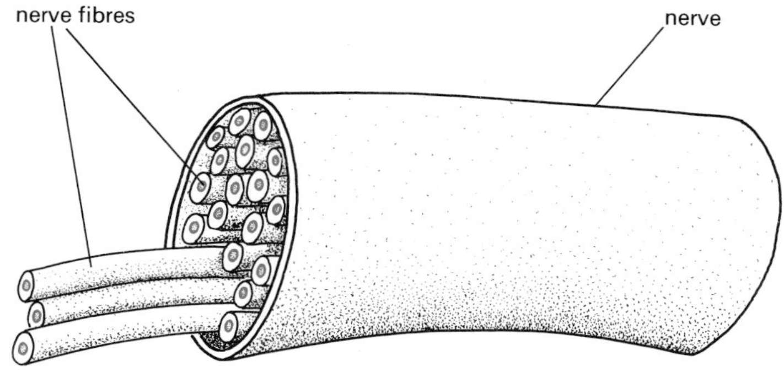

Fig. 4 Nerve fibres grouped into a nerve

Some of the nerve fibres are very long. The nerve fibres to the foot have their cell bodies in the spinal cord and the fibres run inside the nerves, without a break, down to the skin of the toes or the muscles of the foot. Thus a single nerve cell may have a fibre about 1 metre long.

QUESTIONS

1 What is the difference between a nerve and a nerve fibre?
2 In what ways are sensory neurones and motor neurones similar (*a*) in structure, (*b*) in function? How do they differ?
3 Can (*a*) a nerve fibre, (*b*) a nerve, carry both sensory and motor impulses? Explain your answers.

Synapse

Although nerve fibres are insulated, it is necessary for impulses to pass from one neurone to another. An impulse from the finger-tips has to pass through at least three neurones before reaching the brain and so produce a conscious sensation. The regions where impulses are able to cross from one neurone to the next are called **synapses**.

At a synapse, a branch at the end of one fibre is in close contact with the cell body or dendrite of another neurone (Fig. 5). When an impulse arrives at the synapse, it releases a tiny amount of a chemical substance (a chemical transmitter) which sets off an impulse in the next neurone. Sometimes several impulses have to arrive at the synapse before enough chemical is released to cause an impulse to be fired off in the next neurone.

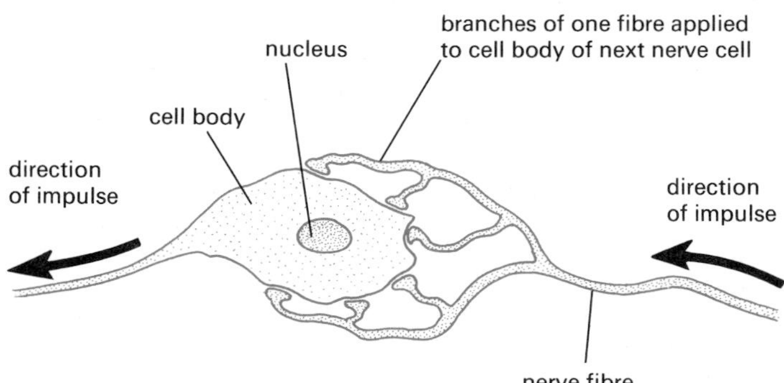

Fig. 5 Diagram of synapses

The nerve impulse

The nerve fibres do not carry sensations like pain or cold. These sensations are felt only when a nerve impulse reaches the brain. The impulse itself is a series of electrical pulses which travel down the fibre. Each pulse lasts about 0.001 second and travels at speeds of up to 100 metres per second. All nerve impulses are similar; there is no difference between nerve impulses from the eyes, ears or hands.

We are able to tell where the sensory impulses have come from and what caused them only because the impulses are sent to different parts of the brain. The nerves from the eye go to the part of the brain concerned with sight. So when impulses are received in this area, the brain recognizes that they have come from the eyes and we 'see' something.

QUESTIONS

4 Look at Fig. 6*b*. (*a*) How many cell bodies are drawn? (*b*) How many synapses are shown?
Look at Figure 8 and answer the same questions.
5 If you could intercept and 'listen to' the nerve impulses travelling in the spinal cord, could you tell which ones came from pain receptors and which from cold receptors? Explain your answer.

The reflex arc

One of the simplest situations where impulses cross synapses to produce action is in the reflex arc. A **reflex action** is an automatic response to a stimulus. When a particle of dust touches the cornea of the eye, you will blink; you cannot prevent yourself from blinking. A particle of food touching the lining of the windpipe will set off a coughing reflex which cannot be suppressed. When a bright light shines in the eye, the pupil contracts (see p. 198). You cannot stop this reflex and you are not even aware that it is happening.

The nervous pathway for such reflexes is called a **reflex arc**. Figure 6 shows the nervous pathway for a well-known reflex called the 'knee-jerk' reflex.

One leg is crossed over the other and the muscles are totally relaxed. If the tendon just below the kneecap of the upper leg is tapped sharply, a reflex arc makes the thigh muscle contract and the lower part of the leg swings forward. Figure 6*b* traces the pathway of this reflex arc. Hitting the tendon stretches the muscle and stimulates a stretch-receptor (p. 195). The receptor sends off impulses in a sensory fibre. These sensory impulses travel in the nerve to the spinal cord.

In the central region of the spinal cord, the sensory fibre passes the impulse across a synapse to a motor neurone which conducts the impulse down the fibre, back to the thigh muscle. The arrival of the impulses at the muscle makes it contract, and jerk the lower part of the limb forward. You are aware that this is happening (which means that sensory impulses must be reaching the brain), but there is nothing you can do to stop it.

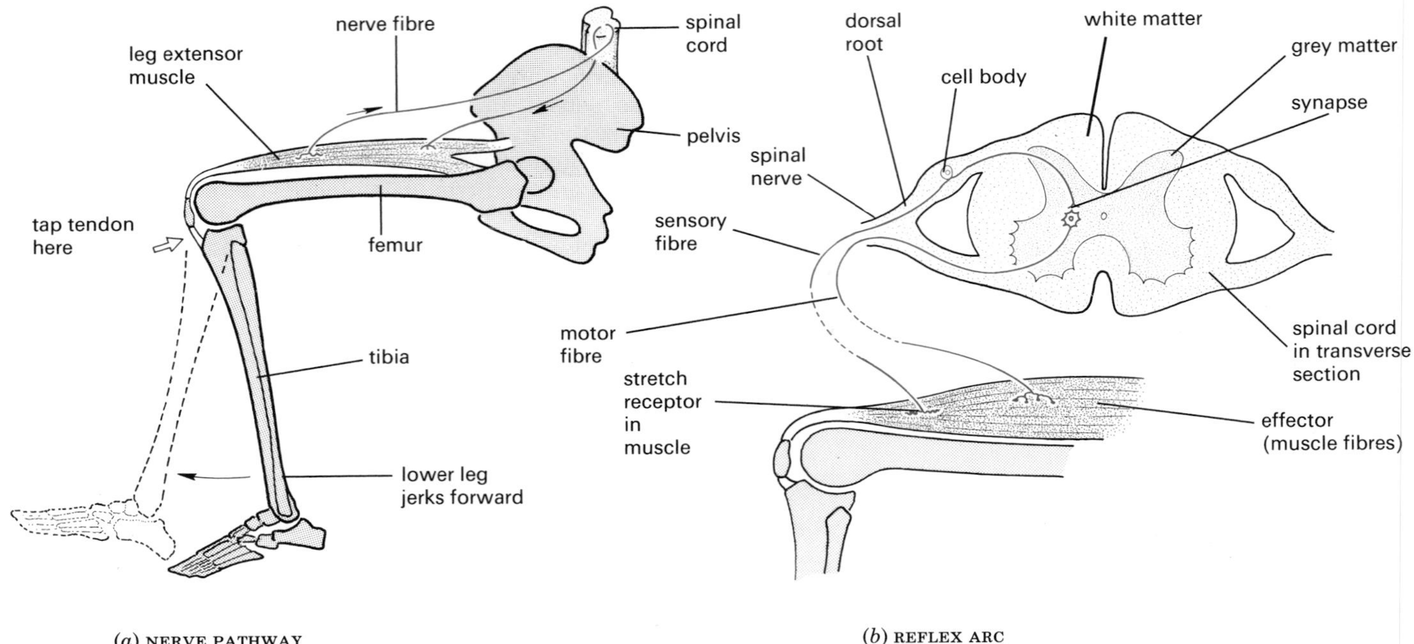

Fig. 6 The reflex knee jerk. This reflex arc needs only one synapse for making the response. Most reflex actions need many more synapses (*a*) to adjust other muscles in the body and (*b*) to send impulses to the brain.

Spinal cord In Fig. 6*b* the spinal cord is drawn in transverse section. The spinal nerve divides into two 'roots' at the point where it joins the spinal cord. All the sensory fibres enter through the **dorsal root** and the motor fibres all leave through the **ventral root**, but both kinds of fibre are contained in the same spinal nerve. This is like a group of insulated wires in the same electric cable. The cell bodies of all the sensory fibres are situated in the dorsal root and they make a bulge called a **ganglion** (Fig. 7).

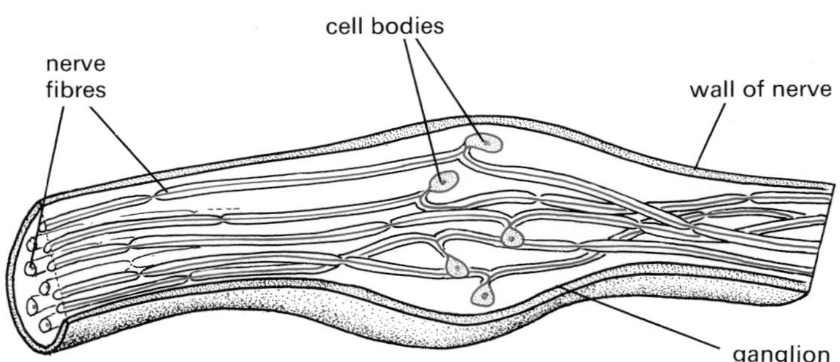

Fig. 7 Cell bodies forming a ganglion

In even the simplest reflex action, many more nerve fibres, synapses and muscles are involved than are described here. Figure 8 shows the reflex arc which would result in the hand being removed from a painful stimulus. On the left side of the spinal cord, an incoming sensory fibre makes its first synapse with a **relay neurone** (or 'intermediate' neurone). This can pass the impulse on to many other motor neurones, although only one is shown in the diagram. On the right side of the spinal cord, some of the incoming sensory fibres are shown making synapses with neurones which send nerve fibres to the brain, thus keeping the brain informed about events in the body. Also, nerve fibres from the brain make synapses with motor neurones in the spinal cord so that 'commands' from the brain can be sent to muscles of the body.

Reflexes The reflex just described is a **spinal reflex**. The brain, theoretically, is not needed for it to happen. Responses which take place in the head, such as blinking, coughing and iris contraction, have their reflex arcs in the brain, but may still not be consciously controlled.

The reflex closure of the iris (p. 198) protects the retina from bright light; the withdrawal reflex removes the hand from a dangerously hot object; the coughing reflex dislodges a foreign particle from the windpipe. Thus, these reflexes have a protective function.

There are many other reflexes going on inside our bodies. We are usually unaware of these, but they maintain our blood pressure, breathing rate, heart beat, etc. and so maintain the body processes.

Inhibition Motor impulses do not always produce action. Some of them inhibit (i.e. suppress the action of) muscle contraction. You will appreciate that, in Fig. 8, a contraction of the biceps muscle will extend the triceps (see Fig. 8, p. 186) and cause its stretch-receptors to fire. These receptors would set off a reflex in the triceps, making it contract and oppose the action of the biceps. However, one of the synapses from the relay neurone in Fig. 8 would send a nerve impulse to the triceps to inhibit it from contracting while the biceps was in action. This kind of inhibition must be going on all the time during co-ordinated movement, or every muscle contraction would be immediately opposed by its antagonistic partner.

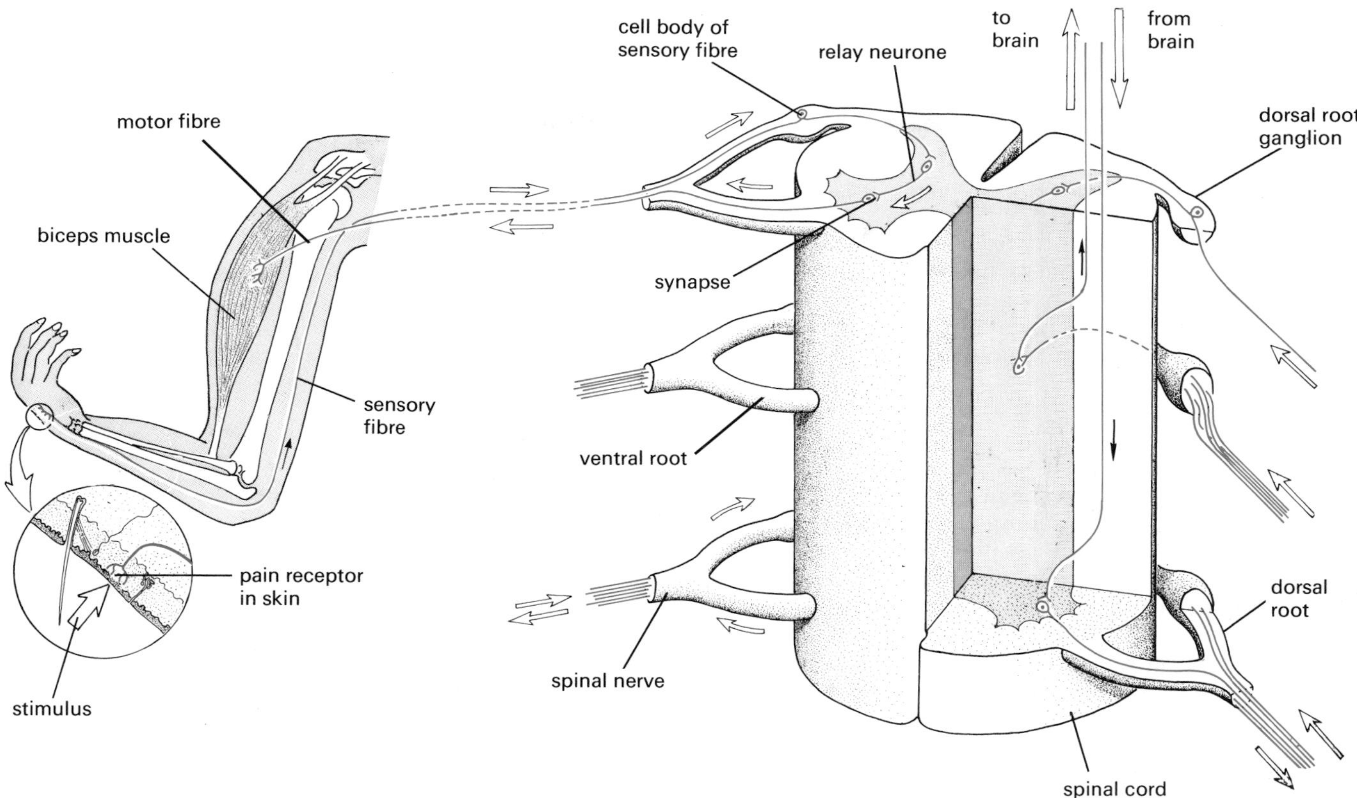

Fig. 8 Reflex arc (withdrawal reflex)

Conditioned reflexes

In most simple reflexes, the stimulus and response are related. For example, the chemical stimulus of food in the mouth produces the reflex of salivation. After a period of learning or training, however, it is possible for a different and often unrelated stimulus to produce the same response. In such a case, a 'conditioned reflex' has been estabished, and the animal is said to be **conditioned** to this stimulus. Pavlov, a Russian biologist of the 1890s, carried out a great many experiments on conditioned reflexes with dogs. One of these experiments is now something of a classic.

The taste of food is a stimulus that activates a dog's salivary glands, making its mouth water. For several days, Pavlov rang a bell at the time the food was given to the dogs. Later, the sound of the bell alone was a sufficient stimulus to cause a dog's mouth to water, without the taste of food. The original chemical stimulus of the food had been replaced by an unrelated stimulus through the ears (Fig. 9).

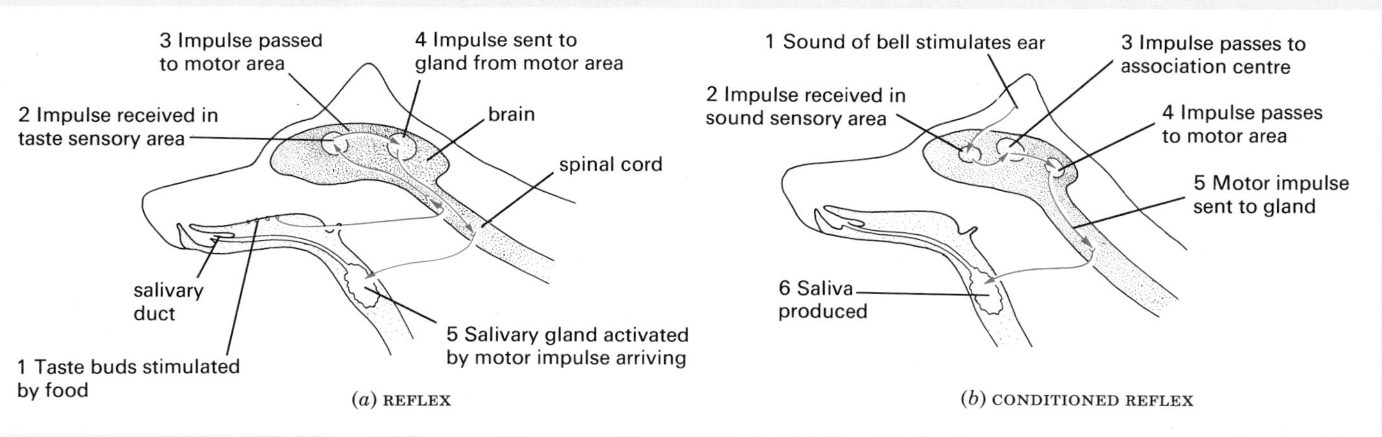

Fig. 9 Possible nervous pathway for conditioning

The training of animals is done largely by conditioning them to respond to new stimuli. Many of our own actions, such as walking and riding a bicycle, involve complicated sets of conditioned reflexes which we acquired in the first place by concentration and practice. Conditioned reflexes thus play a part in learning.

Survival value The ability to learn by conditioned reflex, or by any other process, has survival value (see p. 238). An animal which has learned to avoid danger or to take advantage of a favourable situation, is likely to live longer and to leave more offspring.

Voluntary actions

A voluntary action starts in the brain. It may be the result of external events, such as seeing a book on the floor, but any resulting action, such as picking up the book, is entirely voluntary. Unlike a reflex action it does not happen automatically; you can decide whether or not you carry out the action.

The brain sends motor impulses down the spinal cord in the nerve fibres. These make synapses with motor fibres which enter spinal nerves and make connections to the sets of muscles needed to produce effective action. Many sets of muscles in the arms, legs and trunk would be brought into play in order to stoop and pick up the book, and impulses passing between the eyes, brain and arm would direct the hand to the right place and 'tell' the fingers when to close on the book.

One of the main functions of the brain is to co-ordinate these actions so that they happen in the right sequence and at the right time and place.

QUESTIONS

6 Put the following in the correct order for a simple reflex arc: (*a*) impulse travels in motor fibre, (*b*) impulse travels in sensory fibre, (*c*) effector organ stimulated, (*d*) receptor organ stimulated, (*e*) impulse crosses synapse.
7 Which receptors and effectors are involved in the reflex actions of (*a*) sneezing, (*b*) blinking, (*c*) contraction of the iris?
8 Explain why the tongue may be considered to be both a receptor and an effector organ.
9 Discuss whether coughing is a voluntary or reflex action.

CENTRAL NERVOUS SYSTEM

Spinal cord

Like all other parts of the nervous system, the spinal cord consists of thousands of nerve cells. Figures 6*b*, 8 and 10 show its structure and Figs 1*a* and *b* on p. 182 show how it is protected by the vertebrae.

All the cell bodies, apart from those in the dorsal root ganglion, are concentrated in the central region called the **grey matter**. The **white matter** consists of nerve fibres. Some of these will be passing from the grey matter to the spinal nerves and others will be running along the spinal cord connecting the spinal nerve fibres to the brain. The spinal cord is thus concerned with (*a*) reflex actions involving body structures below the neck, (*b*) conducting sensory impulses from the skin and muscles to the brain and (*c*) carrying motor impulses from the brain to the muscles of the trunk and limbs.

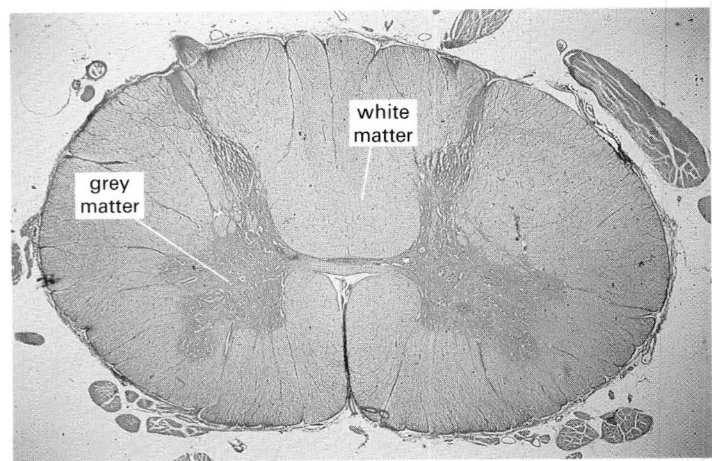

Fig. 10 Section through spinal cord (×7). The light area is the white matter, consisting largely of nerve fibres running to and from the brain. The dark central area is the grey matter, consisting largely of nerve cell bodies.

The spinal cord is enclosed in two membranes. The **pia mater** closely surrounds the cord and contains many blood vessels. The **dura mater** is a tough membrane outside the pia mater and separated from it by a space containing **cerebro-spinal fluid.**

The brain

The brain may be thought of as the expanded front end of the spinal cord. Certain areas are greatly enlarged to deal with all the information arriving from the ears, eyes, tongue, nose and semicircular canals. Figure 11*d* gives a simplified diagram of the main regions of the brain as seen in vertical section. The **medulla** is concerned with regulation of the heart beat, body temperature and breathing rate. The **cerebellum** controls posture, balance and co-ordinated movement. The **mid-brain** deals with reflexes involving the eye. The largest part of the brain, however, consists of the **cerebrum**, made up of two **cerebral hemispheres**. These are very large and highly developed in mammals, especially man, and are thought to be the regions concerned with intelligence, memory, reasoning ability and acquired skills.

In the cerebral hemispheres and the cerebellum, there is an outer layer of grey matter, the **cortex**, with hundreds of thousands of multipolar neurones (Fig. 3*c*) forming the outer layers and making possible an enormous number of synapse connections between the dendrites (Fig. 13).

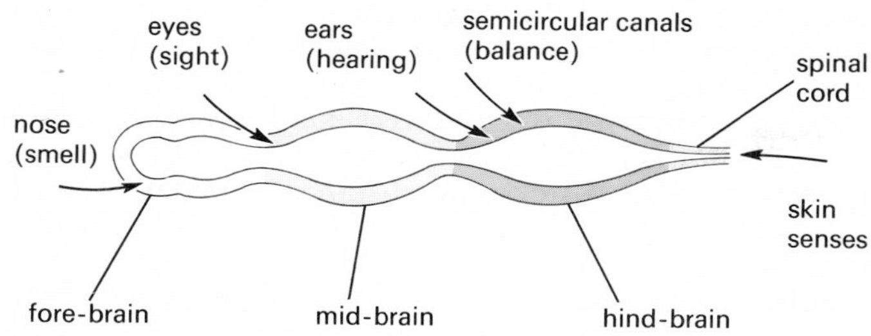

(*a*) The front end of the spinal cord develops three bulges: the fore-, mid- and hind-brain. Each region receives impulses mainly from sense organs in the head.

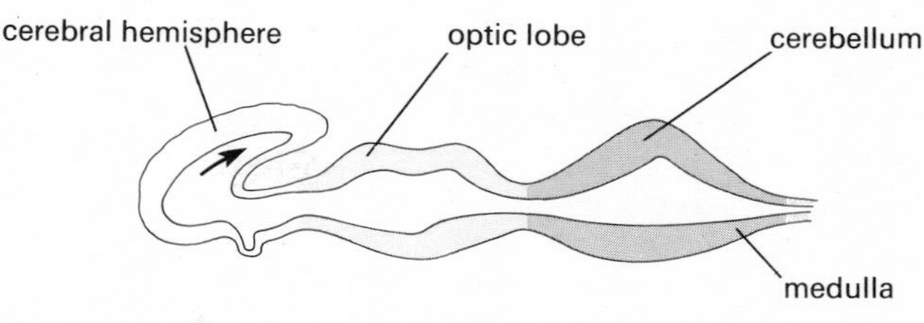

(*b*) The roofs of the fore-, mid- and hind-brain become thicker and form the cerebral hemispheres, optic lobes and cerebellum. The floor of the hind-brain thickens to form the medulla.

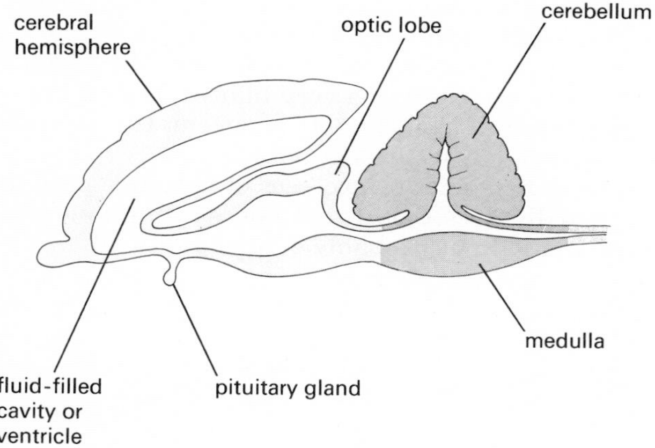

(*c*) A rabbit's brain would look something like this in vertical section.

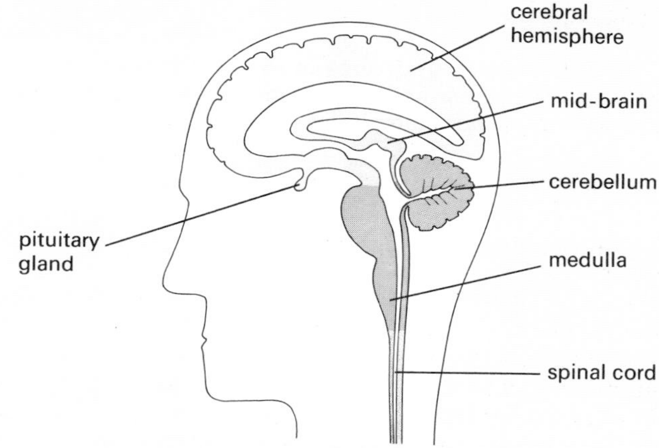

(*d*) The same regions are present in a human brain, but because of our upright position, the brain is bent through 90°.

Fig. 11 Development of the brain of a mammal (vertical sections)

Localization in the cerebral cortex Figure 12 shows the left cerebral hemisphere. The left hemisphere controls the right side of the body; the right hemisphere controls the left side of the body. It is also possible to work out which area of the brain receives impulses from or sends impulses to particular parts of the body.

The region numbered '4' in the motor area on the diagram sends impulses to the hand. If this part of the brain were given a small electrical stimulus, the hand would move, whether the person wanted it to or not.

Similar regions can be mapped out in the sensory area. If the brain centre concerned with hearing were to be stimulated artificially, we would think we were hearing sounds. Stimulation of the 'sight' area would probably cause us to 'see' flashes of light or complete images.

Association centres in the brain are not primarily concerned with any particular sensory or effector system. Association centres receive impulses from many different parts of the brain and relay the impulses to the cortex for further processing or to the motor centres to produce action.

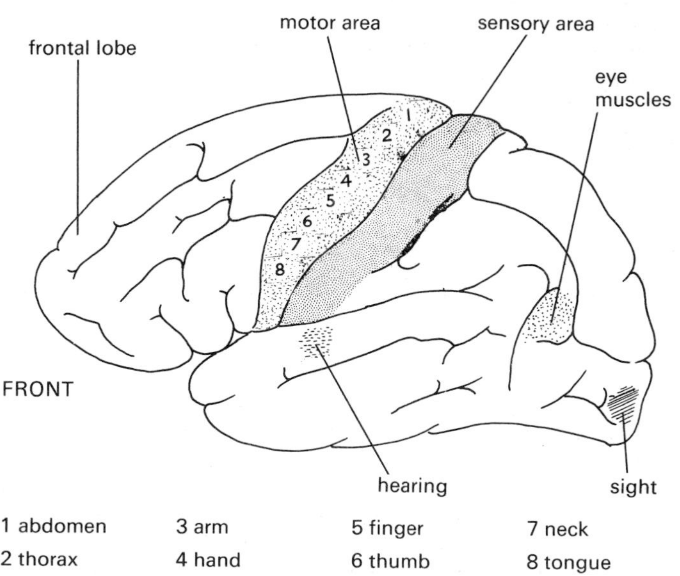

1 abdomen 2 thorax 3 arm 4 hand 5 finger 6 thumb 7 neck 8 tongue

Fig. 12 The left cerebral hemisphere

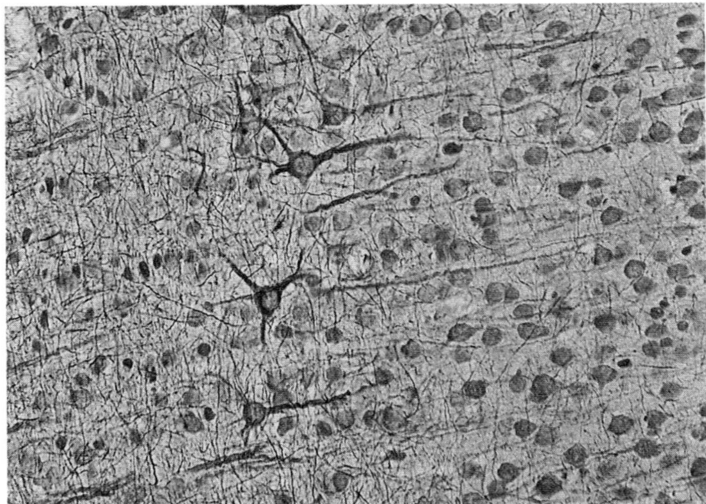

Fig. 13 Multi-polar neurones in the cerebral cortex (×350). The cell bodies and their branching fibres have been darkly stained to make them show up.

Functions of the brain To sum up:

1. The brain receives impulses from all the sensory organs of the body.
2. As a result of these sensory impulses, it sends off motor impulses to the glands and muscles, causing them to function accordingly.
3. In its association centres it correlates the various stimuli from the different sense organs and the memory.
4. The association centres and motor areas co-ordinate bodily activities so that the mechanisms and chemical reactions of the body work efficiently together.
5. It 'stores' information so that behaviour can be modified according to past experience.

QUESTIONS

10 Would you expect synapses to occur in grey matter or in white matter? Explain your answer.
11 Look at Fig. 2. If the spinal cord were damaged at a point about one-third of the way up the vertebral column, what effect would you expect this to have on the bodily functions?
12 (*a*) With which senses are the fore-, mid- and hind-brain mainly concerned? (*b*) Which part of the brain seems to be mainly concerned with keeping the basic body functions going?
13 Describe the biological events involved when you hear a sound and turn your head towards it.
14 If area number 8 of the left cerebral hemisphere (Fig. 12) were stimulated, what would you expect to happen?

THE ENDOCRINE SYSTEM

Co-ordination by the nervous system is usually rapid and precise. Nerve impulses, travelling at up to 100 metres per second, are delivered to specific parts of the body and produce an almost immediate response. A different kind of co-ordination is brought about by the endocrine system. This system depends on chemicals, called **hormones**, which are released from special glands, called **endocrine glands**, into the bloodstream. The hormones circulate round the body in the blood and eventually reach certain organs, called **target organs**. Hormones speed up or slow down or alter the activity of those organs in some way. After being secreted, hormones do not remain permanently in the blood but are changed by the liver into inactive compounds and excreted by the kidneys.

Unlike the digestive glands, the endocrine glands do not deliver their secretions through ducts. For this reason, the endocrine glands are sometimes called the 'ductless glands'. The hormones are picked up directly from the glands by the blood circulation.

Responses of the body to hormones are much slower than responses to nerve impulses. They depend, in the first instance, on the speed of the circulatory system and then on the time it takes for the cells to change their chemical activities. Many hormones affect long-term changes such as growth rate, puberty and pregnancy. Nerve impulses often cause a response in a very limited area of the body, such as an eye-blink or a finger movement. Hormones often affect many organ systems at once.

Serious deficiencies or excesses of hormone production give rise to illnesses. Small differences in hormone activity between individuals probably contribute to differences of personality and temperament.

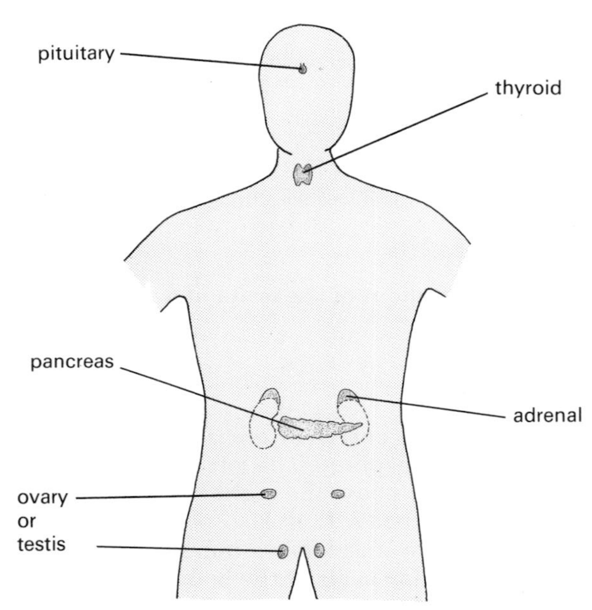

Fig. 14 Position of endocrine glands in the body

The position of the endocrine glands in the body is shown in Fig. 14. Notice that the pancreas and the reproductive organs have a dual function.

Thyroid gland

The thyroid gland is situated in the front part of the neck and lies in front of the windpipe. It produces a hormone called **thyroxine**, which is formed from an amino acid and iodine (p. 116). Thyroxine has a stimulatory effect on the metabolic rate of nearly all the body cells. It controls our level of activity, promotes normal skeletal growth and is essential for the normal development of the brain.

Adrenal glands

These glands are attached to the back of the abdominal cavity, one above each kidney. Each adrenal gland is made up of two distinct regions with different functions. There is an outer layer called the **adrenal cortex** and an inner zone called the **adrenal medulla**. The medulla receives nerves from the brain and produces the hormone **adrenaline**. The cortex has no nerve supply and produces a number of hormones called **corticosteroids**. The corticosteroids help to control the metabolism of carbohydrates, fats, proteins, salts and water.

Adrenaline, from the medulla, has less important but more obvious effects on the body. In response to a stressful situation, nerve impulses are sent from the brain to the adrenal medulla which releases adrenaline into the blood. As adrenaline circulates round the body it affects a great many organs, as shown in the table below.

All these effects make us more able to react quickly and vigorously in dangerous situations that might require us to run away or put up a struggle. However, in many stressful situations, such as taking examinations or giving a public performance, vigorous activity is not called for. So the extra adrenaline in our bodies just makes us feel tense and anxious. You will recognize the sensations described in column 4 of Table 1 as characteristic of fear and anxiety.

Most of the systems affected by adrenaline are also controlled by a section of the nervous system called the **sympathetic nervous system** which affects internal organs other than voluntary muscles. It is difficult to tell whether it is adrenaline or the sympathetic nervous system which is mainly responsible for the stress reactions. It is known, however, that loss of the adrenal medulla seems to cause no ill-effects.

Adrenaline is quickly converted by the liver to a less active compound which is excreted by the kidneys. All hormones are similarly altered and excreted, some within minutes, others within days. Thus their effects are not long-lasting. The long-term hormones, such as thyroxine, are secreted continuously to maintain a steady level.

The pancreas

The pancreas is a digestive gland which secretes enzymes into the duodenum through the pancreatic duct (p. 129). It is also an endocrine (ductless) gland. Most of the pancreas cells produce digestive enzymes but some of them produce hormones. The hormone-producing cells are arranged in small isolated groups called **islets** (Fig. 15) and secrete their hormones directly into the bloodstream. One of the hormones is called **glucagon** and the other is **insulin**.

If the level of sugar in the blood falls, the islets release glucagon into the bloodstream. Glucagon acts on the cells in the liver and causes them to convert some of their stored glycogen into glucose and so restore the blood-sugar level (p. 132).

Insulin has the opposite effect to glucagon. If the concentration of blood sugar increases (e.g. after a meal), insulin is released from the islet cells. When the insulin reaches the liver it stimulates the liver cells to remove glucose from the blood and store it as glycogen. Insulin

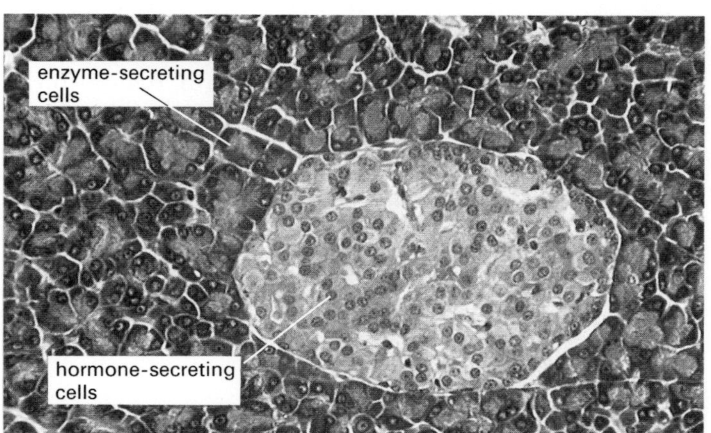

Fig. 15 Section of pancreas tissue showing islets ($\times 250$).

Table 1 Responses to adrenaline

Organ	*Effects of adrenaline*	*Biological advantage*	*Effect or sensation*
Heart	beats faster	sends more glucose and oxygen to the muscles	thumping heart
Breathing centre of the brain	faster and deeper breathing	increased oxygenation of the blood; rapid removal of carbon dioxide	panting
Arterioles of the skin	constricts them (see p. 168)	less blood going to the skin means more is available to the muscles	person goes paler
Arterioles of the digestive system	constricts them	less blood for the digestive system allows more to reach the muscles	dry mouth
Muscles of alimentary canal	relax	peristalsis and digestion slow down; more energy available for action	'hollow' feeling in stomach
Muscles of body	tenses them	ready for immediate action	tense feeling; shivering
Liver	conversion of glycogen to glucose	glucose available in blood for energy production	no sensation
Fat depots	conversion of fats to fatty acids	fatty acids available in blood, for muscle contraction	no sensation

also promotes the conversion of carbohydrates to fat and slows down the conversion of protein to carbohydrate.

All these changes have the effect of regulating the level of glucose in the blood to within narrow limits—a very important example of homeostasis (p. 134).

If anything goes wrong with the production or function of insulin, the person will show the symptoms of **diabetes**.

Diabetes This may result from a failure of the islet cells to produce sufficient insulin or from the reduced ability of the body cells to use it. Two forms of diabetes are recognized, juvenile-onset and adult-onset diabetes.

Juvenile-onset diabetes is the less common form. It mainly affects young people and results from the islets producing too little insulin. There is a slight inherited tendency towards the disease, but it may be triggered off by a virus infection which affects the islets. The patient's blood is deficient in insulin and he or she needs regular injections of the hormone in order to control blood sugar level and so lead to a normal life. This form of the disease is, therefore, sometimes called 'insulin-dependent' diabetes.

Adult-onset diabetes usually affects people after the age of 40. The level of insulin in their blood is often not particularly low but it seems that their bodies are unable to use the insulin properly. This condition can be controlled by careful regulation of the diet and does not usually require insulin injections.

In both forms of diabetes, the patient is unable to regulate the level of glucose in the blood. It may rise to such a high level that it is excreted in the urine or fall so low that the brain cells cannot work properly and the person goes into a coma.

All diabetics need a carefully regulated diet to reduce the intake of carbohydrates and keep the blood sugar within reasonable limits.

Reproductive organs

These produce hormones as well as gametes (sperms and ova) and their effects have been described on page 178.

The hormones from the ovary, **oestrogen** and **progesterone**, both prepare the uterus for the implantation of the embryo, by making its lining thicker and increasing its blood supply.

The hormones **testosterone** (from the testes) and oestrogen (from the ovaries) play a part in the development of the secondary sexual characters as described on page 178.

During pregnancy, the placenta produces a hormone which has effects similar to those of progesterone.

Pituitary gland

This gland is attached to the base of the brain (Fig. 11*d*). It produces many hormones. One of these (anti-diuretic hormone, ADH) acts on the kidneys and regulates the amount of water reabsorbed in the kidney tubules (p. 160). Another pituitary hormone (growth hormone) affects the growth rate of the body as a whole and the skeleton in particular. Several of the pituitary hormones act on the other endocrine glands and stimulate them to produce their own hormones. For example, the pituitary releases into the blood a **follicle-stimulating hormone** (FSH) which, when it reaches the ovaries, makes one of the follicles start to mature and to produce oestrogen. **Luteinizing hormone** (LH) is also produced from the pituitary and, together with FSH, induces ovulation.

A **thyroid-stimulating hormone** (TSH) acts on the thyroid gland and makes it produce thyroxine.

Homeostasis and feedback

Homeostasis The endocrine system plays an important part in maintaining the composition of the body fluids (homeostasis, see p. 134).

A rise in blood sugar after a meal, stimulates the pancreas to produce insulin. The insulin causes the liver to remove the extra glucose from the blood and store it as glycogen (p. 132). This helps to keep the concentration of blood sugar within narrow limits.

The brain monitors the concentration of the blood passing through it. If the concentration is too high, the pituitary gland releases ADH (anti-diuretic hormone). When this reaches the kidneys (the target organs) it causes them to reabsorb more water from the blood passing through them (p. 160). If the blood is too dilute, production of ADH is suppressed and less water is absorbed in the kidneys. Thus ADH helps to maintain the amount of water in the blood at a fairly constant level.

Feedback Some of the endocrine glands are themselves controlled by hormones. For example, pituitary hormones such as LH (luteinizing hormone) affect the endocrine functions of the ovaries. In some cases, the output of hormones is regulated by a process of **negative feedback**.

Figure 16*a* shows that the thyroxine produced by the thyroid gland, suppresses the production of TSH (thyroid-stimulating hormone) from the pituitary. A drop in the level of TSH causes a reduction in the production of thyroxine by the thyroid. Low levels of thyroxine allow the pituitary to produce TSH once again. This feedback causes fluctuations in the production of thyroxine within narrow limits.

The feedback between the pituitary and the ovaries produces a more obvious fluctuation which causes the menstrual cycle (p. 178).

When the level of oestrogen in the blood rises, it affects the pituitary gland, suppressing its production of FSH (follicle-stimulating hormone). A low level of FSH in the blood reaching the ovary will cause the ovary to slow down its production of oestrogen. With less oestrogen in the blood, the pituitary is able to resume its production of FSH which, in turn, makes the ovary start to produce oestrogen again (Fig. 16*b*). This cycle of events takes about 1 month and is the basis of the monthly menstrual cycle.

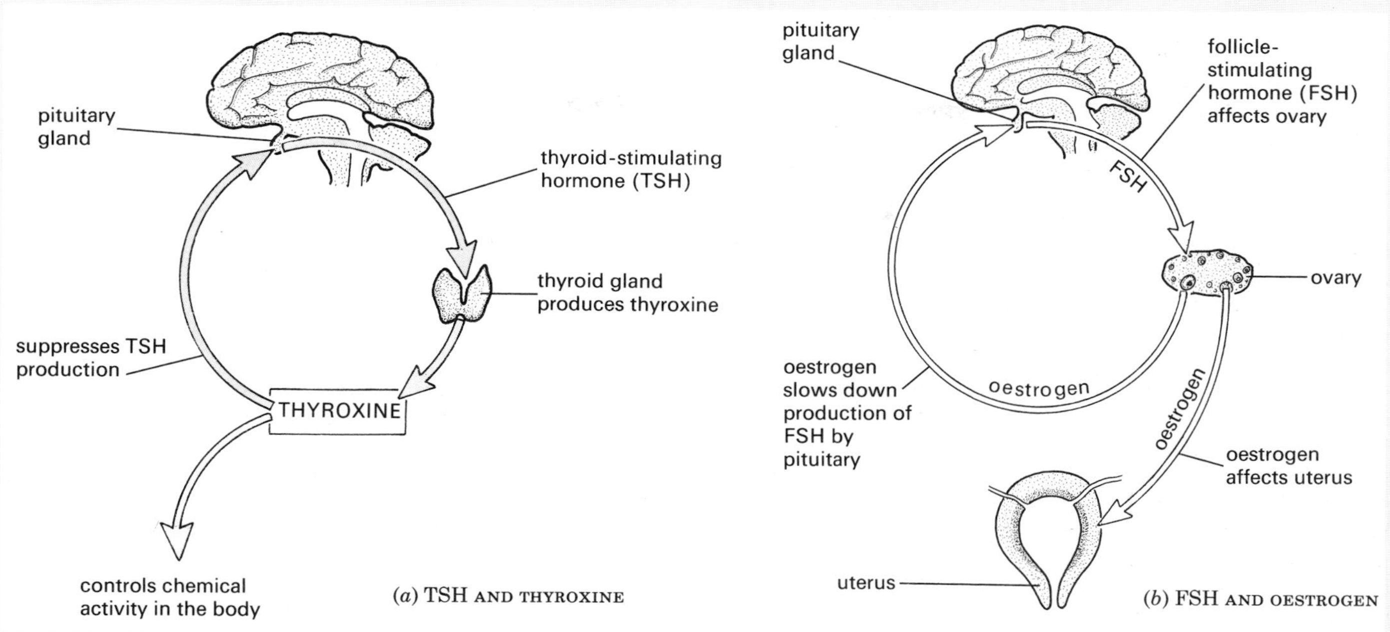

(*a*) TSH AND THYROXINE

(*b*) FSH AND OESTROGEN

Fig. 16 Feedback

The oestrogen and progesterone in the female contraceptive pill act on the pituitary and suppress the production of FSH. If there is not enough FSH, none of the follicles in the ovary will grow to maturity and so no ovum will be released.

QUESTIONS

15 Briefly state the differences between co-ordination by hormones and co-ordination by the nervous system, under the headings 'Routes', 'Speed of conduction', 'Target organs', 'Speed of response', 'Duration of effects'.
16 The pancreas has a dual function in producing digestive enzymes as well as hormones. Which other endocrine glands have a dual function and what are their other functions? (See also p. 178.)
17 What are the effects on body functions of (*a*) too much insulin, (*b*) too little insulin?
18 Why do you think urine tests are carried out to see if a woman is pregnant?

MOOD-INFLUENCING DRUGS

Any substance used in medicine to help our bodies fight illness or disease is called a **drug**. One group of drugs helps to control pain and relieve feelings of distress. These are the mood-influencing drugs.

Events in your life may make you feel excited, depressed, anxious or angry. All these sensations must arise from changes taking place in your nervous and endocrine systems. The chemical substance adrenaline, when released into your blood, makes you feel tense and anxious or excited. In a similar way, it is thought that different chemicals produced by nerve endings in your brain give rise to most of your emotional sensations.

It is not always easy to be sure which is cause and which is effect. Feelings of anxiety may cause the production of adrenaline, or it may be that adrenaline causes feelings of anxiety. Nevertheless, it is known that swallowing or injecting certain substances can give rise to distinct changes of mood. These substances act on the central nervous system but it is often not known how they produce their effect. Even the method of action of alcohol, one of the oldest known mood-influencing drugs, is not known.

Drugs which affect the central nervous system may be classed as stimulants, depressants or analgesics, but these are not rigid headings.

Stimulants

These drugs act, probably, mainly on the cerebral cortex. They increase wakefulness, reduce sensations of fatigue and also depress the appetite. Some of them, such as amphetamines, cause a deterioration in judgement and accuracy. As a result, the drugs often give a false feeling of confidence. It is dangerous for athletes to use amphetamines as they cause excessively high blood pressure and overheating.

Coffee, tea and cocoa contain the substance caffeine, which is a mild stimulant.

Depressants

Depressants, e.g. sedatives, act on the central nervous system to decrease emotional tension and anxiety. Different types of sedative probably affect different areas of the brain but they all lead to relaxation and, in sufficient doses, to sleep or anaesthesia. In excessive doses, they suppress the breathing centre of the brain and cause death.

Tranquillizers Some tranquillizers have been extremely valuable in treating severe mental illnesses such as schizophrenia and mania. Many thousands of mental patients have been enabled to leave hospital and live normal lives as a result of using the tranquillizing drug **chlorpromazine**.

Nowadays, tranquillizers are being prescribed in their millions for the relief of anxiety and tension. Some people think that these drugs are being used merely to escape the stresses of everyday life that could be overcome by a little more will-power and determination. Others think that there is no reason why people should suffer the distress of acute anxiety when drugs are available for its relief. On the other hand, some degree of anxiety is probably needed for mental and physical activity. These activities are unlikely to be very effective in people who tranquillize themselves every time a problem crops up. In some cases, use of certain tranquillizers has led to addiction.

Alcohol The alcohol in wines, beer and spirits is a depressant of the central nervous system. Small amounts give a sense of well-being, with a release from anxiety. However, this is accompanied by a fall-off in performance in any activity requiring skill. It also gives a misleading sense of confidence in spite of the fact that one's judgement is clouded. The drunken driver usually thinks he is driving extremely well.

Alcohol causes vaso-dilation in the skin, giving a sensation of warmth but in fact leading to a greater loss of body heat (see p. 168). A concentration of 500 mg of alcohol in 100 cm^3 of blood results in unconsciousness. More than this will cause death because it stops the breathing centre in the brain.

Some people build up a tolerance to alcohol and this may lead to both emotional and physical dependence (alcoholism). The way alcohol acts on the nervous system is not known, but if taken in excess for a long time, it causes damage to the brain and the liver which cannot be cured (see also p. 120).

Analgesics

Analgesics are drugs which relieve pain. Although the cause of pain may start in the body, the sensation of pain occurs in the brain. Depressants such as alcohol and barbiturates have analgesic effects because they alter the brain's reaction to pain.

Analgesics do not relieve tension or help you to sleep, but they may reduce the pain which is a cause of sleeplessness or anxiety.

Aspirin and paracetamol These are mild analgesics, particularly useful for relief of pain resulting from inflammation of tissues. They are also used to lower the body temperature during a fever.

Tolerance and dependence

If these mood-influencing drugs are used wisely and under medical supervision, they can be very helpful. A person who feels depressed to the point of wanting to commit suicide may be able to lead a normal life with the aid of an anti-depressant drug which removes the sensation of depression. However, if drugs are used for trivial reasons, to produce sensations of excitement or calm, they may be extremely dangerous because they can cause **tolerance** and **dependence**.

Tolerance This means that if the substance is taken over a long period, the dosage has to keep increasing in order to have the same effect. The continuing use of barbiturates in sleeping pills may require the dose to increase from one to two or three tablets in order to get to sleep. People who drink alcohol in order to relieve anxiety may find that they have to keep increasing their intake to reach the desired state. If the dosage continues to increase it will become so large that it causes death.

Dependence This is the term used to describe the condition in which the user cannot do without the substance. Sometimes a distinction is made between emotional and physical dependence. A person with emotional dependence may feel a craving for the substance, may be bad-tempered, anxious or depressed without it, and may commit crimes in order to obtain it. Cigarette-smoking is one example of emotional dependence. Physical dependence involves the same experiences but in addition there are physical symptoms, called **withdrawal symptoms**, when the substance is withheld. These may be nausea, vomiting, diarrhoea, muscular pain, uncontrollable shaking and hallucinations. Physical dependence is sometimes called **addiction**.

Not everyone who takes a mood-influencing drug develops tolerance or becomes dependent on it. There are millions of people who can take alcoholic drinks in moderation with no obvious physical or mental damage. Those who become dependent cannot drink in moderation; their bodies seem to develop a need for permanently high levels of alcohol and the dependent person (an **alcoholic**) gets withdrawal symptoms if alcohol is withheld.

Physical or emotional dependence is a very distressing state. Getting hold of the substance becomes the centre of the addicts' lives, and they lose interest in their persons, their jobs and their families. Because the substances they need cannot be obtained legally or because they need the money to buy them, they resort to criminal activities. Cures are slow, difficult and usually unpleasant. There is no way of telling in advance which person will become dependent and which will not. Dependence is much more likely with some drugs than with others, and these are, therefore, prescribed with great caution. Experimenting with drugs for the sake of emotional excitement is extremely unwise.

QUESTIONS

19 What is the difference between (*a*) becoming tolerant of a drug and (*b*) becoming dependent on a drug? Which of these do you think is meant by being 'hooked' on a drug?
20 Why are amphetamine stimulants unsuitable for improving performance in (*a*) athletics, (*b*) examinations?
21 Why should drinking alcohol cause you to lose heat but make you 'feel' warm?
22 Why is it dangerous to take alcoholic drinks before driving?
23 Why is it dangerous to take an overdose of a depressant drug?
24 What to you think is the difference between an anaesthetic, a sedative and an analgesic?

CHECK LIST

- **The body systems are made to work efficiently together by the nervous system and the endocrine system.**

NERVOUS SYSTEM

- **The nervous system consists of the brain, the spinal cord and the nerves.**
- **The nerves consist of bundles of nerve fibres**
- **Each nerve fibre is a thin filament which grows out of a nerve cell body.**
- **The nerve cell bodies are mostly in the brain and spinal cord.**
- **Nerve fibres carry electrical impulses from sense organs to the brain or from the brain to muscles and glands.**
- **A reflex is an automatic nervous reaction that cannot be consciously controlled.**
- **A reflex arc is the nervous pathway which carries the impulses causing a reflex action.**
- **The simplest reflex involves a sensory nerve cell and a motor nerve cell, connected by synapses in the spinal cord.**
- **The brain and spinal cord contain millions of nerve cells.**
- **The millions of possible connections between the nerve cells in the brain allow complicated actions, learning, memory and intelligence.**

ENDOCRINE SYSTEM

- **The thyroid, adrenal and pituitary are all endocrine glands.**
- **The testes, ovaries and pancreas are also endocrine glands in addition to their other functions.**
- **The endocrine glands release hormones into the blood system.**
- **When the hormones reach certain organs they change the rate or kind of activity of the organ.**
- **Too much or too little of a hormone can cause a metabolic disorder.**
- **Under-production of insulin causes diabetes.**

MOOD-INFLUENCING DRUGS

- **Mood-influencing drugs are valuable for treating mental disorders but dangerous to experiment with.**
- **'Tolerance' means that you have to keep taking greater doses to achieve the same effect.**
- **'Dependence' (addiction) means that you feel physically ill and mentally disturbed if you do not take the drug.**

Examination Questions

Section 3: Human Physiology

Do not write on this page. Where necessary copy drawings, tables or sentences.

1 Carbohydrates and proteins are two important classes of food which are essential in the human diet.

(*a*) Compare carbohydrates with proteins using the following headings:

(i) their chemical structure,

(ii) how they are digested in the human gut,

(iii) their roles in the efficient functioning of the human body.

(*b*) (i) Describe the structure of a **named** seed.

(ii) Describe the nature and distribution of **one** storage material in the seed, how it is made available and then used during germination. (M)

2 The table shows the average number of heart beats per minute for four mammals.

Mammal	*Heart beats per minute while resting*
Mouse	700
Cat	120
Man	72
Blue whale (largest mammal in the world)	20

(*a*) Use the table to state how the number of heart beats per minute is linked to the size of the mammal.

(*b*) Suggest how many beats per minute an elephant has when resting. (W)

3 A student wanted to find out if the digestion of fat by pancreatic extract is speeded up by bile. The student set up the two tubes shown in the diagram below.

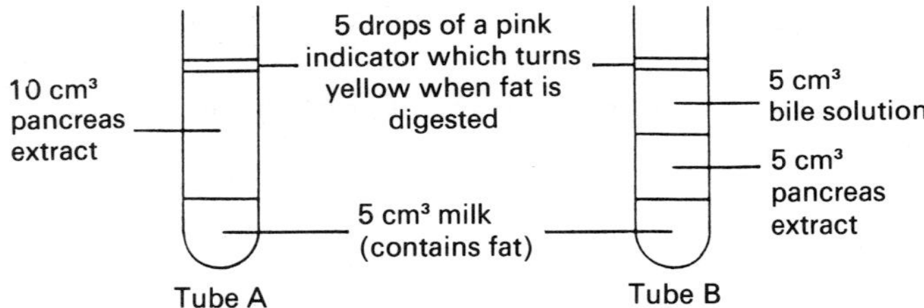

After shaking the tubes the student timed the change from pink to yellow by the indicator. The results are shown below.

Tube	*Time taken for indicator to turn yellow (mins)*
A	10
B	3

The student concluded 'bile helps an enzyme in the extract to break down the fat in the milk'.

(*a*) Give **two** reasons why this is not a valid conclusion.

(*b*) Give **two** ways in which the experiment could be improved so that the student's conclusion might be valid. (N)

4 The diagrams below show a human egg and a human sperm.

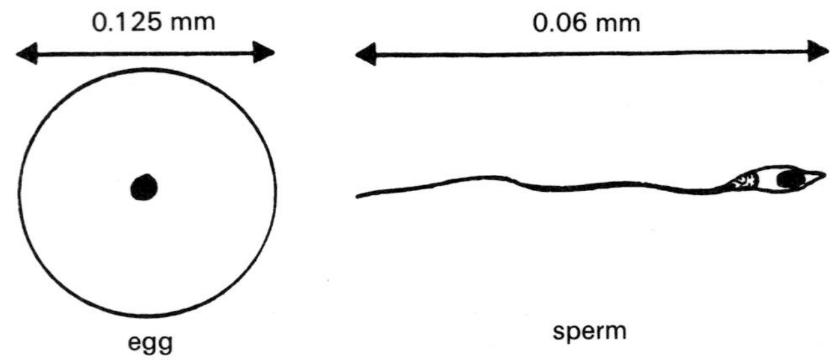

(*a*) (i) How does the sperm move?

(ii) Why is the egg much larger than the sperm?

(*b*) (i) Name an animal whose eggs are fertilized outside its body.

(ii) For the animal you have named, describe **two** features which help to make sure that the eggs are fertilized. (N)

5 (*a*) Explain how bacteria cause tooth decay.

(*b*) Write down how brushing teeth and using a 'fluoride' toothpaste help to prevent tooth decay. (L)

6 (*a*) Match up the letters A–D with the correct boxes in the diagram.

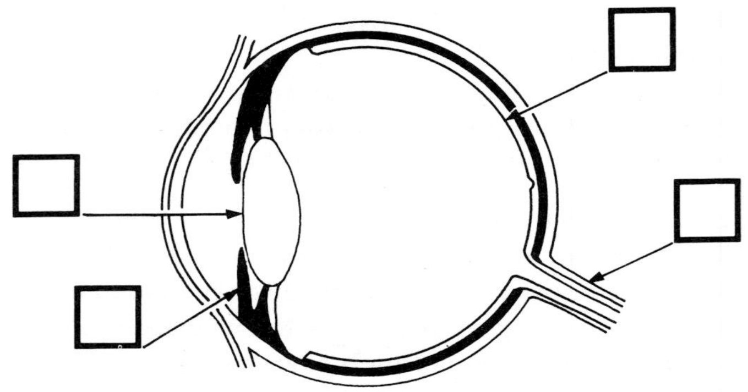

A Light enters the pupil.
B Messages pass along the optic nerve.
C Light reaches the retina.
D The iris reacts.

(*b*) State what the following are used for in the eye:

(i) the sclerotic (outer coat),

(ii) the lens,

(iii) choroid. (W)

7 Which one of the following is a response to an external stimulus?

A peristalsis
B dilation of the pupil
C release of insulin
D production of bile (N)

8 (*a*) What does the term reflex action mean?

(*b*) Describe the pathway of the nerve impulses in a spinal reflex. In your description explain the job carried out by each of the parts you name. This question must **not** be answered by means of a diagram. (M)

SECTION 4

GENETICS AND HEREDITY

23 Cell Division and Chromosomes

HEREDITY AND GENETICS

Explanation of terms.

CHROMOSOMES AND MITOSIS

Mitosis: the movement of chromosomes at cell division. Function of chromosomes: genes on the chromosomes control the cell's physiology and structure. Number of chromosomes: a fixed number for each species.

GAMETE PRODUCTION AND CHROMOSOMES

Meiosis: the chromosomes are shared between the gametes. Meiosis and mitosis compared.

GENES

The structure of the gene. The role of DNA. The genetic code. Genetic engineering: manipulating genes. Mutations: changes in genes and chromosomes.

PRACTICAL WORK

Observing chromosomes in plant cells.

Heredity and genetics

We often talk about people inheriting certain characteristics: 'John has inherited his father's curly hair', or 'Mary has inherited her mother's blue eyes'. We expect tall parents to have tall children. The inheritance of such characteristics is called **heredity** and the branch of biology which studies how heredity works is called **genetics**.

Genetics also tries to forecast what sort of offspring are likely to be produced when plants or animals reproduce sexually. What will be the eye colour of children whose mother has blue eyes and whose father has brown eyes? Will a mating between a black mouse and a white mouse produce grey mice, black-and-white mice or some black and some white mice?

To understand the method of inheritance, we need to look once again at the process of sexual reproduction and fertilization. In sexual reproduction, a new organism starts life as a single cell called a **zygote** (p. 170). This means that you started from a single cell. Although you were supplied with oxygen and food in the uterus, all your tissues and organs were produced by cell division from this one cell. So, the 'instructions' that dictated which cells were to become liver, or muscle, or bone must all have been present in this first cell. The 'instructions' which decided that you should be tall or short, dark or fair, male or female must also have been present in the zygote.

To understand how these 'instructions' are passed from cell to cell, we need to look in more detail at what happens when the zygote divides and produces an organism consisting of thousands of cells. This type of cell division is called **mitosis**. It does not take place only in a zygote but occurs in all growing tissues.

QUESTION

1 (*a*) What are gametes? What are the male and female gametes of (i) plants and (ii) animals called, and where are they produced?
(*b*) What happens at fertilization?
(*c*) What is a zygote and what does it develop into?
(The information needed to answer these questions is given on pages 86 and 170.)

CHROMOSOMES AND MITOSIS

Mitosis

When a cell is not dividing, there is not much detailed structure to be seen in the nucleus even if it is treated with special dyes called stains. Just before cell division, a number of long, thread-like structures appear in the nucleus and show up very clearly when the nucleus is stained (Figs 1*a* and 2). These thread-like structures are called **chromosomes**. Although they are present in the nucleus all the time, they only show up clearly at cell division because at this time they get shorter and thicker.

Each chromosome is seen to be made up of two parallel strands, called **chromatids**. When the nucleus divides into two, one chromatid from each chromosome goes into each daughter nucleus. The chromatids in each nucleus

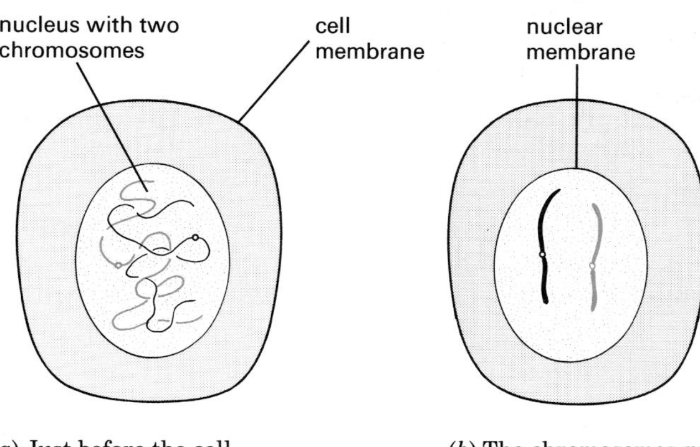

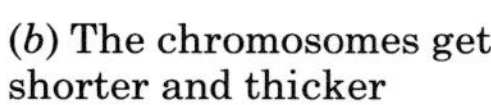

(*a*) Just before the cell divides, chromosomes appear in the nucleus

(*b*) The chromosomes get shorter and thicker

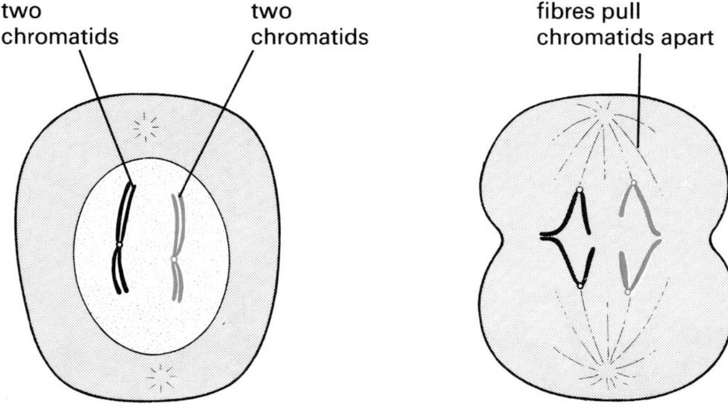

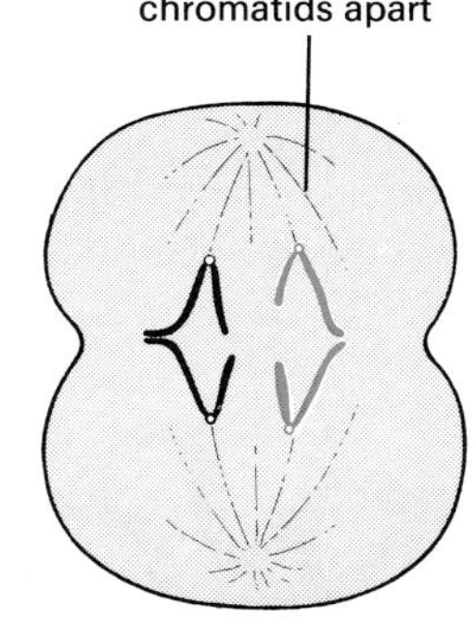

(*c*) Each chromosome is now seen to consist of two chromatids

(*d*) The nuclear membrane disappears and the chromatids are pulled apart to opposite ends of the cell

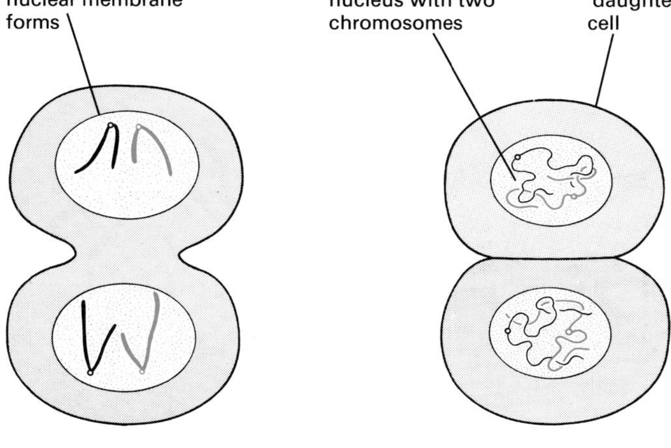

(*e*) A nuclear membrane forms round each set of chromatids, and the cell starts to divide

(*f*) Cell division completed, giving two 'daughter' cells, each containing the same number of chromosomes as the parent cell

Fig. 1 Mitosis. Only one pair of chromosomes is shown. Three of the stages described here are shown in Fig. 2.

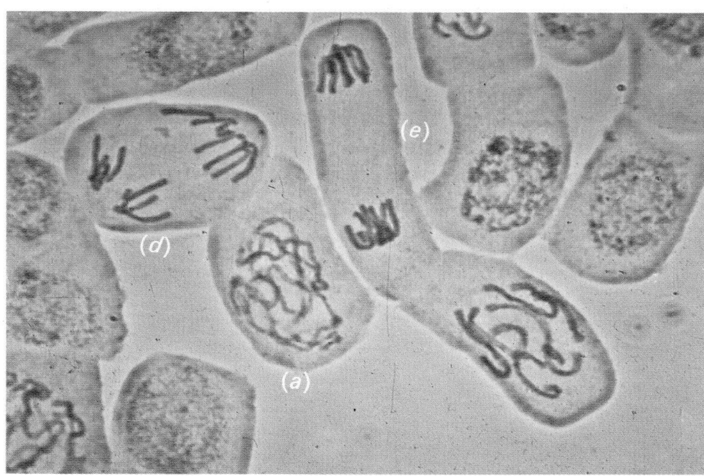

Fig. 2 Mitosis in a root tip (×500). The letters refer to the stages described in Fig. 1. (The tissue has been squashed to separate the cells.)

now become chromosomes and later they will make copies of themselves ready for the next cell division. The process of copying is called **replication** because each chromosome makes a replica (an exact copy) of itself. Figure 1 is a diagram of mitosis, showing only two chromosomes, but there are always more than this. Human cells contain 46 chromosomes.

Mitosis will be taking place in any part of a plant or animal which is producing new cells for growth or replacement. Bone marrow produces new blood cells by mitosis; the epidermal cells of the skin are replaced by mitotic divisions in the Malpighian layer; new epithelial cells lining the alimentary canal are produced by mitosis; growth of muscle or bone in animals, and root, leaf, stem or fruit in plants, results from mitotic cell divisions.

An exception to this occurs in the final stages of gamete production in the reproductive organs of plants and animals. The cell divisions which give rise to gametes are not mitotic but meiotic, as explained on page 223.

Cells which are not involved in the production of gametes are called **somatic cells**. Mitosis takes place only in somatic cells.

QUESTIONS

2 In the nucleus of a human cell just before cell division, how many chromatids will there be?
3 Why can chromosomes not be seen when a cell is not dividing?
4 Look at Fig. 4 on page 167. Where would you expect mitosis to be occurring most often?
5 In which human tissues would you expect mitosis to be going on in (*a*) a five-year old, (*b*) an adult?

The function of chromosomes

When a cell is not dividing, its chromosomes become very long and thin. Along the length of the chromosome is a series of chemical structures called **genes** (Fig. 3). The chemical which forms the genes is called DNA (which is short for deoxy-ribose nucleic acid, p. 17). Each gene controls some part of the chemistry of the cell. It is these genes which provide the 'instructions' mentioned at the beginning of the chapter. For example, one gene may 'instruct' the cell to make the pigment which is formed in

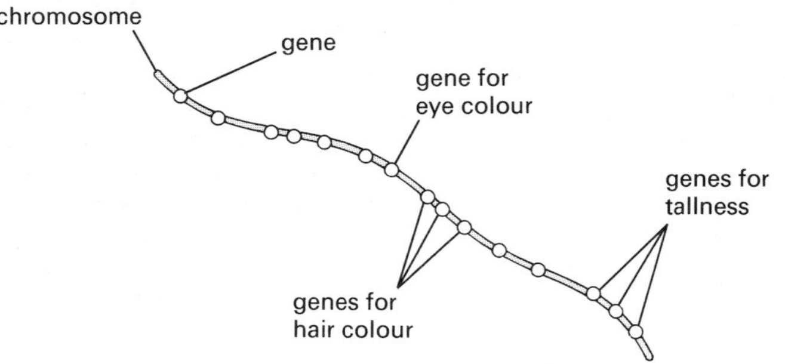

Fig. 3 Relationship between chromosomes and genes. The drawing does not represent real genes or a real chromosome. There are probably thousands of genes on a chromosome.

the iris of brown eyes. On one chromosome there will be a gene which causes the cells of the stomach to make the enzyme pepsin. When the chromosome replicates, it builds up an exact replica of itself, gene by gene (Fig. 4). When the chromatids separate at mitosis, each cell will receive a full set of genes. In this way, the chemical instructions in the zygote are passed on to all cells of the body. All the chromosomes, all the genes and, therefore, all the 'instructions' are faithfully reproduced by mitosis and passed on complete to all the cells.

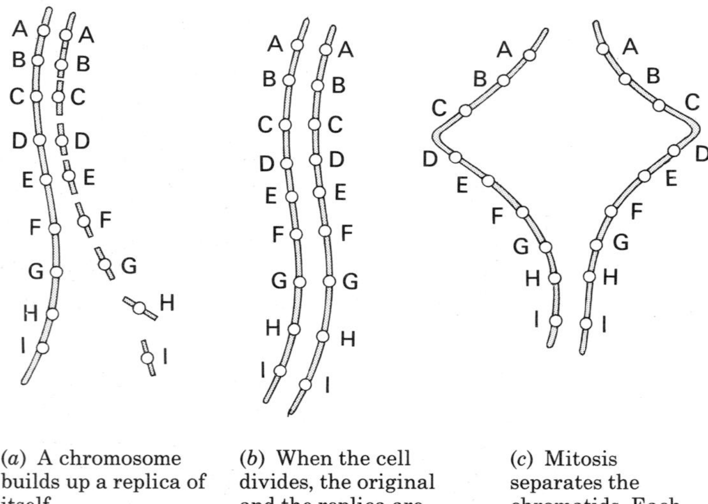

(*a*) A chromosome builds up a replica of itself

(*b*) When the cell divides, the original and the replica are called chromatids

(*c*) Mitosis separates the chromatids. Each new cell gets a full set of genes

Fig. 4 Replication. A, B, C, etc. represent genes.

Which of the 'instructions' are used depends on where a cell finally ends up. The gene which causes brown eyes will have no effect in a stomach cell and the gene for making pepsin will not function in the cells of the eye. So the gene's chemical instructions are carried out only in the correct situation.

Number of chromosomes

(*a*) There is a fixed number of chromosomes in each species. Man's body cells each contain 46 chromosomes, mouse cells contain 40, and garden pea cells 14 (see also Fig. 5).

(*b*) The number of chromosomes in a species is the same in all of its body cells. There are 46 chromosomes in each of your liver cells, in every nerve cell, skin cell and so on.

(*c*) The chromosomes have different shapes and sizes and can be recognized by a trained observer.

(*d*) The chromosomes are always in pairs (Fig. 5), e.g. two long ones, two short ones, two medium ones. This is because when the zygote is formed, one of each pair comes from the male gamete and one from the female gamete. Your 46 chromosomes consist of 23 from your mother and 23 from your father.

(*e*) The number of chromosomes in each body cell of a plant or animal is called the **diploid number**. Because the chromosomes are in pairs, it is always an even number.

The chromosomes of each pair are called **homologous** chromosomes. In Fig. 7*b*, the two long chromosomes form one homologous pair and the two short chromosomes form another.

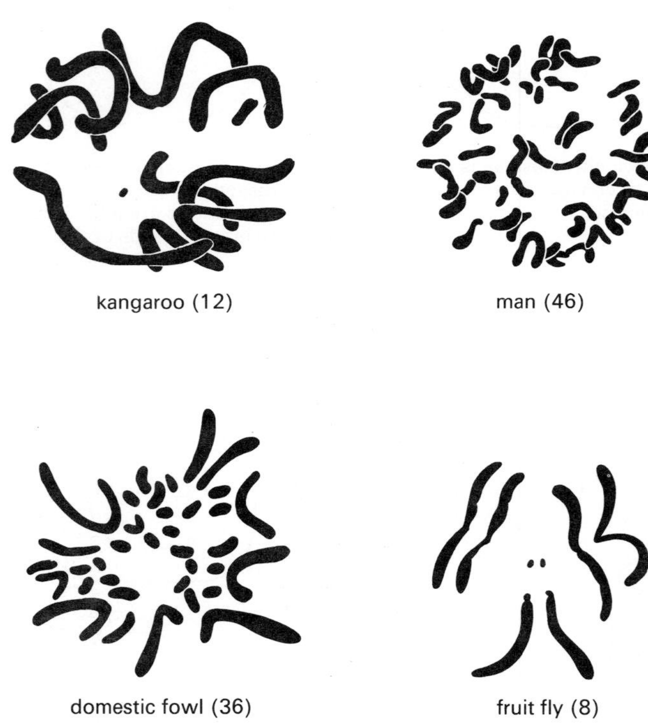

Fig. 5 Chromosomes of different species. Note that the chromosomes are always in pairs.

QUESTIONS

6 How many chromosomes would there be in the nucleus of (*a*) a human muscle cell, (*b*) a mouse kidney cell, (*c*) a human skin cell that has just been produced by mitosis?
7 What is the diploid number in humans?

GAMETE PRODUCTION AND CHROMOSOMES

The genes on the chromosomes carry the 'instructions' which turn a single-cell zygote into a bird, or a rabbit or an oak tree. The zygote is formed at fertilization, when a male gamete fuses with a female gamete. Each gamete brings a set of chromosomes to the zygote. The gametes, therefore, must each contain only half the diploid number of chromosomes, otherwise the chromosome number would double each time an organism reproduced sexually. Each human sperm cell contains 23 chromosomes and each human ovum has 23 chromosomes. When the sperm and ovum fuse at fertilization (p. 170), the diploid number of 46 (23+23) chromosomes is produced (Fig. 6).

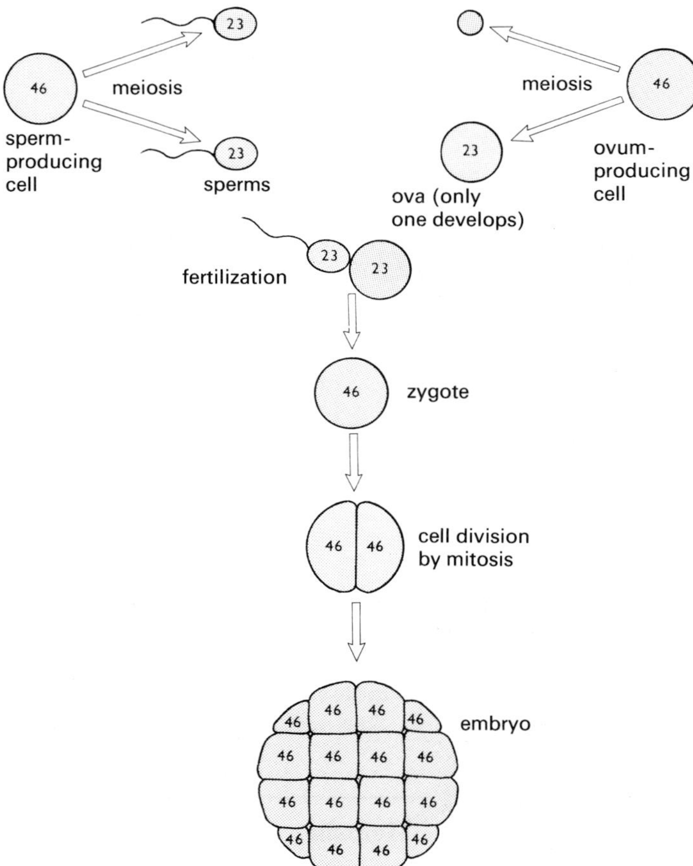

Fig. 6 Chromosomes in gamete production and fertilization

The process of cell division which gives rise to gametes is different from mitosis because it results in the cells containing only half the diploid number of chromosomes. This number is called the **haploid number** and the process of cell division which gives rise to gametes is called **meiosis**.

Meiosis will take place only in reproductive organs.

Meiosis

In a cell which is going to divide and produce gametes, the diploid number of chromosomes shorten and thicken as in mitosis. The pairs of homologous chromosomes, e.g. the two long ones and the two short ones in Fig. 7*b*, lie alongside each other and, when the nucleus divides for the first time, it is the chromosomes and not the chromatids which are separated. This results in only half the total number of chromosomes going to each daughter cell. In Fig. 7*c* the diploid number of four chromosomes is being reduced to two chromosomes prior to the first cell division.

By now (Fig. 7*d*), each chromosome is seen to consist of two chromatids and there is a second division of the nucleus (Fig. 7*e*) which separates the chromatids into four

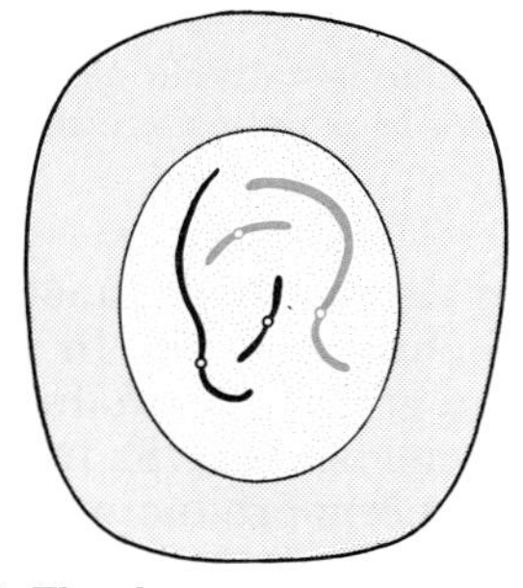

(*a*) The chromosomes appear. Those in colour are from the organism's mother; the black ones are from its father

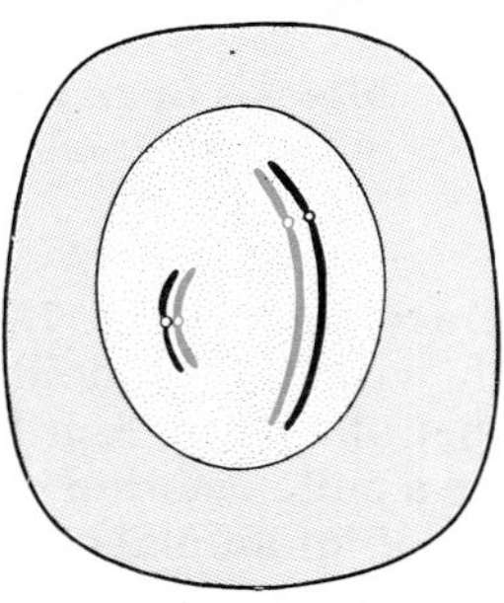

(*b*) Homologous chromosomes lie alongside each other

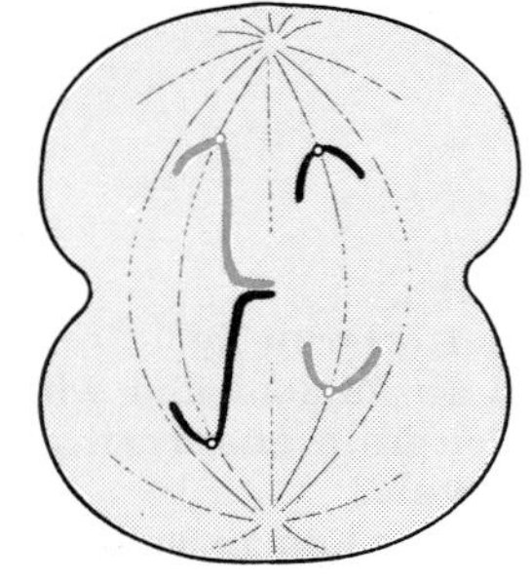

(*c*) The nuclear membrane disappears and corresponding chromosomes move apart to opposite ends of the cell

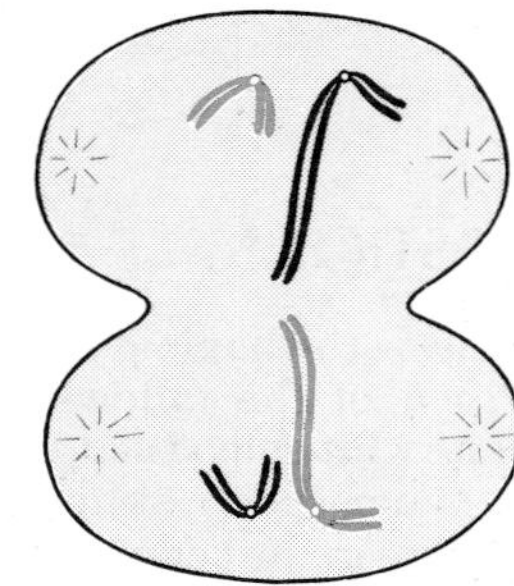

(*d*) By now each chromosome has become two chromatids

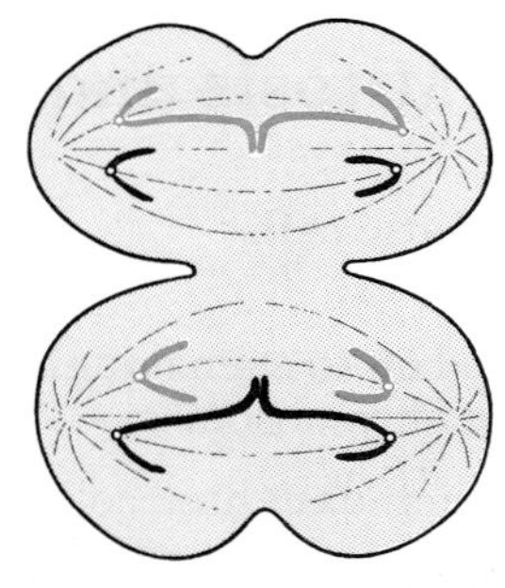

(*e*) A second division takes place to separate the chromatids

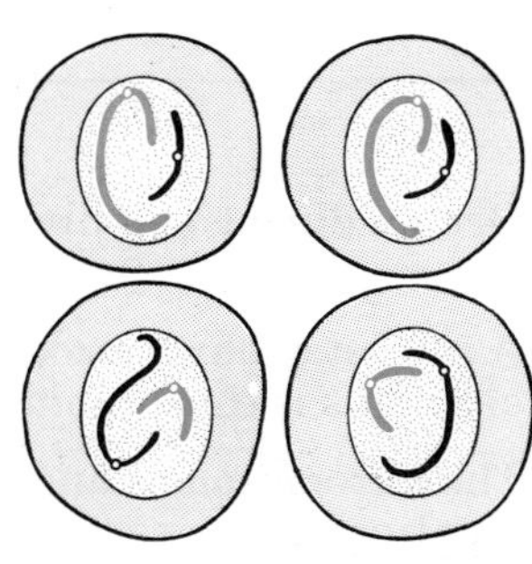

(*f*) Four gametes are formed. Each contains only half the original number of chromosomes

Fig. 7 Meiosis

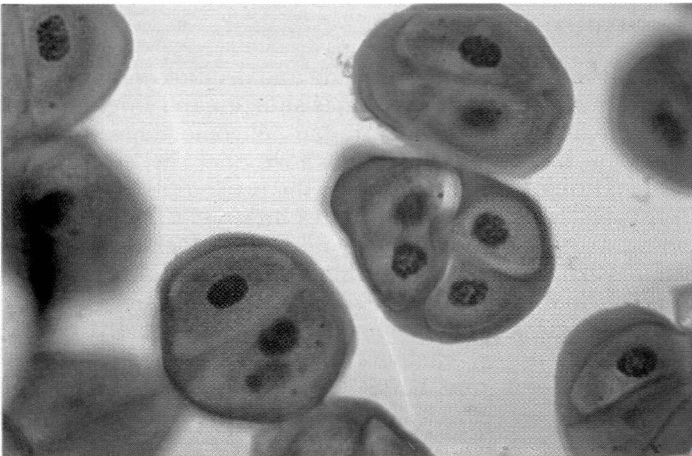

Fig. 8 Meiosis in an anther (×1000). The last division of meiosis in the anther of a flower produces four pollen grains.

distinct nuclei (Fig. 7*f*). This gives rise to four gametes, each with the haploid number of chromosomes. In the anther of a plant, four haploid pollen grains are produced when a pollen mother cell divides by meiosis (Fig. 8). In the testis of an animal, meiosis of each sperm-producing cell forms four sperms. In the cells of the ovule of a flowering plant or in the ovary of a mammal, meiosis gives rise to only one mature female gamete. Although four gametes may be produced initially, only one of them turns into an egg cell which can be fertilized.

QUESTIONS

8 What is the haploid number for (*a*) man, (*b*) fruit fly?
9 Which of the following cells would be haploid and which diploid: white blood cell, male cell in pollen grain, guard cell, root hair, ovum, sperm, skin cell, egg cell in ovule?
10 Where in the body of (*a*) a human male, (*b*) a human female and (*c*) a flowering plant, would you expect meiosis to be taking place?
11 How many chromosomes would be present in (*a*) a mouse sperm cell, (*b*) a mouse ovum?
12 Why are organisms, which are produced by asexual reproduction, identical to each other?

GENES

The structure of the gene

The diagram of genes and chromosomes given in Fig. 3 is greatly over-simplified. Chromosomes consist of a protein framework, with the long DNA molecule coiled round the framework in a complicated way (Fig. 9). It is the DNA part of the chromosome which controls the inherited characters, and it is sections of the DNA molecule which constitute the genes.

On page 17, the structures of nucleotides and nucleic acids were briefly described. DNA is a nucleic acid consisting of a long chain of nucleotides joined

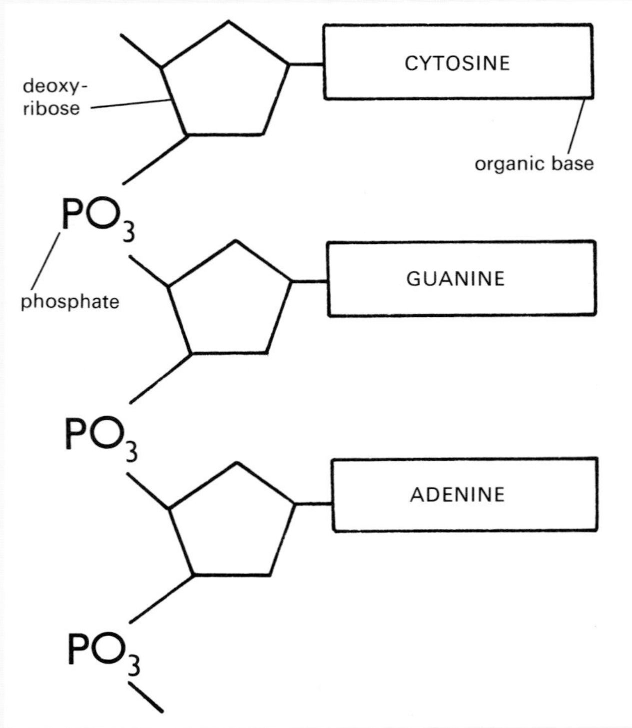

Fig. 10 Part of a DNA molecule

Mitosis and meiosis compared

Mitosis	Meiosis
Occurs during cell division of somatic cells.	Occurs in the final stages of cell division leading to production of gametes.
A full set of chromosomes is passed on to each daughter cell. This is the diploid number of chromosomes.	Only half the chromosomes are passed on to the daughter cells, i.e. the haploid number of chromosomes.
The chromosomes and genes in each daughter cell are identical.	The homologous chromosomes and their genes are randomly assorted between the gametes. (See p. 235 for a fuller explanation of this.)
If new organisms are produced by mitosis in asexual reproduction (e.g. bulbs, p. 101) they will all resemble each other and their parents. They are said to form a 'clone'.	New organisms produced by meiosis in sexual reproduction will show variations from each other and from their parents.

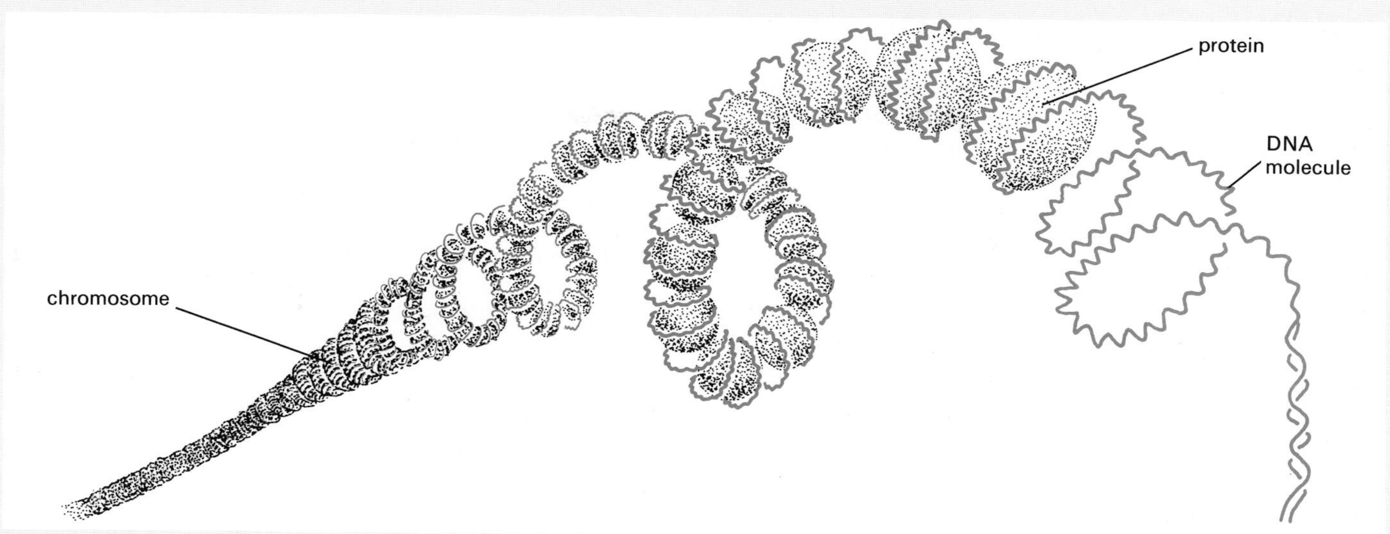

Fig. 9 Simplified model of chromosome structure. This is a '1974 model', which has been superseded by something much more complicated.

together by their phosphate groups (Fig. 10). The phosphate and sugar (deoxyribose) molecules do not vary but the bases may be any one of four kinds; either adenine, guanine, cytosine or thymine.

The sequence of bases down the length of the DNA molecule forms a code which instructs the cell to make particular proteins. Proteins are made from amino acids linked together (p. 14). The type and sequence of the amino acids joined together will determine the kind of protein formed. For example, one protein molecule may start with the sequence *alanine–glycine–glycine* . . . A different protein may start *glycine–serine–alanine* . . .

It is the sequence of bases in the DNA molecule which decides which amino acids are used and in which order they are joined. Each group of three bases stands for one amino acid, e.g. the triplet of bases *cytosine–guanine–adenine* (CGA) specifies the amino acid *alanine*; the base triplet *cytosine–adenine–thymine* (CAT) specifies the amino acid *valine*, and the triplet (CCA) stands for *glycine*. The tripeptide, *valine–glycine–alanine* would be specified by the DNA code CAT–CCA–CGA (Fig. 11).

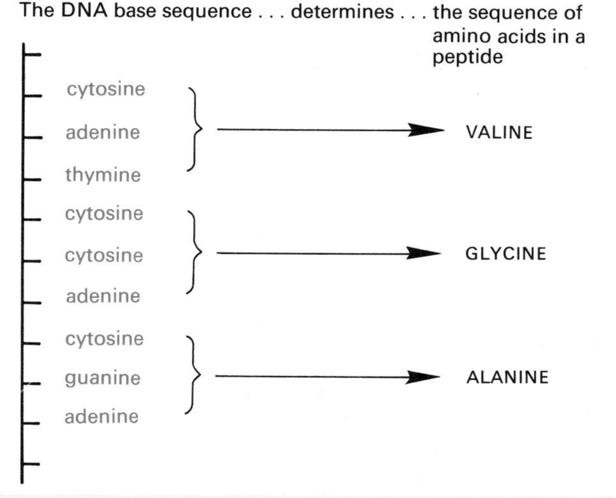

Fig. 11 The genetic code

A gene, then, is a sequence of triplets of the four bases, which specifies an entire protein. Insulin is a small protein with only 51 amino acids. A sequence of 153 (i.e. 3×51) bases in the DNA molecule would constitute the gene which makes an islet cell in the pancreas produce insulin. Most proteins are much larger than this and most genes consist of a thousand or more bases.

The chemical reactions which take place in a cell determine what sort of a cell it is and what its functions are. These chemical reactions are, in turn, controlled by enzymes. Enzymes are proteins. It follows, therefore, that the genetic code of DNA, in determining which proteins, particularly enzymes, are produced in a cell, also determines the cell's structure and function. In this way, the genes also determine the structure and function of the whole organism.

Genetic engineering

The genetic code for a great many proteins has been worked out in the last 20 years or so, and it is possible to synthesize DNA molecules in the laboratory. It is also possible to remove sections of DNA from the nuclei of cells, using special enzymes.

Genetic engineering consists mainly of obtaining lengths of DNA from an organism and inserting them into other organisms, usually bacteria. For example, the gene for human insulin can be inserted into a bacterium. The bacterium is thus made to produce insulin which can be isolated and purified from the bacterial culture and used for treating diabetics (p. 214).

It must be admitted, however, that though these techniques are being successfully developed in the laboratory, they are running into problems when production is stepped up to an industrial scale. Nevertheless, there are many possible applications of genetic engineering in the future. A blood-clotting factor (factor VIII) has been made by genetic engineering. This factor is needed by people suffering from one form of haemophilia, a genetic disease in which the blood fails to clot adequately.

It may prove possible to insert genes for disease resistance into crop plants, or improve the nitrogen-fixing process in root nodule bacteria (p. 251). Enzymes may be commercially produced by genetic engineering and it may be possible, one day, to replace defective human genes with genetically engineered normal ones.

A mutation may be as small as the substitution of one organic base for another in the DNA molecule, or as large as the breakage, loss or gain of a chromosome.

A disease called **sickle-cell anaemia** (p. 239) results from a defective haemoglobin molecule which causes the red blood cells to distort when subjected to a low oxygen concentration. The defective haemoglobin molecule differs from normal haemoglobin by only one amino acid, i.e. *valine* replaces *glutamic acid*. This could be the result of faulty replication at meiosis. When the relevant parental chromosome replicated at gamete formation, the DNA could have produced the triplet –CAT– (specifies *valine*) instead of –CTT– (specifies *glutamic acid*).

Mutations

A mutation is a spontaneous change in a gene or a chromosome. Any change in a gene or chromosome usually has an adverse effect on the cell in which it occurs. If the mutation occurs in a gamete, it will affect all the cells of the individual which develops from the gamete. Thus, the whole organism may be affected. If the mutation occurs in a somatic cell (body cell), it will affect only those cells produced, by mitosis, from the affected cell.

Thus, mutations in gametes may result in genetic disorders in the offspring. Mutations in somatic cells may give rise to cancers of the affected tissues.

An inherited form of mental and physical retardation, known as Down's syndrome, results from a chromosome mutation in which the ovum carries an extra chromosome. The affected child, therefore, has 47 chromosomes in his or her cells instead of the normal 46.

Mutations in bacteria often produce resistance to drugs. Bacterial cells reproduce very rapidly, perhaps as often as once every 20 minutes. Thus a mutation, even if it occurs only rarely, is likely to appear in a large population of bacteria. If a population of bacteria, containing one or two drug-resistant mutants, is subjected to that particular drug, the non-resistant bacteria will be killed, but the drug-resistant mutants survive. Mutant genes are inherited in the same way as normal genes, so when the

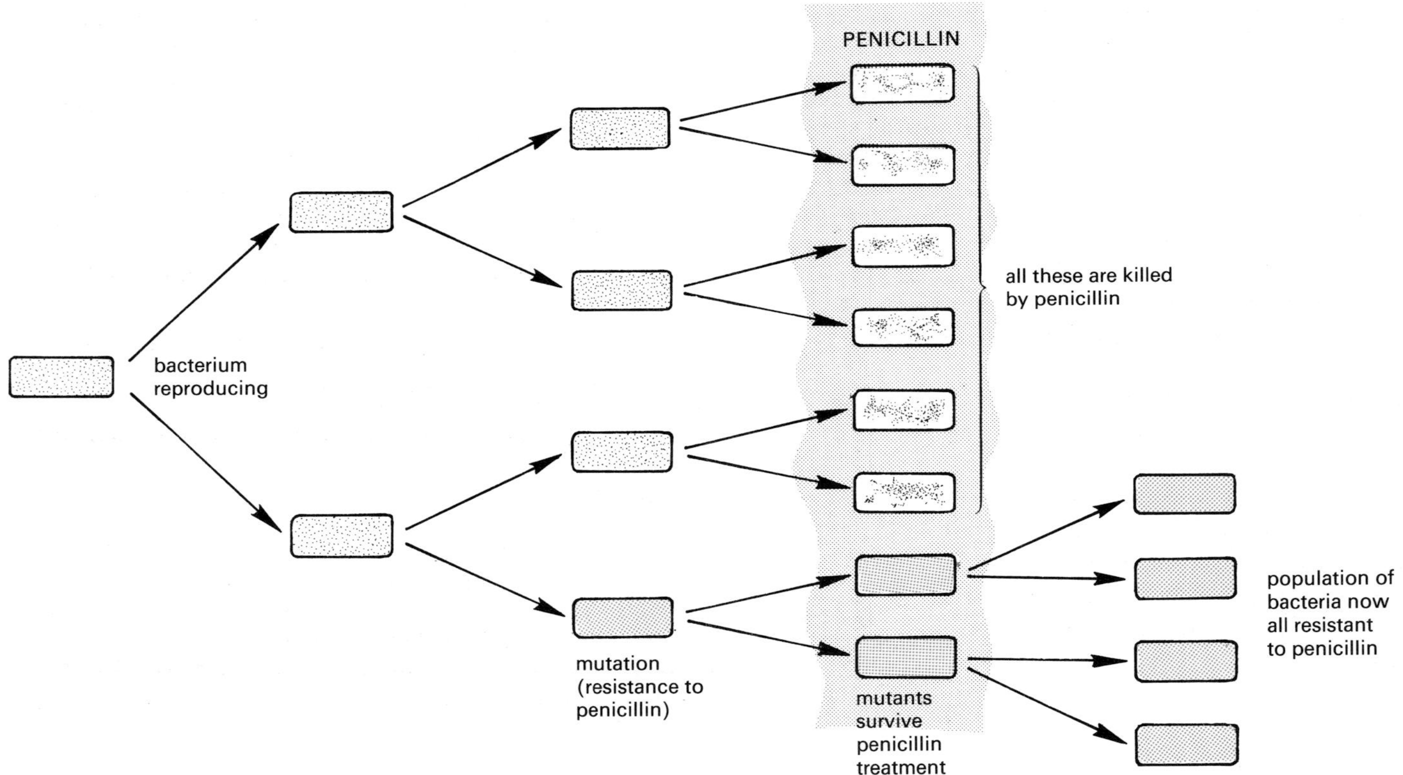

Fig. 12 Mutation in bacteria can lead to drug resistance

surviving mutant bacteria reproduce, all their offspring will be resistant to the drug (Fig. 12).

Mutations are comparatively rare events; perhaps only one in every 100 000 replications results in a mutation. Nevertheless they do occur all the time. (There are bound to be some mutants in the 500 million sperms produced in a human ejaculate.) It is known, however, that exposure to ultra-violet light, X-rays, atomic radiation and certain chemicals does increase the rate of mutation.

PRACTICAL WORK

Squash preparation of chromosomes using acetic orcein

Material *Allium cepa* (onion) root tips. Support onions over beakers or jars of water. Keep the onions in darkness for several days until the roots growing into the water are 2–3 cm long. Cut off about 5 mm of the root tips, place them in a watch glass and

1. cover them with 9 drops acetic orcein and 1 drop molar hydrochloric acid;
2. heat the watch glass gently over a very small Bunsen flame till the steam rises from the stain, but do not boil;
3. leave the watch glass covered for at least 5 minutes;
4. place one of the root tips on a clean slide, cover with 45 per cent ethanoic (acetic) acid and cut away all but the terminal 1 mm;
5. cover this root tip with a clean cover slip and make a squash preparation as described below.

Making the squash preparation Squash the softened, stained root tips by lightly tapping on the cover slip with a pencil: hold the pencil vertically and let it slip through the fingers to strike the cover slip (Fig. 13). The root tip will spread out as a pink mass on the slide; the cells will separate and the nuclei, many of them with chromosomes in various stages of mitosis (because the root tip is a region of rapid cell division), can be seen under the high power of the microscope (× 400).

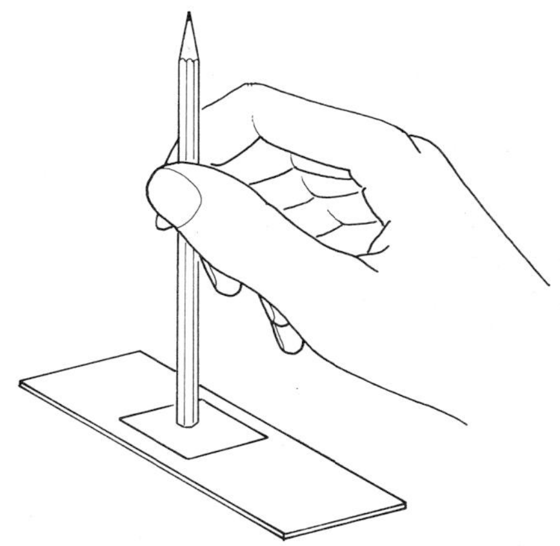

Fig. 13 Tap the cover slip gently to squash the tissue

QUESTIONS

13 What peptide is specified by the DNA sequence CGACGACATCCACAT?

14 A mutation in the DNA sequence in question 13 produces a valine in place of the glycine. What change in the genetic code could have produced this result?

15 State briefly the connection between genes, enzymes and cell structure.

16 Why is it particularly important to prevent radiation from reaching the reproductive organs?

CHECK LIST

- **In the nuclei of all cells there are thread-like structure called 'chromosomes'.**
- **The chromosomes are in pairs; one of each pair comes from the male and one from the female parent.**
- **On these chromosomes are carried the genes.**
- **The genes control the chemical reactions in the cells and, as a result, determine what kind of organism is produced.**
- **Each species of plant or animal has a fixed number of chromosomes in its cells.**
- **When cells divide by mitosis, the chromosomes and genes are copied exactly and each new cell gets a full set.**
- **At meiosis, only one chromosome of each pair goes into the gamete.**
- **The DNA molecule is coiled along the length of the chromosome.**
- **Genes consist of particular lengths of DNA.**
- **Most genes control the type of enzyme that a cell will make.**
- **Genetic engineering involves transferring lengths of DNA from one species to another.**
- **A mutation is a spontaneous change in a gene or chromosome. Most mutations produce harmful effects.**

24 Heredity

PATTERNS OF INHERITANCE

Single-factor inheritance: dominant and recessive genes. Breeding true: homozygous and heterozygous individuals. Genotype and phenotype. Alleles. The three to one ratio: breeding experiments with mice. The recessive back-cross: testing for a heterozygote. Co-dominance and incomplete dominance. Determination of sex: X and Y chromosomes.

APPLIED GENETICS

Improving crop plants and farm animals.

PATTERNS OF INHERITANCE

A knowledge of mitosis and meiosis allows us to explain, at least to some extent, how heredity works. The gene in a mother's body cells which causes her to have brown eyes may be present on one of the chromosomes in each ovum she produces. If the father's sperm cell contains a gene for brown eyes on the corresponding chromosome, the zygote will receive a gene for brown eyes from each parent. These genes will be reproduced by mitosis in all the embryo's body cells and when the embryo's eyes develop, the genes will make the cells of the iris produce brown pigment and the child will have brown eyes.

In a similar way, the child may receive genes for curly hair. Figure 1 shows this happening, but it does not, of course, show all the other chromosomes with thousands of genes for producing the enzymes, making different types of cell and all the other processes which control the development of the organism.

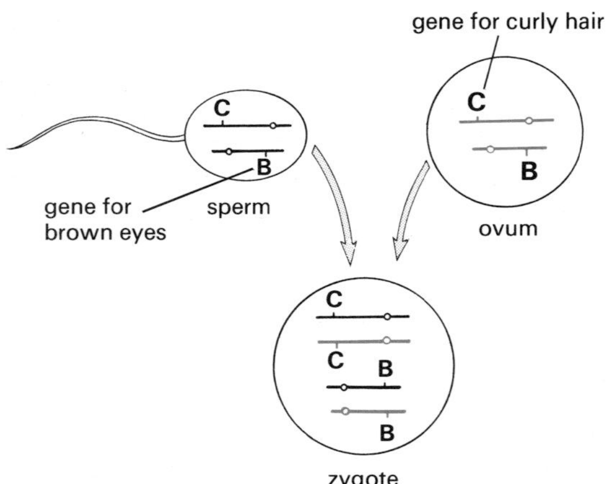

Fig. 1 Fertilization. Fertilization restores the diploid number of chromosomes and combines the genes from the mother and father.

Single factor inheritance

Because it is impossible to follow the inheritance of the thousands of characteristics controlled by genes, it is usual to start with the study of a single gene which controls one characteristic. We have used eye colour as an example so far. Probably more than one gene pair is involved, but the simplified example will serve our purpose. It was explained above how a gene for brown eyes from each parent would result in the child having brown eyes. Suppose, however, that the mother has blue eyes and the father brown eyes. The child might receive a gene for blue eyes from its mother and a gene for brown eyes from its father (Fig. 2). If this happens, the child will, in fact, have brown eyes. The gene for brown eyes is said to be **dominant** to the gene for blue eyes. Although the gene for blue eyes is present in all the child's cells, it does not contribute to the eye colour. It is said to be **recessive** to brown.

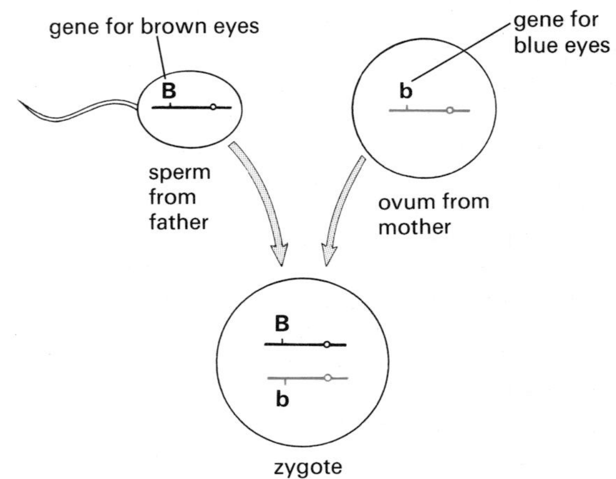

Fig. 2 Combination of genes in the zygote (only one chromosome is shown). The zygote has both genes for eye colour; the child will have brown eyes.

This example illustrates the following important points:

1. There is a pair of genes for each characteristic, one gene from each parent.
2. Although the gene pairs control the same character, e.g. eye colour, they may have different effects. One tries to produce blue eyes, the other tries to produce brown eyes.
3. Often one gene is dominant over the other.
4. The genes of each pair are on corresponding chromosomes and occupy corresponding positions. For example, in Fig. 1 the genes for eye colour are shown in the corresponding position on the two short chromosomes and the genes for hair curliness are in corresponding positions on the two long chromosomes.

In diagrams and explanations of heredity:

(*a*) genes are represented by letters;
(*b*) genes controlling the same characteristic are given the same letter; and
(*c*) the dominant gene is given the capital letter.

For example, in rabbits, the dominant gene for black fur is labelled **B**. The recessive gene for white fur is labelled **b** to show that it corresponds to **B** for black fur. If it were labelled **w**, we would not see any connection between **B** and **w**. **B** and **b** are obvious partners. In the same way **L** could represent the gene for long fur and **l** the gene for short fur.

QUESTIONS

1 Some plants occur in one of two sizes, tall or dwarf. This characteristic is controlled by one pair of genes. Tallness is dominant to shortness.
Choose suitable letters for the gene pair.
2 Why are there two genes controlling one characteristic? Do the two genes affect the characteristic in the same way as each other?
3 The gene for red hair is recessive to the gene for black hair. What colour hair will a person have if he inherits a gene for red hair from his mother and a gene for black hair from his father?

Breeding true

A white rabbit must have both the recessive genes **b** and **b**. If it had **B** and **b**, the dominant gene for black (**B**) would override the gene for white (**b**) and produce a black rabbit. A black rabbit, on the other hand, could be either **BB** or **Bb** and, by just looking at the rabbit, you could not tell the difference. When the male black rabbit **BB** produces sperms by meiosis, each one of the pair of chromosomes carrying the **B** genes will end up in different sperm cells. Since the genes are the same, all the sperms will have the **B** gene for black fur (Fig. 3*a*).

The black rabbit **BB** is called a true-breeding black and is said to be **homozygous** for black coat colour ('homo-' means 'the same'). If this rabbit mates with another black (**BB**) rabbit, all the babies will be black because all will receive a dominant gene for black fur. When all the offspring have the same characteristic as the parents, this is called 'breeding true' for this characteristic.

When the **Bb** black rabbit produces gametes by meiosis, the chromosomes with the **B** genes and the chromosomes with the **b** genes will end up in different gametes. So 50 per cent of the sperm cells will carry **B** genes and 50 per cent will carry **b** genes (Fig. 3*b*). Similarly, in the female, 50 per cent of the ova will have a **B** gene and 50 per cent will have a **b** gene. If a **b** sperm fertilizes a **b** ovum, the offspring, with two **b** genes (**bb**), will be white. The black **Bb** rabbits are not true-breeding because they may produce some white babies as well as black ones. The **Bb** rabbits are called **heterozygous** ('hetero-' means 'different').

The black **BB** rabbits are homozygous dominant.
The white **bb** rabbits are homozygous recessive.

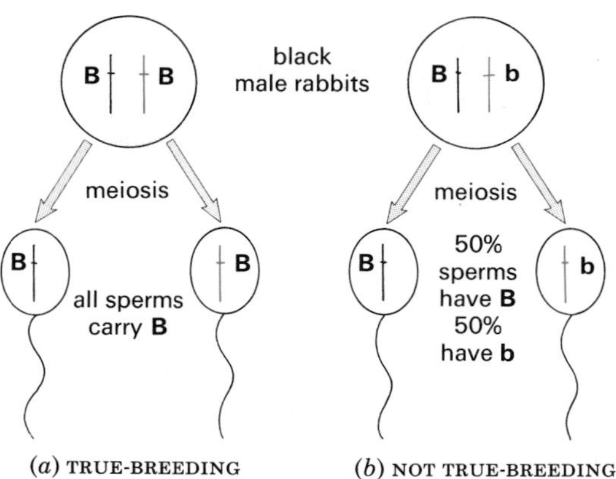

Fig. 3 Breeding true

QUESTIONS

4 (*a*) Read question 3 again. Choose letters for the genes for red hair and black hair and write down the gene combination for having red hair.
(*b*) Would you expect a red-haired couple to breed true?
(*c*) Could a black-haired couple have a red-haired baby?
5 Use the words 'homozygous', 'heterozygous', 'dominant' and 'recessive' (where suitable) to describe the following gene combinations: **Aa**, **AA**, **aa**.
6 A plant has two varieties, one with red petals and one with white petals. When these two varieties are cross-pollinated, all the offspring have red petals. Which gene is dominant? Choose suitable letters to represent the two genes.

Genotype and phenotype

The two kinds of black rabbit **BB** and **Bb** are said to have the same **phenotype**. This is because their coat colours look exactly the same. However, because they have different gene pairs for coat colour they are said to have different **genotypes**, i.e. different combinations of genes. One genotype is **BB** and the other is **Bb**.

You and your brother might both be brown-eyed phenotypes but your genotype could be **BB** and his could be **Bb**. You would be homozygous dominant for brown eyes; he would be heterozygous for eye colour.

Alleles

The genes which occupy corresponding positions on homologous chromosomes and control the same character are called **allelomorphic genes** or **alleles**. The word 'allelomorph' means 'alternative form'. The genes **B** and **b** are alternative forms of a gene for eye colour. **B** and **b** are alleles.

There are often more than two alleles of a gene. The human ABO blood groups (p. 147) are controlled by three alleles, $\mathbf{I}^A$, $\mathbf{I}^B$ and **i**, though only two of these can be present in one genotype.

The three to one ratio

Figure 4*a* shows the result of a mating between a true-breeding (homozygous) black mouse (**BB**), and a true-breeding (homozygous) brown mouse (**bb**). The illustration is greatly simplified because it shows only one pair of the 20 pairs of mouse chromosomes and only one pair of alleles on the chromosomes.

Because black is dominant to brown, all the offspring from this mating will be black phenotypes, because they all receive the dominant allele for black fur from the father. Their genotypes, however, will be **Bb** because they all receive the recessive **b** allele from the mother. They are heterozygous for coat colour. The offspring resulting from this first mating are called the $\mathbf{F_1}$ **generation**.

Figure 4*b* shows what happens when these heterozygous, F_1 black mice are mated together to produce what is called the $\mathbf{F_2}$ **generation**. Each sperm or ovum produced by meiosis can contain only one of the alleles for coat colour, either **B** or **b**. So there are two kinds of sperm cell, one kind with the **B** allele and one kind with the **b** allele. There are also two kinds of ovum with either **B** or **b** alleles. When fertilization occurs, there is no way of telling whether a **b** or a **B** sperm will fertilize a **B** or a **b** ovum, so we have to look at all the possible combinations as follows:

1. a **b** sperm fertilizes a **B** ovum. Result: **bB** zygote.
2. a **b** sperm fertilizes a **b** ovum. Result: **bb** zygote.
3. a **B** sperm fertilizes a **B** ovum. Result: **BB** zygote.
4. a **B** sperm fertilizes a **b** ovum. Result: **Bb** zygote.

There is no difference between **bB** and **Bb**, so there are three possible genotypes in the offspring—**BB**, **Bb** and **bb**. There are only two phenotypes—black (**BB** or **Bb**) and brown (**bb**). So, according to the laws of chance, we would expect three black baby mice and one brown. Mice usually have more than four offspring and what we really expect is that the **ratio** (proportion) of black to brown will be close to 3:1.

If the mouse had 13 babies, you might expect nine black and four brown, or eight black and five brown. Even if she had 16 babies you would not expect to find exactly 12 black and four brown because whether a **B** or **b** sperm fertilizes a **B** or **b** ovum is a matter of chance. If you spun ten coins, you would not expect to get exactly five heads and five tails. You would not be surprised at six heads and four tails or even seven heads and three tails. In the same way, we would not be surprised at 14 black and two brown mice in a litter of 16.

To decide whether there really is a 3:1 ratio, we need a lot of results. These may come either from breeding the same pair of mice together for a year or so to produce many litters, or from mating 20 black and 20 brown mice, crossing the offspring and adding up the number of black and brown babies in the F_2 families (see also Fig. 5).

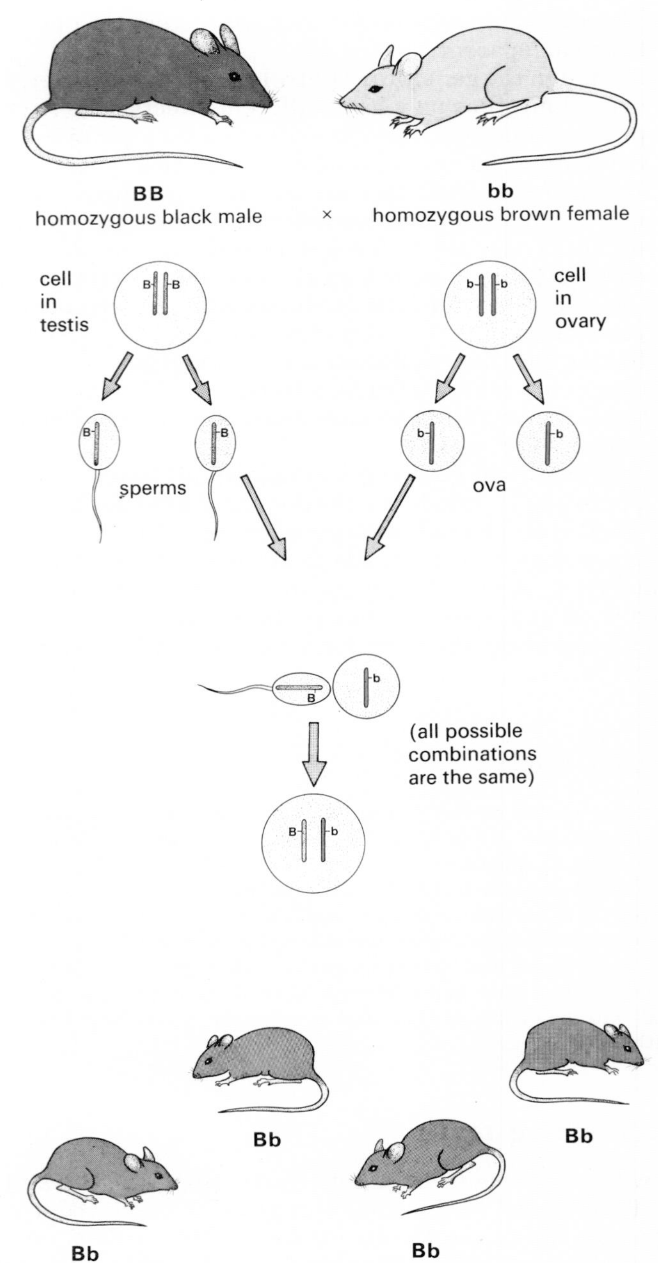

(*a*) All the F_1 generation are heterozygous black

Fig. 4 Inheritance of coat colour in mice

QUESTIONS

7 Look at Fig. 4*a*. Why is there no possibility of getting a **BB** or a **bb** combination in the offspring?

8 In Fig. 4*b* what proportion of the F_2 black mice are true-breeding?

9 Two black guinea-pigs are mated together on several occasions and their offspring are invariably black. However, when their black offspring are mated with white guinea-pigs, half of the matings result in all black litters and the other half produce litters containing equal numbers of black and white babies.

From these results, deduce the genotypes of the parents and explain the results of the various matings, assuming that colour in this case is determined by a single pair of alleles.

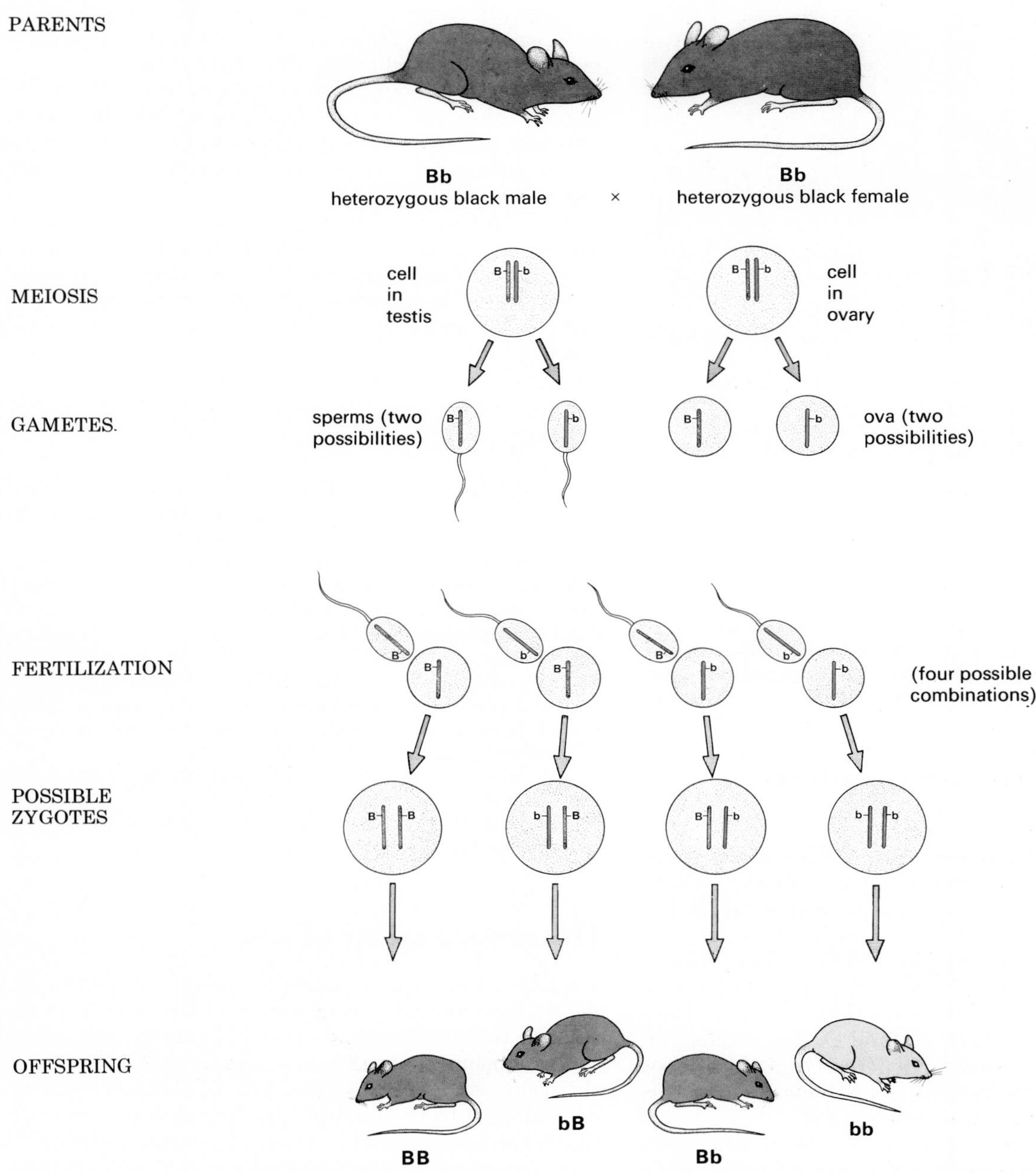

(*b*) The probable ratio of coat colours in the F_2 generation is 3 black : 1 brown

The recessive test-cross (back-cross)

A black mouse could have either the **BB** or the **Bb** genotype. One way to find out which, is to cross the black mouse with a known homozygous recessive mouse, **bb**. The **bb** mouse will produce gametes with only the recessive **b** allele. A black homozygote, **BB**, will produce only **B** gametes. Thus, if the black mouse is **BB**, all the offspring from the cross will be black heterozygotes, **Bb**.

Half the gametes from a black, **Bb**, mouse would carry the **B** allele and half would have the **b** allele. So, if the black mouse is **Bb**, half of the offspring from the cross will, on average, be brown homozygotes, **bb**, and half will be black heterozygotes, **Bb**.

The term 'back-cross' refers to the fact that, in effect, the black, mystery mouse is being crossed with the same genotype as its brown grandparent, the **bb** mouse in Fig. 4*a*. Mouse ethics and speed of reproduction makes the use of the actual grand-parent quite feasible!

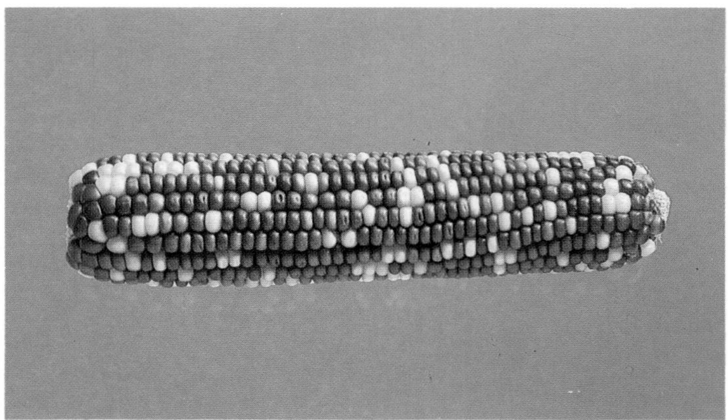

Fig. 5 F_2 maize cobs. These are the F_2 offspring from a breeding experiment using maize instead of mice. One of the gene pair for colour gives yellow grains, the other gives dark grains. What was the colour of the seeds which produced the plants with these cobs?

QUESTION

10 Two black rabbits thought to be homozygous for coat colour were mated and produced a litter which contained all black babies. The F_2, however, resulted in some white babies which meant that one of the grandparents was heterozygous for coat colour. How would you find out which parent was heterozygous?

Co-dominance and incomplete dominance

Co-dominance If both genes of an allelomorphic pair produce their effects in an individual (i.e. neither allele is dominant to the other) the alleles are said to be co-dominant.

The inheritance of the human ABO blood groups includes an example of co-dominance. On page 147 it was explained that, in the ABO system, there are four phenotypic blood groups, A, B, AB and O. The alleles for groups A and B are co-dominant. If a person inherits alleles for group A and group B, half his red cells will carry the antigen *A* and half will carry the antigen *B*.

However, the alleles for groups A and B are both completely dominant to the allele for group O. (Group O people have neither *A* nor *B* antigens on their red cells.)

Table 1 shows the genotypes and phenotypes for the ABO blood groups. (Note that the allele for group O is sometimes represented as **i** and sometimes as $\mathbf{I^O}$).

Table 1. The ABO blood groups

Genotype	*Blood Group (phenotype)*
$\mathbf{I^AI^A}$ or $\mathbf{I^Ai}$	A
$\mathbf{I^BI^B}$ or $\mathbf{I^Bi}$	B
$\mathbf{I^AI^B}$	AB
ii	O

Since the genes for groups A and B are dominant to that for group O, a group A person could have the genotype $\mathbf{I^AI^A}$ or $\mathbf{I^Ai}$. Similarly a group B person could be $\mathbf{I^BI^B}$ or $\mathbf{I^Bi}$. There are no alternative genotypes for groups AB and O.

Incomplete dominance This term is sometimes taken to mean the same as 'co-dominance' but, strictly, it applies to a case where the effect of the recessive allele is not completely masked by the dominant allele.

An example occurs with sickle-cell anaemia (p. 239). If a person inherits both recessive genes ($\mathbf{Hb^SHb^S}$) for sickle-cell haemoglobin, then he or she will show manifest signs of the disease, i.e. distortion of the red cells leading to severe bouts of anaemia.

A heterozygote ($\mathbf{Hb^AHb^S}$), however, will have a condition called 'sickle-cell trait'. Although there may be mild symptoms of anaemia the condition is not serious or life-threatening. In this case, the normal haemoglobin allele ($\mathbf{Hb^A}$) is not completely dominant over the recessive ($\mathbf{Hb^S}$) allele.

QUESTIONS

11 What are the possible blood groups likely to be inherited by children born to a group A mother and a group B father? Explain your reasoning.

12 A woman of blood group A claims that a man of blood group AB is the father of her child. A blood test reveals that the child's blood group is O. Is it possible that the woman's claim is correct? Could the father have been a group B man? Explain your reasoning.

13 A red cow has a pair of alleles for red hairs. A white bull has a pair of alleles for white hairs. If a red cow and a white bull are mated, the offspring are all 'roan', i.e. they have red and white hairs equally distributed over their body.

(*a*) Is this an example of co-dominance or incomplete dominance?

(*b*) What coat colours would you expect among the offspring of a mating between two roan cattle?

Determination of sex

Whether you are a male or female depends on one particular pair of chromosomes called the 'sex chromosomes'. In females, the two sex chromosomes, called the X chromosomes, are the same size as each other. In males, the two sex chromosomes are of different sizes. One corresponds to the female sex chromosomes and is called the X chromosome. The other is smaller and is called the Y chromosome. So the female genotype is XX and the male genotype is XY.

When meiosis takes place in the female's ovary, each ovum receives one of the X chromosomes, so all the ova are the same for this. Meiosis in the male's testes results in 50 per cent of the sperms getting an X chromosome and 50 per cent getting a Y chromosome (Fig. 6). If an X sperm fertilizes the ovum, the zygote will be XX and will grow into a girl. If a Y sperm fertilizes the ovum, the zygote will be XY and will develop into a boy. There is an equal chance of an X or Y chromosome fertilizing an ovum, so the numbers of girl and boy babies are more or less the same.

QUESTION

14 A married couple has four girl children but no boys. This does not mean that the husband produces only X sperms. Explain why not.

APPLIED GENETICS

It is possible for biologists to use their knowledge of genetics to produce new varieties of plants and animals. For example, suppose one variety of wheat produces a lot of grain but is not resistant to a fungus disease. Another variety is resistant to the disease but has only a poor yield of grain. If these two varieties are cross-pollinated (Fig. 7), the F_1 offspring should be disease-resistant and give a good yield of grain (assuming that the useful characteristics are controlled by dominant genes).

R represents a dominant allele for resistance to disease, and **r** is the recessive allele for poor resistance. **H** is a dominant allele for high yield and **h** is the recessive allele for low yield. The high-yield/low-resistance variety (**HHrr**) is crossed with the low yield/high-resistance variety (**hhRR**). Each pollen grain from the **HHrr** plant will contain one **H** and one **r** allele (**Hr**). Each ovule from the **hhRR** plant will contain an **h** and an **R** allele (**hR**). The seeds will, therefore, all be **HhRr**. The plants which grow from these seeds will have dominant alleles for both high yield and good disease resistance.

The offspring from crossing two varieties are called **hybrids**. If the F_1 hybrids from this cross bred true, they could give a new variety of disease-resisting, high-yielding wheat. As you learned on page 230, the F_1 generation from a cross does not necessarily breed true. The F_2 generation of wheat may contain:

1. high yield, disease resistant
2. low yield, disease prone
3. low yield, disease resistant } parental
4. high yield, disease prone } types

This would not give such a successful crop as the F_1 plants.

With some commercial crops, the increased yield from the F_1 seed makes it worth while for the seedsman to make the cross and sell the seed to the growers. The hybrid corn (maize) grown in America is one example. The F_1 hybrid gives nearly twice the yield of the standard varieties.

In other cases it is possible to work out a cross-breeding programme to produce a hybrid which breeds true (Fig. 8). If, instead of the **HhRr** in Fig. 7 an **HHRR** could be produced, it would breed true.

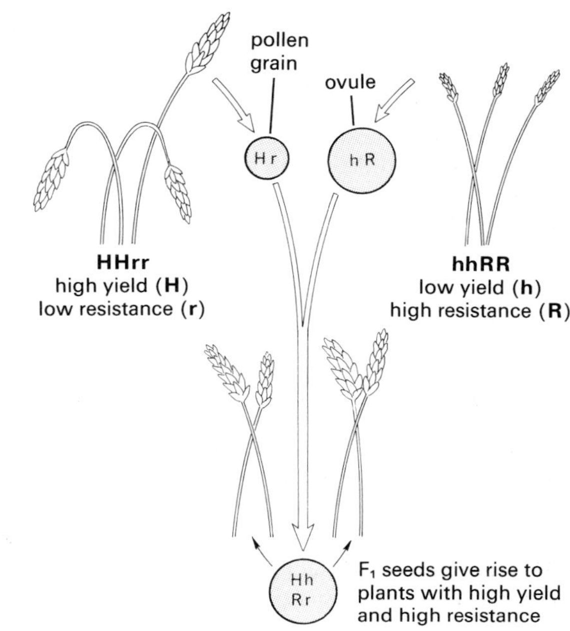

Fig. 7 Combining useful characteristics

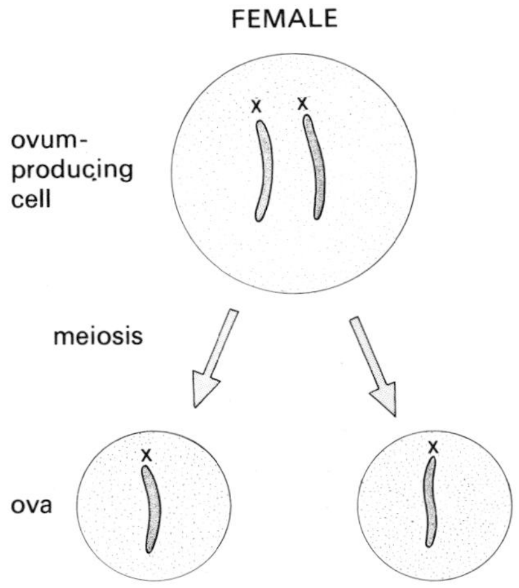

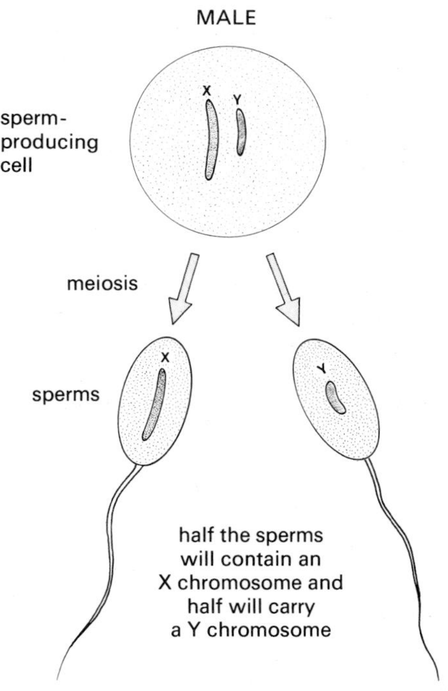

Fig. 6 Determination of sex. Note that:
(i) Only the X and Y chromosomes are shown
(ii) The Y chromosome is not smaller than the X in all organisms
(iii) Details of meiosis have been omitted
(iv) In fact, four gametes are produced in each case, but two are sufficient to show the distribution of X and Y chromosomes

An important part of any breeding programme is the selection of the desired varieties. The largest fruit on a tomato plant might be picked and its seeds planted next year. In the next generation, once again only seeds from the largest tomatoes are planted. Eventually it is possible to produce a true-breeding variety of tomato plant which forms large fruits. Figure 8 (p. 239) shows the result of such selective breeding.

The same principles can be applied to farm animals. Desirable characteristics, such as high milk yield and resistance to disease, may be combined. Stock-breeders will select calves from cows which give large quantities of milk. These calves will be used as breeding stock to build a herd of high yielders. A characteristic such as milk yield is probably under the control of many genes. At each stage of selective breeding the farmer, in effect, is keeping the beneficial genes and discarding the less useful genes from his animals.

Selective breeding in farm stock can be slow and expensive because the animals often have small numbers of offspring and breed only once a year.

QUESTIONS

15 Suggest some good characteristics that an animal-breeder might try to combine in sheep by mating different varieties together.

16 A variety of barley has a good ear of seed but has a long stalk and is easily blown over. Another variety has a short, sturdy stalk but a poor ear of seed.

Suggest a breeding programme to obtain and select a new variety which combined both of the useful characteristics.

Choose letters to represent the genes and show the genotypes of the parent plants and their offspring.

(*a*) (*b*) (*c*) (*d*) (*e*)

Fig. 8 The genetics of bread wheat. A primitive wheat (*a*) was crossed with a wild grass (*b*) to produce a better-yielding hybrid wheat (*c*). The hybrid wheat (*c*) was crossed with another wild grass (*d*) to produce one of the varieties of wheat (*e*) which is used for making flour and bread.

CHECK LIST

- **In breeding experiments, the effect of only one or two genes (out of thousands) is studied, e.g. colour of fur in rabbits or mice.**
- **The genes are in pairs (allelomorphic pairs), because the chromosomes are in pairs.**
- **Although each pair of alleles controls the same character, the alleles do not necessarily have the same effect. For example, of a pair of alleles controlling fur colour, one may try to produce black fur and the other may try to produce white fur.**
- **Usually, one allele is dominant over the other, e.g. the allele (B) for black fur is dominant over the allele (b) for white fur.**
- **This means that a rabbit with the alleles Bb will be black even though it has a gene for white fur.**
- **Although BB rabbits and Bb rabbits are both black, only the BB rabbits will breed true.**
- **Bb black rabbits mated together are likely to have some white babies.**
- **The expectation is that, on average, there will be one white baby rabbit to every three blacks.**
- **Meiosis is the kind of cell division that leads to production of gametes.**
- **Only one of each chromosome pair goes into a gamete.**
- **A Bb rabbit would produce two kinds of gametes for coat colour; 50 per cent of the gametes would have the B gene and 50 per cent would have the b gene.**
- **In some cases, neither one of a pair of alleles is fully dominant over the other. This may be called incomplete dominance or co-dominance.**
- **Sex, in mammals, is determined by the X and Y chromosomes. Males are XY; females are XX.**
- **A knowledge of genetics enables breeders to produce new varieties of plants or animals.**
- **Cross-breeding two varieties enables beneficial genes from both to be brought together in a new variety.**

25 Variation and Selection

VARIATION

New gene combinations and mutations. Meiosis and new gene combinations. Discontinuous variation: distinct differences. Continuous variation: graduated differences. Interaction of genes and environment.

NATURAL SELECTION

The selection of more efficient varieties. Peppered moth: selection of the different forms. Sickle-cell anaemia: selection in different environments. Artificial selection for improved breeds.

VARIATION

The term 'variation' refers to observable differences within a species. All domestic cats belong to the same species, i.e. they can all interbreed, but there are many variations of size, coat colour, eye colour, fur length, etc.

Those variations which can be inherited are determined by genes. They are genetic or heritable variations.

There are also variations which are not heritable, but determined by factors in the environment. A kitten which gets insufficient food will not grow to the same size as its litter mates. A cat with a skin disease may have bald patches in its coat. These conditions are not heritable. They are caused by environmental effects.

Similarly, a fair-skinned person may be able to change the colour of his skin by exposing it to the sun, so getting a tan. The tan is an **acquired characteristic**. You cannot inherit a sun tan. The dark skin of a negro, on the other hand, is an **inherited characteristic**.

Many features in plants and animals are a mixture of acquired and inherited characteristics (Fig. 1). For example, some fair-skinned people never go brown in the sun, they only become sunburned. They have not inherited the genes for producing the extra brown pigment in their skin. A fair-skinned person with the genes for producing pigment will only go brown if he exposes himself to sunlight. So his tan is a result of both inherited and acquired characteristics.

Heritable variation may be the result of mutations (p. 226), or new combinations of genes in the zygote.

New combinations of genes If a grey cat with long fur is mated with a black cat with short fur, the kittens will all be black with short fur. If these offspring are mated together, in due course the litters may include four varieties: black–short, black–long, grey–short and grey–long. Two of these are different from either of the parents. (See 'Meiosis and new combinations of characteristics' below.)

Mutations Many of the coat variations mentioned above may have arisen, in the first place, as mutations in a wild stock of cats. A recent variant produced by a mutation is the 'rex' variety, in which the coat has curly hairs.

Many of our high-yielding crop plants have arisen as a result of mutations in which the whole chromosome set has been doubled.

Meiosis and new combinations of characteristics

On page 223 it was explained that, during meiosis, homologous chromosomes pair up and then at the first nuclear division, separate again.

One of the homologous chromosomes comes from the male parent and the other from the female parent. The alleles for a particular characteristic occupy identical positions on the homologous chromosomes but they do not necessarily control the characteristic in the same way.

Fig. 1 Acquired characteristics. These apples have all been picked from different parts of the same tree. All the apples have the same genotype, so the differences in size must have been caused by environmental effects.

The allele for eye colour will be in the same position on the maternal and paternal chromosome but, in one case, it may be the allele for brown eyes and in the other case, for blue eyes. Separation of homologous chromosomes at meiosis means that the alleles for blue and brown eyes will end up in different gametes (Fig. 2).

On a second pair of homologous chromosomes there may be allelomorphic genes for hair curliness (**C** = curly; **c** = straight). These chromosomes and their alleles will also be separated at the first division of meiosis.

Suppose a father has brown eyes and straight hair, and a mother has blue eyes and curly hair. Also suppose that the father is heterozygous for eye colour (**Bb**) and the mother is heterozygous for hair curliness (**Cc**).

In the mother's ovary, the **b** and **b** alleles will be separated at the first division of meiosis and so will the **C** and **c** alleles. The **b** allele could finish up in the same gamete as either the **C** or **c** allele. So, the genotype of the ovum could be either **bC** or **bc**.

Similarly, meiosis in the father's testes will produce equal numbers of **Bc** and **bc** gametes.

At fertilization, it is a matter of chance which of the two types of sperm fertilizes which of the two types of ovum. (Although usually only one ovum is released, there is a 50:50 chance of its being **bC** or **bc**.)

The grid in Fig. 2*c* shows the possible genotypes of the children in the family.

Offspring (2) and (3) would have the same combination of characteristics as their parents: (2) has brown eyes and straight hair (father's phenotype), and (3) has blue eyes and curly hair (mother's phenotype). Offspring (1) and (4), however, would have different combinations of these two characteristics; combinations which are not present in either parent, namely, (1) brown eyes and curly hair, (4) blue eyes and straight hair.

The separation of parental chromosomes at meiosis and their recombination at fertilization has thus introduced the possibility of new combinations of characteristics. It occurs because the homologous chromosomes derived from one parent do not all go into the same gamete, but move independently of each other.

This recombination of characteristics as a result of meiosis is important for plant and animal breeding programmes as described on page 233. It is also important as a source of variation for natural selection to act on, as described on page 237.

Meiosis takes place only at gamete formation and this is an essential feature of sexual reproduction. It can be claimed, therefore, that one of the important biological advantages of sexual reproduction is the production of new varieties which might be more successful than existing varieties.

sperm mother cell

ovum mother cell

(*a*) Parental genotypes

(*b*) Possible gene combinations in gametes

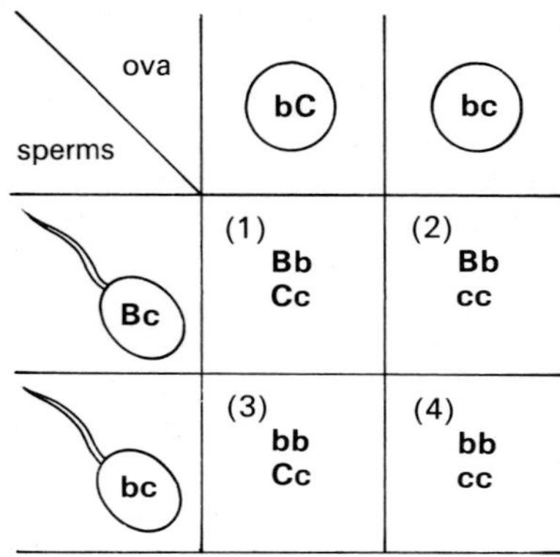

(*c*) Possible combinations of genes in the children

Fig. 2 New combinations of characteristics

QUESTIONS

1 Which of the following do you think are (*a*) mainly inherited characteristics, (*b*) mainly acquired characteristics or (*c*) a more or less equal mixture: manual skills, facial features, body build, language, athleticism, ability to talk?

2 (*a*) Bearing in mind the role of the sex chromosomes, suggest two other new variations which might occur among the children in the example given above.

(*b*) In the example given above, if the mother had been homozygous for hair curliness and the father had been homozygous for eye colour, is there a possibility of new combinations of those characters in the offspring?

3 What new combinations of characters are possible as a result of crossing a tall plant with yellow seeds (**TtYy**) with a dwarf plant with green seeds (**ttyy**)?

4 What are the environmental effects which might have caused the variation in apple size in Fig. 1?

Discontinuous variations

These are variations under the control of a single pair of alleles or a small number of genes. The variations take the form of distinct, alternative phenotypes with no intermediates. The mice in Fig. 4 on page 230 are either black or brown; there are no intermediates. You are either male or female. Apart from a small number of abnormalities, sex is inherited in a discontinuous way.

Discontinuous variations cannot usually be altered by the environment. You cannot change your blood group or eye colour by altering your diet. A genetic dwarf cannot grow taller by eating more food.

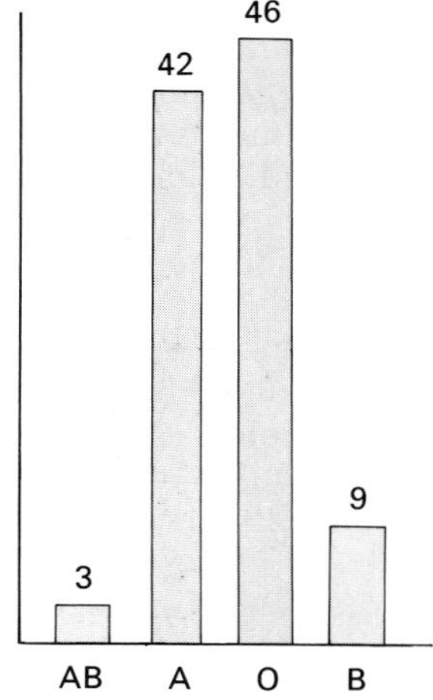

(*a*) Discontinuous variation. Frequencies of ABO blood groups in Britain. The figures could not be adjusted to fit a smooth curve because there are no intermediates.

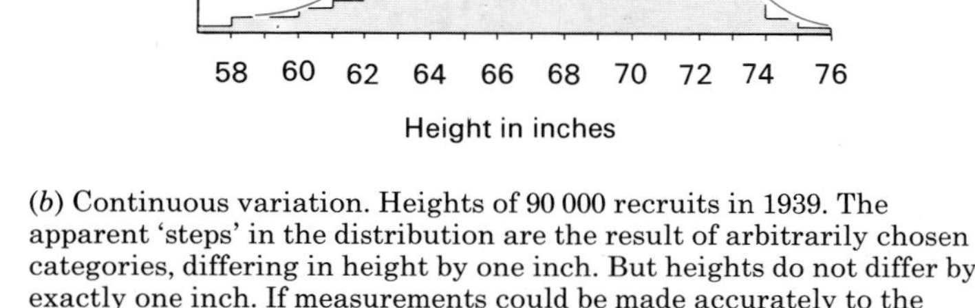

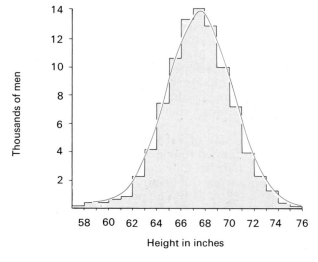

(*b*) Continuous variation. Heights of 90 000 recruits in 1939. The apparent 'steps' in the distribution are the result of arbitrarily chosen categories, differing in height by one inch. But heights do not differ by exactly one inch. If measurements could be made accurately to the nearest millimetre, there would be a smooth curve like the one shown in colour.

Fig. 3 Discontinuous and continuous variation

Continuous variation

There are no distinct categories of height; people are not either tall or short. There are all possible intermediates between very short and very tall. This is a case of continuous variation (Fig. 3*b*).

Continuously variable characteristics are usually controlled by several pairs of alleles. There might be five pairs of alleles for height—(**Hh**), (**Tt**), (**Ll**), (**Ee**) and (**Gg**)—each dominant allele adding 4 cm to your height. If you inherited all ten dominant genes (**HH**, **TT**, etc.) you could be 40 cm taller than a person who inherited all ten recessive genes (**hh**, **tt**, etc.).

The actual number of genes which control height, intelligence and even the colour of hair and skin, is not known.

Continuously variable characteristics are greatly influenced by the environment. A person may inherit genes for tallness and yet not get enough food to grow tall. A plant may have the genes for large fruits but not get enough water, minerals or sunlight to produce large fruits. Continuous variations in human populations, such as height, physique and intelligence are always the result of interaction between the genotype and the environment.

There are many characteristics which are difficult to classify as either wholly continuous or discontinuous variations. Human eye colour has already been mentioned. People can be classified roughly as having blue eyes or brown eyes, but there are also categories described as grey, hazel or green eyes. Probably there is a small number of genes for eye colour and a dominant gene for brown eyes which overrides all the others when it is present. Similarly, red hair is a discontinuous variation but it is masked by genes for other colours and there is a continuous range of hair colour from blond to black.

QUESTION

5 Which of the following would you expect to be inherited in (*a*) a continuous way and (*b*) a discontinuous way: presence of horns in sheep, black colour in cats, number of seeds in an ear of wheat, size of apples, colour of roses, ability to sing, physical strength?

NATURAL SELECTION

Theories of evolution have been put forward in various forms for hundreds of years. In 1858, Charles Darwin and Alfred Russel Wallace published a theory of evolution by natural selection which is still an acceptable theory today.

The theory of natural selection suggests that

1. Individuals within a species are all slightly different from each other (Fig. 4). These differences are called **variations**.
2. If the climate or food supply changes, some of these variations may be better able to survive than others. A variety of animal that could eat the leaves of shrubs as well as grass would be more likely to survive a drought than one which fed only on grass.
3. If one variety lives longer than others, it is also likely to leave behind more offspring. A mouse that lives for 12 months may have ten litters of five babies (50 in all). A mouse that lives for 6 months may have only five litters of five babies (25 in all).
4. If some of the offspring inherit the variation that helped the parent survive better, they too will live longer and have more offspring.
5. In time, this particular variety will outnumber and finally replace the original variety.

Thomas Malthus, in 1798, suggested that the increase in the size of the human population would outstrip the rate of food production. He predicted that the number of people would eventually be regulated by famine, disease and war. When Darwin read the Malthus essay, he applied its principles to other populations of living organisms.

He observed that animals and plants produce vastly more offspring than can possibly survive to maturity and he reasoned that, therefore, there must be a 'struggle for survival'.

For example, if a pair of rabbits had eight offspring which grew up and formed four pairs, eventually having eight offspring per pair, in four generations the number of rabbits stemming from the original pair would be 512 (i.e. 2 → 8 → 32 → 128 → 512). The population of rabbits, however, remains more or less constant. Many of the offspring in each generation must, therefore, have failed to survive to reproductive age.

Competition and selection There will be **competition** between members of the rabbit population for food, burrows and mates. If food is scarce, space is short and the number of potential mates limited, then only the healthiest, most vigorous, most fertile and otherwise well-adapted rabbits will survive and breed.

The competition does not necessarily involve direct conflict. The best adapted rabbits may be able to run faster from predators, digest their food more efficiently, have larger litters or grow coats which camouflage them better or more effectively reduce heat losses. These rabbits will survive longer and leave more offspring. If the offspring inherit the advantageous characteristics of their parents, they may give rise to a new race of faster, different coloured, thicker furred and more fertile rabbits which gradually replace the original, less well-adapted varieties. The new variations are said to have **survival value**.

This is natural selection; the better adapted varieties are 'selected' by the pressures of the environment (**selection pressures**).

For natural selection to be effective, the variations have to be heritable. Variations which are not heritable are of no value in natural selection. Training may give athletes more efficient muscles, but this characteristic will not be passed on to their children.

Evolution Most biologists believe that natural selection, among other processes, contributes to the evolution of new species and that the great variety of living organisms on the Earth is the product of millions of years of evolution, involving natural selection.

The peppered moth

An example of natural selection is provided by a species of moth called the peppered moth. The common form is speckled but there is also a variety which is black. The black variety was rare in 1850, but by 1895 in the Manchester area its numbers had risen to 98 per cent of the population of peppered moths. Observation showed that the light variety was concealed better than the dark

Fig. 4 Variation. The garden tiger moths in this picture are all from the same family. There is a lot of variation in the pattern on the wings and abdomen.

Fig. 5 Selection for varieties of the peppered moth
(*a*) A light moth and a dark one are resting on a lichen-covered tree trunk. Which is better hidden?

(*b*) The two forms are resting on a tree trunk which has no lichen and whose bark is darkened with soot. Which form is more likely to be taken by a bird?

variety when they rested on tree-trunks covered with lichens (Fig. 5). In the Manchester area, pollution had caused the death of the lichens and the darkening of the tree-trunks with soot. In this industrial area the dark variety was the better camouflaged (hidden) of the two and was not picked off so often by birds. So the dark variety survived better, left more offspring and nearly replaced the light form.

The selection pressure, in this case, was presumed to be mainly predation by birds. The adaptive variation which produced the selective advantage was the dark colour.

(In fact, the story is more complex than this and some of the evidence has been challenged.)

Sickle-cell anaemia

This condition has already been mentioned on page 232. A person with sickle-cell disease has inherited both recessive alleles (**Hb^{S}Hb^S**) for defective haemoglobin. The distortion and destruction of the red cells which occurs in low oxygen concentrations leads to bouts of severe anaemia (Fig. 6).

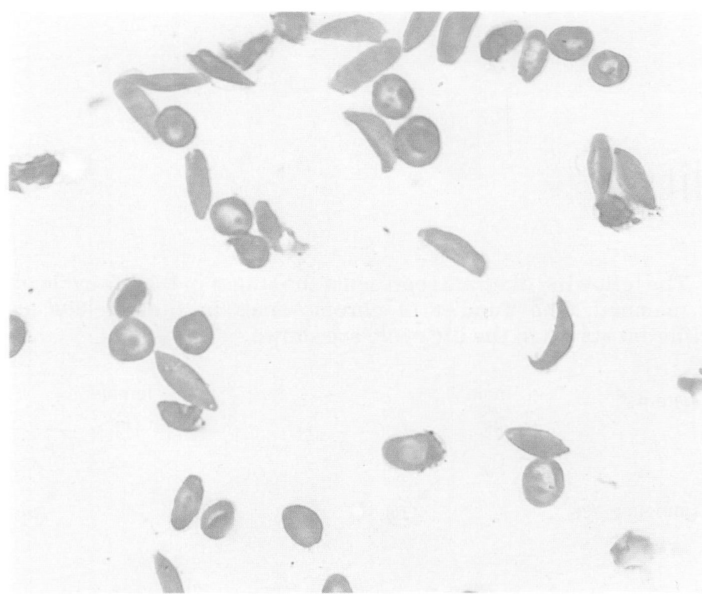

Fig. 6 Sickle-cell anaemia. At low oxygen concentration the red cells become distorted.

In many African countries, sufferers have a reduced chance of reaching reproductive age and having a family. There is thus a selection pressure which tends to remove the homozygous recessives from the population. In such a case, you might expect the harmful **Hb^S** allele to be selected out of the population altogether. However, the heterozygotes (**Hb^{A}Hb^S**) have virtually no symptoms of anaemia but do have the advantage that they are more resistant to malaria than the homozygotes **Hb^{A}Hb^A**.

The selection pressure of malaria, therefore, favours the heterozygotes over the homozygotes and the potentially harmful **Hb^S** allele is kept in the population (Fig. 7).

When Africans migrate to countries where malaria does not occur, the selective advantage of the **Hb^S** allele is lost and the frequency of this allele in the population diminishes.

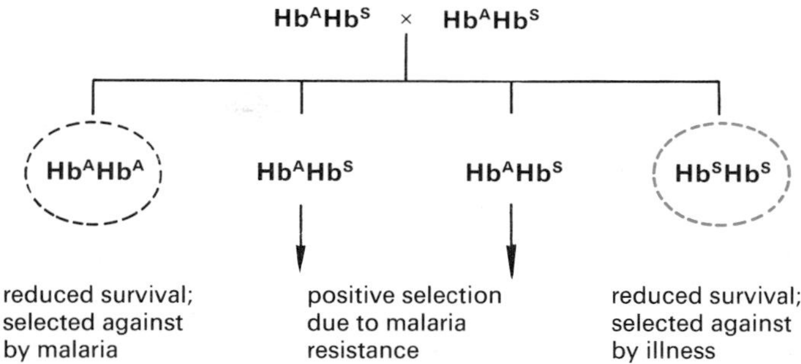

Fig. 7 Selection in sickle-cell disease

Artificial selection

Human communities practise a form of selection when they breed plants and animals for specific characteristics. The many varieties of dog that you see today have been produced by selecting individuals with short legs, curly hair, long ears, etc. One of the puppies in a litter might vary from the others by having longer ears. This individual, when mature, is allowed to breed. From the offspring, another long-eared variant is selected for the next breeding stock, and so on, until the desired or 'fashionable' ear length is established in a true-breeding population.

Fig. 8 Selective breeding in tomatoes. Different breeding programmes have selected different genes for fruit size, colour and shape. Similar processes have given rise to most of our cultivated plants and domesticated animals.

More important are the breeding programmes to improve agricultural livestock or crop plants. Animal-breeders will select cows for their high milk yield, sheep for their wool quality, pigs for their long backs (more bacon rashers). Plant-breeders will select varieties for their high yield and resistance to fungus diseases (Fig. 8). (See p. 233.)

QUESTIONS

6 What features of a bird's appearance and behaviour do you think might help it compete for a mate?
7 What selection pressures do you think might be operating on the plants in a lawn?

CHECK LIST

- **Variations within a species may be inherited or acquired.**
- **Inherited variations arise from different combinations of genes or from mutations.**
- **At meiosis the maternal and paternal chromosomes are randomly distributed between the gametes.**
- **Because the gametes do not carry identical sets of genes, new combinations of genes may arise at fertilization.**
- **Discontinuous variation results, usually, from the effects of a single pair of alleles, and produces distinct and consistent differences between individuals.**
- **Discontinuous variations cannot be changed by the environment.**
- **Continuous variations are usually controlled by a number of genes affecting the same characteristic.**
- **Continuous variation can be influenced by the environment.**
- **Members of a species compete with each other for food and mates.**
- **Some members of a species may have variations which enable them to compete more effectively.**
- **These variants will live longer and leave more offspring.**
- **If the beneficial variations are inherited, the offspring will also survive longer.**
- **The new varieties may gradually replace the older varieties.**
- **Natural selection involves the elimination of the less well-adapted varieties by the pressure of the environment.**
- **Artificial selection is used to improve commercially useful plants and animals.**

Examination Questions

Section 4: Genetics and Heredity

Do not write on this page. Where necessary copy drawings, tables or sentences.

1 In humans the body cells of a girl have

A two X chromosomes
B one X chromosome only
C one X chromosome and one Y chromosome
D two Y chromosomes (N)

2 In humans, eye colour is controlled by a particular gene. This gene has one form, **B**, which produces brown eyes and is dominant to another form, **b**, which produces blue eyes.

(*a*) Use the symbols **B** and **b** to write down the possible genotypes of
(i) a person with blue eyes,
(ii) a person with brown eyes.

(b) If two blue-eyed parents had a child what colour eyes would the child have?

(*c*) (i) An embryo was formed from two blue-eyed parents. The embryo was transplanted and allowed to develop inside a brown-eyed woman. What colour eyes would this baby have?
(ii) Into which part of the reproductive system of the woman should the embryo have been placed when it was transplanted? (L)

3 When 'test tube babies' are formed, eggs are fertilized in Petri dishes by sperms from the father. Some scientists have evidence to suggest that sperms with X sex chromosomes separate towards the bottom of test tubes which are spun at high speed in a machine called a centrifuge. If this is true, explain concisely how it could be used to control the sex of a baby. (W)

4 The following diagram represents the stages in the life cycle of a mammal. The number of chromosomes in the nucleus at different stages in the life cycle are shown.

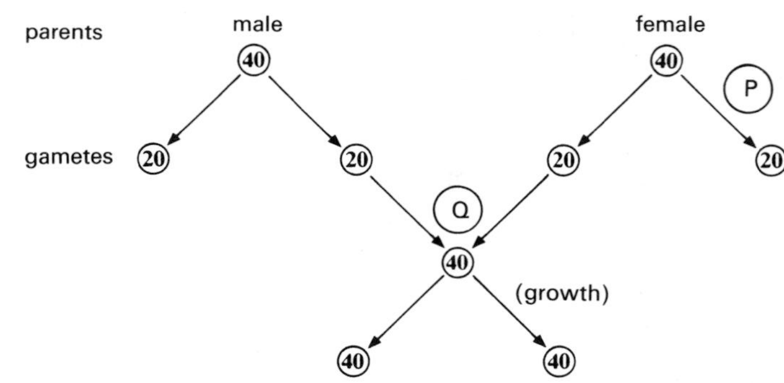

(*a*) Name (i) the type of cell division taking place at P,
(ii) the process taking place at Q.

(*b*) Name the places in the female body where P and Q take place. (W)

5 When the first child of two normal parents is born, it is found to have a serious genetic disease. The parents are told that the disease is caused by a recessive gene, and that only one pair of genes is involved. If they have a second child

A the second child is certain to have the disease
B there is a 1 in 2 chance of the second child having the disease
C there is a 1 in 4 chance of the second child having the disease
D there is no chance of the child having the disease (N)

SECTION 5
ORGANISMS AND THEIR ENVIRONMENT

26 The Interdependence of Living Organisms

FOOD CHAINS AND FOOD WEBS
Pyramids of numbers, recycling.

THE CARBON CYCLE AND NITROGEN CYCLE

THE WATER CYCLE

MANURING AND CROP ROTATION

TYPES OF NUTRITION
Autotrophs, heterotrophs, parasites, symbionts.

ENERGY FLOW IN AN ECOSYSTEM

'Interdependence' means the way in which living organisms depend on each other in order to remain alive, grow and reproduce. For example, bees depend for their food on pollen and nectar from flowers. Flowers depend on bees for pollination (p. 83). Bees and flowers are, therefore, interdependent.

FOOD CHAINS AND FOOD WEBS

One important way in which organisms depend on each other is for their food. Many animals, such as rabbits, feed on plants. Such animals are called **herbivores**. Animals called **carnivores** eat other animals. A **predator** is a carnivore which kills and eats other animals. A fox is a predator which preys on rabbits. **Scavengers** are carnivores which eat the dead remains of animals killed by predators. These are not hard and fast definitions. Predators will sometimes scavenge for their food and scavengers may occasionally kill living animals.

Food chains

Basically, all animals depend on plants for their food. Foxes may eat rabbits, but rabbits feed on grass. A hawk eats a lizard, the lizard has just eaten a grasshopper but the grasshopper was feeding on a grass blade. This relationship is called a **food chain** (Fig. 1).

The organisms at the beginning of a food chain are usually very numerous while the animals at the end of the chain are often large and few in number. The **food**

Fig. 1 A food chain. The caterpillar eats the leaf; the blue tit eats the caterpillar but may fall prey to the kestrel.

pyramids in Fig. 2 show this relationship. There will be millions of microscopic, single-celled green plants in a pond (Fig. 3*a*). These will be eaten by the larger but less numerous water-fleas and other crustacea (Fig. 3*b*), which in turn will become the food of small fish, like minnow and stickleback. The hundreds of small fish may be able to provide enough food for only four or five large carnivores, like pike or perch.

The organisms at the base of the food pyramids in Fig. 2 are plants. Plants produce food from carbon dioxide, water and salts (see 'Photosynthesis', p. 44), and are, therefore, called **producers**. The animals which eat the plants are called **primary consumers**, e.g. grasshoppers. Animals which prey on the plant-eaters are called **secondary consumers**, e.g. shrews, and these may be eaten by **tertiary consumers**, e.g. weasels or kestrels (Fig. 4).

The position of an organism in a food pyramid is sometimes called its **trophic level**.

Pyramids of numbers and biomass

The width of the bands in Fig. 2 is meant to represent the relative number of organisms at each trophic level. So, the diagrams are sometimes called **pyramids of numbers**.

However, you can probably think of situations where a pyramid of numbers would not show the same effect. For example, a single sycamore tree may provide food for thousands of greenfly. One oak tree may feed hundreds of caterpillars. In these cases the pyramid of numbers is upside-down.

The way round this problem is to consider not the single tree, but the mass of the leaves that it produces in the growing season, and the mass of the insects which can live on them. **Biomass** is the term used when the mass of living organisms is being considered, and pyramids of biomass can be constructed as in Fig. 18, p. 252.

An alternative is to calculate the energy available in a year's supply of leaves and compare this with the energy needed to maintain the population of insects which feed on the leaves. This would produce a **pyramid of energy**, with the producers at the bottom having the greatest amount of energy. Each successive trophic level would show a reduced amount of energy.

Later in the chapter, the recycling of matter is discussed. The elements which make up living organisms are recycled, i.e. they are used over and over again. This is not the case with energy which flows from producers to consumers and is eventually lost to the atmosphere as heat. (See 'Energy Flow', p. 251.)

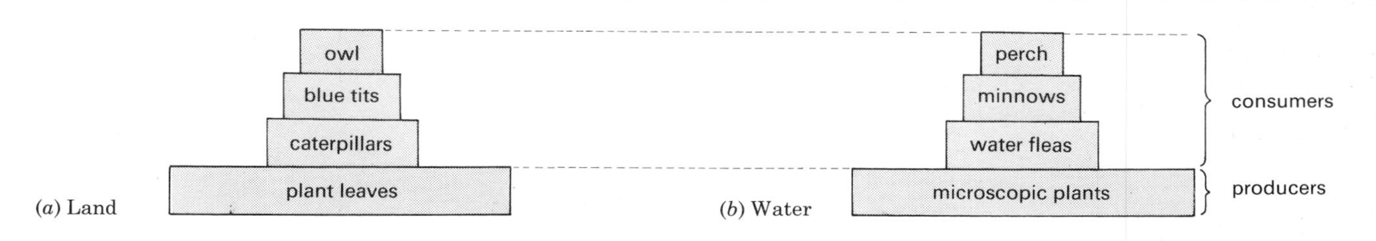

Fig. 2 Examples of food pyramids

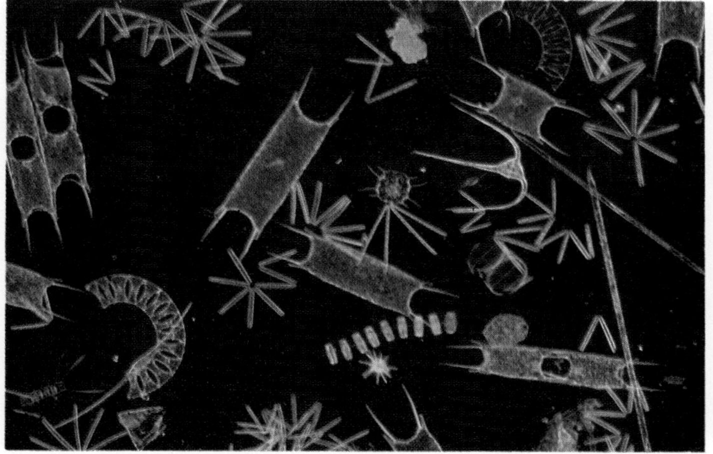

(*a*) Phytoplankton (× 100). These microscopic plants form the basis of a food pyramid in the water.

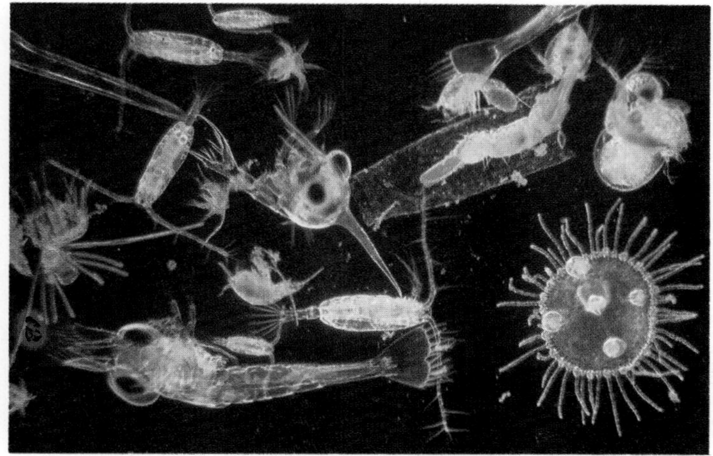

(*b*) Zooplankton (× 20). These crustacea will eat microscopic plants.

Fig. 3 Plankton. The microscopic organisms which live in the surface waters of the sea or fresh water are called, collectively, plankton. The single-celled plants (see p. 327) are the phytoplankton. They are surrounded by water, salts and dissolved carbon dioxide. Their chloroplasts absorb sunlight and use its energy for making food by photosynthesis. The phytoplankton is eaten by small animals in the zooplankton, mainly crustacea (see p. 336). Small fish will eat the crustacea.

Fig. 4 The Kestrel; a secondary or tertiary consumer

Food webs

Food chains are not really as straightforward as described above, because most animals eat more than one type of food. A fox, for example, does not feed entirely on rabbits but takes beetles, rats and voles in its diet. To show these relationships more accurately, a **food web** can be drawn up (Fig. 5 and Fig. 1, p. 265).

The food webs for land, sea and fresh water, or for ponds, rivers and streams will all be different. Food webs will also change with the seasons when the food supply changes.

If some event interferes with a food web, all the organisms in it are affected in some way. For example, if the rabbits in Fig. 5 were to die out, the foxes, owls and stoats would eat more beetles and rats. Something like this happened in 1954 when the disease myxomatosis wiped out nearly all the rabbits in England. Foxes ate more voles, beetles and blackberries, and attacks on lambs and chickens increased. Even the vegetation was affected because the tree seedlings which the rabbits used to nibble off were allowed to grow. As a result, woody scrub-land started to develop on what had been grassy downs. Figure 6 shows a similar effect.

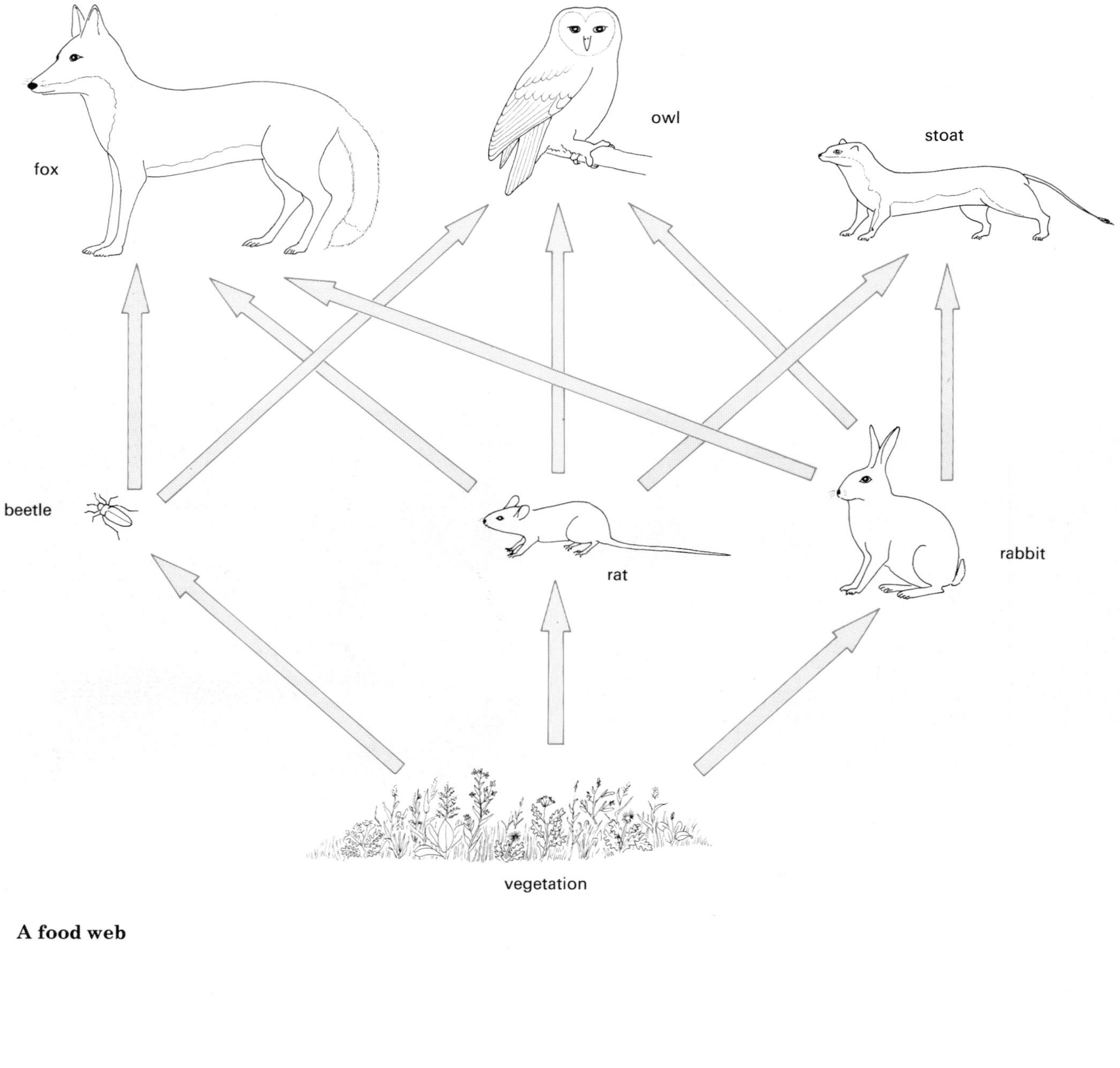

Fig. 5 A food web

(a)

(b)

Fig. 6 Effect of grazing. (*a*) Sheep have eaten any seedlings which grew under the trees. (*b*) Ten years later, the fence has kept the sheep off and the tree seedlings have grown.

Dependence on sunlight

If you take the idea of food chains one step further you will see that all living organisms depend on sunlight and photosynthesis (p. 44). Green plants make their food by photosynthesis, which needs sunlight. Since all animals depend, in the end, on plants for their food, they therefore depend indirectly on sunlight. A few examples of our own dependence on photosynthesis are given below.

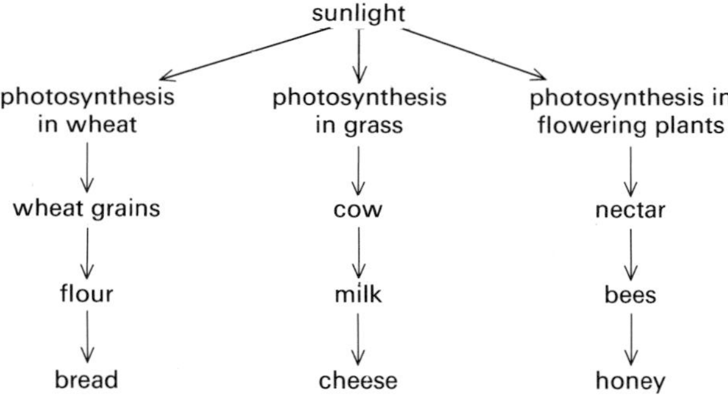

Nearly all the energy released on the Earth can be traced back to sunlight. Coal comes from tree-like plants, buried millions of years ago. These plants absorbed sunlight for their photosynthesis when they were alive. Petroleum was formed, also millions of years ago, from the partly decayed bodies of microscopic plants which lived in the sea. These, too, had absorbed sunlight for photosynthesis.

Today it is possible to use mirrors and solar panels to collect energy from the sun directly, but the best way, so far, of trapping and storing energy from sunlight is to grow plants and make use of their products, such as starch, sugar, oil, alcohol and wood, for food or as energy sources for other purposes. For example, sugar from sugar-cane can be fermented (p. 24) to alcohol and used as motor fuel instead of petrol.

Recycling

There are a number of organisms which have not been fitted into the food webs or food chains described so far. Among these are the **saprophytes**. Saprophytes do not obtain their food by photosynthesis, nor do they kill and eat living animals or plants. Instead they feed on dead and decaying matter such as dead leaves in the soil or rotting tree-trunks (see Fig. 7). The most numerous examples are

Fig. 7 Saprophytes. These toadstools are getting their food from the rotting tree stump.

the fungi, such as mushrooms, toadstools or moulds (see Fig. 4, p. 257), and the bacteria, particularly those which live in the soil. They produce extra-cellular enzymes (p. 19) which digest the decaying matter and then they absorb the soluble products back into their cells. In so doing, they remove the dead remains of plants and animals which would otherwise collect on the Earth's surface. They also break these remains down into substances which can be used by other organisms. Some bacteria, for example, break down the protein of dead plants and animals and release nitrates which are taken up by plant roots and there built into new amino acids and proteins (p. 14). This use and re-use of materials in the living world is called **recycling**.

Figure 8 shows the general idea of recycling. The green plants are the producers, and the animals which eat the plants and each other are the consumers. The bacteria and

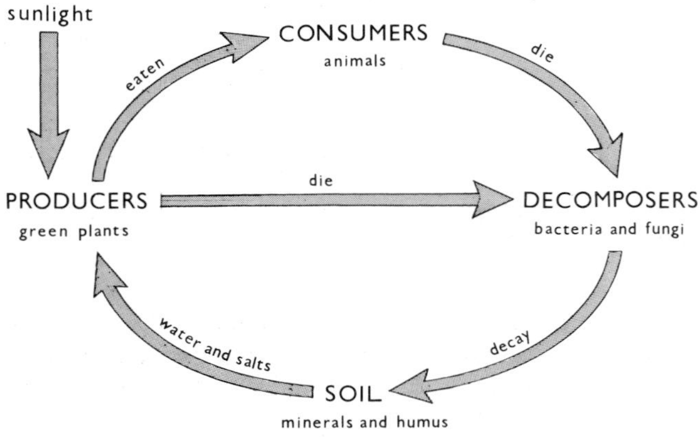

Fig. 8 Recycling in an ecosystem

fungi, especially those in the soil, are called the **decomposers** because they break down the dead remains and release the chemicals for the plants to use again. Two examples of recycling, one for carbon and one for nitrogen, are described below.

QUESTIONS

1 Try to construct a simple food web using the following: sparrow, fox, wheat seeds, cat, kestrel, mouse.
2 Describe briefly all the possible ways in which the following might depend on each other: grass, earthworm, blackbird, oak tree, soil.
3 Explain how the following foodstuffs are produced as a result of photosynthesis: wine, butter, eggs, beans.
4 An electric motor, a car engine and a race-horse can all produce energy. Show how this energy could come, originally, from sunlight. What forms of energy on the Earth are *not* derived from sunlight?
5 How do you think evidence is obtained in order to place animals such as a fox and a pigeon in a food web?

THE CARBON CYCLE

Carbon is an element which occurs in all the compounds which make up living organisms. Plants get their carbon from carbon dioxide in the atmosphere and animals get their carbon from plants. The carbon cycle, therefore, is mainly concerned with what happens to carbon dioxide (Fig. 9).

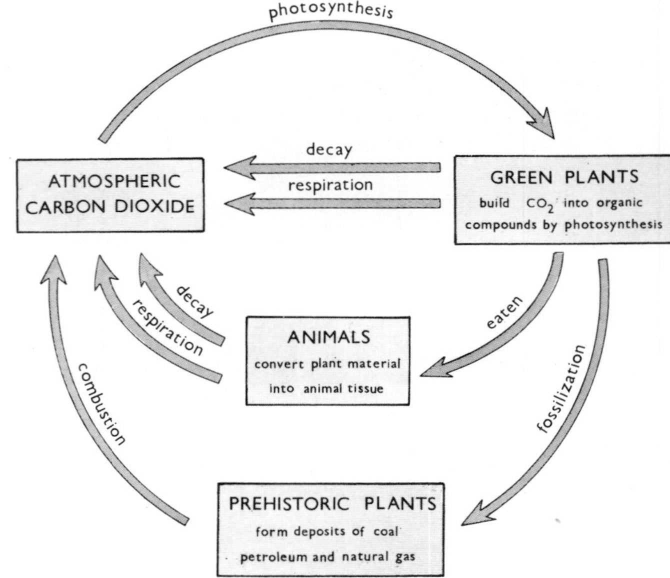

Fig. 9 The carbon cycle

Removal of carbon dioxide from the atmosphere

Green plants remove carbon dioxide from the atmosphere as a result of their photosynthesis. The carbon of the carbon dioxide is built first into a carbohydrate such as sugar. Some of this is changed into starch or the cellulose of cell walls, and the proteins, pigments and other compounds of a plant. When the plants are eaten by animals, the organic plant material is digested, absorbed and built into the compounds making up the animals' tissues. Thus the carbon atoms from the plant become part of the animal.

Addition of carbon dioxide to the atmosphere

Respiration Plants and animals obtain energy by oxidizing carbohydrates in their cells to carbon dioxide and water (p. 24). The carbon dioxide and water are excreted and so the carbon dioxide returns once again to the atmosphere.

Decay The organic matter of dead animals and plants is used by saprophytes, especially bacteria and fungi, as a source of energy. These micro-organisms decompose the plant and animal remains and turn the carbon compounds into carbon dioxide.

Combustion (burning) When carbon-containing fuels such as wood, coal, petroleum and natural gas are burned, the carbon is oxidized to carbon dioxide ($C + O_2 \longrightarrow CO_2$). The hydrocarbon fuels, such as coal and petroleum, come from ancient plants which have only partly decomposed over the millions of years since they were buried.

So, an atom of carbon which today is in a molecule of carbon dioxide in the air may tomorrow be in a molecule of cellulose in the cell wall of a blade of grass. When the grass is eaten by a cow, the carbon atom may become part of a glucose molecule in the cow's bloodstream. When the glucose molecule is used for respiration, the carbon atom will be breathed out into the air once again as carbon dioxide. (See p. 269 for the 'greenhouse effect'.)

The same kind of cycling applies to nearly all the elements of the Earth. No new matter is created, but it is repeatedly rearranged. A great proportion of the atoms of which you are composed will, at one time, have been part of other organisms.

QUESTIONS

6 (*a*) Why do living organisms need a supply of carbon?
(*b*) Give three examples of carbon-containing compounds which occur in living organisms (see pp. 14–16).
(*c*) Where do (i) animals, (ii) plants get their carbon from?

7 Write three chemical equations (*a*) to illustrate that respiration produces carbon dioxide (see p. 24), (*b*) to show that burning produces carbon dioxide, (*c*) to show that photosynthesis uses up carbon dioxide (see p. 44).

8 Outline the events that might happen to a carbon atom in a molecule of carbon dioxide which entered the stoma in the leaf of a potato plant, and became part of a starch molecule in a potato tuber which was then eaten by a man. Finally the carbon atom is breathed out again in a molecule of carbon dioxide.

9 Large areas of tropical forest, particularly in South America, are being cut down to make way for roads, cities and agriculture. What effect might this have on the carbon dioxide level in the Earth's atmosphere?

10 Construct a diagram, on the lines of the carbon cycle (Fig. 9) to show the cycling process for hydrogen (starting from the water used in photosynthesis).

THE NITROGEN CYCLE

When a plant or animal dies, its tissues decompose, mainly as a result of the action of saprophytic bacteria. One of the important products of this decay is **ammonia** (NH_3, a compound of nitrogen), which is washed into the soil.

The excretory products of animals contain nitrogenous waste-products such as ammonia, urea and uric acid (p. 159). The organic matter in their droppings is also decomposed by soil bacteria.

Processes which add nitrates to soil

Nitrifying bacteria These are bacteria living in the soil which use the ammonia from excretory products and decaying organisms as a source of energy (as we use glucose in respiration). In the process of getting energy from ammonia, the bacteria produce **nitrates**.

The 'nitrite' bacteria oxidize ammonium compounds to nitrites ($NH_4^- \longrightarrow NO_2^-$).

'Nitrate' bacteria oxidize nitrites to nitrates ($NO_2^- \longrightarrow NO_3^-$).

Although plant roots can take up ammonia in the form of its compounds, they take up nitrates more readily, so the nitrifying bacteria increase the fertility of the soil by making nitrates available to the plants.

Nitrogen-fixing bacteria This is a special group of nitrifying bacteria which can absorb nitrogen as a gas from the air spaces in the soil (p. 255), and build it into

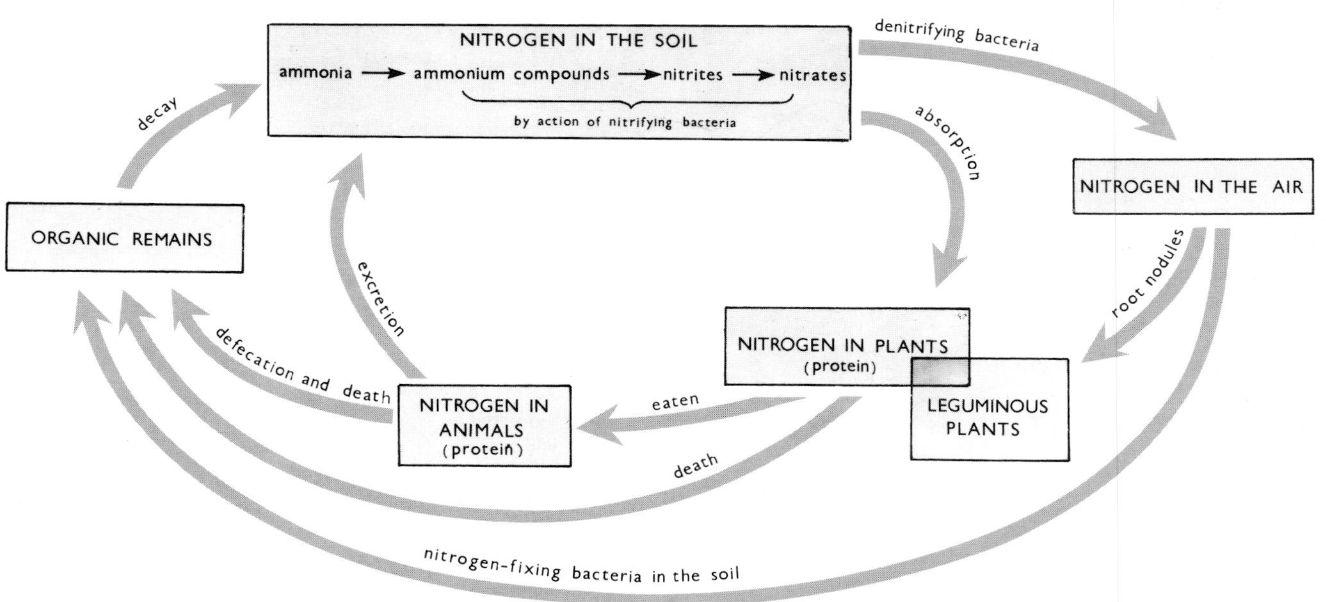

Fig. 10 The nitrogen cycle

compounds of ammonia. Nitrogen gas cannot itself be used by plants. When it has been made into a compound of ammonia, however, it can easily be changed to nitrates by other nitrifying bacteria. The process of building the gas, nitrogen, into compounds of ammonia is called **nitrogen fixation**. Some of the nitrogen-fixing bacteria live freely in the soil. Others live in the roots of **leguminous plants** (peas, beans, clover), where they cause swellings called **root nodules** (Fig. 15). These leguminous plants are able to thrive in soils where nitrates are scarce, because the nitrogen-fixing bacteria in their nodules make compounds of nitrogen available for them (see p. 251). Leguminous plants are also included in crop rotations (see p. 249) to increase the nitrate content of the soil.

Blue-green algae Another important group of nitrogen-fixing organisms are the blue-green algae. These organisms need sunlight for photosynthesis and nitrogen-fixation. Consequently they are effective nitrogen-fixers mainly in the sea and fresh water, and in the soil surface when it does not have a dense cover of vegetation.

Lightning The high temperature of lightning discharge causes some of the nitrogen and oxygen in the air to combine and form oxides of nitrogen. These dissolve in the rain and are washed into the soil as weak acids, where they form nitrates. Although several million tonnes of nitrate may reach the Earth's surface in this way each year, this forms only a small fraction of the total nitrogen being recycled.

Processes which remove nitrates from the soil

Uptake by plants Plant roots absorb nitrates from the soil and combine them with carbohydrates to make proteins (p. 52).
Leaching Nitrates are very soluble (i.e. dissolve easily in water), and as rain-water passes through the soil it dissolves the nitrates and carries them away in the run-off or to deeper layers of the soil. This is called **leaching**.
Denitrifying bacteria These are bacteria which obtain their energy by breaking down nitrates to nitrogen gas which then escapes from the soil into the atmosphere.

These processes are summed up in Fig. 10.

THE WATER CYCLE

All the elements which make up living organisms are recycled, not just carbon and nitrogen. It would be possible to trace out cycles for hydrogen, oxygen, phosphorus, sulphur, iron, etc. The 'water cycle', however, is somewhat different because only a tiny proportion of the water which is recycled passes through living organisms.

Animals lose water by evaporation (p. 168), defecation (p. 131), urination (p. 162) and exhalation (p. 154). They gain water from their food and drink. Plants take up water from the soil (p. 75) and lose it by transpiration (p. 71). Millions of tonnes of water is transpired, but only a tiny fraction of it has taken part in the reactions of respiration (p. 24) or photosynthesis (p. 44).

The great proportion of water is recycled without the intervention of animals or plants. The sun shining and the

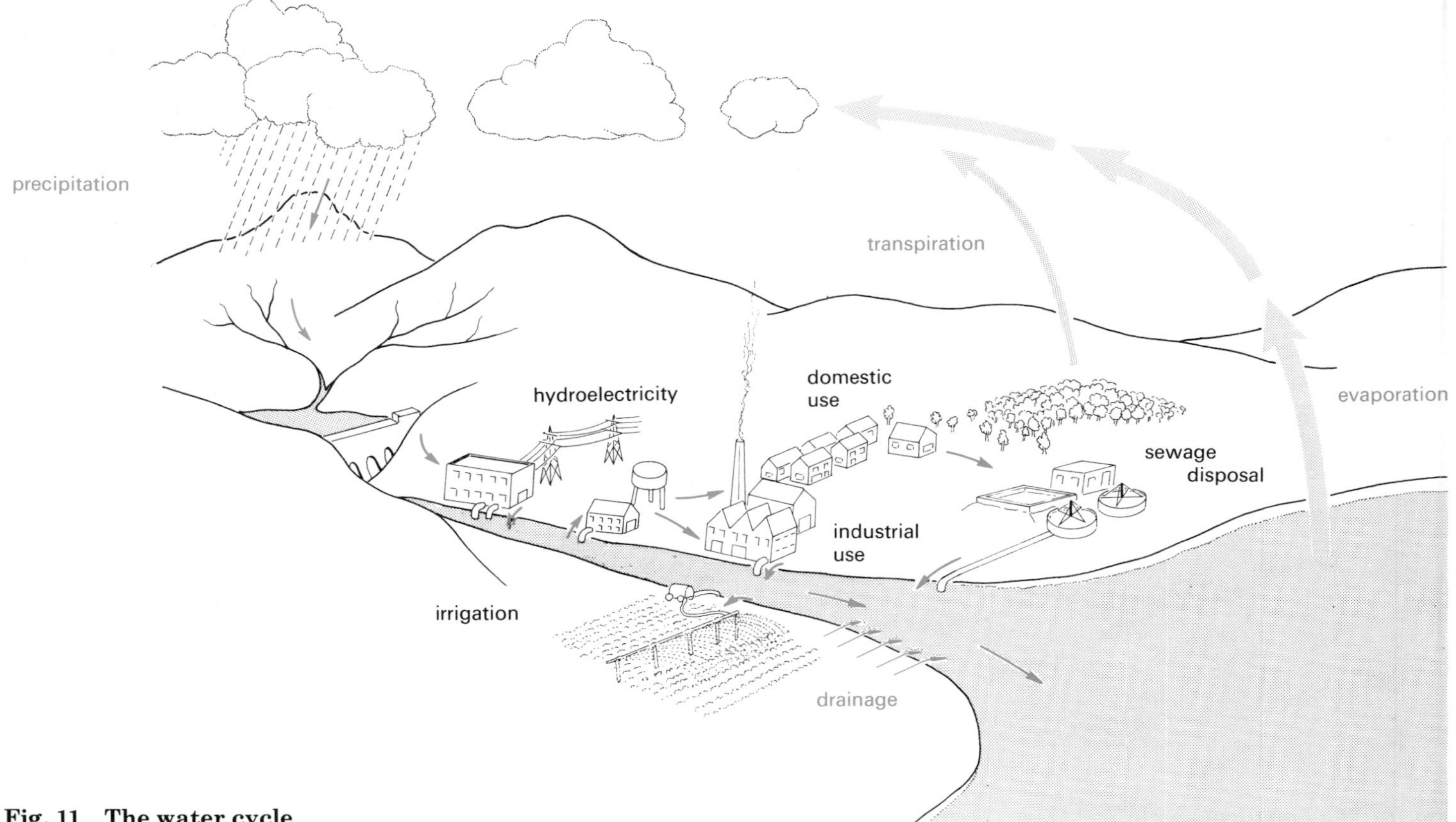

Fig. 11 The water cycle

wind blowing over the oceans evaporate water from their vast, exposed surfaces. The water vapour produced in this way enters the atmosphere and eventually forms clouds. The clouds release their water in the form of rain or snow (precipitation). The rain collects in streams, rivers and lakes and ultimately finds its way back to the oceans. The human population diverts some of this water for drinking, washing, cooking, irrigation, hydro-electric schemes and other industrial purposes, before allowing it to return to the sea (Fig. 11).

QUESTIONS

11 On a lawn growing on nitrate-deficient soil, the patches of clover often stand out as dark green and healthy against a background of pale green grass. Suggest a reason for this contrast. (See also p. 54.)
12 Very briefly explain the difference between nitrifying, nitrogen-fixing and denitrifying bacteria.

MANURING AND CROP ROTATION

Manuring

In a natural community of plants and animals, the processes which remove nitrates from and add nitrates to the soil are in balance. In agriculture, most of the crop is usually removed so that there is little or no organic matter left on the soil for the nitrifying bacteria to act on. In a farm with animals, the animal manure, mixed with straw, is ploughed back into the soil or spread on the pasture. The manure thus replaces the nitrates and other minerals removed by the crop. It also gives the soil a good structure and improves its water-holding properties (p. 255).

When animal manure is not available in large enough quantities, artificial fertilizers are used. These are mineral salts made on an industrial scale. Examples are ammonium sulphate (for nitrogen and sulphur), ammonium nitrate (for nitrogen) and compound NPK fertilizer for nitrogen, phosphorus and potassium (see p. 53). These are spread on the soil in carefully calculated amounts to provide the minerals, particularly nitrogen, phosphorus and potassium that the plants need. Figures 12 and 13 show how these artificial fertilizers increase the yield of crops from agricultural land, but they do little to maintain a good soil structure because they contain no humus (see pp. 255 and 269).

Fig. 12 Experimental plots of wheat. The rectangular plots have been treated with different fertilizers.

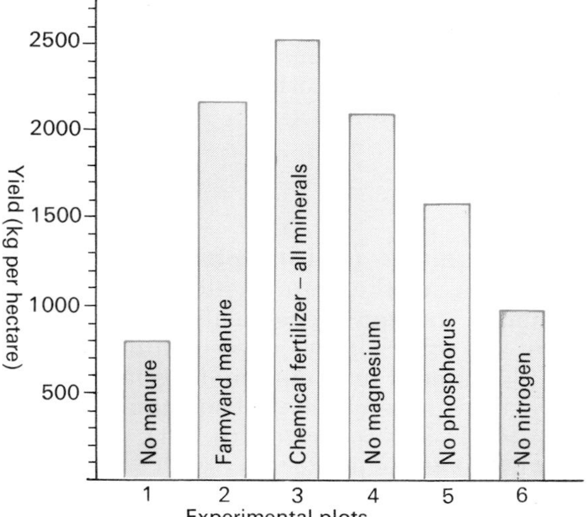

Fig. 13 Average yearly wheat yields from 1852 to 1925. Broadbalk field, Rothamsted Experimental Station.
Plot 1 received no mature or chemical fertilizer for 73 years.
Plot 2 received an annual application of farmyard manure.
Plot 3 received chemical fertilizer with all necessary minerals.
Plots 4 to 6 received chemical fertilizer lacking one element.

Crop rotation

Different crops make differing demands on the soil; potatoes and tomatoes use much potassium, for example. By changing the crop grown from year to year, no single group of minerals is continuously removed from the soil. Leguminous crops such as clover and beans may help to replace the nitrogen content of the soil because their root nodules contain nitrogen-fixing bacteria. The nitrates are released into the soil when the plant dies and decays.

The use of artificial fertilizers has made crop rotation, at least for the reasons above, largely unnecessary. However, turning arable land over to grass for a year or two does improve the crumb structure of the soil (p. 256), its drainage and other properties. Rotation also reduces the chances of infectious diseases that can enter the crop through the soil. For example, repeated crops of potatoes in the same field will increase the population of the fungus causing the disease 'potato blight'. If potatoes are not planted for a few years, the crop will suffer less from this disease when potatoes are grown again.

QUESTIONS

13 To judge from Fig. 13, which mineral element seems to have the most pronounced effect on the yield of wheat? Explain your answer.
14 Draw up two columns headed A and B. In A list the processes which add nitrates to the soil and in B list those which remove nitrates from the soil. How might the activities of man alter the normal balance between these two processes?
15 Suggest (*a*) some advantages, (*b*) some disadvantages of (i) organic manures (e.g. compost or farmyard manure) and (ii) chemical fertilizers.

TYPES OF NUTRITION AND LIFE-STYLES

A great many different terms are used to describe the ways in which living organisms get their food and how they depend on each other. The following account describes a few of them.

Autotrophic

Autotrophic organisms (**autotrophs**) are those which can build up all the organic substances they need from simple inorganic chemicals. All plants, some bacteria and other single-celled organisms are autotrophs. Plants need only carbon dioxide, water and salts to make all their essential substances (see Chapter 5, p. 44.) In any ecosystem (p. 281), it is the autotrophs which are the producers.

Heterotrophic

Heterotrophic nutrition refers to organisms (**heterotrophs**) which must use ready-made, complex organic compounds as a source of food. These organic compounds will have been made by the autotrophs. The heterotrophs digest the organic compounds and absorb the products into their bodies. In an ecosystem, the heterotrophs are the consumers or the decomposers.

Heterotrophs may be herbivores, carnivores, omnivores, saprophytes or parasites.

Herbivores are animals which feed exclusively on plants.

Carnivores are animals which eat other animals.

Omnivores are animals, like ourselves, which include plants and animals in their diet.

Saprophytes are mainly fungi and bacteria. They feed by secreting enzymes into the dead and decaying remains of animals and plants and absorbing the soluble products back into their bodies (see pp. 321–2).

Parasites A parasite is an organism which derives its food from another organism, called the **host**, while the host is still alive. **Ectoparasites** live, for most of their life cycle, on the surface of their host; **endoparasites** live inside the host.

A flea is an ectoparasite which lives in the fur or feathers of mammals or birds and sucks blood from the skin. An aphid (greenfly) is a parasite on plants and sucks food from the veins in their leaves or stems. A tapeworm is an endoparasite living in the intestine of a vertebrate host and absorbing the host's digested food. In some cases the host may be weakened by the presence of the parasite, or its metabolism may be upset by the parasite's excretory products. In many cases, however, the host appears to suffer no serious disadvantage. Often, this will depend on how many parasites the host is carrying.

Many bacteria and fungi (Fig. 14) are parasitic and may cause diseases in their hosts. Bacteria often live on or in their hosts without apparently causing any harm. In this case they might be called **commensals** rather than parasites (see below). Sometimes the host is thought to benefit from the activities of the bacteria which may then be called **symbionts**.

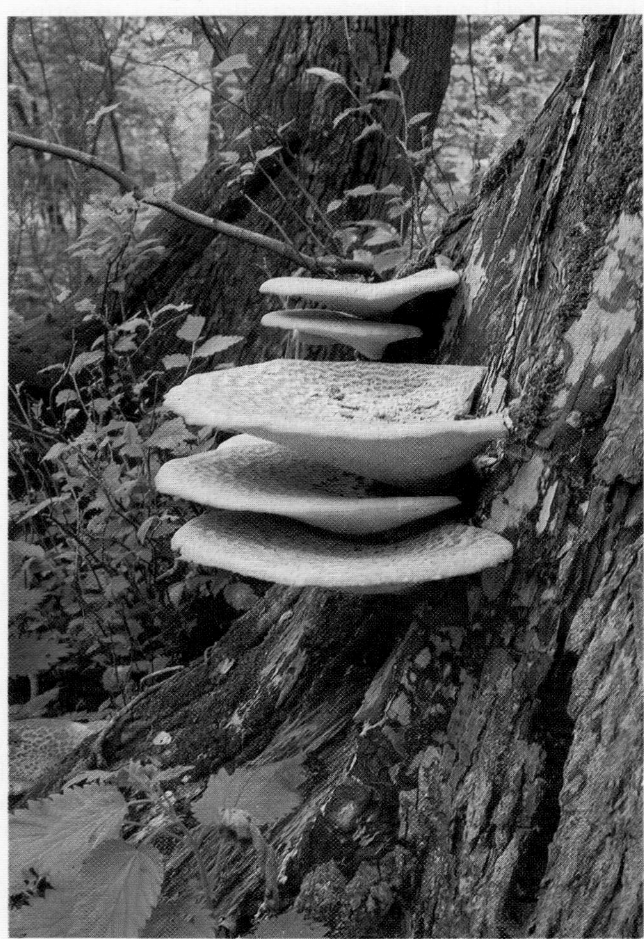

Fig. 14 A parasitic fungus on an elm tree

Symbiosis and commensalism

Symbiosis The term 'symbiosis' sometimes refers to two unrelated organisms living more or less permanently together, irrespective of whether one of them is a parasite. More commonly, however, it implies that both organisms derive some benefit from the association.

For example, in the stomachs of cattle and sheep there live large numbers of bacteria. They cause no symptoms of illness and they are thought to be of value to the animal because they help to digest the cellulose in its food (see p. 129). The cow benefits from the relationship because it is better able to digest grass. The bacteria are thought to benefit by having an abundant supply of food, though they are themselves digested when the grass moves along the cow's digestive tract.

Fig. 15 Root nodules on pea plant. The nodules contain bacteria which fix nitrogen.

The nitrogen-fixing bacteria in root nodules (Fig. 15) provide a further example of symbiosis. The plant benefits from the extra nitrates that the bacteria provide, while the bacteria are protected in the plant's cells and can also use the sugars made by the plant's photosynthesis.

Sometimes the terms **mutualism** and **commensalism** are used to describe a permanent relationship between two unrelated species.

Mutualism The term 'mutualism' means the same as symbiosis, in the sense that it is used in this chapter, i.e. both partners are presumed to derive some benefit from the association.

Commensalism This is an association between two species in which one of the partners is thought to benefit and the other is not harmed. It most often refers to an association in which one of the organisms feeds on the surplus or discarded food of the other. Many of the bacteria living in our intestine may be commensals, feeding on the undigested or unabsorbed food in the ileum or colon (p. 131) and doing us no harm. There is evidence, however, that some of the bacteria produce a substance called vitamin K, which we need. On the other hand there is little or no evidence that we actually absorb the vitamin from this source.

Harmless commensal bacteria living on our skin may become harmful parasites if the skin is damaged.

It is often difficult to produce good evidence that one or both organisms are getting some benefit or suffering damage from these relationships. Consequently, it is not always easy to make a distinction between parasitism, commensalism and symbiosis. The argument is about words rather than biology. It is far more important to try and find out what the relationship between organisms really is rather than to argue about what word to apply to the relationship.

QUESTIONS

16 What type of nutrition is likely to be characteristic of (*a*) an apple tree, (*b*) a mushroom, (*c*) a human, (*d*) a goat?

17 Different types of bacteria exhibit almost every kind of nutrition. Which kind do they *not* exhibit (see p. 310)?

18 Nitrifying bacteria and green plants both use carbon dioxide to make carbohydrates. How do they differ from each other in the way they do it?

19 Some species of wood-eating termites have large populations of protozoa in their intestines. If the insect is kept at 36 °C it is unharmed but the protozoa are killed. How could you use this knowledge to try and find out if the protozoa were symbionts, commensals or parasites? What controls would you carry out?

ENERGY FLOW IN AN ECOSYSTEM

An **ecosystem** is a community of living organisms and the habitat in which they live (p. 281). A pond is an ecosystem consisting of plants, animals, water, dissolved air, minerals and mud. The input of energy from the sun and a supply of water from rain is all that the pond community needs to maintain its existence. A forest is an ecosystem, so is an ocean. The whole of the Earth's surface may be considered as one vast ecosystem.

With the exception of atomic energy and tidal power, all the energy released on Earth is derived from sunlight. The energy released by animals comes, ultimately, from plants that they or their prey eat and the plants depend on sunlight for making their food (p. 245). The energy in organic fuels also comes ultimately from sunlight trapped by plants. Coal is formed from fossilized forests and petroleum comes from the cells of ancient marine plants.

Use of sunlight To try and estimate just how much life the Earth can support it is necessary to examine how efficiently the sun's energy is used. The amount of energy from the sun reaching the Earth's surface in 1 year ranges from 2 million to 8 million kilojoules per 1 m^2 ($2–8 \times 10^9$ J m^{-2} y^{-1}) depending on the latitude. When this energy falls on to grassland, about 20 per cent is reflected by the vegetation, 39 per cent is used in evaporating water from the leaves (transpiration), 40 per cent warms up the plants, the soil and the air, leaving only about 1 per cent to be used in

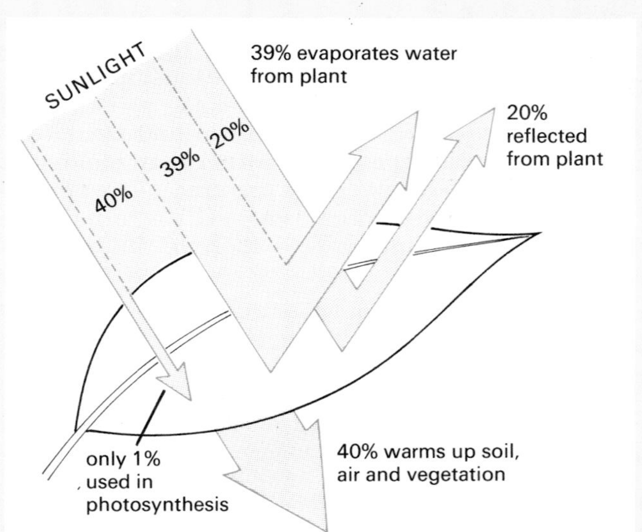

Fig. 16 Absorption of sun's energy by plants

photosynthesis for making new organic matter in the leaves of the plants (Fig. 16).

This figure of 1 per cent will vary with the type of vegetation being considered and with climatic factors such as availability of water and the soil temperature. Sugar-cane grown in ideal conditions can convert 3 per cent of the sun's energy into photosynthetic products; sugar-beet at the height of its growth has nearly a 9 per cent efficiency. Tropical forests and swamps are far more productive than grassland but it is difficult, and probably undesirable, to harvest and utilize their products (p. 269).

In order to allow crop plants to approach their maximum efficiency they must be provided with sufficient water and mineral salts. This can be achieved by irrigation and the application of fertilizer.

Energy transfer between organisms Having considered the energy conversion from sunlight to plant products the next step is to study the efficiency of transmission of energy from plant products to primary consumers. On land, primary consumers eat only a small proportion of the available vegetation. In a deciduous forest only about 2 per cent is eaten; in grazing land, 40 per cent of the grass may be eaten by cows. In open water, however, where the producers are microscopic plants (phytoplankton) and are swallowed whole by the primary consumers in the zooplankton, 90 per cent or more may be eaten. In the land communities, the parts of the vegetation not eaten by the primary consumers will eventually die and be used as a source of energy by the decomposers.

A cow is a primary consumer; over 60 per cent of the grass it eats passes through its alimentary canal (p. 125) without being digested. Another 30 per cent is used in the cow's respiration to provide energy for its movement and other life processes. Less than 10 per cent of the plant material is converted into new animal tissue to contribute to growth (Fig. 17). This figure will vary with the diet and the age of the animal. In a fully grown animal all the digested food will be used for energy and replacement and none will contribute to growth. Economically it is desirable to harvest the primary consumers before their rate of growth starts to fall off.

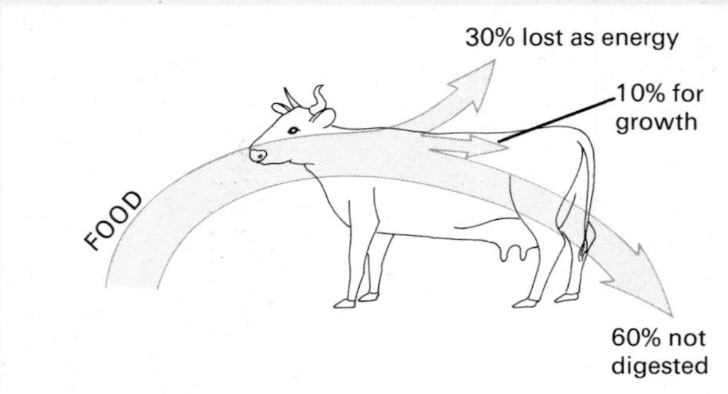

Fig. 17 Energy transfer from plants to animals

The transfer of energy from primary to secondary consumers is probably more efficient since a greater proportion of the animal food is digested and absorbed than is the case with plant material. The transfer of energy at each stage in a food chain may be represented by classifying the organisms in a community as producers, or primary, secondary or tertiary consumers, and showing their relative masses in a pyramid such as the one shown in Fig. 2 but on a more accurate scale. In Fig. 18 the width of the horizontal bands is proportional to the masses (dry weight) of the organisms in a shallow pond.

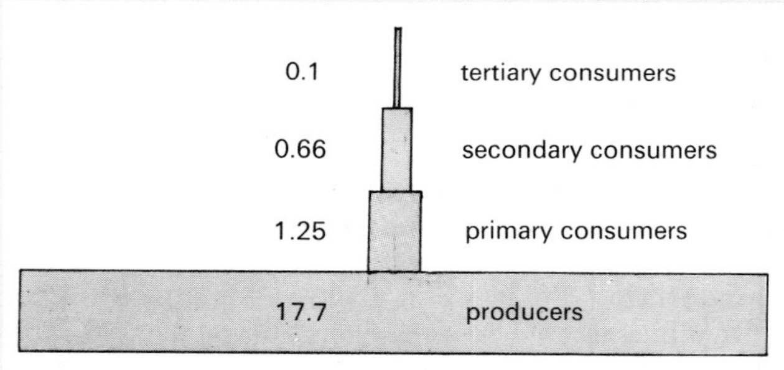

Fig. 18 Biomass (dry weight) of living organisms in a shallow pond (in grams per square metre).
(After R. H. Whittaker, *Communities and Ecosystems*, Macmillan, N.Y. ©1975)

Energy transfer in agriculture In human communities, the use of plant products to feed animals which provide meat, eggs and dairy products is wasteful, because only 10 per cent of the plant material is converted to animal products. It is more economical to eat bread made from the wheat than to feed the wheat to hens and then eat the eggs and chicken meat. This is because eating the wheat as bread avoids using any part of its energy to keep the chickens alive and active. Energy losses can be reduced by keeping hens indoors in small cages where they lose little heat to the atmosphere and cannot use much energy in movement. The same principles can be applied in 'intensive' methods of rearing calves. However, many people feel that these methods are less than humane, and the saving of energy is far less than if the plant products were eaten directly by man.

Consideration of the energy flow in a modern agricultural system reveals other sources of inefficiency. To produce 1 tonne of nitrogenous fertilizer takes energy equivalent to burning 5 tonnes of coal. Calculations show that if the energy needed to produce the fertilizer is added to the energy used to produce a tractor and to power it, the energy derived from the food so produced is less than that expended in producing it.

QUESTIONS

20 It can be claimed that the sun's energy is used indirectly to produce a muscle contraction in your arm. Trace the steps in the transfer of energy which would justify this claim.

21 A group of explorers is stranded on a barren island where there is no soil and no vegetation. From their stores they have salvaged some living hens and some wheat. To make these resources last as long as possible should they

(*a*) eat the wheat and when it is finished, kill and eat the hens, or

(*b*) feed the wheat to the hens, collect and eat the eggs laid and when the wheat is gone, kill and eat the hens, or

(*c*) kill and eat the hens first and when they are finished, eat the wheat? Justify your answer.

22 Discuss the advantages and disadvantages of human attempts to exploit a food chain nearer to its source, e.g. the plankton in Fig. 3.

CHECK LIST

- **All animals depend, ultimately, on plants for their source of food.**
- **Since plants need sunlight to make their food, all organisms depend, ultimately, on sunlight for their energy.**
- **Plants are the producers in a food web; animals may be primary, secondary or tertiary consumers.**
- **The materials which make up living organisms are constantly recycled.**
- **Plants take up carbon dioxide during photosynthesis; all living organisms give out carbon dioxide during respiration; the burning of carbon-containing fuels produces carbon dioxide.**
- **The uptake of carbon dioxide by plants balances the production of carbon dioxide from respiration and combustion.**
- **Soil nitrates are derived naturally from the excretory products of animals and the dead remains of living organisms.**
- **Nitrifying bacteria turn these products into nitrates which are taken up by plants.**
- **Nitrogen-fixing bacteria can make nitrogenous compounds from gaseous nitrogen.**
- **Autotrophs are organisms (e.g. plants) which build up their food from simple substances.**
- **Heterotrophs are organisms (e.g. animals) which have to take in food derived from other organisms.**
- **Parasites derive their food from another living organism while living on it or in it.**
- **Symbiosis and commensalism are terms used to describe a close permanent association between two unrelated organisms from which one, at least, derives some benefit and the other is unharmed.**
- **An ecosystem is a self-contained community of organisms.**
- **Only about 1 per cent of the sun's energy which reaches the Earth is trapped by plants during photosynthesis.**
- **At each step in a food chain, only a small proportion of the food is used for growth. The rest is used for energy to keep the organism alive.**

27 The Soil

SOIL FORMATION

SOIL COMPONENTS

Mineral particles, organic matter, air and water, salts, micro-organisms.

SOIL FERTILITY

Composition, crumb structure, acidity.

LIFE IN THE SOIL

Micro-organisms, invertebrates, plant roots, mycorrhiza.

PRACTICAL WORK

Presence of particles, permeability to and retention of water, pH, soil animals, air, water and humus content, capillary attraction, micro-organisms.

Soil is a layer of material covering parts of the land. **Topsoil** is the crumbly, dark material which gardeners dig and farmers plough. It contains a good deal of organic matter and many living organisms (Fig. 1). **Subsoil** is the lighter-coloured material between the topsoil and the underlying stratum. The subsoil can usually be penetrated by plant roots but it contains little organic matter.

The depth, colour and composition of soil vary greatly according to the climate, the geology and the vegetation of the region. All soils consist primarily of mineral particles, e.g. sand, silt or clay particles. These particles are formed from rocks by the various processes of erosion such as wind, rain, heat and cold which cause the rocks to crumble into smaller and smaller fragments. The fragments may be washed off hillsides by rain and accumulate in valleys. They may have been carried by glaciers millions of years ago or they may simply build up on top of the rocks from which they are being formed.

Accumulations of particles are colonized by plants and micro-organisms which help to hold the particles together. The dead remains of these organisms are added to the mineral particles to form the material we call soil.

Soils will differ from each other in the proportion of sand, silt and clay particles, the amount of organic matter, the soluble salts present and the organisms living on or in the soil.

The soil forms an ecosystem (p. 281) in which a large number of organisms live (see p. 257). In addition, soil is an essential medium for most land plants. Plant roots anchor the plant in the soil and help to hold the plant upright (Fig. 3). It is from the soil that land plants obtain the water and mineral salts they need for photosynthesis and production of food (Chapter 5).

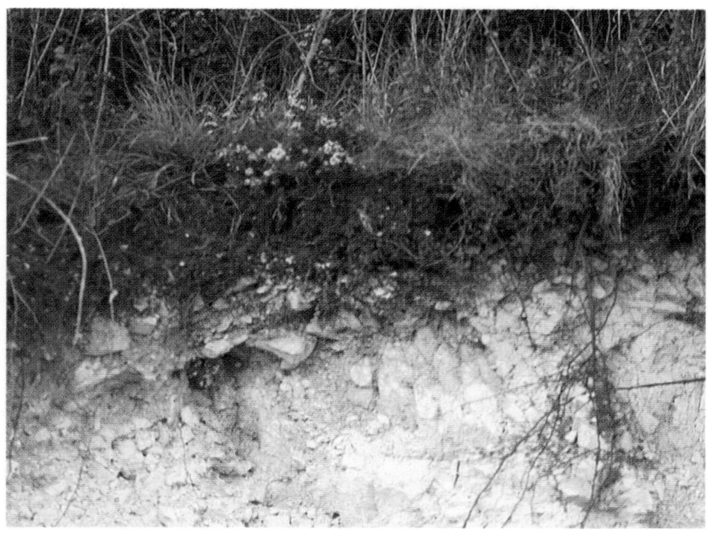

Fig. 1 A soil profile. Thin, chalky soil. The soil layer varies from a few centimetres to 30 or 40 cm. In this case there is a 25 cm layer of soil over limestone rock.

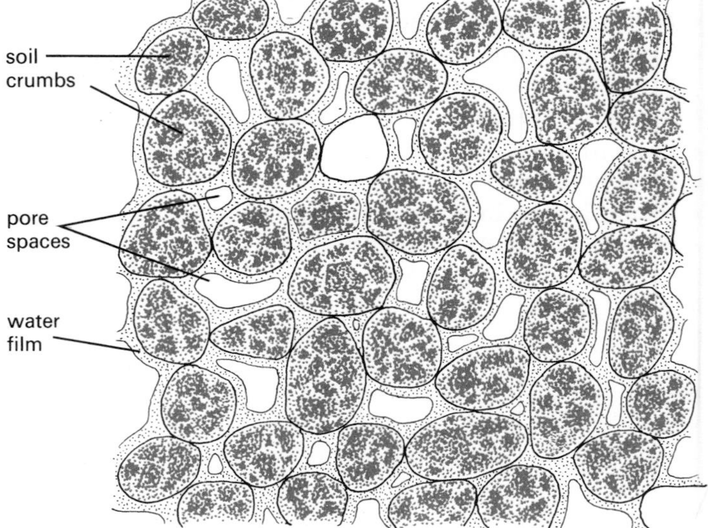

Fig. 2 Soil structure

SOIL COMPONENTS

Sand, silt and clay

Sand particles are formed from quartz (silicon oxide or silica). If the particles are from 0.02 to 2.0 mm across, they are described as 'sand particles'. Particles between 0.002 and 0.02 mm are called silt.

The silicon oxide of sand and silt is inert; that is, it cannot be decomposed or contribute to the supply of minerals needed for the nutrition of plants. The sand particles do, however, form the 'skeleton' of the soil. Their irregular shapes (Fig. 2) leave spaces between them, to be occupied by air and water, and their surfaces become coated with clay particles, bacteria, humus and salts.

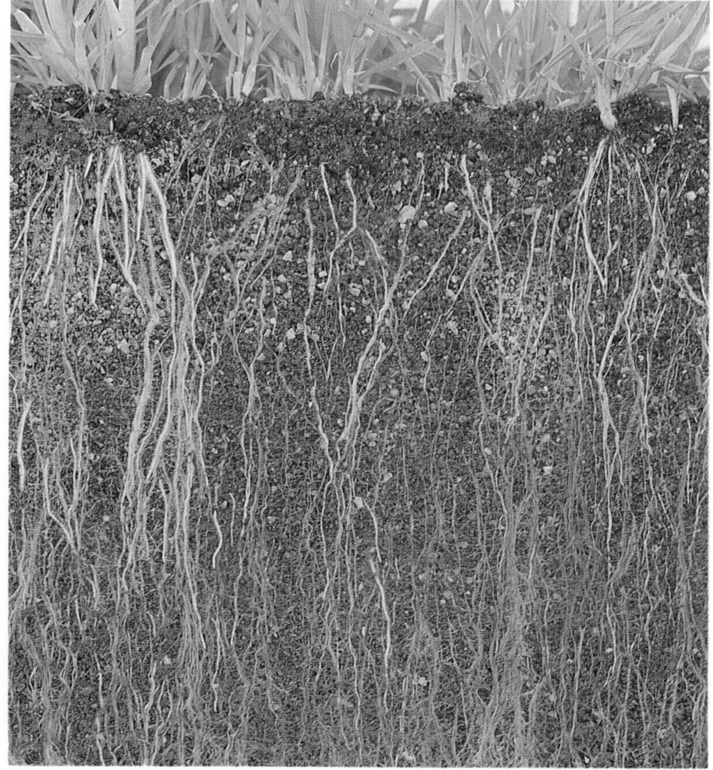

Fig. 3 Deep agricultural top soil. The soil is deeper than in Fig. 1 and the plant roots can be seen.

Clay particles are less than 0.002 mm across. Apart from their much smaller size, they also differ from sand particles in being made up of silicon oxide and aluminium oxide forming microscopic hexagonal plates. The clay particles also carry negative charges on their surface. These charges hold positively charged metallic ions such as potassium, calcium and magnesium. Clay is, therefore, a valuable source of these ions which are needed by plants (p. 52).

The countless billions of tiny, flat, clay particles present an enormous surface area. (1 g of clay could have a surface area of 10 m^2.) This helps to retain water, held by capillary attraction between the particles, but it also makes the particles stick together. Consequently a clay soil is sticky when wet and may form hard clods when dry.

Organic matter

The organic matter in a soil is derived from the dead and decaying remains of plants, animals and micro-organisms. Saprophytic bacteria and fungi in the soil digest wood, dead leaves and animal remains. This results in the production of **humus**. Humus is a structureless, black, jelly-like material which forms a coating round the sand particles and helps to 'glue' sand and clay particles together to form soil **crumbs**.

Vegetable compost, leaf-mould and animal manure mixed with straw are also sometimes described as 'humus' but they are really just coarse organic matter which provides the source of the humus.

The organic matter, with its spongy open texture, helps to hold water in the soil. It also makes the soil texture 'lighter' and easier to dig or plough.

As the organic matter breaks down, it releases some of the salts (as ions) which plants need for their growth. Organic matter, therefore, improves the fertility of the soil.

Crumb structure

In a mature, relatively undisturbed soil, the sand and clay particles become cemented together to form aggregates which eventually make soil **crumbs** up to 3 mm across. How the crumbs form is not fully understood but the humus and jelly-like hydroxides of iron and aluminium seem to form the 'glue'. The presence of organic matter is essential and the activities of bacteria, fungi and plant roots seem to play an important part.

The crumbs hold the clay and silt particles together and prevent them from being washed away and from blocking the drainage pores. Between the soil crumbs, air and water can penetrate, giving a well-drained and well-aerated soil. The crumbs themselves are porous and hold water.

Pore spaces

The irregular sand grains and crumbs do not pack tightly together. The spaces between them are called **pore spaces**. The pore spaces may occupy from 40 to 70 per cent of the soil volume. In a waterlogged soil, the pore spaces may be filled with water but, normally, air fills most of the space and water is confined to a thin film round the particles, being held there by capillary attraction. The air provides the oxygen for the respiration of soil organisms and plant roots. The water and its dissolved salts are taken up by plant roots.

Water

Rain-water falling on the soil, drains through under the influence of gravity, carrying soluble salts with it. The rate at which water passes through the soil will depend on the size and number of pore spaces, the crumb structure and the proportion of clay. In addition to being retained on the sand particles by capillary attraction, water may also be held to the clay particles by their surface charges.

Capillary attraction may help to distribute water from waterlogged regions to dryer regions. Apart from this, there is not much evidence of water movement by capillarity in the soil.

Mineral salts

Salts of potassium, iron, magnesium, phosphates, sulphates and nitrates are present in solution in the soil water. They come from rock particles and from the action of bacteria on the organic matter in the soil (see 'The nitrogen cycle', p. 247). These salts are taken up as ions by the roots of plants and used to build up the substances needed for their cells (see p. 52). Negative ions, such as sulphate (SO_4), nitrate (NO_3) and phosphate (PO_4) are likely to be in solution in the soil water and, therefore, easily washed out by rain. Positive metallic ions, such as calcium, magnesium and iron may be held close to the clay particles by the negative charges on these particles.

Micro-organisms

The bacteria and fungi in the soil play a vital part in converting organic matter to humus, making mineral salts available to plants and cementing soil particles together to form crumbs. They are given further consideration on pages 257–8.

QUESTIONS

1 What is the difference between a soil 'particle' and a soil 'crumb'?
2 Make a list of all the things you would expect to find if you carefully analysed a sample of soil.
3 What forces tend (*a*) to remove water from the soil, (*b*) to retain water in the soil?
4 What are the main differences (*a*) in structure, (*b*) in properties between sand and clay particles?

SOIL FERTILITY

In natural conditions, soil fertility is maintained by the activities of the organisms living on it or in it. For example, plant roots maintain the soil's crumb structure, the burrows of earthworms improve its drainage, and nitrogen-fixing bacteria keep up the supply of nitrates. Although plants remove mineral salts, these are replaced by the death and decomposition of plant and animal bodies.

The practice of agriculture interrupts natural cycles by removing the crops at harvest but not returning the dead remains of either the plants or the animals which eat them. If this practice is continued for more than a year or two the soil loses much of its fertility.

In agricultural terms, a naturally fertile soil is one which produces a large yield of crop plants with a minimum expenditure of energy and cash. This means that the soil, initially, has ideal proportions of sand and clay particles (i.e. 15–20 per cent clay), adequate humus and mineral salts, a stable crumb structure, is well drained and aerated and is not too acid or alkaline.

Proportion of clay and sand

There is not much a farmer can do to alter the proportions of clay and sand in the soil but some of their effects can be changed.

Adding organic manure to a heavy clay soil will improve the crumb structure and increase the pore spaces, making the soil better aerated and also easier to plough or dig. Organic manure on a light, sandy soil will help it to retain water and reduce the washing out (leaching) of soluble salts by the rain.

Humus and mineral salts

Addition of farmyard manure or other organic matter provides both humus and mineral salts. The humus helps to maintain crumb structure and the salts provide the ions needed by the crop plants.

For large arable farms, however, there is insufficient animal manure, and artificial fertilizers have to be used (p. 53). Commercial fertilizers can greatly increase yields but they do nothing to maintain a stable crumb structure. After many years of intensive arable farming using artificial fertilizer, some soils become dry and powdery and are liable to be blown away in strong winds (Fig. 16, p. 271).

Crop rotation (p. 249) helps to maintain both the crumb structure and the supply of nitrate.

Crumb structure, aeration and drainage

A good crumb structure helps to keep the soil well drained and aerated. Grass roots seem to have a positive effect on crumb structure, so a rotation which includes grass will help to maintain fertility. Repeated application of artificial fertilizers and compaction of soil with heavy machinery tends to destroy the crumb structure and so reduce the pore spaces and the drainage.

Earthworm burrows assist drainage. If pesticides are used which affect the earthworm population, drainage may be impaired.

Ploughing increases aeration but does not help to conserve soil structure. Direct drilling is a technique in which weeds are killed off with herbicides and the crop seeds planted in unploughed soil. This helps to maintain the structure of some types of soil.

If the subsoil is prone to waterlogging, pipes made from clay or plastic can be laid in the soil to carry off the excess water.

Acid and alkaline soils

It is not always clear what makes a soil acid or alkaline. Soil in a chalky or limestone district will probably be slightly alkaline because of the particles of calcium carbonate in the soil. A light sandy soil may be slightly acid because the rain washes away soluble compounds that would otherwise neutralize the acids. Most British soils are slightly acid, with pH values of 6.4–6.9. A chalky soil could have a pH as high as 8 and the soil of heaths and moorland can be as low as 4.5.

Acid soils are usually infertile because the acidity makes the mineral salts very soluble and they are easily washed away by rain. Alkaline soils are usually quite fertile, with plenty of mineral salts, but in some cases, the alkalinity may be so high that the salts are insoluble and cannot be taken up by the plants. Soil acidity is usually reduced by adding calcium carbonate or calcium hydroxide, both described as 'lime'.

QUESTIONS

5 Make a list of all the effects that addition of organic manure can have on the fertility of a soil.
6 List the ways in which a soil's crumb structure might be (*a*) destroyed, (*b*) maintained.
7 Study the information on pages 53 and 249. Then name two artificial fertilizers and say (*a*) what mineral elements they supply and (*b*) why plants need these elements.

LIFE IN THE SOIL

Except to the keen gardener, soil may look dull and lifeless, but in fact it contains millions of organisms moving, growing, reproducing and competing with each other for food (Fig. 5) At the bottom of the food pyramid (p. 243) are the micro-organisms. Other abundant soil organisms are nematodes, arthropods and earthworms.

Soil micro-organisms

Soil micro-organisms (Fig. 5) are the bacteria (p. 310), blue-green algae (p. 300), actinomycetes, fungi (p. 321) and protozoa (p. 327). The bacteria, actinomycetes and protozoa inhabit the water film adhering to the soil particles. The blue-green algae, being photosynthetic, will be in the top few millimetres of the soil where they can receive light. The fungal hyphae will spread throughout the soil penetrating between the particles.

The blue-green algae and some bacteria have pigments similar to chlorophyll and can use sunlight to build their food substances by photosynthesis. However, most bacteria, actinomycetes and fungi are saprophytes (p. 322). They secrete digestive enzymes into the organic matter of

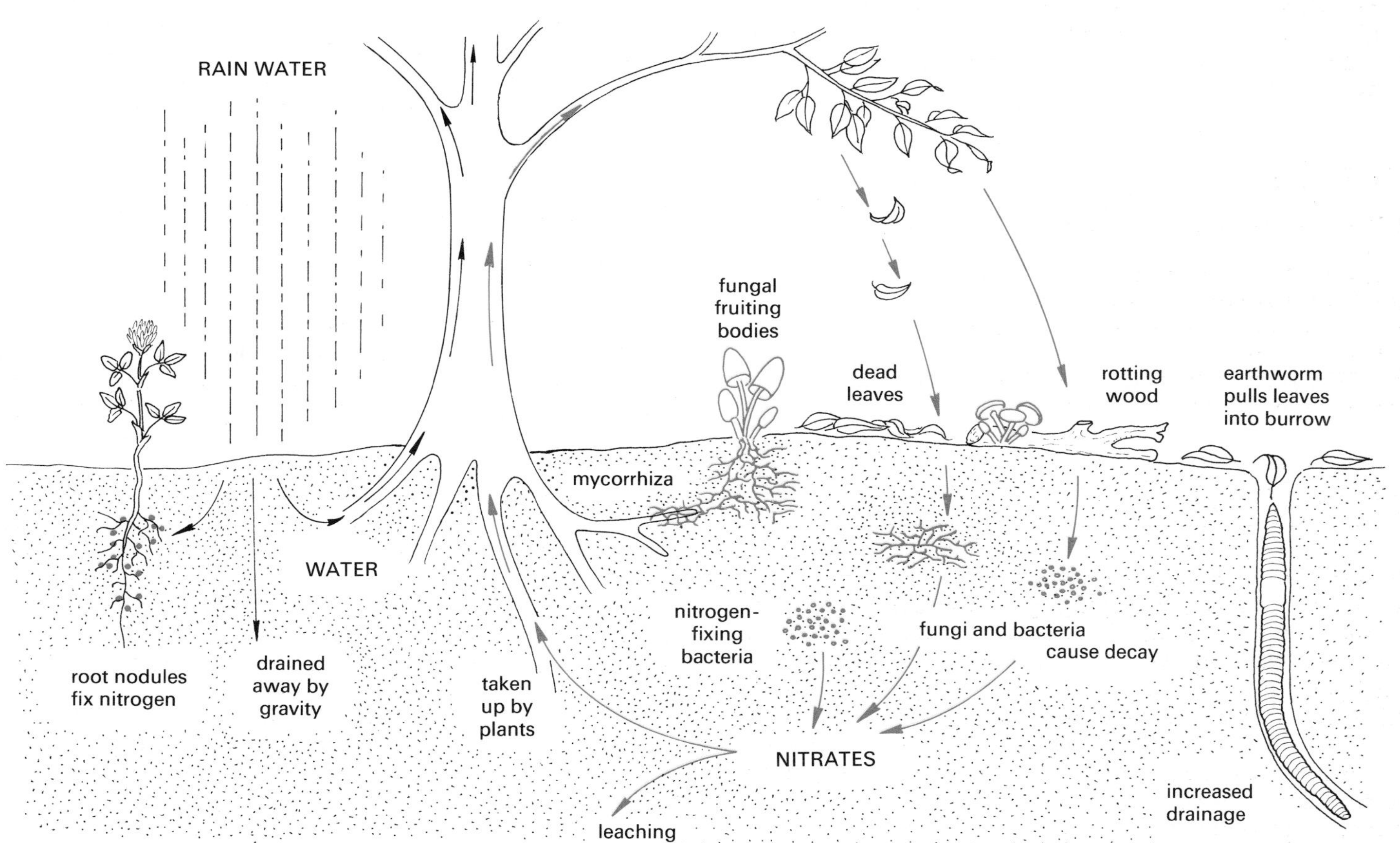

Fig. 4 Some of the interactions in the soil

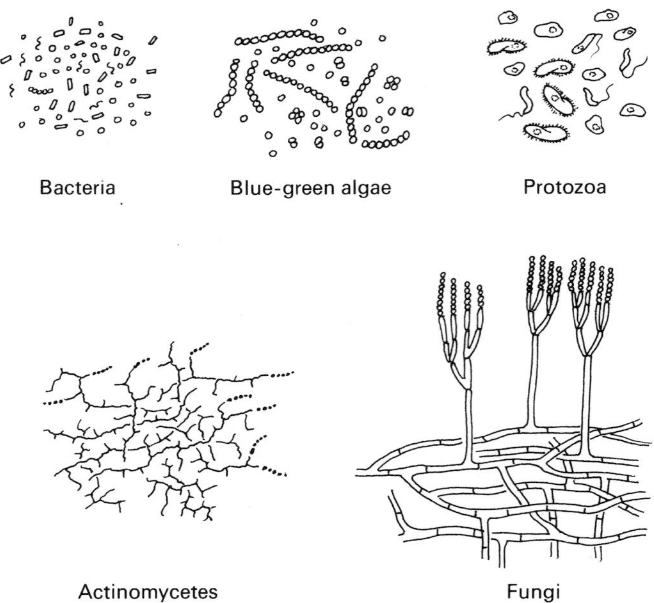

Fig. 5 Some soil micro-organisms

the soil and absorb the soluble products back into their cells or hyphae.

There is considerable competition between these organisms for the limited organic matter. One aspect of this competition is the production by some actinomycetes and fungi of chemicals which inhibit the growth of competing bacteria. It is these chemicals which can be extracted and purified to make the antibiotic drugs used to treat bacterial diseases in humans. The antibiotic **streptomycin** comes from the actinomycete *Streptomyces*; **penicillin** is produced by the fungus *Penicillium*.

The protozoa are single-celled organisms. These include *Amoeba*, ciliates such as *Paramecium* and flagellates like *Euglena* (p. 327). Many protozoa feed by ingesting bacteria though some of the flagellates have chloroplasts and can photosynthesize.

Soil micro-organisms and plant roots

The rhizosphere It has been found that soil micro-organisms, particularly bacteria and fungi, are far more numerous in the vicinity of plant roots than in the rest of the soil. This area round the root is called the **rhizosphere**. There is evidence, at least from laboratory experiments, that substances such as sugars, amino acids and vitamins escape from plant roots. These substances could promote the growth of the micro-organism population. In addition, there is the dead material left behind by growing roots, e.g. the defunct root hairs and the cells rubbed off from the root cap (p. 65). This dead material may form a source of food for the bacteria and fungi. Numerous micro-organisms grow on the root surface as well as in the rhizosphere.

In some cases, roots may exude chemicals which inhibit the growth of specific bacteria or fungi.

Mycorrhiza Most plant roots are very closely associated with fungal hyphae (p. 321). This association may be parasitic or symbiotic (p. 250). The symbiotic fungi form a **mycorrhiza** (Fig. 5). This is a dense network of fungal hyphae round the lateral roots, usually in certain trees. Some of the hyphae penetrate into the root cortex, others spread out into the soil. It seems that the fungal hyphae digest organic matter in the soil and that some of the digestion products, e.g. nitrates, pass along the hyphae and into the plant. The mycorrhiza may completely replace the root hair system.

The fungus is thought to benefit by receiving a supply of carbon-containing substances from the tree, e.g. carbohydrates or simpler compounds. Experiments with radioactive carbon (p. 31) have shown that such substances do pass from the tree to the fungus.

Some of the toadstools you see growing in woods are the fruiting bodies (p. 322) of the mycorrhizal fungi associated with the tree roots.

Nematodes

These worm-like creatures occur in vast numbers in the soil. They are from 0.5 to 1.5 mm long and live in the pore spaces, feeding on bacteria, actinomycetes, algae, protozoa and other nematodes. Some nematodes are ectoparasites (p. 250), piercing the roots of plants and absorbing the cell contents. Others are endoparasites of plant roots and cause serious damage to crops, e.g. the potato root 'eelworm'.

Arthropods

These include insect larvae, springtails, centipedes, millipedes and mites (Fig. 6). It is not possible to generalize about the life-styles and feeding habits of so large and diverse a group but, between them, they feed on living and dead plant material, faeces, fungi, nematodes, small worms and each other. Some of the insect larvae, such as the 'wireworms', 'leather-jackets' and 'cutworms' can become serious agricultural pests by eating the roots of crop plants.

Earthworms

About ten different species of earthworm commonly live in agricultural and garden soils. They make burrows in the soil (Fig. 7) partly by pushing between the particles and partly by swallowing the soil. The organic matter in the soil is digested and the remaining soil is passed out. Some species pass out this soil at the surface forming 'worm-casts'. One common species pulls leaves into its burrow and ingests them a little at a time.

It is generally believed that earthworms improve soil fertility, though it would be difficult to demonstrate that they actually improved crop yields.

As soil passes through the earthworm's intestine, the material is finely ground and made more alkaline.

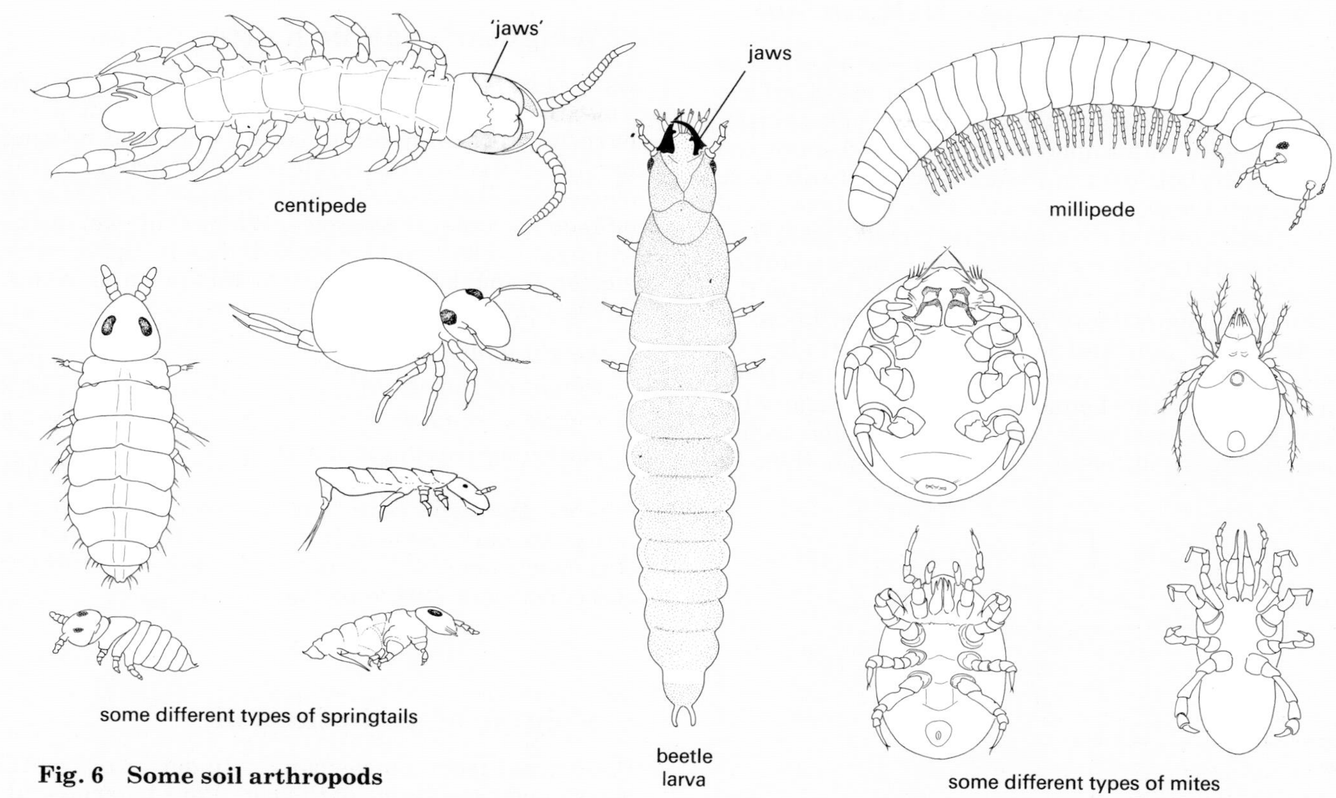

Fig. 6 Some soil arthropods

Secretions from the intestine help to bind the organic and mineral particles together. The raised pH favours the growth of soil bacteria; the grinding of coarse humus and the mixing and binding with mineral particles should contribute to a good crumb structure.

The species which cast at the surface have been estimated to add a layer of about 5 mm fine topsoil to the soil surface each year. Most species help to mix the soil layers, taking organic matter to the deeper layers and bringing mineral particles, with associated ions, to the surface layers. Earthworm burrows also improve the drainage and aeration of the soil.

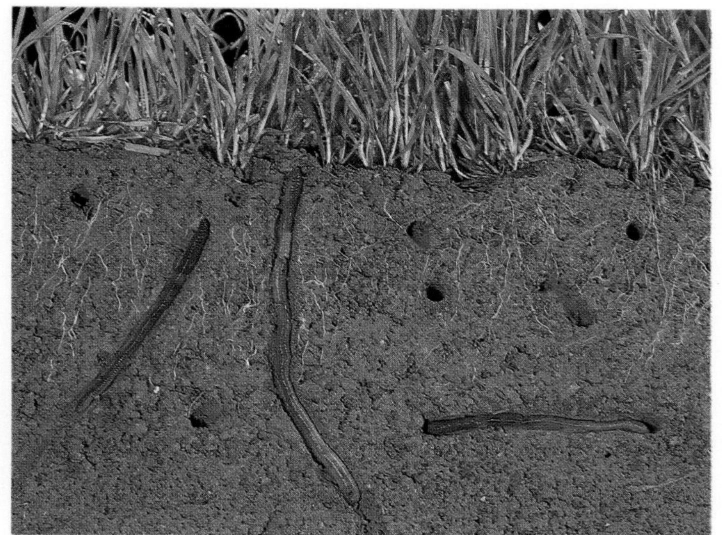

Fig. 7 Earthworms in their burrows

QUESTIONS

8 Make three lists of the living organisms that you might expect to find in a sample of topsoil, under the headings, 'Useful', 'Harmful' and 'Harmless'.
9 What is the main source of food for soil organisms?
10 Make a list of the activities of (*a*) soil bacteria, (*b*) soil fungi which might have an effect on the growth of plants.
11 List the ways in which the activities of earthworms might help to increase the fertility of the soil.
12 Try to draw up a food chain or food web (see p. 242) using only soil organisms.
13 What might be the disadvantage of adding an insecticide to the soil to combat insect pests such as wireworms?
14 Suggest some reasons why it is likely to be difficult to study the life-style and feeding habits of organisms in the soil.

PRACTICAL WORK

It is important to remember that as soon as you take up a trowel full of soil and put it in a container, you have disorganized its natural structure. Earthworm burrows will collapse, soil crumbs will fall apart, plant root systems will be disrupted, extra air may enter or some may be expelled. This means that the results of experiments designed to measure soil properties may have little bearing on what actually happens in an intact, undisturbed soil.

Nevertheless, it is possible to derive some information from soil samples, particularly if comparisons are made between contrasting soil types.

1. Observation of mineral particles in soil

Sieve some dry soil to remove stones and particles larger than about 3 mm. Crush the soil lightly to break up the crumbs and place 50 grams of it in a small, flat-sided bottle. Fill the bottle with water almost to the top and screw the cap on. Shake the bottle for at least 30 seconds to disperse the soil through the water. Then allow the bottle to stand for 10–15 minutes so that the soil settles down (Fig. 8). The large particles will fall most rapidly and form the bottom layer. Smaller particles will remain suspended in the water. The larger particles of organic matter will float to the top. As the soil particles settle they may form layers. The layers may not be very distinct, but use your judgement to mark the boundaries between them with a marker pen and measure the depth of each layer. The composition of different soil samples can then be compared.

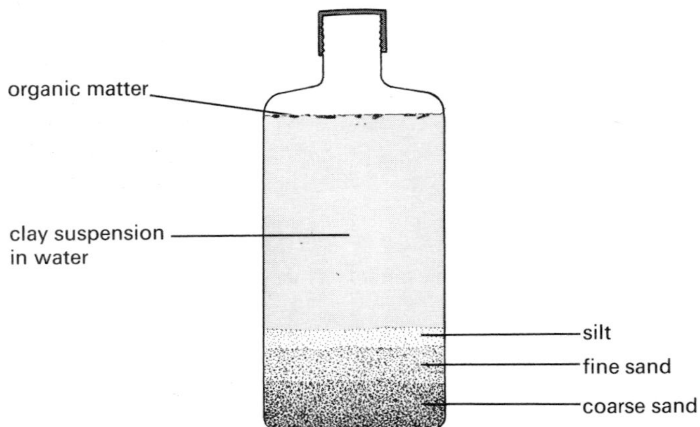

Fig. 8 Mineral particles in the soil

2. Weight of water in the soil

Place a sample of soil in a weighed evaporating basin and then weigh it again. Heat the basin in an oven at 100 °C for 2 days to drive off the soil water. Weigh the basin and soil again. Strictly speaking you should heat and reweigh the soil and basin until two weighings give the same result, showing that all the soil water has been evaporated. However, this is time-consuming and tedious and may be omitted, at the cost of some accuracy.

The difference between the second and final weighings will give the weight of water that was present at first. Temperatures higher than 100 °C must not be used because they will cause the organic matter to burn away and so give an extra loss of weight. Here is a sample calculation:

weight of basin	200 g
weight of basin + moist soil	250 g
∴ weight of moist soil	250 − 200 = 50 g
final weight of basin + dry soil	240 g
∴ loss in weight	250 − 240 = 10 g
percentage water in moist soil	$\frac{10 \times 100}{50} = 20\%$

3. Weight of organic matter in soil

Use dry soil from the previous experiment. Place it in a metal tray and heat strongly over a Bunsen flame to burn off all the organic matter. Continue heating until smoke stops coming off and the charred organic matter has disappeared. This will leave only the grey or reddish mineral particles. When cool, weigh the soil again. The loss in weight is due to the organic matter being burnt to carbon dioxide and water. Sample calculation:

weight of dry soil	40 g
weight of 'burnt' soil	38 g
∴ weight of humus	40 − 38 = 2 g
percentage humus in dry soil	$\frac{2 \times 100}{40} = 5\%$

Since the residue of 'burnt' soil consists of the inorganic particles (sand and clay), it follows that, in this experiment, 95 per cent of the dry weight of the soil consists of sand and clay.

4. Volume of air in the soil

Pour water from a measuring cylinder into a metal can to find the volume of the can. Pour two cans full of water into a large, straight-sided jar (e.g. a 200 g or 8 oz coffee jar), and mark the water level with a marker pen on the outside of the jar.

Pour enough water from the jar back in the can to just fill it and mark the new water level in the jar. The two marks on the jar now represent the volume of one and two cans. The jar should be left with water up to the 1-can mark (Fig. 9).

Empty the can and press it, open end down, into some undisturbed soil which is soft enough to admit it. Stamp the can down until its base is just level with the soil surface. Then dig the can out carefully. Cut away any surplus soil from the mouth of the can.

Now use a stick or a spoon to dig the soil out and let it fall into the jar of water. Since the volume of soil in the can is the same as the volume of water you removed from the jar, you might expect the water level to go up to the 2-can mark. However, there was air in the soil and this will have escaped when you dislodged the soil into the jar. The new water level will, therefore, be below the 2-can mark.

Fill the measuring cylinder to the top with water and pour some of the water into the jar until the level reaches the 2-can mark. Note the reading in the measuring cylinder so that you can calculate the volume of water you have added. This is the same as the volume of air which escaped from the soil. Sample calculation:

volume of the metal can	200 cm^3
volume of water added to jar	28 cm^3
∴ volume of air in 200 cm^3 soil	28 cm^3
percentage air in soil	$\frac{28 \times 100}{200} = 14\%$

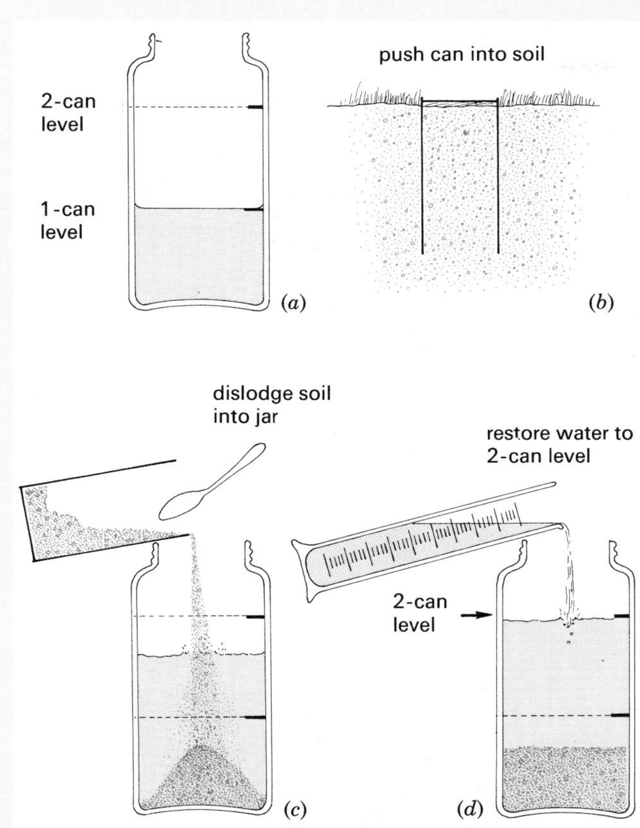

Fig. 9 Measuring the volume of air in soil

5. A comparison of the water-holding properties of soils

Two equal-sized plastic cups have their bases perforated and lined with glass fibre. Partly pack one cup with dry sand. Pack the other to the same depth with dry soil, rich in organic matter. Support the cups, in turn, over a beaker and pour a known volume of water through the soil from a measuring cylinder (Fig. 10). When all the water has drained through, return it to the measuring cylinder to see how much has been retained by the soil. Repeat the experiment with the other sample, using the same volume of water as before.

It is expected that the soil rich in organic matter will hold a greater proportion of water than the sand, thus showing one of the advantages of organic matter in the soil.

6. The permeability (porosity) of soil

The permeability of a soil means how easily water will pass through it. Plug the stems of two glass funnels with glass wool and half-fill one with sandy soil and the other with an equal volume of clay soil. Cover both lots of soil with water and keep the levels the same by topping up during the experiment. Thus, there will be no difference in the water pressure in the two funnels (Fig. 11). Collect the water that runs through in a given time in a measuring cylinder. The sandy soil will probably be more permeable to water than the clay soil and will allow far more water to run through in a fixed time.

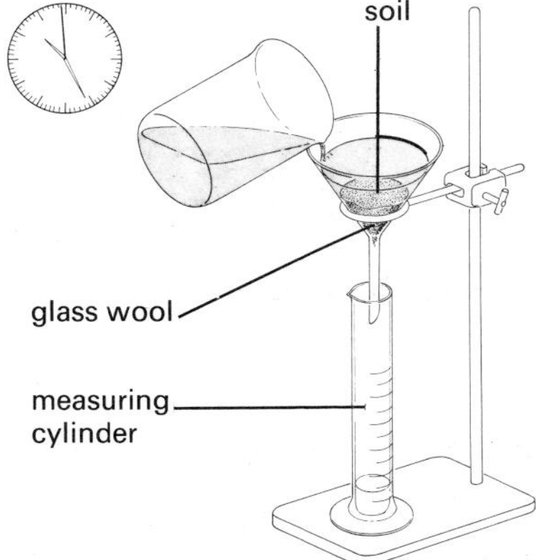

Fig. 11 Comparing the permeability of soils to water. (Only one of the two funnels is shown.)

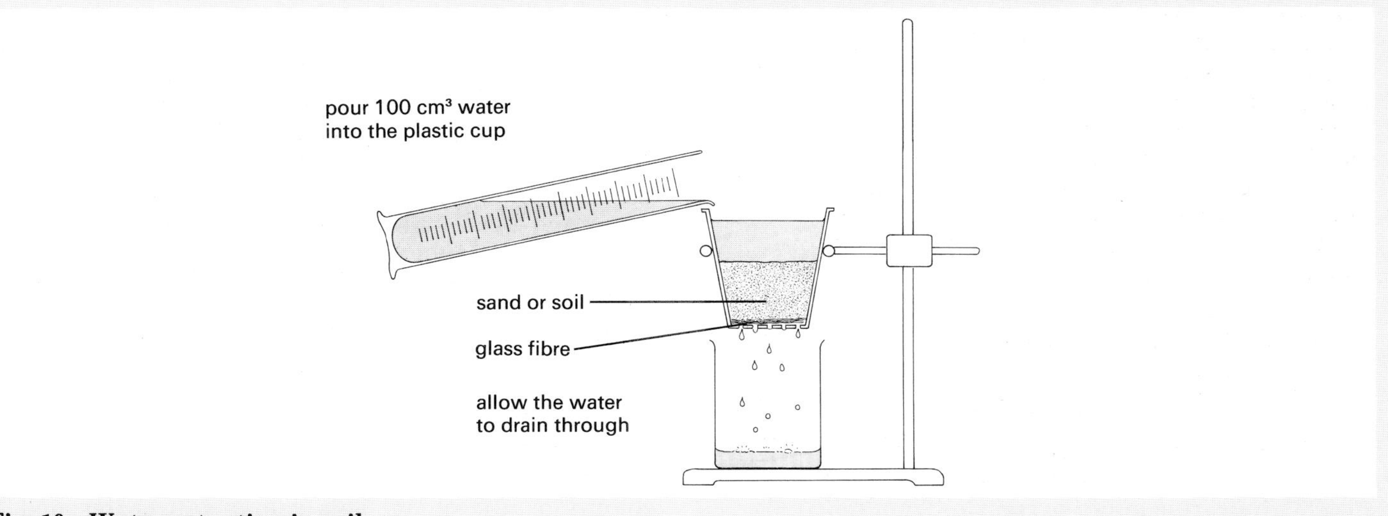

Fig. 10 Water retention in soil

7. Capillary attraction in soil

Pack two glass tubes 1 cm or more wide and about 50 cm long with dry sand. Fill one tube with fine sand and the other with coarse sand. Plug the ends with glass wool and clamp the tubes upright in a beaker of water (Fig. 12). The water will travel up through the sand by capillary attraction, and its level can be seen by the darker colour of the sand. Measure and compare the levels after one day. The results will show that water travels further in the fine sand. The smaller the particles in a soil, the greater is the capillary attraction.

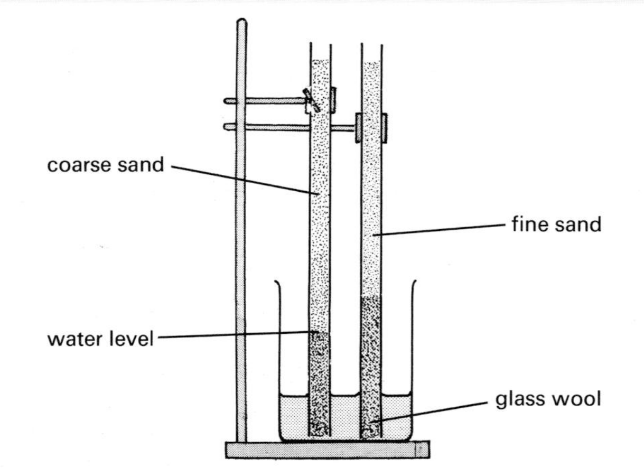

Fig. 12 Capillary attraction in sand

8. The presence of micro-organisms in the soil

Two Petri dishes of sterile, vegetable agar are prepared as described on page 263. In one of them sprinkle some particles of soil (Fig. 13) and in the other scatter some particles of sand that have been sterilized by heating. This dish is the control (p. 30). If bacteria or fungi are present in the soil, they will appear as colonies on the surface of the agar in a few days. The absence of any colonies from the control will prove that the micro-organisms were in the soil and did not come from the air, from the dish, from the agar or from the instruments used to scatter the particles.

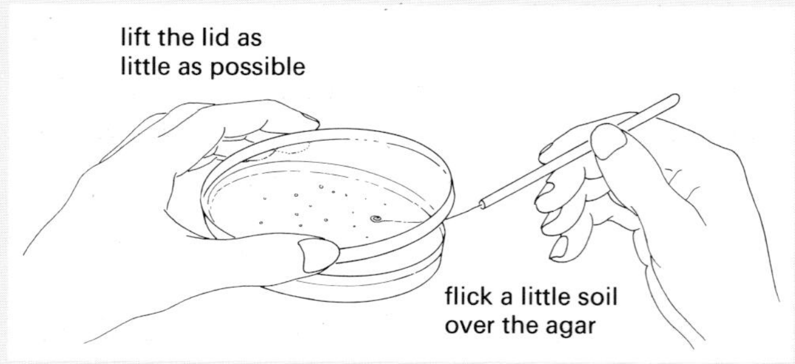

Fig. 13 Culturing micro-organisms from soil

9. Extraction of some of the larger soil animals

An apparatus such as the one in Fig. 14 is set up with a soil sample. The heat from the lamp gradually dries out the soil from the top and so drives some of the insects, nematodes and mites down into the Petri dish where they are trapped in a preserving fluid. After a few days, the 'catch' is examined under the low power of the microscope (Fig. 6).

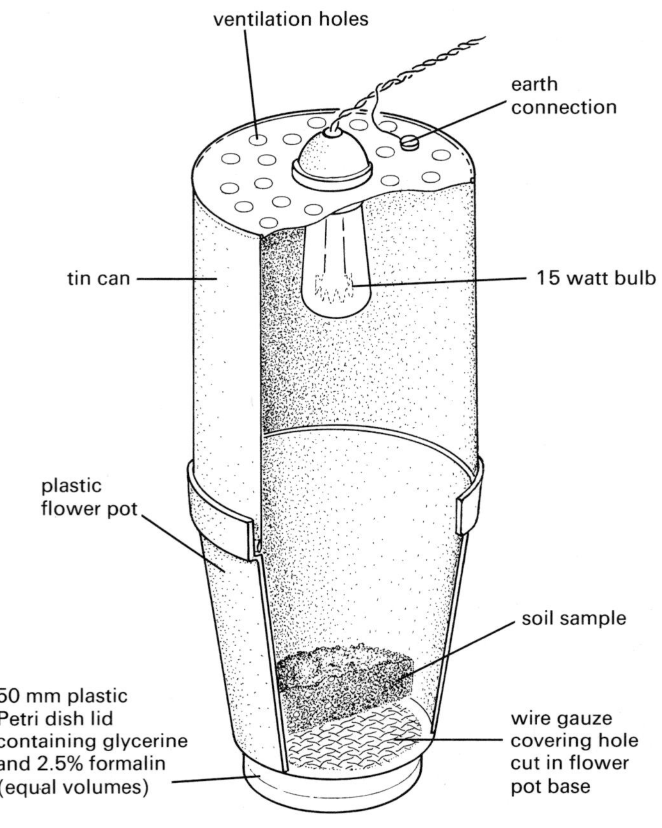

Fig. 14 Extraction of small animals from the soil

10. Measuring the pH of the soil

Place about 10 mm soil in the bottom of a test-tube. Add about 10 mm barium sulphate powder. This is a neutral salt which will precipitate the clay in the soil so that, later on, the colour of the solution can be seen. Add 10 cm^3 distilled water and 2 cm^3 soil indicator to the test-tube. Close the mouth of the tube with a bung and shake the tube vigorously to mix the contents.

Let the tube stand for a minute and a clear coloured solution should appear at the top as the soil particles settle down. The colour is produced by the soil's acidity or alkalinity acting on the indicator dye.

Hold the tube against the colour chart. (This may be on the label of the indicator bottle or on a separate card.) Try to match the colour of the liquid in the tube to one of the colours on the chart. This colour will correspond to a particular pH or pH range (Fig. 15).

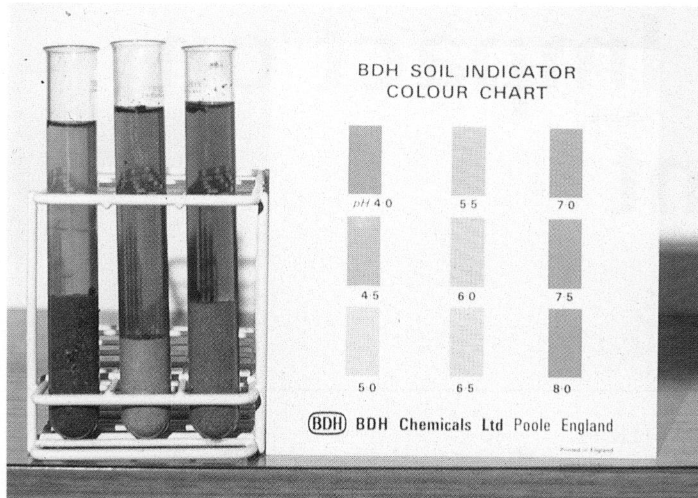

Fig. 15 Testing the pH of a soil. Compare the colours in the test-tubes with the colours on the chart.

11. Culturing soil protozoa

Cut up some hay with scissors and boil it in a beaker with some rain-water for about 5 minutes. Filter the liquid when it is cool enough and leave it in an open jar. Bacterial spores from the hay will germinate and create a population of bacteria. Add some soil to this liquid and leave it for a few days. The protozoa in the soil will feed on the bacteria and their numbers will increase. If you take small samples of water from the surface of the liquid, and study it under the microscope, you will probably find a large number of ciliates and some rotifers (which are not protozoa).

Note Fuller details for these and other experiments can be found in *Experimental Work in Biology*, Combined Edition (see p. 364).

QUESTIONS

15 Because you omitted the heating and reweighing in Experiment 2, would you expect your figure for the percentage of soil water to be too large or too small? Explain.
16 Why is it necessary to start with dry soil in order to measure the weight of organic matter in the soil (Experiment 3)?
17 Suggest some causes of inaccuracy in the results of Experiment 4.
18 Suggest some reasons why you would expect the permeability of an intact soil to differ from the permeability of the soil as measured in Experiment 6.
19 Would you expect a soil which is very permeable to water (Experiment 6) to also retain a lot of water (Experiment 5)? Explain your answer.

Preparation of culture plates for soil micro-organisms

This should be done by the teacher or technician.

The Petri dishes are sterilized by super-heated steam in an autoclave or pressure cooker for 15 minutes at a pressure of 1 kg per cm^2. This kills any bacteria that are already on the glassware. The culture medium is made by adding 2 g agar (p. 319) and 25 cm^3 vegetable juice (e.g. tomato juice) to 100 cm^3 hot distilled water. This mixture is stirred until all the substances are dissolved, and then sterilized in an autoclave. When it is fairly cool but still liquid, the agar is poured into the Petri dishes which are covered at once. The agar sets to a jelly when cool. After the soil particles have been added to the dishes, the lids should be sealed on with adhesive tape and not opened again. When the experiment is finished, the dishes must be sterilized before opening them up and disposing of the cultures.

CHECK LIST

- **Soil is a mixture of sand, silt and clay particles with humus.**
- **Soil has a 'structure'; the particles form 'crumbs' and there are pore spaces between the particles and crumbs.**
- **The larger pore spaces contain air, the smaller pore spaces contain water.**
- **Mineral salts are present, in solution, in soil water.**
- **Bacteria, fungi, protozoa and other micro-organisms live in the soil and affect its structure and fertility.**
- **There is also a population of arthropods and nematodes in the soil. Some of them are agricultural pests.**
- **A fertile soil contains adequate humus and mineral salts. It is well drained and aerated, has a stable crumb structure and is not too acid or alkaline.**
- **Organic matter in the soil improves its structure, helps to retain water and is a source of mineral ions.**
- **Plant roots penetrate the soil and extract water and mineral salts.**
- **Soil water is held round the particles by capillary attraction, a force which also resists the uptake of water by roots.**
- **There is often an intimate relationship between plant roots and soil fungi which benefits the plant and the fungus.**

28 The Human Impact on the Environment

FOOD WEBS

Hunting, agriculture, pesticides, eutrophication.

FORESTS

Erosion, flooding, greenhouse effect.

SOIL

Erosion, pesticides.

WATER

Sewage, pollution, radioactive waste.

AIR

Sulphur dioxide, oxides of nitrogen, lead, smog, carbon monoxide, chlorofluorocarbons.

A few thousand years ago, most of the humans on the Earth probably obtained their food by gathering leaves, fruits or roots and by hunting animals. The population was probably limited by the amount of food that could be collected in this way.

Human faeces, urine and dead bodies were left on or in the soil and so played a part in the nitrogen cycle (p. 247). Life may have been short, and many babies may have died from starvation or illness, but humans fitted into the food web and nitrogen cycle like any other animal.

Once agriculture had been developed, it was possible to support much larger populations and the balance between humans and their environment was upset.

An increasing population has three main effects on the environment.

1. Intensification of agriculture Forests and woodland are cut down and the soil is ploughed up in order to grow more food. This destroys important wildlife habitats and may even affect the carbon dioxide levels in the atmosphere.

Tropical rainforest is being cut down at the rate of 43 000 square miles per year. Since 1950, between 30 and 50 per cent of British deciduous woodlands have been felled to make way for farmland or conifer plantations.

The application of chemical fertilizers can cause deterioration of the soil structure and, in some cases, results in pollution of rivers and streams. Application of pesticides often kills beneficial creatures as well as pests.

2. Urbanization The development of towns and cities makes less and less land available for wildlife. In addition, the crowding of growing populations into towns leads to problems of waste-disposal. The sewage and domestic waste from a town of several thousand people can cause disease and pollution in the absence of effective means of disposal.

When fuels are burned for heating and transport, they produce gases which pollute the atmosphere.

3. Industrialization In some cases, an increasing population is accompanied by an increase in manufacturing industries which produce gases and other waste products which damage the environment.

The effects of the human population on the environment are complicated and difficult to study. They are even more difficult to forecast. In their ignorance, humans have destroyed many plants and animals and great areas of natural vegetation. Unless we control our consumption of the Earth's resources, limit our own numbers and treat our environment with more care and understanding, we could make the Earth's surface impossible to live on and so cause our own extinction.

The account which follows, mentions just some of the ways in which our activities damage the environment.

THE HUMAN IMPACT ON FOOD WEBS

The hunting of animals

One obvious way to upset a food web is to remove some of the animals or plants which form part of it. If the tawny owls were removed from the food web in Figure 1, we would expect the numbers of shrews to increase because fewer were being eaten by the owls. The numbers of woodlice and earthworms might then go down because there were more shrews to eat them. The effect of the rabbit disease, myxomatosis, on a food web, has been described on page 244.

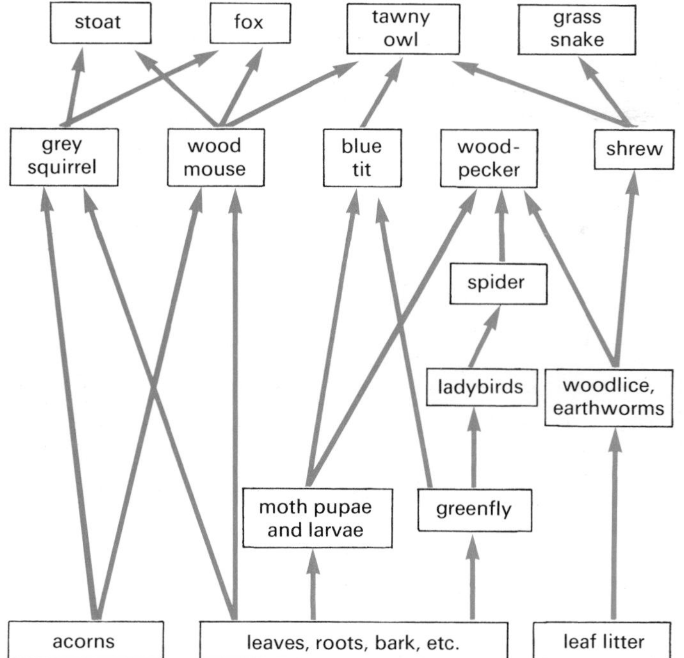

Fig. 1 The food web of an oak tree (only a small sample of animals is shown).
(From P. W. Freeland, *School Science Review*, 1973.)

In 1910, in the Grand Canyon National Game Reserve (USA), an attempt was made to protect the deer population by shooting the animals which ate them. These were cougars, wolves, bobcats and coyotes. After fourteen years, the deer population had increased from about 4000 to 100 000, and the environment could not support them. The grass was overgrazed, the trees and young shrubs were destroyed by browsing and the deer were dying from starvation in large numbers. Ignorant human interference with the food web had not only destroyed hundreds of cougars, wolves and coyotes but threatened to lead to the destruction of the environment and the deer which lived there (Fig. 2).

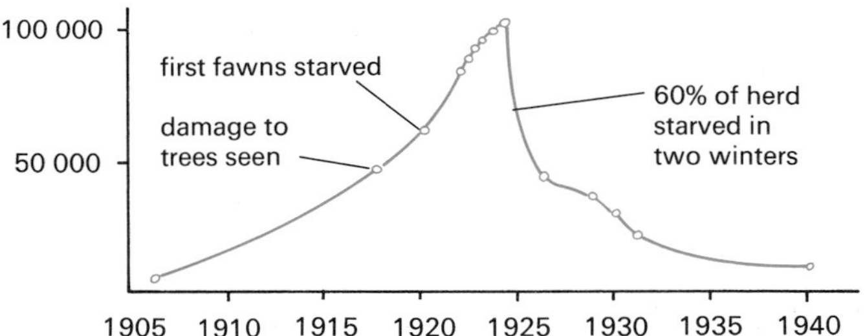

Fig. 2 The result of human interference with a food web: changes in deer population after predators were killed.
(From J. Barker, *In the Balance*, Evans, 1975.)

It is, of course, easy to be 'wise after the event', but it is not always obvious what should be done to conserve a population or its habitat. Arguments are currently taking place about the elephant population in some of Africa's game parks. Some people think the elephants are too numerous and could destroy their own habitat. Others disagree and feel that the population is self-regulating.

Many animal populations are threatened because humans kill them for food, profit or 'sport'. Over-fishing has reduced some fish stocks to the point where they cannot reproduce fast enough to keep up their numbers.

Animals like the leopard and tiger have been reduced to dangerously low levels by hunting, in order to sell their skins (Fig. 3). The blue whale's numbers have been reduced from about 2 000 000 to 6000 as a result of intensive hunting.

Fig. 3 One of the threats to wild animals. So long as people are willing to buy these products, other people will be prepared to kill the endangered species.

Agriculture

Monoculture The whole point of crop farming is to remove a mixed population of trees, shrubs, wild flowers and grasses and replace it with a dense population of only one species such as wheat or beans (Figs 4 and 5). This is called a **monoculture**.

Figure 1 is a simplified diagram of a food web which can be supported by a single oak tree. Similar food webs could be constructed for grasses, wild flowers and shrubs. Clearly, a field of wheat could not support such a mixed population of creatures. Indeed, every attempt is made to destroy any organisms such as rabbits, insects or pigeons, which try to feed on the crop plant.

So, the balanced life of a natural plant and animal

Fig. 4 Natural vegetation. Uncultivated land carries a wide variety of plant species.

Fig. 5 A monoculture. Only wheat is allowed to grow. All competing plants are destroyed.

community is displaced from farmland and left to survive only in small areas of woodland, heath or hedgerow. We have to decide on a balance between the amount of land to be used for agriculture, roads or building and the amount of land left alone in order to keep a rich variety of wildlife on the Earth's surface.

Pesticides This is a general name for any chemicals which destroy agricultural pests. For a monoculture to be maintained, plants which compete with the crop plant for root space, soil minerals and sunlight are killed by chemicals called **herbicides** (Figs 6 and 7). The crop

Fig. 6 Weed control by herbicide spraying. A young wheat crop is sprayed with herbicide to suppress weeds.

Fig. 7 Effect of a herbicide spray. The area in the foreground has been treated with a herbicide which has prevented the growth of the yellow-flowered charlock.

Fig. 8 Control of fungus disease. The tree bearing the apples on the right has been sprayed with a fungicide. The apples on the unsprayed tree have developed apple scab.

plants are protected against fungus diseases by spraying them with chemicals called **fungicides** (Fig. 8). To destroy insects which eat and damage the plants, the crops are sprayed with **insecticides**.

Pesticide	*kills*
insecticide	insects
fungicide	parasitic fungi
herbicide	'weed' plants

The trouble with nearly all these pesticides is that they kill the harmless or beneficial organisms as well as the harmful ones.

In about 1960, a group of chemicals, including one called **dieldrin**, were used as insecticides to kill wireworms and other insect pests in the soil. Dieldrin was also used as a seed dressing. If seeds were dipped in the chemical before planting, it prevented certain insects from attacking the seedlings. This was thought to be better than spraying the soil with dieldrin which would have killed all the insects in the soil. Unfortunately pigeons, rooks, pheasants and partridges dug up and ate so much of the seed that the dieldrin poisoned them. Thousands of these birds were poisoned and, because they were part of a food web, birds of prey and foxes, which fed on them, were also killed.

The concentration of insecticide often increases as it passes along a food chain (Fig. 9). Clear Lake in California was sprayed with DDT to kill gnat larvae. The insecticide made only a weak solution of 0.015 parts per million (ppm) in the lake water. The microscopic plants and animals which fed in the lake water built up concentrations of about 5 ppm in their bodies. The small fish which fed on the microscopic animals had 10 ppm. The small fish were eaten by larger fish, which in turn were eaten by birds called grebes. The grebes were found to have 1600 ppm of DDT in their body fat and this high concentration killed large numbers of them.

These new insecticides had been thoroughly tested in the laboratory to show that they were harmless to humans and other animals when used in low concentrations. It had not been foreseen that the insecticides would become more and more concentrated as they passed along the food chain.

Insecticides like this are called persistent because they last a long time without breaking down. This makes them good insecticides but they also persist for a long time in the soil, in rivers, lakes and the bodies of animals, including humans. This is a serious disadvantage.

Eutrophication

On page 52 it was explained that plants need a supply of nitrates for making their proteins, and a source of phosphates for many chemical reactions in their cells. The rate at which plants grow is often limited by how much nitrate and phosphate they can obtain. In recent years, the amount of nitrate and phosphate in our rivers and lakes has been greatly increased. This leads to an accelerated process of **eutrophication**.

Eutrophication is the enrichment of natural waters with nutrients which allow the water to support an increasing amount of plant life. This process takes place naturally in many lakes but usually very slowly. Human activities have greatly speeded up the eutrophication of rivers and lakes and the following processes are the main causes.

Discharge of treated sewage In a sewage treatment plant, human waste is broken down by bacteria (p. 277) and made harmless, but the breakdown products include phosphates and nitrates. When the water from the sewage treatment is discharged into rivers it contains large quantities of phosphate and nitrate which allows the microscopic plant-life to grow very rapidly (Fig. 10).

Fig. 10 Growth of algae in a canal. Abundant nitrate and phosphate from treated sewage and from farmland make this growth possible.

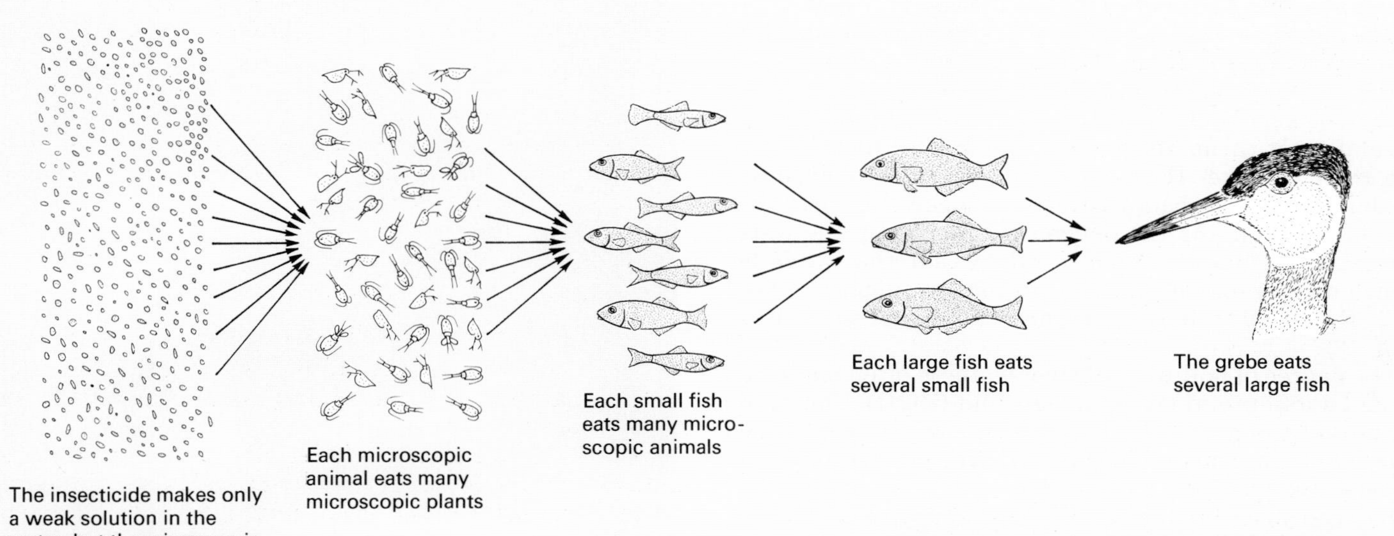
The insecticide makes only a weak solution in the water, but the microscopic plants take up the DDT
Each microscopic animal eats many microscopic plants
Each small fish eats many micro-scopic animals
Each large fish eats several small fish
The grebe eats several large fish

Fig. 9 Pesticides may become more concentrated as they move along a food chain. The intensity of colour represents the concentration of DDT.

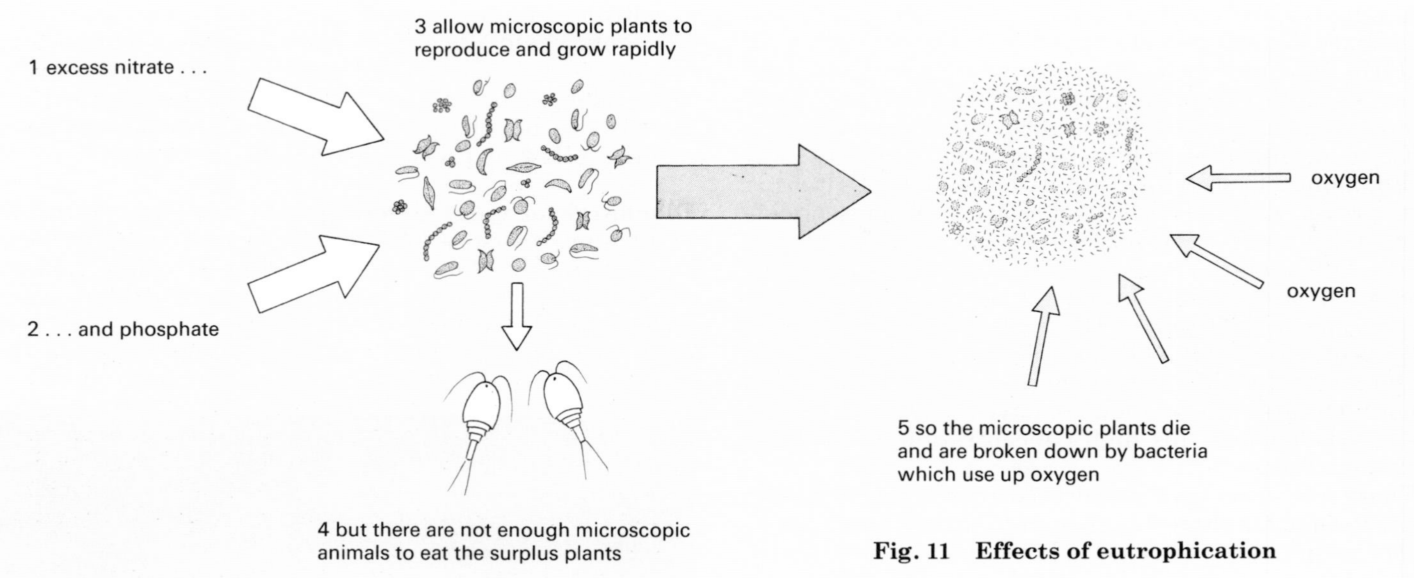
1 excess nitrate . . .
2 . . . and phosphate
3 allow microscopic plants to reproduce and grow rapidly
4 but there are not enough microscopic animals to eat the surplus plants
5 so the microscopic plants die and are broken down by bacteria which use up oxygen
oxygen
oxygen

Fig. 11 Effects of eutrophication

Use of detergents Some detergents contain a lot of phosphate. This is not removed by sewage treatment and is discharged into rivers. The large amount of phosphate encourages growth of microscopic plants (algae).

Agriculture Intensification of agriculture has led to nitrates being washed out of the soil by rain. The nitrates find their way into streams and rivers where they allow the overgrowth of the microscopic algae. In some cases, the level of nitrates getting into drinking-water reaches dangerous levels, particularly for babies and young children.

'Factory farming' Chickens, calves and pigs are often reared in large sheds instead of in open fields. Their urine and faeces are washed out of the sheds with water. If this mixture gets into streams and rivers it supplies an excess of nitrates and phosphates for the microscopic algae.

The microscopic algae are at the bottom of a food chain. The extra nitrates and phosphates from the processes listed above enable them to increase so rapidly that they cannot be kept in check by the microscopic animals which normally eat them. So they die and fall to the bottom of the river or lake. Here, their bodies are broken down by bacteria. The bacteria need oxygen to carry out this breakdown and the oxygen is taken from the water (Fig. 11). So much oxygen is taken that the water becomes deoxygenated and can no longer support animal life. Fish and other organisms die from suffocation (Fig. 12).

The degree of pollution of river water is often measured by its **Biochemical Oxygen Demand (BOD)**. This is the amount of oxygen used up by a sample of water in a fixed period of time. The higher the BOD, the more polluted the water is likely to be.

It is possible to reduce eutrophication by using
1. detergents with less phosphates;
2. agricultural fertilizers that do not dissolve so easily;
3. animal wastes on the land instead of letting them reach rivers.

QUESTIONS

1 What might be the effect of the removal of earthworms from the food web in Fig. 1 on p. 265?
2 Give five examples of a monoculture.
3 What might be the effect on the food web of Fig. 1 of spraying the tree with an insecticide?
4 At one time, elm trees were sprayed with DDT to kill the beetles which carried Dutch elm disease. In the autumn, the sprayed leaves fell to the ground and were eaten by earthworms. From the food web in Fig. 1, suggest what effects this might have had on other organisms.
5 Explain briefly why too much nitrate could lead to too little oxygen in river water.

Fig. 12 Fish killed by pollution. The water may look clear but is so short of oxygen that the fish have died from suffocation.

Fig. 13 Cutting a road through a tropical rain forest. The road not only destroys the natural vegetation, it also opens up the forest to further exploitation.

HUMANS AND FORESTS

Forests have a profound effect on climate, water supply and soil maintenance. They have been described as environmental buffers. For example, they intercept heavy rainfall and release the water steadily and slowly to the soil beneath and to the streams and rivers that start in or flow through them. The tree roots hold the soil in place.

At present, we are destroying forests, particularly tropical forests, at a prodigious rate (*a*) for their timber, (*b*) to make way for agriculture, roads (Fig. 13) and settlements, and (*c*) for firewood. At the current rate of destruction, it is estimated that all tropical rainforests will have disappeared in the next 85 years.

Removal of forests allows soil erosion, silting up of lakes and rivers, devastating floods and the loss for ever of thousands of species of animals and plants.

Trees can grow on hillsides even when the soil layer is quite thin. When the trees are cut down and the soil is ploughed, there is less protection from the wind and rain. Heavy rainfall washes the soil off the hillsides into the rivers. The hillsides are left bare and useless and the rivers become choked up with mud and silt which can cause floods (Figs 14 and 15*a*). For example, Argentina spends 10 million dollars a year on dredging silt from the River Plate estuary to keep the port of Buenos Aires open to shipping. It has been found that 80 per cent of this sediment comes from a deforested and overgrazed region 1800 km upstream which represents only 4 per cent of the river's total catchment area. Similar sedimentation has halved the lives of reservoirs, hydro-electric schemes and irrigation programmes. The disastrous floods in India and Bangladesh in recent years can be attributed largely to deforestation.

The soil of tropical forests is usually very poor in nutrients. Most of the organic matter is in the leafy canopy of the tree tops. For a year or two after felling and burning, the forest soil yields good crops but the nutrients are soon depleted and the soil eroded. The agricultural benefit from cutting down forests is very short-lived, and the forest does not recover even if the impoverished land is abandoned.

The 'greenhouse effect' of carbon dioxide In the last 100 years, the atmospheric carbon dioxide concentration has risen from about 0.027 per cent to over 0.033 per cent as a result of the increased combustion of coal and petroleum, in our industries and motor vehicles.

Carbon dioxide is removed from the atmosphere by being dissolved in the sea and being taken up by photosynthesis. Destruction of large areas of tropical forest could significantly reduce the proportion of carbon dioxide removed by photosynthesis.

An increasing concentration of atmospheric carbon dioxide may have the effect of 'trapping' the sun's radiant energy in a similar way to a greenhouse. This could result in a warming of the Earth's atmosphere, the melting of the polar ice-caps and a rise in sea level. There could also be climatic changes which would affect the important food-growing areas of the world

The evidence for global warming is not conclusive, but many scientists now believe that significant climatic changes are inevitable.

Fig. 14 Soil erosion. Removal of forest trees from steeply sloping ground has allowed the rain to wash away the topsoil.

AGRICULTURE AND THE SOIL

Soil erosion

Bad methods of agriculture lead to soil erosion. This means that the soil is blown away by the wind, or washed away by rainwater. Erosion may occur for a number of reasons.

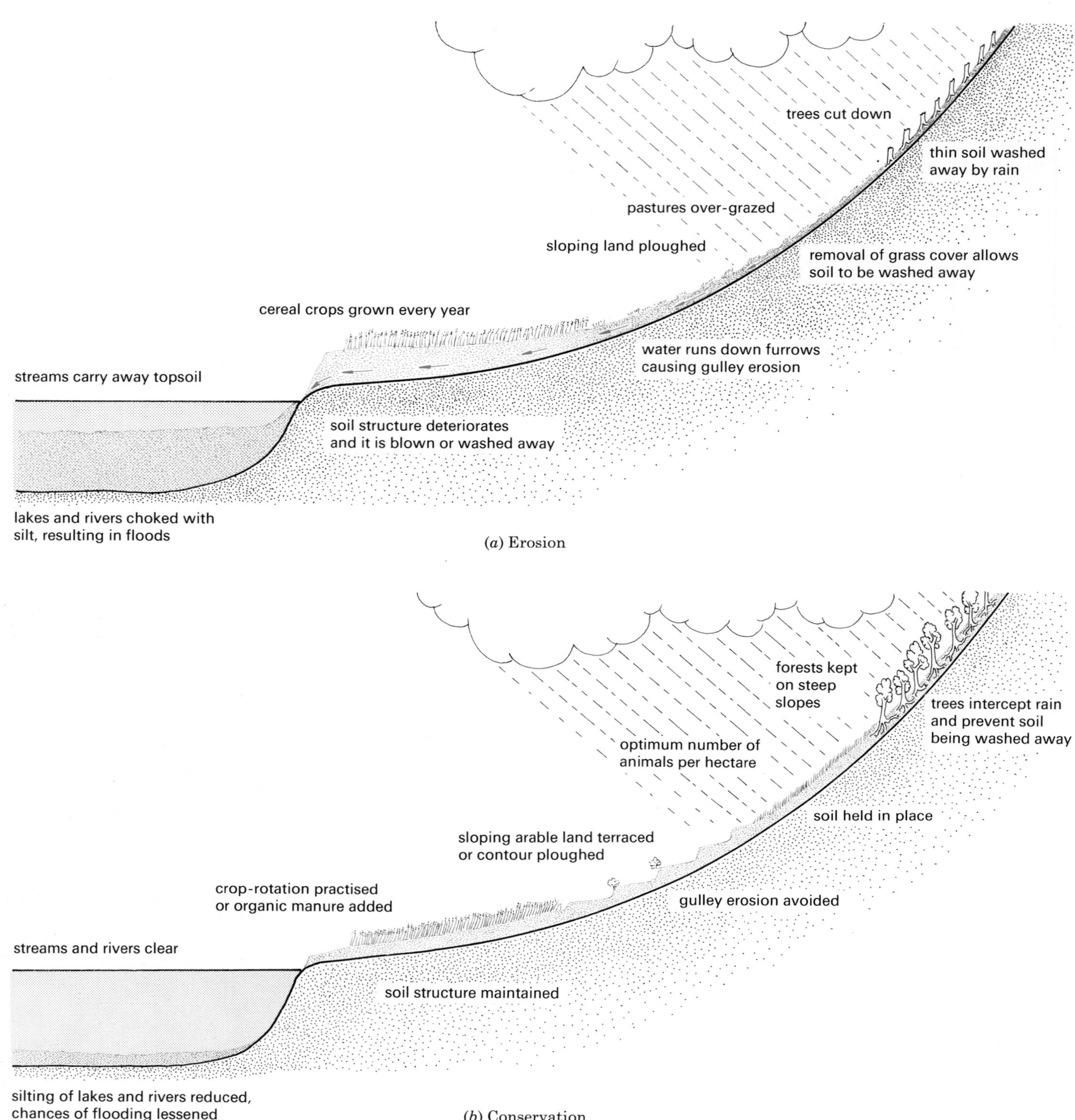

trees cut down
thin soil washed away by rain
pastures over-grazed
sloping land ploughed
removal of grass cover allows soil to be washed away
cereal crops grown every year
water runs down furrows causing gulley erosion
streams carry away topsoil
soil structure deteriorates and it is blown or washed away
lakes and rivers choked with silt, resulting in floods
(a) Erosion
forests kept on steep slopes
trees intercept rain and prevent soil being washed away
optimum number of animals per hectare
soil held in place
sloping arable land terraced or contour ploughed
crop-rotation practised or organic manure added
gulley erosion avoided
streams and rivers clear
soil structure maintained
silting of lakes and rivers reduced, chances of flooding lessened
(b) Conservation

Fig. 15 Soil erosion and conservation (From W. E. Shewell-Cooper, *The ABC of Soils*, English Universities Press.)

Deforestation The soil cover on steep slopes is usually fairly thin but can support the growth of trees. If the forests are cut down to make way for agriculture, the soil is no longer protected by a leafy canopy from the driving rain. Consequently, some of the soil is washed away eventually reaching streams and rivers (Figs 14 and 15).

Bad farming methods If land is ploughed year after year and treated only with chemical fertilizers, the soil's structure (see p. 255) may be destroyed and it becomes dry and sandy. In strong winds it can be blown away as dust (Fig. 16), leading to the formation of 'dust bowls', as in central USA in the 1930s, and even to deserts.

Fig. 16 Topsoil blowing in the wind. A dry sandy soil can easily be eroded by the wind.

Over-grazing If too many animals are kept on a pasture, they eat the grass down almost to the roots, and their hooves trample the surface soil into a hard layer. As a result, the rainwater will not penetrate the soil and so it runs off the surface, carrying the soil with it.

Use of pesticides

The effect of insecticides on food webs was described on page 266. When insecticides get into the soil, they kill the insect pests but they also kill other organisms. The effects of this on the soil's fertility are not very clear. An insecticide called **aldrin** was found to reduce the number of species of soil animals in a pasture to half the original number. Ploughing up a pasture also reduces the number of species to the same extent, so the harm done by the insecticide is not obvious.

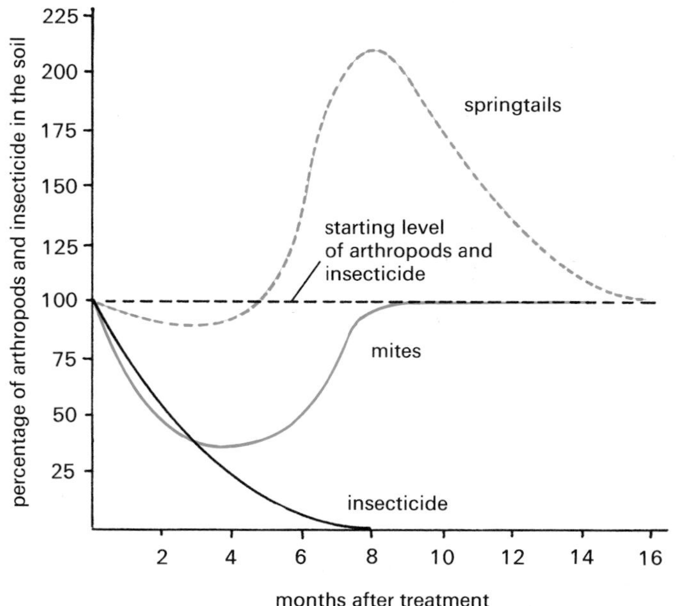

Fig. 17 The effect of insecticide on some soil organisms (From Clive A. Edwards, *Soil Pollutants and Soil Animals*, © 1969 Scientific American Inc.)

QUESTIONS

6 Read pp. 256 and 270 and say why the continuous use of chemical fertilizers may destroy the soil's crumb structure.
7 In what ways might trees protect the soil on a hillside from being washed away by the rain?
8 If a farmer ploughs a steeply sloping field, in what direction should the furrows run to help cut down soil erosion?
9 The graph in Fig. 17 shows the change in the numbers of mites and springtails (see Fig. 4 on p. 259) in the soil after treating it with an insecticide. Mites eat springtails. Suggest an explanation for the changes in numbers over the 16-month period.

WATER POLLUTION

Human activity sometimes pollutes streams, rivers (Fig. 18), lakes and even coastal waters. This affects the living organisms in the water and sometimes poisons humans or infects them with disease.

Fig. 18 River pollution. The river is badly polluted by the effluent from a paper mill.

Sewage Diseases like typhoid and cholera are caused by certain bacteria when they get into the human intestine. The faeces passed by people suffering from these diseases will contain the harmful bacteria. If the bacteria get into drinking-water they may spread the disease to hundreds of other people. For this reason, among others, untreated sewage must not be emptied into rivers. It is treated at the sewage works so that all the solids are removed and the water discharged into rivers is free from harmful bacteria and poisonous chemicals (but see 'Eutrophication' on p. 267).

Eutrophication When nitrates and phosphates from farmland and sewage escape into water they cause excessive growth of microscopic green plants. This may result in a serious oxygen shortage in the water as explained on page 268.

Chemical pollution Many industrial processes produce poisonous waste products. Electroplating, for example, produces waste containing copper and cyanide. If these chemicals are released into rivers they poison the animals and plants and could poison humans who drink the water. It is estimated that the River Trent receives 850 tonnes of zinc, 4000 tonnes of nickel and 300 tonnes of copper each year from industrial processes (see Fig. 4, p. 277).

In 1971, 45 people in Minamata Bay in Japan died and 120 were seriously ill as a result of mercury poisoning. It was found that a factory had been discharging a compound of mercury into the bay as part of its waste. Although the mercury concentration in the sea was very low, its concentration was increased as it passed through the food chain (see p. 267). By the time it reached the people of Minamata Bay, in the fish and other sea food which formed a large part of their diet, it was concentrated enough to cause brain damage, deformity and death.

High levels of mercury have also been detected in the Baltic Sea and in the Great Lakes of North America.

Oil pollution of the sea is becoming a familiar event. In 1967, a tanker called the *Torrey Canyon* ran on to the rocks near Land's End and 100 000 tonnes of crude oil spilled into the sea. Thousands of sea birds were killed by the oil (Fig. 19). The detergents, which were used to try and disperse the oil, killed many more birds and sea creatures living on the coast. Since 1967, there have been even greater spillages of crude oil from tankers and off-shore oil wells.

Fig. 19 Oil pollution. Oiled sea birds like this Guillemot cannot fly to reach their feeding grounds. They also poison themselves by trying to clean the oil from their feathers.

Radioactive waste

Radioactive products, unless very carefully controlled, could pollute land, sea, air and inland waters. In 1983, an accidental discharge of radioactive waste from the nuclear processing plant at Sellafield, polluted the beaches and the sea in the area.

Nuclear (atomic) power stations and other industries use or process radioactive materials. The radiation from these materials can cause cancers such as leukaemia. The radioactivity cannot be destroyed by burning or any other means of disposal and many of the compounds remain radioactive for thousands of years. The compounds have to be stored or transported in containers which do not allow the radiation to penetrate.

The wastes can be disposed of at sea or by burying on land or under the sea bed but, at present, there is considerable opposition to both these methods.

QUESTIONS

10 What are the possible dangers of dumping and burying poisonous chemicals on the land?

11 Before most water leaves the waterworks, it is exposed for some time to the poisonous gas, chlorine. What do you think is the point of this?

AIR POLLUTION

Some factories and all motor vehicles release poisonous substances into the air. Factories produce smoke and sulphur dioxide; cars produce lead compounds, carbon monoxide and the oxides of nitrogen which lead to smog (Fig. 20).

Fig. 20 Photochemical 'smog' over Paris

Smoke This consists mainly of tiny particles of carbon and tar which come from burning coal either in power stations or in the home. The tarry drops contain chemicals which may cause cancer. When the carbon particles settle, they blacken buildings and damage the leaves of trees. Smoke in the atmosphere cuts down the amount of sunlight reaching the ground. For example, since the Clean Air Act of 1956, London has received 70 per cent more sunshine in December.

Smoke also caused the dense 'pea-soup' fogs of industrial districts. When the water droplets in these fogs were inhaled, they contributed to illness and death from bronchitis. The Clean Air Acts of 1956 and 1968 have effectively stopped these lethal fogs in Britain, but they have not stopped atmospheric pollution by sulphur dioxide and nitrogen oxides.

Sulphur dioxide and oxides of nitrogen Coal and oil contain sulphur. When these fuels are burned, they release sulphur dioxide (SO_2) into the air. Although the tall chimneys of factories (Fig. 21) send smoke and sulphur dioxide high into the air, the sulphur dioxide dissolves in rainwater and forms an acid. When this acid falls on buildings, it slowly dissolves the limestone and mortar. When it falls on plants, it reduces their growth and damages their leaves.

This form of pollution has been going on for many years and is getting worse. In North America, Scandinavia and

Scotland, forests are being destroyed (Fig. 22) and fish are dying in lakes, at least partly as a result of 'acid rain'.

Oxides of nitrogen from power stations and vehicle exhausts also contribute to atmospheric pollution and acid rain. The nitrogen oxides dissolve in raindrops and form nitric acid.

Oxides of nitrogen also take part in reactions with other atmospheric pollutants and produce ozone. It may be the ozone and the nitrogen oxides which are largely responsible for the damage observed in forests.

There is still some argument about the causes of acidification of lakes and damage to forests but there is a mass of circumstantial evidence which points very clearly to the industrial areas of America, Britain and Central Europe as the principal sources of the sulphur dioxide and nitrogen oxides which make acid rain.

One effect of acid rain is that it dissolves out the aluminium salts in the soil. These salts eventually reach toxic levels in streams and lakes.

Smog This is a thin fog which occurs in cities in certain climatic conditions (see Fig. 20). Smog is irritating to the eyes and lungs and also damages plants. It is produced when sunlight and ozone (O_3) in the atmosphere, act on the oxides of nitrogen and unburnt hydrocarbons released from vehicle exhausts. This type of smog is called 'photochemical smog' to distinguish it from the smoke plus fog that used to afflict British cities.

Carbon monoxide This gas is also a product of combustion in the engines of cars and trucks. When inhaled, carbon monoxide combines with haemoglobin in the blood to form a fairly stable compound, carboxyhaemoglobin. The formation of carboxyhaemoglobin reduces the oxygen-carrying capacity of the blood and this can be harmful, particularly in people with heart disease or anaemia.

A smoker is likely to inhale far more carbon monoxide from cigarettes than from the atmosphere. Nevertheless, the carbon monoxide levels produced by heavy traffic in towns can be harmful.

Chlorofluorocarbons (CFCs) These are gases which readily liquefy when compressed. This makes them useful as refrigerants, propellants in aerosol cans and in plastic foams. Chlorofluorocarbons are very stable and accumulate in the atmosphere, where they react with ozone (O_3).

Ozone is present throughout the atmosphere but reaches a peak at about 25 km, where it forms what is called the 'ozone layer'. This layer filters out much of the ultraviolet radiation in sunlight.

The fear is that chlorofluorocarbons will deplete the ozone layer and allow more ultraviolet (UV) radiation to reach the Earth's surface. An increased level of UV radiation could cause more skin cancer, affect crops, interfere with the oxygen cycle and even distort weather patterns.

The reactions involved are very complex. There are also natural processes which destroy atmospheric ozone and some which generate ozone. The balance between destruction and creation is not known.

In July 1990 nearly 100 countries, including Britain, agreed to cut production of CFCs by 50 per cent by 1995, by 75 per cent by 1997 and to phase them out entirely by the year 2000. In 1992 tougher measures will probably be put forward.

Fig. 21 Air pollution by industry. Tall chimneys keep pollution away from the immediate surroundings but the atmosphere is still polluted.

Lead Compounds of lead are mixed with petrol to improve the performance of motor cars. The lead is expelled with the exhaust gases into the air. In some areas of heavy traffic it may reach levels which are dangerous and may cause damage to the brain in children.

Although there are other sources of lead pollution, such as some canned food, or water from lead pipes, the main source of lead entering the body is leaded petrol.

Laws have been passed to reduce the level of lead in petrol and the results of such legislation in America are shown in Figure 22. In 1985 in Britain, the lead content in petrol was reduced from 0.4 to 0.15 grams per litre by law, but the best step is to remove lead from petrol altogether. Claims that cars would not run so well on unleaded petrol, or that it would cause increased engine wear are not soundly based. It is neither difficult nor expensive to provide alternatives to leaded petrol.

The cost of cleaning up

Most of the forms of pollution described in this chapter could be prevented provided we were prepared to pay the cost of the necessary measures. Removal of sulphur dioxide from the waste gases of power stations might increase our electricity bills by 5 per cent. Lead-free petrol may cost a little more than leaded petrol (unless subsidised by a reduced tax). It is probably essential to bear these extra costs if we are to preserve our environment. Furthermore, when the costs of reducing pollution are compared with the costs of environmental damage and human ill-health, the difference may not be all that great.

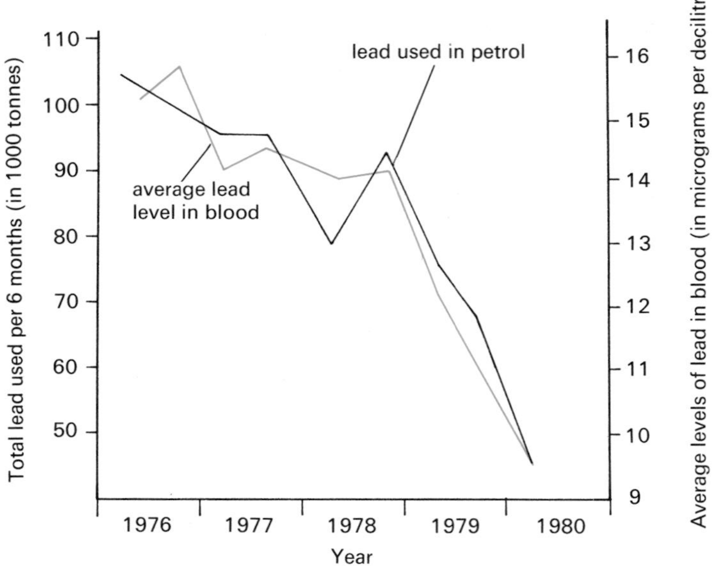

Fig. 22 The effect of reducing lead in petrol. In 1975 the U.S. government began to phase out the use of lead in petrol. This was subsequently matched by a fall in the levels of lead in peoples' blood. This suggests (but does not prove) a close connection between lead in exhaust fumes and the lead in the body. (First published in *New Scientist.*)

Correlation and causation

Figure 22 shows what is called a correlation. As the percentage of lead in petrol falls, so does the lead level in people's blood. A correlation between two things, however, cannot usually prove that one of the effects is caused by the other.

Nevertheless, a strong correlation often arouses suspicions of a 'cause-and-effect' relationship and leads to further investigation. A correlation between deaths from lung cancer and number of cigarettes smoked led to investigations which now point convincingly to cigarette smoking being a major cause of lung cancer (see pp. 155 and 157).

In the studies on lead pollution (Fig. 22), it would be necessary to make sure, for example, that there had been no decrease in the use of other lead-containing substances over the same period which might have accounted for the decrease in blood levels of lead. An Italian study between 1977 and 1979 used special isotopes of lead in petrol. No other substances contained those particular isotopes. A correlation between the levels of the lead isotopes in petrol and in blood, is strong evidence for causation (sometimes described as a 'causal relationship').

QUESTIONS

12 To what extent do tall chimneys on factories reduce atmospheric pollution?
13 What are thought to be the main causes of 'acid rain'?
14 If compounds of lead and mercury get into the body, they are excreted only very slowly. Why do you think this makes them dangerous poisons even when they are in low concentrations in the air or the water?
15 It costs money to prevent harmful chemicals escaping into the air from factories and cars. The effects of pollution also cost a great deal of money. List some of the ways in which the effects of pollution (*a*) affect our health and (*b*) cost us money.

CHECK LIST

- **The plants and animals in a food web are so interdependent that even a small change in the numbers of one group has a far-reaching effect on all the others.**
- **Hunting activities and farming upset the natural balance between other living organisms.**
- **Pesticides kill insects, weeds and fungi that could destroy our crops.**
- **Pesticides help to increase agricultural production but they kill other organisms as well as pests.**
- **A pesticide or pollutant which starts off at a low, safe level can become dangerously concentrated as it passes along a food chain.**
- **Eutrophication of lakes and rivers results in the excessive growth of algae followed by an oxygen shortage when the algae die and decay.**
- **Soil erosion results from removal of trees from sloping land, use of only chemical fertilizers on ploughed land and putting too many animals on pasture land.**
- **The conversion of tropical forest to agricultural land usually results in failure because forest soils are poor in nutrients.**
- **Removal of forests can lead to erosion, silting-up of lakes and rivers and to flooding.**
- **We pollute our lakes and rivers with industrial waste and sewage effluent.**
- **We pollute the sea with crude oil and factory wastes.**
- **We pollute the air with smoke, sulphur dioxide and nitrogen oxides from factories, and lead and nitrogen oxides from motor vehicles.**
- **The acid rain resulting from air pollution leads to poisoning of lakes and possibly destruction of trees.**

29 Conservation and the Reduction of Pollution

CONTROL OF AIR POLLUTION

Sulphur dioxide, nitrogen oxides, lead.

CONTROL OF WATER POLLUTION

1973 Water Act, sewage disposal.

CONSERVATION

Habitats, reclamation, renewable energy sources, endangered species, gene banks.

REDUCTION OF POLLUTION

Pollution can be reduced voluntarily or by passing laws (legislation) which restrict the emission of pollutants. Pressure of public opinion has sometimes been effective in persuading a factory voluntarily to cut down its emission of polluting gases or liquid effluent. However, to reduce pollution, on a national scale, it is necessary to introduce legislation, i.e. pass laws which set limits on the amount of pollutants that may be released into the environment. The laws include penalties that may be applied if these levels are exceeded. One example of legislation is the Clean Air Act.

Control of air pollution

The Clean Air Acts of 1956 and 1968 These Acts designated certain city areas as 'smokeless zones'. The use of coal for domestic heating was prohibited and factories were not allowed to emit black smoke. This was effective in abolishing dense fogs in cities but did not stop the discharge of sulphur dioxide and nitrogen oxides in the country as a whole (Fig. 1). The effect of legislation on lead pollution can be seen in Fig. 22 on p. 274.

Emission of sulphur dioxide and nitrogen oxides The concern over the damaging effects of acid rain has led several countries to press for regulations to reduce emissions of these acid gases. European Commission regulations require Britain to reduce SO_2 discharges by 20 per cent (from 1980 levels) by 1993, 40 per cent by 1998 and 60 per cent by 2003. Britain has made plans to reduce emissions from three power stations by 1997. This will cut SO_2 emissions by about 14 per cent from the 1980 level.

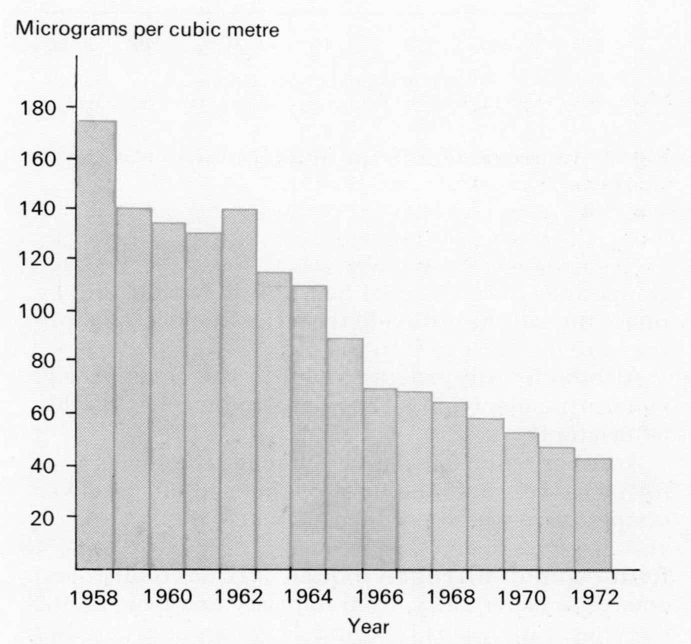

Fig. 1 Average smoke concentration near ground level in the United Kingdom 1958–72
(From John Barker, *Breathing Space*, Evans 1975.)

Reduction of sulphur dioxide This can be achieved by removing a proportion of the sulphur

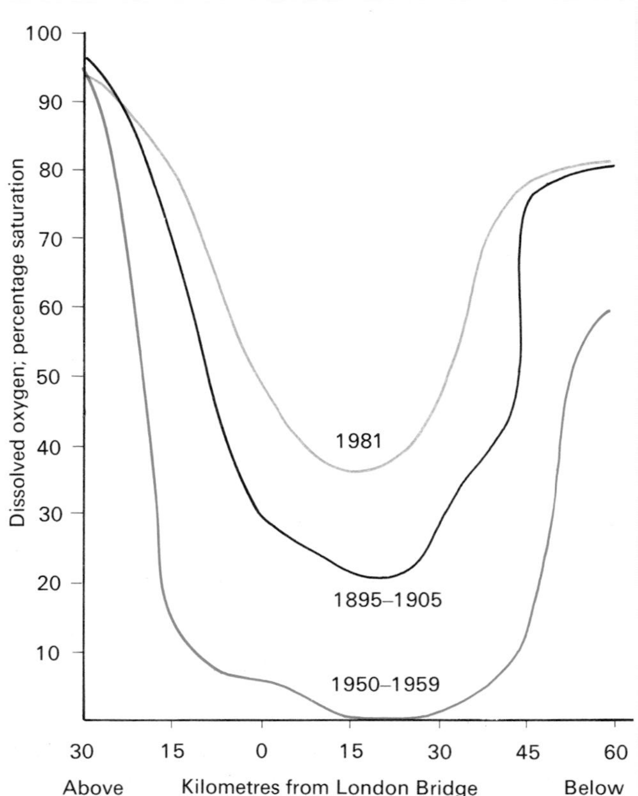

Fig. 2 Improvement in the quality of water in the Thames (Thames Water Authority.)

compounds from the coal before it is burned and by removing sulphur dioxide from the flue gases before they are discharged into the air.

Although both processes add to the costs of, e.g. generating electricity, they can produce marketable by-products.

Another solution is to change the design of furnaces so that the fuel is burned at a lower temperature which produces less acid gas.

Reduction of nitrogen oxides Oxides of nitrogen come, almost equally, from industry and from motor vehicles. Flue gases from industry can be treated to remove most of the nitrogen oxides. Vehicles can have catalytic converters fitted to their exhaust systems. These converters remove most of the oxides of nitrogen, carbon monoxide and unburned hydrocarbons, but they could add £300 to the cost of a car and will only work if lead-free petrol is used.

Another solution is to redesign car engines to burn petrol at lower temperatures and so produce smaller amounts of nitrogen oxides. However, this will not reduce emissions of hydrocarbons.

In 1988, 24 countries agreed to reduce nitrogen oxide emissions to 1987 levels by 1994. Twelve of the countries (not including Britain) have agreed to a further 30 per cent reduction by 1998.

Lead in petrol The European Community would like to move rapidly towards lead-free petrol but the member countries cannot agree on the timing of the measures. Britain has reduced the lead in petrol from 0.4 to 0.15 g/litre and has introduced lead-free petrol.

Lead-free petrol has been available in the USA for some years. Lead emissions have been reduced by 55 per cent and the levels of lead in the blood have dropped by 36.7 per cent across the country (p. 274).

Control of water pollution

It is an offence under the Control of Pollution Act 1974 to discharge water that has been used for any industrial purpose into a sewer or directly into a river without the consent of the appropriate water authority.

In practice, this may reduce pollution but does not stop it. The industries concerned reach agreement with the water authority about the levels of pollution that are acceptable. These levels are usually too high and even if they are exceeded the fines imposed are very small.

The Thames Water Authority has been successful in cleaning up the Thames, to the extent that Atlantic salmon have returned to the river after an absence of 60 years (Figs 2 and 3). However, there are still 2800 km of other rivers classified as unfit for fish (Fig. 4). Pollution control was only extended to tidal rivers, estuaries and the coast in 1987.

Fig. 3 The Thames Bubbler. Heavy storms sometimes wash extra sewage effluent into the river. The bubbler injects oxygen into the water to stop the oxygen level falling to a point where fish would suffocate.

When the ten water authorities were 'privatized' in 1989, an independent National Rivers Authority (NRA) was set up to control water use, conservation and pollution nationwide.

Fig. 4 River pollution. Untreated industrial effluent being discharged into a tributary of the Mersey.

Sewage disposal Inland towns have to treat their sewage to make it harmless. Paper, plastic and other debris are removed by passing the raw sewage through metal grids. As the flow slows down, gravel settles out and, later, the organic solids settle in sedimentation tanks and form sewage sludge. The remaining liquid is treated in one of two ways, the biological filter or the activated sludge method. Only the former will be described.

The liquid sewage is sprayed over a circular bed of porous bricks (Fig. 5). As it trickles over the bricks, the liquid becomes well aerated and this encourages the growth of aerobic bacteria (p. 311) which feed on the dissolved organic matter in the liquid. The bacteria themselves are eaten by single-celled creatures which, in turn, are eaten by aquatic worms and insect larvae. Thus a food chain is set up which removes harmful bacteria and nitrogenous waste from the liquid sewage.

Fig. 5 Sewage treatment. Micro-organisms in the filter beds remove organic matter from the effluent.

The dead remains of the organisms are washed out of the filter beds and allowed to settle as sludge. The remaining liquid, now free of bacteria and harmful substances, can be discharged into a river. Effective sewage treatment, by destroying intestinal bacteria, prevents the spread of infectious diseases such as typhoid and cholera.

However, there are often nitrates and phosphates still present in the clear sewage effluent and these can cause problems of pollution (see p. 267).

The sewage sludge can be dried and used as fertilizer provided it does not contain heavy metals (e.g. mercury and cadmium) from industrial processes. In some cases, the sewage sludge is fed into a 'digester' where anaerobic bacteria (p. 311) act on it and produce enough methane gas to drive all the machinery at the sewage works.

It is important that sewage should not contain industrial chemicals which could kill the organisms in the filter beds. It is also desirable that any chemicals, e.g. detergents, should be **biodegradable**, i.e. capable of being digested by bacteria. If they are not biodegradable the chemicals will pass unchanged into the river.

Many coastal towns discharge untreated sewage into the sea where it is considered to become sufficiently diluted to be harmless. In practice, adverse winds and tides sometimes deposit most of it back on the shore and it may become a health hazard.

QUESTIONS

1 Why have the Clean Air Acts not prevented atmospheric pollution?
2 How can (*a*) sulphur dioxide emissions, (*b*) nitrogen oxide emissions be reduced?
3 What do the graphs in Fig. 2 show?
4 In what ways does sewage treatment (*a*) reduce pollution, (*b*) cause pollution?
5 In the long run, who pays the cost of reducing pollution?

CONSERVATION

Conservation involves preserving habitats and protecting individual species of plants and animals.

Conservation of habitats

In 1949, the *Nature Conservancy* (later the *Nature Conservancy Council*), was established by the National Parks and Countryside Act. The Council's job was to establish, manage and maintain nature reserves, protect threatened habitats and to conduct research into matters relevant to conservation.

The NCC had established 195 Nature Reserves (Fig. 6) but, in addition, it had responsibility for no-

Fig. 6 A National Nature reserve at Scolt Head, Norfolk. The reserve attracts shore birds and provides a nesting site for large numbers of terns.

Fig. 7 Site of Special Scientific Interest. This land in Bedfordshire is privately owned but protected by a management agreement with the landowner.

Fig. 8 Destruction of a hedgerow. Once an area has been declared an SSSI, damage such as this can be prevented. But there are still loopholes in the law and some landowners have destroyed SSSI's.

tifying planning authorities of **Sites of Special Scientific Interest** (SSSIs). These are privately owned lands which include important habitats or rare species (Fig . 7). The NCC established management agreements with the owner so that the site was not destroyed by felling trees, ploughing land or draining marshes or fens (Fig. 8). There are now 4150 SSSIs.

There are several other, non-governmental organizations which have set up reserves and which help to conserve wildlife. The Nature Conservation Trust Reserve has about 1400 reserves; the Royal Society for Protection of Birds (RSPB) has 93; the Woodland Trust has 102, and there are about 160 other reserves managed by other organizations.

The National Parks Commission has set up 10 National Parks, covering some 9 per cent of England and Wales, e.g. Dartmoor, Snowdonia and the Lake District. Although the land is privately owned, the Park Authorities are responsible for protecting the landscape and wildlife, and for planning public recreation such as walking, climbing or gliding.

Reclamation There are probably more than 10 000 hectares of derelict land in Britain, resulting from abandoned industrial sites, excavations and dumping of waste from the coal, limestone and steel industries. Government grants are made available, in some cases, for reclamation of these derelict areas. Spoil heaps are levelled and excavations are filled in. Topsoil and sewage sludge are added; grass and trees are planted.

In this way, the land may become a public amenity or be restored to agriculture.

The margins of clay and gravel pits may be replanted and restored to attractive aquatic habitats or developed for recreations such as sailing.

Despite reclamation schemes, it is estimated that the area of derelict land is increasing at the rate of 1400 hectares per year.

Renewable resources and recycling

Our supplies of coal, oil and metal ores are running out. These are non-renewable resources because we cannot produce these materials from other substances. As the supplies of non-renewable resources dwindle, we are compelled to use alternative or renewable sources of energy and to recycle our metals.

Alternative sources of energy are, for example, atomic energy, hydro-electric power, wind generators and solar panels (Fig. 9).

For most of these, there is an environmental price to pay. Nuclear power stations produce radioactive waste; hydro-electric schemes involve flooding valleys, damming rivers or building tidal barrages.

Damage to the environment can be reduced, but not avoided, by cutting down on our demands for

Fig. 9 Wind generators in California, USA. On otherwise unproductive land, these generators can make a small but significant contribution to the electricity supply.

energy. Using surplus heat from power stations for domestic heating; improved insulation in buildings; more efficient combustion of our remaining fuels; these are just some of the ways of conserving energy.

Renewable sources of energy are, for example, wood, alcohol and vegetable oil. These are all the products of photosynthesis. Sugar produced by sugar-cane, can be fermented by yeasts to produce alcohol. This forms an acceptable fuel for driving vehicles and produces fewer pollutants than petrol. Sunflower oil has been tried as an alternative to diesel oil, but this has run into technical problems. In fact, there are technical and economic problems to be overcome with any renewable source of energy. Land which is used for producing sugar-cane for alcohol cannot be used for producing food and there is already enormous pressure on land resources.

Recycling By reusing scrap metals, glass from bottles and jars, and wastepaper, we help to conserve raw materials and timber. In some cases there may also be savings in energy costs. Recycling metals and glass also reduces some of the pollution due to dumping of these wastes.

Straw-burning at harvest time can cause pollution with smoke and has often damaged hedges. Ways are being sought to recycle the straw either by ploughing it back, treating it to make it suitable for animal fodder or even compressing it to make fuel.

Recovery of metals such as mercury and cadmium from industrial effluents, reduces the pollution of rivers and oceans and also helps conserve supplies of these non-renewable materials.

Domestic waste can be mechanically sorted into metals, organic waste, fabrics, paper and plastic. The metals and paper are recycled, the organic waste can be fermented to produce methane which drives the plant, and even the plastic can be processed into fuel pellets.

QUESTIONS

6 Explain why some renewable energy sources depend on photosynthesis.

7 Why should recycling wastepaper help to conserve our deciduous woodland?

8 In what ways does recycling of metals help to conserve the environment?

9 Explain why some of the alternative and renewable energy resources are less likely to cause pollution than coal and oil.

10 What schemes exist in your locality for (*a*) recycling waste materials, (*b*) conservation of habitats, (*c*) reclamation of waste land?

Conservation of species

The world is losing one species every day and, within 20 years, at least 25 per cent of all forms of wildlife could be extinct. There are laws in Britain which protect wildlife. For example, it is an offence to capture or kill almost all species of wild birds or to take eggs from their nests; wild flowers in their natural habitat may not be uprooted; badgers, otters and bats are three of the protected species of mammal (Fig. 10).

Fig. 10 Badger. One of a number of species protected by law.

Animals and plants cannot be conserved unless their habitats are protected. Thus, the organizations mentioned above are concerned with habitats as well as with endangered species. In addition, other organizations are particularly concerned with endangered species. Three examples are given below.

CITES (Convention on International Trade in Endangered Species) gives protection to about 1500 animals and thousands of plants, by persuading governments to restrict or ban trade in endangered species or their products, e.g. snake skins or rhino horns. There are about 70 countries who are party to the Convention.

The World Wide Fund for Nature (WWF) operates on a global scale and is represented in 25 countries. The WWF raises money for conservation projects in all parts of the world, with particular emphasis on endangered species and habitats (Fig. 11).

The WWF calls on advice from the International Union for the Conservation of Nature (IUCN), a representative group of experts from governments and conservation agencies in over 100 countries.

The International Whaling Commission (IWC) was set up to try and avoid the extinction of whales as a result of uncontrolled whaling, and has 40 members.

The IWC allocates quotas of whales that member countries may catch but, having no powers to enforce its decisions, cannot prevent countries from exceeding their quotas. For example, a ban on catching sperm whales in 1982 was ignored by Japan. In 1985, the IWC declared a moratorium (i.e. a complete ban) on all whaling. The position is to be reviewed in 1990. It is not yet certain whether countries such as Japan, Iceland, Norway and South Korea will comply with the moratorium.

Conservation of genes

On p. 234 it was explained that crossing a wild grass with a strain of wheat, produced an improved variety. This is only one example of many successful attempts to try to improve yield, drought resistance and disease resistance in food plants. Some 25 000 plant species are threatened with extinction at the moment. This could result in a devastating loss of hereditary material (genes, see p. 224) and a reduction of about 10 per cent in the genes available for crop improvement. 'Gene banks' have been set up to preserve a wide range of plants but these banks are vulnerable to accidents, disease and human error. The only secure way of preserving the full range of genes is to keep the plants growing in their natural environments.

Fig. 11 Trying to stop the trade in endangered species. A representative of the World Wildlife Fund checks an illegal cargo impounded at an Indian customs post.

QUESTIONS

11 (*a*) What are the differences between the Nature Conservancy Council and the National Parks Commission?

(*b*) In what ways do both organizations contribute to conservation?

12 What is the difference between an SSSI and a nature reserve?

13 Discuss whether habitat conservation would automatically result in species conservation.

CHECK LIST

- **Sulphur dioxide emissions can be reduced by removing sulphur compounds from coal.**
- **Sulphur dioxide and nitrogen oxides can be removed from flue gases.**
- **Nitrogen dioxide, carbon monoxide and hydrocarbons can be removed by fitting catalytic converters to vehicle exhausts.**
- **Removal of lead from petrol reduces lead pollution of the air.**
- **The EEC is pressing for regulations to reduce discharge of all these pollutants.**
- **Water pollution of rivers is regulated by Regional Water Authorities.**
- **Estuaries and tidal rivers are not included in these regulations.**
- **It is urgently necessary to stop the destruction of wildlife and their habitats.**
- **The Nature Conservancy Council has responsibility for this in Britain.**
- **Many voluntary organizations help to conserve wildlife and habitats.**
- **Alternative sources of energy must be used as coal and oil stocks run out.**
- **No form of energy production can entirely avoid causing some environmental damage or problems of waste disposal.**
- **Recycling of waste helps reduce pollution and conserve our natural reserves.**

30 Ecology

DEFINITIONS

Environment, habitat, community, ecosystem, competition, population, niche.

ECOLOGY OF FRESH WATER

Physical aspects, plant and animal communities, adaptations, succession and colonization.

FIELD TECHNIQUES

Measuring physical factors, sampling techniques.

Ecology is the study of living organisms in relation to their natural environment, as distinct from in the laboratory. This does not mean that laboratory work is ruled out, but its object is always to explain how the organism survives, how it relates to other organisms and why it is successful in its particular environment.

The following are some of the terms used in any discussion of ecology.

Environment

This means everything in the surroundings of an organism that could possibly influence it. The environment of a tadpole consists of water. The temperature of the water will influence the tadpole's rate of growth and activity. The watery environment contains plants and animals on which the tadpole will feed, but it also contains fish and insects which may eat the tadpole. The water contains dissolved oxygen which the tadpole breathes by means of its gills. The water, the oxygen, the food and the predators are all part of the tadpole's environment.

Habitat

A habitat is where an organism lives, i.e. where it obtains its food and shelter, and where it reproduces. The habitat of a limpet is a rocky shore. The environment includes air, sea water and sunlight but the habitat is the shore. The habitat of the tapeworm is the intestine of a mammal. Its environment, however, is the warm digested food and digestive juices of its host. The habitat of an aphid may be a bean plant, but its environment will include sun, wind, rain, ladybirds, ants and bacteria.

Population

In biology, this term always refers to a single species. A biologist might refer to the population of sparrows in a farmyard or the population of carp in a lake. In each case this would mean the total numbers of sparrows or the total numbers of carp in the stated area.

Community

A community is made up of all the plants and animals living in a habitat. In the soil (p. 254) there is a community of organisms which includes earthworms, springtails and other insects, mites, fungi and bacteria. In a lake, the animal community will include, fish, insects, crustacea, molluscs and protozoa.

The plant community will consist of rooted plants with submerged leaves, rooted plants with floating leaves, reed-like plants growing at the lake margin, plants floating freely on the surface, filamentous algae like *Spirogyra* (p. 329) and single-celled plants (p. 243) in the surface waters.

Ecosystem

The community of organisms in a habitat, plus the non-living part of the environment (air, water, soil, light, etc.) make up an ecosystem. A lake is an ecosystem which consists of the plant and animal communities mentioned above, and the water, minerals, dissolved oxygen, soil and sunlight on which they depend. An ecosystem is self-supporting (Fig. 1).

individuals of the same species } = POPULATION + populations of other species } = COMMUNITY + non-living part of environment } = ECOSYSTEM

In a woodland ecosystem, the plants absorb light and rainwater for photosynthesis, the animals feed on the

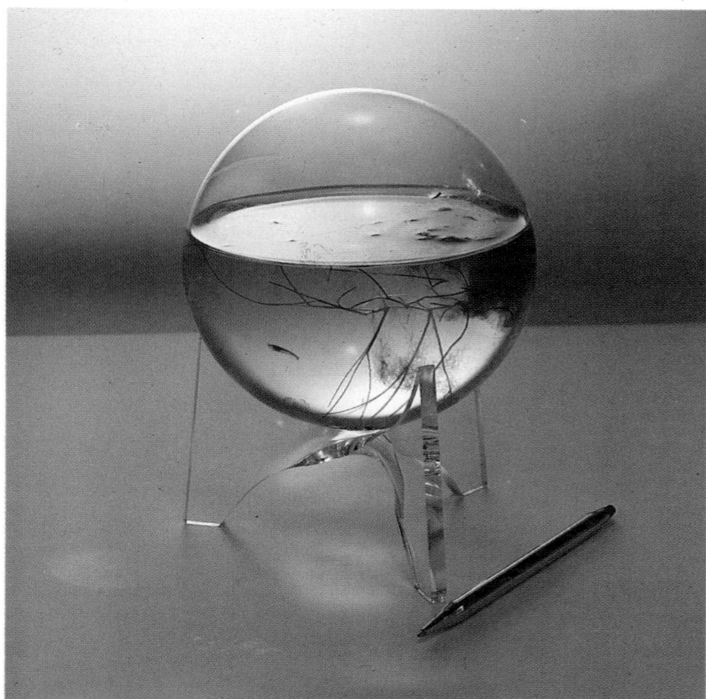

Fig. 1 The 'Ecosphere'. The 5-inch globe contains sea-water, bacteria, algae and a few Pacific shrimps. Given a source of light it is a self-supporting system and survives for several years (at least). The shrimps live for up to 5 years but do not reproduce.

plants and on each other. The dead remains of animals and plants, acted upon by fungi and bacteria, return nutrients to the soil.

Lakes and ponds are clear examples of ecosystems. Sunlight, water and minerals allow the plants to grow and support animal life. The recycling of materials from the dead organisms maintains the supply of nutrients.

All ecosystems will contain producers, consumers and decomposers as described on p. 243.

It is quite reasonable to regard a single oak tree as an ecosystem. The tree grows as a result of photosynthesis. Lichens may grow on its branches; ivy may grow up its trunk, squirrels will eat its acorns, and aphids and caterpillars feed on its leaves. Birds eat the caterpillars and may nest in the branches. The tree, the community of plants and animals which live on it, and the air, water and sunlight which nourish it, all form an ecosystem.

On the other hand, the whole of that part of the Earth's surface which contains living organisms (called the **biosphere**), may be regarded as one vast ecosystem.

Competition

Living organisms compete with each other for resources such as food, light, rooting space and breeding partners. As explained on p. 238, all organisms produce more offspring than can possibly survive, so competition is unavoidable. **Intra-specific competition** takes place between members of the same species for territory, food, nesting sites or mates (Fig. 2).

Inter-specific competition occurs between different species. On page 293 there is a description of how competition between two species of *Paramecium* leads to a decline in the population of one of them.

Niche

A niche is the position which a particular species occupies in an ecosystem. This 'position' may refer to its habitat, its place in the food web, or both. A caterpillar and an aphid may both live in the same habitat, e.g. on a cabbage leaf, but they occupy different ecological niches because the caterpillar bites off pieces of leaf, whereas the aphid sucks sap from the veins.

An organism may be very closely adapted to its niche and would have difficulty in occupying a different one. A rabbit's teeth and digestive system are adapted to eating short vegetation such as grass and clover. Its behaviour is adapted to running for cover and burrowing. A rabbit could not survive in an exclusively woodland habitat or change to a carnivorous diet.

A badger, on the other hand, is not closely adapted to an ecological niche. It will eat a wide variety of food, including small mammals, insects, earthworms, acorns and grass, and it ranges over many different habitats.

So, a *population* of carp forms part of the animal *community* living in a *habitat* called a lake. The communities in this habitat, together with their watery *environment*, make up a self-supporting *ecosystem*.

A carp is a *secondary consumer*, occupying an ecological *niche* at the top of a *food chain*, where it is in *competition* with other species of fish for food and with other carp for food and mates.

Fig. 2 Intra-specific competition in red deer. The 'pushing contest' will determine who has access to the females.

QUESTIONS

1 What communities might be present in an area of woodland?
2 (*a*) What is the habitat of an earthworm?
(*b*) What makes up the environment of the earthworm?
3 Describe the ecological niche occupied by the grey squirrel (see Fig. 1 on p. 265).
4 Name some of the producers, consumers and decomposers that might be present in a grassland ecosystem.

THE ECOLOGY OF FRESH WATER

Ponds, lakes and rivers form clearly defined examples of ecosystems. Some of the properties of a pond can be reproduced by setting up a balanced aquarium in the laboratory.

Physical aspects of a freshwater environment

Density Water is far more dense than air. It offers resistance to moving animals but it also physically supports the animals and plants. Plants and animals living in rivers and streams must be able to withstand or avoid the force of the flowing water.

Surface tension The surface of water behaves as if it had a thin elastic skin on it. There is, in fact, no skin; the effect is due to the attraction of water molecules at the surface, downwards into the bulk of the liquid. Nevertheless, the surface 'film' is used by animals above and below the surface.

Pond skaters (Fig. 12) can walk over the surface film without getting wet; pond snails and flatworms crawl along under the surface film; insects such as water boatmen (Fig. 16) and mosquito larvae (Fig. 14) can hang from it while they take in air at the surface.

Temperature Water can absorb a good deal of heat from the sun without its temperature rising much. Similarly, when water loses heat, its temperature does not fall much. A very small or shallow pond might heat up during the day and cool down at night but in most freshwater habitats, the temperature remains fairly constant (Fig. 3).

When the temperature of water rises, it can hold less dissolved oxygen. Thus, a rise in temperature, well below the lethal level, might impair the breathing of some aquatic animals.

Water also has an unusual property. Like most liquids, water contracts when it cools, but at temperatures below 4 °C, it expands, becomes less dense and rises. So, ice will form only at the surface of a pond. Water at lower levels will not cool below 4 °C.

Light Except for places where trees and shrubs grow on the banks, the surface water will receive a high light intensity. The small particles suspended in the water absorb the light so that at a depth of a metre or two, there may not be enough light to allow plants to photosynthesize and grow. The changes in light intensity and other conditions are shown in Fig. 4.

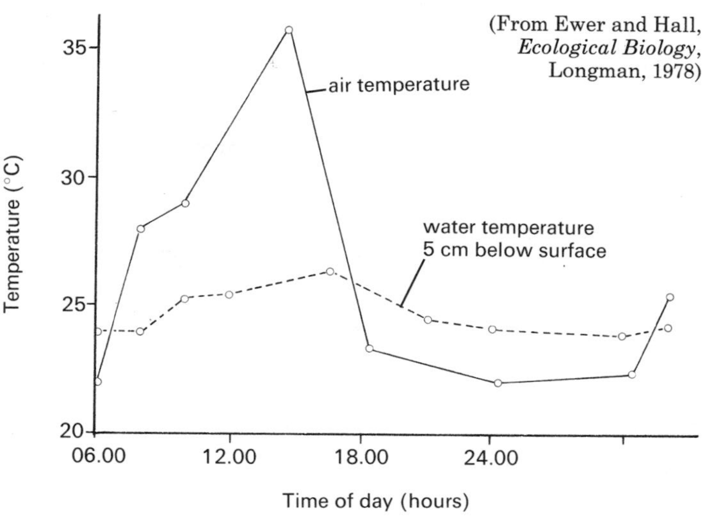

(From Ewer and Hall, *Ecological Biology*, Longman, 1978)

Fig. 3 Daily temperature change in air and water in a small tropical lake

Oxygen Although water is H_2O, the oxygen in its molecule is not available for respiration. The oxygen that plants and animals use for respiration is dissolved in the water. It comes from the plants' photosynthesis during the day and also diffuses continuously through the water surface from the air.

There is much less oxygen in water than there is in air. At 0 °C, 100 cm^3 water can hold only about 1 cm^3 dissolved oxygen. (100 cm^3 air contains about 20 cm^3 oxygen.) This means that a stationary animal or plant in still water quickly uses all the oxygen from the water immediately around it. Anything, such as eutrophication (p. 267) which reduces this small oxygen concentration in fresh water puts the animals at risk of suffocation. Any event which breaks up the surface of the water, e.g. waterfalls or breaking waves, helps to introduce more oxygen from the air.

Minerals The water flowing in a river or into a lake brings a supply of minerals. Water draining from heavily fertilized farmland may even bring too many minerals (see

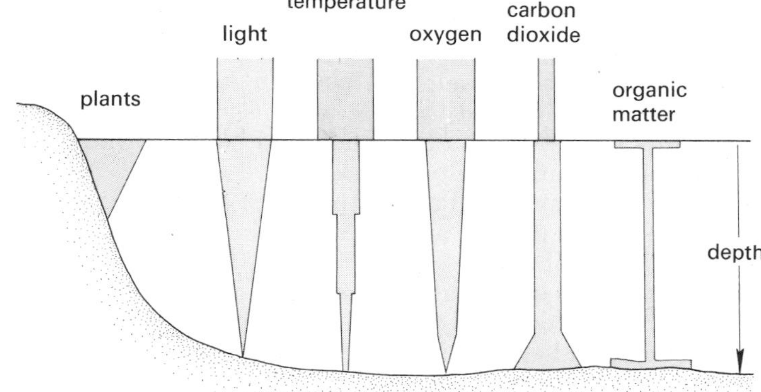

Fig. 4 How conditions change with depth. The width of each column is proportional to the factors named.
(From Bennet and Humphries, *Introduction to Field Biology*, Arnold, 1974)

p. 268). In enclosed lakes and ponds, the supply of minerals will be maintained by normal recycling processes (p. 245).

Substrate This is the mud or silt on the bottom. It allows plants to root and animals to burrow in it. If it contains a lot of decaying organic matter, it may be very short of oxygen. This is because the bacteria which break down the organic matter use up oxygen in their respiration. Streams which flow into a lake may bring down deposits of silt which collect on the bottom and gradually fill the lake.

QUESTIONS

5 What processes might (*a*) increase, (*b*) decrease the amount of oxygen dissolved in water?
6 What factors are likely to limit the growth of plants in a lake?
7 What properties of water make it a good environment for living organisms?

Fig. 5 ***Elodea canadensis*; Canadian pondweed** **Fig. 6** ***Ceratophyllum*; hornwort**

The plant community

The surface waters of ponds, lakes and rivers contain the microscopic algae which form the phytoplankton (p. 243). There will be diatoms and blue-green algae, and in small ponds there may be filamentous algae like *Spirogyra* (p. 329). The high light intensity of these surface waters allows rapid photosynthesis.

Floating freely on the surface of still waters in lakes and ponds are plants such as duckweed (*Lemna*). Since these plants receive direct sunlight and can reproduce rapidly by vegetative propagation (p. 100) they may spread to cover a large surface of the water and restrict the light reaching the submerged plants. The roots and lower parts of the leaves contain air spaces which enable the plants to float, and the waxy cuticle on the leaves repels water if the leaves are temporarily submerged by waves.

Submerged plants Plants such as Canadian pondweed (*Elodea canadensis*) (Fig. 5) are rooted in the substrate and can only grow where the water is shallow enough or clear enough to allow plenty of light to penetrate. The many leaves of *E. canadensis* and the thin, branching leaves of hornwort (*Ceratophyllum*, Fig. 6) and the river crowfoot (*Ranunculus fluitans*) present a large surface area to the water. This probably helps to speed up gaseous exchange between the plant and the water. In *Ceratophyllum*, the cuticle is very thin and the epidermal cells, unlike most land plants, contain chloroplasts.

Inside the leaves and stems there are air spaces which keep the plant shoots or leaves floating as near to the surface as possible.

The thin leaves of plants such as river crowfoot also offer very little resistance to water flow. They stream out with the current and are not likely to be pulled off (Fig. 7).

Because water is so much denser than air, it buoys up and supports the plants submerged in it. Consequently the plants do not need as much strengthening tissue in their stems and leaves as do land plants. For this reason aquatic plants removed from their environment are limp and floppy.

Fig. 7 ***Ranunculus fluitans*; river crowfoot.** The thin flexible leaves stream out in the water current.

Plants with leaves floating at the surface The leaves of plants such as the water-lily (*Nymphaea*) receive direct sunlight. Their stomata are on the upper surface and so exchange gases directly with the atmosphere rather than with the water. The upper surface of the leaf has a waxy cuticle which allows wave splashes and rainwater to run off.

There is very little oxygen in the mud at the bottom of lakes and rivers and it is not always clear how plant roots can respire in these conditions. In some cases, such as in the water-lily, there are air spaces running from leaves to roots, which would allow diffusion of oxygen (Fig. 8).

The flowers of these plants and of those in the next group are brought above the surface and pollinated probably by insects or the wind.

Growing on the underwater parts of all these plants there is a community, called the **periphyton**, of small organisms such as single-celled creatures, filamentous algae (p. 329) and blue-green algae. These are grazed by fish, tadpoles, pond snails and insect larvae.

Submerged and aerial leaves In plants such as water crowfoot (*Ranunculus aquatilis*) (Fig. 9) and arrow head (*Sagittaria sagittifolia*) the submerged leaves show the

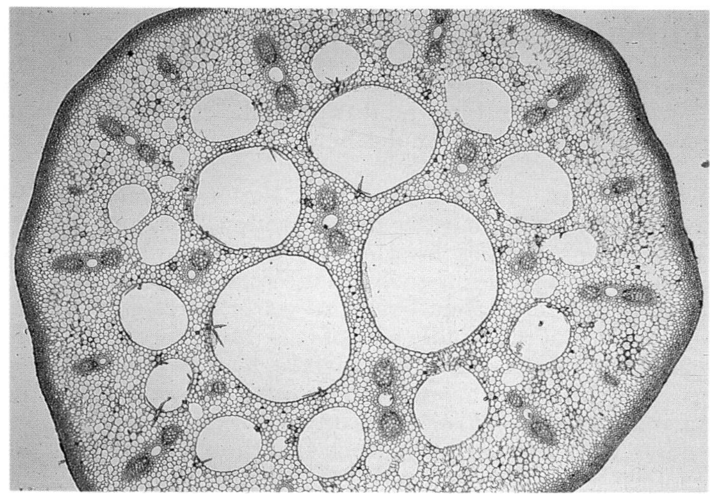

Fig. 8 ***Nymphaea* (water lily) leaf stalk, transverse section** (× 12). The large air spaces may allow diffusion of carbon dioxide and oxygen between the leaves and roots.

adaptations described for an aquatic environment, while the aerial leaves are characteristic of land plants.

Plants growing at the edge If the banks are shelving, there is usually a gradual change-over from water plants to land plants (Fig. 11).

Fig. 9 ***Ranunculus aquatilis*; water crowfoot.** Notice the difference between the aerial and submerged leaves.

The roots and lower stems of plants such as the reedmace (*Typha latifolia*) or reed (*Phragmites communis*) are totally submerged in shallow water. The abundant air spaces in the lower stem permit oxygen to diffuse into the roots. The types of plants represented in the change from aquatic to terrestrial vegetation probably reflect the extent to which they can tolerate having their roots submerged and deprived of oxygen.

Succession and colonization

Succession The stems and submerged leaves of aquatic plants interfere with the free flow of water and cause it to deposit part of the sediment it is carrying. This gradually makes the water shallower and so favours the growth of plants which thrive better in these new conditions. The wholly submerged plants are gradually replaced by partly submerged plants and then by the waterside plants. So the natural tendency for any lake, or area of slowly moving water, is to fill with sediment and change over to a land ecosystem. The process by which one community is gradually replaced by another is called succession (Fig. 11 and p. 241).

The progression from a shallow lake to a terrestrial ecosystem will depend on the depth of water, rate of flow, pH and other conditions. In non-acid conditions, with a slow water flow, the shallow edges of the lake are colonized by reedmace and reeds, forming a **reed-swamp**. As the water becomes shallower, the reeds and reedmace colonize areas nearer the centre of the lake and the drier margins are taken over by sedges (e.g. *Carex paniculata*), and flowering plants such as the hairy willow herb (*Epilobium hirsutum*) and purple loosestrife (*Lythrum salicaria*), forming a narrow **marsh zone**.

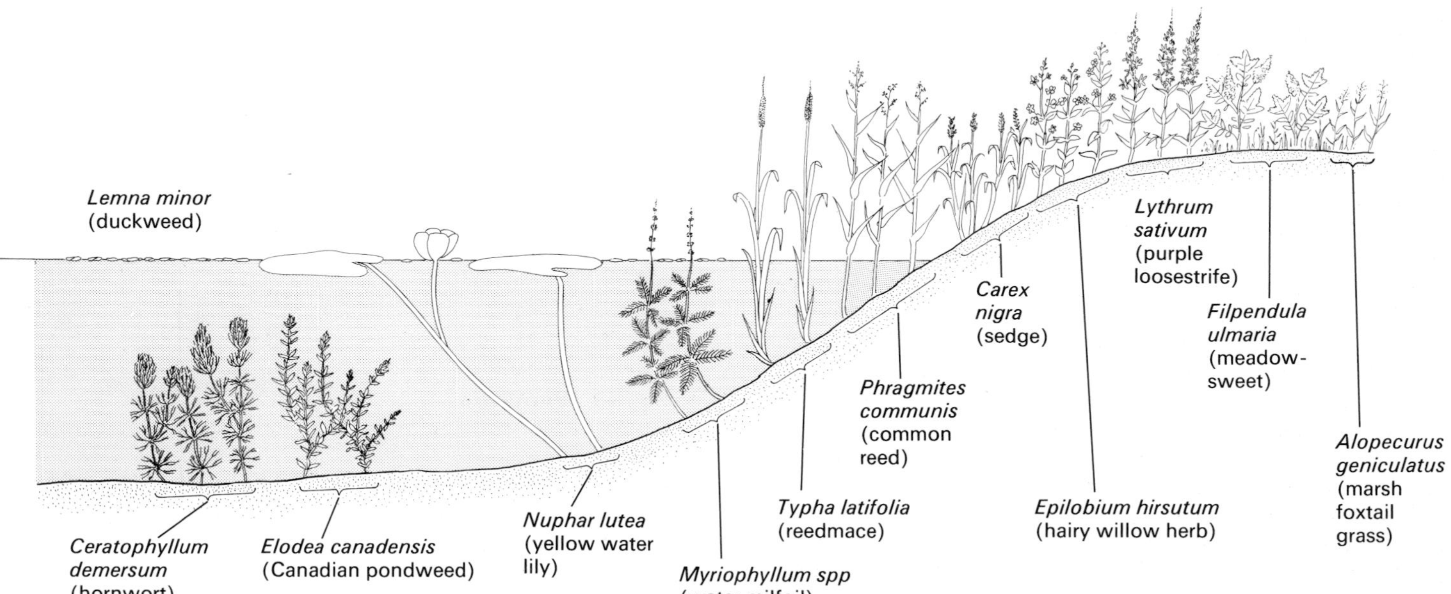

Fig. 10 **Transition from water to land.** This is just one example; there are many other sequences and combinations of plants.

Fig. 11 Zonation of vegetation at a lake side. The succession does not correspond exactly to Fig. 10 but the transition from submerged plants (Nuphar) to swamp carr (Rhamnus) can be seen.

Seedlings of trees such as the sallow (*Salix cinerea*), alder (*Alnus glutinosa*) and buckthorn (*Rhamnus catharticus*) become established, making a swampy scrub called a **swamp-carr**. The trees which colonize a swamp-carr are those which can grow in a waterlogged or nearly waterlogged soil.

Eventually, as the soil level continues to rise, oak (*Quercus robur*) and birch (*Betula pubescens*) may take over to produce a **woodland carr**.

The final, stable stage of a plant succession is called the **climax vegetation** for that region. Grassland left to itself (i.e. not cut or grazed) gradually becomes colonized by scrub, mainly hawthorn bushes, and progresses to woodland. Beech woods are the climax on clay soil and birch-pine woods on dry sandy soils.

Progression towards a climax vegetation can be deflected by human activities such as grazing sheep on grassy downs (Fig. 6, p. 245) or coppicing a wood (i.e. cutting down chestnut and hazel trees to the stumps every 10 years or so).

Colonization If an area of soil is dug over and left bare, it will soon be colonized by plants and animals. Some organisms will already be present, e.g. bacteria, fungi, mites, springtails, earthworms and the seeds of flowering plants. Other seeds will arrive as a result of dispersal (p. 89), particularly wind-borne seeds of plants such as groundsel, dandelion and thistle.

In favourable conditions of warmth and moisture, the seeds will germinate and the ground will become colonized, first by annual plants which grow quickly and produce seeds in a single season. Later, the perennials such as couch grass, nettles, willow herb and brambles take over. These plants have perennating and storage organs (p. 100), so that they gradually spread over the ground, filling the spaces previously occupied by the annuals.

Seeds of trees will arrive at the same time but grow more slowly. The wind-dispersed ash and sycamore may be the first to arrive. The tree seedlings grow taller than the herbaceous vegetation, the stems and branches persist throughout the winter, and the roots penetrate deep into the soil. Thus, the tree seedlings compete successfully for light, water and minerals, and eventually dominate the area, giving rise to a climax vegetation as described above.

The animal community

There are a great many different species of animals living in freshwater habitats and it is possible to mention only a few representative examples here.

Surface film Pond skaters (Fig. 12*a*) are insects which move about on the surface film of static or sluggish water. The water-repelling bristles on the tips of their legs prevent them breaking through the surface film and so the pond skater can glide over the water surface by brisk movements of the middle pair of legs. Most species feed on insects and crustacea at the surface, by sucking the fluids from their bodies through the piercing mouth parts. Tiny springtails (Fig. 12*b*) also live on the surface film. They are simple, wingless insects and are thought to feed on bacteria and algae in the surface film.

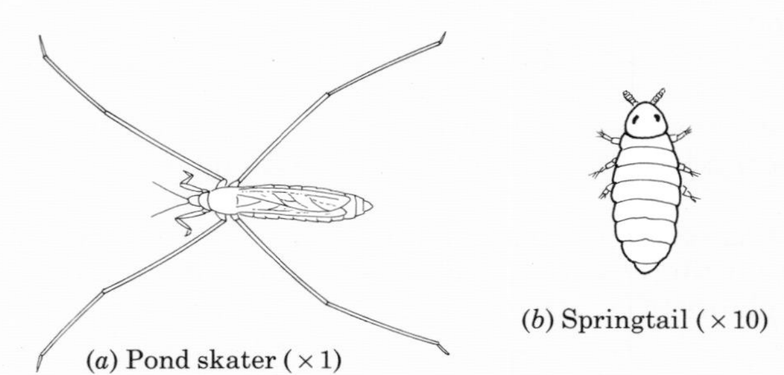

(*a*) Pond skater (× 1)

(*b*) Springtail (× 10)

Fig. 12 Animals of the surface film

Plankton In the top few centimetres of water, there is usually a dense population of zooplankton (p. 243). This includes protozoans such as *Paramecium* (p. 327) and small crustaceans such as water fleas and *Cyclops* (Figs 13*a* and *b*). *Paramecium* feeds on the microscopic plants and bacteria in the phytoplankton. The water fleas have a filter-feeding mechanism. They draw a current of water between the two halves of their body covering and filter out the edible plankton.

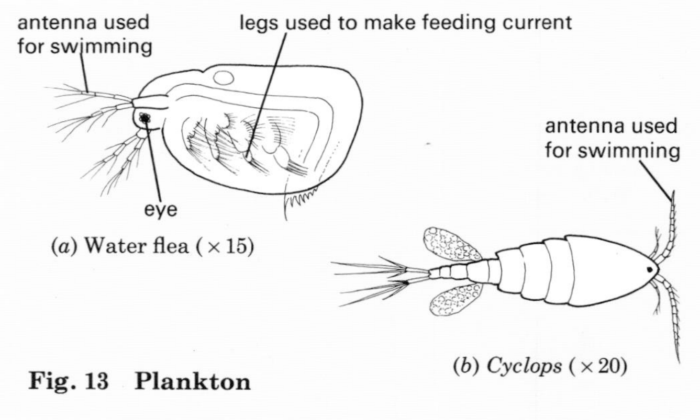

(*a*) Water flea (× 15)

(*b*) *Cyclops* (× 20)

Fig. 13 Plankton

Surface feeders Mosquito larvae hang from the surface film and filter out plankton by the flicking movements of their 'mouth brushes' (Fig. 14*a*). Some species of flatworm may be found gliding along under the surface film by means of their cilia. They are probably feeding on insects and crustacea in the surface film. Flatworms and pond snails are also to be found on the underside of floating leaves. The snails feed by rasping off the algae covering the leaf, with their 'tongues'. The phantom midge larva (*Chaoborus*, Fig. 14*b*) floats at the surface where it captures its prey.

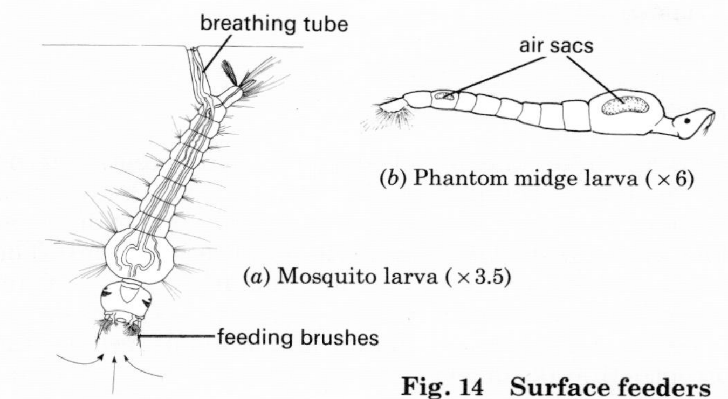

Fig. 14 Surface feeders

Bottom-living animals Dragonflies and mayflies have larvae (Fig. 15*a*) which live in fresh water. When the larvae reach full size, they emerge from the water and change into the adult form. There are many different types of mayfly and their larvae may be adapted to burrowing in mud, clinging to stones in fast streams, or swimming about fairly freely in sluggish waters. Caddis fly larvae (Fig. 15*b*) also live in the bottom waters of well-aerated streams and lakes. They make themselves a tubular casing of pebbles, sand or vegetation and some species spin a silk net for catching their food. Also living on the substrate or in the vegetation is the water louse, *Asellus* (Fig. 15*c*). This appears to be a scavenger in its feeding. Another crustacean, the freshwater shrimp, *Gammarus* (Fig. 15*d*) prefers well-oxygenated water.

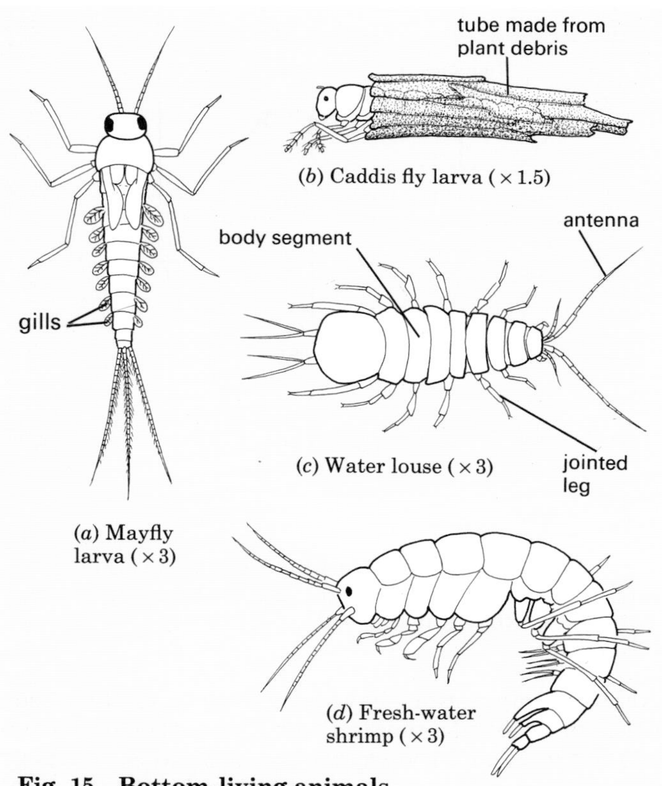

Fig. 15 Bottom-living animals

Freshwater mussels (*Anodonta cygnea*) and pea shells (*Pisidium amnicum*) are molluscs (p. 334) which live in the mud at the bottom of lakes and rivers. Inside their two shells are net-like gills, covered with cilia. The beating of the cilia draws water into the shells and the gills filter out small organisms from the water. These organisms are then trapped in sticky mucus and swallowed by the clam.

Free-swimming animals Water beetles (Fig. 16*a*) and water boatmen move about freely in ponds and lakes though they do cling to water weed to stop themselves floating to the surface when they stop swimming. There are many different species of water beetle which may be either carnivorous or plant eaters. There are two common families of water boatmen, *Notonecta* (Fig. 16*b*), which swims on its back, and *Corixa* (Fig. 16*c*), which swims the right way up. They have short, tubular mouth parts which they use to suck up particles of plant and animal debris.

Fish such as the minnow, stickleback and roach may be found swimming in any part of ponds, lakes and rivers through they have preferred areas for obtaining their food.

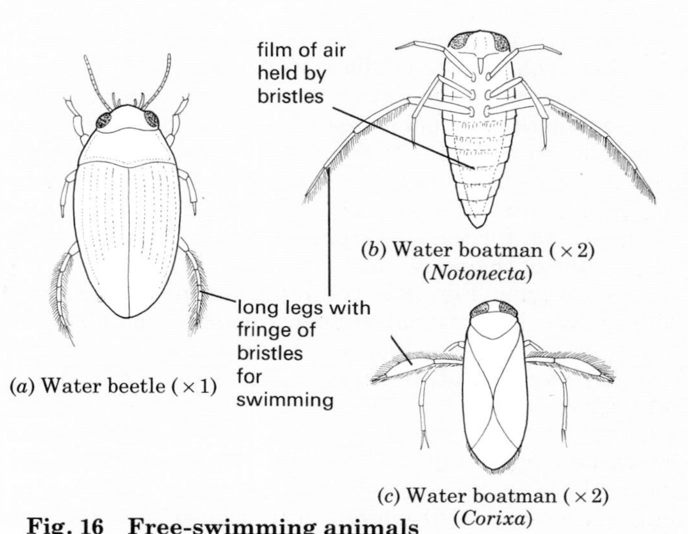

Fig. 16 Free-swimming animals

Adaptations of aquatic animals

Adaptations for movement Rapid movement through a dense medium like water requires a streamlined shape. This is seen in most fish and in some mayfly larvae. The latter fold their legs and gills flat against their bodies and swim with rapid wriggling movements.

Water beetles and water boatmen have legs which are adapted for propulsion through water. The last pair of legs is particularly long, slightly flattened and fringed with bristles. The long legs enable the insect to 'row' through the water, the bristles offering maximum resistance during the driving stroke and very little in the recovery stroke.

Some of the aquatic animals have adaptations to make them buoyant. Thus they do not sink to the bottom when they stop swimming. Fish have a swim bladder (p. 340); the transparent phantom midge larvae (*Chaoborus*, Fig. 14*b*) which floats, almost invisible, near the surface has two pairs of air sacs. The abundant bristles on the antennae of the water flea offer a lot of water resistance and so help it to swim, but they also slow down its rate of sinking.

In swiftly moving streams and rivers, some animals may have adaptations which prevent them being swept away in the current. *Gammarus* and some mayfly larvae are flattened and so offer little resistance to the water flow. Leeches (Fig. 17) have suckers which attach firmly to their host animal but also prevent them being carried off in water currents.

Fig. 17 Leech (×1)

Adaptations for breathing There are two options open to freshwater animals; either they extract the dissolved air from the water or they go to the surface and breathe air directly from the atmosphere. Those which use the dissolved oxygen usually have some form of gills which present a large surface to the water. They also have some method of changing the water in contact with the gills. The breathing method of fish is described on p. 340. The gills of the mayfly larvae (Fig. 15*a*) have air tubes running into them. A rhythmic flicking movement of the gills keeps fresh supplies of water moving past them. The caddis larva has gills and forces a stream of water over them by undulations of the body inside the tube.

Tubifex worms (Fig. 18*a*) live in the mud at the bottom of ponds where the oxygen concentration is usually low. They build tubes in the mud and their bodies project from them and wave about, so renewing the supply of water in contact with them. The lower the oxygen concentration, the more of their body protrudes from the tube. Their blood contains haemoglobin which helps to absorb what little oxygen there is. Some species of the midge *Chironomus* have larvae (Fig. 18*b*) which, unlike most insects, have haemoglobin in their blood. These, too, live in tubes in poorly oxygenated mud.

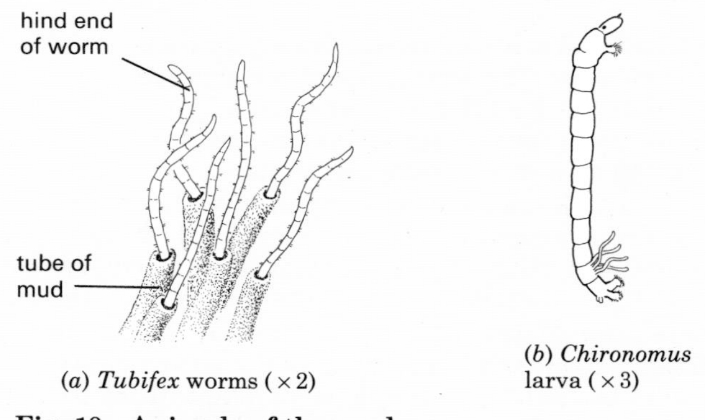

(*a*) *Tubifex* worms (×2)

(*b*) *Chironomus* larva (×3)

Fig. 18 Animals of the mud

Mosquito larvae (Fig. 14*a*) and pupae hang by a tube ('siphon') from the surface film (p. 283). This siphon allows a gaseous exchange to take place between the air in the breathing system and the atmosphere. Water beetles carry a store of air under their wing cases and rise to the surface from time to time to replenish the supply. Water boatmen carry a film of air trapped in bristles covering the surface of their bodies. Pond snails have a simple 'lung' under their shells (Fig. 4, p. 335). They fill the lung with air at the water surface at intervals.

Interrelationships

Ponds and lakes may be self-sufficient ecosystems with no new material being brought into them. The oxygen produced by the photosynthesis of green plants in daylight replaces the oxygen used up by the plants and animals. The carbon dioxide produced by plant and animal respiration is used by plants during their daytime photosynthesis.

The dead remains of plants and animals fall to the bottom of the lake or river and decay as a result of bacterial action; the nitrates and phosphates so released into the water are used by the plants for their growth (p. 52).

The plants and animals of fresh water form a complex food web (p. 244). For example, protozoa eat single-celled plants; water fleas eat protozoa; fish and *Hydra* (p. 350) eat water fleas; big fish eat the smaller fish, and herons may eat some of the big fish (Fig. 19).

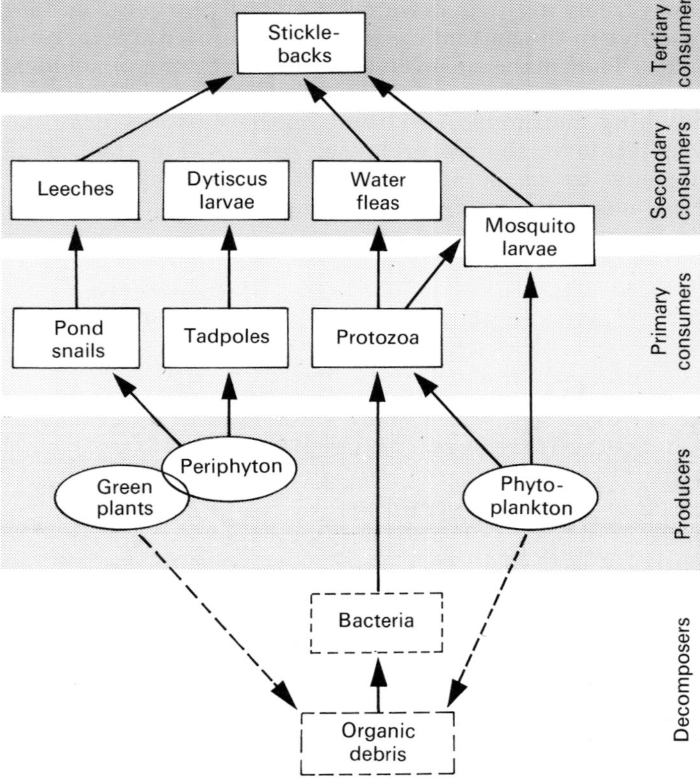

Fig. 19 Part of a fresh-water food web. Many other organisms could be added. Sticklebacks are themselves eaten by perch and pike. The animals also contribute to the organic debris.

Although suggestions have been made in the text about the food of the pond animals, there is a great deal still to be found out about what they eat. In many cases, it depends on the species. One species of water beetle may eat plants while another species may be carnivorous. It is not always easy to find out exactly what some of the small animals are eating. It is not clear, for example, whether pond snails eat the leaves of water plants or just scrape off the algae which are growing on the leaves. Some authors claim that water boatmen are carnivorous and pursue their prey but others say they only suck up plant and animal debris. Only by careful observation and by examining the gut contents is it possible to decide what is the main food source of an animal.

QUESTIONS

8 Give one example of an animal and one of a plant which is adapted in a way that prevents it being swept away in a flow of water. Say what the adaptive feature is.
9 In what two ways do aquatic animals obtain their oxygen? Give an example of each.
10 Give three examples of animals which exploit the surface tension of water and say how they use it.
11 What disadvantages might there be for plants and animals living at the bottom of a pond?

FIELD TECHNIQUES

In order to make a study of a pond or stream ecosystem, you will need to measure the physical factors which could influence it and to find out the variety and numbers of species that live there.

Physical factors

The factors most likely to affect a freshwater ecosystem are light, temperature, pH, flow rate, dissolved minerals and oxygen. Measurement of the last two requires either special equipment or rather elaborate methods.

Temperature The simplest way to find the temperature is to place a mercury thermometer in the water and wait for the reading to become steady. However, the temperature in the surface layers is likely to be higher than that at the bottom. In this case, the thermometer bulb can be coated with warm wax and lowered to the bottom for about 10 minutes. It can then be pulled up and the temperature read. The layer of wax slows down the exchange of heat so that the bottom temperature can be read before the mercury starts to rise again.

Light To compare the depths to which light penetrates, a white disc (Fig. 20) is lowered into the water until it just disappears from view. By noting the length of string let out, the depth of water which is needed to obscure the disc can be measured. The clearer the water, the greater the depth to which the disc can be lowered before it disappears.

This method does not give a value for the light intensity but is useful for comparing the light penetration at different times of the year or in different situations.

pH Some pond or stream water is placed in a test-tube and a few drops of Universal Indicator added. The colour is then compared with the colour chart on the label of the bottle or with a special colour chart supplied with the indicator (Fig. 15, p. 263).

Flow rate This can be estimated by timing a floating object over a measured distance. To avoid distortions caused by the wind, it is best to use a partly submerged object such as a plastic bottle, two thirds filled with water.

You will need to bear in mind that the flow rate near the bank is likely to be slower than in the middle. There will also be variations in rate where the stream narrows or becomes deeper.

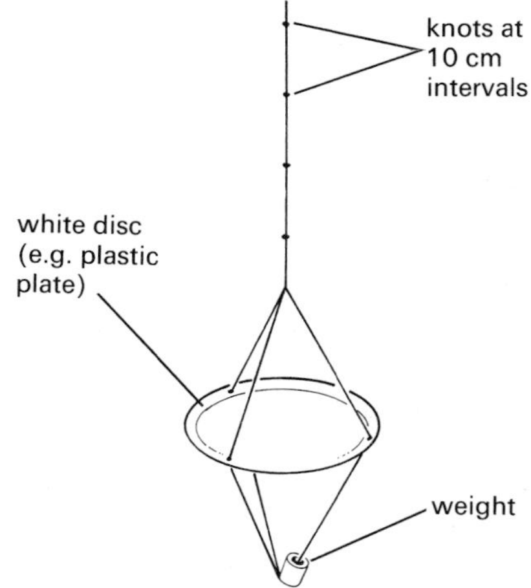

Fig. 20 Absorption of light by water

Sampling techniques

The object of these techniques is to find out what plants and animals are present in the ecosystem and, if possible, their numbers. It is rarely possible to count all the organisms in a pond or stream and so a sample has to be taken and the total population calculated. The figures obtained are likely to be inaccurate but, provided the sampling methods are consistent, can be used to compare the populations at different sites or seasons.

Sweep net A fixed number of sweeps is made with a strong net such as that in Fig. 21*a*, through an area of water. The net is then inverted into a tray of water and the

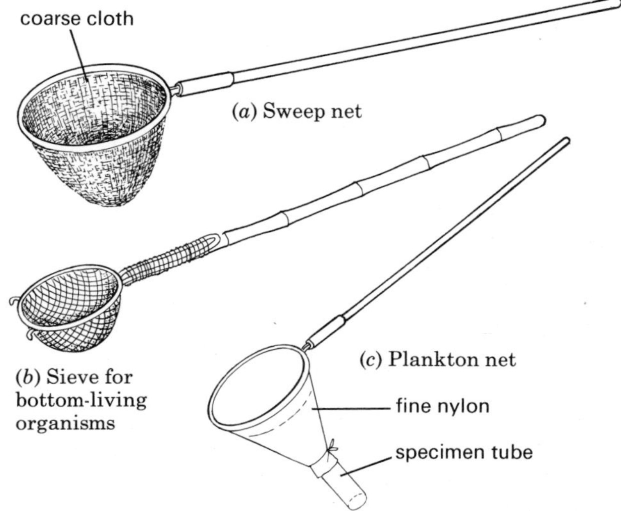

Fig. 21 Nets for collecting fresh-water organisms

organisms can be identified and recorded or transferred to a carrying jar. The same number of sweeps should be made in a different area if a comparison is to be made.

Plankton net This is similar to a sweep net but the very fine nylon mesh traps small crustacea and some algae and protista from the plankton. The net must be moved slowly through the surface water at about 50 cm per second, to allow the water to pass through the fine mesh. Small organisms are collected in the specimen tube at the end (Fig. 21*c*). From the diameter of the net opening and the length and number of sweeps, the volume of water sampled can be calculated.

For example, if the diameter of the net is 15 cm and two sweeps of 1 metre long are made, the volume of water passing through the net will be

$$\pi \times \left(\frac{15}{2}\right)^2 \times 2 \times 100 \text{ cm}^3 = 35343 \text{ cm}^3$$

or approximately 35 dm^3.

The organisms in the tube may be too numerous to count, so smaller samples may have to be taken. It would be sensible to take several samples, e.g. 5 samples of 1 cm^3. This will reduce the sampling error. Suppose you counted 1, 6, 3, 5, 5 water fleas in the successive samples, the average count is 4 water fleas per cm^3. If the tube from the plankton net holds 50 cm^3 water, there is a population of $4 \times 50 = 200$ water fleas in the tube.

This number was collected from 35 dm^3 of water, so the population density of water fleas in that part of the pond is

$$\frac{200}{35} = 5.7 \text{ per dm}^3.$$

This figure, on its own, does not have much meaning but it can be used to compare variations in population density at different depths, in different habitats or at different times of the year.

An alternative method, is to collect a known volume of water in a large container, pour it through the plankton net and count the organisms in the tube.

Search technique Organisms living under stones or crawling in the vegetation are unlikely to be captured in a sweep net. The stones have to be picked up to identify and count e.g. caddis fly larvae or stone fly larvae. Water-lily leaves are turned over to observe and count pond snails or flatworms crawling on the under-surface. In all cases, any stones, twigs or leaves must be restored to their original position.

If different habitats are to be compared, it is necessary to search over a fixed time, e.g. 10 minutes, so that comparisons of numbers are valid. Alternatively, the same number of stones or leaves can be examined in each survey.

Sampling the mud A kitchen strainer, tied to a stick (Fig. 21*b*), can be used to scoop up a mud sample. The mud is washed through the wire mesh, leaving organisms such as *Tubifex* and *Chironomus* to be identified and counted. The volume of the strainer can be found by lining it with a plastic bag and pouring in water from a measuring cylinder.

Sampling the plant community A sample of the submerged plants can be obtained by means of a grapnel (Fig. 22). This is thrown into the pond or stream and pulled back on a line. The bent wires will pick up samples of the submerged plants.

This method will not give accurate information about the population density but some idea of relative abundance can be obtained.

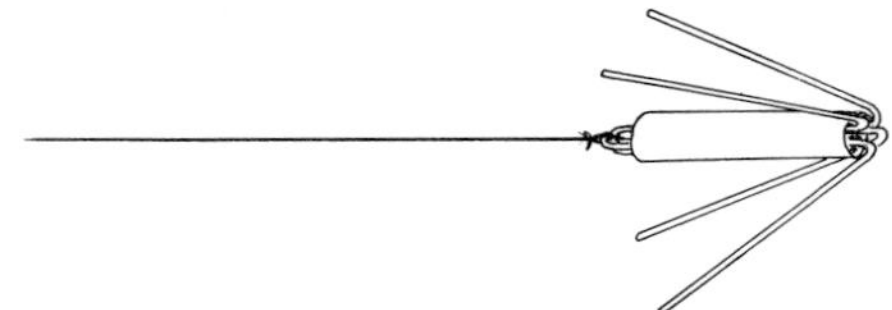

Fig. 22 A grapnel. When pulled through the water, the hooks collect pieces of plant.

The species and relative numbers of plants at the edge of the water can often be found simply by careful observation and recording. If an objective method is needed, a line transect can be used.

A length of plastic clothes-line has ink marks at 5 or 10 cm intervals, or a piece of rope or string is knotted at similar points. The line is laid across the zone to be studied, e.g. from shallow water, up the bank to a meadow (Fig. 23). Any plant whose position corresponds to one of the marks, is identified and recorded. The depth of water or height above the water should be measured and recorded for each

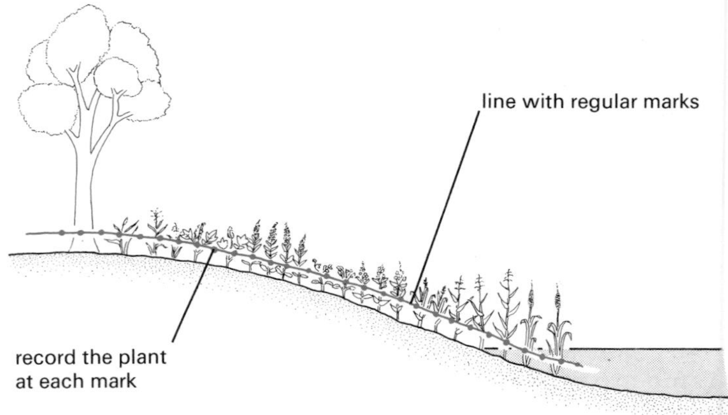

Fig. 23 A line transect

Height above or below water level (cm)

60
40
0
20
40

Filipendula
Lythrum
Epilobium
Carex
Phragmites
Alopecurus
Typha
Nuphar
Elodea
Ceratophyllum

0 1 2 3 4 5 6 7 8
Metres

Fig. 24 Result of a series of line transects at the edge of a pond

plant. Any plant which occurs between two marks is ignored.

It is desirable to take several transects fairly close together in order to get a representative picture of the vegetation. The positions should be selected systematically, e.g. at 1 metre intervals. They should not be chosen deliberately to include particularly interesting (or empty) parts of the bank.

The results can be recorded as a chart, similar to Fig. 24.

QUESTIONS

12 What are the possible sources of error in estimating population densities from sweep net samples?

13 In the calculation of numbers of water fleas, given on p. 290, how would the results have been affected if you had taken only the first sample of 1 cm^3 from the tube?

14 If a plant community is sampled by only a single line transect, in what ways might this give an inaccurate picture of the vegetation?

CHECK LIST

- **A habitat is where an organism lives, feeds and breeds.**
- **A community is all the organisms living in a habitat.**
- **An ecosystem is a self-supporting community of organisms plus the physical features of their environment.**
- **A population is the number of a given species in a defined habitat.**
- **There is competition within and between species for food, light, space and mates.**
- **A niche is the position occupied by a particular species in an ecosystem.**
- **A freshwater environment is affected by light, temperature, oxygen, pH and minerals.**
- **A plant community in fresh water consists of floating plants, submerged plants, rooted plants with floating leaves and rooted plants with leaves and flowers above water.**
- **The numerous, thin or finely divided leaves of submerged plants help speed up gaseous exchange.**
- **Most ecosystems, if left alone, gradually change to a climax vegetation of woodland.**
- **Freshwater animals may live on or just under the surface film, in the mud at the bottom, on the leaves of aquatic plants, on or under stones or swimming freely in the water.**
- **The activities of decomposers (mainly bacteria) in the mud, may lower the oxygen concentration.**
- **Aquatic animals obtain oxygen either from the air at the surface or from dissolved oxygen in the water.**
- **Many aquatic animals have adaptations which make them buoyant or enable them to move effectively in water.**
- **It is fairly easy to compare temperature, light penetration, and pH in different aquatic environments.**
- **Populations of animals and plants may be estimated by sampling techniques. It is important to ensure that sufficient samples are taken and that the samples are not biased.**

31 Populations

POPULATION GROWTH AND FLUCTUATION

Population dynamics, pest control.

HUMAN POPULATION

Birth and death rates, world population, population patterns.

POPULATION CHANGES

On p. 281 it was explained that a biological population is defined as the total number of individuals of any one species in a particular habitat. Such a population will not necessarily be evenly spread throughout the habitat, nor will its numbers remain steady. The population will also be made up of a wide variety of individuals: adults (male and female), juveniles, larvae, eggs or seeds, for example. In studying populations, these variables often have to be simplified.

Population growth

In the simplest case, where a single species is allowed to grow in laboratory conditions, the population develops more or less as shown in Fig. 1.

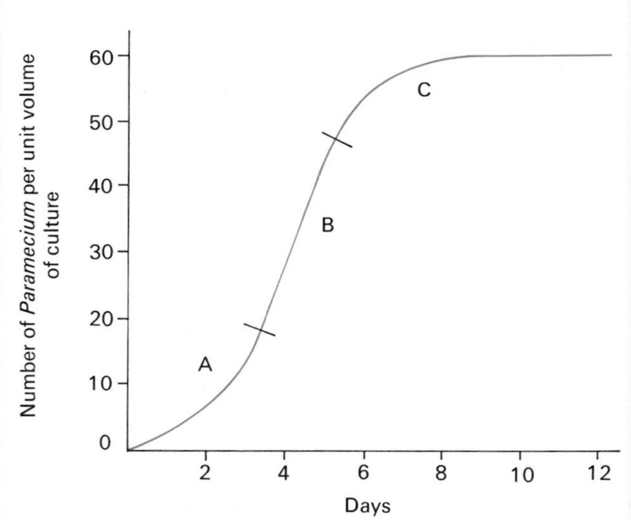

Fig. 1 The sigmoid curve (*Paramecium caudatum*). This is the characteristic growth pattern of a population when food is abundant at first and there are no other factors limiting growth and reproduction.

(From Lewis and Taylor, *Introduction to Experimental Ecology*, Academic Press, 1966.)

The population might be of yeast cells (p. 324) growing in a sugar solution, flour beetles in wholemeal flour or weevils in a grain store. The experiment illustrated by Fig. 1 uses a single-celled organism called *Paramecium* (see p. 327), which reproduces by dividing into two (binary fission). The **sigmoid** (S-shaped) form of the graph can be explained as follows:

A The population increase is exponential or logarithmic, i.e. it does not increase 2–4–6–8, etc. but 2–4–8–16–32, etc. One *Paramecium* divides into two, the two offspring each divide, producing 4, the 4 divide to give 8 cells, and so on. In other words, the population doubles at each generation. In 10 generations it would reach 1024. When a population of 4 organisms doubles, it is not likely to strain the resources of the habitat, but when a population of 1024 doubles, there is likely to be considerable competition for food and space.

B The population continues to grow but at a steady rate. This may be because the food resources are limiting the rate of growth and reproduction: or the effect of crowding may itself reduce the reproduction rate. Also, some of the mature organisms may be dying.

C At this point the population ceases to grow. The **reproduction rate** equals the **mortality rate** (death rate). The number of offspring produced will still be greater than the number of adults which die, but fewer of these offspring will live long enough to reproduce themselves.

After this stage, the population may start to decline. This can happen because the food supply is insufficient, waste products contaminate the habitat or disease spreads through the dense population.

QUESTIONS

1 In Fig. 1, how many days does it take for the mortality rate to equal the replacement rate?

2 From the graph in Fig. 1, what is the approximate increase in the population of *Paramecium* (*a*) between day 0 and day 2, (*b*) between day 2 and day 4, and (*c*) between day 8 and day 10?

3 In section B of the graph in Fig. 1, what is the approximate reproduction rate of *Paramecium* (i.e. the number of new individuals per day)?
4 In 1937, 2 male and 6 female pheasants were introduced to an island off the N.W. coast of America. There were no other pheasants and no natural predators. The population for the next 6 years increased as follows:

1937 – 24	1940 – 563
1938 – 65	1941 – 1122
1939 – 253	1942 – 1611

Plot a graph of these figures and say whether it corresponds to any part of the sigmoid curve.

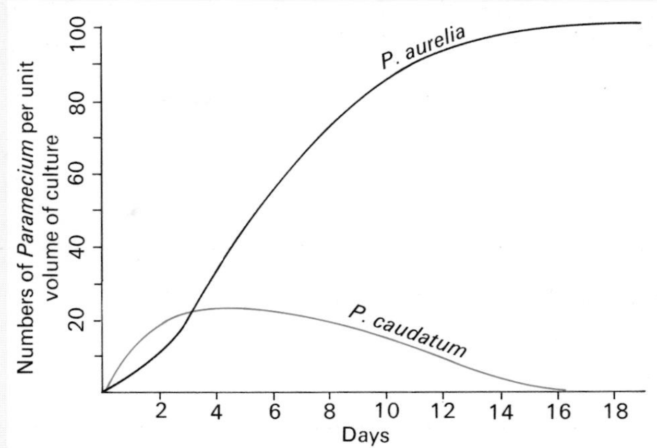

Fig. 2 The effect of competition. *Paramecium aurelia* and *P. caudatum* eat the same food but *P. aurelia* can capture and ingest it faster than *P. caudatum*.
(From Lewis and Taylor, *Introduction to Experimental Ecology*, Academic Press, 1966.)

Population fluctuations

The sigmoid curve is a very simplified model of population growth. Few organisms occupy a habitat on their own, and the conditions in a natural habitat will be changing all the time. The steady state of the population in part C of the sigmoid curve is rarely reached in nature. In fact, the population is unlikely to reach its maximum theoretical level because of the many factors limiting its growth.

If, in the laboratory, two species of *Paramecium* (*P. aurelia* and *P. caudatum*) are placed in an aquarium tank, the population growth of *P. aurelia* follows the sigmoid curve but the population of *P. caudatum* soon declines to zero because *P. aurelia* takes up food more rapidly than *P. caudatum* (Fig. 2).

This example of competition for food is only one of many factors in a natural environment which will limit a population or cause it to change.

Plant populations will be affected by **abiotic** (non-biological) factors such as rainfall, temperature and light intensity. The population of small annual plants may be greatly reduced by a period of drought; a severe winter can affect the numbers of more hardy perennial plants. **Biotic** (biological) factors affecting plants include their leaves being eaten by browsing and grazing animals or by caterpillars and other insects, and the spread of fungus diseases.

Animal populations, too, will be limited by abiotic factors such as seasonal changes (Fig. 3). A cold

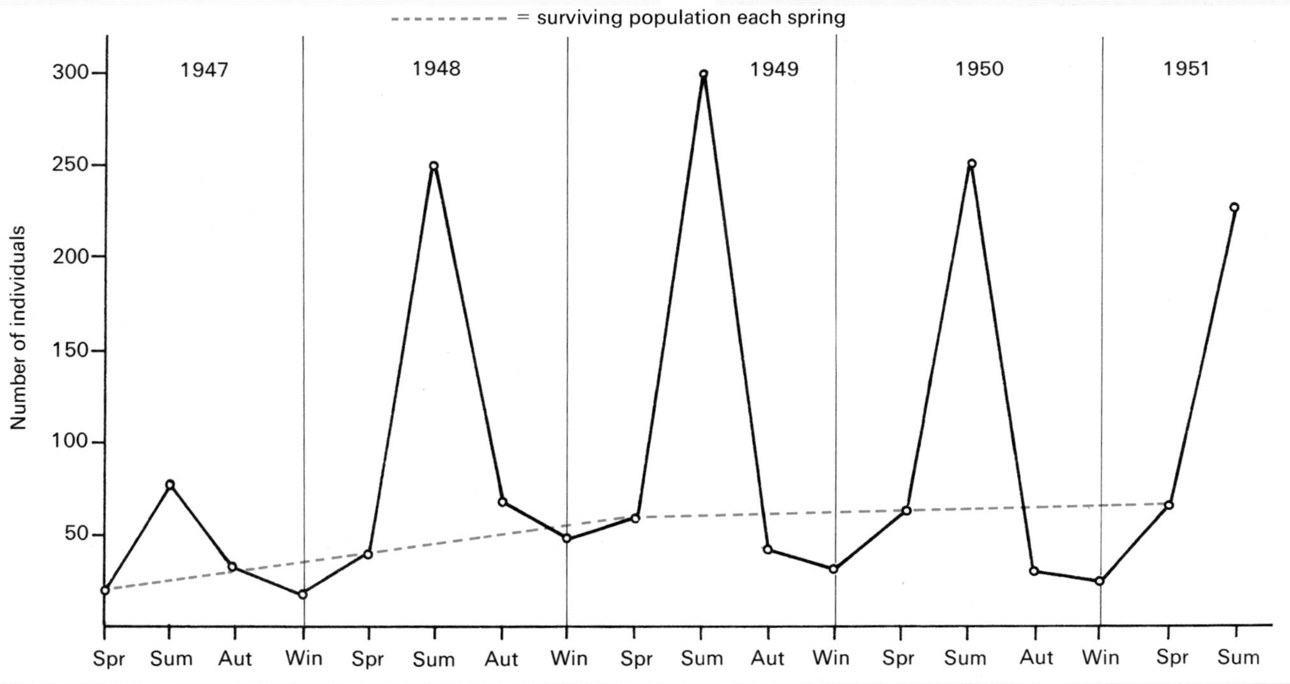

Fig. 3 Seasonal changes in the population of great tits in a wood. Since each pair rears about 10 young, there is a surge in numbers each summer but the average number of adults increases steadily from about 10 to 30 breeding pairs. The winter of 1947 was severe.

(From David Lack, *Natural Regulation of Animal Numbers*, Oxford, 1954.)

winter can severely reduce the populations of small birds. However, animal populations are also greatly affected by biotic factors such as the availability of food, competition for nest sites (Fig. 4), predation (i.e being eaten by other animals), parasitism and diseases.

Fig. 4 Colony of nesting gannets. Availability of suitable nest sites is one of the factors which limits the population.

Figure 5 shows the effect of predation in a laboratory experiment using *P. aurelia* and yeast (*Paramecium* eats yeast cells). It can be seen that as the population of *P. aurelia* increases, the yeast population declines. When the population of *P. aurelia* is low, the yeast population recovers.

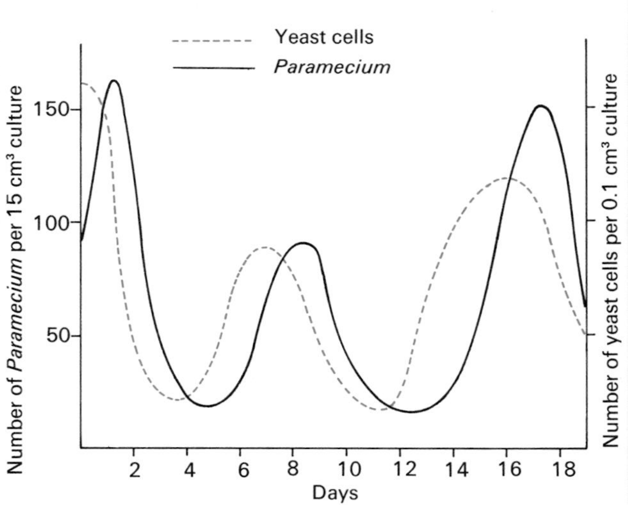

Fig. 5 Effects of predation. Yeast cells multiply rapidly in a culture solution containing sugar, but *Paramecium* eats the yeast cells and multiplies at their expense.

(From Lewis and Taylor, *Introduction to Experimental Ecology*, Academic Press, 1966.)

In this simple experimental situation, it is reasonable to deduce that when the population of *P. aurelia* is high, yeast cells are eaten faster than they can reproduce. When the yeast population falls, there is insufficient food to maintain the reproduction rate of *P. aurelia*.

In a natural environment, it is rarely possible to say whether the fluctuations observed in a population are mainly due to one particular factor because there are so many factors at work. In some cases, however, key factors can be identified as mainly responsible for limiting the population.

QUESTIONS

5 In Fig. 2, which part of the curve approximately represents the exponential growth of the *P. aurelia* population? Give the answer in days.

6 What forms of competition might limit the population of sticklebacks in a pond?

7 From Fig. 3 it seems that about 175 new individuals were added to the population of great tits each summer but, by next spring, the numbers were back to about 50. What might have happened to the other 125 birds?

8 From Fig. 5 say (*a*) approximately how long it takes for the *Paramecium* population to recover from its low levels, (*b*) what is the approximate time lag between the maximum population of yeast and the maximum population of *Paramecium*? Suggest reasons for this lag.

9 Suggest (*a*) some abiotic factors, (*b*) some biotic factors that might prevent an increase in the population of sparrows in a farmyard.

Population dynamics

The study of population changes and their causes is called population dynamics. It is an important study for trying to understand interrelationships between living creatures and it has numerous applications. For example, trying to decide the maximum number of herrings that can be taken from the North Sea without persistently reducing the population. Ideally, one needs to catch fish that have just passed their maximum growth rate, and avoid taking fish that still have a growth potential. It is also important not to remove fish at a rate which exceeds the maximum reproduction rate.

The same principles apply to whaling and similar activities which exploit natural populations.

Population dynamics also helps in deciding how to control populations of plants and animals which are regarded as pests. For example, population dynamics can help to discover the density of aphids which makes it necessary to spray an entire crop of wheat. Another example is the control of rats or mice. There is little point in setting traps when the reproduction rate of these rodents greatly exceeds the rate of successful trapping. A better strategy is to cut down their reproduction rate by denying them access to food and breeding sites.

Pest control

Cultivation Ploughing the soil, planting crop plants at a high density and keeping down weeds is a widespread method of controlling plant populations.
Biological control In some situations, predators or parasites have been used to control populations of pest species. A moth, *Cactoblastis*, was used to control the spread of the prickly pear cactus in Australia. The moth's caterpillars ate the leaves of the prickly pear and reduced the population to manageable proportions.

A small wasp is sold to market gardeners to use against the greenhouse whitefly. The wasp lays its eggs in the immature whiteflies. When the eggs hatch they feed on the internal organs of the larvae and eventually kill them.

The myxomatosis virus was used to control the rabbit population in Australia and Britain (see p. 244).
Pesticides The main method of controlling pest populations is by chemicals called pesticides. These are usually very effective but they are often not selective. Thus they may kill a wide variety of insects and other organisms, including beneficial species, as well as the pests (see p. 266). Many of the pesticides are potentially very poisonous to humans.

QUESTIONS

10 Suggest why shooting is unlikely to be an effective way of controlling a population of wood pigeons.
11 In the summer, when trees are in full leaf, it is impossible to count the number of great tits in a wood. In the study illustrated in Fig. 3. the great tits used nesting boxes for breeding. How would this help to estimate the summer population?

HUMAN POPULATION

Birth and death rates

The birth rate is the number of live births per 1000 people in the population per year. For example, if there were 500 live births in a population of 25 000, the birth rate would be 500 ÷ 25 = 20.
The death rate is the number of deaths per 1000 people per year. For example, if 300 people died, out of a population of 25 000, the death rate would be 300 ÷ 25 = 12.

Obviously, if the birth rate is higher than the death rate, the population will grow. If the rates are the same, you might expect the population to be stable, but this would be true only if all the children survived to reproductive age.

Population growth

As a very simple model, imagine 5 couples, aged 20–30; each couple has 4 children. When these children grow up, they, too, pair off and have 4 children per pair. The following table shows how the population of this little community would grow in 60 years.

Table 1 Population growth

Year	*Number of individuals*	*Generation*	*Age*	*Total population*
1	10 (5 pairs) 20	parents children	20–30 0–10	30
20	10 20 40	parents children grand-children	40–50 20–30 0–10	70
40	10 20 40 80	parents children grand-children great grand-children	60–70 40–50 20–30 0–10	150
60	10 20 40 80 160	parents children grand-children great grand-children great, great grand-children	dead 60–70 40–50 20–30 0–10	300

In the next 3 generations, the number of deaths will increase from 10 to 20, 40 and 80 but the number of births will increase from 160 to 320, 640 and 1280. This is the exponential growth described on p. 292 and must eventually be limited by the availability of food.

If you try drawing up a similar table, assuming that each couple has only 2 children, you will see that the population stops increasing after the third generation because the birth rate and death rate are the same.

If the population growth is plotted on a graph, it would appear as in Fig. 6.

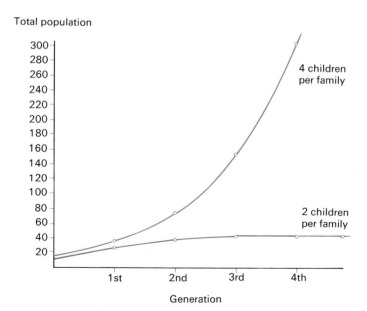

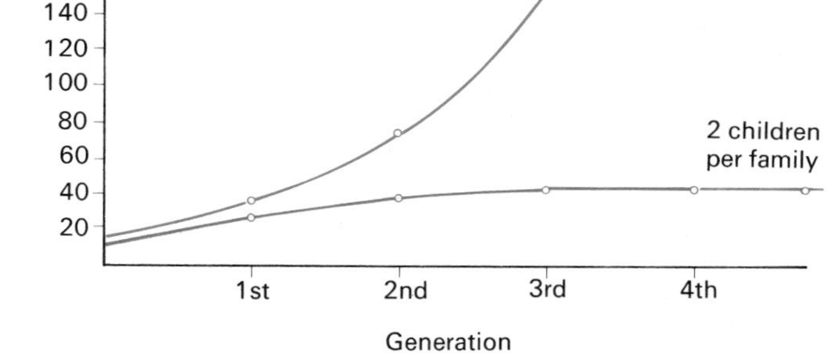

Fig. 6 Population growth; 2 or 4 children per family

Of course, these two models are greatly simplified. Children may die before reaching reproductive age; couples may be infertile or have more or less children than in the models.

To keep the population stable, the replacement rate needs to average out at just over 2 children per family.

Population patterns

Fig. 7*a* shows the population of England in 1841, divided into 5-year groups. The tapering pattern indicates that the bulk of the population was below 35 years old and a relatively small proportion of people survived into their seventies. About 130 years later (Fig. 7*b*), the pattern has changed. Improved standards of nutrition and health care have resulted in many more people surviving beyond the ages of 30–70.

The tapering pattern of Fig. 7*a* is still characteristic of some Third World countries where the infant mortality is fairly high and the average length of life is low. Half the population of Nigeria, for example, is below the age of 14.

In the industrialized countries, the birth rate is diminishing and the average length of life is increasing. You can probably see that this could lead eventually to a pyramid with a relatively narrow base and a bulge in the middle.

This pattern raises the problem of how a decreasing population of young people can support and care for an increasing population of the elderly.

World population

This topic was introduced on p. 180. The world population is estimated to be 4.7 billion, and 10 years ago it was increasing at the rate of 2 per cent per year. This may not sound very much, but it means that the population doubles every 35 years. This doubles the demand for food, space and other resources.

Recently, the rate of growth has slowed to 1.7 per cent and the population is expected to stabilize at about 10 billion by the end of the next century, i.e. the year 2100 (Fig. 8, and Fig. 20 on p. 180).

However, the growth rate is not the same in all parts of the world. Kenya's population, for example, is growing at the rate of 3.9 per cent, while western Europe's grows by only 0.1 per cent.

The effects of high growth rates are complicated by climate, resources, economics and politics, but it does seem

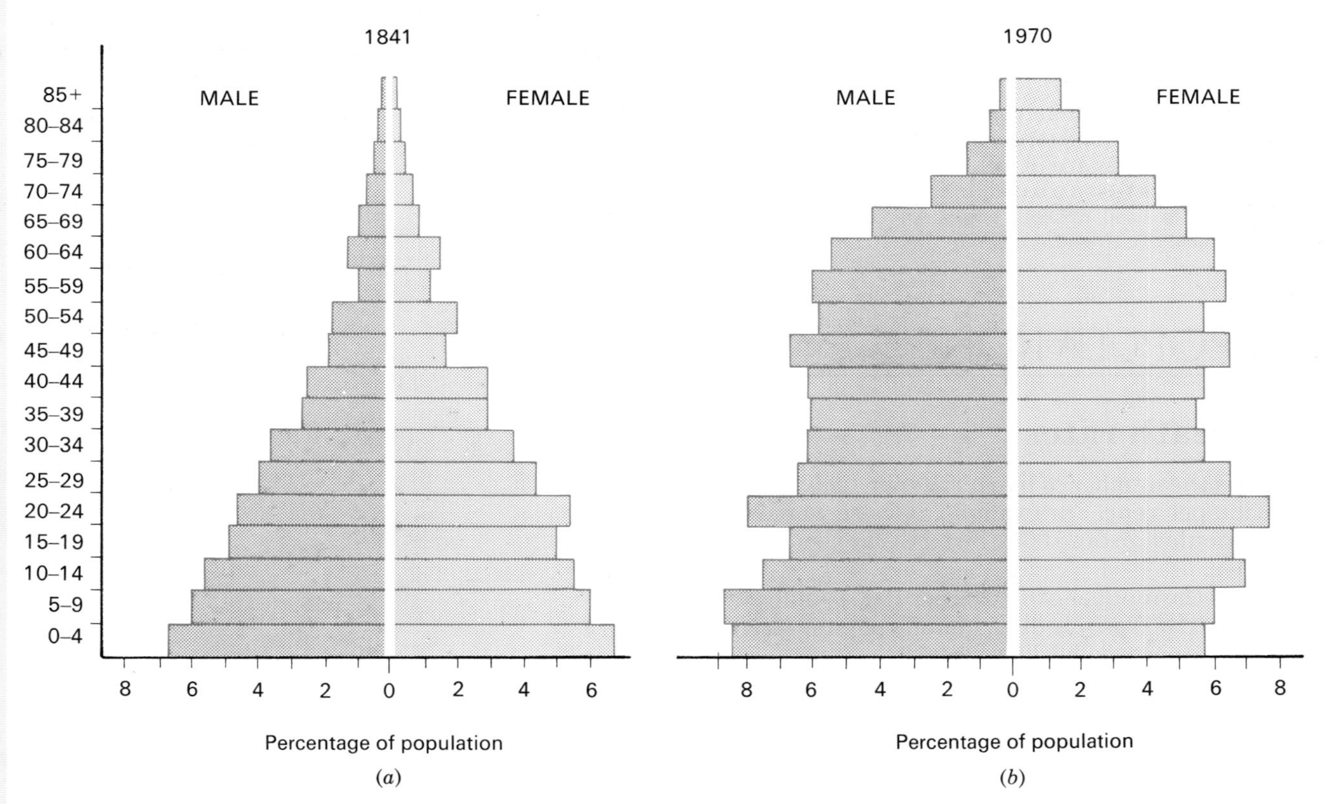

Fig. 7 Age pyramids for England in 1841 and 1970. The tapering pattern of (*a*) is characteristic of a population with a high birth rate and a low average life expectation. The almost rectangular pattern of (*b*) is characteristic of an industrial society with a steady birth rate and a life expectancy of about 70.

(From David Hay, *Human Populations*, Penguin Education, 1972.)

Fig. 8 Human population. It may theoretically be possible to feed 10 billion people but is this the population density at which you would choose to live?

rational to try to limit the rate of population increase so that, eventually, everyone can enjoy an adequate standard of living. This could be achieved, at least in theory, by widespread use of one or other of the family-planning methods outlined on pp. 179–80.

QUESTIONS

12 If there are 12 000 live births in a population of 400 000 in one year, what is the birth rate?
13 If you have not already done so, draw up a table similar to Table 1 but with each couple having only 2 children.
14 Try to explain why, on average, couples need to have just over 2 children if the population is to remain stable.
15 Study Fig. 7*b* and then comment on (*a*) the relative number of boy and girl babies, (*b*) the relative number of men and women of reproductive age (20–40), and (*c*) the relative numbers of the over-70s.

CHECK LIST

- **A population is the total number of individuals of a species in a habitat.**
- **The growth of plant populations is limited by competition for, e.g. light, water, minerals, rooting space and the abundance of herbivores.**
- **The growth of animal populations is limited by competition for food, water, breeding space and the abundance of predators.**
- **Abiotic (non-living) factors affecting population size include temperature, rainfall and soil conditions.**
- **Biotic (biological) factors affecting population size include predation, disease, and competition within and between species for food, etc.**
- **In the fishing industry, a detailed study of populations is desirable in order to achieve economic returns without depleting the breeding stock.**
- **In agriculture, we manipulate populations by cultivation and pest control.**
- **The world population is growing at the rate of 1.7 per cent each year. At this rate, the population more than doubles every 50 years.**
- **In theory, it may be possible to feed this increasing population but at present, we seem unable to distribute the food effectively.**
- **In western Europe, an average replacement rate of 2.1 children per family would maintain a stable population.**

Examination Questions

Section 5: Organisms and their Environment

Do not write on this page. Where necessary copy drawings, tables or sentences.

1 All of the energy which passes along a food chain comes originally from

A oxygen
B carbon dioxide
C glucose
D light (N)

2 (*a*) The diagram below shows parts of the nitrogen cycle.

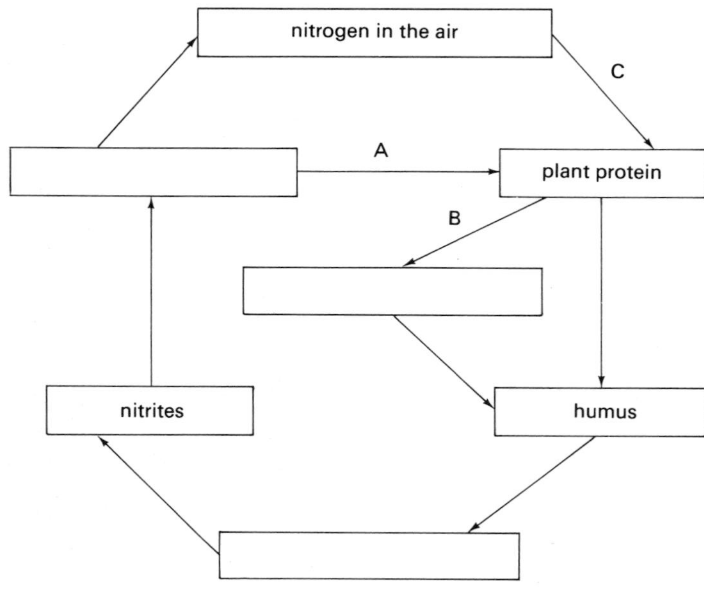

(i) Fill in the **three** empty boxes by choosing the correct words from the following list: ammonia; animal protein; nitrates.

(ii) What process takes place at A?

(iii) What process takes place at B?

(iv) In some plants, nitrogen in the air can be changed into plant protein. This is shown by arrow C. Explain how this happens.

(*b*) Why is it important to recycle elements, such as nitrogen, in nature?

(*c*) (i) Suggest **three** ways in which a good woodland soil is a suitable habitat for an earthworm.

(ii) Describe how earthworms improve the soil for the plants and animals living there. (L)

3 Either **A**

(*a*) What is the importance of the recycling of elements in nature?

(*b*) Draw a labelled diagram to show the nitrogen cycle.

(*c*) Explain the role of the micro-organisms in any **three** stages of the nitrogen cycle.

Or **B**

(*a*) Explain how the structure and properties of soil make it a suitable habitat for living organisms.

(*b*) Explain the need for conservation and, using a suitable example, suggest ways in which conservation could be carried out. (L)

4 The diagram below shows part of a food web in a pond.

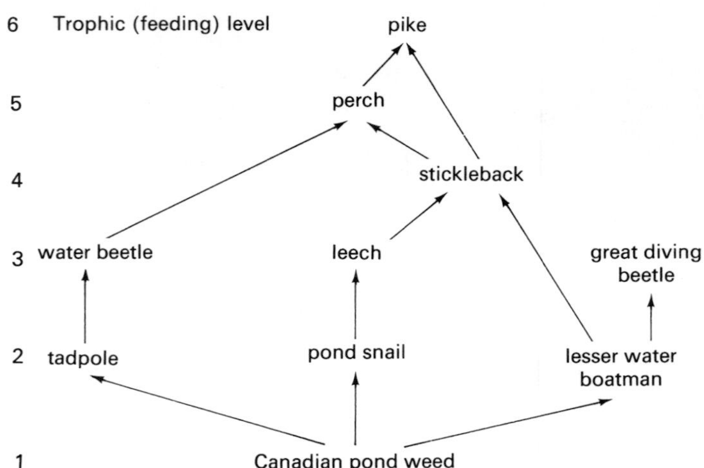

(*a*) Name an example from this food web of (i) a producer, and (ii) a secondary consumer.

(*b*) Write down a food chain, from the web, with at least **four** organisms including the perch.

(*c*) Write down **two** ways in which energy is lost from the web at each feeding level.

(*d*) Metal ions in the water, such as cadmium, can accumulate in the pike. Explain how this might happen. (L)

5 The Earth's atmosphere and much of its water (present in rivers, lakes and seas) is polluted.

(*a*) With reference to **five** named examples of pollutants which have increased markedly in the air or water during the present century, describe, for each of these, how this has come about.

(*b*) (i) Explain the importance of reducing the level of a **named** pollutant in the environment.

(ii) Describe **two** ways in which attempts are being made to reduce pollution in the environment. (M)

6 The map below shows an island off the coast of Britain. Most of the island is covered by grassland. There is a small wood in one part of the island. No grazing animals live on the island and no people live there.

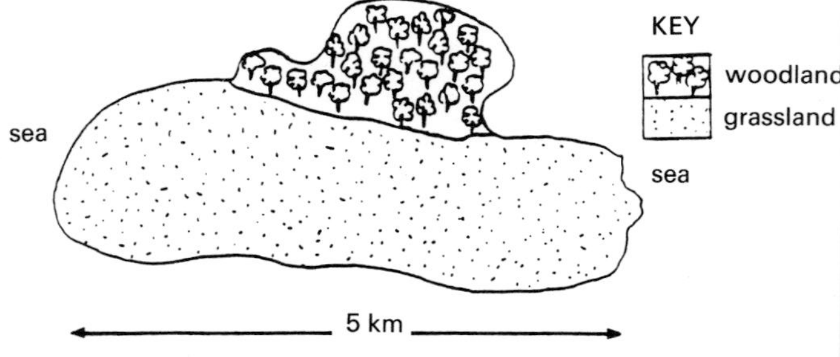

A small herd of deer, with both males and females, is put on the island. The deer can feed on all the plants found on the island.

Describe how the **populations** of the deer and the plants would change if the island were left alone for many years.

You should write in full sentences. Marks will be given for clear explanations. (N)

SECTION 6
DIVERSITY OF ORGANISMS

32 Classification

KINGDOMS
Monera, Protista, Fungi, Plants, Animals.

SUBDIVISIONS
Phylum, class, genus, species.

BINOMIAL NOMENCLATURE

ANIMAL CLASSIFICATION

PLANT CLASSIFICATION

INCREASE IN COMPLEXITY

KEYS FOR IDENTIFICATION

You do not need to be a biologist to realize that there are millions of different organisms living on the Earth, but it takes a biologist to sort them out into some kind of meaningful order, i.e. to **classify** them.

There are many possible ways of classifying organisms. You could group all aquatic organisms together or put all black and white creatures into the same group. However, these do not make very meaningful groups; a seaweed and a porpoise are both aquatic organisms; a magpie and a zebra are both black and white but neither of these pairs has much in common apart from being living organisms and the latter two being animals.

A biologist looks for important features which are shared by as large a group as possible. In some cases it is easy. Birds all have wings, beaks and feathers; there is rarely any doubt about whether a creature is a bird or not. In other cases it is not so easy. For example, some single-celled organisms are not obviously either plants or animals. As a result, biologists change their ideas from time to time about how living things should be grouped. New groupings are suggested and old ones abandoned.

Kingdoms

The largest group of organisms recognized by biologists is the kingdom. But how many kingdoms should there be? Most biologists used to favour the adoption of two kingdoms, namely **Plants** and **Animals**. This, however, caused problems in trying to classify fungi, bacteria and single-celled organisms which do not fit obviously into either kingdom. A scheme now favoured by many, but by no means all, biologists is the Whittaker 5-kingdom scheme comprising the **Monera**, **Protista**, **Fungi**, **Plants** and **Animals**.

Kingdom Monera These are the bacteria (p. 310) and the blue-green algae. They consist of single cells but differ from other single-celled organisms because their chromosomes are not organized into a nucleus.

Kingdom Protista These are single-celled (unicellular) organisms which have their chromosomes enclosed in a nuclear membrane to form a nucleus.

Some of them, e.g. *Euglena* (p. 327), possess chloroplasts and make their food by photosynthesis. These protista are often referred to as unicellular 'plants' or **protophyta**.

Organisms such as *Amoeba* and *Paramecium* (p. 327), take in and digest solid food and thus resemble animals in their feeding. They may be called unicellular 'animals' or **protozoa**.

Kingdom Fungi Most fungi are made up of thread-like hyphae (p. 321), rather than cells, and there are many nuclei distributed throughout the cytoplasm in their hyphae.

Kingdom Plants These are made up of many cells – they are multicellular. Plant cells have an outside wall made of cellulose. Many of the cells in plant leaves and stems contain chloroplasts with photosynthetic pigments, e.g. chlorophyll. Plants make their food by photosynthesis.

Kingdom Animals Animals are multicellular organisms whose cells have no cell walls or chloroplasts. Most animals ingest solid food and digest it internally.

It is still not easy to fit all organisms into this scheme. For example, many protista with chlorophyll (the protophyta) show important resemblances to some members of the Algae, but the Algae are classified into the plant kingdom. The viruses are not included in the scheme because, in many respects, viruses are not independent living organisms.

This kind of problem will always occur when we try to devise rigid classificatory schemes with distinct boundaries between groups. The process of evolution would hardly be expected to result in a tidy scheme of classification for biologists to use.

QUESTIONS

1 Which kingdoms contain organisms with (*a*) many cells, (*b*) nuclei in their cells, (*c*) cell walls, (*d*) hyphae, (*e*) chloroplasts?
2 State the main differences between animals and plants?
3 If it was decided to use a 2-kingdom classification (plants and animals), in which kingdoms would you place the bacteria, protista and fungi? Suggest reasons for your decision.

Subdivisions of the kingdoms

Phylum A kingdom can be divided into smaller groups called phyla (singular, phylum). For example, the animal kingdom is divided into about 23 different phyla; the actual number depends on whose scheme of classification you adopt. Members of a phylum have some major features in common.

Two familiar phyla are the **annelids** (segmented worms such as earthworms and lugworms), and the **arthropods** (animals with a hard external skeleton, e.g. crustaceans, insects, spiders and millipedes).

Class The phylum can be further divided into classes. The vertebrate phylum includes all the animals with vertebral columns, and is divided into 5 classes: **fishes**, **amphibia**, **reptiles**, **birds** and **mammals**.

Order Within each class there are groups called orders. Some of the orders in the Class mammals are the **rodents** (e.g. rats and mice), the **carnivores** (e.g. lions, wolves), **insectivores** (e.g. shrews, moles, hedgehogs) and the **primates** (lemurs, monkeys, apes and humans).

Genus When organisms within an order share many features in common they are classified into a genus. For example, in the carnivore order the genus *Mustelus* includes stoats, weasels and polecats.

Species The smallest natural group of organisms is the species. Robins, blackbirds and sparrows are three different species of bird. Apart from small variations, members of a species are almost identical in their anatomy, physiology and behaviour.

Members of a species also often resemble each other very closely in appearance, though there are some notable exceptions where humans have taken a hand in the breeding programmes. All dogs belong to the same species but there are wide variations in the appearance of different breeds.

One of the main features which determines whether organisms belong to the same species is whether they can successfully breed together. A spaniel and a labrador retriever may look very different but they have no problems in breeding together to produce a litter of 'mongrel' puppies.

Binomial nomenclature

Species must be named in such a way that the name is recognized all over the world.

The 'cuckoo flower' and the 'lady's smock' are two common names for the same wild plant. If you are not aware that these are alternative names this could lead to confusion. If the botanical name, *Cardamine pratensis*, is used, however, there is no chance of error. The Latin form of the name allows it to be used in all the countries of the world irrespective of language barriers.

Binomial means two names; the first name gives the genus and the second gives the species. For example, the stoat and weasel are both in the genus *Mustela* but they are different species; the stoat is *Mustela erminea* and the weasel is *Mustela nivalis*.

The name of the genus (generic name) is always given a capital letter and the name of the species (specific name) always starts with a small letter.

Frequently, the specific name is descriptive, e.g. *hirsutum* = hairy, *aquatilis* = living in water, *bulbosus* = having a bulb.

Classification of plants

The classification of plants follows the same lines as those given above for animals. However, botanical classification uses the term division instead of phylum.

An outline classification of plants and animals is given below and illustrated in Figs 1–4.

Note that subdivisions of phyla are given only for groups whose names are likely to be familiar.

ANIMAL KINGDOM

PHYLUM (only 8 out of 23 listed here)

- **Coelenterates** (sea anemones, jellyfish)
- **Flatworms**
- **Nematode worms**
- **Annelids** (segmented worms)
- **Arthropods**
  - CLASS
  - [1] Crustacea[2] (crabs, shrimps, water fleas)
  - Insects
  - Arachnids (spiders and mites)
  - Myriapods (centipedes and millipedes)
- **Molluscs** (snails, slugs, mussels, octopuses)
- **Echinoderms** (starfish, sea urchins)

- **Vertebrates**[3]
  - CLASS
  - Fishes[4]
  - Amphibia (frogs, toads, newts)
  - Reptiles (lizards, snakes, turtles)
  - Birds
  - Mammals
    - ORDER (only 4 out of about 26 are listed)
    - Insectivores
    - Carnivores
    - Rodents
    - Primates

[1] All the organisms which do not have a vertebral column are often referred to as invertebrates. Invertebrates are not a natural group, but the term is convenient to use.
[2] Crustacea may be classified as a sub-phylum.
[3] Vertebrates are, in fact, a sub-phylum of the chordate phylum.
[4] Fishes, strictly speaking, comprise three classes.

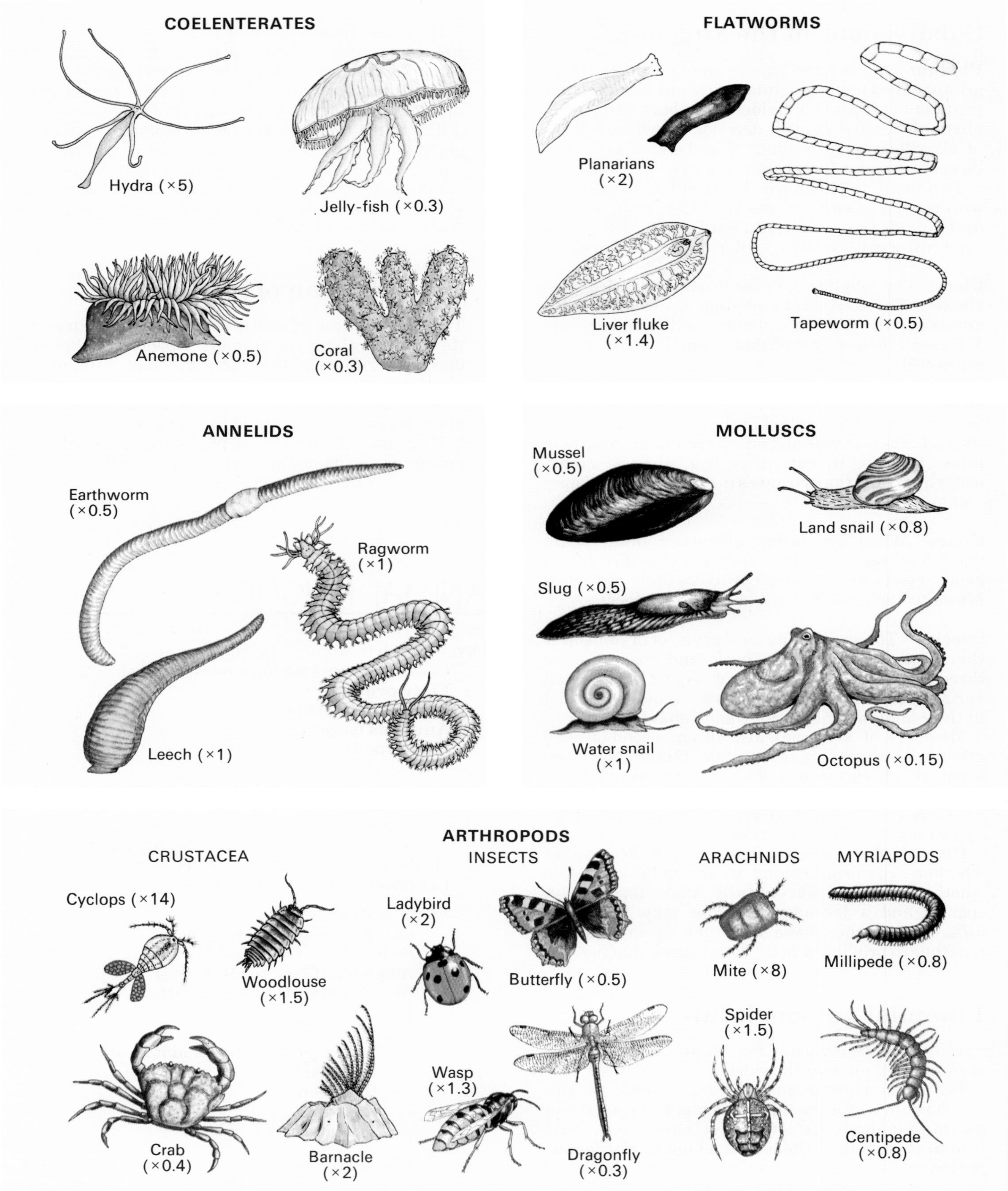

Fig. 1 The animal kingdom; 5 invertebrate phyla

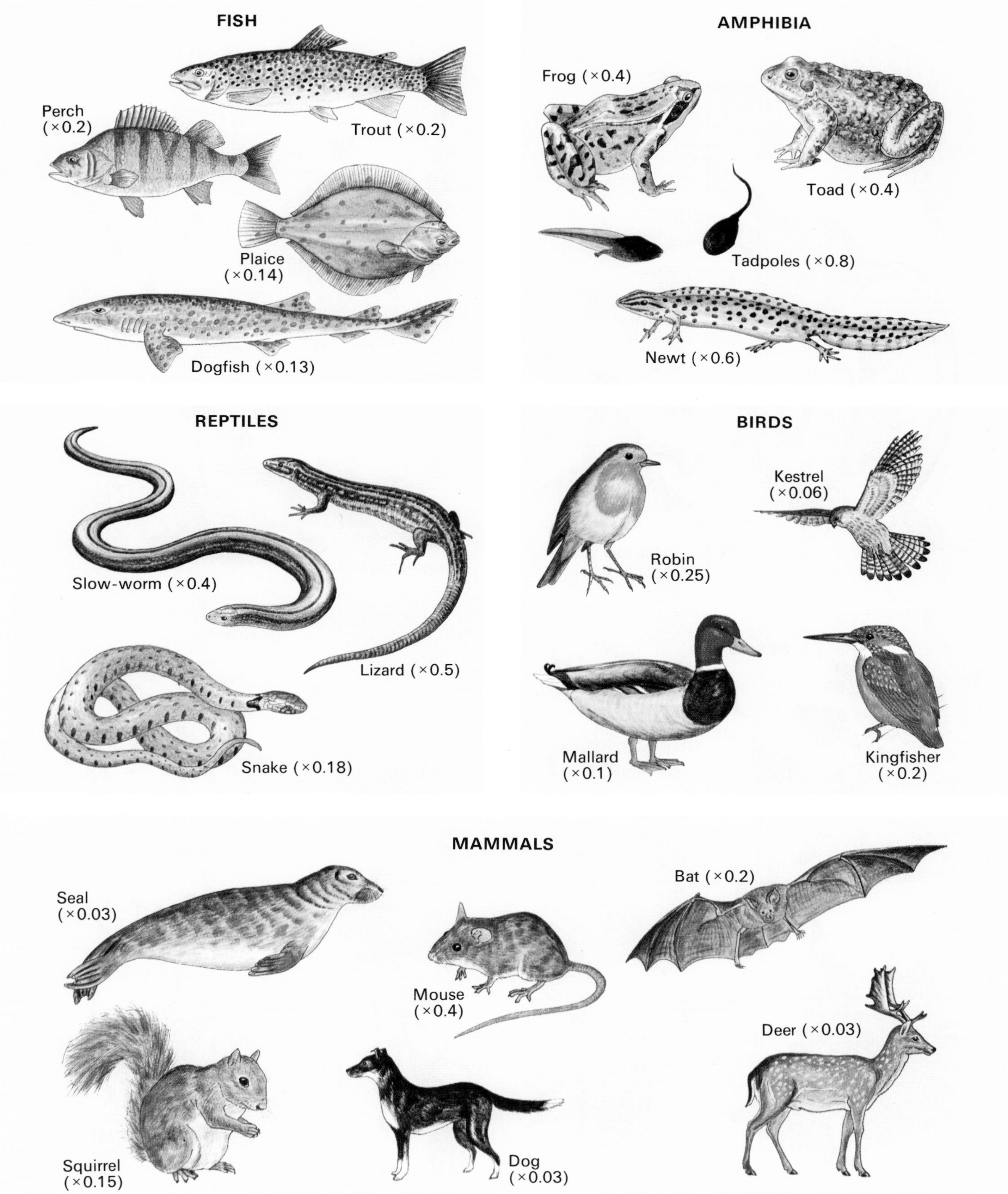

Fig. 2 The animal kingdom; the vertebrate classes

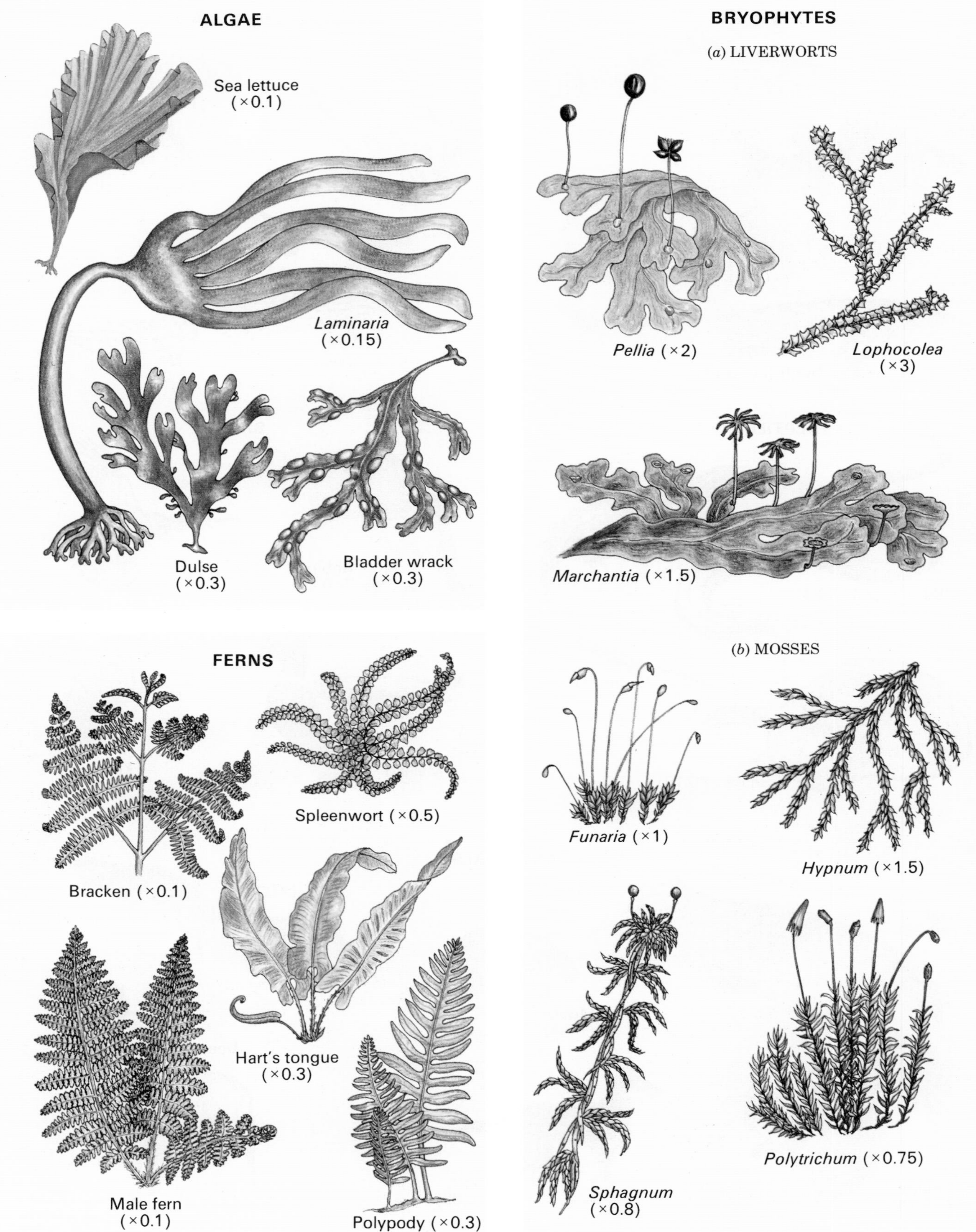

Fig. 3 The plant kingdom; plants that do not bear seeds

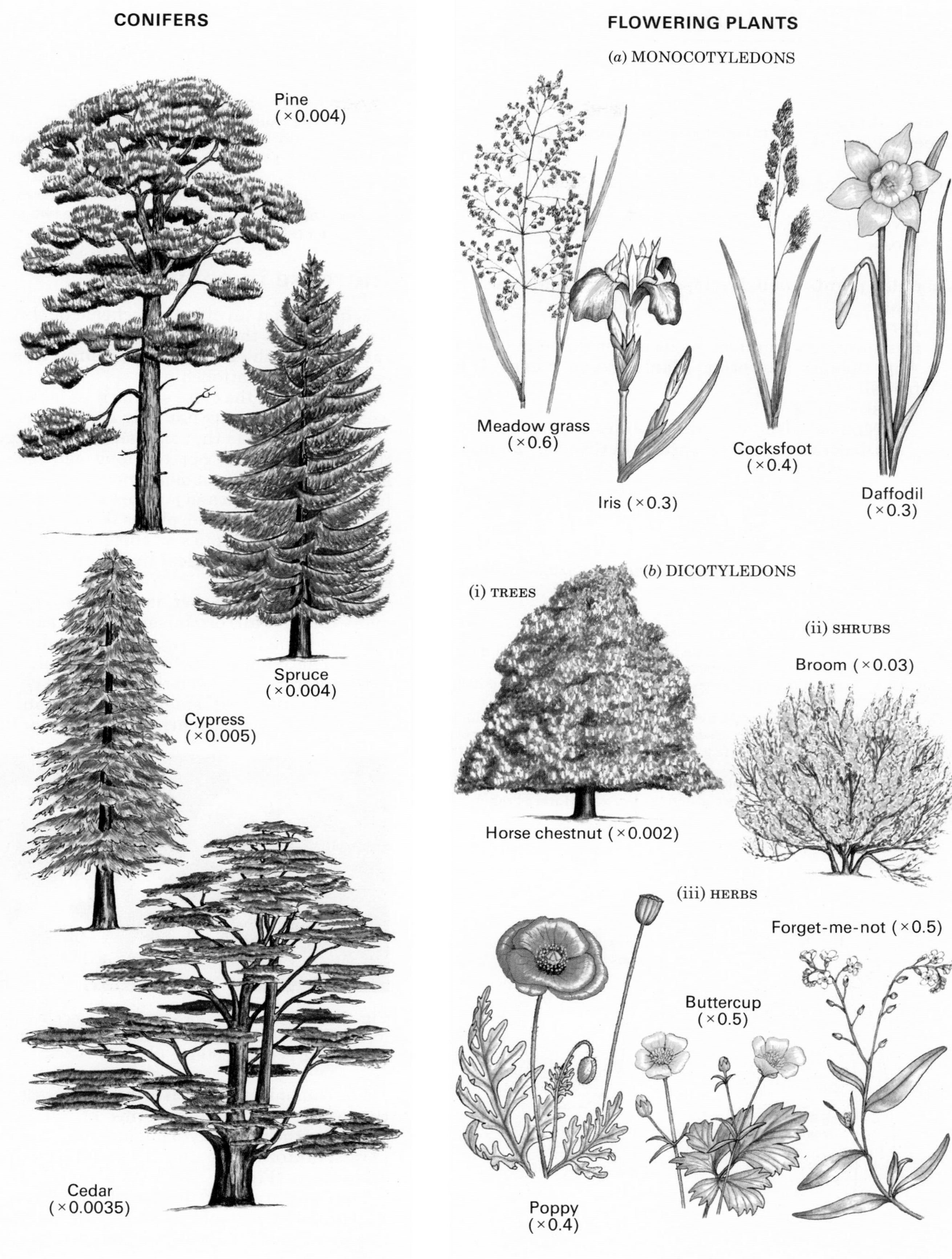

Fig. 4 The plant kingdom; seed-bearing plants

PLANT KINGDOM

DIVISION

Red algae
Brown algae
Green algae
} seaweeds and filamentous[1] forms; mostly aquatic

Bryophytes (no specialized conducting tissue)

- CLASS
  - Liverworts
  - Mosses

Vascular plants (well-developed xylem and phloem)

- CLASS
  - Ferns
  - 2 { Gymnosperms (conifers, seeds not enclosed in fruits)
  - Angiosperms (flowering plants, seeds enclosed in fruits) }
    - SUB-CLASS
      - Monocotyledons[3] (grasses, lilies)
      - Dicotyledons (trees, shrubs, herbaceous plants)
        - FAMILY
          - e.g. Ranunculaceae (one of about 70 families)
            - GENUS
              - e.g. *Ranunculus*
                - SPECIES
                  - e.g. *Ranunculus bulbosus* (bulbous buttercup)

[1] In some classificatory schemes, some of the Algae are classified into a sub-kingdom and some into divisions.
[2] The gymnosperms and angiosperms are sometimes called collectively 'the seed-bearing plants'.
[3] The monocotyledons (monocots for short), are flowering plants which have only one cotyledon in their seeds (p. 92). Most, but not all, monocots also have long, narrow leaves (e.g. grasses, daffodils, bluebells) with parallel leaf veins (Fig. 5).
The Dicotyledons (dicots for short), have two cotyledons in their seeds (p. 92). Their leaves are usually broad and the leaf veins form a branching network.

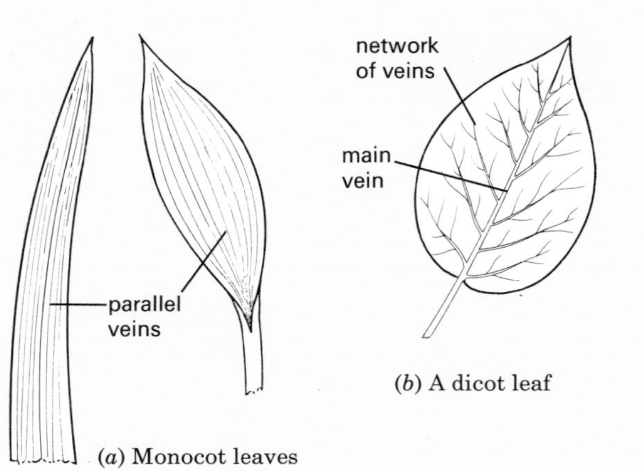

Fig. 5 Leaf types in flowering plants

QUESTIONS

4 Put the following terms in order, starting with those containing the largest groups of organisms: phylum, species, kingdom, genus, order, class.
5 Try to classify the following organisms: beetle, sparrow, weasel, gorilla, pine tree, buttercup, moss. E.g. Butterfly: Kingdom, animal; Phylum, arthropod; Class, insect.
In some cases you can get only as far as the class; in others you may be able to get to order, genus or species.
6 The white deadnettle is *Lamium album*; the red deadnettle is *Lamium purpureum*. Would you expect these two plants to cross-pollinate successfully?

Increase in complexity

It is hoped that an effective scheme of classification will do more than just sort organisms into tidy categories. It should also help to throw some light on evolutionary relationships.

For example, the close similarities between stoats, weasels and polecats not only determines that they should be placed in the same genus, but suggests that they have evolved, not too long ago (geologically speaking) from a common ancestor. Similarly, the fact that their teeth and jaws are very similar to those of cats, dogs, otters, etc. places them in the carnivore order and also implies a common ancestry, a bit further back in geological time, with other members of the order.

It is also assumed that, in the course of evolution, many (but not all) organisms became more complex. The first forms of life were probably single cells. These evolved into simple, multicellular organisms but with little difference between the individual cells in their bodies. In the next stages of evolution, the cells became specialized to carry out particular functions.

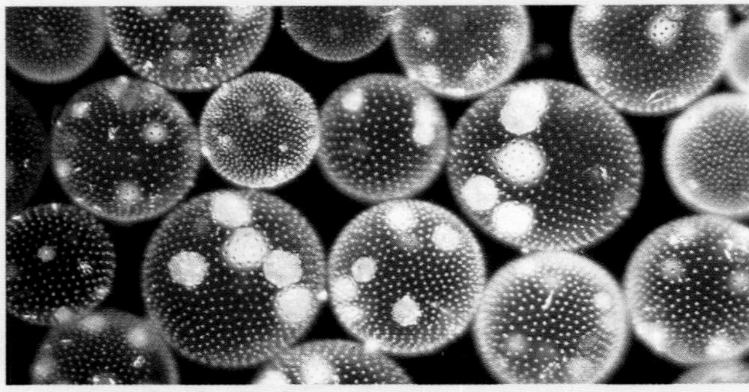

Fig. 7 Volvox. This is a simple, many-celled alga, in the form of a ball of cells. The earliest multicellular creatures might have been something like this.

For example, the protista might represent some of the early, single-celled forms of life. Organisms similar to the protophyta might have given rise to the plants; protozoa-like unicells could have started the animal line (Fig. 6).

The evolution of multicellular forms can be illustrated by a simple green alga which consists of a ball of cells. All the cells are the same as each other, but the alga moves around and reproduces rather like a multi-cellular organism (Fig. 7).

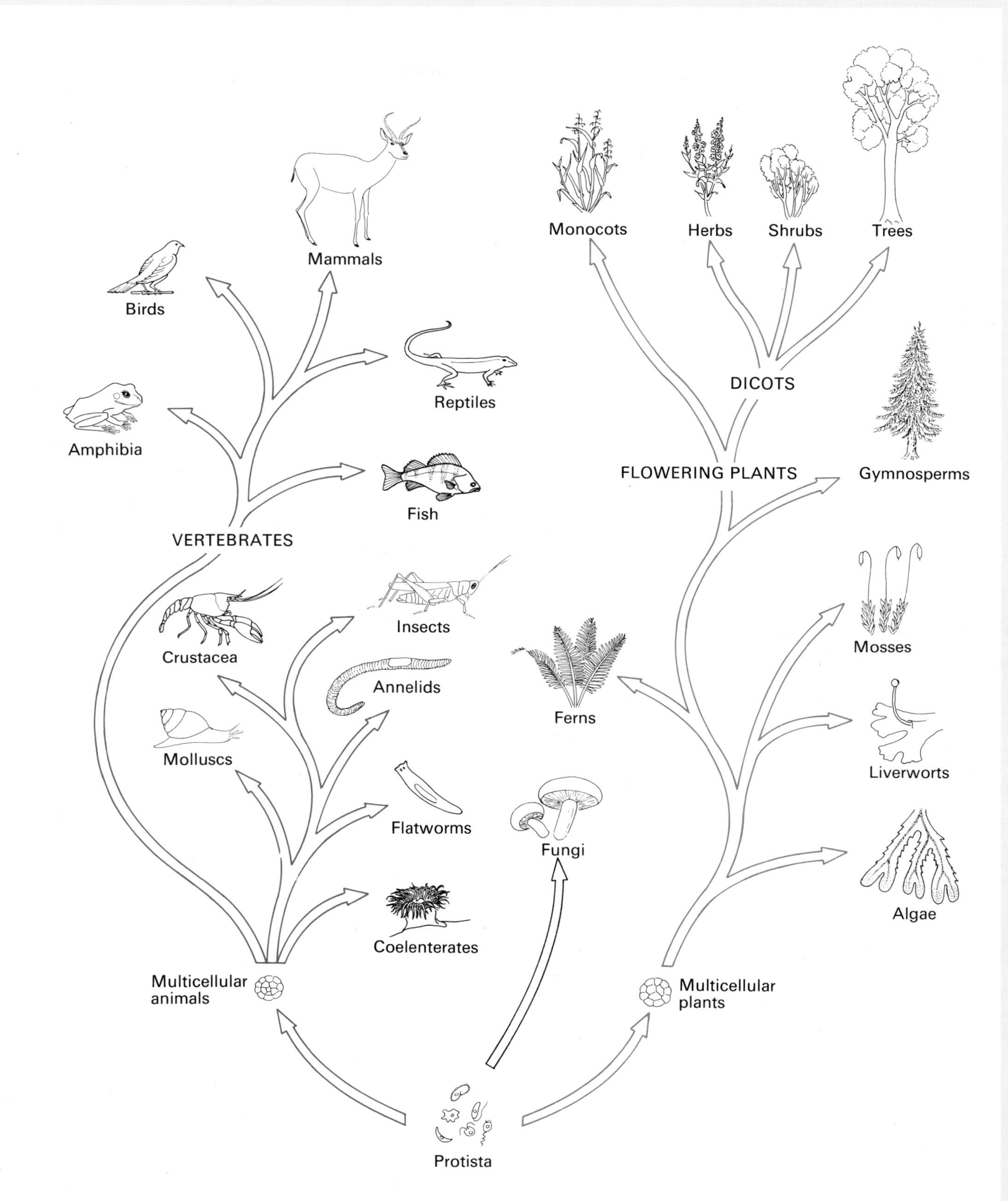

Fig. 6 Classification and evolution. The kingdoms, phyla, sub-phyla and some classes can be arranged to show a possible evolutionary sequence.

The liverworts are fairly uncomplicated green plants but some of their cells are specialized to absorb water, and others become specialized to form reproductive structures.

The mosses have stems and leaves (Fig. 9 on p. 331), but there is no specialized conducting tissue as seen in the ferns and the seed-bearing plants.

The seed-bearing plants are very much more complex than the ferns in their structure and method of reproduction. Ferns, mosses and liverworts reproduce by means of single-celled spores, whereas the seed-bearing plants have cones or elaborate flower structures and produce multicellular seeds.

A similar increase in complexity can be seen in the animal kingdom. The protozoa (p. 327) are single-celled organisms; all the living processes are carried out in one cell. The coelenterates are multicellular with some cells specialized for digestion, conduction of nerve impulses or contraction. In the flatworms, the body plan is still simple but there is a recognizable nervous system. The annelids (p. 333) have a circulatory system and a well-organized nervous system.

The vertebrates (p. 340) have a skeleton, a central nervous system with a brain, a blood circulatory system and highly developed sense organs.

It must be remembered that the increase in complexity shown by our classification systems, may well illustrate a similar series of changes during evolution but is not sound evidence that this is how evolution actually occurred. The present-day forms of algae and mosses are not the same organisms that gave rise to the higher plants some 400 million years ago. Algae and mosses are also likely to have changed during 400 million years of evolution.

QUESTION

7 Figure 9*b* on p. 331 shows a section through a moss leaf. Figure 2 on p. 59 shows a section through the leaf of a flowering plant. In what ways is the leaf of the flowering plant more complex than that of the moss?

Keys for identification

Once you know the main characteristics of a group, it is possible to draw up a systematic plan for identifying an unfamiliar organism. One such plan is shown below.

An alternative form of key is the **dichotomous key**. Dichotomous means two branches, so you are confronted with two possibilities at each stage.

Fig. 8 is an example of a dichotomous key that could be used to place an unknown vertebrate in the correct class. Item 1 gives you a choice between 2 alternatives. If the animal is 'cold-blooded' you move to item 2 and make a further choice. If it is a 'warm-blooded' animal you move to item 4 for your next choice

DICHOTOMOUS KEY FOR VERTEBRATE CLASSES

1	'Cold-blooded'	2
	'Warm-blooded'	4
2	Has fins but no limbs	FISH
	Has 4 limbs	3
3	Has no scales on body	AMPHIBIAN
	Has scales	REPTILE
4	Has feathers	BIRD
	Has fur	MAMMAL

Fig. 8

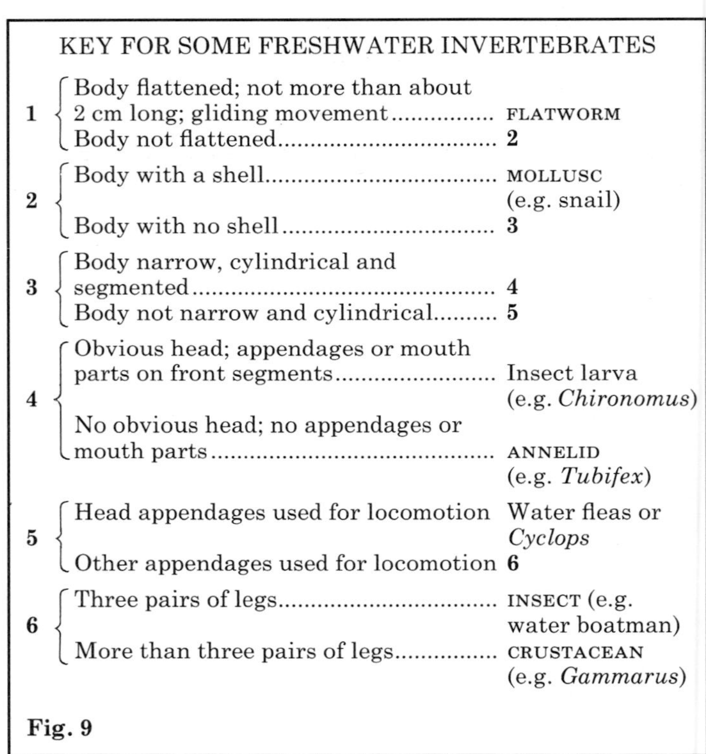
KEY FOR SOME FRESHWATER INVERTEBRATES

1	Body flattened; not more than about 2 cm long; gliding movement	FLATWORM
	Body not flattened	2
2	Body with a shell	MOLLUSC (e.g. snail)
	Body with no shell	3
3	Body narrow, cylindrical and segmented	4
	Body not narrow and cylindrical	5
4	Obvious head; appendages or mouth parts on front segments	Insect larva (e.g. *Chironomus*)
	No obvious head; no appendages or mouth parts	ANNELID (e.g. *Tubifex*)
5	Head appendages used for locomotion	Water fleas or *Cyclops*
	Other appendages used for locomotion	6
6	Three pairs of legs	INSECT (e.g. water boatman)
	More than three pairs of legs	CRUSTACEAN (e.g. *Gammarus*)

Fig. 9

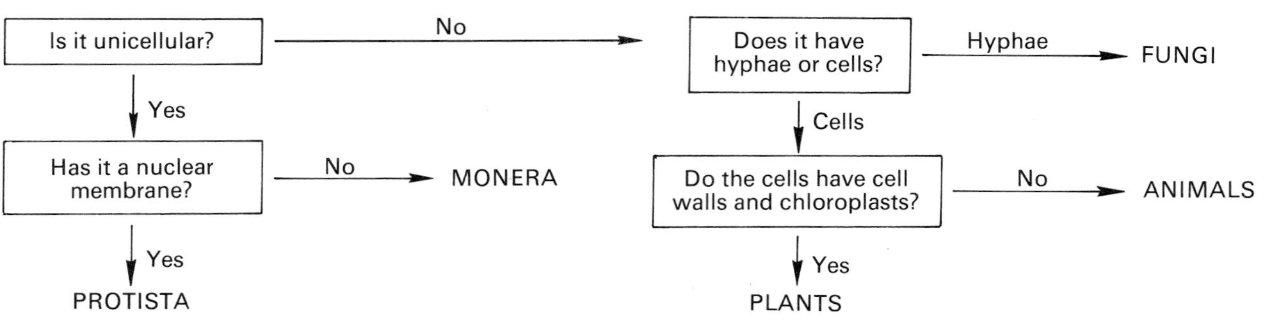

The same technique may be used for assigning an organism to its order, genus or species. However, the important features may not always be easy to see and you have to make use of less fundamental characteristics.

Fig. 9 is a key for assigning pond organisms to their correct phylum or sub-phylum. It is not based on the fundamental characteristics of the phylum, e.g. 'not more than 2 cm long' is not a feature of flatworms, but it is a useful guide to those flatworms that live in ponds. A key such as this is sometimes called an artificial key.

QUESTION

8 Figure 1 on p. 327 shows some protista. Using only the features shown in the drawings, construct a dichotomous key that could be used to identify these organisms.

CHECK LIST

- **There is no universally accepted system of classification.**
- **One system groups all living organisms into 5 kingdoms: Monera, Protista, Fungi, Plants and Animals.**
- **Kingdoms are subdivided into phyla, which are themselves divided into classes.**
- **The smallest natural group is the species.**
- **Members of a species are alike in all important respects and can interbreed.**
- **When a group of species are very similar to each other, they are classed as a genus.**
- **Each species is given a 2-part (binomial) name. The first part names the genus and the second part names the species.**
- **The binomial system of naming organisms is used throughout the world.**
- **A classification system can be used to show an evolutionary sequence.**
- **The sequence of kingdoms and the order of phyla and classes usually show an increase in the complexity of organisms.**
- **A dichotomous key is a way of identifying organisms by posing alternative questions or choosing between alternative statements.**

33 Bacteria and Viruses

BACTERIA
Structure and physiology.

VIRUSES
Structure and multiplication.

USEFUL BACTERIA
In natural cycles, in biotechnology.

DISEASES
Methods of transmission, virus diseases, bacterial diseases, sexually transmitted diseases.

CONTROL OF DISEASES
Water and sewage, food-handling, resistance and immunity.

PRACTICAL WORK
Culturing bacteria, effect of antibiotics.

BACTERIA

Bacterial structure

Bacteria (singular = bacterium) are very small organisms consisting of single cells rarely more than 0.01 mm in length. They can be seen only with the higher powers of the microscope.

Their cell walls are made, not of cellulose, but of a complex mixture of proteins, sugars and lipids. Some bacteria have a **slime capsule** outside their cell wall. Inside the cell wall is the cytoplasm which may contain granules of glycogen, lipid and other food reserves (Figs 1 and 2).

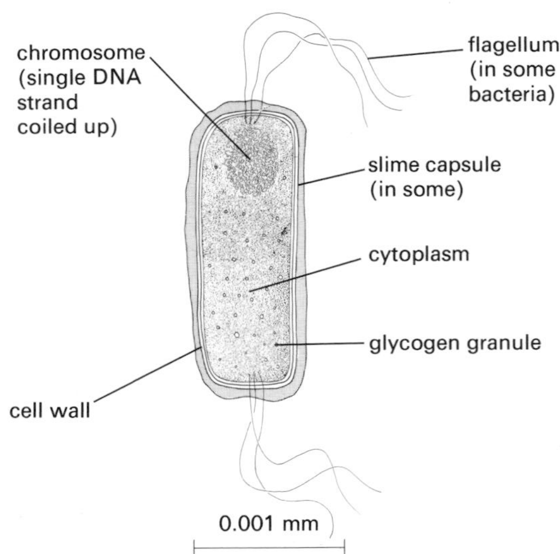

Fig. 1 Generalized diagram of a bacterium

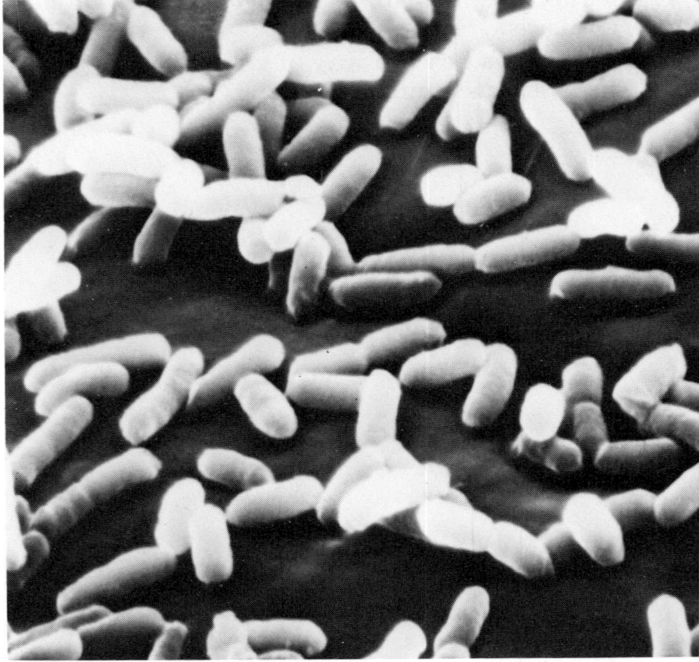

Fig. 2 Bacteria as seen by the scanning electron microscope (× 11 000).

Each bacterial cell contains a single chromosome, consisting of a circular strand of DNA (p. 224). The chromosome is not enclosed in a nuclear membrane but is coiled up to occupy part of the cell (Fig. 3).

Individual bacteria may be spherical, rod-shaped or spiral and some have filaments, called **flagella**, projecting from them. The flagella can flick, and so move the bacterial cell about.

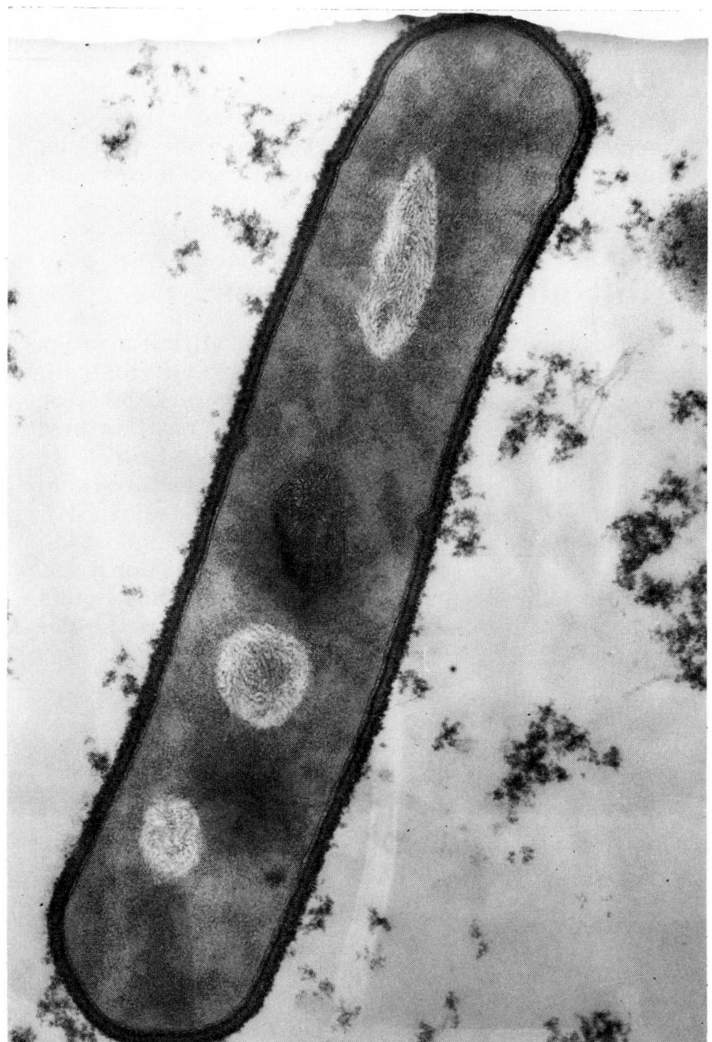

Fig. 3 Longitudinal section through a bacterium (×45 000). The light areas are the coiled DNA strands. There are three of them because the bacterium is about to divide twice (see Fig. 4).

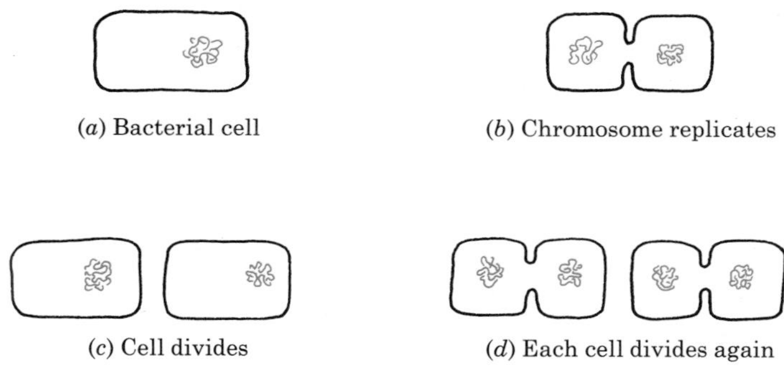

Fig. 4 Bacterium reproducing. This is asexual reproduction by cell division.

Bacterial physiology

Nutrition There are a few species of bacteria which contain a photosynthetic pigment like chlorophyll, and can build up their food by photosynthesis. Most bacteria, however, live in or on their food. They produce and release enzymes which digest the food outside the cell. The liquid products of digestion are then absorbed back into the bacterial cell.

Respiration The bacteria which need oxygen for their respiration are called **aerobic bacteria.** Those which do not need oxygen for respiration are called **anaerobic bacteria.** The bacteria used in the filter beds of sewage plants (p. 277) are aerobic, but those used to digest sewage sludge and produce methane, are anaerobic.

Reproduction Bacteria reproduce by cell division or **fission.** Any bacterial cell can divide into two and each daughter cell becomes an independent bacterium (Fig. 4).

In some cases, this cell division can take place every 20 minutes so that, in a very short time, a large colony of bacteria can be produced. This is one reason why a small number of bacteria can seriously contaminate our food products.

This kind of reproduction, without the formation of gametes, is called **asexual reproduction** (p. 350).

Effect of heat Bacteria, like any other living organisms, are killed by high temperatures. The process of cooking destroys any bacteria in food, provided high enough temperatures are used. If drinking-water is boiled, any bacteria present are killed.

However, some bacteria can produce spores which are resistant to heat. When the cooked food or boiled water cools down, the spores germinate to produce new colonies of bacteria, particularly if the food is left in a warm place for many hours. For this reason, cooked food should be eaten at once or immediately refrigerated. (Refrigeration slows down bacterial growth and reproduction.) After refrigeration, food should not merely be warmed up but either eaten cold or heated to a temperature high enough to kill any bacteria that have grown, i.e. to 90°C or more (p. 121).

VIRUSES

Most viruses are very much smaller than bacteria and can be seen only with the electron microscope at magnifications of about ×30 000. Figure 5 shows virus particles inside a bacterial cell.

Virus structure

There are many different types of virus and they vary in their shape and structure. All viruses, however, have a central core of RNA or DNA (p. 17) surrounded by a protein coat. Viruses have no nucleus, cytoplasm, cell organelles, or cell membrane, though some forms have a membrane outside their protein coats.

Virus particles, therefore, are not cells. They do not feed, respire, excrete or grow and it is debatable whether they can be classed as living organisms. Viruses do reproduce, but only inside the cells of living organisms, using materials provided by the host cell.

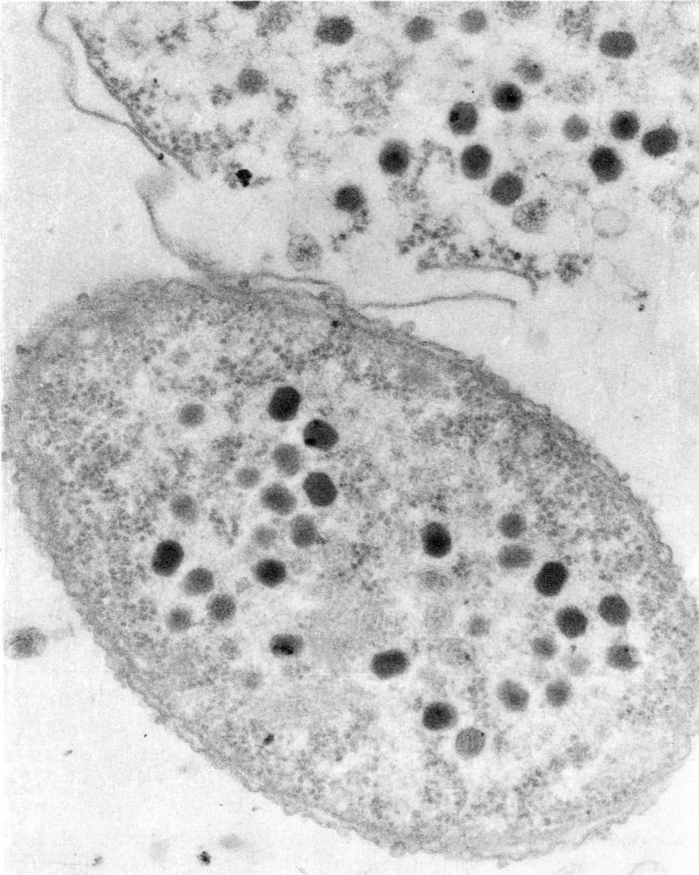

Fig. 5 Viruses attacking a bacterium (×60 000). The viruses have invaded the bacterial cell and can be seen as black blobs. The viruses reproduce inside the cell. The cell at the top has burst open and the viruses are escaping.

Figure 6 shows a generalized virus particle. The nucleic acid core is a coiled single strand of RNA. The coat is made up of regularly packed protein units called **capsomeres** each containing many protein molecules. The protein coat is called a **capsid**. Outside the capsid, in the influenza and some other viruses, is an envelope which is probably derived from the cell membrane of the host cell.

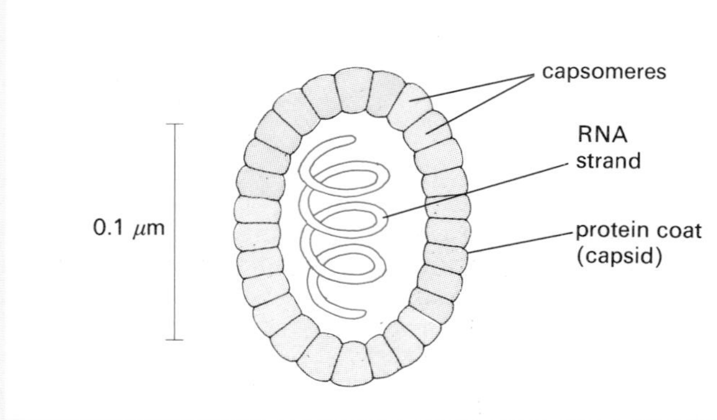

Fig. 6 Structure of a virus

Figure 7 is an electronmicrograph of the tobacco mosaic virus (TMV) which invades the cells of a wide variety of plants, including tomatoes. Figure 8 is a schematic drawing of part of a TMV particle, showing the regularly arranged capsomeres surrounding the coiled RNA core.

Multiplication of viruses

Viruses can survive outside the host cell, but in order to reproduce they must penetrate into a living cell. How they do this, in many cases, is not known for certain. In most instances, the virus particle first sticks to the cell membrane. It may then 'inject' its DNA or RNA into the cell's cytoplasm or the whole virus may be taken in by a kind of endocytosis (p. 34).

Once inside the host cell, the virus is 'uncoated', i.e. its capsid is dispersed, exposing its DNA or RNA. The DNA or RNA then takes over the host cell's physiology. It arrests the normal syntheses in the cell

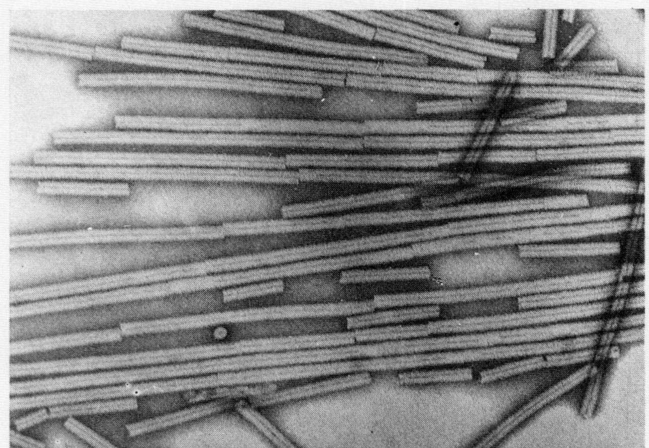

Fig. 7 Tobacco mosaic virus (×86 000). The virus causes disease in the plants of tobacco and tomato.

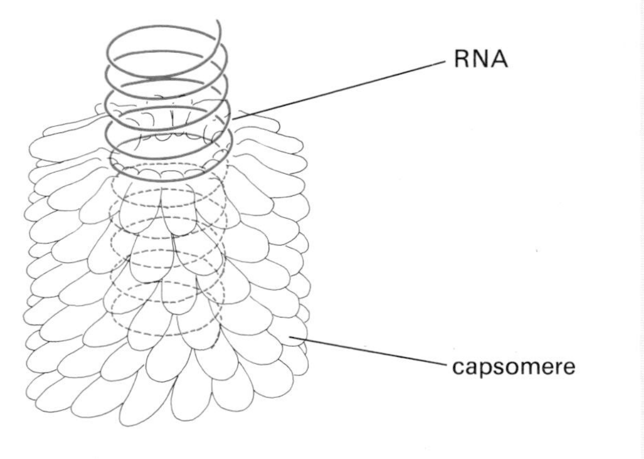

Fig. 8 Structure of the tobacco mosaic virus

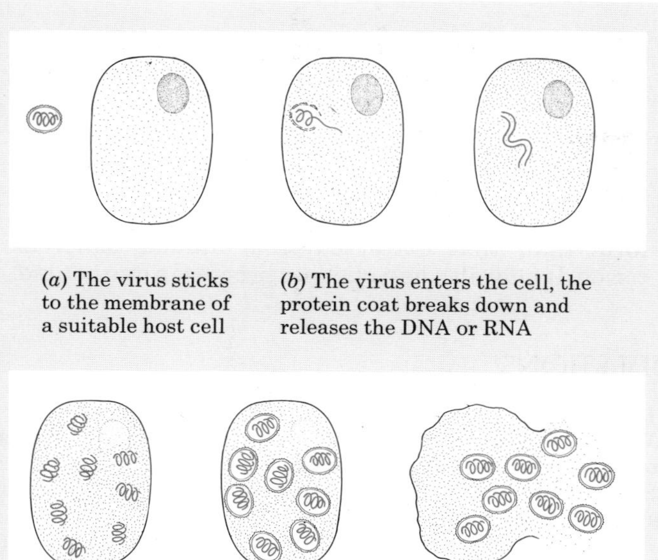

(*a*) The virus sticks to the membrane of a suitable host cell

(*b*) The virus enters the cell, the protein coat breaks down and releases the DNA or RNA

(*c*) The viral DNA or RNA replicates and directs the host cell to make new protein coats

(*d*) The new viruses escape from the host cell

Fig. 9 Reproduction of a virus

and makes the cell produce new viral DNA or RNA and new capsomeres. The nucleic acid and the capsomeres are assembled in the cell to make new virus particles which escape from the cell (Fig. 9).

The cell may be destroyed in this process or the viruses may escape, wrapping themselves in pieces of the host's cell membrane as they do so. These activities give rise to the signs and symptoms of disease.

QUESTIONS

1 Which of the following structures are present in both bacterial cells and plant cells: cytoplasm, cellulose, DNA, cell wall, nucleus, chromosome, vacuole, glycogen?
2 If 5 bacteria landed in some food and reproduced at the maximum possible rate, what would be the population of bacteria after 4 hours?
3 (*a*) Why is a virus particle not considered to be a cell?
(*b*) Why are viruses not easy to classify as living organisms?
4 Why do you think that commercial growers of tomatoes forbid smoking in their greenhouses?
5 How does the reproduction of a virus differ from that of a bacterium?

USEFUL BACTERIA

Natural cycles

When people talk about bacteria, they are usually thinking about those which cause disease or spoil our food. In fact, only a tiny minority of bacteria are harmful. Most of them are harmless or extremely useful.

Bacteria which feed saprophytically (p. 250), bring about decay. They secrete enzymes into dead organic matter and liquefy it. This may be a nuisance if the organic matter is our food but, in most cases, it consists of the excreta and dead bodies of organisms. If it were not for the activities of the decay bacteria (and fungi), we should be buried in ever-increasing layers of dead vegetation and animal bodies.

The decay bacteria also release essential elements from the dead remains. For example, proteins are broken down to ammonia and the ammonia is turned into nitrates by nitrifying bacteria (p. 247). The nitrates are taken up from the soil by plants, which use them to build up their proteins. In a similar way, sulphur, phosphorus, iron, magnesium and all the elements essential to living organisms are recycled in the course of bacterial decomposition.

Biotechnology

Biotechnology can be defined as the application of biological organisms, systems or processes to manufacturing and service industries.

Although biotechnology is 'hot news', we have been making use of it for hundreds of years. Wine-making, beer-brewing, the baking of bread and the production of cheese all depend on fermentation processes brought about by yeasts, other fungi and bacteria, or enzymes from these organisms.

Antibiotics, such as penicillin, are produced by mould-fungi or bacteria. The production of industrial chemicals such as citric acid or lactic acid needs bacteria or fungi to bring about essential chemical changes.

Sewage disposal (p. 277) depends on bacteria in the filter beds to form the basis of the food chain which purifies the effluent.

Biotechnology is not concerned solely with the use of micro-organisms. Cell cultures and enzymes also feature in modern developments. In this chapter, however, there is space to consider only one or two instances of the use of bacteria.

The role of bacteria in biotechnology

Biological conversions *Gluconbacteria* is used to convert ethanol to ethanoic acid (e.g. vinegar). Species of *Streptococcus* and *Lactobacillus* convert milk to yoghurt; other members of the same genera are used in cheese-making.

An increasingly important conversion process exploits bacteria in extracting metals from low-grade ores. Spoil heaps of residues from, for example, copper-mines are sprayed with acidified water. This promotes the growth of a species of *Thiobacillus*, which can use the sulphur in the ore and produce sulphuric acid. The acid dissolves out the copper as copper sulphate, which can be collected from the bottom of the spoil heap and the copper reclaimed.

Similar bacteriological processes may be used to convert toxic wastes to less harmful compounds.

An increasing number of biotechnological processes are using enzymes extracted from micro-organisms instead of the micro-organisms themselves. This often increases the efficiency of the process. Microbial enzymes are used in some washing powders to digest away stains caused by organic materials.

Single-cell protein (SCP) The principle is that bacteria are grown in a 'feedstock', which may be a petroleum product, an agricultural product or waste material from agriculture or the food industry, e.g. whey from cheese-making.

In the right conditions of temperature, aeration, etc., the bacteria grow very rapidly and can be separated from the feedstock, dried and used as food for animals or, possibly, humans.

Before 1985 ICI manufactured a single-cell protein called 'Pruteen' (Fig. 10). The feedstock was methanol produced from the methane in natural gas, and the bacterium used was *Methylophilus methylotrophus*. The bacteria were allowed to grow in a dilute solution of methanol with air, ammonia and essential elements added. In the fermentation tanks, the bacterial cells multiplied and were then separated from the liquid medium by centrifugation and drying. The product was a creamy white powder with no smell and little taste. It was added, in powder form, to animal feedstuffs or made into pellets for cattle.

Fig. 10 ICI's 'Pruteen' factory. The tall building in the centre houses the fermentation tanks; on the left are the storage containers; in between is the drying unit.

Since the single cells consist mainly of protoplasm, the product is very rich in protein. Whether the project will be a financial success is not certain.

Production of drugs Bacteria in the *Streptomyces* genus are used to produce antibiotics such as streptomycin and tetracycline. A *Pseudomonas* bacterium is used in the preparation of vitamin B_{12}.

Some of the most promising developments in biotechnology are coming from the genetic engineering (p. 225) of bacteria. It has proved possible to insert pieces of human DNA, carrying specific genes, into bacterial cells. For example, the gene which controls production of human insulin has been inserted into bacterial cells. The genetically engineered bacteria produce insulin when they are cultured and the insulin can be separated and purified from the culture medium. This process produces insulin in a purer form than that extracted from animal pancreas tissue. It may also be a cheaper process in due course.

QUESTIONS

6 In what ways are decay bacteria (*a*) useful, (*b*) harmful?

7 Why, do you think, is it usually only the hard parts of animals (bones, teeth, shells) which are preserved as fossils?

8 Suggest why it is that paper and wood will rot, but glass and many plastic materials will not.

9 In producing single-cell protein why is it essential to ensure that the feedstock and industrial plant do not contain any bacteria other than the one it is intended to culture?

DISEASES CAUSED BY BACTERIA AND VIRUSES

All the bacteria described above are saprophytic (p. 250). They obtain their food from dead or non-living organic material. The bacteria which cause disease are parasites (p. 250). They live in the cells of plants or animals and feed on the cytoplasm. Parasitic organisms which cause disease are called **pathogens**. The organism in which they live and reproduce is called the **host.** All viruses are pathogens but only a small proportion of bacteria are pathogenic.

Viruses and pathogenic bacteria may cause diseases because of the damage they do to the host's cells, but most bacteria also produce poisonous waste products called **toxins**. The toxin produced by the *Clostridium* bacteria (which cause tetanus) is so poisonous that as little as 0.00023 g is fatal. Viruses do not produce toxins but they do destroy the cells in which they reproduce.

Methods of disease transmission

Droplet infection When we cough, sneeze or merely talk, we produce a fine spray of droplets of saliva. If we are carrying an infection, these drops may contain virus particles or bacteria. The droplets may remain suspended in the air for long periods or they may evaporate leaving the infectious particles floating in the air or falling to the ground in dust.

If a healthy person inhales the infected droplets, he or she may catch the disease. Diseases which are spread by droplets are often those affecting the respiratory passages (nose, throat, bronchi and lungs). Examples are influenza and colds (viruses), and tuberculosis (bacteria).

Food and drink The diseases spread by food or drink are those which affect the alimentary canal, e.g. cholera, typhoid and salmonella food poisoning (bacteria), and poliomyelitis (virus).

The faeces of an infected person will contain the harmful micro-organisms. These may be transferred to food or utensils by houseflies which walk on the faeces (Fig. 11). The micro-organisms may contaminate the fingers of the infected person if he or she is not scrupulous about scrubbing the hands after using the toilet. If this person handles food that is to be eaten by others, the disease-causing organisms can be spread.

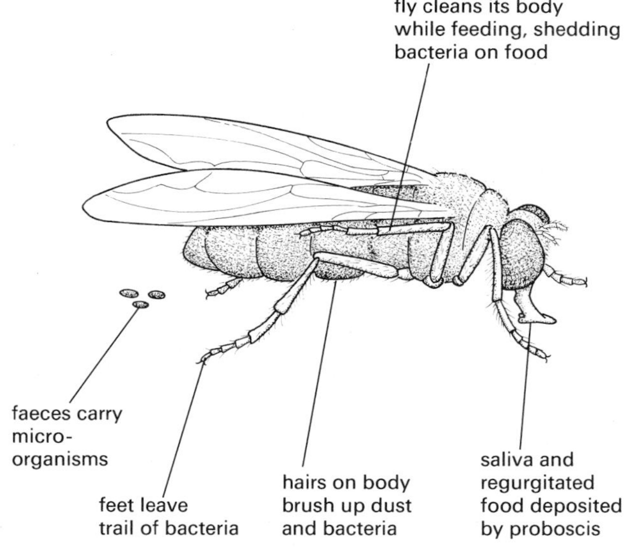
fly cleans its body while feeding, shedding bacteria on food
faeces carry micro-organisms
feet leave trail of bacteria
hairs on body brush up dust and bacteria
saliva and regurgitated food deposited by proboscis

Fig. 11 Transmission of bacteria by houseflies
(After Brian Jones, *Human and Social Biology*)

In countries which do not have the advantage of efficient sewage disposal or piped water, intestinal diseases can be spread in drinking-water if infected faeces reach streams, rivers or lakes. This can cause widespread outbreaks of typhoid and cholera.

Contagion Contagious diseases are spread by physical contact. Tinea, a fungus infection (e.g. athlete's foot, ringworm) can be spread by contact with an infected area of skin or, more likely, by infected clothing or towels.

Some of the most serious contagious diseases are those spread by sexual contact; the sexually transmitted diseases (STDs). Examples are syphilis and gonorrhoea (bacterial) and genital herpes (viral). These are described in more detail below.

Some virus diseases

The common cold This is caused by a **rhinovirus.** It is spread by droplet infection and contact. The symptoms of the disease develop within 12–78 hours after infection and are very familiar: dry throat, watering of the eyes, a copious secretion of watery mucus from the nose, swollen (congested) nasal membranes making it difficult to breathe through the nose.

These symptoms last for a few days but the damage done by the virus to the nose and throat membranes often allows *Streptococcus* bacteria to invade. This secondary bacterial infection may give rise to a sore throat, a cough and catarrh.

Although the body develops immunity to the 'cold' virus, there are least 80 different strains of rhinovirus, as well as other species of virus which cause colds. Immunity to one of these strains does not extend to the others, and so colds can follow in swift succession.

There is no cure for a cold. Antibiotics are ineffective against rhinoviruses, but they may be prescribed to combat a secondary, bacterial infection if it is acute or persistent.

Influenza ('flu) is caused by a virus which exists in 3 strains, A, B and C. The virus attacks the lining of the throat and respiratory passages giving rise to inflammation of the trachea, bronchi and bronchioles. The patient will have a raised temperature, headache, dry cough and a mild sore throat and will feel generally 'rotten'. The symptoms subside in 2–4 days but the damage to the respiratory linings may allow *Streptococci* to invade, causing a secondary bacterial infection.

There are no specific drugs, but aspirin helps to lower the temperature, and antibiotics may be used against any secondary infection.

A bout of infection confers immunity for several years but the 'A' strain of the virus undergoes mutations (p. 226) very readily and immunity to one form is not effective against other mutants.

There are sometimes severe epidemics of influenza. If it is known which mutant is responsible, it is possible to prepare a vaccine which gives protection for a few months to people most at risk and to doctors and nurses dealing with the epidemic.

Herpes One variety of the herpes virus causes the 'cold sore'. The virus remains in the skin in 'dormant' condition, but may be activated by sunlight or by catching a cold. It produces small blisters which eventually break and form a dry scab under which the skin heals spontaneously, usually in a few days.

One strain of the herpes virus causes **genital herpes**. The blisters and 'sores', in this case, appear on the penis or scrotum in men and on the labial lips of the vulva in women. Genital herpes is spread mainly by sexual contact.

The sores usually heal in a week or two and there are drugs which alleviate the symptoms. The drugs, however, do not destroy the viruses, which remain dormant in the tissues and cause further outbreaks from time to time.

If a woman catches genital herpes in the late stages of pregnancy, the infection can harm the foetus, sometimes fatally.

Another form of herpes virus causes chicken-pox in the young, and shingles in older people.

AIDS The initials stand for Acquired Immune Deficiency Syndrome. (A 'syndrome' is a pattern of symptoms associated with a particular disease.) AIDS is caused by the human immunodeficiency virus (HIV). It attacks certain kinds of lymphocyte (p. 138), and thus weakens the body's immune responses.

As a result, the patient succumbs to illnesses, such as pneumonia and rare forms of blood disease to which he or she would normally be resistant.

The virus can enter the body directly through the bloodstream or during sexual intercourse. The people most at risk are promiscuous male homosexuals and drug addicts who share needles. Some infections have occurred from transfusions of infected blood.

AIDS cannot be spread by droplet infection or by casual contact such as touching or kissing.

The disease has appeared only recently in the industrialized countries and there are not yet any effective drugs or vaccines against it, though intensive work is taking place on both fronts.

Some bacterial diseases

Tuberculosis (TB) The bacterium *Mycobacterium tuberculosis* causes this disease. It usually enters the body through the nose or mouth and affects the lungs. The bacteria damage the lung tissue and give rise to a raised temperature, loss of weight and a persistent cough.

The disease is associated with poverty and has declined greatly in the more wealthy nations. In the Third World, however, there are probably 15 million people with the disease and about 3 million deaths per year.

The disease can be prevented largely by inoculation with BCG vaccine. This consists of a harmless form of the bacteria, which promote the formation of antibodies. Young people, who are the most susceptible to the disease, usually receive the vaccine when they are about 12 years old. Immunity lasts for 3–7 years.

The antibiotic streptomycin and the drug Isoniazid are among several effective treatments for TB.

Salmonella food poisoning One of the commonest forms of food poisoning is caused by the bacterium *Salmonella typhimurium*. This bacterium lives in the intestines of cattle, pigs, chickens and ducks. Humans may develop food poisoning if they drink milk or eat meat or eggs which are contaminated with *Salmonella* bacteria from the alimentary canal of an infected animal (Fig. 12).

The symptoms are diarrhoea and vomiting, which occur from 12 to 24 hours after eating the contaminated food. Although these symptoms are unpleasant, the disease is not serious except in the case

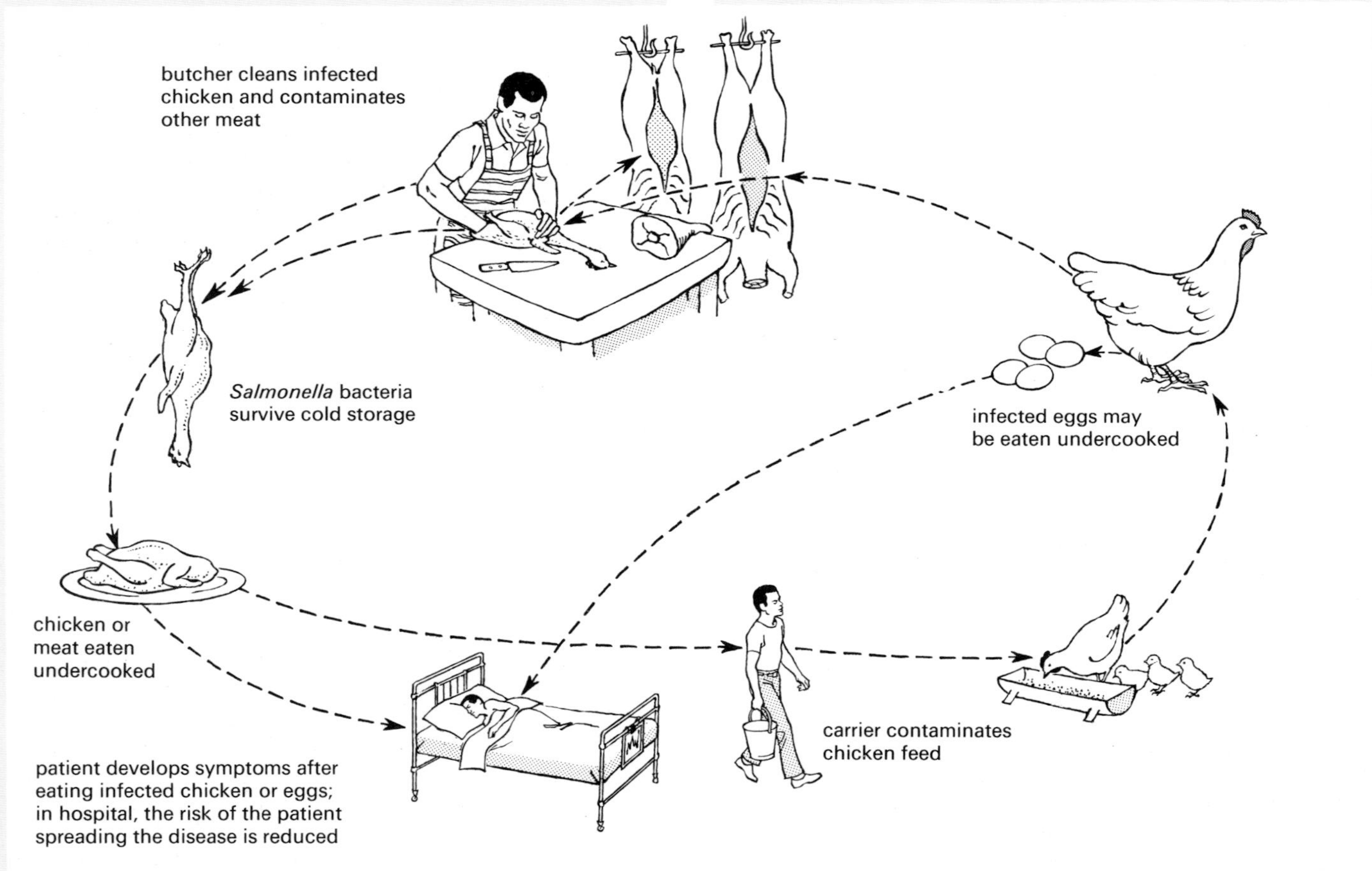

Fig. 12 Transmission of *Salmonella* food poisoning (After Brian Jones, *Human and Social Biology*)

of the elderly or the very young and does not need treatment with drugs.

The *Salmonella* bacteria are killed when meat is cooked or milk is pasteurized (heated for 30 minutes at 62 °C). Infection is most likely if untreated milk is drunk, meat is not properly cooked, or cooked meat is contaminated with bacteria transferred from raw meat.

It follows that to avoid the disease all milk should be pasteurized, meat should be thoroughly cooked and shop-assistants, and cooks, etc. should not handle cooked food at the same time as they handle raw meat unless they wash their hands thoroughly between the two activities.

In the past few years there has been an increase in the outbreaks of salmonella food poisoning in which the bacteria are resistant to antibiotics. Some scientists suspect that the practice of feeding antibiotics to farm animals to increase their growth rate, also causes populations of drug-resistant *Salmonellae* to develop (see p. 226).

The sexually transmitted diseases

These are diseases caught, almost exclusively, by having sexual intercourse with an infected person. It follows that the diseases can be avoided by not having sexual contact with infected people. The symptoms are often not obvious, but the people most likely to carry the disease are prostitutes and individuals who have sexual relationships with many other people.

Gonorrhoea is caused by a bacterium called *Neisseria gonorrhoea*. The first symptoms in men are pain and a discharge of pus from the urethra. In women, there may be similar symptoms or no symptoms at all.

In men, the disease leads to a blockage of the urethra and to sterility. In a woman, the disease can be passed to her child during birth. The bacteria in the vagina invade the infant's eyes and cause blindness.

The disease can be cured with penicillin but some strains of *Neisseria* have become resistant to this antibiotic. There is no immunity to gonorrhoea; having had the disease once does not prevent you catching it again.

Syphilis This is caused by a spirochaete bacterium called *Treponema pallidum*. In the first stage of the disease, a lump or ulcer appears on the penis or the vulva one week to three months after being infected. The ulcer usually heals without any treatment after about 6 weeks. By this time the spirochaetes have entered the body and may affect any tissue or organ. There may be a skin rash, a high temperature and swollen lymph nodes, but the symptoms are variable and the infected person may appear to be in good health for many years. However, if the disease is not treated in the early stages, the spirochaetes will eventually cause inflammation almost anywhere in the body and do permanent damage to the blood vessels, heart or brain, leading to paralysis and insanity.

In a pregnant woman, the spirochaetes can get across the placenta and infect the foetus.

Penicillin will cure syphilis but unless it is used in the early stages of the disease, the spirochaetes may do permanent damage.

Herpes Genital herpes is a virus disease and has been described on p. 315.

AIDS is also a virus disease and was discussed on p. 316.

If a person suspects that he or she has caught a sexually transmitted disease, treatment must be sought at once. Information about treatment can be obtained by phoning one of the numbers listed under 'Venereal disease' in the telephone directory; treatment is always confidential. The patients must, however, ensure that anyone with whom they have had sexual contact also gets treatment. There is no point in one partner being cured if the other is still infected.

CONTROL OF DISEASES

Clean water and efficient sewage disposal

The water that comes to our homes has been passed through filters which remove any bacteria. Chlorine has been added to destroy bacteria that might have escaped the filters. This treatment prevents the distribution of disease organisms in water.

Water from lakes, streams or rivers should be regarded with suspicion unless you are quite sure that the source is not contaminated. If in doubt, always boil water before drinking. This will kill any pathogenic bacteria present.

Sewage After sewage has been treated (p. 277), the effluent from the sewage works should be free from any intestinal bacteria and is safe to discharge into rivers.

In estuaries, however, untreated sewage, containing pathogenic micro-organisms, may be discharged and so estuaries could become a source of infection.

Hygienic food-handling

Intestinal diseases can be passed on from the hands of a person carrying the disease. People employed in the food industry may be medically examined to ensure that they are not carriers of intestinal disease. Apart from that, great care must be taken by food-handlers, whether in the home, in shops (Fig. 13) or in factories, to ensure that their hands and equipment do not become contaminated with bacteria that are present in faeces.

Food must be prepared, stored and displayed in such a way that flies cannot walk on it, or infected droplets fall on it (Fig. 14).

Fig. 13 Hygienic handling of food. A shop assistant avoids handling food with her fingers.

Fig. 14 Protection of food on display. The glass barrier stops customers from touching the food and helps stop bacteria from falling on the food.

Natural resistance

Barriers to entry The skin forms the first line of defence against the invasion of the blood and body cavities by micro-organisms. This is achieved partly by the physical barrier of the dead, cornified, outer layer of the skin and partly by production of chemicals which destroy bacteria. The eyes, for example, are protected by an enzyme in normal tears. The lining of the alimentary canal and respiratory passages also resist the entry of bacteria.

The mucus in the respiratory passages traps dust and bacteria which are then carried away by the action of the cilia (p. 9).

Defences in the blood The white cells, called **phagocytes**, engulf and digest bacteria. Other white cells, **lymphocytes**, produce antibodies which make the bacteria harmless, or easier for the phagocytes to ingest (see p. 146). Lymphocytes also produce anti-toxins which neutralize the poisonous toxins released by some bacteria.

Once having recovered from a disease, we are better able to combat a subsequent attack because the antibodies are present in the blood or can be made rapidly by the lymphocytes. We are then said to be **immune** to the disease. Not all diseases produce immunity. There seems to be no immunity to syphilis or gonorrhoea and immunity to some other diseases lasts only a short time.

Artificial immunity If you are inoculated with dead or harmless forms of micro-organisms or their products, your lymphocytes will make the appropriate antibodies. If the real micro-organisms get into your body later on, the antibodies will be ready and waiting to destroy them. This process was discussed more fully on p. 147.

Drug therapy The ideal drug for curing disease would be a chemical that destroyed the pathogen without harming the tissues of the host. In practice, modern antibiotics such as penicillin come pretty close to this ideal, for bacterial infections. Antibiotics seem to interfere with chemical processes in the bacterial cell wall and so cause it to break down. Since these processes do not occur in animal cells, the host tissues are unaffected. Nevertheless, many antibiotics cause some side-effects such as allergic reactions.

Not all bacteria are susceptible to antibiotics and most bacteria have a nasty habit of mutating to forms which are resistant to these drugs.

There are no very effective drugs against viruses. This is because the virus is so intimately involved in its host cell's physiology that any chemical which harms the virus, also harms the cell. In most cases virus diseases have to take their course.

Interferon When a virus invades an animal cell, the cell produces proteins called interferons. Interferons combat the viruses in several ways, though these have not been fully worked out. They probably block the action of viral DNA or RNA and, in some cases prevent the entry of the virus particles into healthy cells. A great deal of time and money is currently being spent in trying to manufacture interferon cheaply enough (by biotechnology) to try it out as a defence against virus diseases and some forms of cancer.

QUESTIONS

10 Why should people who sell, handle and cook food be particularly careful about their personal hygiene?
11 Coughing or sneezing without covering the mouth and nose with a handkerchief is thought to be inconsiderate behaviour. Why is this?
12 Inhaling cigarette smoke can stop the action of the cilia in the trachea and bronchi for about 20 minutes. Why should this increase a smoker's chances of catching a respiratory infection?
13 People living in closed, isolated communities, e.g. an Arctic expedition, often get a cold or two at first and then have several months free from colds. However, when the supply ship arrives, the colds start again. Try to explain this phenomenon.
14 After a disaster, such as an earthquake, the survivors are urged to boil all drinking-water. Why do you think this is so?
15 Antibiotics are ineffective against virus diseases. Why, then, are antibiotics sometimes given to people suffering from a virus infection such as influenza?

PRACTICAL WORK

Bacteria culture

In order to identify and investigate bacteria, it is necessary to culture them. This is done by mixing fruit juice, meat extract or other nutrients with agar jelly[1] to form a medium in which the bacteria can grow and reproduce. The bacteria multiply and form visible colonies (Fig. 15) whose colour and chemical reactions make them identifiable. By including or excluding certain substances from the culture medium, the essential conditions for the growth and reproduction of a particular bacterial species can be found. Also the effects of drugs, antibiotics and disinfectants can be investigated.

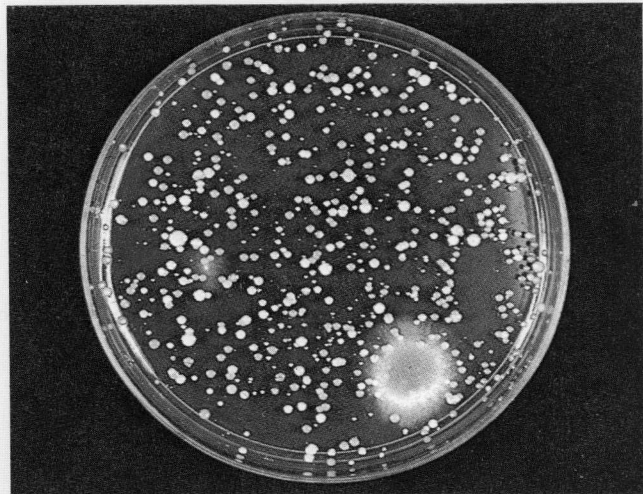

Fig. 15 Micro-organisms from the air. The nutrient agar was exposed to the air for a few minutes. Subsequently, these bacterial (and one fungus) colonies grew.

Bacteria are everywhere, in the air, on surfaces, on clothing and the skin. If you want to culture one particular species of bacteria, it is important to ensure that all other species are excluded. This means using **aseptic techniques**. The glassware and the culture media must be sterilized, and precautions must be taken to avoid allowing bacteria from the air or from the experimenter to contaminate the experiments.

The glassware and media are sterilized by heating them to temperatures which destroy all bacteria and their spores. Contamination is avoided by handling the apparatus in a way which reduces the chances of unwanted bacteria entering the experiments.

[1] Agar jelly, derived from seaweed, is used because bacteria cannot grow on it unless nutrients are added, and it is not liquefied by bacterial enzymes.

Culturing bacteria

The glassware must be sterilized by super-heated steam in an autoclave or pressure cooker for 15 minutes at 1 kg/cm^2 so that the unwanted bacteria on the glassware are destroyed.

1.5 g agar is stirred into 100 cm^3 of hot distilled water and 1 g beef extract, 0.2 g yeast extract, 1 g peptone and 0.5 g sodium chloride added. The mixture is sterilized in an autoclave and poured into sterile petri dishes which are covered at once and allowed to cool.

Precautions Culture methods can give rise to very dense colonies of potentially harmful bacteria. All bacteria should be treated as if they were harmful. When the petri dishes have been inoculated with the bacterial samples,the lids should be sealed in place with adhesive tape and not removed until the plates have been sterilized again at the end of the experiment. Any colonies which appear must be examined with the lids still on the dishes.

Inoculating the plates If a little cooked potato or other vegetable matter is allowed to rot for a few days in water, bacteria will arrive and grow in the liquid. A drop of this liquid is picked up in a sterile wire loop[2] and streaked lightly across the surface of the agar jelly, lifting the lid of the dish just far enough to admit the wire loop, but not enough to let bacteria from the air fall on to the agar (Fig. 16). The lid is then sealed on the dish with adhesive tape, and the dish is kept, upside down,[3] in an incubator or similar warm place for about 2 days. A 'control' dish is left unopened.

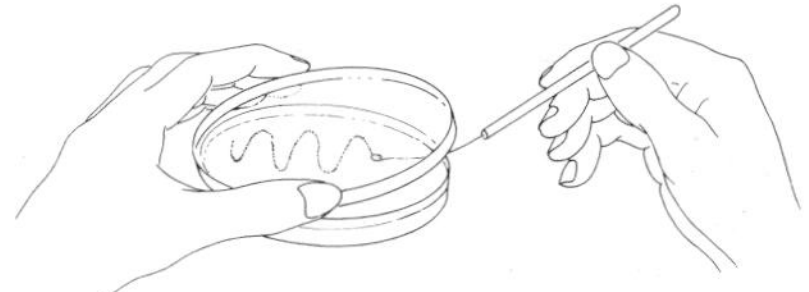

Fig. 16 Lift the lid as little as possible

Bacterial colonies should grow on the streaked dish, following the path taken by the loop (Fig. 17). There should be no growth of bacteria in the control dish, showing that the bacteria in the experiment came from the inoculating liquid and not from the glassware or the medium.

All the dishes, still sealed, are then sterilized once more in the autoclave before they are washed up.

The effect of antibiotics

The culture medium is made as before. When it is cool but before it sets, a few drops of a pure bacteria

[2] The wire loop is sterilized by heating it to redness in a bunsen flame.

[3] If water condenses in the petri dish, it will fall on the lid and not spread over the surface of the agar.

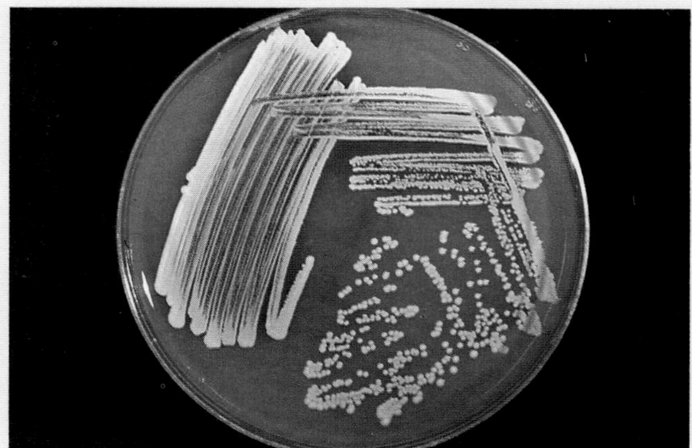

Fig. 17 A 'streak plate'. Notice how the bacterial colonies have grown where the plate was streaked.

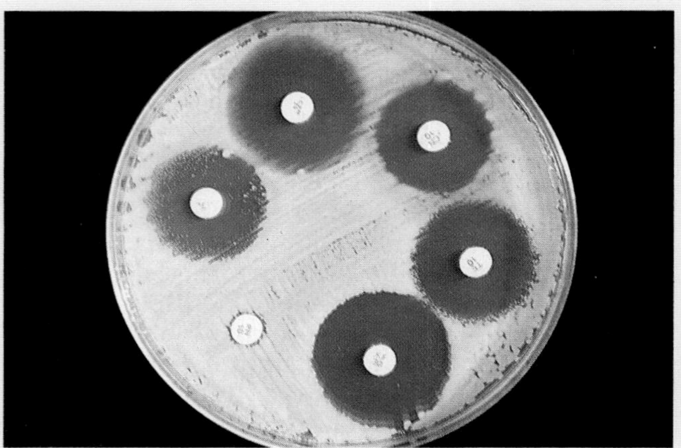

Fig. 18 Testing antibiotics. Antibiotics have diffused out from the discs and suppressed the growth of bacteria.

culture[4] are added and thoroughly mixed in with the medium. The medium is then poured into sterile petri dishes as before and allowed to set.

Using sterile forceps, some discs containing antibiotics are placed on the surface of the agar and the lids sealed on. The dishes are then incubated at about 35 °C for 24 hours.

The bacteria, which are dispersed in the agar, will grow and make the medium look cloudy, but in regions surrounding some of the antibiotic discs, bacterial growth will have been suppressed and the agar will look clear (Fig. 18).

If the bacteria are able to grow right up to the edge of a disc, it means that the antibiotic in that disc is unable to suppress the growth of the species of bacteria used in the trial.

[4] e.g. *Escherichia coli* or *Staphylococcus albus* from a reputable supplier.

QUESTION

16 Suggest reasons for the variation in extent of the clear areas round the discs in Fig. 18.

CHECK LIST

- **Bacteria are single cells; they have a cell wall, cytoplasm and a single chromosome.**
- **Bacteria produce enzymes which digest the surrounding medium.**
- **Bacteria reproduce by cell division.**
- **Bacteria are killed by heat but their spores survive.**
- **Viruses are smaller than bacteria and cannot, strictly, be classed as living organisms.**
- **Each virus particle consists of a DNA or RNA core enclosed in a protein coat.**
- **Viruses can reproduce only inside a living cell.**
- **Viruses take over the host cell's physiology and make it produce new virus particles.**
- **Most bacteria are saprophytes and bring about decay.**
- **Bacterial decay releases essential substances for recycling.**
- **Biotechnology exploits bacteria and other micro-organisms by making them produce materials useful to us.**
- **Viruses and bacteria can cause disease.**
- **They are spread by droplet infection, or in food and drink or by contact.**
- **Influenza, colds, herpes and AIDS are virus diseases.**
- **Tuberculosis, salmonella food poisoning, gonorrhoea and syphilis are bacterial diseases.**
- **Sexually transmitted diseases include syphilis, gonorrhoea, herpes and AIDS.**
- **You can become immune to many diseases by developing antibodies either after an infection or as a result of inoculation with a vaccine.**
- **Antibiotics are effective against many bacterial diseases but there are few drugs effective against virus diseases.**

34 Fungi

BIOLOGY OF FUNGI

Structure and nutrition.

EXAMPLES OF FUNGI

Mucor, rust fungus, mushroom, yeast, brewing and baking.

ECONOMIC IMPORTANCE OF FUNGI

PRACTICAL WORK

Growing mould-fungi, starch digestion, yeast in bread dough.

THE BIOLOGY OF FUNGI

The kingdom of the fungi includes fairly familiar organisms such as mushrooms, toadstools, puffballs, and the bracket fungi that grow on tree-trunks. There are also the less obvious, but very important, mould-fungi which grow on stale bread, cheese, fruit or other food. Many of the mould-fungi live in the soil or in dead wood. The yeasts are single-celled fungi similar to the moulds in some respects.

Some fungal species are parasites and live in other organisms, particularly plants, where they cause diseases which can affect crop plants.

Structure

Many fungi are not made up of cells but of microscopic threads called **hyphae**. The branching hyphae spread through the material on which the fungus is growing, and absorb food from it. The network of hyphae that grow over or through the food material is called the **mycelium** (Fig. 1).

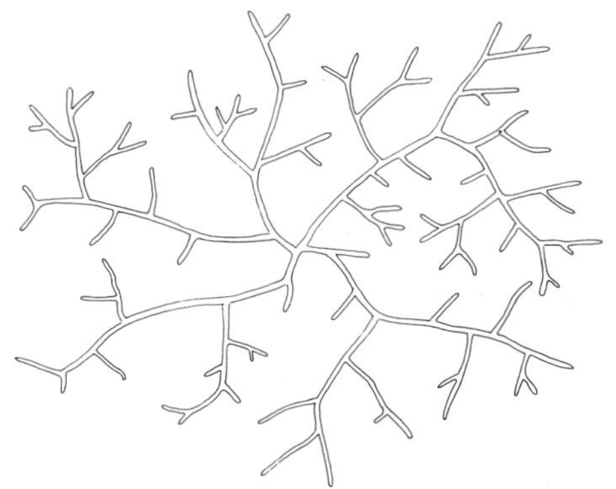

Fig. 1 The branching hyphae form a mycelium

Mushrooms and toadstools are the reproductive structures, 'fruiting bodies', of an extensive mycelium that spreads through the soil or the dead wood on which the fungus is growing.

The hyphae are like microscopic tubes lined with cytoplasm. In the centre of the older hyphae there is a vacuole, and the cytoplasm contains organelles and inclusions (p. 6). The inclusions may be lipid droplets or granules of glycogen but, unlike plants, there are no chloroplasts or starch grains (Fig. 2).

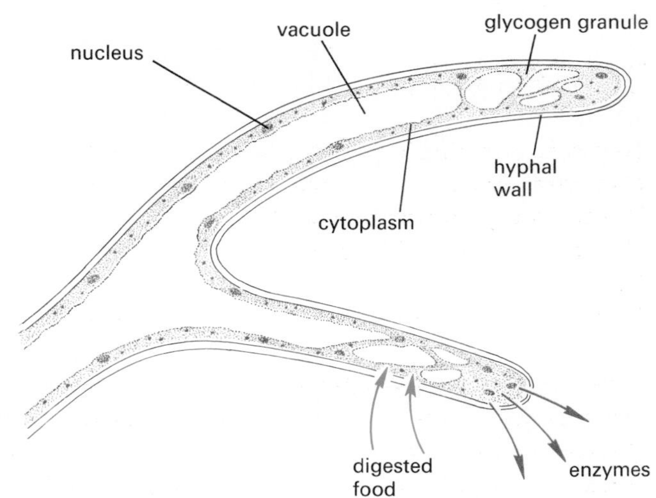

Fig. 2 Structure of a fungal hypha

The hyphal wall may contain cellulose or chitin or both, according to the species. Chitin is similar to cellulose but the chitin molecule contains nitrogen atoms.

In some species of fungi, there are incomplete cross-walls dividing the hyphae into cell-like regions, but the cytoplasm is free to flow through large pores in these walls. In the species which do have cross-walls (septa), there may be one, two or more nuclei in each compartment. In the species without septa, the nuclei are distributed throughout the cytoplasm.

Nutrition

Saprophytic fungi Most fungi are saprophytes (p. 250), living on dead organic matter (Fig. 3).

The hyphae secrete enzymes into the organic material and digest it to liquid products. The digested products are then absorbed back into the hyphae and used for energy or for the production of new protoplasm or hyphal walls.

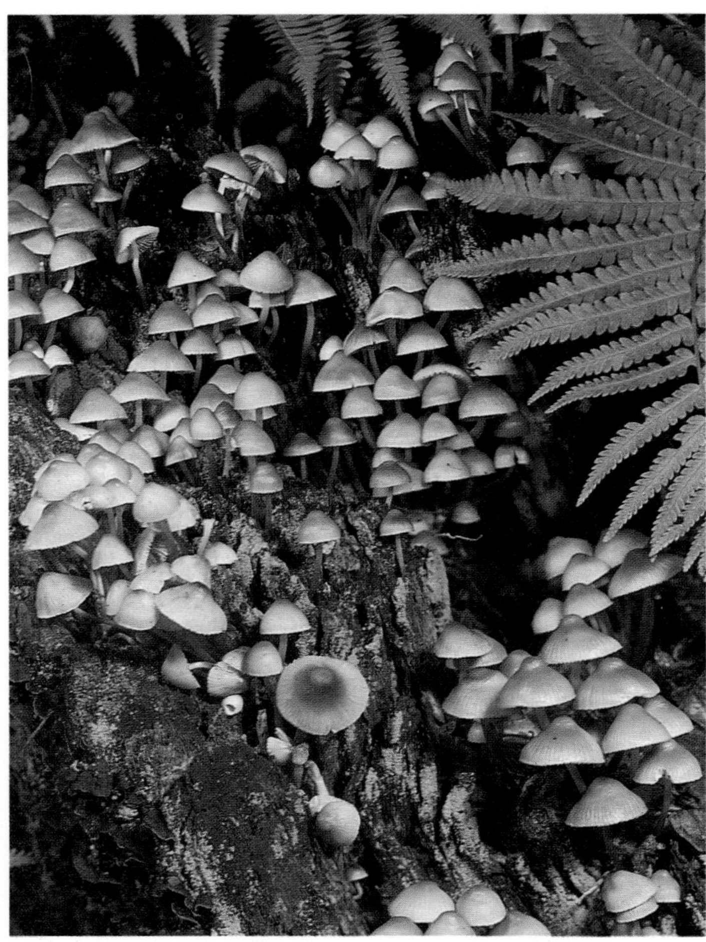

Fig. 3 Toadstools growing on a tree stump. The toadstools are the reproductive structures which produce spores. The feeding hyphae are inside the tree stump, digesting the wood.

Fig. 4 Mould fungus growing on an over-ripe orange

The type of enzymes produced by the hyphae will depend on the species of fungus and the material it is growing on. Those species which produce a **cellulase**, i.e. an enzyme which digests cellulose, are important in helping to break down plant remains (Fig. 4). Saprophytic fungi and bacteria are the 'decomposers' in most food webs (p. 244), and are largely responsible for the recycling of essential nutrients in any ecosystem (p. 281).

Parasitic fungi The hyphae of parasitic fungi penetrate the tissues of their host plant and digest the cells and their contents. If the mycelium spreads extensively through the host, it usually causes the death of the plant. The bracket fungus shown in Fig. 5 is the fruiting body of a mycelium that is spreading through the tree and will eventually kill it.

Fig. 5 A parasitic fungus. The 'brackets' are the reproductive structures. The mycelium in the trunk will eventually kill the tree.

Fungus diseases such as blight, mildews or rusts (Fig. 6) are responsible for causing considerable losses to arable farmers, and there is a constant search for new varieties of crop plants which are resistant to fungus disease, and for new chemicals (fungicides) to kill parasitic fungi without harming the host.

A few parasitic fungi cause diseases in animals, including humans. One group of these fungi cause tinea or ringworm. The fungus grows in the epidermis of the skin and causes irritation and inflammation. One form of tinea is athlete's foot, in which the skin between the toes becomes infected.

Tinea is very easily spread by contact with infected towels or clothing, but can usually be cured quickly with a fungicidal ointment.

Reproduction

Fungi reproduce by releasing microscopic, single-celled **spores**. You can see a cloud of these spores if you squeeze a mature puffball (Fig. 7).

The spores are budded off from the tips of special hyphae. Each spore contains a little cytoplasm and one or more nuclei, depending on the species. The spores are dispersed in air currents or by other methods. When a spore lands on suitable organic matter or on a new host, it germinates to produce a mycelium.

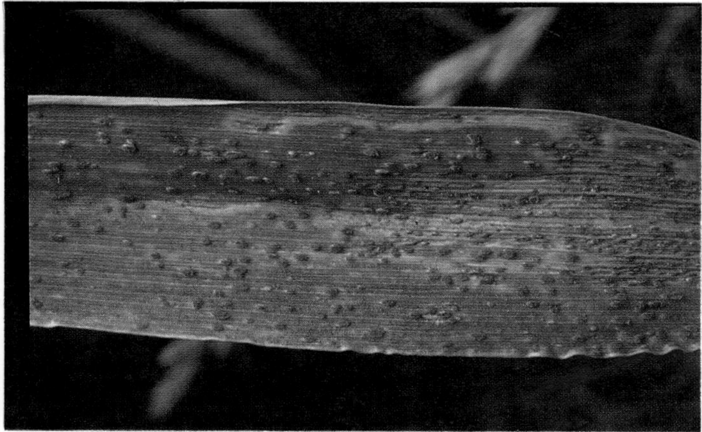

Fig. 6 'Rust' disease in wheat. The hyphae are in the leaf tissue digesting the cells. The brown spots are caused by spore-bearing hyphae breaking out of the leaf.

Fig. 7 Puff-ball dispersing spores. When a rain drop hits the ripe puff-ball, a cloud of spores is ejected.

The spores are produced from a single mycelium without the involvement of gametes or any sexual process, and so this is an example of **asexual reproduction** (p. 350).

Most fungi, however, do have a sexual process in their life cycle and this may precede the production of asexual spores.

QUESTIONS

1 When something goes mouldy what is actually happening to it?
2 Suggest why bread, wood and leather may go mouldy, while glass and plastic do not.
3 Suggest why toadstools may be found growing in very dark areas of woodland where green plants cannot flourish.

SOME EXAMPLES OF FUNGI

Mucor

Mucor is a common mould-fungus that grows on stale, damp food such as bread or cooked apple. The mycelium grows over and into the food, digesting and absorbing nutrients. Vertical hyphae grow up from the mycelium, giving it a 'fluffy' appearance. The tips of the vertical hyphae expand and produce **sporangia** (Figs 8 and 9). Each sporangium contains hundreds of spores, which are released when the sporangium wall breaks down.

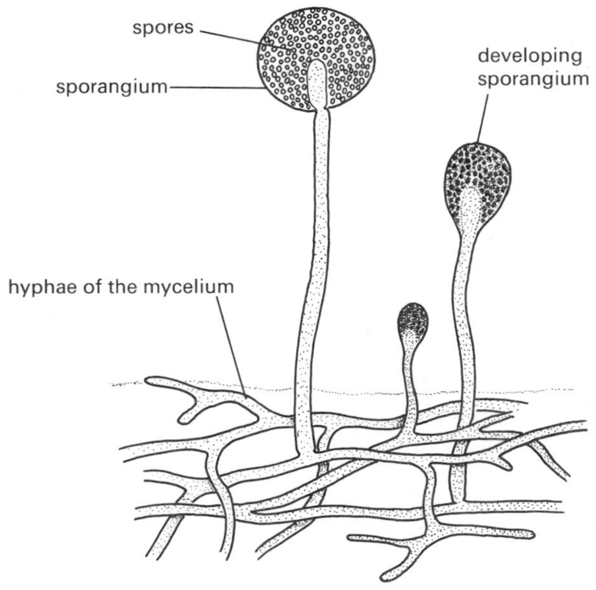

Fig. 8 Spore production in *Mucor*

Fig. 9 A mould fungus ($\times 100$). The white threads are the hyphae; the black knobs are the sporangia.

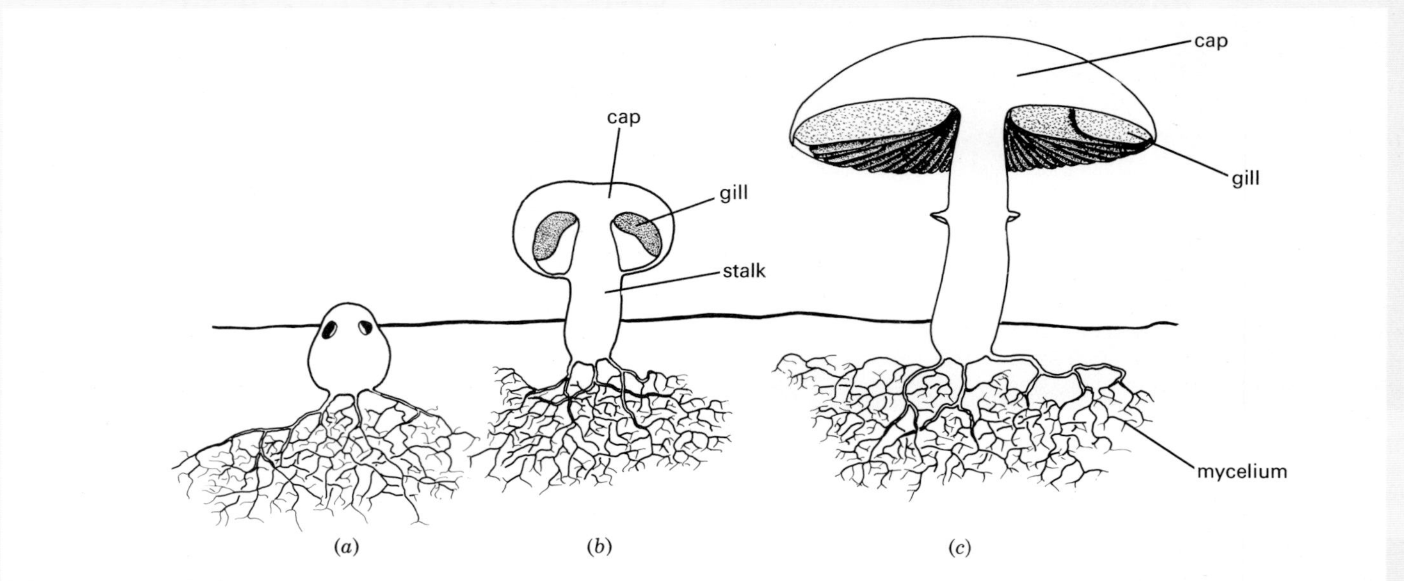

Fig. 10 Development of a mushroom

Mushrooms

The mycelium of a mushroom is a diffuse, branching network of hyphae which may spread over a radius of many feet in the surface layers of the soil. At the time of reproduction, usually in the autumn, some of the hyphae form a dense mass which emerges from the soil as a compact 'fruiting body'. The growth and structure of this fruiting body is shown in Fig. 10.

The gills are formed on the underside of the cap and they consist of special hyphae which bud off spores at their tips. The spores are flicked into the spaces between the gills and are carried away in air currents (see Experiment 6).

Yeast

The yeasts are a rather unusual family of fungi. Only a few of the several species can form true hyphae. The majority of them consist of separate, spherical cells, which can be seen only under the microscope. They live in situations where sugar is likely to be available, e.g. in the nectar of flowers or on the surface of fruits.

The thin cell wall encloses the cytoplasm, which contains a nucleus and a vacuole. In the cytoplasm are granules of glycogen and other food reserves (Fig. 11*a* and Fig. 11, p. 6).

The cells reproduce by budding. An outgrowth from the cell appears, enlarges and is finally cut off as an independent cell. When budding occurs rapidly, the individuals do not separate at once and, as a result, small groups of attached cells may sometimes be seen (Fig. 11*b*).

Fermentation Yeasts are of economic importance in promoting alcoholic fermentation. Yeast cells contain many enzymes, some of which can break down sugar into carbon dioxide and alcohol. This chemical change provides energy for the yeast cells to use in their vital processes.

$$C_6H_{12}O_6 \rightarrow 2CO_2 + \underset{\text{alcohol}}{2C_2H_5OH} + 118\ \text{kJ}$$

Alcoholic fermentation is a form of anaerobic respiration (p. 24) but, if the yeast is supplied with carbohydrates other than sugar, the yeast needs oxygen to convert these substances to sugar first.

In brewing, barley is allowed to germinate. During germination, the barley grains convert their starch reserves into maltose (p. 16). The germinating barley is then killed by heat and the sugars are dissolved out with water. Yeast is added to this solution and ferments it to carbon dioxide and alcohol.

In making beer, hops are added to give the brew a bitter flavour and the liquid is sealed into casks or bottles so that the carbon dioxide is under pressure. When the pressure is released, the dissolved carbon dioxide escapes from the liquid giving it a 'fizz'.

In making spirits such as whisky, the fermentation is allowed to go on longer and the alcohol is distilled off. Wine is made by extracting the juice from fruit, usually grapes, and allowing the yeasts, which live on the surface of the fruit, to ferment the sugar to alcohol.

If too much oxygen is admitted or fungi and bacteria are allowed to enter, the alcohol may be oxidized to ethanoic (acetic) acid, i.e. vinegar.

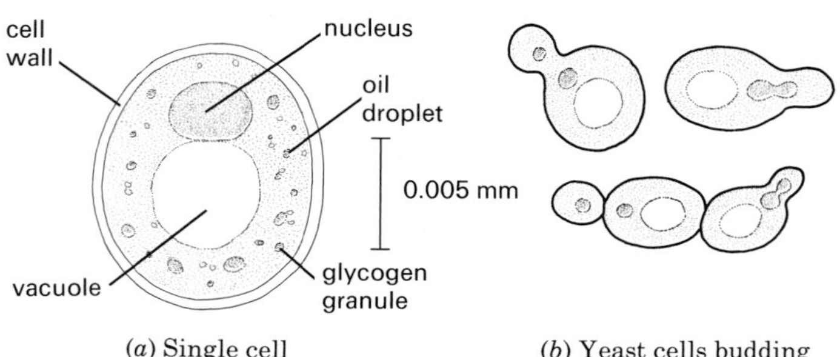

Fig. 11 Yeast

Baking In baking, flour, salt, fat and water are mixed to produce 'dough'. Enzymes in the flour start to convert the flour-starch to sugar. When yeast is added, it ferments the sugar and produces carbon dioxide. The bubbles of carbon dioxide make the dough 'rise'. The bubbles also expand when the dough is baked and so give the bread a 'light' texture.

Yeast as food Yeasts may become important as a source of single-cell protein (p. 314) for humans and their farm animals. Given only sugar and inorganic salts, these micro-organisms will grow and reproduce very rapidly, converting the sugar and salts into protein.

An eminent biologist once said that, 'In 24 hours, half a tonne of bullock will make a pound of protein; half a tonne of yeast will make 50 tonnes and needs only a few square metres to do it on.'

Yeasts contain most of the essential amino acids and vitamins needed by humans and, at present, the yeasts produced on a commercial scale are used to supplement diets which are inadequate in these respects. Yeast is also used to remedy vitamin B deficiency diseases.

ECONOMIC IMPORTANCE OF FUNGI

Harmful fungi

These are the fungi which cause diseases in our crop plants and ourselves, and which cause food spoilage or rotting of timber and leather.

Beneficial fungi

The saprophytes These fungi play a part in recycling essential nutrients (p. 245).

Biotechnology This exploits fungi which bring about chemical changes, e.g. in the manufacture of citric acid. Biotechnology includes the production of single-cell protein and the long-standing industries of brewing and baking (p. 313).

Antibiotics Many of the fungi which live in the soil compete with bacteria and each other for supplies of organic material to use as food. One form of competition involves the production of chemicals which inhibit the growth of competitors. We have learned to extract these chemicals, now called antibiotics, and to use them in combating some of the bacteria which cause human diseases.

Pure cultures of the fungi are grown on nutrient solutions on a large scale and the antibiotics are extracted and purified from the growth medium. A well-known example is the production of penicillin from the mould *Penicillium*.

There are also soil bacteria which produce similar antibiotics, e.g. *Streptomyces* from which we obtain streptomycin. Some antibiotics are now made synthetically rather than from culturing micro-organisms.

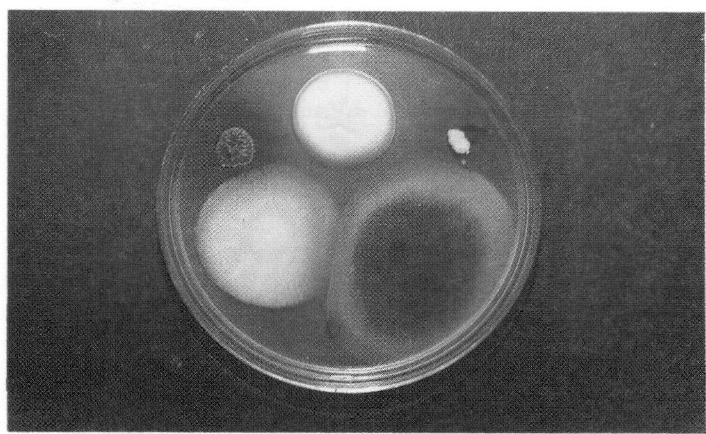

Fig. 12 Different species of mould fungus growing on vegetable agar

QUESTION

4 Fungi do more good than harm. Put some points for and against this statement.

PRACTICAL WORK

1. Growing mould-fungi

Moisten a piece of stale bread, place it in a petri dish base and leave it exposed to the air. After a day, moisten it again if it has dried out, and then cover it with a beaker or small jar. In the humid atmosphere inside the jar, colonies of mould-fungi should grow on the bread within a few days.

If the moulds *Mucor* or *Rhizopus* grow, the sporangia will appear as black dots on the ends of the vertical hyphae (Fig. 9 on p. 323).

The experiment can be duplicated by using different food material, e.g. pieces of banana or cooked apple.

2. Culturing fungi

Mould-fungi or even small toadstools can be cultured on dishes of agar as described for bacteria on p. 319. Follow the same techniques as described for bacteria, but use vegetable agar in place of nutrient agar. The vegetable agar is made simply by dissolving 2 g agar and 25 cm^3 tomato juice in 100 cm^3 water. The vegetable agar and 5 petri dishes are sterilized in an autoclave. The cool, but still liquid agar is poured into the sterile dishes and allowed to set.

The dishes are then exposed to the air or inoculated with fungus spores or hyphae using a sterile wire loop as shown in Fig. 16 on p. 319. The fungi may be selected from one of the colonies that grew in Experiment 1.

The dishes are incubated for 24 hours at about 35 °C or left at room temperature for 2 or 3 days (Fig. 12).

3. Starch digestion by a mould-fungus

Prepare starch agar by dissolving 0.3 g starch powder and 1 g agar in 100 cm^3 water. If the experiment can be examined 1 or 2 days after setting up, there is no need to sterilize the agar or the glassware.

Pour the agar, when cool, into petri dishes, replace the lids and allow the medium to set.

Inoculate the agar with fungal spores or hyphae, by making one or two strokes across it with a sterile wire loop that has been dipped into a mature fungal colony, as described for Experiment 2. Replace the lid and leave the dish for 1 or 2 days.

After this time, the fungus should be seen growing along the lines of the streaks.

Remove the lid, pour iodine solution into the dish to cover the agar and leave for a minute or two. Finally wash away the iodine with water.

Iodine solution will turn starch-agar blue. So, any areas which remain clear have no starch in them and it is reasonable to assume that the fungus has secreted an enzyme into the agar and digested the starch.

A control should be set up at the same time as the experiment, by streaking a starch-agar plate with a wire loop that has been heated to redness in a bunsen flame. This plate should also be tested with iodine solution.

4. Anaerobic respiration in yeast

An experiment to investigate anaerobic respiration in yeast has been described on p. 29.

5. The function of yeast in bread dough*

One day before the experiment make a yeast-suspension 'starter' by mixing a level teaspoonful of dried yeast with half a teaspoonful of sugar and 20 cm^3 water, in a small flask or bottle. Plug the mouth of the flask with cotton wool and leave it in a warm place for 24 hours.

Mix the following in measures of level teaspoons: plain white flour, 2; sugar, $\frac{1}{2}$; yeast suspension, $\frac{1}{2}$; water, 2. This should make a sticky dough.

Withdraw the plunger of a 1 cm^3 syringe to the 0.5 mark and then draw up about 0.3 cm^3 dough into the syringe. Note the level of the dough and plug the nozzle of the syringe with a piece of sharpened matchstick. Place the syringe in a water bath kept at about 35 °C and note the level of the dough at 5-minute intervals.

A control should be set up using the same mixture but made with yeast suspension that has been boiled for a minute to kill the yeast cells.

The results should indicate what part is played by yeast in making bread. The experiment also demonstrates anaerobic respiration, since there is little or no air present in the bread dough.

The rate of anaerobic respiration at different temperatures may be compared if similar syringes are placed in water baths at 15, 25 and 45 °C. The rate at which the dough rises is a measure of the rate of carbon dioxide production.

*After P. W. Freeland, *School Science Review*, No. 194.

6. A spore map from a mushroom

Remove the stalk from a mature mushroom (one with dark brown gills). Place the cap on a piece of paper, so that the gills are touching the paper. One day later, carefully lift the cap off the paper. A pattern of dark lines will show where the spores have been ejected from the gill surface.

QUESTION

5 Explain the purpose of the controls in Experiments 3 and 4.

CHECK LIST

- **Fungi are formed from thread-like hyphae rather than cells.**
- **The branching hyphae produce a network called a mycelium.**
- **The hyphae have walls, cytoplasm, nuclei, vacuoles, organelles and inclusions.**
- **Fungi secrete enzymes into their food and absorb the digested products.**
- **Saprophytic fungi digest dead organic matter.**
- **Parasitic fungi digest living tissues.**
- **The saprophytic bacteria are important as decomposers; they release essential nutrients from dead organic matter.**
- **Parasitic fungi cause plant diseases some of which affect our crops.**
- **Fungi reproduce asexually by releasing spores which can grow into new mycelia.**
- **Mushrooms and toadstools are the reproductive bodies of a hidden mycelium.**
- **Attempts are made to control parasitic fungi by using fungicides and by breeding resistant strains of crop plants.**
- **Yeasts are single-celled fungi, which are important in brewing, baking and other forms of biotechnology.**
- **Yeasts respire anaerobically, causing a fermentation process which produces CO_2 and alcohol.**
- **Fungi and bacteria produce chemicals called antibiotics, which can be used to suppress the growth of disease bacteria in humans.**

35 Protista and some Lower Plants

PROTISTA

Paramecium, organization within a single cell, useful and harmful protistans, surface area and volume.

ALGAE

Plankton, *Spirogyra*, seaweeds.

MOSSES AND FERNS

PROTISTA

Most living things have bodies made up of thousands of cells but there are also large numbers of tiny creatures which consist of only a single cell. These are the protista. Most of them are microscopic but a few can be seen as specks with the naked eye.

Nearly all protista live in water; in the sea, rivers and lakes, in puddles and in the body fluids of animals. Protista differ from bacteria because they have a definite nucleus in their cells and because they feed in a variety of different ways. Some take in solid food and some photosynthesize.

Protista may also be referred to as 'unicellular organisms' or 'unicells'. Those protista which take in solid or liquid food from their surroundings may be called **protozoa** (animal-like unicells); those which possess chlorophyll may be called unicellular plants, unicellular algae or **protophyta**.

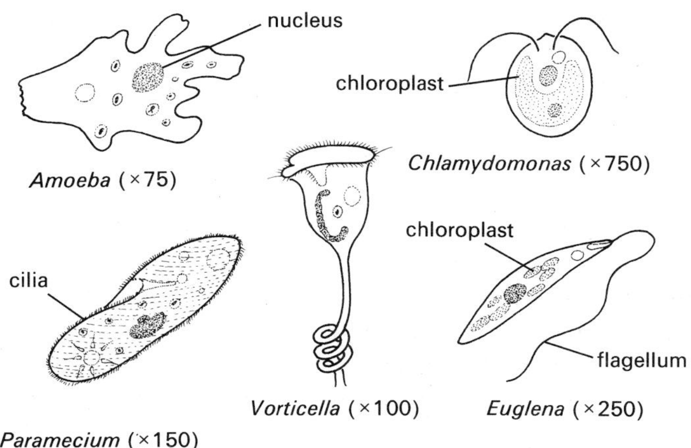

Fig. 1 Protista. *Chlamydomonas* and *Euglena* have chloroplasts and can photosynthesize. The others are protozoa and ingest solid food.

Amoeba (Fig. 1) is a protozoan which moves by a flowing movement of its cytoplasm. It feeds by picking up bacteria and other microscopic organisms as it goes. *Vorticella* has a contractile stalk and feeds by creating a current of water with its cilia (p. 9). The current brings particles of food to the cell. *Euglena* and *Chlamydomonas* have chloroplasts in their cells and feed, like plants, by photosynthesis.

Many of the protista can move about by means of structures such as **cilia** or **flagella**. Cilia are like microscopic hairs which can flick to and fro. Protista which have flagella are called **flagellates**; those which move by means of cilia are called **ciliates**.

Paramecium

Paramecium is a fairly typical ciliate. It is abundant in fresh water where it feeds on bacteria and other microscopic organisms.

Figure 2 shows its structure in more detail. The cilia covering its body are arranged in rows, and they beat in a regular rhythm, driving the *Paramecium* through the

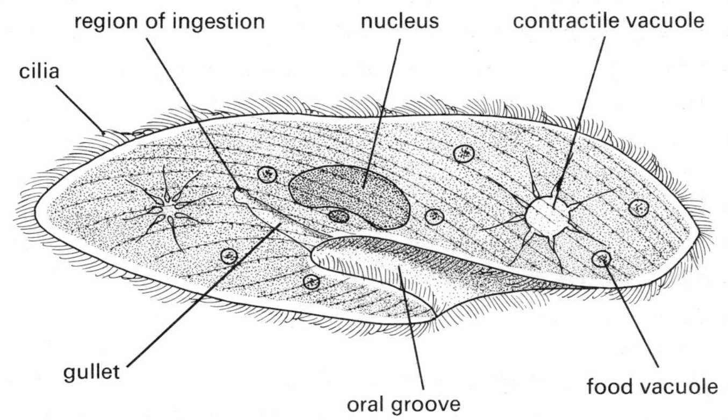

Fig. 2 ***Paramecium.*** A ciliate protozoan.

water. The ciliary beat can be reversed, so the *Paramecium* can go backwards.

In one region of its body, a special band of cilia creates a current of water which wafts small particles into the **oral groove.** At the end of the oral groove, the particles are taken into the cytoplasm in small vacuoles (endocytosis, p. 34). The cytoplasm secretes enzymes into the vacuole and these enzymes digest the food. The soluble products are then absorbed back into the cytoplasm, and the undigested remains are expelled at a point on the cell surface not far from the oral groove (exocytosis, p. 34).

In addition to the **food vacuoles**, there are two larger vacuoles which can be seen to expand and contract alternately. These are the **contractile vacuoles** which are responsible for expelling excess water. *Paramecium*'s cytoplasm has a lower water potential (p. 37) than the surrounding water, so water enters the cell by osmosis. This water collects in the contractile vacuoles and is expelled when they contract. The contractile vacuoles thus have an osmo-regulatory function (see p. 162).

Paramecium reproduces by **binary fission**. This is an asexual process in which, first the nucleus divides, and then the cytoplasm, to produce two independent *Paramecia* (Fig. 3). There is also a sexual process in the life cycle of *Paramecium.*

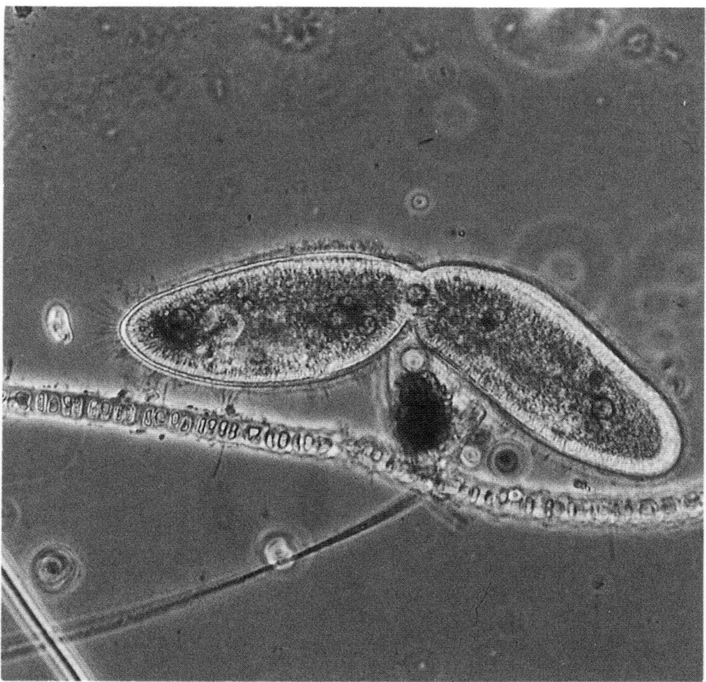

Fig. 3 ***Paramecium*** **dividing (binary fission)** (×350). The nucleus has already divided and now the cytoplasm is pinching off in the middle to produce two organisms.

Life processes in the protista

The animals and plants studied so far in this book, have been multicellular, i.e. their bodies are made up of thousands of millions of cells. Some of these cells are specialized to carry out particular functions, and the specialized cells are grouped into organs and systems (p. 10).

A protistan, however, can carry out all the vital processes within a single cell. If there is any specialization, it is specialization of a region of cytoplasm. For example, the oral groove in *Paramecium* is specialized to take in food; the cilia are specialized filaments of cytoplasm which can flick to and fro.

Digestion In most multicellular animals, digestion is carried out by an alimentary canal with specialized regions and glands. In *Paramecium,* almost any part of the cytoplasm can form a food vacuole. There are no glands; enzymes are produced by the cytoplasm which surrounds the food vacuole.

Osmo-regulation In higher animals this involves the complex structure of the kidneys and is effected, in *Paramecium,* by the specialized regions of cytoplasm which form the contractile vacuoles.

Gaseous exchange In the vertebrates, there are elaborate structures, such as lungs or gills, for gaseous exchange, and a blood circulation to carry oxygen and dissolved food to all parts of the body.

In a protistan, the surface area of its single-celled body is very large in comparison with its volume (see below). This allows absorption of oxygen from the surrounding water to take place rapidly by diffusion. No special absorptive surfaces are needed.

Transport Similarly, the distance across a protistan is so small, that diffusion is rapid enough to distribute oxygen from the water, and food from the food vacuoles, to all parts of the cytoplasm. There is no need for a specialized transport system.

Nitrogenous excretory products and carbon dioxide diffuse rapidly out of the cytoplasm and into the water. There is no specialized excretory system.

The same principles apply to those protistans that have chloroplasts, and produce food by photosynthesis. Carbon dioxide diffuses into the cell quickly, and oxygen diffuses out. Food made in the chloroplasts diffuses to all parts of the cytoplasm.

Protista and humans

The protista form the basis of most food webs in the sea and fresh water. Their activities are exploited in the filtration of drinking-water and the purification of sewage. The ciliates, in particular, digest the bacteria present in untreated water and sewage effluent. The protists are themselves eaten by small multicellular organisms such as worms and insect larvae.

Some protists are human parasites. An Amoeba-like species causes amoebic dysentery; a flagellate called *Trypanosoma* causes sleeping sickness; another *Amoeba*-like protist, called *Plasmodium,* invades the red blood cells and causes malaria.

The malarial parasite digests the red cell's cytoplasm and also reproduces inside it. The newly formed parasites break out of the cell and invade other red cells (Fig. 4). This gives rise to a malarial fever and causes anaemia. The parasites are carried from person to person by blood-sucking mosquitoes.

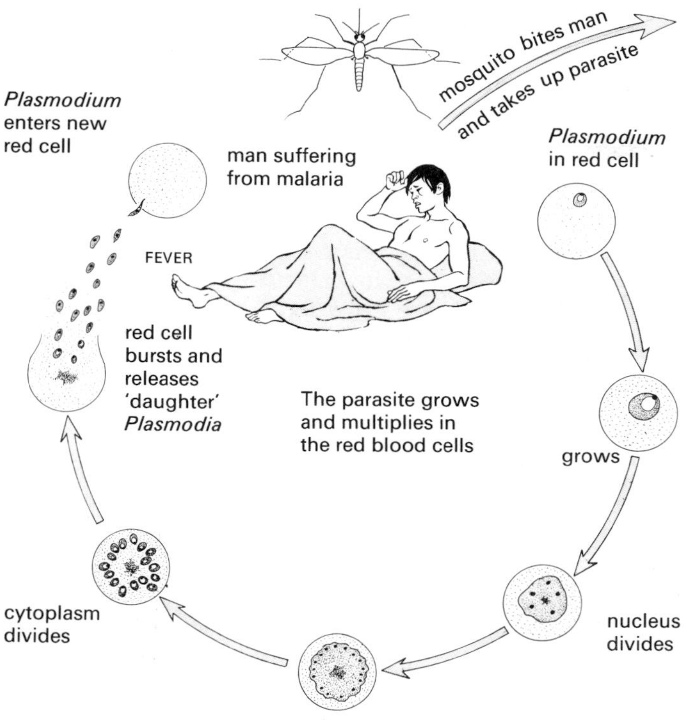

Fig. 4 ***Plasmodium*, the malarial parasite**

Surface area and volume

Imagine an unlikely animal which consists of a 2 cm cube of cytoplasm (Fig. 5*a*). The animal's volume is $2 \times 2 \times 2 = 8$ cm^3. Its surface area is 6×4 cm$^2 = 24$ cm^2. Suppose that 1 cm^3 of cytoplasm needs 5 cm^3 oxygen per hour. The cubic animal will need 5×8 cm^3 oxygen hourly and it will absorb the gas through its surface.

Now the animal divides, by binary fission (Fig. 5*b*). The volume of each offspring is half that of its parent, namely, 4 cm^3, but its surface area is 16 cm^2, i.e. more than half. Each offspring will need half the oxygen its parent needed, i.e. 5×4 cm^3 oxygen per hour, but will have 16 cm^2 surface, instead of 12 cm^2 to absorb it.

This crude example illustrates the fact that, as you move down the scale of size, the ratio, surface area/volume, becomes greater as the size diminishes, and the relatively increased surface area makes it easier to exchange gases with the environment.

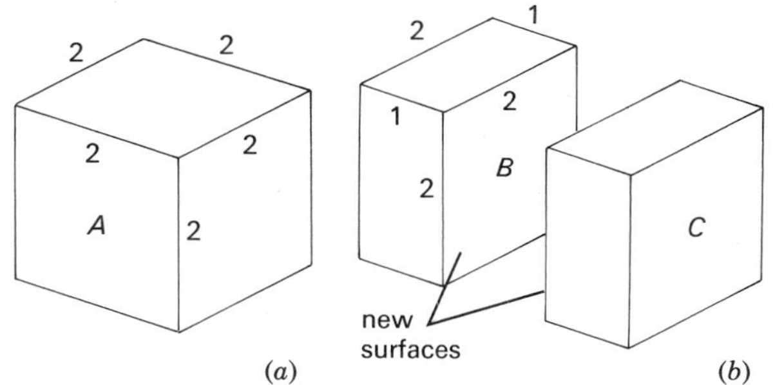

Fig. 5 Surface area and volume. When *A* divides into *B* and *C* two new surfaces are formed.

QUESTIONS

1 In what ways are feeding and digestion in *Paramecium* similar to the same processes in humans?

2 If a single cell from your body was placed in pond water it could not survive for long. *Paramecium* is a single cell and can survive in pond water. What can *Paramecium* do that your body cell cannot do?

ALGAE

The algae are plants which range from very simple filaments, consisting of a single row of identical cells, to large seaweeds. The protistans with chloroplasts (the protophyta) may also, with good reason, be classified as algae.

Most members of the algae have cellulose cell walls and chloroplasts. The chloroplasts contain chlorophyll, but other pigments may be present in brown and red algae.

The algae, for the most part, grow in water and have little need of specialized supporting or water-conducting tissue. Compared with vascular plants, therefore, their cells and tissues are relatively unspecialized. The aquatic algae cannot survive prolonged desiccation when out of water.

Plankton

In open water, it is the single-celled algae which form the basis of the food web. These algae, particularly the forms called **diatoms**, are abundant in the surface waters of ponds, lakes and the sea. Usually they are invisible to the naked eye, but occasionally, in a pond or neglected swimming-pool, the unicellular plants are so numerous that the water looks green.

The protophyta make their food by photosynthesis and reproduce rapidly. These microscopic plants are eaten by small animals, particularly crustaceans such as water fleas and *Cyclops* (p. 286) and their marine counterparts.

The protophyta and the small animals which eat them are called, collectively the **plankton**. The microscopic plants are the **phytoplankton**; the small animals are the **zooplankton** (Figs 3*a* and *b* on p. 243). Larger, shrimp-like crustacea and small fish feed on the plankton and are themselves eaten by larger animals.

The plankton, in effect, is the 'pasture of the sea' and other open waters.

Spirogyra

Spirogyra is a filamentous alga. It consists of hair-like green threads. When the threads are studied under the microscope, they are seen to consist of single cells joined end to end (Fig. 6). All the cells are the same; there is no specialization and any one cell is capable of surviving on its own.

Each cell has a cellulose cell wall, a lining of cytoplasm and a central vacuole. There is a ribbon-like chloroplast which runs through the cytoplasm in a helix. The nucleus is in the centre of the cell, supported by strands of cytoplasm (Fig. 7).

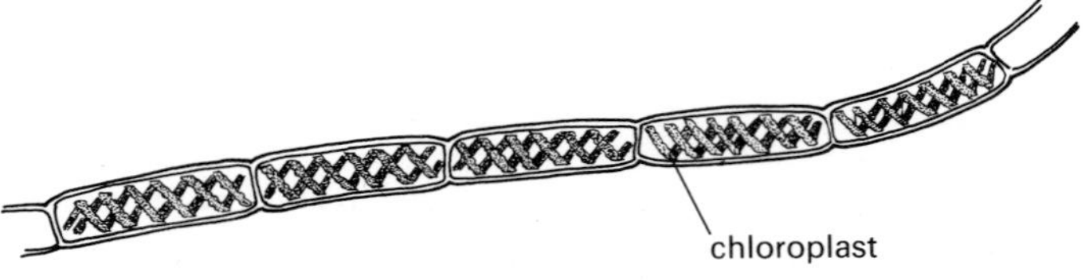

Fig. 6 ***Spirogyra*, part of a filament**

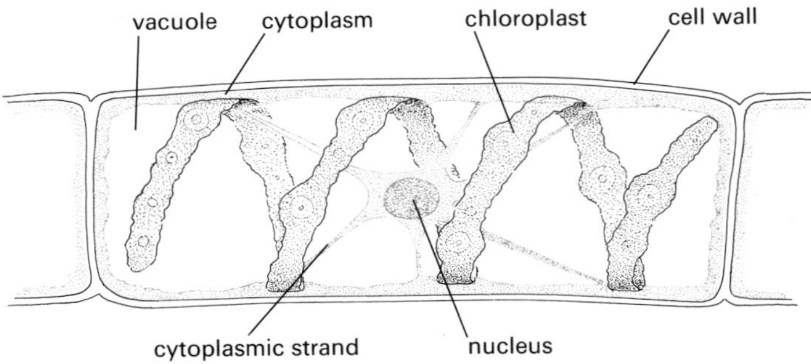

Fig. 7 ***Spirogyra*, one cell from a filament**

Each *Spirogyra* cell can photosynthesize. The cells absorb water, carbon dioxide and salts from their surroundings and build up their food with the aid of sunlight (p. 44). The chloroplast absorbs the sunlight and makes its energy available for chemical processes within the cell.

The colonies of *Spirogyra* filaments are often seen as a cloudy green mass in shallow water. When photosynthesis is rapid, the oxygen given off as a waste product often makes the *Spirogyra* colonies float to the surface as a kind of green scum.

There are many different kinds of filamentous algae. The different species can usually be identified by the shape of the chloroplast when seen under the microscope.

Seaweeds

Seaweeds are larger and more complex algae than *Spirogyra* (Fig. 3, p. 304), with many more cells in their bodies. The cells are arranged to form a flat, ribbon-like *thallus* which, at one end, is attached firmly to a rock (Fig. 8). At the other end are actively dividing cells which cause the thallus to grow in length.

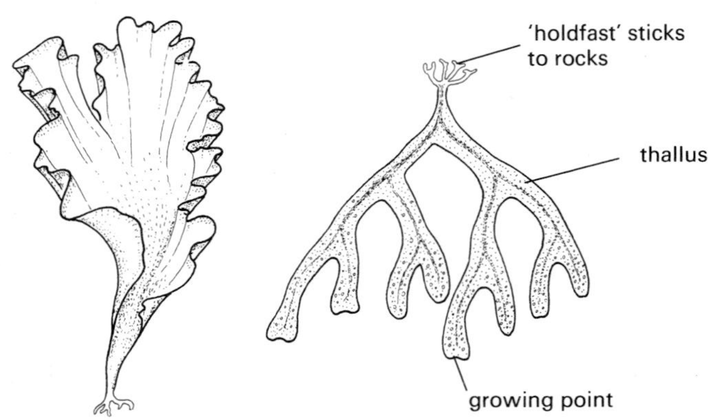

Fig. 8 **Two types of seaweed**

There is not much obvious specialization in the cells. The outer cells of the thallus contain chloroplasts and make food by photosynthesis. The cells in the region of the midrib are often elongated and may have a conducting function. There are generally no roots, leaves or stem and no conducting tissues like the xylem and phloem of the flowering plants. There is no supporting tissue in the thallus of seaweeds, so at low tide they collapse over the rocks on which they are growing. At high tide they are buoyed up and supported by the surrounding water.

QUESTIONS

3 Compare a cell of *Spirogyra* with the palisade cell shown in Fig. 7 on p. 4. List the structures that are present in both cells. In what ways is the *Spirogyra* cell different from the palisade cell?
4 What do the unicellular algae in the phytoplankton need in order to produce food? Where do they get these things from?
5 Look at the picture of *Laminaria* in Fig. 3 on p. 304. What specialization does there appear to be in the body of this seaweed?

MOSSES AND FERNS

These are non-flowering plants with stems and leaves. They do not produce seeds but reproduce by means of unicellular spores.

Mosses

The mosses are simple, but successful land plants. Each plant is quite small, on average about 1–5 cm long, but they usually grow in dense tufts which probably helps to support them and to conserve moisture.

Each plant has a slender stem with numerous small leaves arising from it (Fig. 9). Sometimes the stem is horizontal and has many branches. Most moss leaves are only one cell thick, so there is no specialization of cells like there is in the leaves of flowering plants. In the stem, the innermost cells conduct water and food, but apart from being longer than other cells, they are not particularly specialized in their structure. The outer cells have thick walls and probably give the stem strength.

There are no proper roots, though structures called **rhizoids**, rather like multicellular root hairs, grow into the soil from the base of the stem.

Mosses reproduce sexually but do not form seeds. When the male and female gametes fuse at the tip of a moss plant, they produce a spore capsule on the end of a long stalk (Fig. 10). The spore capsule opens to release single-celled spores which are scattered and grow into new moss plants.

Four different species of moss are illustrated in Fig. 3 on p. 304.

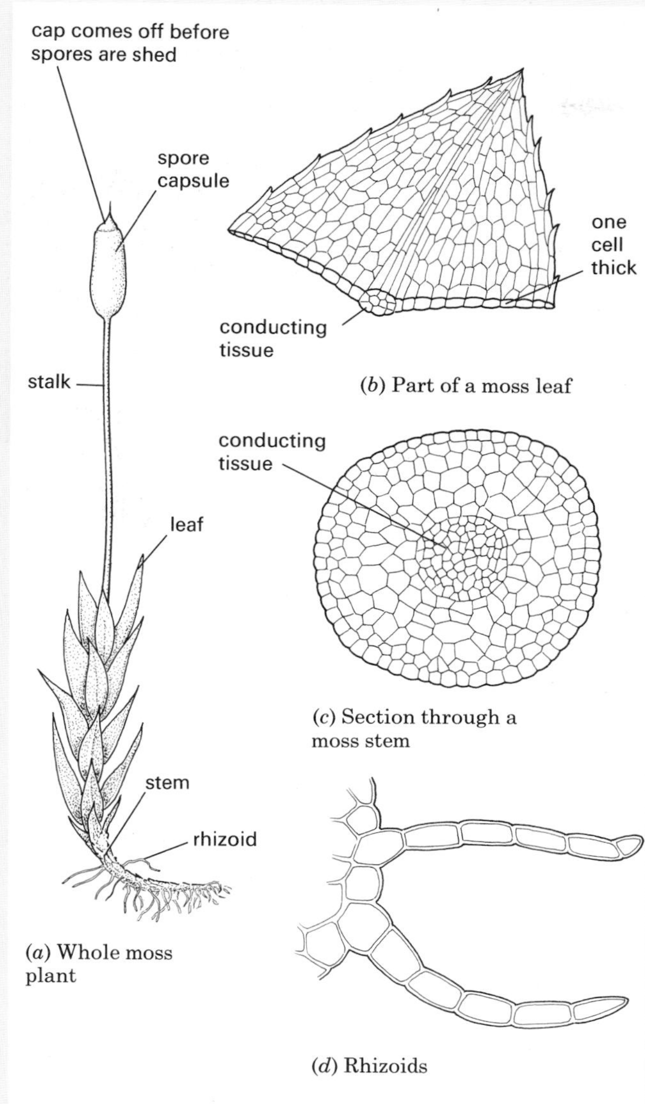

(*a*) Whole moss plant

(*b*) Part of a moss leaf

(*c*) Section through a moss stem

(*d*) Rhizoids

Fig. 9 Structure of a moss

Fig. 10 Moss plants growing in a wood. Some of the plants have produced spore capsules.

Ferns

Ferns are land plants, usually much larger than mosses, with more highly developed structures. Their stems, leaves and roots are very similar to those of the flowering plants.

The stem is usually entirely below ground and takes the form of a **rhizome** (p. 100). In bracken, the rhizome grows horizontally below ground, sending up leaves at intervals (Fig. 11). The roots which grow from the rhizome are called adventitious roots. This is the name given to any roots which grow directly from the stem rather than from other roots.

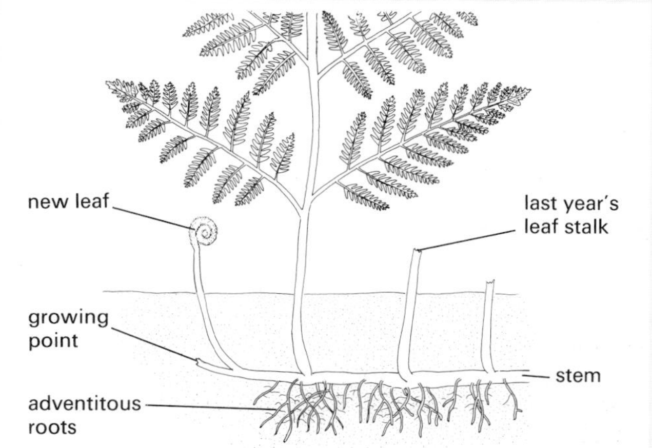

Fig. 11 Bracken rhizome. The stem is horizontal and below the ground.

The stem and leaves have sieve-tubes and water-conducting cells similar to those in the xylem and phloem of a flowering plant (p. 62). For this reason, the ferns and seed-bearing plants are sometimes referred to as vascular plants, because they all have vascular bundles or vascular tissue. Ferns also have multicellular roots with vascular tissue.

The leaves of ferns vary from one species to another (Fig. 12 and Fig. 3 on p. 304), but they are all much

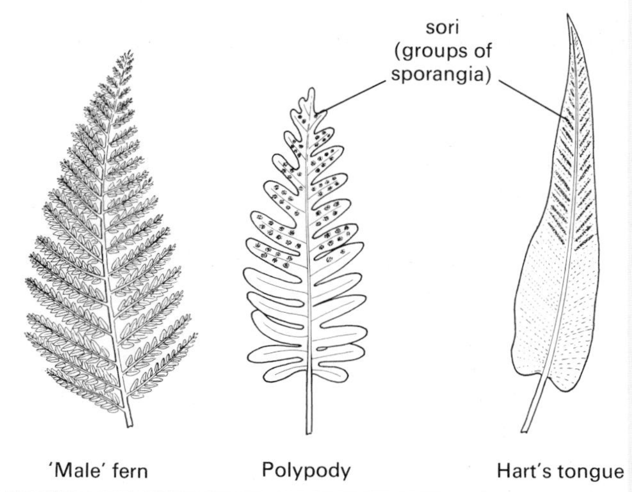

Fig. 12 Three species of fern

larger than those of mosses and are several cells thick. Most of them have an upper and lower epidermis, a layer of palisade cells and a spongy mesophyll similar to the leaves of a flowering plant (Fig. 13).

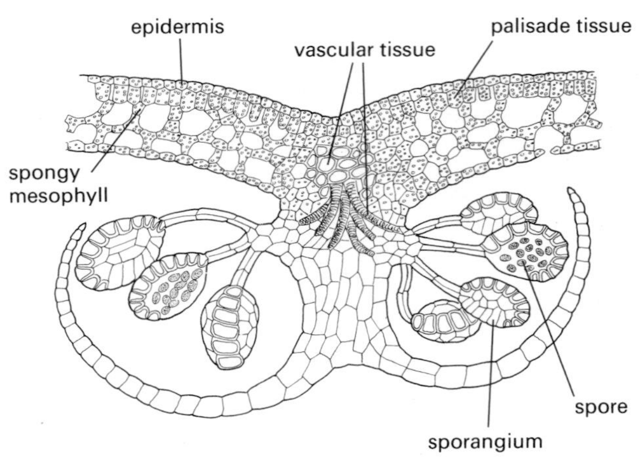

Fig. 13 Section through a fern leaf with sporangia

Ferns produce gametes but no seeds. The zygote gives rise to the fern plant, which then produces single-celled spores from numerous sporangia (spore capsules) on its leaves. The sporangia are formed on the lower side of the leaf but their position depends on the species of fern. The sporangia are usually arranged in compact groups called **sori** (singular = sorus) (Fig. 14).

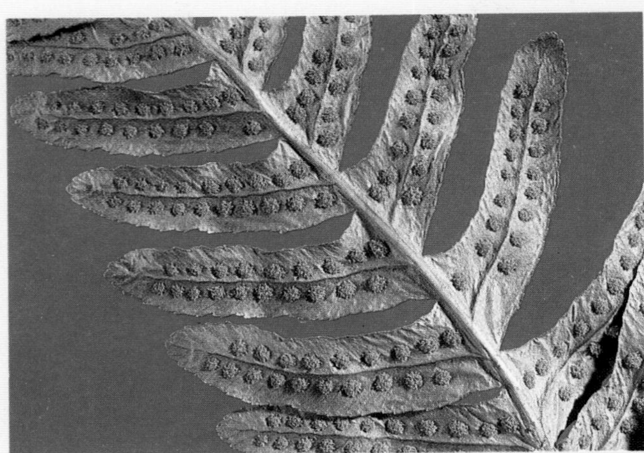

Fig. 14 Polypody fern. Each brown patch on the underside of the leaf is made up of many sporangia.

When the sporangia break down, the spores are scattered and grow to form an inconspicuous, leaf-like structure only a few millimetres across. This leaf-like structure produces the gametes which fuse and grow into a fern.

QUESTIONS

6 If a heath fire destroys all the above-ground vegetation, the bracken will still grow well next season. Suggest why this should be so.

7 Which groups of plants are sometimes called vascular plants and why are they so called?

CHECK LIST

- **Protistans are organisms whose bodies consist of one cell only.**
- **Some protistans take in solid food particles and are called protozoa.**
- **Some protistans have chloroplasts and feed by photosynthesis; they are called protophyta.**
- **Regions of cytoplasm in a protistan may be specialized to carry out particular functions.**
- **Cilia and flagella are filaments of cytoplasm specialized for locomotion.**
- **Contractile vacuoles are specialized areas of cytoplasm which are osmo-regulatory.**
- **Protista reproduce by binary fission; the cell divides into two.**
- **Unicells have a large surface area in relation to their volume and so diffusion is rapid enough to meet their needs for gaseous exchange.**
- **Some protista are parasites and cause disease in humans, e.g. malaria.**
- **The algae are simple plants with little specialization of tissues.**
- **Algae are nearly all aquatic and may be single cells, multicellular filaments or seaweeds.**
- **Plankton is the name given to microscopic plants and animals living in the surface waters of the sea, lakes and ponds.**
- **The phytoplankton consists of unicellular algae and is the basis of food webs in the sea and other open waters.**
- ***Spirogyra* is a filamentous alga made up of chains of identical cells.**
- **Seaweeds are multicellular algae but without much specialization in their bodies.**
- **A moss plant has a stem and simple leaves but no proper roots.**
- **Ferns have roots, leaves and stems, with well-developed vascular tissue.**
- **Mosses and ferns reproduce by means of unicellular spores.**

36 Some Invertebrates

ANNELIDS
Earthworms.

MOLLUSCS
Snails.

ARTHROPODS
Crustacea, insects, metamorphosis, arachnids and myriapods.

The term 'invertebrate' refers to an animal which does not have a vertebral column. Fish, amphibia, reptiles, birds and mammals all have vertebral columns and are called 'vertebrates'; all other animals are invertebrates. The 'invertebrates' do not constitute a classificatory group because the phyla included in this heading are all very different from each other. However, it is a convenient, if not a strictly scientific term.

All animals can be classified into **phyla** (p. 301). There are about 20 invertebrate phyla. Some of these will be fairly familiar, e.g. the **coelenterates**, which include sea anemones, corals and jellyfish. Other phyla include animals which are less familiar because of their rarity, limited distribution or small size.

Representatives of three familiar phyla are described below.

ANNELIDS

Annelids are worms (see Fig. 1 on p. 302). Most of them have elongated, cylindrical bodies which are divided into segments. All the segments have identical sets of organs, though those at the front end may have specialized structures. Some organs, e.g. the alimentary canal, the nerve cord and the main blood vessels run the whole length of the body.

Earthworms are annelids, but there are many more annelid species living in fresh water and the sea. Lugworms, bristle-worms and ragworms, for example, are annelids which burrow in the sand on the sea shore. *Tubifex* (Fig. 18*a*, p. 288) is a freshwater annelid living in the mud at the bottom of ponds.

Earthworms

There are about 15 common species of earthworm in Britain. Figure 1 shows the external features of the species, *Lumbricus terrestris*. Each segment has two pairs of fine bristles, called **chaetae**. The chaetae can be retracted into the body, or protruded to help the earthworm get a grip on the walls of its burrow (Fig. 7, p. 259).

The mouth, at the front end, takes in soil, dead leaves and other organic matter. The organic material is digested and the remaining soil is expelled

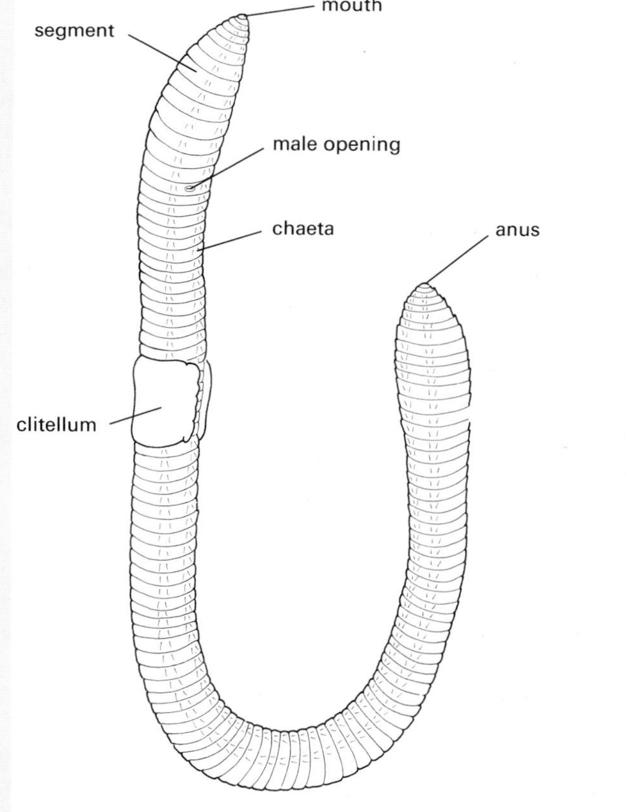

Fig. 1 ***Lumbricus terrestris*, an earthworm**

from the anus. In some species, the expelled soil forms the familiar worm casts on lawns.

The **clitellum**, or 'saddle', is a glandular region between segments 32 and 36. It plays a part in the reproduction of the earthworm and makes a cylindrical egg-case (cocoon) into which the eggs are laid.

A pair of openings is visible on segment 15. These are the male openings from which sperms are released during mating. The female openings on segment 14 are less obvious. (Earthworms are bisexual, i.e. each worm has male and female reproductive organs.)

The egg-case made by the clitellum is moved forwards by muscular contractions of the body and, as it passes segment 14, the eggs are laid into it.

The earthworm takes in oxygen and eliminates carbon dioxide through its skin. The skin is thin and supplied with many blood capillaries. Both these features make the gaseous exchange rapid enough to meet the needs of the earthworm. Glands in the skin produce mucus which keeps the skin moist. The moist skin may help the worm to absorb oxygen from the air but also makes it vulnerable to desiccation (drying up).

MOLLUSCS

Molluscs include snails, whelks, slugs, mussels, oysters and (perhaps surprisingly) squids and octopuses (see Fig. 1 on p. 302).

Many of the molluscs have a shell. In snails, the shell is usually a coiled, tubular structure. In mussels and clams (the bivalves), the shell consists of two halves which can be partially open or tightly closed. In squids the shell is a plate-like structure enclosed in the body. In other molluscs, the shell is reduced or absent, e.g. slugs, octopuses.

All molluscs have a muscular **foot**. In the snails and slugs it forms a flattened structure which protrudes from the shell during locomotion. In bivalves, the foot can protrude from between the halves of the shell and burrow in the sand (e.g. cockles). In the squids and octopuses, the foot has become the array of tentacles.

Snails

The common garden snail is *Helix aspersa*, but you are probably familiar with the pink, brown, yellow or banded snails that occur in hedgerows. These are different species and varieties of the genus *Cepea*. If you have made a study of fresh water, you will have seen the pond snail, *Limnea*, and the ram's horn snail, *Planorbis*, with its shell in the form of a flat spiral.

Helix aspersa is illustrated in Fig. 2. As the snail grows, it adds new material to the opening of the shell, so the shell increases in size to accommodate the snail. Contraction of a muscle in the snail's body, pulls the whole body inside the shell, where it is protected, to some extent, from predators. The snail's skin is permanently moist and retraction into the shell does protect the snail from desiccation in dry weather.

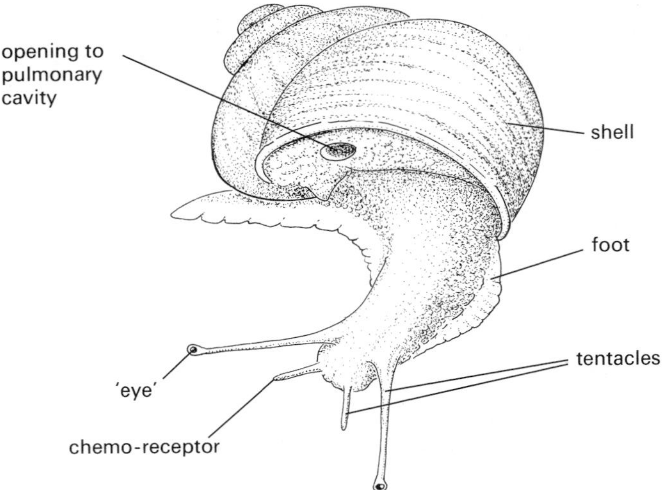

Fig. 2 ***Helix aspersa*, the garden snail**
(From Diana Kershaw, *Animal Diversity*, University Tutorial Press, 1983)

A rippling muscular movement along the base of the foot moves the snail along. Glands in the foot secrete a layer of sticky mucus. This forms a track over which the snail moves.

At the head end are two pairs of tentacles. The first, shorter pair are **chemo-receptors**, i.e. they are sensitive to chemicals, and function as organs of taste or smell. The second, longer pair of tentacles have a light-sensitive region ('eyes') at their tips. These 'eyes' cannot form a detailed image but can detect the presence and direction of light.

An opening just below the upper lip of the shell, leads into a **pulmonary cavity**. This cavity is lined with a thin membrane, well supplied with blood vessels which can exchange oxygen and carbon dioxide with the air. The pulmonary cavity, therefore, functions like a simple lung.

Snails can protrude a firm, tongue-like structure, called a **radula**, from their mouths. The radula is covered with microscopic teeth, which give it a surface like sandpaper. Using the radula, the snail rasps at vegetation and swallows the particles it scrapes off.

Pond snails (Figs 3 and 4) are similar in structure to the land snails but they have only one pair of tentacles, which

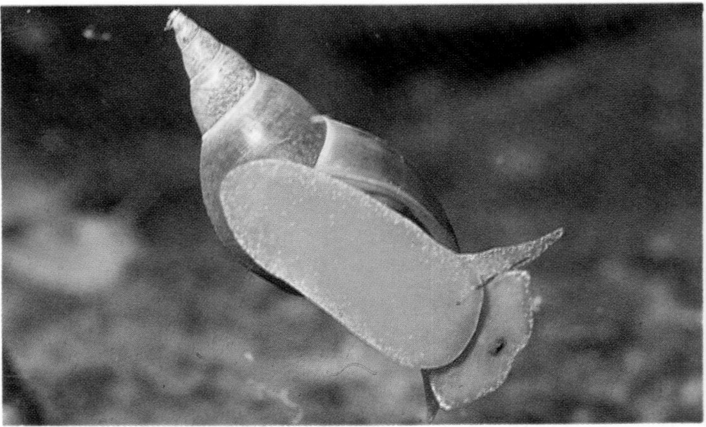

Fig. 3 The pond snail, *Limnea*. The snail is moving over the glass wall of the aquarium by means of its foot.

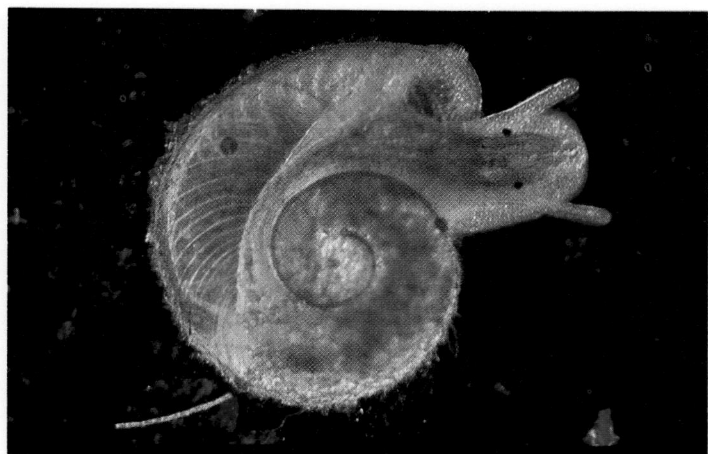

Fig. 4 The pond snail, *Planorbis*. In this young specimen, the pulmonary cavity can be seen under the shell.

are probably the chemo-receptors. The 'eyes' are found at the base of the tentacles. The rasping radula scrapes algae from the leaves and stems of water plants. The snails in an aquarium keep the glass walls free from algae, and you can see the radula at work as the snail crawls up the glass.

Although pond snails are aquatic, many species breathe air. If you have an aquarium, you will see the snails at the surface, exchanging the air in their pulmonary cavities through an aperture which opens at the water surface. Because these snails breathe air, they are able to survive in stagnant water where the level of dissolved oxygen is too low for many organisms.

QUESTIONS

1 What position do (*a*) earthworms, (*b*) snails occupy in their respective food webs?
2 An earthworm can withdraw into its burrow; a snail can retract into its shell. What is the likely benefit of this behaviour to these two organisms?

ARTHROPODS

The arthropod phylum includes the crustacea, insects, centipedes and spiders (Fig. 5, and Fig. 1 on p. 302). The name arthropod means jointed limbs and this is a feature common to them all. They also have a hard, firm external skeleton, called a **cuticle**, which encloses their bodies. Their bodies are segmented and, between the segments, there are flexible joints which permit movement. In most arthropods, the segments are grouped together to form distinct regions, e.g. the head, thorax and abdomen.

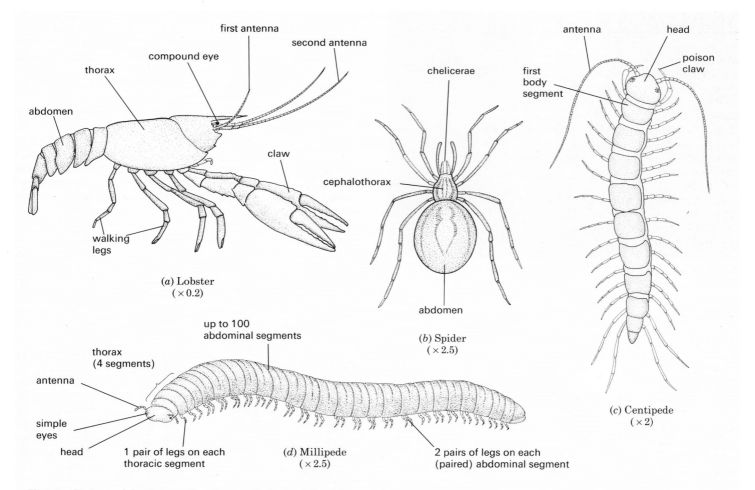

Fig. 5 External features of arthropods (other than insects)

Ecdysis The cuticle is formed from a substance which will not stretch. Although new cells and tissues may be formed inside the exoskeleton, increase in size can take place only by splitting open the cuticle and wriggling out of it. The new cuticle is soft, and the arthropod expands its body, usually by swallowing air or water, before the cuticle hardens (Fig. 6).

This moulting of the cuticle from time to time is called **ecdysis**.

Fig. 6 The final ecdysis of a locust. The old cuticle (above) is left attached to the branch. The newly emerged locust will expand its wings by blood pressure, the wings will then harden.

Crustacea

Marine crustacea are crabs, prawns, lobsters, shrimps and barnacles. Freshwater crustacea are water fleas, *Cyclops*, the freshwater shrimp (*Gammarus*) and the water louse (*Asellus*). Woodlice are land-dwelling crustacea. Some of these crustacea are illustrated on pages 286, 287 and 302.

Like other arthropods, crustacea have an exoskeleton and jointed legs. They also have two pairs of antennae which are sensitive to touch and to chemicals, and they have **compound eyes** (Fig. 7, and Fig. 12 on p. 6).

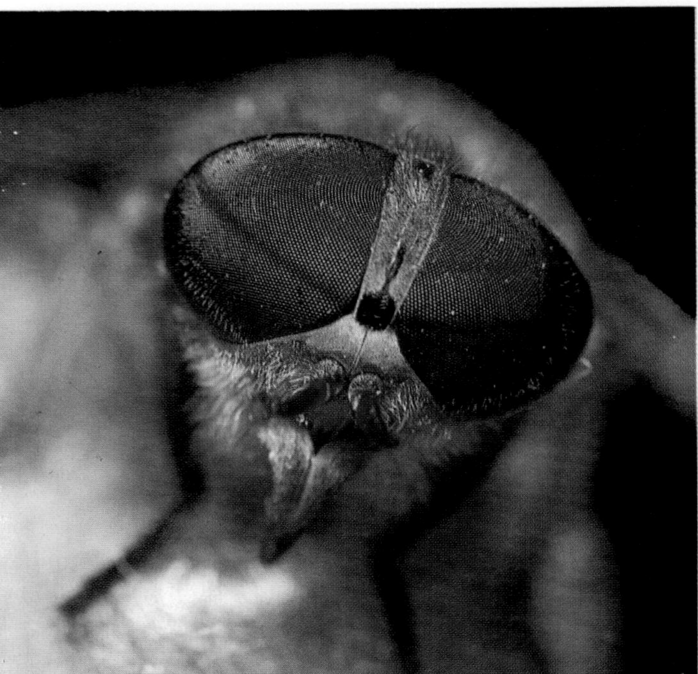

Fig. 7 Compound eyes of a horsefly (×20)

Compound eyes are made up of tens or hundreds of separate lenses with light-sensitive cells beneath. They are able to form a crude image and are very sensitive to movement.

Typically, crustacea have a pair of jointed limbs on each segment of the body, but those on the head segments are modified to form antennae or specialized mouth parts for feeding.

Insects

The insects form a very large class of arthropods. Bees, butterflies, mosquitoes, houseflies, earwigs, greenfly and beetles are just a few of the sub-groups (orders) in this class.

Insects have segmented bodies with a firm exoskeleton, three pairs of jointed legs, compound eyes and, typically, two pairs of wings. The segments are grouped into distinct head, thorax and abdomen regions (Fig. 8).

Insects differ from crustacea in having wings, only one pair of antennae and only three pairs of legs. There are no limbs on the abdominal segments.

The insects have very successfully colonized the land. One reason for their success is the relative impermeability of their cuticles which prevent desiccation even in very hot, dry climates.

Insects such as dragonflies, butterflies, bees and beetles have two pairs of wings on their thoracic segments. In the beetles, the first pair of wings is modified, forming hard wing-cases (Fig. 9). Flies and mosquitoes have only one pair of wings, and some insects have no wings at all, either because they are 'primitive', e.g. springtails (Fig. 6, p. 259), or because the wings have been lost in the course of evolution, e.g. fleas.

The appendages on the head segments form antennae and mouth parts. The mouth parts vary widely according

Fig. 9 Larva of the great diving-beetle (*Dytiscus*). Notice the 3 pairs of legs on the thorax and the circle of 6 simple eyes on the side of the head. The jaws have impaled a tadpole.

Fig. 10 The great diving-beetle (*Dytiscus*) adult. The body form is quite different from the larva, there are compound eyes on the head, and wings covered by wing cases. Notice the third pair of legs modified for swimming.

to the feeding habits of the species. Beetles have jaws which bite off pieces of vegetation; mosquitoes have sharp, needle-like stylets which can pierce the skin and suck blood; bees and butterflies have tubular mouth parts which suck up nectar. The water-beetle larva in Fig. 9 has sharp jaws which impale the prey and inject digestive enzymes.

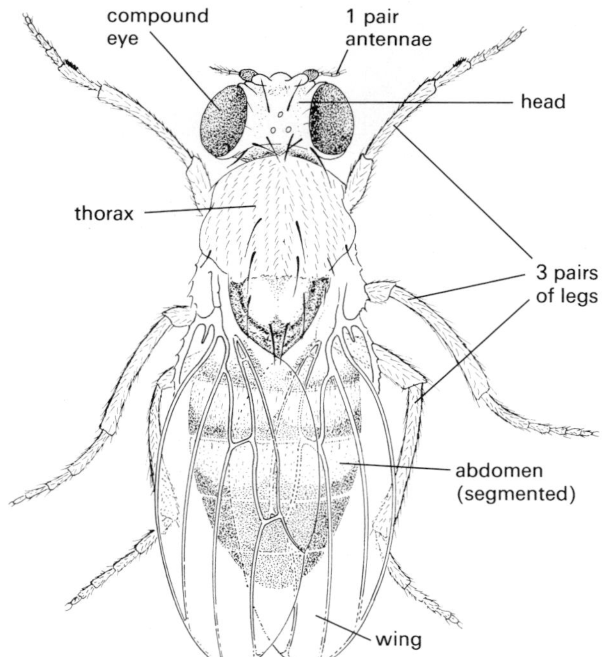

Fig. 8 External features of an insect (fruit fly)

Growth and metamorphosis As in all arthropods, insects have to undergo ecdysis in order to increase in size, but only the larval forms do this.

Insects lay eggs which hatch to **larvae**. These larvae often bear little resemblance to the adult form. (Compare the water-beetle larva and adult water beetle in Figs 9 and 10).

Figures 12 and 13 show the changes which take place in the life cycle of the large white butterfly. The eggs hatch into larvae, called **caterpillars** (Fig. 11), which feed on vegetation, e.g. cabbage leaves. They grow rapidly, shedding their cuticle four times. When the caterpillar is fully grown, its last ecdysis reveals a change. The active larva has become a **pupa**, which does not move or feed. Inside the pupa, the larval tissues are digested and replaced by adult organs. When the pupa splits open, it is a butterfly which emerges (Fig. 14). The butterfly does not grow or shed its cuticle.

The change from larva to adult is called **metamorphosis**. In insects such as butterflies, with a pupal stage and a drastic change at the final ecdysis, the process is called **complete metamorphosis**. In certain other insects, the larval form that hatches from the egg bears some resemblance to the adult and becomes more like it at each ecdysis. The metamorphosis, in this case, is described as 'incomplete'. **Incomplete metamorphosis** is seen in grasshoppers, locusts, cockroaches, dragonflies and mayflies.

In both types of metamorphosis, the larval forms often live in an environment quite different from that of the adult and have different methods of feeding, moving and, in some cases, breathing.

Caterpillars live on vegetation and use their jaws to bite

Fig. 11 Caterpillars (large white butterfly) hatching. The caterpillars will eat their egg shells later on.

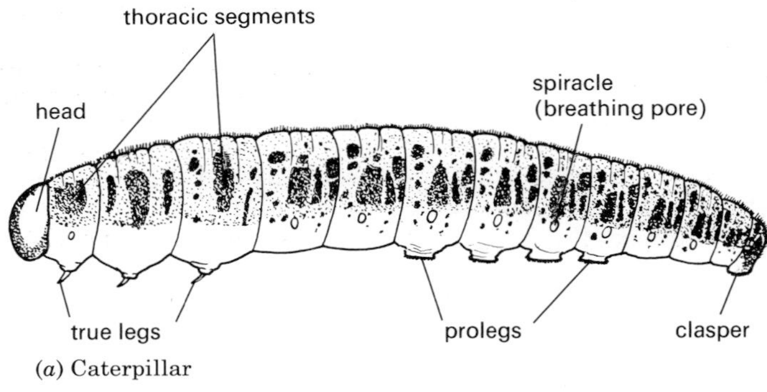

(*a*) Caterpillar

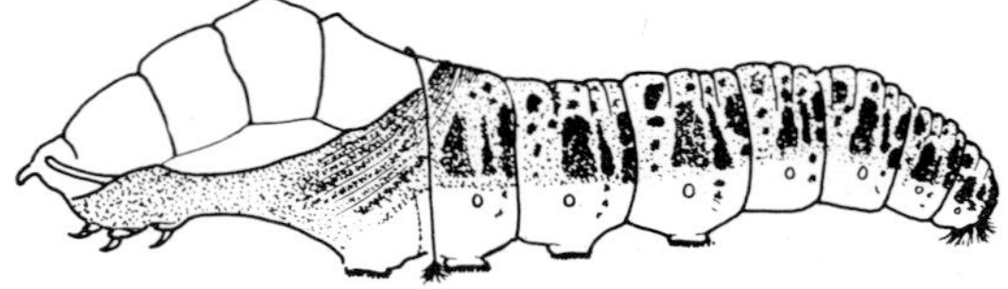

(*b*) The last ecdysis. The cuticle splits above the thorax and is 'shrugged' off by rhythmic contractions.

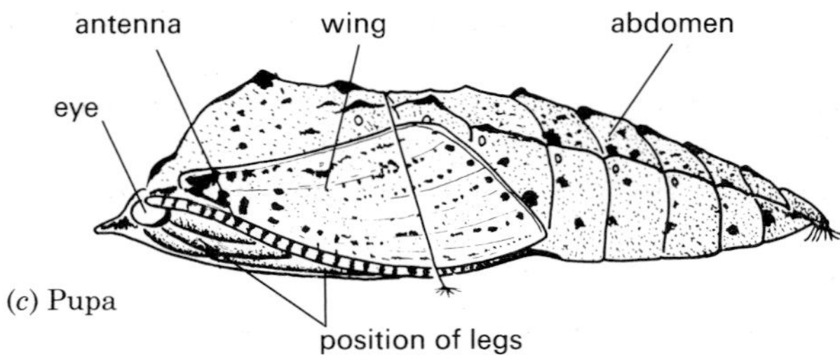

(*c*) Pupa

Fig. 12 Large white butterfly, stages in metamorphosis

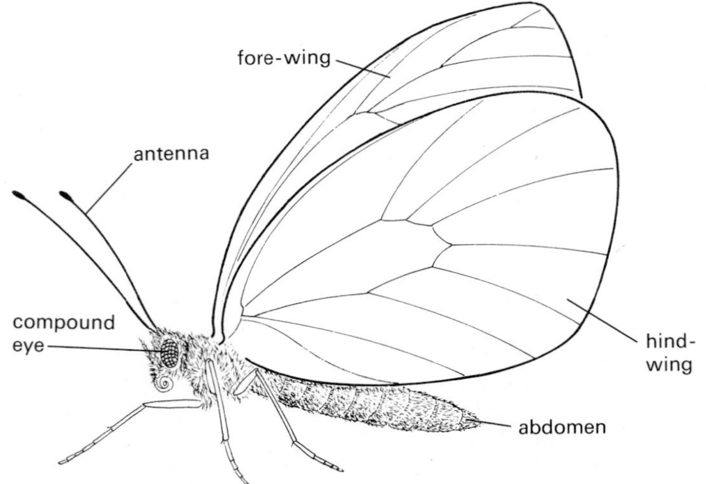

Fig. 13 Large white butterfly, adult

Fig. 14 Metamorphosis of the small tortoiseshell butterfly. The pupa's cuticle splits and the adult butterfly emerges.

off pieces of leaf. They move by means of three pairs of 'true-legs' and five pairs of 'pro-legs'. Butterflies feed on nectar with their specialized, tubular mouth parts. They have no pro-legs and their principal method of locomotion is by flying.

Some larval forms of insects are aquatic and their methods of breathing differ from those of the adult. Mayfly larvae (Fig. 15*a*, p. 287) live in fresh water and obtain dissolved oxygen by means of gills on the abdomen. After metamorphosis, the gills are lost and the adult mayfly breathes air.

ARACHNIDS

Arachnids are spiders, mites, scorpions and harvestmen.

A spider's body is not obviously segmented but is divided into two regions, a **cephalothorax** and an abdomen (Fig. 4*b*). The cephalothorax carries four pairs of walking legs and two pairs of appendages for feeding. There are no antennae and, although there are eight eyes, they are not compound eyes like those of insects but they can detect movement (Fig. 15).

The first pair of appendages on the cephalothorax are the **chelicerae**. These carry poison fangs which seize the prey and inject poison, followed by digestive enzymes.

There are over 70 000 species of spiders and many of them produce silk threads which they use to make webs and trap prey.

Fig. 15 Jumping spider. Notice the array of 4 simple eyes.

MYRIAPODS

Centipedes and millipedes are myriapods (Figs 4*c* and *d*). They each have a head and a segmented body. Centipedes have 18 body segments and millipedes may have up to 100, depending on the species. There is no obvious thorax and each abdominal segment bears a pair of walking legs (two pairs in millipedes). Myriapods have one pair of antennae and two clusters of simple eyes on the head.

Centipedes are carnivores. The first pair of legs is modified to form poison claws which impale and kill the prey. Millipedes are herbivores, feeding mainly on decaying organic matter. Both types of myriapod may be found in leaf litter and loose soil.

QUESTIONS

3 Figures 12–18 on pages 286–8 show some freshwater invertebrates. Try to assign these organisms to their correct phyla.
4 Cockles, shrimps, mussels, whelks, prawns and crabs are often referred to as shellfish. Say (*a*) why they are not fish and (*b*) which ones do not have shells.
5 If you were shown an unfamiliar animal which was segmented and had legs, what features would you look for to help decide whether it was a crustacean, an insect or a myriapod?
6 Draw up a dichotomous key (p. 308) that would enable a biologist to decide whether an organism was an insect, a myriapod, a spider or a crustacean.
7 What is the difference between ecdysis and metamorphosis?

CHECK LIST

- **Invertebrates are animals which do not have a vertebral column.**
- **Annelids are segmented worms; earthworms are annelids.**
- **Molluscs are snails, bivalves and octopuses.**
- **All molluscs have a 'foot' which is used for locomotion or feeding. Many molluscs have a shell.**
- **Snails have a creeping foot, a coiled shell, a pulmonary cavity for breathing, a rasping 'tongue' and one or two pairs of tentacles.**
- **Arthropods have segmented bodies with jointed legs or appendages; they all have an exoskeleton called a cuticle.**
- **The cuticle is inextensible and has to be shed (ecdysis) before the arthropod can increase in size.**
- **Crustacea are arthropods with jointed legs on most segments, compound eyes and two pairs of antennae. Examples are crabs, shrimps and water fleas.**
- **Insects are arthropods with three pairs of legs, one or two pairs of wings, one pair of antennae, compound eyes and bodies divided into head, thorax and abdomen.**
- **Insect eggs hatch into larvae which may be different from the adult in appearance and mode of life.**
- **The change from the larval to the adult insect is called metamorphosis.**
- **The larvae may move, feed and breathe in ways quite different from the adult.**
- **Arachnids are spiders, mites and scorpions; they have four pairs of legs, no antennae and bodies in two sections (cephalothorax and abdomen).**
- **The myriapods are millipedes and centipedes, with segmented bodies and legs on each segment.**

37 Vertebrates

FISH
Characteristics, breathing, reproduction.

AMPHIBIA
Characteristics, frog reproduction.

REPTILES
Characteristics, reproduction.

BIRDS
Characteristics, reproduction.

MAMMALS
Characteristics, reproduction.

Vertebrates are animals which have a vertebral column. The vertebral column is sometimes called the spinal column or just spine and consists of a chain of cylindrical bones (vertebrae) joined end to end (Fig. 1 on p. 182).

Each vertebra carries an arch of bone on its dorsal (upper) surface. This arch protects the spinal cord (p. 210), which runs most of the length of the vertebral column. The front end of the spinal cord is expanded to form a brain which is enclosed and protected by the skull.

The skull carries a pair of jaws which, in most vertebrates, have rows of teeth.

The five classes of vertebrates are fish, amphibia, reptiles, birds and mammals.

FISH

Fish are 'cold-blooded' (poikilothermic) vertebrates. Many of them have streamlined bodies, which make it possible to move rapidly in water.

Figure 1 shows the external features of the three-spined stickleback, which inhabits fresh water; Fig. 2 on page 303 illustrates four other species.

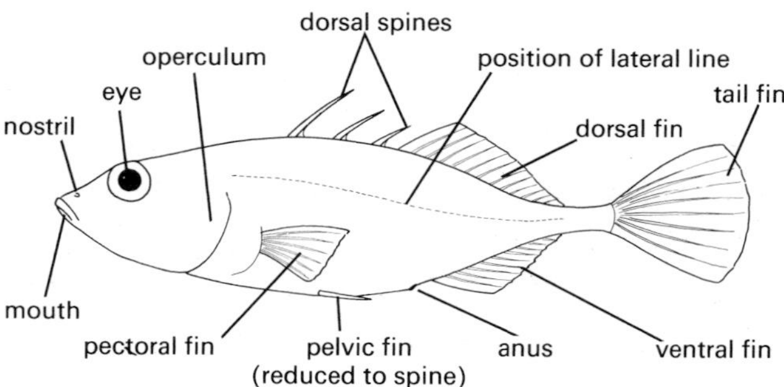

Fig. 1 External features of the 3-spined stickleback

The bodies of fish are covered with overlapping scales (not particularly obvious in the stickleback), which themselves are covered by a thin layer of skin.

The fins are either median, e.g. **dorsal** and **ventral**, or paired, e.g. **pectoral** and **pelvic**. In the stickleback, the pelvic fins have become spines. The fins are formed from skin supported by bony fin rays. Three of the fin rays of the stickleback's dorsal fin are modified to form spines. These can be raised during threat displays and may also discourage some predators from swallowing the stickleback.

The tail fin is important in propelling the fish through the water. The median fins help to reduce rolling and assist in turning movements. The paired fins help to steer the fish up or down.

Some fish have in their bodies a structure called a **swim-bladder**. This is an air-filled sac in the upper part of the body cavity. It makes the fish buoyant so that it can remain at any level in the water without having to keep swimming.

The **lateral line** is a sensory organ. It consists of a fine tube with sensory nerve endings and runs just beneath the skin. It is sensitive to movements and vibrations in the water.

The **operculum** is a bony plate which covers and protects the gills and also acts as a valve in the breathing movements. It allows water to escape after it has passed over the **gills**, but prevents it from entering when the next lot of water is sucked in through the mouth.

The fish absorbs dissolved oxygen from the water by means of the gills. There are usually four gills on each side, under the operculum. Each gill consists of a curved bony bar with branched filaments radiating from it (Fig. 2). The **gill filaments** have a rich blood supply which brings carbon dioxide from the tissues and takes oxygen back to them.

A current of water is maintained over the gills by rhythmic movements of the floor of the mouth. Water is sucked in through the mouth and forced out through the operculum on each side (Fig. 3).

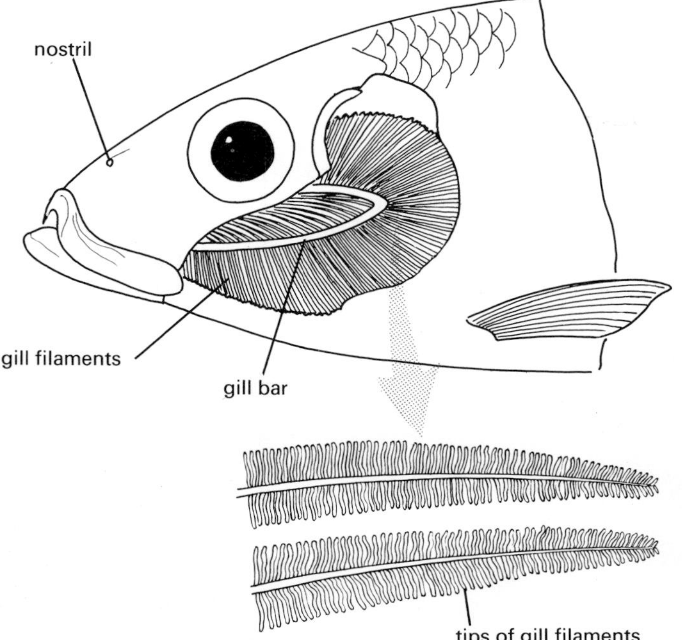

Fig. 2 Gills; the operculum has been cut away.

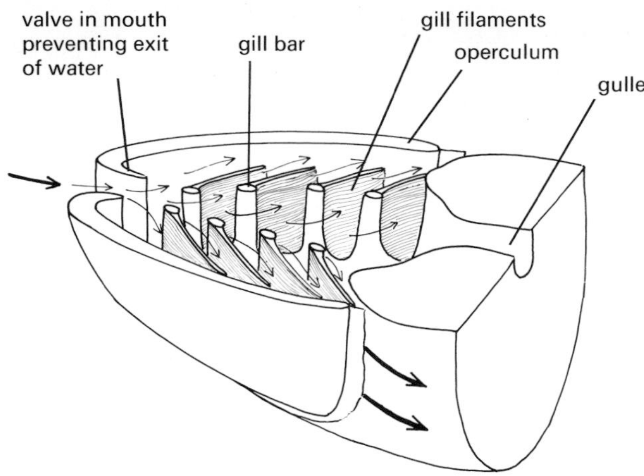

Fig. 3 Diagram to show breathing current

Reproduction Fish reproduce sexually but fertilization takes place externally, i.e. the female lays eggs and the male sheds sperms on them after they are laid.

There is often a 'courtship' which brings the pair together and ensures that sperms are shed at the right time and place. Sometimes a simple nest is made in which the eggs are laid and where they hatch eventually.

The male three-spined stickleback digs a hollow at the bottom of the pond and covers it with a dome of vegetation (Fig. 4). He then induces a female to lay eggs in the nest, after which he enters the nest and fertilizes the eggs with his sperms.The female leaves the nest area but the male remains, chasing off intruders who come near the nest and fanning currents of water through the nest with his pectoral fins. This improves the supply of oxygen to the eggs.

Fig. 4 Male stickleback at the nest. The male develops blue eyes and a red belly in the breeding season. This may attract females and warn off rival males. The female is in the nest, laying eggs.

In the trout, the female uses her tail to scrape a hollow in the gravel at the bottom of a stream. After a courtship pattern with the male, she lays eggs in the hollow and he sheds sperm on them. The female covers the fertilized eggs with gravel and they hatch in about four months.

There is no parental care as there is in the stickleback but the young fish remain attached to their large **yolk sacs** (Fig. 5) which provide their food for another two or three weeks. After this they start to feed themselves.

Fig. 5 Newly hatched trout (×4). The young trout are still attached to their yolk sacs. They will absorb food from the yolk for several weeks.

Both the stickleback and the trout are carnivores and, therefore, secondary consumers in their food webs, eating insect larvae and small crustaceans.

QUESTIONS

1 A mammal's lungs and a fish's gills both have large absorbing surfaces. How is this achieved in each case?
2 In what way does the construction of even a crude nest improve the chances of the eggs being fertilized?

AMPHIBIA

Amphibia are 'cold-blooded' vertebrates with four limbs and no scales. The class includes frogs, toads and newts. The name, amphibian, means double life and refers to the fact that the organism spends part of its life in water and part on the land. In fact, most frogs, toads and newts spend much of their time on the land, albeit in moist situations, and return to ponds, etc. only to lay eggs.

Figure 6 shows the external features of the common frog: Fig. 2 on p. 303 shows the toad and the newt.

The toad's skin is drier than that of the frog and it has glands which can exude an unpleasant-tasting chemical which discourages predators. Newts differ from frogs and toads in having a tail. All three groups are carnivorous.

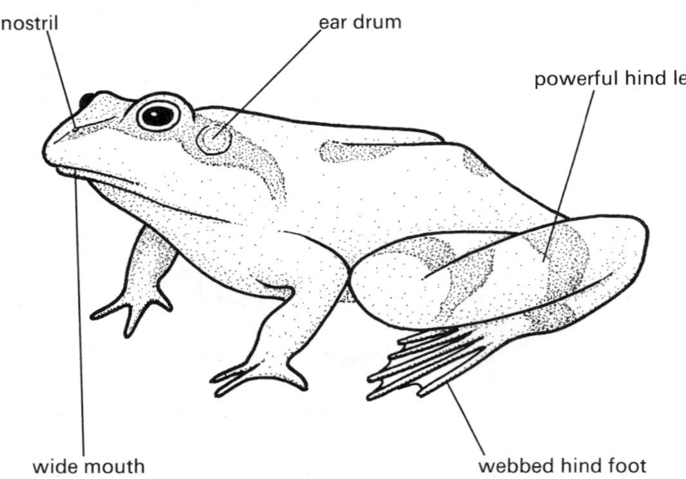

Fig. 6 External features of the frog

Amphibians have four limbs. In frogs and toads, the hind feet have a web of skin between the toes. This offers a large surface area to thrust against the water when the animal is swimming. Newts swim by a wriggling, fish-like movement of their bodies and make less use of their limbs for swimming.

Amphibia have moist skins with a good supply of capillaries which can exchange oxygen and carbon dioxide with the air or water. They also have lungs which can be inflated by a kind of swallowing action. They do not have a diaphragm or ribs.

Reproduction The moist skin of the amphibian restricts it to humid situations were it can avoid desiccation. The same restriction applies to the eggs which do not have a shell, and quickly dry out if they are not surrounded with water. Consequently, all amphibia have to return to water to breed.

Frogs migrate to ponds where the males and females pair up. The male climbs on the female's back and grips firmly with his front legs (Fig. 7). When the female lays eggs, the male simultaneously releases sperms over them. Fertilization, therefore, is external even though the frogs are in close contact for the event.

Fig. 7 Frogs pairing. The male's thumbs develop special, non-slip skin pads which help to maintain his grip on the female.

Frog and toad eggs hatch to tadpoles which spend the first 12 weeks or so of their lives in water. They swim by wriggling movements of their tails and they feed at first on algae; later on they eat animal matter such as small crustacea or dead fish.

During this time the tadpoles breathe water, exchanging gases by means of gills. As they mature, the tadpoles lose their gills and develop lungs. They also lose their tails and develop legs, after which they are ready to emerge on to the land as miniature frogs or toads.

In the early stages, tadpoles are primary consumers and scavengers in the pond. They nibble at algae growing on the surface of water weeds but will also eat the remains of dead animals on the bottom of the pond. Later, when they change to a carnivorous diet, they become secondary consumers. Adult frogs and toads eat insects, slugs and worms which they capture by extruding a sticky tongue. They are, therefore, secondary or tertiary consumers in their land-based food webs.

QUESTIONS

3 By what means might a frog breathe (*a*) when it is on land and (*b*) when swimming under water?
4 Why are frogs restricted to damp habitats and why must they return to water in order to breed?
5 How does the pairing behaviour of frogs and toads improve the chances of fertilization?

REPTILES

The reptiles include lizards, snakes, turtles, tortoises and crocodiles. In Britain we have only three species of lizard and three species of snake (Fig. 8, and Fig. 2 on p. 303).

Fig. 8 Grass snake. Notice the scales on the head and body. The flicking tongue picks up chemicals in the air and carries them to a sense organ in the roof of the mouth which 'tastes' them.

Reptiles are land-living vertebrates. Their skins are dry and the outer layer of epidermis forms a pattern of scales. This dry, scaly skin resists water loss. Also the eggs of most species have a tough, parchment-like shell. Reptiles, therefore, are not restricted to damp habitats, nor do they need water in which to breed.

The reptiles are cold-blooded but they can regulate their temperature to some extent. They do this by basking in the sun until their bodies warm up. This enables them to move about rapidly in pursuit of insects and other prey.

Reproduction Fertilization of the eggs takes place internally. There is a behaviour pattern which leads to copulation during which the male introduces sperms into the female's reproductive tracts.

In the common lizard, the adder and the slow worm, the eggs are retained in the body of the female where they develop to tiny lizards or snakes. These are then born, fully formed with perhaps just a thin egg membrane surrounding them. This method of giving birth to young rather than laying eggs, is called **viviparity**.

The grass snake and the sand lizard lay eggs which take several weeks to hatch. The sand lizard buries its eggs, and the grass snake lays hers in a heap of rotting vegetation, but there is no parental care.

QUESTION

6 Suggest what might be the advantages of (*a*) internal fertilization and (*b*) viviparity.

BIRDS

Birds are warm-blooded (homoiothermic) vertebrates: that is, they keep their body temperature more or less constant and, for the most part, above that of their surroundings.

The external features of a chaffinch are illustrated in Fig. 9, and four other species are shown in Fig. 2 on p. 303.

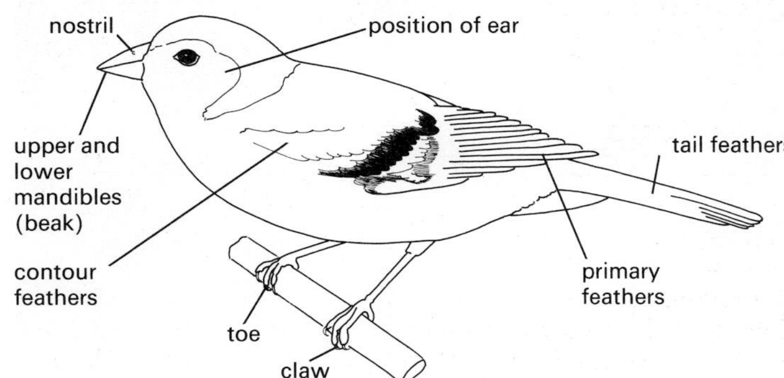

Fig. 9 External features of a chaffinch

The vertebral column in the neck is flexible but the rest of the vertebrae are fused to form a rigid structure. This is probably an adaptation to flight, as the powerful wing muscles need a rigid air-frame to work against.

The epidermis over most of the body, produces a covering of feathers but, on the legs and toes, the epidermis forms scales. The feathers are of several kinds. The fluffy down feathers form an insulating layer close to the skin; the contour feathers cover the body and give the bird its shape and coloration; the large quill feathers on the wing are essential for flight.

Birds have four limbs, but the forelimbs are modified to form wings. The feet have four toes with claws which help the bird to perch, scratch for seeds or capture prey, according to the species.

The upper and lower jaws are extended to form a beak which is used for feeding in various ways.

Reproduction Birds lay eggs with hard shells. Fertilization is internal. The male mates with the female and passes sperms into her oviducts to fertilize the eggs before the shell is formed. The female lays the eggs in a nest and then incubates them, i.e. keeps them at body temperature by sitting on them.

There is always some form of parental care. Birds which nest on the ground keep the young together. The chicks respond to warning cries by crouching motionless, and they imitate the parents' feeding behaviour. Birds which nest above ground have chicks which are naked, blind and helpless at first. The parents have to bring food to the nest (Fig. 10) and brood the chicks, i.e. keep them warm, until they are old enough to fly.

QUESTIONS

7 What do you think are the possible advantages of nest-building?

8 Suggest some of the different ways that birds use their beaks. Consider, for example, eagles, ducks, thrushes and sparrows.

Fig. 10 Parental care in birds. The song thrush has brought a beakful of worms to feed the chicks in the nest. The bright colour on the inside of their beaks stimulates the parents to put food into them.

MAMMALS

Mammals are warm-blooded vertebrates with four limbs. They differ from birds in having hair rather than feathers.

Fig. 11 Mammalian features. The furry coat, the external ear pinnae and the facial whiskers (vibrissae) are visible mammalian features in this gerbil.

Unlike the other vertebrates they have a diaphragm which plays a part in breathing (p. 152). They also have mammary glands and suckle their young on milk.

Figure 2 on p. 303 shows a sample of mammals and Figs 11 and 12 illustrate some of the mammalian features.

Humans are mammals and most of the physiology described in Section 3 applies to all mammals.

Reproduction Mammals give birth to fully formed young instead of laying eggs. The eggs are fertilized internally (see p. 173) and undergo a period of development in the uterus.

The young may be blind and helpless at first, e.g. dogs and cats (Figs 11 and 12), or they may be able to stand up and move about soon after birth, e.g. sheep and cows. In either case, the youngster's first food is the milk which it sucks from the mother's teats (Fig. 12). The milk is made in the mammary glands and contains all the nutriments that the offspring need for the first few weeks or months, depending on the species.

Fig. 12 Suckling. All mammals suck milk from their mother's mammary glands. The kittens will be suckled for about 8 weeks. In the wild state, the mother would then start to bring mice, etc. for them.

As the youngsters get older, they start to feed on the same food as the parents. In the case of carnivores, the parents bring the food to the young until they are able to fend for themselves.

QUESTIONS

9 Which vertebrate classes (*a*) are warm-blooded, (*b*) have four legs, (*c*) lay eggs, (*d*) have internal fertilization and (*e*) have some degree of parental care?

10 Draw up a key (see p. 308) which a biologist could use to place an animal in its correct phylum.

CHECK LIST

- **Vertebrates have a vertebral (spinal) column, formed from a series of vertebrae.**
- **They also have a spinal cord and brain. The spinal cord is enclosed in the vertebral column; the brain is enclosed in the skull.**
- **Most vertebrates have teeth and jaws.**
- **Fish, amphibia, reptiles, birds and mammals are vertebrates.**
- **Fish are cold-blooded, aquatic vertebrates with scales and fins; they breathe by means of gills; their eggs are fertilized externally.**
- **Amphibia are frogs, toads and newts.**
- **Amphibia are cold-blooded vertebrates with four limbs. Amphibia can breathe through their moist skins or with lungs, and can live on land or in water.**
- **In amphibia, fertilization is external and takes place in water. The early stages of development are entirely aquatic.**
- **The moist skin of amphibia restricts them to damp habitats.**
- **Reptiles are lizards, snakes, turtles and crocodiles.**
- **Reptiles are cold-blooded vertebrates with four limbs and dry, scaly skins which resist desiccation.**
- **Fertilization in reptiles is internal and they lay eggs with soft shells.**
- **Snakes and lizards are not restricted to damp habitats; they lay their eggs on land, but some are viviparous.**
- **Birds are warm-blooded vertebrates with four limbs; the forelimbs are wings and the jaws are extended to form a beak.**
- **The bird's epidermis produces feathers, and scales over the legs and feet.**
- **Birds make nests and lay eggs which are fertilized internally.**
- **There is a period of incubation followed by parental care when the young hatch. Some species bring food for their young.**
- **Mammals are warm-blooded vertebrates with four limbs: their bodies are covered with hair.**
- **In mammals, fertilization is internal and they give birth to fully formed young and suckle them on milk produced by the mammary glands.**

38 Characteristics of Living Organisms

RESPIRATION
Anaerobic and aerobic.

FEEDING

BREATHING
Animals and plants, transport systems.

EXCRETION
Animals and plants, nitrogenous and other excretory products.

SEXUAL REPRODUCTION
Internal and external fertilization, parental care, new gene combinations.

ASEXUAL REPRODUCTION
Clones.

GROWTH
Measurement, growth curves.

MOVEMENT

SENSITIVITY
Receptors and effectors, tropisms, survival value.

PRACTICAL WORK
Choice chamber.

The last six chapters have concentrated on the diversity of living organisms. This chapter is concerned with the characteristics shared by all organisms. All living organisms, whether they are bacteria, fungi, plants or animals:

RESPIRE	REPRODUCE
FEED	GROW
BREATHE	MOVE
EXCRETE	RESPOND TO STIMULI

RESPIRATION

All organisms need a supply of energy. This energy is used sometimes for obvious activities such as muscle contraction and movement. In most cases, however, the energy is used to drive chemical reactions in the cells, e.g. building up proteins from amino acids (p. 14), or generating nerve impulses.

In all cases, the energy is derived from the process of respiration. In this process, carbohydrates or fats are broken down to carbon dioxide. In the course of the breakdown, energy is released and transferred to other chemical substances such as ATP (p. 25). ATP can store the energy and make it available for a wide variety of reactions in the cell.

Plants can use energy from sunlight but only to build up their food by photosynthesis. All other processes in their cells are driven by energy from respiration.

Anaerobic and aerobic respiration

In the process of anaerobic respiration, energy is released from food substances by breaking them down to simpler compounds, in the absence of oxygen (p. 24).

In aerobic respiration, the simpler, intermediate compounds are oxidized completely to carbon dioxide and water with the aid of oxygen. Aerobic respiration releases more energy than anaerobic respiration.

Some bacteria can respire only anaerobically; oxygen is harmful to them. Some bacteria must have a supply of oxygen in order to respire aerobically and others can use either aerobic or anaerobic respiration according to the circumstances.

In vertebrates, the first stages of respiration are usually anaerobic. Carbohydrates are broken down to carbon dioxide and intermediate substances, e.g. lactic acid, without the involvement of oxygen. The final stages of respiration are aerobic, and the lactic acid is oxidized completely to carbon dioxide and water with the aid of oxygen.

RESPIRATION	
Releases energy from food; carbon dioxide is produced	
ANAEROBIC	*AEROBIC*
Does not need oxygen	Needs oxygen
Incomplete breakdown of food	Complete breakdown of food to carbon dioxide and water
Not so much energy as from aerobic	More energy than anaerobic

QUESTIONS

1 Which of the following are needed for aerobic respiration: food, proteins, oxygen, carbon dioxide, energy, carbohydrates, sunlight, enzymes?
2 List the stages involved in the transfer of energy from sunlight to a protein molecule in a cow.
3 Alcohol is produced industrially by the anaerobic respiration of yeast. In what way would aerobic respiration ruin this industrial process?

FEEDING

The energy released during respiration comes from food. All living organisms, therefore, must feed.

Plants can make their food from carbon dioxide, water and mineral salts by the process of photosynthesis, (p. 44). Blue-green algae and some bacteria can photosynthesize but all other creatures must obtain their food directly or indirectly from plants or animals.

The food must be digested before it can be absorbed and used in the cells. Saprophytes digest their food externally and absorb the digested products (p. 250). Holozoic organisms take in solid food and digest and absorb it internally, e.g. *Paramecium* (p. 327) or humans (p. 125).

Parasites which live in their host are called **endoparasites**. They may digest the host's tissues, e.g. parasitic fungi (p. 322), and the malarial parasite (p. 328), or they may absorb digested food from the host's blood or digestive system, e.g. a tapeworm in a rabbit's intestine.

Parasites which live outside their host are called **ectoparasites**. They pierce the host's skin or epidermis in order to suck up the body fluids. Aphids (greenflies) push their sharp, tubular mouth parts through the epidermis of a leaf and suck food from the phloem. Mosquitoes pierce the skin of vertebrates and suck blood from the capillaries.

No matter how the food is obtained, it will be used either for energy (respiration) or for making new cytoplasm or other cell products (growth, replacement or repair). Food which is not used at once may be stored. Plants store food mostly as starch, sucrose or oil, either in their fruits or in special storage organs (p. 102). Animals store excess food as glycogen or fat. Glycogen may be stored in the liver or the muscles. Fat is stored either in special fat bodies or fat depots, or under the skin.

QUESTIONS

4 Make a list of the different ways in which food may be used by a living organism.
5 Revise pages 242–5 and then list the different ways in which organisms obtain their food.

BREATHING

Aerobic respiration needs a supply of oxygen, and both types of respiration produce carbon dioxide. The process by which organisms take in oxygen and get rid of carbon dioxide is called **gaseous exchange**. At some stage in the process of gaseous exchange, oxygen diffuses into the organism from the environment and carbon dioxide diffuses out.

It was explained on p. 329 that very small organisms have a large surface area compared with their volume. In bacteria, protista and small multicellular organisms, therefore, diffusion of gases through their surface is rapid enough to meet their needs.

Animals

The surface area of larger animals is relatively small and their skins or cuticles are not very permeable to gases. In these animals there are special 'respiratory' organs, e.g. lungs or gills, which carry out gaseous exchange.

Although these respiratory organs occupy a fairly small space, they have a very large absorptive surface. This is achieved in the lungs by thousands of microscopic pockets, called alveoli (p. 152). In the gills it is the hundreds of branched gill filaments (p. 340) which produce the large absorbing surface.

Two other features of respiratory surfaces are their thin lining and their rich supply of blood capillaries. The thin epithelium offers little resistance to diffusion of gases and the blood supply maintains a steep diffusion gradient (p. 34). The blood does this by (*a*) removing oxygen as fast as it enters the absorbing surface and (*b*) replacing the carbon dioxide as fast as it diffuses out.

Most respiratory organs have a means of exchanging the water or air in contact with the absorbing surface. This **ventilation**, as it is called, also helps to establish a steep diffusion gradient. The water or air in contact with the respiratory surface is exchanged before it can become saturated with carbon dioxide or depleted of oxygen.

In mammals, the processes of inhaling and exhaling (p. 152) ventilate the lungs. In fish, a current of water is forced past the gills by movements of the mouth and operculum (p. 340).

Transport systems In a protistan such as *Paramecium* or *Amoeba*, no part of the cytoplasm is more than a fraction of a millimetre away from the absorbing surface (Fig. 1). Over such short distances, diffusion is rapid enough to carry oxygen from any part of the surface to any part of the cytoplasm which needs it.

In larger animals, oxygen can be absorbed only by the respiratory organ and this may be a long distance from the parts of the body which need the oxygen. In these animals, a circulatory system has evolved. The blood circulatory system (p. 137) carries oxygen rapidly from the lungs or gills to all parts of the body.

The same circulatory system, of course, carries dissolved food and excretory products to or from the specialized organs to other parts of the body which are some distance away.

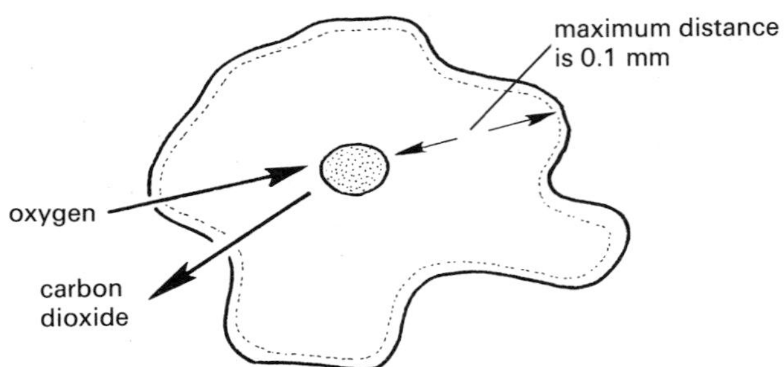

Fig. 1 Diffusion in protista. The distances are so small that diffusion is rapid enough to meet the organism's needs.

Plants

Plants, for the most part, are static and their energy demands are less than those of animals. There are no special respiratory organs but the leaves may be regarded as adapted for gaseous exchange as well as for absorbing sunlight. There are numerous stomata (p. 60) and the extensive intercellular spaces in the mesophyll present a large absorbing surface.

Diffusion is rapid enough to meet the plant's demand for carbon dioxide during daylight hours, when it is photosynthesizing. Diffusion also supplies enough oxygen for respiration when photosynthesis slows down or stops (see p. 61).

In young stems, gaseous exchange takes place through stomata. In older stems this function is taken over by lenticels.

QUESTIONS

6 What gaseous exchange will be taking place in (*a*) a plant in sunlight, (*b*) a plant in darkness, (*c*) an animal and (*d*) an anaerobic bacterium?
7 What are the main features of a respiratory organ which contribute to its efficiency?
8 Criticize the use of the term respiratory in the context 'respiratory organ'.

EXCRETION

The term, **metabolism**, refers to all the chemical reactions which take place in cells and which keep an organism alive. These reactions produce waste products which, though harmless in low concentrations, could poison the cells if allowed to accumulate.

Excretion is the process by which living organisms get rid of the waste products of metabolism (p. 159) and any excess water and salts taken in with the diet.

Animals

Nitrogenous excretion The nitrogen produced from the breakdown of proteins and amino acids appear first as ammonia, NH_3. Ammonia is a toxic substance but can be excreted as a very dilute solution by organisms, such as freshwater fish, which can afford to lose a great deal of water without becoming dehydrated. In mammals, the ammonia is converted to urea, which is less toxic, before excretion. In birds, the ammonia is changed to uric acid which is not very soluble in water and can be excreted with very little loss of water.

Ammonia, urea and uric acid are called nitrogenous excretory products because of the nitrogen they contain.

Other excretory products The waste products of respiration are carbon dioxide and water. Carbon dioxide is excreted by the lungs, gills or other respiratory organs. The water enters the circulatory system and is eventually excreted by the kidneys. Water which is produced from respiration is sometimes called **metabolic water**, though once it has been produced, it is indistinguishable from water from any other source, e.g. drinking or osmo-regulation.

The process of excretion also gets rid of substances which are harmful or which are taken in excess with the food, e.g. excess salts and water.

Expulsion of excess water is osmo-regulation rather than excretion, except in so far as some of the water is metabolic water derived from respiration.

In the protista, e.g. *Paramecium* (p. 327), the contractile vacuole is osmo-regulatory rather than excretory. Excretory products diffuse out from the large surface as effectively as oxygen diffuses in.

Egestion The expulsion of faeces is not usually considered to be excretion because faeces consist mainly of undigested food, dead bacteria and water which have not been involved in cellular metabolism. However, in the vertebrates at least, the faeces usually contain bile pigments which are derived from the breakdown of haemoglobin in the red cells of the blood.

Plants

When a plant is photosynthesizing, it will be producing oxygen, which escapes into the intercellular spaces of the leaf and out of the stomata. Oxygen, therefore, is an excretory product. The plant also produces carbon dioxide as a waste product of respiration. In daylight, however, photosynthesis uses up this carbon dioxide as fast as it is produced, so it is not excreted. Only when the rate of respiration exceeds the rate of photosynthesis does carbon dioxide diffuse out of the leaf (p. 49).

There are no excretory organs; diffusion is fast enough to remove oxygen and carbon dioxide as fast as they accumulate.

There is no nitrogenous excretion. A plant uses nitrogenous compounds, e.g. nitrates, in building up amino acids and proteins from carbohydrates (p. 52).

QUESTIONS

9 (*a*) Name 2 nitrogenous excretory products and 2 non-nitrogenous excretory products.
(*b*) Why do plants not excrete nitrogenous products?
10 Revise page 133 and then say what part the liver plays in nitrogenous excretion.

REPRODUCTION

Most organisms die because they are eaten by another creature or they succumb to disease or they just die of old age.

In an annual plant, for example, a seed germinates, the plant grows and produces flowers and seeds. The parent then becomes senescent (old) and dies. Only the seeds persist to start a new generation. There are many variations on this theme in all living organisms but, for the most part, reproduction is the way in which the species is perpetuated, even though the individuals must eventually die.

Sexual reproduction

Gametes Sexual reproduction involves the production of gametes which fuse to form a zygote. The zygote then develops into a new organism (p. 170).

Gametes are single cells produced in the reproductive organs of the higher, multicellular organisms. In some protistans, the whole cell may become a gamete and fuse with another cell. In *Spirogyra* (p. 329) any one of the cells may act as a gamete.

In the animal kingdom, the male gametes are sperms and the female gametes are ova (p. 170). In seed-bearing plants, the male gamete is a cell in the pollen grain and the female gamete is a cell in the ovule (p. 86).

Male gametes are generally very small, consisting of a nucleus and very little cytoplasm. The female gametes are larger, with a nucleus and more abundant cytoplasm and, in some cases, a large amount of stored food in the form of yolk, e.g. bird's eggs or trout eggs (p. 341).

Female gametes are not mobile. They may be placed in a nest (trout, stickleback), or retained in the body of the female (birds, mammals) or in the female part of the organism (plants), before they are fertilized.

Male gametes are usually motile, e.g. sperms have tails which propel them towards the ova. Pollen grains are not self-propelled but they are carried long distances by insects or the wind. When they reach the female organ they grow pollen tubes and the male gamete travels down the pollen tube to reach the egg cell.

Both male and female gametes are produced by a form of cell division known as **meiosis** (p. 223). In this process, each gamete receives only half the chromosome complement of the adult cell. The gamete is said to be haploid or monoploid. Human cells contain 46 chromosomes ($2n$), but the gametes contain only 23 (n). The $2n$, or diploid number of chromosomes, is restored at fertilization.

Fertilization Fertilization occurs when the male and female gametes meet and their nuclei fuse to form a zygote (p. 170). In animals, fertilization may be external or internal.

External fertilization must take place in water. The female releases the ova and the male sheds sperms on them after they have left the female's body. Often, one of the pair makes a nest of some kind, which keeps the ova in one place. A pattern of courtship behaviour ensures that the male releases sperms at the right time and place. Nest building and courtship, therefore, improve the chances of the ova being fertilized.

External fertilization occurs in fish, amphibia and many aquatic invertebrates (see Chapter 37).

Internal fertilization occurs when the female retains the ova in her body and the male, in a process of copulation, introduces sperms into her oviducts. The sperms swim to the ova and fertilize them inside the female's body. The eggs may then be laid (e.g. birds and some reptiles), or retained in the female's body until the young are more or less fully formed (e.g. mammals and some reptiles).

Placing the sperms in the body of the female increases the chances of successful fertilization. In general, animals which exhibit internal fertilization lay a small number of eggs or have a small number of offspring. Animals which fertilize their ova externally, usually lay a large number of eggs. This may compensate for those which remain unfertilized or are eaten by predators.

Plants In the majority of flowering plants, individual flowers have male and female organs. The egg cell is retained in the ovule, inside the ovary. The pollen cell reaches the egg cell with the aid of insects or wind, and the growth of the pollen tube. The fertilized ovule is retained in the plant as it develops to a seed.

Development In all living organisms, the **zygote** is a single cell. Its nucleus contains chromosomes from both parents.

The zygote undergoes cell division and the cells become tissues and organs. The changes which turn a new cell into a skin cell, a mesophyll cell, a nerve cell, etc. are called **differentiation**, i.e. the cells become specialized and different from each other.

Once cell division begins, the zygote is called an **embryo**. In the flowering plants, the embryo is released as a seed and development ceases until the seed germinates.

In the animals, the embryo continues to develop in the egg or the uterus until it is fully formed.

If the zygote is retained in the mother's body, her blood circulation supplies the embryo (via the placenta) with all the food it needs for its growth and development (p. 175).

If the zygote is in an egg outside the body of the female, the egg often contains a food store in the form of yolk. The embryo uses this food store to form its tissues and supply its energy.

The embryo in the seed of a flowering plant draws on the food reserves in its cotyledons (or endosperm) during the early stages of germination (p. 94).

Parental care and dispersal In many cases, once the ova have been fertilized and the eggs laid, the embryos and young organisms are left to fend for themselves. Usually, in such cases, a large number of eggs are produced, and this compensates for the heavy losses due to predation.

Some of the vertebrates exhibit parental care. The young may be protected in a nest, fed and kept warm. Parental care increases the chances of survival for the young animals. In those species where parental care is practised, the numbers of offspring are usually fairly low.

Even where there is parental care, the young usually disperse once they can obtain their own food, though there are many variations, e.g. herds of deer, lion families hunting together, wolf packs, etc. Dispersal of individuals or groups probably helps to reduce the competition between members of the same family for food, space and mates.

Dispersal of fruits and seeds has been discussed on page 89.

New gene combinations One of the effects of sexual reproduction is to produce new variations in the offspring.

The process of meiosis (p. 223) which precedes gamete formation, shares out the parental chromosomes between the gametes. However, the chromosomes are not shared out in the same way as they are in mitosis (Fig. 2). In a human cell there are 46 chromosomes, 23 from the mother and 23 from the father. When these chromosomes are shared between two gametes, one gamete may receive e.g. 10 maternal and 13 paternal chromosomes, or 17 maternal and 6 paternal chromosomes. There are 2^{23} possible combinations.

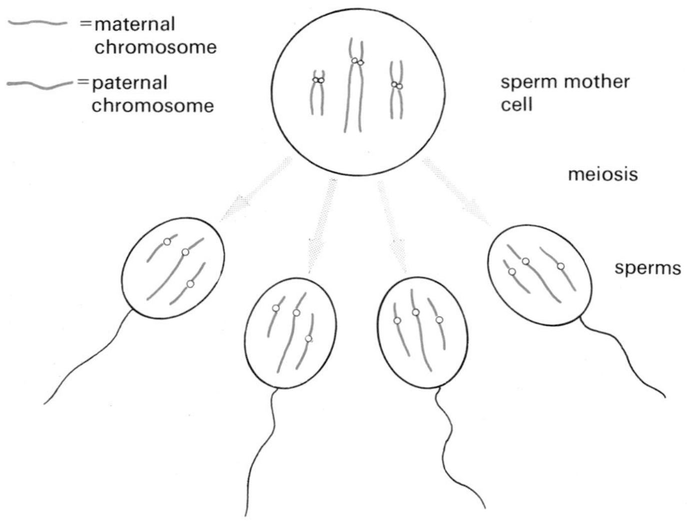

Fig. 2 New gene combinations. Some of the different chromosome combinations in gametes. How many other combinations are possible?

When the sperm and ovum fuse, there is the possibility of many new combinations of maternal and paternal characteristics. Moreover, in animals and in many plants, the male and female gametes come from different individuals with different characteristics. This increases, even further, the chances of new combinations of characteristics (p. 235) and, therefore, new variations in the offspring. These are some of the variations on which natural selection may act (p. 237).

Asexual reproduction

Asexual means without sex and this method of reproduction does not involve gametes.

The simplest form of asexual reproduction is the cell division (binary fission) of bacteria and protista (pp. 311 and 328). A single individual may reproduce many offspring without involving a second individual or any fusion of nuclei.

Some of the simpler invertebrates, e.g. some coelenterates (p. 302), can reproduce asexually.

Hydra is a small, freshwater coelenterate. It lives in ponds and feeds on small crustacea which it traps with its tentacles. It does have a sexual method of reproduction but also reproduces asexually by the process of budding depicted in Fig. 3. The hydra-like organisms which make up coral reefs, reproduce by a form of budding and produce the vast colonies you will have seen in television nature programmes.

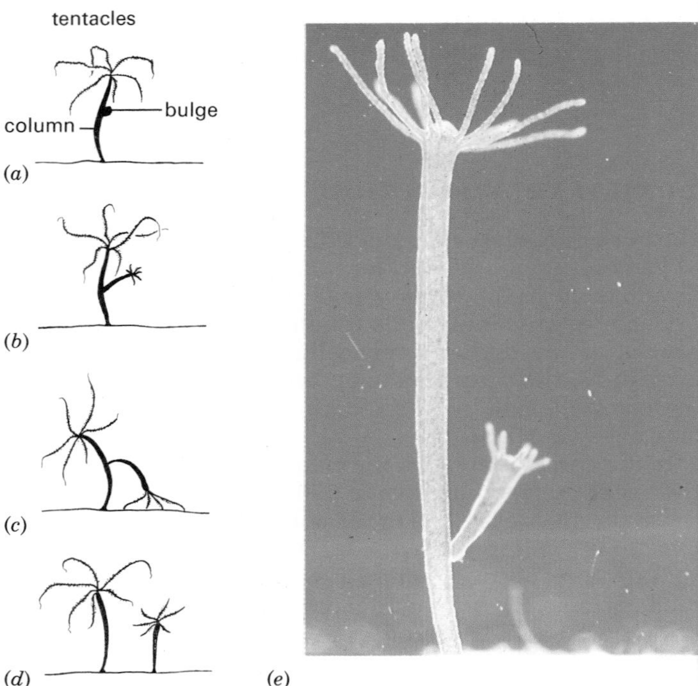

Fig. 3 Asexual reproduction in *Hydra*
(*a*) A group of cells on the column divides rapidly and produces a bulge.
(*b*) The bulge develops tentacles.
(*c*) The daughter *Hydra* pulls itself off the parent
(*d*) and becomes an independent animal.
(*e*) *Hydra* with bud (×10).

Fungi, mosses and ferns reproduce asexually by spores (p. 323). Although spores are mostly single cells, they can grow into a new individual without fusing with another spore.

Many flowering plants are experts at asexual reproduction. Bulbs, corms, runners and rhizomes (p. 100) produce buds which grow into independent plants.

Most of the organisms which can reproduce asexually also have a sexual method. Daffodils, bluebells and strawberries, for example, produce flowers and seeds by sexual reproduction.

Clones Asexual reproduction is brought about by cell division. This may be the division of a single cell, as in bacteria, or the division of many cells as in bulbs. In either case, the cell division is mitotic (p. 220), and the 'daughter' cells receive identical sets of chromosomes and genes.

All the offspring from asexual reproduction, therefore, are genetically identical and form what are known as clones. Each colony of bacteria on the culture plate in

Fig. 15 on p. 319 is a clone derived from a single parent. Each clump of daffodils in Fig. 6 on p. 102 is a clone derived originally from one bulb.

Asexual reproduction, therefore, does not produce any varieties on which natural selection can act. However, it can be exploited when we want to perpetuate the desirable characteristics of a crop plant (p. 103).

You may come across the term clone in connection with genetic engineering (p. 225). After a biologist succeeds in introducing a gene for, say, insulin into the chromosome of a bacterium, the bacterium is cloned. This means that it is allowed to reproduce asexually in a culture medium and all the offspring will carry the gene for making insulin.

REPRODUCTION	
Perpetuation of the species	
SEXUAL	*ASEXUAL*
Two parents needed	Only one parent
Gametes produced	No gametes
New variations in the offspring	Offspring all identical

QUESTIONS

11 Which of the following do not play a part in asexual reproduction: mitosis, gametes, meiosis, cell division, chromosomes, zygote?

12 Revise pages 100–104 and then say how we exploit the process of asexual reproduction in plants.

13 A fish may lay hundreds of eggs; a bird may lay only five or six. Despite this difference in the number of eggs the numbers of birds and fish do not change much. Suggest reasons for this.

GROWTH

In its simplest terms growth is an increase in size, or an increase in mass, or both. Growth takes place by cell division followed by cell enlargement. In a growing embryo, there will also be cell differentiation as cells become specialized to form tissues and organs.

In the early stages of growth, all the cells are able to divide. In the later stages, the dividing cells are restricted to certain regions, e.g. root tips (p. 65), buds (p. 67), Malpighian layer of the skin (p. 167). Specialized cells lose their ability to divide.

Growth does not take place uniformly in all parts of an organism. In young humans, the head grows relatively little compared with the limbs (Fig. 4).

Measurement of growth To investigate growth, it is possible to measure increase in length, mass, volume or area. The apparent pattern of growth revealed by these measurements will differ according to which method is used. For example, when a zygote undergoes cell division it does not increase in size or mass; a relatively large single cell simply divides to form a number of smaller cells (Fig. 5).

If growth is defined as increase in the total amount of protoplasm, then the only reliable indicator of growth is

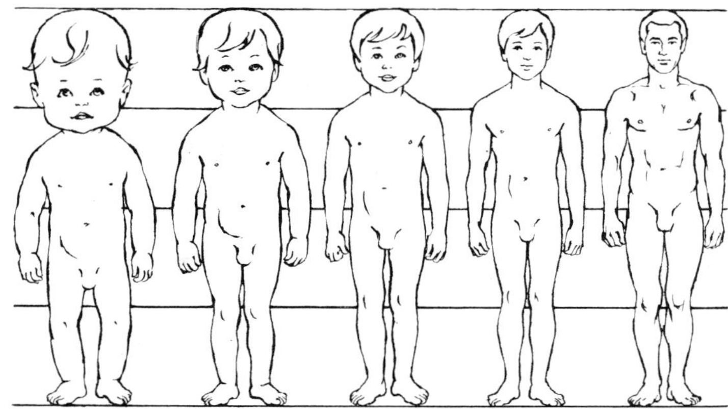

Fig. 4 Human growth. All the figures are drawn to the same height to show how body proportions change with age.
(After C. M. Jackson)

the increase of **dry mass** (p. 26). An increase of living mass may be the result of a temporary intake of water by osmosis, which is not necessarily a feature of growth.

However, measuring dry mass involves killing the organism and heating it in an oven at 110 °C until it loses no more weight. This is often neither desirable nor feasible, so one of the other measurements has to be used, but with regard to its possible limitations.

Growth curves Any of the measurements listed above can be plotted against time to produce a graph called a growth curve. The idealized growth curve is the sigmoid curve (p. 292) seen in Fig. 6. Growth is slow at first (cell division without cell enlargement), then becomes rapid and finally slows down and stops when the organism reaches full size.

Very few organisms conform exactly to the sigmoid pattern in their growth, and the growth curve will vary according to which part of the organism is measured and what kind of measurement is made.

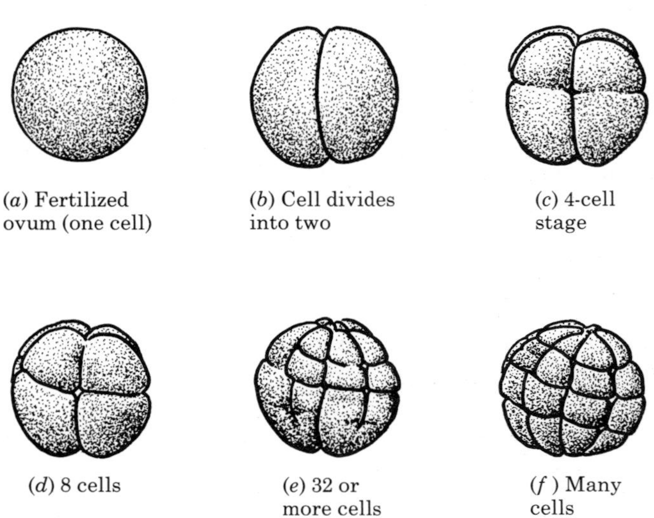

Fig. 5 Cell division in a frog's egg. There is no increase in size at first.

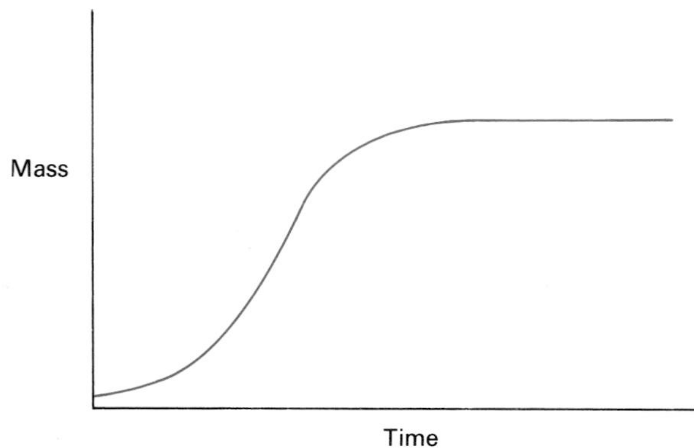

Fig. 6 Sigmoid growth curve

Figure 7 shows a fairly typical human growth curve based on increase in mass (not dry mass of course). It has an approximately sigmoid shape but shows two growth spurts; one in the first year or two and another at adolescence.

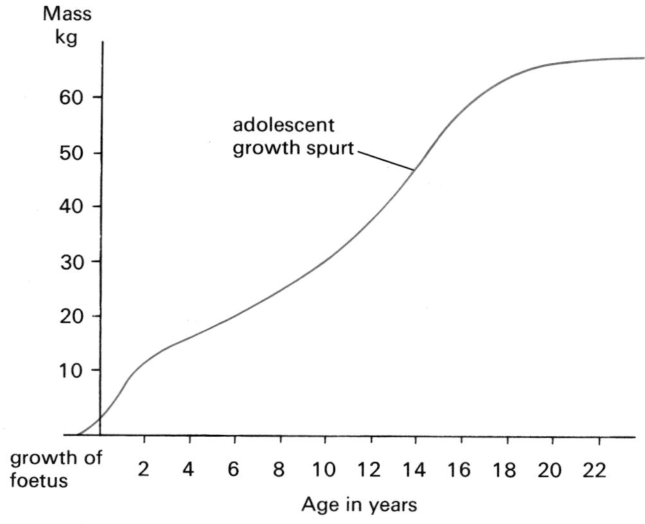

Fig. 7 Human growth curve

Many fish and some invertebrates never stop growing, although their growth rate does slow down.

The growth curve for an insect, which can increase its size only at ecdysis (p. 336), would look something like that in Fig. 8. However, this pattern of discontinuous growth relates only to increase in length. The growth rate as measured by the increase in mass, would approximate to the sigmoid curve.

QUESTIONS

14 Which of the following would not be included in the term growth: extension of root cells by osmosis (p. 65), production of branches on a tree, *Paramecium's* contractile vacuole enlarging (p. 328), binary fission in bacteria (p. 311), metamorphosis of an insect (p. 337)? Explain your answers.

15 The following figures show the growth of an insect over 200 days.
(*a*) Plot a graph of the figures with the days on the horizontal axis.
(*b*) Say whether you think the graph is sigmoid.
(*c*) Why is it so different from the graph in Fig. 8 which also represents insect growth?

Mass in mg	10	20	40	120	240	400	640	940	950	960
Time in days	20	40	60	80	100	120	140	160	180	200

MOVEMENT

Most animals can be seen to move about or to move parts of their bodies. Even a static creature, such as a mussel, can be seen to open its shell when it is submerged in water; many animals which lead a sedentary life have actively swimming larvae.

Movement in plants is less obvious and may be absent altogether. Leaves and flower petals may make quite obvious movements (p. 105) and the growth movements of seedlings may be included under this heading. In some of the lower plants, the male gametes are motile.

SENSITIVITY

Sensitivity (sometimes called irritability) is the ability of living organisms to respond to stimuli. A **stimulus** is a change in the external or internal environment of an organism.

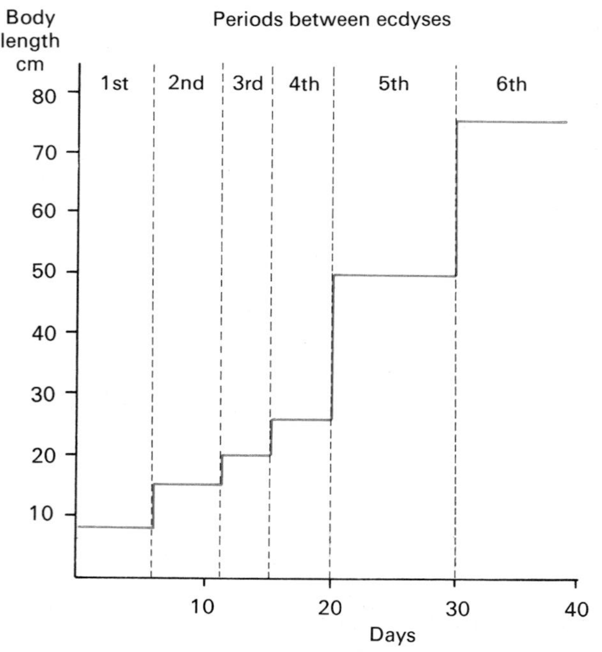

Fig. 8 Discontinuous growth in the locust. The discontinuity applies only to increase in body size and is a result of ecdysis. Mass increases continuously.

An external stimulus might be a touch, a change of light intensity or temperature, a sound or a movement. An internal stimulus might be food arriving in the stomach, a rise in blood pressure or a fall in body temperature.

An organism has the means of detecting these stimuli and responding to them in appropriate ways. In animals, stimuli are detected by structures called **receptors**. Receptors may be distributed over the surface of an animal (e.g. touch-sensitive bristles on an insect, pressure receptors in the human skin), or they may be concentrated into specialized sensory organs such as the antennae, the eyes or the ears (p. 196).

The organs which make a response to the stimuli are called **effectors**. An effector could be a muscle which contracts or a gland which produces a secretion.

The growing roots and shoots of plants respond to the stimuli of one-sided light and gravity. These responses are called **tropisms** (p. 106). There are no specialized receptors or effectors but the ability to detect and respond to the stimuli may be restricted to certain regions of the root and shoot.

In some plants, the petals and leaves respond to light or touch (p. 105) and this is particularly obvious in some insectivorous plants such as the Venus fly-trap.

Survival value of behaviour Appropriate responses to stimuli are essential to the survival of an organism.

A unicellular alga that moved away from the light would be unable to photosynthesize: a male frog which failed to release sperms when his partner laid her eggs would not father any offspring.

The chicks of a ground-nesting bird such as a pheasant, respond to their mother's warning cries by 'freezing', i.e. crouching and remaining motionless. They are thus less likely to be seen by a predator. A chick which does not respond in this way is likely to be caught and eaten and, therefore, will not leave any offspring. The chick's inappropriate behaviour has been 'weeded out' by natural selection.

QUESTIONS

16 A fly lands on your hand; you attempt to swat it but the fly gets away. What are the possible stimuli to which the fly responded?
17 Which of your sense organs respond to the stimuli of (*a*) light, (*b*) temperature, (*c*) chemicals?
18 What effectors are involved when you unexpectedly touch a rose thorn, cry out and remove your hand?
19 In what way do shoots show a directional response to a stimulus other than light?
20 What directional responses do wasps appear to make (*a*) when they pester you at a picnic, (*b*) when they try to escape from a room?

PRACTICAL WORK

A choice chamber

Figure 9 shows a choice chamber made from a plastic petri dish. The air on the left side will be moist (humid) because of the water in the lower compartment. On the right, the air will be drier because the silica gel absorbs water vapour.

Some woodlice are kept in a dry container for 3 hours before the experiment, and ten of them are placed in the top chamber and the lid replaced. The woodlice move about at random but on the dry side they will move rapidly and on the moist side they will move more slowly and eventually stop.

It looks as if they have moved directionally towards the humid side but this is not the case. Their final positions result from the fact that the stimulus of dry air makes them move rapidly in any direction. When, by chance, a movement happens to bring them into the humid air, they stop moving.

If you try this experiment, you must take certain precautions if your results are to be reliable.
1. You must make sure that there is no stimulus other than humidity to which the woodlice could be responding. If light was coming from the right side, it could be claimed that the woodlice were moving away from the light rather than responding to the humidity.

You would have to place the petri dish so that light came from above and not from one side. Alternatively you could repeat the experiment several times with light coming from each side in turn to show it made no difference to the results; or you could cover the choice chamber with a box to exclude all light, and lift the box at one-minute intervals to record the position of the woodlice.
2. You must use at least 10 woodlice. If you used only 5 and ended up with 2 on the dry side and 3 on the humid side, this is just as likely to have happened by chance. For the results to be significant, the woodlice must end up with at least 9 on the humid side and one or none on the dry side. Even 8:2 would not be accepted as very much better than chance.

You must do the experiment several times. The more often you repeat the experiment, and the more often you get a 9:1 or 10:0 distribution of woodlice, the more confident you can be that the animals are responding to humidity.

QUESTIONS

21 How would you modify the choice chamber experiment to see if woodlice made a non-directional response to light intensity?
22 What would be a good control experiment (p. 30) for the choice chamber, to show that it was humidity and not some other stimulus that affected the woodlice? What results would you expect in the control experiment?

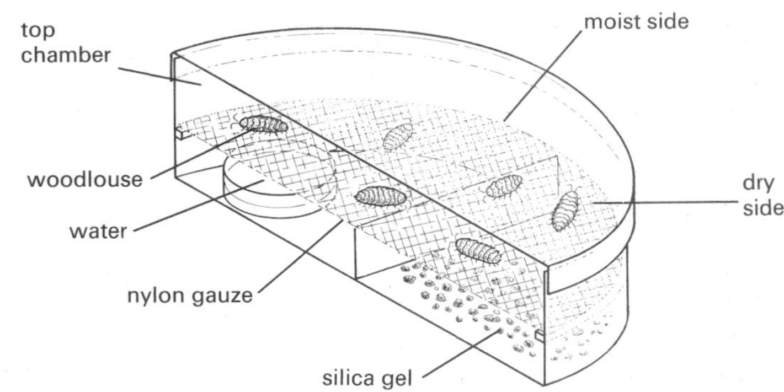

Fig. 9 A 'choice chamber' (only one half shown)

CHECK LIST

- **All living organisms do the following things.**
- **Respire: they break down food to obtain energy. Most organisms need oxygen for this.**
- **Feed: food is needed for energy and for making new cells. Plants make their food but animals, fungi and most bacteria must obtain it from other living organisms.**
- **Breathe: nearly all organisms take in oxygen from their surroundings and give out carbon dioxide as a result of respiration.**
- **Excrete: respiration and other chemical processes in cells, produce waste products which have to be expelled.**
- **Reproduce: all organisms eventually die. Species persist only as a result of reproduction. Reproduction may be sexual or asexual.**
- **Grow: the cells of an organism increase in number and size. The cells also become specialized and the organism increases in complexity.**
- **Move: most organisms move, or make movements at some stage in their lives.**
- **Respond to stimuli: living organisms exhibit sensitivity and respond to changes in their external and internal environment.**

Examination Questions

Section 6: Diversity of Organisms

Do not write on these pages. Where necessary copy drawings, tables or sentences.

1 Diagram A shows the external features of a bony fish and Diagram B shows the external features of a fern plant.

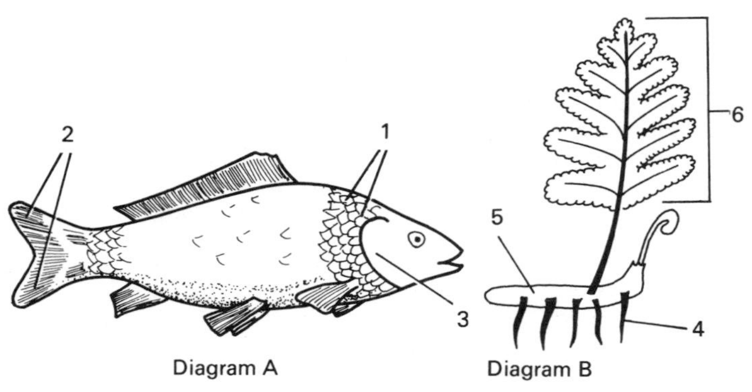

(*a*) Name the parts numbered 1–6.

(*b*) Use diagram A to help you explain TWO ways in which the fish is suited for life in water. (NI)

2 List **four** ways in which the structure of a bony fish helps it to live in water. (L)

3 (*a*) (i) In the diagram below fill in the boxes A, B and C to show the missing stages in the life cycle of a butterfly or moth.

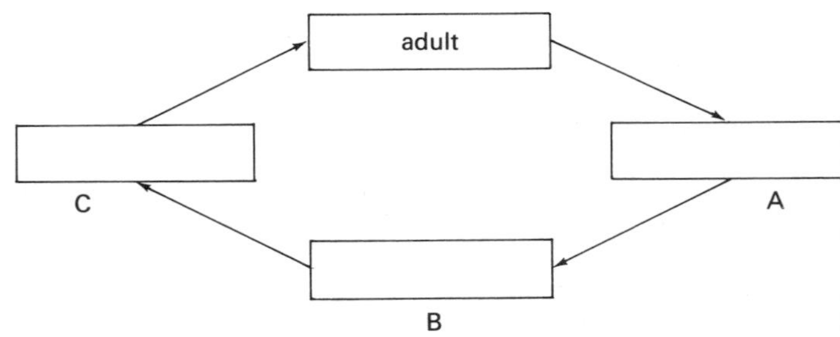

(ii) Which **one** of the following processes is carried out only by the adult butterfly or moth: excretion, feeding, growth, locomotion, reproduction, respiration, response to stimuli?

(*b*) State **four** external features of the adult butterfly or moth which help us to know that the adult is an insect. (L)

4 Gonorrhoea is a sexually transmitted disease.

(*a*) Write down **two** of the early symptoms of gonorrhoea in a man.

(*b*) State **two** ways which help stop the disease gonorrhoea from spreading. (L)

5 Three samples of milk were used in an experiment. They were spread across the surface of sterile nutrient agar in a Petri dish in an 'S' shape. The lids were replaced and sealed. The dishes were then placed in an incubator at 25 °C for 48 hours. The results are shown below.

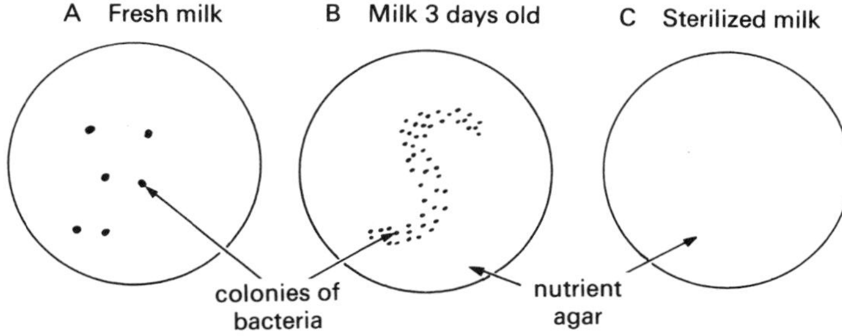

Explain why
(i) **sterile** agar was used,
(ii) **nutrient** agar was used,
(iii) the dishes were **sealed**,
(iv) an **incubator** was used,
(v) a few colonies of bacteria were found in A,
(vi) large numbers of colonies were found in B,
(vii) no colonies were present in C. (W)

6 The graph below shows the death rate from a bacterial disease, typhoid, in an underdeveloped country over a period of 15 years. In 1908 chlorine was added to the country's drinking water for the first time.

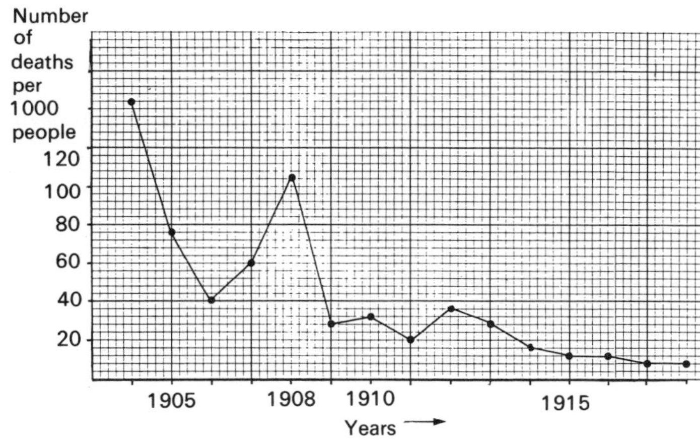

(*a*) From the graph state the death rate from typhoid in the year (i) 1905, (ii) 1910.
(*b*) Suggest reasons why the death rate in 1910 was (i) lower than 1905, (ii) higher than 1915.
(*c*) From the information in the graph, what can be deduced about the way in which typhoid bacteria enter the body?
(*d*) Suggest a way by which deaths could have been reduced before 1908. (W)

7 Increasingly Man is able to apply technology in order to mass produce certain useful products of biological processes.
Illustrate the truth of this statement with reference to the processes involved in the production of any **four** of the following products: (*a*) yoghurt, (*b*) cheese, (*c*) single-cell protein (SCP), (*d*) ethanol (alcohol), (*e*) bread. (M)

8 Malaria is caused by a microscopic organism.
(*a*) Why is the organism which causes malaria called a parasite?
(*b*) There are many things which can be done to control the spread of the malarial parasite. Name **four** different control methods and describe briefly how each works. (M)

9 An investigation was carried out into the effect of different antibiotics on a disease-causing bacterium.
Equal-sized samples of bacteria were placed on four dishes of agar.
The agar in three of the dishes contained different antibiotics A, B and C. The fourth dish had no antibiotic in it. The dishes were incubated at 35 °C for two days.
The diagrams below show the number of colonies of bacteria which grew on each dish. Each dot represents one bacterial colony.

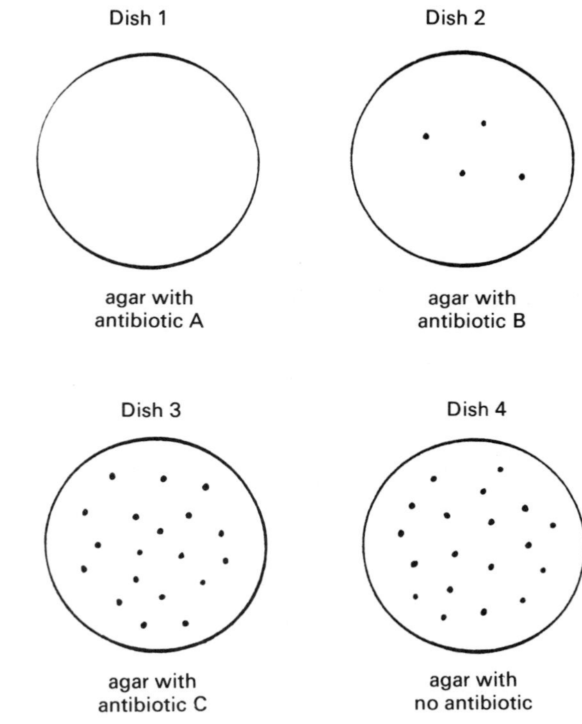

(*a*) Describe what the antibiotic did to the bacteria in (i) Dish 1, (ii) Dish 2, (iii) Dish 3.
(*b*) Which antibiotic would be the best for treating the disease caused by this bacterium?
(*c*) Describe **one** way in which our blood system protects us against bacteria.
(*d*) We can be protected against typhoid disease by an injection of dead typhoid bacteria. Explain how this injection will protect the body against typhoid. (N)

10 Some types of disease-causing bacteria have become resistant to some antibiotics. This is a result of
A bacteria becoming used to the antibiotics.
B the antibiotics causing mutations in the bacteria.
C the antibiotics not being powerful enough.
D survival of varieties of bacteria which are not affected by the antibiotic. (N)

Further Examination Questions

Do not write on the following pages. Where necessary copy drawings, tables or sentences.

Section 1: Some Principles of Biology

1 Match the numbers of the reactions with the correct letter describing them.

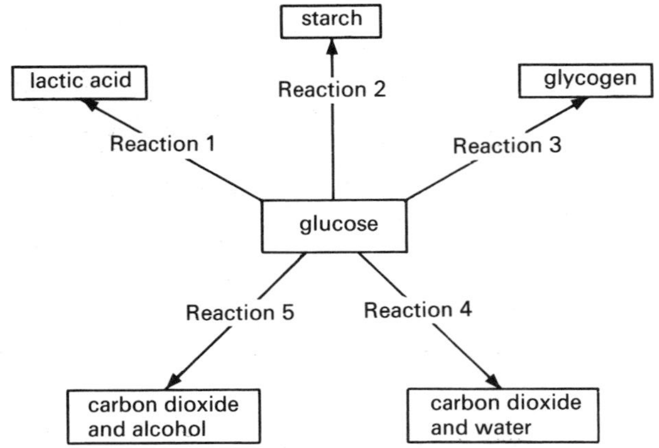

A Aerobic respiration
B Change taking place in the liver
C Anerobic respiration in yeast
D Change taking place in a plant storage organ, e.g. potato. (W)

2 An investigation was carried out to find the effect of different amounts of fertilizer on the amount of grain produced by five varieties of rice. These were an old variety and four new ones, A, B, C and D. The results are shown in the table below.

Amount of fertilizer applied (kg per hectare)	*Amount of grain produced* (tonnes per hectare)				
	Old variety	*Variety A*	*Variety B*	*Variety C*	*Variety D*
0	4.8	3.9	4.4	4.6	5.2
10	5.0	4.2	5.0	5.0	5.3
20	5.1	4.4	5.6	5.2	5.5
30	4.9	4.7	6.2	5.0	5.1
40	4.8	5.1	6.8	4.7	4.9
50	4.4	4.6	7.4	4.5	4.6
60	4.3	4.4	8.0	4.4	4.0

(*a*) Which variety produced the most grain when 20 kg per hectare of fertilizer was applied?
(*b*) Which TWO varieties produced the same amount of grain when 50 kg per hectare of fertilizer was applied?
(*c*) What is the effect of increasing the amount of fertilizer on the amount of grain produced by the old variety?
(*d*) (i) In India most farmers cannot afford to buy fertilizer. Which variety should they grow?
(ii) Explain how you decided this.
(iii) Suggest **one** other factor which should be considered when deciding which variety to grow. (N)

3 (*a*) Two sets of apparatus (A and B below) were used to investigate the process of photosynthesis. Before the apparatus was set up both plants were kept in the dark for 48 hours.

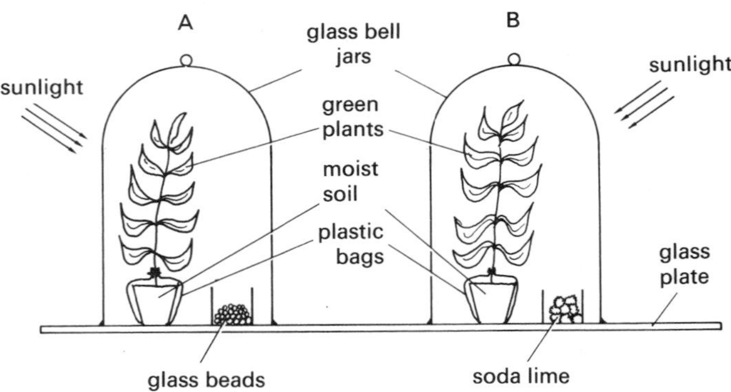

(i) Why were the plants kept in the dark for 48 hours before starting the investigation?
(ii) Which environmental factor necessary for photosynthesis was missing from one of the bell jars?
(iii) Why were two sets of apparatus used in the investigation?
(iv) What chemical substance in the leaves do you usually test for to show that photosynthesis has taken place?
(v) How would the results of this test show the importance of the missing environmental factor to the process of photosynthesis?
(*b*) Name **one** important element taken up by plants from the soil which is required for the production of (i) chlorophyll, (ii) cell walls.
(*c*) State **two** different ways in which a leaf is well adapted to carry out photosynthesis.
(*d*) Horticulturalists sometimes introduce extra lighting and carbon dioxide into their greenhouses to increase photosynthesis. Explain why **each** action increases photosynthesis. (NI)

4 The diagram below shows part of a section through a leaf.

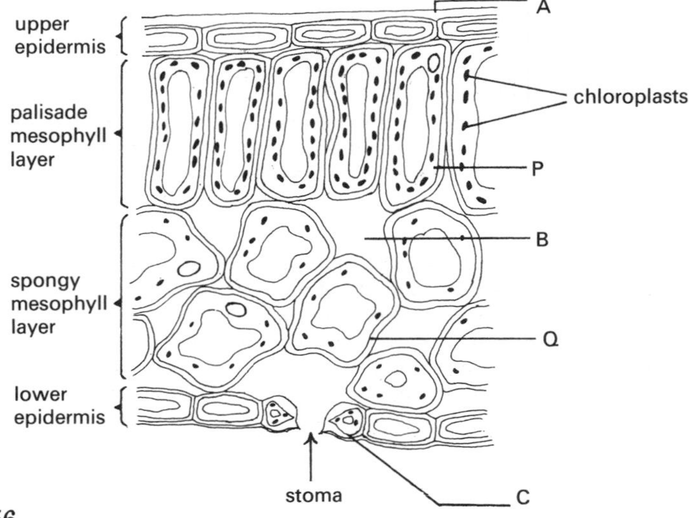

(*a*) Name A, B and C.

(*b*) (i) Photosynthesis takes place in the chloroplasts. Count the number of chloroplasts in the cells labelled P and Q.

Cell	*Number of chloroplasts*
P	
Q	

(ii) Which of the cells, P or Q, would produce more sugar in the daylight?

(iii) Which surface of the leaf has more stomata?

(*c*) Explain how the palisade and spongy mesophyll layers help the leaf to carry out photosynthesis.

(*d*) What are the functions of the upper and lower epidermis? (L)

Section 2: Flowering plants

5 Two chrysanthemum cuttings were treated as shown in the diagrams below.

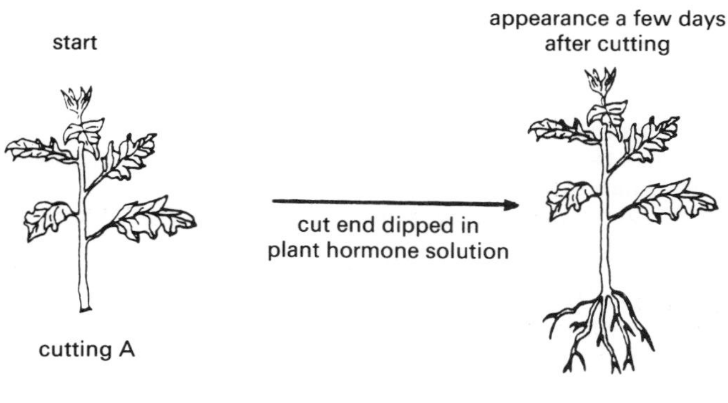

(*a*) (i) Give **one** difference between the cuttings a few days after planting.

(ii) How would this difference help cutting A to grow better?

(*b*) Explain why cutting B was dipped in distilled water.

(*c*) The hormone does not keep for long in bright light or when it is dissolved in water. Suggest how a manufacturer could overcome these problems when packaging the hormone for sale to gardeners. (N)

6 An experiment was set up to find out the effect of the hormone auxin on the growth of coleoptiles (shoots).

Ten oat seedlings with straight coleoptiles were chosen when the coleoptiles were 2 cm long. Some lanolin paste was mixed with the auxin. This mixture was spread along one side of each of five coleoptiles as shown in the diagram below of seedling A at day 0. The other five coleoptiles were used as a control. Both sets were then left in the dark for two days.

The right-hand diagram shows an incomplete drawing of seedling A after 2 days.

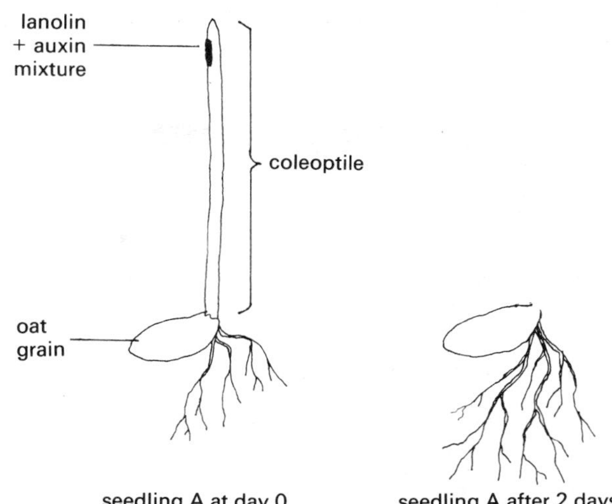

(*a*) Complete the drawing of seedling A as it might appear after two days.

(*b*) What is the magnification of the drawing of seedling A at day 0?

(*c*) (i) What should have been smeared along the side of the control coleoptiles?

(ii) Explain why the treatment in (i) would act as a control.

(*d*) Why is it better to use five coleoptiles in each set rather than one? (L)

7 List **three** factors which are needed for seeds to start to grow (germinate). (L)

Section 3: Human Physiology

8 Some food companies have developed non-meat foods which look and taste similar to meat. One food called mycoprotein is produced from a fungus which is closely related to the mushroom.

The table below compares some of the classes of food in beef and mycoproteins as percentages of their dry mass.

Class of food	*Beef % dry mass*	*Mycoprotein % dry mass*
Protein	68.2	44.3
Lipid	30.2	13.8
Dietary fibre	0.0	37.6

(*a*) Using information in the table write down **two** differences between mycoprotein and beef which make mycoprotein a healthier food to eat. In each case give an explanation for your answer.

(*b*) Some people find mycoprotein unpleasant to eat and it goes bad very quickly. Food additives can overcome these problems. Name **two** examples of food additives and in each case explain the reason for adding it.

(*c*) Write down **two** ways in which fungi are different from green plants. (L)

9 Exercise causes the heart to beat faster. The simplest way of detecting this is by taking a person's pulse.

(*a*) Describe how you would find a person's pulse and measure its rate.

(*b*) Describe how you could investigate whether the effect of exercise on the pulse rate of fourteen-year-old pupils was the same for boys and girls.

In your account you must make clear how you would be sure to obtain a 'typical' response for boys and girls and how you would make sure that the comparison was a fair one. (S)

10 The diagram below shows the breathing organs inside the human chest.

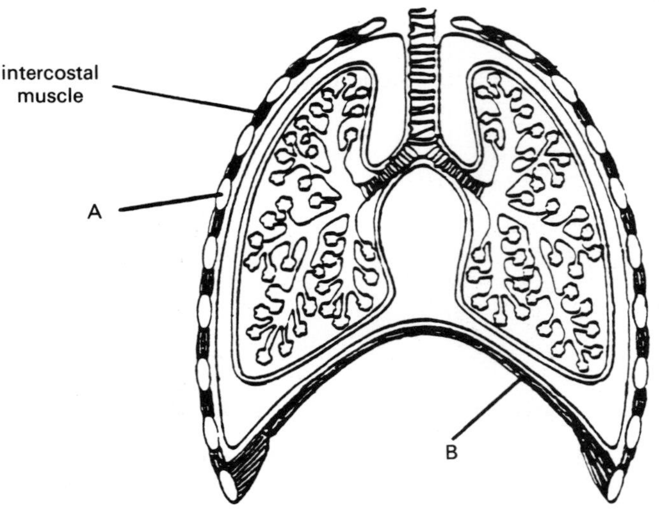

(*a*) Name the parts marked A and B.
(*b*) (i) What makes part A move?
(ii) Which way does part B move when we breathe in?
(iii) Why does air move into the lungs when part B moves?
(*c*) In the table below, give two differences between the air we breathe in and the air we breathe out.

Air breathed in	*Air breathed out*

(*d*) Give **two** ways in which breathing changes during exercise. (N)

11 We produce an enzyme in our mouth which digests starch. A scientist measured the amount of this enzyme in three groups of people who eat different foods. The results were:

Group	*Food eaten*	*Amount of enzyme in mouth (in units per cm³)*
A	Mixed diet of meat, vegetables, fruit and cereals	101
B	Mainly meat	22
C	Mainly starchy cereals	248

(*a*) How does the type of food eaten affect the amount of enzyme produced?
(*b*) The following information was obtained from an experiment involving the digestion of starch by this enzyme.

Temperature °C	*Amount of starch digested in one minute*
5	32
25	164
35	216
45	204
65	36

(i) Plot a graph to show these results.
(ii) What does the graph tell you about the effect of temperature on starch digestion? (W)

12 The diagram below shows the urinary system of the human body.

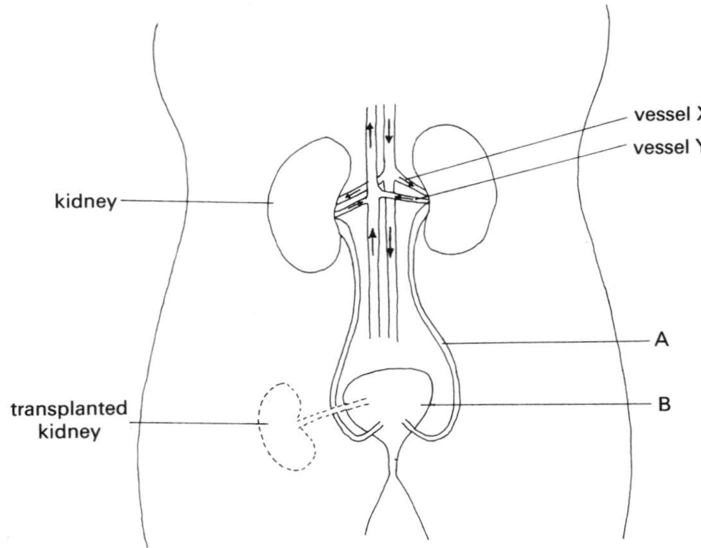

(*a*) Name the parts A and B.
(*b*) Name **two** substances which would be less concentrated in vessel Y than in vessel X.
(*c*) In some people the kidneys stop working. When this happens, another kidney may be transplanted into the position shown in the diagram. Tube A from the transplanted kidney is joined to part B.
(i) What other link must be made between the transplanted kidney and the body to allow the kidney to work?
(ii) Suggest **one** advantage of transplanting the kidney in the position shown rather than in the normal position. (L)

13 The diagrams show the condition of the teeth of two twelve-year-old pupils. One visits the dentist regularly for fluoride treatment, the other does not.

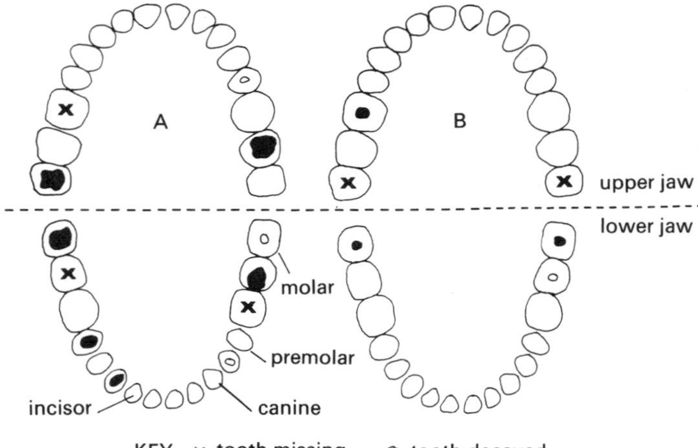

(*a*) How many healthy teeth are present in A?
(*b*) (i) Which diagram represents the person who received the fluoride treatment?
(ii) State a reason for your answer.
(*c*) (i) One type of tooth in A is perfectly healthy. Name this type.
(ii) Explain why this is so.
(*d*) (i) Which type of teeth appear to have decayed the most?
(ii) Suggest **two** reasons for this. (W)

Section 4: Genetics and Heredity

14 (*a*) Discuss whether or not each of the following statements is accurate.

(i) Although a human sperm and a human ovum are very different in size, both contain an equal amount of genetic material.

(ii) Cell division in the human body can either be of the type which ensures exact replicas of the original nucleus or of the type which results in variation of the original composition of the nucleus.

(iii) The appearance (phenotype) of an organism is not just a result of its genotype.

(*b*) A red-flowered plant of species A was crossed with a white-flowered plant of the same species. All the offspring were red-flowered.

A red-flowered plant of species B was crossed with a white-flowered plant of species B. All the offspring of this cross produced pink flowers.

Using **this information**, explain the difference between each of the following pairs of biological terms.

(i) complete dominance and incomplete dominance

(ii) homozygous and heterozygous

(iii) phenotype and genotype (M)

15 The drawings below show two forms of the Peppered Moth.

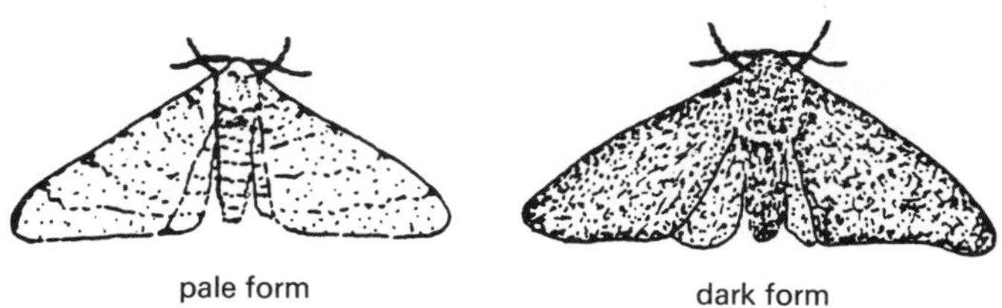

When one of the pale Peppered Moths is crossed with one of the dark moths, all of the young moths have dark wings and body.

The wing and body colour is determined by one pair of genes only.

(*a*) Which wing and body colour is dominant? Explain your answer.

(*b*) Complete the genetic diagram below to explain the results of the cross described. Use suitable symbols for each gene.

Symbols for each gene	Pale..................	Dark..................
Parental genotypes	....	....
Gamete genotypes	....	....
Genotype of young moths	....	

(*c*) The pale form of the moth is well camouflaged on lichen-covered tree-trunks in unpolluted woodland. In polluted areas the lichen is killed and tree-trunks are dark, so the dark form has better camouflage. The moths are preyed on by birds.

(i) What would you expect to happen to dark moths in an unpolluted wood?

(ii) If an area in which most of the moths were pale became polluted, what would you expect to happen to the proportions of pale and dark moths in the population? Explain your answer.

(iii) Explain how the gene for pale colour might be preserved in a population of dark moths. (N)

Section 5: Organisms and their Environment

16 Some of the most fertile waters in the world lie just off the coast of Peru. Plankton multiply rapidly and support great shoals of fish. The major consumers of the plankton are small fish called anchovetas. These, in turn, are eaten by larger fish such as sea bass and tunny and by vast numbers of birds such as gulls, pelicans and guanays. Fifty years ago there were over five million guanays nesting on the bare rocks of the offshore islands. The guanays' droppings were rich in nitrogenous waste. Some of these droppings fell in the sea and promoted the growth of the plankton. Some of these droppings fell on the rocks and accumulated to make a layer which was once fifty metres deep. The guano, as it was called, made a good fertilizer and was sold all over the world. Fleets of fishing boats harvested the tunny and sea bass to provide food for people all over Peru.

Then, thirty years ago, chemical fertilizers caused the price of guano to fall. The people who lived by the sea decided to catch anchovetas instead of digging guano. The anchovetas were not to be eaten by people but made into fertilizer. The fishing was not controlled and within a few years the shoals of fish had all but disappeared. Most of the guanays then starved. Now the sea is not fertilized by their droppings and little guano is made.

(*a*) Draw a food web, including humans, for the coastal part of Peru as it was thirty years ago.

(*b*) What destroyed the food web?

(*c*) Suggest **one** thing that could be done to restore the populations of fish and birds to their previous levels. (N)

17 The diagram below shows the carbon cycle.

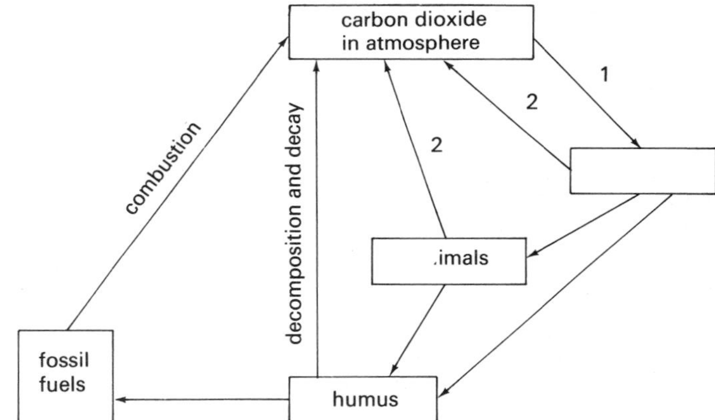

(*a*) (i) Name the process which is shown by arrow 1.

(ii) Name the raw material, other than carbon dioxide, which is needed for process 1.

(*b*) Name the process shown by both the arrows labelled 2.

(*c*) Name **two** groups of organisms which bring about decomposition and decay.

(*d*) Name **two** fossil fuels in common use.

(*e*) Suggest **one** effect the burning of fossil fuels has on the composition of the air.

(*f*) In some parts of the world large areas of forest are being cleared. Suggest **two** ways in which this might affect the carbon cycle. (L)

Appendix 1: First Aid and Emergency Treatment

OBJECTIVES

The object of First Aid is to save life and prevent an injured person's condition getting any worse while waiting for expert medical help.

Attempts by an unskilled person to treat an injury may do more harm than good. On the other hand a very simple, common-sense act may save a person's life. For example, trying to move an injured person or clumsy attempts to put a splint on a broken limb may cause additional damage. In contrast, simply turning an unconscious person on his side may save his life by keeping his mouth and wind-pipe clear so that he is not suffocated by his own relaxed tongue or choked by his own saliva or vomit.

PRIORITIES

Non-medical priorities

If you are the first or only person at the scene of an accident you will need to make quick decisions about what to do first. Your first action may have nothing to do with 'First Aid'. If there has been a motoring accident, the most important action may be to try and stop oncoming traffic so that the accident is not worsened by further collisions. If other people are present, your first act might be to send someone to get help while you attend to the injured people. If you are the only able-bodied person present, you should not go in search of assistance until you have attended to the injured people.

Medical priorities

If there are several injured people, you must decide who to treat first. Somebody who is standing, or sitting up and crying out with pain or fright, is certainly conscious and breathing and unlikely to be in immediate danger unless he or she is bleeding profusely. A person who is lying still and silent is probably unconscious and may not be breathing. You can check whether a person is unconscious simply by speaking to him or her. If there is no response or movement, the person is probably unconscious. This person must be examined first and your priorities are:

(*a*) Restore breathing.
(*b*) Stop any bleeding.
(*c*) Reduce the effects of shock.
(*d*) Do not leave an unconscious person lying on his back.

(*a*) **Restore breathing.** If a person is not breathing, the heart may still be beating but the blood it is pumping to the brain will be deoxygenated (p. 138). If the brain cells are deprived of oxygen for more than four minutes they die. Since brain cells cannot be replaced, this will result in permanent brain damage even if the person recovers. It is important to restore breathing at once by the method of mouth-to-mouth resuscitation as described on the next page.

(*b*) **Stop any bleeding.** If the injured person is breathing, your next task is to look for any signs of bleeding. Small cuts and scratches are not likely to need urgent treatment but if an artery or vein is cut, the casualty may lose blood rapidly and be in serious danger after only a few minutes. You may have to try and stop serious bleeding while still trying to restore the casualty's breathing. Resuscitation may take an hour or more, during which time the casualty could bleed to death if serious bleeding is not stopped.

(*c*) **Reduce the effects of shock.** 'Shock', in the medical sense, does not mean a fright but the drastic fall in blood pressure which often follows injury. This fall in blood pressure may cause death even if the injuries themselves are not fatal. The first-aider cannot treat an injured person for shock but he can do much to prevent it getting worse.

(*d*) **Do not leave an unconscious person lying on his or her back.** When you have restored the person's breathing, stopped serious bleeding and taken precautions against shock, it is then reasonable to go for help, provided you do not leave unconscious persons lying on their backs. In this position, the relaxed tongue, saliva, vomit or blood may block the wind-pipe and suffocate them. Turn the person very carefully on his side as shown in Fig. 1, pulling the jaw forward and tilting the head back so that the airway is clear and fluids can escape from the mouth. Apart from this, you should not move injured persons more than necessary.

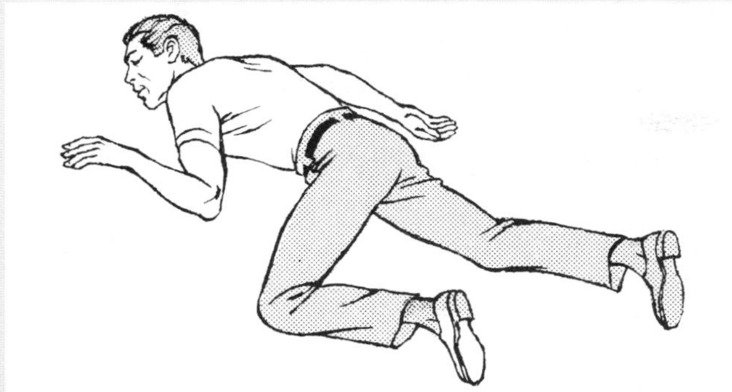

Fig. 1 Recovery position

FIRST AID TECHNIQUES

Resuscitation

Your exhaled breath contains about 16 per cent oxygen which is quite enough to oxygenate the casualty's blood. Turn the person carefully on to his back, tilt the head back as far as possible and pull the jaw forward, opening the mouth (Fig. 2). This will open the wind-pipe. Open your mouth wide and place it over that of the casualty; close his nostrils by pinching his nose with your fingers. Breathe air gently into the casualty's lungs till you see his chest rise. If there is no chest movement, examine the mouth and pharynx and remove any material that might be blocking the air passage. Then tilt the casualty's head back again as far as possible, pull his jaw well forward and try again. Inflate his lungs about ten times per minute until he starts to breathe or until a doctor tells you to stop. When breathing is restored, turn the patient on his side as described in (*d*), p. 360.

Arrest of bleeding

Severe bleeding, with the blood flowing out rapidly, must be stopped at once by pressing with the fingers directly on the wound or pressing the edges of the wound together for at least ten minutes. There will be no time to search for sterile dressings or bandages. Lay the casualty down and if the wound is in a limb, raise it, provided it is not fractured. When the bleeding has slowed down press a pad of material over the wound and tie it firmly in place. If blood oozes through, apply more material on top of the original pad.

It may sometimes be impossible to get the edges of the wound together because so much tissue has been removed by the injury. In this case, if bleeding is severe, a dressing must be placed in the wound to plug it. The 'dressing' may have to be a vest or a shirt if you are not carrying a First Aid kit. It is more important to stop the bleeding than to worry about infection. Rapid loss of blood may lead to death in as little time as 20 minutes or so and the casualty may not reach hospital alive. If the blood loss is stopped and the casualty gets to hospital, the medical staff will be in a position to treat any infection introduced by the dressing.

Reducing the effects of shock

Shock is a drop in blood pressure which may result from heart failure, loss of blood or poisoning. Blood may be lost from an external wound, or bleeding may be taking place internally and so not be noticed. At the site of a severe burn, there is a rapid seepage of blood plasma which can cause a fall in blood volume.

If the blood pressure falls, the brain does not work so well and does not send out the right impulses to the heart and circulatory system. As a result, the blood

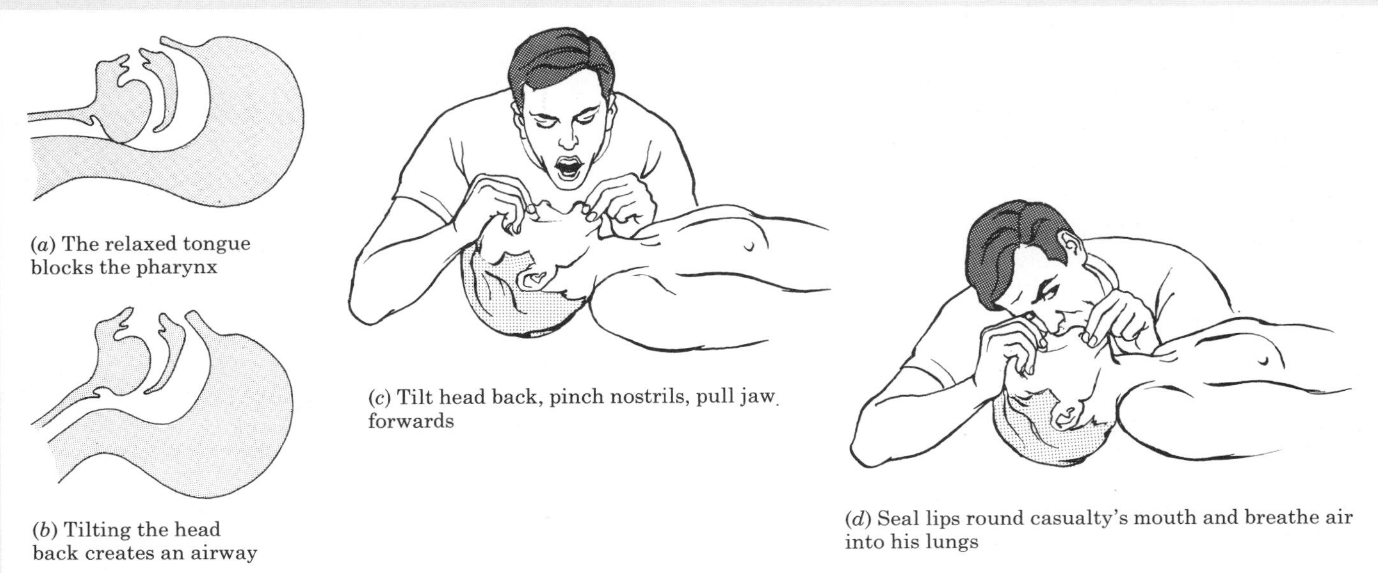

(*a*) The relaxed tongue blocks the pharynx

(*b*) Tilting the head back creates an airway

(*c*) Tilt head back, pinch nostrils, pull jaw forwards

(*d*) Seal lips round casualty's mouth and breathe air into his lungs

Fig. 2 Mouth-to-mouth resuscitation

pressure falls even more and, unless the first-aider acts promptly, the victim may die.

The symptoms of shock are a rapid, weak pulse, cold, clammy skin and shallow, rapid breathing. You should assume that the patient is shocked in all cases of electric shock, drowning, burns and serious injury.

Keep the casualty lying down, with legs raised, if possible, to maintain the blood pressure in the brain. Do not give anything to drink and do not try to warm the casualty but do loosen any tight clothing round the neck, chest and waist. Do not move the casualty more than is absolutely necessary for his or her safety.

The treatment for shock is to restore the blood volume by transfusing blood (p. 147) or plasma, so the casualty must be sent to hospital as soon as possible.

Electric shock

A severe electric shock will cause loss of consciousness. On no account should you touch or even go close to an unconscious person who is likely to be still in contact with an electrical supply, because the electricity passing through his body may reach you too. The current must first be switched off by you or somebody who has the necessary skill before touching the unconscious person.

It is likely that the casualty will not be breathing and so resuscitation must be started at once, together with precautions against shock and possibly treatment for burns.

Burns and scalds

A *burn* is caused by dry heat, e.g. a flame or hot object. A *scald* is caused by steam or boiling water. The treatment is the same for both except that, in the case of scalds, any wet clothing should be carefully removed, while burned clothing should be left in place. The treatment is to flood the burned or scalded area with clean, cold water for at least ten minutes or, in the case of hands and arms, immerse the limb in running cold water.

Do not apply any ointments or lotions but lightly cover the injured area with clean, dry cloth to keep bacteria from falling on to the damaged tissue. Remember that shock is likely to result from burns which affect a large area of skin.

Fainting

This is caused by a temporary fall in the blood supply to the brain, sometimes brought on by emotional shock or prolonged standing. Treat the casualty as if he or she were suffering from shock, but if the unconsciousness was not the result of injury or poisoning, there is no need for the person to go to hospital.

Drowning

A person pulled unconscious from the water may not be breathing. In this case, mouth-to-mouth resuscitation must be started at once, followed by efforts to reduce shock. Do not worry about water in the lungs. It is very difficult for water to get into the lungs and even if it does, the first-aider can do nothing about it. Bear in mind that the casualty is quite likely to be suffering from hypothermia (see below).

Accidental poisoning

If you have reason to suspect that a person, especially a child, has swallowed a harmful substance, ask at once what it was or get hold of a sample so that a suitable antidote can be prepared at the hospital. If the casualty is conscious, he can be given water or milk to drink, to dilute the poison. Do not give solutions of salt or any other mixture intended to induce vomiting. If the casualty is conscious and the poison is not corrosive, i.e. has not caused burns to the mouth and throat, he may be persuaded to vomit by getting him to put a finger to the back of the throat.

Fractures or dislocations

If a limb or joint is distorted, badly bruised or swollen and the casualty cannot move it, it is probably fractured or dislocated and needs expert attention. The limb should not be moved unless the casualty has to be moved to safety, in which case, support the limb by tying it to the casualty's body, e.g. one leg to the other, or the arm to the chest.

Particles in the eye

In the eye really means 'on the cornea or sclera'. The eyes will water and the person will blink. These two actions may wash the particle into one corner of the eye or on to the lower lid. If the object is on the lower lid, in the corner of the eye or free to move on the sclera (white of the eye) it can usually be brushed gently away with the corner of a clean handkerchief or piece of cotton wool soaked in water. If this does not work, the casualty should put his or her face in a bowl of clean water and open and close the eyelids. The object will usually float away.

If the particle is on the cornea or if it appears to be embedded in the surface of the eye and does not move, or cannot be seen clearly, no attempt should be made to dislodge it. The casualty must be dissuaded from rubbing the eye. The eyelid should be kept closed and covered with a soft pad held lightly in place with a bandage, while the casualty is taken to a hospital or surgery.

Hypothermia

The causes of, and treatment for, this condition have been described on p. 169.

Appendix 2: Reagents

(Only teachers or technicians should prepare these.)

Acetic orcein It is simplest to buy the concentrated solution (3.3% orcein in glacial acetic acid). Dilute 10 cm^3 with 12 cm^3 water just before use. The diluted stain does not keep.

Adenosine triphosphate Purchase 2 cm^3 ampoules from a supplier (Expensive; over £20 for 5 ampoules. Store in a refrigerator.)

Ammonia solution (2M) Dilute 11 cm^3 0.880 ammonia with 89 cm^3 water.

Benedict's solution Dissolve 170 g sodium citrate and 100 g sodium carbonate in 800 cm^3 distilled water. Add a solution of copper sulphate made from 17 g copper sulphate in 200 cm^3 distilled water.

Bicarbonate indicator (*See* Hydrogencarbonate indicator.)

Biuret test 10% sodium hydroxide; 1% copper sulphate.

Cobalt chloride paper Dissolve 1 g cobalt chloride in 20 cm^3 distilled water. Soak filter paper in the solution and allow to dry.

Ethanoic (acetic) acid (M/10) Place 6 cm^3 glacial ethanoic (acetic) acid in a graduated flask and make up the volume to 1 litre with distilled water.

Formalin Dilute 1 part of 40% formaldehyde with 39 parts of water.

Gelatin Make a 10% solution by dissolving the crystals in tap water and heating to boiling point. Keep the solution moving about or it will burn on the bottom. Stopper the solution while hot to reduce the chance of bacterial contamination.

Glucose phosphate Make a 5% solution in distilled water. (Glucose phosphate costs over £2 per gram. Make only enough to meet your needs.)

Hydrochloric acid (M/10) Dilute 10 cm^3 conc. acid with 990 cm^3 distilled water.

Hydrogencarbonate indicator (bicarbonate indicator). Dissolve 0.2 g thymol blue and 0.1 g cresol red powders in 20 cm^3 ethanol. Dissolve 0.84 g 'Analar' sodium hydrogencarbonate (sodium bicarbonate) in 900 cm^3 distilled water. Add the alcoholic solution to the hydrogencarbonate and make the volume up to 1 litre with distilled water. Just before use, dilute the appropriate amount of this solution 10 times, i.e. add 9 times its own volume of distilled water.

To bring the solution into equilibrium with atmospheric air, bubble air from outside the laboratory through the diluted indicator using a filter pump or aquarium pump. After 10 minutes, the dye should be red.

Hydrogen peroxide Use a 20-volume solution, from suppliers or pharmacists.

Hormone rooting powder From hardware shops or garden centres. Powders suitable for woody or soft-stemmed varieties are available.

Iodine solution Grind 1 g iodine and 1 g potassium iodide in a mortar with distilled water. Make up to 100 cm^3 and dilute 5 cm^3 of this solution with 100 cm^3 water for experiments.

Lime water Shake distilled water with an excess of calcium hydroxide and allow the lime to settle. Decant off the clear liquid. Before use, test the liquid by bubbling exhaled air through it.

Macerating fluid Mix 10% nitric acid and 10% chromic acid in equal volumes just before the fluid is needed. Incubate plant tissues at 35 °C for 12–24 hours. Wash well with water before giving the tissues to students. (Purchase the 10% chromic acid. Make 10% nitric acid by adding 86 cm^3 water to 14 cm^3 conc. nitric acid.)

Manometer liquid (Experiment 3, p. 28). Try water with a few drops of liquid detergent plus a cooking dye such as cochineal substitute.

Methylene blue Dissolve 0.5 g methylene blue in 30 cm^3 ethanol and dilute with 100 cm^3 distilled water.

PIDCP (Phenol-indo-2,6-dichlorophenol). Make a 0.1% solution in distilled water.

Pyrogallic acid Dissolve 10 g pyrogallic acid in 100 cm^3 2N sodium hydroxide solution just before use. Alternatively, place 1 g pyrogallic acid in each flask and add 10 cm^3 2N sodium hydroxide.

Ringer's solution Dissolve 0.3 g calcium chloride (anhydrous), 0.25 g potassium chloride and 8.5 g sodium chloride in 1 litre (1 dm^3) distilled water.

Soda-lime Use the self-indicating form which changes colour when it loses its activity.

Sodium carbonate (M/20) Dissolve 5.3 g anhydrous sodium carbonate in 1 litre distilled water.

Sodium hydrogencarbonate Use a 10% solution.

Sugar solution (osmosis experiments). Dissolve sucrose in its own weight of tap water.

Starch agar Mix 1 part agar, 0.3 parts starch and 100 cm^3 water. Mix thoroughly and heat gently till the mixture starts to boil. When the liquid cools to about 45 °C, dispense it into Petri dishes. If the experiment is to be left for more than a day or two, it is advisable to sterilize the agar and glassware in an autoclave as described on p. 319.

Starch solution (1%) Shake 1 g starch powder with 100 g water and heat the mixture gently, with stirring, until the liquid just starts to boil.

Taste, solutions for Sweet: 5% sucrose. Sour: 0.5% citric acid. Salt: 2% sodium chloride. Bitter: 1 cm^3 tincture of quinine in 100 cm^3 water, or boil 3 g dried hops in 200 cm^3 water for 30 min. Strain the mixture and make up to 200 cm^3 with water.

Water cultures 2 g calcium nitrate, 0.5 g each potassium nitrate, magnesium sulphate and potassium phosphate (KH_2PO_4) in 2 litres distilled water. Add a few drops of iron(II) chloride solution.

No nitrate; use potassium and calcium chlorides instead of the nitrates.

No calcium; use potassium nitrate instead of calcium nitrate.

No phosphate; use potassium sulphate instead of the phosphate.

Water Unless distilled water is specified in this list of reagents, tap water is adequate.

Appendix 3: Book List

Section 1: Some Principles of Biology

Living cells (British Museum/Cambridge University Press, 1981). A short, well illustrated account.

Eukaryotic cell, M.R. Ingle (Blackwell, 1985). For reference; a concise overview.

Cell Biology, B. King (Allen and Unwin, 1983). Advanced text; teachers' reference.

Section 2: Flowering Plants

Organization in plants, W.M.M. Baron (Arnold, 1979). Plant physiology for 'A' level; some suggested practical work.

The physiology of flowering plants, H.E. Street and H. Opik (Arnold 1984). An up-to-date reference book on plant physiology.

Anatomy and activities of plants, C.J. Clegg and G. Cox (Murray, 1978). A concise, well-illustrated account of plant anatomy.

Plants in action, A. Hibbert and J. Brooks (BBC Publications, 1981). A general introduction to the life of plants; good for applied aspects.

Green plants and their allies, T.J. King (Nelson, 1983). Up-to-date reference source.

Section 3: Human Physiology

Success in Nutrition, Magnus Pyke (Murray, 1975). A useful reference for food and diet.

E for additives, M. Hanssen (Thorsons, 1984). A list of food additives, their purpose and possible effects.

Human Biology (British Museum/ Cambridge University Press, 1981). Based on the exhibition; well illustrated simple account.

Teaching about the senses, Centre for Life Studies (ILEA, 1984). A series of five booklets designed for teachers; full of useful ideas.

Biology of the mammal, A.G. Clegg and P.C. Clegg (Heinemann, 1975). Clear, readable teachers' reference.

Co-ordination, C. Morgan (Macdonald, 1977). Useful for pupils' background reading.

The life of mammals, J.Z. Young (Oxford, 1975). Advanced reference book on mammalian physiology.

Good mouthkeeping, J. Besford (Oxford, 1984). A non-technical practical guide to oral hygiene.

Section 4: Genetics and Heredity

Origin of species (British Museum/ Cambridge University Press, 1981). Based on the museum's exhibition but valuable in its own right.

Discovering genetics, N. Cohen (Longman, 1982). Simple, well-illustrated account.

Illustrated 'Origin of species', C. Darwin, edited by R.E. Leakey (Faber and Faber, 1979). Parts of Darwin's text selected and annotated by Richard Leakey. Many illustrations.

Darwin for beginners, J. Miller (Writers and Readers Publishing Co-op, 1982). Cartoon approach to understanding Darwin's work.

Evolution, R. Moore and Editors of *Time-Life* (*Time-Life*, 1962). Readable and well illustrated account of the classical story.

Evolution, C. Patterson (Routledge and Kegan Paul, 1978). Advanced reference account.

Looking at Genetics, N. Sully (Batsford, 1985). A very simple introduction to a difficult topic.

Section 5: Organisms and their Environment

Introduction to field biology, D.P. Bennett and D.A. Humphries (Arnold, 1974). Teachers' reference to fieldwork.

Introducing ecology: nature at work (British Museum/ Cambridge University Press, 1978). Well illustrated introduction.

Ecology: principles and practice, W.H. Dowdeswell (Heinemann, 1984). For reference; a sound introduction to the subject.

Wild flowers of Britain, R. Phillips (Ward Lock and Pan Books, 1977). For reference; good colour photographs.

Discovering ecology, T. Shreve (Longman, 1982). Broad, simple account.

Life in the soil, R.M. Jackson and F. Raw (Arnold, 1966). A brief account of soil organisms and methods of studying them.

Natural communities, O. Bishop (Murray, 1973). Useful account of habitats with lists of organisms one would expect to find.

Acid rain, Steve Elsworth (Pluto Press, 1984). A campaigning book but soundly based.

How to save the world, R. Allen (Kogan Page, 1980). A very readable account of world conservation strategy.

Key to pond organisms, Nuffield Foundation (Longman, 1974). Illustrated key to the common species.

Section 6: Diversity of Organisms

Adventures with small animals, O. Bishop (Murray, 1982). Simple introduction with suggestions for investigations.

Adventures with small plants, O. Bishop (Murray, 1983). Simple introduction with suggestions for practical work.

Adventures with micro-organisms, O. Bishop (Murray, 1984). Good background reading for pupils; some interesting practical exercises.

Classification (British Museum, 1983). A beginner's guide to some of the main systems in use.

Plant types: 1. Algae, fungi and lichens; 2. Mosses, ferns, conifers and flowering plants, R.N. Miller (Hutchinson, 1982/85). Reference books for the plant kingdom.

Behaviour, C. Morgan (Macdonald, 1978). Background reading for pupils; well illustrated.

Animal types 1, 2, M.A. Robinson and J. Wiggins (Hutchinson, 1970/71). Useful reference for basic details of the main forms of an animal life.

Discovery of animal behaviour, J. Sparks (BBC Publications, 1983). Introduction to animal behaviour by a historical approach.

Life of vertebrates, J.Z. Young (Oxford University Press, 1981). For reference; an advanced standard text.

Biotechnology, J.E. Smith (Arnold, 1981). Brief survey of the scope of this subject.

Microbiology: An HMI guide for schools, D.E.S. (H.M.S.O., 1985). Guidelines for handling micro-organisms; suggestions for suitable experiments and organisms; safe procedures.

Practical Work

Experimental Work in Biology, D.G. Mackean (Murray, 1983). About 140 tested experiments, with detailed instructions to the students. Expected results are not given but students are asked questions to test their understanding of the experimental design and their ability to make critical interpretations of the results.

Appendix 4: Resources

The **Biology Resource Pack** by D. G. Mackean contains 117 pages of photocopiable self-assessment questions (with answers) and questions for use as exercises or for discussion. There are also 55 pages devoted to lists of teaching resources for each topic, including films, videos, slides, wall-charts, computer software, teaching packs and suggestions for practical work.

Films, videos, slides and software

AVP (Audio-Visual Productions) School Hill Centre, Chepstow, Gwent, NP6 5PH
Slide sets, overhead projector transparencies, computer software.

BBC Publications, 35 Marylebone Street, London, W1X 4AA
Filmstrips associated with radio vision programmes.

Boulton Hawker Films Ltd., Hadleigh, Ipswich, Suffolk, IP7 5BG
Films and videos for hire or sale. Mostly from USA.

Concord Video and Film Council, 201 Felixstowe Road, Ipswich, Suffolk, IP3 9PJ
Videos and 16 mm films for hire. Mostly concerned with social aspects of biology.

Focal Point Audio Visual, 251 Copnor Road, Portsmouth, Hants., PO3 5EE
Wide range of slides and videos especially in ecology.

GBI Laboratories, Northgate, Pontefract, N. Yorks., WF8 1HJ
Slide material by Gene Cox covering a wide range.

Guild Sound and Vision, 6 Royce Road, Peterborough, PE1 5YB
Videos and 16 mm films for hire. Good source of films released by television companies.

International Centre for Conservation Education, Greenfield House, Guiting Power, Cheltenham, Glos., GL54 5TZ
A long list of slides, filmstrips and tape/slide programmes on aspects of ecology and conservation.

National Audio Visual Aids Library, The George Building, Normal College, Bangor, Gwynedd, LL57 2PZ
Extensive catalogue of films and videos for hire.

Philip Harris Biological Ltd, Oldmixon, Weston-super-Mare, Avon, BS24 9BJ
Wide range of slides and filmstrips, videos and software.

Scottish Central Film and Video Library, 74 Victoria Crescent Road, Glasgow, G12 9JN
A source of health education films and videos. Distributors for Unilever films.

Slide Centre, (Rickitt Educational Media) Ilton, Ilminster, Somerset, TA19 9HS
Slide folios covering ecology and wildlife.

Viewtech Audio-Visual Media, 161 Winchester Road, Brislington, Bristol, BS4 3NJ
Videos and 16 mm films for hire or sale.

Charts, booklets, teaching packs

Cancer Research Campaign, 2 Carlton House Terrace, London, SW1 5AR
Charts on cell structure.

Health Education Authority, Hamilton House, Mabledon Place, London, WC1H 9TX
Leaflets, booklets, charts and resource lists on health education topics.

Nature Conservancy Council, Interpretive Branch, Attingham Park, Shrewsbury, SY4 4TW
Posters and attractively illustrated, inexpensive booklets on habitats and wildlife.

Philip Harris Biological Ltd (see above)
Wide range of charts on human anatomy, plants and animals, life histories, habitats.

Pictorial Charts Educational Trust, 27 Kirchen Road, London, W13 0UD
Wide range of attractive, well-produced material such as:
Birds in the school grounds
Food chain in a pond
Life before birth
Spiders, snails and woodlice

Project Icarus Ltd, Raglan House, 4 Clarence Parade, Southsea, Hants., PO5 3NU
Health education posters including drugs, smoking and sexually transmitted diseases.

Roopers Company, 20 Ridgewood Industrial Park, Uckfield, E. Sussex, TN22 5SX
Wall charts on human anatomy and health education.

TACADE (Teachers' Advisory Council on Alcohol and Drugs Education), Furness House, Third Floor, Trafford Road, Salford, M5 2XJ
Teaching packs on health education, alcohol, drugs and smoking.

Thames Water, Marketing Services Department, Room 1503, Nugent House, Vastern Road, Reading, RG1 8DB.
Charts and pamphlets on topics such as:
Sewage treatment
Wildlife of the tidal Thames
The water cycle

Other sources of information

ASH (Action on Smoking and Health), 5–11 Mortimer Street, London, W1N 7RH
Information packs, fact sheets, posters.

The Conservation Trust, George Palmer Site, Northumberland Avenue, Reading, RG2 7PW
Study notes, cards, kits and packs on e.g.
Alternative technology
Conservation of resources
Food and agriculture

Council for Environmental Education, School of Education, University of Reading, London Road, Reading, RG1 5AQ
Lists of resources on environmental topics such as:
Acid rain
Lead pollution
Pond and freshwater studies

Friends of the Earth, 377 City Road, London, EC1V 1NA
Newsletters, books, pamphlets on issues such as:
Radioactive waste
Acid rain
Pesticides

General Dental Council, 37 Wimpole Street, London, W1M 8DQ
A range of material concerned with dental health.

Institute of Biology, 20 Queensberry Place, London, SW7 2DZ
Advice on careers, courses, resources.

London Centre for Biotechnology, Room E255, South Bank Polytechnic, Borough Road, London, SE1 0AA
Posters, slides, teaching packs.

MISAC (Microbiology in Schools Advisory Committee), c/o Institute of Biology (see above)
Arranges courses, speakers and advice on teaching microbiology.

National Centre for School Biotechnology, Department of Microbiology, University of Reading, Reading, RG3 1JL
Newsletter and other resources.

Glossary

(A) SCIENTIFIC TERMS

Acid A sharp-tasting chemical, often a liquid. Some acids can dissolve metals and turn them into soluble salts. Nitric acid acts on copper and turns it into copper nitrate, which dissolves to form a blue-coloured solution. Acids of plants and animals (amino acids, fatty acids) are weaker and do not dissolve metals. Amino acids and fatty acids are organic acids. Hydrochloric, sulphuric and nitric acids are called mineral acids or inorganic acids.

Agar A clear jelly extracted from one kind of seaweed. On its own it will not support the growth of bacteria or fungi, but will do so if food substances (e.g. potato juice or Bovril) are dissolved in it. Agar with different kinds of food dissolved in it is used to grow different kinds of micro-organism.

Alcohol Usually a liquid. There are many kinds of alcohol but the commonest is ethanol (or ethyl alcohol) which occurs in wines, spirits, beer, etc. It is produced by fermentation of sugar. Ethanol vaporizes quickly and easily catches fire.

Alkali The opposite of an acid. An alkali can neutralize an acid and so remove its acid properties. Sodium hydroxide (NaOH) is an alkali. It neutralizes hydrochloric acid (HCl) to form a salt, sodium chloride.

$$Na\boxed{OH + H}Cl \rightarrow \underset{\text{salt}}{NaCl} + \underset{\text{water}}{H_2O}$$

Atom The smallest possible particle of an element. Even a microscopic piece of iron would be made up of millions of iron atoms. When we write formulae, the letters represent atoms. So H_2O for water means an atom of oxygen joined to two atoms of hydrogen.

Calorie Just as a centimetre is a unit of length, a caloıie is a unit of heat or energy. It is the amount of heat that would raise the temperature of one gram of water one degree Celsius. The energy value of food is measured in kilocalories (kcal). 1000 calories = 1 kilocalorie. In scientific studies, calories have been replaced by joules. For the energy in food, however, calories are still used. (1 calorie = 4.2 joules.)

Capillary attraction The tendency of water to fill small spaces is called capillary attraction. If a narrow bore tube is placed in water, the water will rise up it for several centimetres. In a similar way, water will creep into the spaces between the fibres in a piece of blotting paper or between the particles in soil.

Carbon A black, solid non-metal which occurs as charcoal or soot, for example. Its atoms are able to combine together to make ring or chain molecules (see p. 15). These molecules make up most of the chemicals of living organisms (see 'Organic'). One of the simplest compounds of carbon is carbon dioxide (CO_2).

Carbon dioxide A gas which forms 0.03 per cent (by volume) of the air. It is produced when carbon-containing substances burn ($C + O_2 \rightarrow CO_2$). It is also produced by the respiration of plants and animals. It is taken up by green plants to make food during photosynthesis.

Catalyst A substance which makes a chemical reaction go faster but does not get used up in the reaction. Platinum is a catalyst which speeds up the rate at which nitrogen and hydrogen combine to form ammonia, but does not get used up. Enzymes are catalysts for chemical reactions inside living cells.

Caustic A caustic substance can damage the skin and clothing and therefore should be handled with great care.

Compound Two or more elements joined together form a compound. Carbon dioxide, CO_2, is a compound of carbon and oxygen. Potassium nitrate, KNO_3, is a compound of potassium, nitrogen and oxygen.

Cubic centimetre (cm^3) This is a unit of volume. A tea-cup holds about 200cm^3 liquid. One thousand cubic centimetres are called a cubic decimetre (dm^3) but this volume is also called a litre. Some measuring instruments are marked in millilitres (ml). A millilitre is a thousandth of a litre and therefore the same volume as a cubic centimetre. So 1 cm^3 = 1 ml.

Density This is the weight (mass) of a given volume of a substance. Usually it is the weight in grams of one cubic centimetre of the substance, e.g. 1 cm^3 lead weighs 11 grams, so its density is 11 grams per cm^3.

The density of water at 4°C is 1 g per cm^3.

Diffusion The random movement of molecules by which gases or dissolved substances move from a region of high concentration to a region of low concentration.

Dissolve A substance which mixes with a liquid and seems to 'disappear' in the liquid is said to dissolve. Sugar dissolves in water to make a solution.

Element An element is a substance which cannot be broken down into anything else. Sulphur is a non-metallic element. Iron is a metallic element. Oxygen and nitrogen are gaseous elements. Water (H_2O) is not an element because it can be broken down into hydrogen and oxygen.

Energy This can be heat, movement, light, electricity, etc. Anything which can be harnessed to do some kind of work is energy. Food consists of substances containing chemical energy. When food is turned into carbon dioxide and water by respiration, energy is released to do work such as making muscles contract.

Filtrate The clear solution which passes through a filter; e.g. if a mixture of copper sulphate solution and sand is filtered, the blue copper sulphate solution which passes through the filter paper is called the filtrate.

Formula A way of showing the chemical composition of a substance. Letters are chosen to represent elements, and numbers show how many atoms of each element are present. The letter for carbon is C and for oxygen is O. A molecule of carbon dioxide is one atom of carbon joined to two atoms of oxygen and the formula is CO_2. There are more elements than letters in the alphabet, so some of the elements have two letters, e.g. Mg for magnesium. Other elements have letters standing for the latin name, e.g. sodium is Na (= natrium).

Gram (g) A unit of weight in the metric system.

A penny weighs $3\frac{1}{2}$ grams.
A pack of butter is 225 grams.
1000 grams is a kilogram (kg).
One thousandth of a gram is a milligram (mg).

Hydrogen Hydrogen is a gas which burns very readily. It is present in only tiny amounts in the air but forms part of many compounds such as water (H_2O), and organic compounds like carbohydrates (e.g. $C_6H_{12}O_6$ glucose) and fats.

Inorganic Substances like iron, salt, oxygen and carbon dioxide are inorganic. They do not have to come from a living organism. Salt is in the sea, iron is part of a mineral in the ground, oxygen is in the air. Inorganic substances can be made by industrial processes or extracted from minerals.

Insoluble An insoluble substance is one which will not dissolve. Sugar is soluble in water but insoluble in petrol.

Lime water A weak solution of lime (calcium hydroxide) in water. When carbon dioxide bubbles through this solution, it reacts with the calcium hydroxide to form calcium carbonate (chalk) which is insoluble and forms a cloudy suspension. This makes lime water a good test for carbon dioxide.

$$Ca(OH)_2 + CO_2 \rightarrow CaCO_3 + H_2O$$

Manometer An instrument which measures pressure by the displacement of a liquid in a U-tube.

Mass This is the amount of matter in an object. The more mass an object has, the more it weighs, so mass can be measured by weighing something. However, if the force of gravity becomes less, as on the Moon, the same object will weigh less even though the amount of matter in it (its mass) has not changed. So mass and weight are related, but are not the same.

Molecule The smallest amount of a substance which you can have. For example, the water molecule is H_2O, that is, two atoms of hydrogen joined to one atom of oxygen. A drop of water consists of countless millions of molecules of H_2O moving about in all directions and with a lot of space between them.

Organic This usually refers to a substance produced by a living organism. Organic chemicals are things like carbohydrates, protein and fat. They have very large molecules and are often insoluble in water. Inorganic chemicals are usually simple substances like sodium chloride (salt) or carbon dioxide (CO_2).

```
    H   H   H   H
    |   |   |   |       O
H — C — C — C — C — C ⟨
    |   |   |   |       OH
    H   H   H   H
```

molecule of a fatty acid
C_4H_9COOH (organic)

O=C=O

molecule of carbon dioxide
CO_2 (inorganic)

Oxygen Oxygen is a gas which makes up about 20 per cent (by volume) of the air. It combines with other substances and oxidizes them, sometimes producing heat and light energy. In plants and animals it combines with food to release energy.

Permeable Allows liquids or gases to pass through. A cotton shirt is permeable to rain but a PVC mackintosh is impermeable. Plant cell walls are permeable to water and dissolved substances.

pH This is a measure of how acid or how alkaline a substance is. A pH of 7 is neutral. A pH in the range 8–11 is alkaline; pH's in the 6–2 range are acid; pH 6 is slightly acid; pH 2 is very acid.

PIDCP The initials of an organic chemical called phospho-indo-dichlorophenol. It changes from blue to colourless in the presence of certain chemicals, including Vitamin C.

Pigment A chemical which has a colour. Haemoglobin in blood is a red pigment; chlorophyll in leaves is a green pigment. A black pigment called melanin may give a dark colour to human skin, hair and eyes.

Pipette A glass tube designed to deliver controlled amounts of liquid. A bulb pipette has a plastic squeezer on one end so that it can deliver a drop at a time. A graduated pipette has marks on the side to show how much liquid has run out.

Reaction (chemical) A change which takes place when certain chemicals meet or are acted on by heat or light. The change results in the production of new substances. When paper burns, a reaction is taking place between the paper and the oxygen in the air.

Salt A salt is a compound formed from an acid and a metal. Salts have double-barrelled names like sodium chloride (NaCl) and potassium nitrate (KNO_3). The first name is usually a metal and the second name is the acid. Potassium (K) is a metal, and the nitrate (NO_3) comes from nitric acid (HNO_3).

Sodium hydrogencarbonate At one time this was called sodium bicarbonate. It is a salt which is used to make carbon dioxide in experiments. Its formula is $NaHCO_3$.

Sodium hydroxide (NaOH) An alkali with caustic properties, i.e. its solution will dissolve flesh, wood and fabrics.

Soluble A soluble substance is one which will dissolve in a liquid. Sugar is soluble in water.

Solution When something like sugar or salt dissolves in water it forms a solution. The molecules of the solid become evenly spread through the liquid.

Volume The amount of space something takes up, or the amount of space inside it. A milk bottle has an internal volume of one pint. Your lungs have a volume of about 5 litres; they can hold up to 5 litres of air. This cube has a volume of 8 cubic centimetres (8 cm^3).

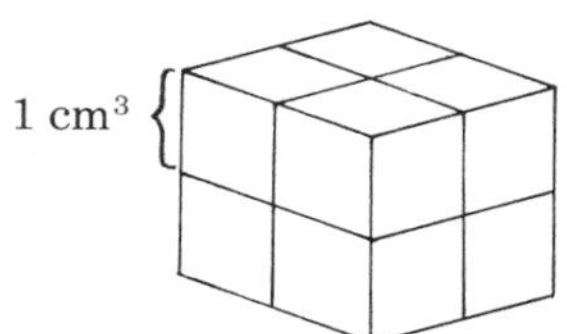

(B) BIOLOGICAL TERMS

(References in brackets are to pages.)

Abdomen (127, 335) The part of the body below the diaphragm which contains stomach, kidneys, liver, etc.; in insects it refers to the third region of the body.

Accommodation (198) Changing the shape (and focal length) of the eye lens to focus on near or distant objects.

Active transport (35) The transport of a substance across a cell membrane with expenditure of energy, often against a concentration gradient.

Adaptation (50) The development, during evolution of an organism, of structures or processes which make it more efficient in its environment.

Alleles (229) Alternative forms of a gene, occupying the same place on a chromosome and affecting the same characteristics but in different ways.

Anabolism (26) The building up of complex substances from simpler ones.

Angiosperms (306) Flowering plants with seeds enclosed in an ovary.

Aseptic technique (319) Method of handling materials or apparatus so that unwanted micro-organisms are excluded.

Asexual reproduction (350) Reproduction without the involvement of gametes.

Assimilation (131) Absorption of substances which are built into other compounds in the organism.

Autotroph (250) An organism which can build up its organic materials from inorganic substances.

Auxin (108) A chemical which affects the rate of growth in plants.

Basal metabolism (26) The minimum rate of chemical activity needed to keep an organism alive.

Biodegradable (277) Able to be broken down to simple inorganic substances by the action of bacteria and fungi.

Biomass (243) The weight of all the organisms in a population, community or habitat.

Biosphere (282) The part of the Earth which contains living organisms.

Biotechnology (313) The use of living organisms or biological processes for industrial, agricultural or medical processes.

Biotic factors (393) The environmental effects of other living things on an organism.

Cardiac To do with the heart.

Catabolism (26) The breakdown of complex substances to simpler substances in the cell, with a release of energy.

Climax vegetation (286) The plant community which finally colonizes a particular habitat.

Clone (103) A population of organisms derived by asexual reproduction from a single individual.

Commensalism (250) Two unrelated organisms living in a close association which benefits one of them and does not harm the other.

Control (30) An experiment which is set up to ensure that only the condition being investigated has affected the results.

Co-ordination (205) The process which makes the different systems in an organism work effectively together.

Cortex (160) An outer layer.

Denature (15) Destroy the structure of a protein by means of heat or chemicals.

Detoxication (133) The process by which the liver makes poisonous chemicals harmless.

Dialysis (39) The separation of small molecules from large molecules in solution by a selectively permeable membrane.

Differentiation (66) The process by which a cell becomes specialized during the course of development.

Ecdysis (336) The periodic shedding of the cuticle during the growth of arthropods.

Ecosystem (281) A community of interdependent organisms and the environment in which they live.

Emulsify (129) Break-up of oil or fat into tiny droplets which remain suspended in water as an emulsion.

Eutrophic (267) An aquatic environment well supplied with nutrients for plant growth.

Fermentation (24) A form of anaerobic respiration in which carbohydrate is broken down to carbon dioxide and, in some cases, alcohol.

Gastric (128) To do with the stomach.

Gene (224) A sequence of chemicals in a chromosome which controls the development of a particular characteristic in an organism.

Genetic code (225) The sequence of bases in a molecule of DNA which specifies the order of amino acids in protein.

Genetic engineering (225) Altering the genetic constitution of an organism by introducing new DNA into its chromosomes.

Genotype (229) The combination of genes present in an organism.

Genus (301) One of the categories in classification; a group of closely related species.

Gestation (176) The period of growth and development of a foetus in the uterus of a mammal.

Hepatic (131) To do with the liver.

Heterotroph (250) An organism which feeds by taking in organic substances made by other organisms.

Heterozygous (229) Carrying a pair of contrasted genes for any one heritable characteristic; will not breed true for this characteristic.

Homeostasis (134) Keeping the composition of the body fluids the same.

Homoiothermic (343) Animals whose body temperature is maintained at a constant level, usually above that of their surroundings. Sometimes called 'warm-blooded'.

Homologous chromosomes (222) A pair of corresponding chromosomes of the same shape and size; one from each parent.

Homozygous (229) Possessing a pair of identical genes controlling the same characteristic; will breed true for this characteristic.

Hypothesis (45) A provisional explanation for an observation; it can be tested by experiments.

Immunity (145) Ability of an organism to resist infection, usually because it carries antibodies in its blood.

Implantation (174) The process in which an embryo becomes attached to the lining of the uterus.

Incubate (343) Maintain at a raised temperature, e.g. birds' eggs or bacteria cultures.

Inflorescence (81) A group of flowers on the same stalk.

Inhibit (208) Slow down a process, or prevent its happening.

Inoculation (147) Deliberate infection with a mild form of disease to stimulate the formation of antibodies. (In the case of culture methods for bacteria or fungi, 'inoculation' means introducing the organism to the culture medium.)

Interferon (318) A group of proteins produced by animal cells as a result of infection by viruses. Interferons inhibit the multiplication of viruses.

Laparoscopy (180) A method of examining the inside of the abdomen by inserting an optical instrument (an endoscope) through the abdominal wall.

Limiting factor (50) A condition which limits the rate of a process, e.g. shortage of light for photosynthesis.

Metabolism (26) All the chemical changes going on in the cells of an organism which keep it alive.

Metamorphosis (337) The relatively sudden change by which the larval form of an insect or amphibian becomes an adult.

Monoculture (265) Growing a single species of crop plant, usually in the same ground for successive years.

Mutation (226) A spontaneous change in a gene or chromosome, which may affect the appearance or physiology of an organism.

Parasite (250) An organism living in or on another organism (the host). The parasite derives its food from the host.

Pathogen (314) A parasite which causes disease or harms its host in other ways.

Phenotype (229) The observable characteristics of an organism which are genetically controlled.

Photoperiodism (105) Response by plants to change in day length. Flowering is the most obvious response.

Phylum (301) A major classificatory group.

Plankton (243) The community of small plants and animals floating in the surface waters of an aquatic environment.

Poikilothermic (340) Having a body temperature which fluctuates with that of the environment. Sometimes referred to (inaccurately) as 'cold-blooded'.

Predator (242) An animal which kills and eats other animals.

Proprioceptor (195) A sense organ which detects changes within the body.

Protista (327) Single-celled organisms which have a proper nucleus.

Protophyta (300) Those protista which have chlorophyll and make their food by photosynthesis.

Protozoa (300) Those protista which take in solid food and digest it.

Puberty (178) The period of growth during which humans become sexually mature.

Receptor (193) A sense organ which detects a stimulus.

Recessive (228) A gene which, in the presence of its contrasting allele, is not expressed in the phenotype.

Recycling (245, 178) As a biological term this means the return of matter to the soil, air or water, and its re-use by other organisms. In daily life, it means the re-use of manufactured materials such as paper, glass and metals.

Renal (160) To do with the kidneys.

Replication (221) Production of a duplicate set of chromosomes prior to cell division.

Scion (103) In a graft, this is the cutting or bud which is grafted on to the stock.

Section (3) A thin slice of tissue which can be examined under the microscope.

Sensitivity (352) The ability to detect and respond to a stimulus.

Sphincter (186) A band of circular muscle which can contract to constrict or close a tubular organ.

Spore (330) A cell or small group of cells which can grow into a new organism.

Stimulus (105) An event in the surroundings or in the internal anatomy of an organism, which provokes a response.

Stoma (60) A structure, in the epidermis of a plant, which consists of a pore enclosed by two guard cells. It permits gaseous exchange with the atmosphere.

Succession (285) Changes, primarily in vegetation, which take place from the time a habitat is first colonized to the establishment of climax vegetation.

Symbiosis (250) A very close association between two unrelated organisms. Each organism derives some benefit from the association. May also be used to describe any close association, whether beneficial or harmful.

Toxin (121) A poisonous protein produced by pathogenic bacteria.

Toxoid (147) A toxin which has been treated to make it harmless, but can still cause the body to make antibodies.

Trophic level (243) An organism's position in a food chain, e.g. primary or secondary consumer.

Turgor (40) The pressure built up in a plant cell as a result of taking in water by osmosis.

Vascular To do with vessels; blood vessels or xylem and phloem.

Viviparous (343) Giving birth to fully formed young rather than laying eggs.

(C) SOME DERIVATIONS

A great many terms used in Biology are derived from Greek or Latin words. In some cases, knowing the general meaning of a prefix or suffix helps to understand or recognize the term. Unless indicated as Latin (L), all the words below are Greek.

autos = self, e.g. autotroph
bios = life, e.g. biology, biomass
bis (L) = twice, e.g. binary fission, bicuspid valve
chloros = pale green, e.g. chlorophyll
chroma = colour, e.g. chromosome (takes up coloured stains)
dia = across, e.g. diaphragm, dialysis
dis = twice, e.g. dipeptide, diploid
ektos = outside, e.g. ectoparasite
epi = upon (above), e.g. epidermis, epicotyl
exo = outside, e.g. exoskeleton, exocytosis
haima = blood, e.g. haemoglobin, haemophilia
heteros = other (i.e. different), e.g. heterozygous
homos = same, e.g. homozygous, homologous
hypo = under, e.g. hypothermia, hypocotyl
inter (L) = between, e.g. inter-cellular, intercostal
intra (L) = within, e.g. intra-cellular, intra-uterine
kytos = vessel (a cell), e.g. cytoplasm, leucocyte
lipos = fat, e.g. lipid, lipase
lysis = dissolution, e.g. lysozyme, dialysis
mesos = middle, e.g. mesophyll, mesenteric
meta = after (change), e.g. metamorphosis
mikros = little, e.g. microvilli, micropyle
morphe = form (shape), e.g. metamorphosis, morphology
phagein = to eat, e.g. phagocyte, oesophagus
phyllon = leaf, e.g. mesophyll, chlorophyll
phyton = plant, e.g. phytoplankton, saprophyte
polys = many, e.g. polypeptide, polysaccharide
protos = first formed, e.g. protista, protoplasm
rhiza = root, e.g. rhizoid, rhizosphere
semi (L) = half, e.g. semi-lunar valve, semicircular canal
sub (L) = under, e.g. subsoil, subclavian
sym-, syn- = together, e.g. symbiosis, synapse
treis = three, e.g. tripeptide, tricuspid valve
trophe = food, e.g. autotroph, trophic level
unus (L) = one, e.g. unicellular, unisexual
vas (L) = vessel, e.g. vascular bundle, vaso-dilation
zoion = animal, e.g. zooplankton, zoology
zygon = yolk, e.g. zygote, homozygous

Index

(Page numbers in bold type show where a subject is introduced or most fully explained)

abdomen (insect) 335–8
abiotic factors 293
ABO blood groups 147–8, 230, 232, 237
abortion 177
absorption 125, 130–31
accommodation 198
acid rain 272–3, 275
acid soil 256–7
acquired characteristics 235
actinomycete 257–8
active transport 35, 76, 131
adaptation 50
 of aquatic animals 287–8
 of aquatic plants 284–5
 of flowers 85
 of leaves 50
addiction 216
additives (food) 121–2
adenine 17, 25, 224–5
adenosine diphosphate (ADP) 25
adenosine monophosphate (AMP) 17, 25
adenosine triphosphate (ATP) **25**, 30
ADH (anti-diuretic hormone) 162, 214
adipose tissue 115, 132, 166–7
ADP (adenosine diphosphate) 25–6, 35
adrenal cortex 213
adrenal gland 212–13
adrenaline 213, 215
adrenal medulla 213
adventitious root 104, 331
aerobic bacteria 311
aerobic respiration **24–5**, 26, 346–7
aerosol cans 273
'after-birth' 176
agar 319, 325
age pyramids 296
agriculture 264–6, 269–70
 energy transfer 252–3
AIDS (acquired immune deficiency syndrome) 316
air pollution 272–3
 control 275–6
air spaces
 in aquatic plants 285
 in leaf 48, 50, 59, 61, 71–2
 in soil 225, 260–61
alanine 15, 115, 225
albumen 15
alcohol 24, 216
 in diet 120–21
 in pregnancy 175
alcoholism 216
aldrin 271
algae 268, 304, 306–7, **329–30**
alimentary canal 125–31
alkaline soil 256–7
allele 229–30
allelomorphic genes 229
alveoli 151–2, 154–6
amino acids **14–15**, 19, 52, 115–16, 126, 129, 225
 absorption and use 131–3
 essential 115–16
ammonia 133, 247–8, 348
 in urine 348
ammonium nitrate 53
amnion 174–5
Amoeba 327
AMP (adenosine monophosphate) 17, 25
amphetamine 215
amphibia 301, 303, 307, 342
ampulla 201
amylase 19, 21, 126–7, 129, 134
anabolism 26
anaemia 116–17
 sickle-cell 232, 239
anaerobic bacteria 311
anaerobic respiration 29, 30, **24–5**, 29–30, 346–7
analgesic 216
angiosperm 306
animal cells 3–5, 8, 13
 osmotic effects 37
animal dispersal (seeds) 90
animal kingdom 300–303
ankle 183, 188
annelid 301–2, 307–8, 333
antagonistic muscles 186–7
antenna 335–8
anther 80, **81**–6
antibiotic 258, 314, 319–20, 325
antibodies 145–**147**, 148, 318
 in milk 178
antidiuretic hormone (ADH) 162, 214
antigens 147–8
anti-toxin 318
anus 127, 131
anxiety 213, 215–16
aorta 139, 141–2
aphid 250
appendix 127, 129
apple
 flower and fruit 88–9
 pollination 87
applied genetics 233–4
aqueous humour 196
arachnids 301–2, 339
arm (skeleton) 182–4, 186, 188
artefacts 7
arterial system 142
arteries 11, **140–42**, 149
arterioles 142–3
 in skin 166, 168
arthropods 301–2, **335–8**
 in soil 258–9
artificial fertilizer 249, 256
artificial immunity 147, 318
'artificial kidney' 162–3
artificial propagation 103–4
artificial selection 239
ascorbic acid 117
Asellus (water louse) 287
aseptic techniques 319
asexual reproduction 100, 323 350–51
 in bacteria 311
aspirin 216
assimilation 131
association centre 209, 211
atheroma 149–50
athlete's foot 322
ATP (adenosine triphosphate) **25**–6, 30, 35, 187
atrium (heart) 139–41
autoclave 319
autotroph 250
autotrophic nutrition 250
auxin 108–9

'backbone' 182
back-cross 231
bacteria 226, **310**, 311, **313–14**, 315, **316–19**
 and eutrophication 268
 in soil 256–8
 and tooth decay 191
bacteria culture 319–20
bacterial diseases 316–18
baking 325
balance (sense of) 201, 211
balanced diet 118–20
ball and socket joint 184
barbiturate 216
barium sulphate 128
bark 62
barnacle 202
basal metabolism 26, 119
base (in DNA) 224–5
bat 303
BCG vaccine 147, 316
beak 343
beans (in diet) 115–16, 118
bees (in pollination) 84
beetle 337
behaviour 353
Benedict's solution 122–3, 363
biceps 186–8, 208–9
bicuspid valve 139–40
bile 129, 133
bile duct 127, 129
bile pigments 129, 133, 159
bile salts 116, 129, 131, 133
bilirubin 133, 138, 159
binary fission 328
binocular vision 199
binomial nomenclature 301
biodegradable 277
biological control 295
biological oxygen demand (BOD) 268
biomass 243, 252
biosphere 282
biotechnology 313–14, 325
biotic factors 393–4
birds 301, 303, 307, **343**
 and seed dispersal 90
birth 176–7
birth control 179–80
birth rate 295
bisexual 80
bitter (taste) 194, 203–4
Biuret test 122–3
bivalve 334
blackberry, flower and fruit 88
bladder 160–61
blanching 121
blind spot 196–7
blinking 207
blood 137–50
 circulation 140, 142
 composition 137–8
 functions 145
blood clot 146, 149
blood donor 148
blood groups 147–8, 230, 237
blood pressure 142–3, 150
blood sugar 132, 213–14
blood transfusion 147
blue-green algae 248, 257–8
BOD (biological oxygen demand) 268
body temperature 167–9
body weight 132
bolus 127
bone 185, 188
bone marrow 138
bottom-living animals 287
bracken 304, 331
 rhizome 101
bracket fungus 322
bracts 83, 90
brain 11, 202, 206–8, **210–12**
 control of homeostasis 164
bread 118
breast bone (sternum) 183
breast-feeding 178–80
breathing **151**–4, 157, 347
breathing rate 154
breech delivery 177
brewing 324
broad bean
 germination 93, 95
 seed structure 93
bronchi (bronchus) 151–2, 156
bronchioles 151–2, 155
bronchitis 156
brushing (teeth) 191–2
bryophytes 304, 306
buccal cavity 127
bud 58, **66–7**, 100–101
budding
 Hydra 350
 yeast 324
bud graft 104
bud scales 66–7
bulbs 101–2
burdock (seed dispersal) 90
butterflies 302, 337–8
 in pollination 84–5

caddis fly larva 287–8
caecum 127, 129
caesarian section 176
caffeine 215
calciferol 117
calcium 54
 in diet 116–17
calcium salts
 in bone 185, 188
 in teeth 190–91
calculus 191
Canadian pondweed (*Elodea canadensis*) 47, 284–5
cancer (lung) 155, 157
canine tooth 189–90
capillaries 130–31, 141, **142**–3, 151–2, 154–5
 in kidney 160
 in skin 166–8
 in teeth 190
capillary attraction (in soil) 256, 262
capillary bed 143
capsid 312
capsomere 312
capsule 90
carbohydrates **15–16**, **114**–15, 118
carbon cycle 246
carbon dioxide
 in carbon cycle 246–7
 diffusion from cells 34
 diffusion into leaf 61
 expelled by lungs 154
 'greenhouse effect' 269

in photosynthesis 45–9, 51
from respiration 24–7, 30–31
transport in blood 145
transport in plants 76
carbon-14 31
carbon monoxide 273
carboxy-haemoglobin 273
caries 191
carnivore 242, 250, 301
carotid artery 142
carpels 80, **81**, 88
cartilage 182, 184, **185**
catabolism 26
catalase 20
catalyst 17
catalytic converter 276
caterpillar 337–8
catkins 85
cell **2–10**, 13
animal 3–5, 8, 13
differentiation 66
diffusion in 33–4, 37
osmosis in 36–8
plant **4**, 5, 8, 12
physiology 14
structure 2–7
surface area 34
cell body **206**–8, 212
cell division **7–8**, 65–6, 170, 174, 221–3, 311, 351
cell extension 108
cell membrane **3–4**, 5, 34–5, 37–8, 43
cell sap 5, 38
cell specialization 7, **9–10**
cellular respiration 24
cellulase 322
cellulose 4, 8, **16**, 18, 114–15, 117
digestion 129, 250
cell wall **4**–5, 8, 38
of bacteria 310
cement (of teeth) 190
centipedes 258–9, 302, 335, 339
central nervous system 206, 210–11
centrifuge 7
cephalothorax 335, 339
Ceratophyllum (hornwort) 284–5
cereals 83
in diet 115, 117
seed germination 95
cerebellum 210–11
cerebral cortex 211–12
cerebral hemispheres 210–11
cerebro-spinal fluid 210
cerebrum 210
cervix 171, 173–4, 176, 179
chaetae 333
chaffinch 343
Chaoborus (phantom midge larva) 287
characteristics of organisms **346**–53
chelicerae 335, 339
chemical senses 194
chemoreceptors 194–5, 334–5
chewing 127, 189
chewing muscles 183
Chironomus 288
chitin 321
Chlamydomonas 327
chlorofluorocarbons 273
chlorophyll 5–6, **44**–5, 47–8, 52
chloroplast 4–**6**, 9, 12, 47–8, 50, 60–61, 330
chlorpromazine 216
choice chamber 353
cholera 315
cholesterol 119–20, 149–50
choroid 196
chromatid **220**–23
chromosome 4, **220–23**, 224–7, 229, 232–3, 310
chrysanthemums 106
cigarette smoking
and bronchitis 156
and heart disease 149, 156
and lung cancer 155, 157
and pregnancy 175
cilia 9, 155–6, 172, 327
ciliary body 196, 198
ciliary muscle 196, 198
ciliated cell 9, 151
ciliates 327
circular muscle 125–6, 128, 130, 186
circulation 140
circulatory system 11, 140, **141–3**, functions 145
CITES (Convention on International Trade in Endangered Species) 279
class 301, 306, 308
classification 300–309
clavicle 183
clay particle 254–6, 260
Clean Air Act 272, 275
cleft graft 104
climax vegetation 286
clinostat 106–7
clitellum 333–4
clone 103, 350–51
clotting (blood) 146
coal 245–6
coat colour (inheritance) 229–31
cochlea 200–201
co-dominance 232
coelenterate 301–2, 307–8, 333
colds 315
cold-sensory endings 193
'cold sore' 315
coleoptile 95
response to light 108, 110
collar bone (clavicle) 182–3
collecting duct (kidney) 160–1
colon 127, 131
colonization 285–6
colour vision 197
combustion 246
commensalism 250–51
common cold 315
community **281**–2
of aquatic animals 286
of aquatic plants 284–5
compensation point 49
competition 238, 282, 293
complete metamorphosis 337
composite flowers 82–3
compound eye 335–8
concentration gradient 34–5
conditioned reflex 209–10
conditioning 209–10
condom 179
cones (of eye) 197
conifers 305–6
conjunctiva 196
conservation 277, 279
constipation 117
consumer 243, 246, 250, 252
contagion 315
continuous variation 237
contraception 179–80
contraceptive pill 180, 215
contractile vacuole 327–8
control (experimental) 27–8, **30–31**, 45
controlled diffusion 34–5
controlled experiments **30–31**, 98
Convention on International Trade in Endangered Species (CITES) 279
cooking 121, 311
co-ordination 205–15
in plants 109
copulation 171–3
coral 302
core temperature 169
Corixa (water boatman) 287
corms 101–2
cornea 117, 196–8
cornified layer 166–7
coronary artery 139–40, 149
coronary heart disease 119, 149–50, 156
coronary thrombosis 149
corpus luteum 173, **178**–9
correlation 150, 157, 274
cortex
adrenal gland 213
brain 210–12
kidney 160–61
root 64
stem 62–4
corticosteroids 213
cotyledons 87, **92**–6
cover slip 12
couch grass rhizome 101
coughing 154, 207–8
crab 302
crop rotation 249, 256
cross breeding 233
cross linkage 15
cross pollination **83**, 233
crown (tooth) 189–90
crumb (soil) 254–6
crumb structure 255–6
crustacea 301–2, 307, 335, **336**
crypt 129–31
cultivation 295
culture plates 263
culturing bacteria 319–20
culturing fungi 325
cupula 201
cusps 189
cuticle
insect 335–6
plant 59–61, 73
cuttings 103–4, 110
Cyclops 286, 302
cysteine 15, 115
cytoplasm **3**–5, 9
cytosine 17, 224–5

daffodil
bulb 101–2
flower 81
daisy 82–3
'sleep movements' 105
dandelion 82–3
fruit dispersal 90
Darwin 237–8
day length (flowering) 105–6
DDT 267
deamination 132–3
death rate 292, 295
decay
organic matter 246
teeth 191
decay bacteria 313
decomposers 246, 250
deer 265, 282
defecation 131
deforestation 270
dehydrogenase 19
delivery (birth) 177
denaturing 15, 19, 37
dendrite 206
denitrifying bacteria 248
dental caries (decay) 191
dental floss 192
dentine 190–91
deoxygenated blood 138–9, 141, 143, 145, 175
deoxy-ribose 17, 224–5
deoxy-ribose-nucleic acid (DNA) 17, 222, 224–5
dependence (on drugs) 216
depressant 216
dermis 165–7
destarching 45
detergent 268
detoxication 133
development 349
diabetes 214
diageotropism 107
dialysis 39–40
kidney machine 162–3
dialysis tubing 135
diaphragm 127, 152–3, 344
diaphragm (contraceptive) 180
diatoms 329
dichotomous key 308
dicotyledons 92, 305–7
dieldrin 266–7
diet 116–20
dietary fibre 117–18, 120
differentiation 66, 349
diffusion **33–4**, 39, 347–8
experiments 39
of gases 39
in liquids 39
in lungs 154–5
rates 34
of water 36–7
diffusion gradient **34**, 155
digestion 125–30
in protista 328
digestive enzymes 125–6, 128–9
experiments 134–5
dipeptide 15
diploid number 222–3
directional stimuli 105–6
disaccharides 16
disc (vertebral) 182
disclosing tablets 192
discontinuous growth 352
discontinuous variation 236–7
disease 314–18
disease resistance 233–4
disease transmission 314–17
dispersal of fruits and seeds 89–90
distance judgement 199
distant vision 198
division of labour 10
DNA (deoxy-ribose-nucleic acid) **17**, 222, **224–5**, 310–13
dominant 228–9
dormancy 99
dorsal fin 340
dorsal root 208–9
dorsal root ganglion 209
dough 326
Down's syndrome 226
dragonfly 302
drinking water 317
droplet infection 314
drug resistance 226–7
drugs (mood-influencing) 215–16
drug therapy 318
drying (food preservation) 121
dry mass (weight) 26, 351
seedlings 96–7
duckweed (*Lemna*) 284–5
ductless glands 212–13
duodenum 127–30
dura mater 210
'dust bowls' 270
Dytiscus 337

ear 199–201
ear bones 199
ear drum 199, 200, 342
earthworm 256–9, 302, **333–4**
ecdysis 336–8, 352
echinoderms 301
ecology 281–90
freshwater 284–91
ecosystem 251, **281–2**
ectoparasite 250, 258, 347
effector 206, 208, 353
effort 188
egestion 125, 131, 348

egg (ovum) 171–3
egg cell (plants) 86–7
egg white 15
  digestion 134–5
ejaculation 172–3
elbow joint 183–4, 186, 188
electron microscope 3, **5**–6
Elodea canadensis (Canadian pondweed) 284–5
embryo 170–71, 174–6, 349
  plant 86–7, 92–3, 96
emphysema 156
emulsification (of fats) 129
enamel (tooth) 190–92
encapsulated nerve ending 193
endocrine system 205, **212**–**15**
endocytosis 34–5
endoparasite 250, 258, 347
endoplasmic reticulum 5–6
endosperm 87, 95
energy **24**–**5**, 26, 29–30
  flow 251–2
  from food 114, 118–19
  for photosynthesis 44, 47
  sources 278–9
  store 26
  from sunlight 44, 47
  transfer 25, 252
enterokinase 129
environment 281–2
enzymes 14–15, **17**–**22**, 25, 94, 225, 322
  digestive 125–6, 128–9, 134–5
  extra-cellular 19
  intra-cellular 19
epicotyl 93, 95–6
epidermis
  leaf 59–61
  human skin 165–7
  plant 12
  stem 62–3, 68
epididymis 171–2
epiglottis 127–8, 151
epithelium 10, 13, 125–6, 128–31, 151, 154–5
erectile tissue 172–3
erosion 269–71
essential amino acids 115–16
Euglena 327
eustachian tube 127, 200
eutrophication 267–8
evaporation (from leaf) 71–2, 78
evergreen plants 73
evolution 237–8, 306–7
excretion **159**–**62**, 164, 348
  in plants 348
excretory products 348
exercise 25
  and breathing rate 154
  and heart disease 150
exhaling 152–3
exocytosis 34–5
exoskeleton 336
experiment design 45
expiration 154
'explosive' fruits 91
extension (limb) 186–7, 208
external fertilization 341–2, 349
external respiration 155
extra-cellular enzymes 19
eye **196**–**8**, 203
eye colour 228, 236
eye-piece 2

F_1 generation 230, 233
F_2 generation 230, 232–3
factor VIII 226
'factory farming' 268
faeces 117, 131
fallopian tube 171
family 306
family planning 179–80
famine 122
fat cell 132
fat depot 132
fatness 120
fats 15
  in diet 114, **115**, 118–19
  digestion 129
  and heart disease 149–50
  storage 132
  test for 123
  use in body 131
fat-soluble vitamins 117
fatty acid **15**, 126, 129, 131
  saturated and unsaturated 119
fear 213
feathers 343
feed-back 195, 214–15
feeding 347
femoral artery and vein 142
femur 11, 182–4, 187, 208
fermentation **24**, 30, 324
ferns 304, 306–8, **331**–**2**
fertility
  human 179
  of soil 256–7
fertilization 349
  flowering plants 86–7
  human 170, **173**–4, 179
fertilizers 53, 249, 268
fetus 174, 176, 185
fibre (in diet) 117–18, 120
fibres (in plants) 63
fibrin 146, 149
fibrinogen 133, 139, **146**, 149
fibrous root system 66
fibula 183
field techniques 289–90
filament 80, **81**–4, 86
filamentous algae 329–30
filter bed 277
finger bones 183
fins 340
fish 301, 303, 307, **340**–**41**
  as food 118
fission 311
fixative 7
fixed joint 184
flaccid 38
flagella 310, 327
flagellates 327
flatworm 301–2, 307
flavour 195
flea 250
flexion (limb) 186
flies (and disease) 315
flight (birds) 343
floods 269–70
floret 82–3
flowering (day length) 105–6
flowering plants 305–7
flowers
  pollination 83–6
  structure **80**–**81**, 82–3, 88–9
flow rate (stream) 289
fluoridation 192
fluoride 116
  and teeth 190–92
focus 195, 197–8
foliar feeding 76
folic acid 117
follicle 172–3, 179
follicle-stimulating hormone (FSH) 180, 214–15
food 114–23
food additives 121–2
food chain **242**–**5**, 267, 272, 277
food-handling 317–18
food poisoning 316–17
food preservation 121
food production 122
food pyramid 243
food storage 347
  in bulb 101–2
  in seed 94, 96
food tests 122–3
food transport (plants) 74
food vacuole 327–8
food web 244, 264–5, 267
  in freshwater 288
foot 182, 187–8
foot (mollusc) 334
fore-brain 211
foreskin 172
forests 269–70
fovea 196–7
foxglove (pollination) 85
fraternal twins 176
free nerve endings 193
free-swimming animals 287
freeze-etched section 6
freezing (food) 121
French bean
  seed germination 94
  seed structure 92
freshwater environment 283–90
freshwater invertebrates 286–8, 308
freshwater shrimp (*Gammarus*) 287–8
frog 303, **342**
fructose 114
fruit 80–81
  formation 86–9
'fruiting bodies' 321–2, 324
fruit and seed dispersal 89–90
FSH (follicle-stimulating hormone) 180, 214–15
fulcrum 188
fungi 300, 307, **321**–**6**
  in soil 256–8
fungicide 266
fungus diseases 322–3

gall bladder 127, 129, 133
gamete 86, 170, 349
  production 223–4
Gammarus (freshwater shrimp) 287–8
ganglion 208
gaseous exchange 154, 334, 347
  in leaf 48–9, 76, 79
  in lung 154
  in protista 328
gastric juice 128
gastrin 129
gene 222, **224**–6, 228–9, 280
gene bank 280
gene combinations 350
genetic code 225
genetic engineering 225–6
genetics 220–25, 228–37
genital herpes 315
genotype 229–30
genus 301, 306
geotropism 106–10
German measles 176
germination 92–8
  conditions for 97–9
gestation 176
gill bar 341
gill filaments 340–41
gills 287–8, 340–41
gingivitis 191–2
girdles (skeletal) 182
glands 10–11, 206
  digestive 125–6, 128–9
  endocrine 212–15
globulin 139
glomerulus 160–61
glottis 127–8, 154
glucagon 213
glucose **15**–**16**, 18, 20, 24–5, 114, 126–7, 129, 131–3
  in blood 213–14
  reabsorption 160–61
  test for 122–3
glucose phosphate 20–21
glutamic acid 226
glutamine 15, 115
glycerol **15**, 126, 129, 131
glyceryl tristearate 15
glycine 14, 115, 225
glycogen 5, **16**, 114–15, 131–3
gonorrhoea 317
grafting 103–4
granular layer (skin) 166–7
grapnel 290
grass 305
  flower 83
  pollination 85
  seed germination 95
  vegetative reproduction 100
grass snake 343
gravity (plant response) 106–7, 109–10
grazing 245
'greenhouse effect' 269
grey matter 208, 210
growth 7–8, 114, 119, 214, 351–2
  of bud 67
  of insects 337
  of root 65–6, 68–9, 97
  of stem 66, 69
growth curve 351–2
growth hormone 214
growth movements 106
growth substances 108–9
guanine 224–5
guard cell 9–10, 59, **60**–**61**
gullet 11, 126–8
gum 190–91
  disease 191–2
gymnosperms 306–7

habitat 281–2
haemoglobin 34, 116, 133 **137**–**8**, 154, 226, 239
haemophilia 226
hair 166–7
  erector muscle 166–7
  follicle 166–7
  plexus 193
half-flower 81
haploid number 223–4
hart's-tongue fern 304, 331
hawkweed 82
hearing 199, 200, 211
heart 11, **139**–**40**, 141–2
heart attack 149–50
heart beat 140
heart disease 149–50
  and smoking 156
heat
  effect on bacteria 311
  lost from body 167–9
  produced by body 167–8, 187
  produced by liver 133
  produced by respiration 24, 29
  regulation 168–9
  transport 146
heat-sensory endings 193
hedgerows 278
height
  genetics 237
  and weight 120
Helix aspersa (snail) 334
hepatic artery and vein 141
hepatic-portal vein 131, 141–2
herbicide 266
herbivore 242, 250
herbs 305, 307
heredity 220–25, 228–33, 235–7
heritable variation 235, 238
hermaphrodite 80
herpes 315
heterotrophic nutrition 250
heterozygous 229
hilum 92–3
hind-brain 211
hinge joint 182–4

hip girdle 182–3
hip joint 184
'holdfast' 330
homeostasis 134, 145, 163–4, 214
homoiothermic 343
homologous chromosomes 222–3
homozygous 229
'honey' guides 85
honeysuckle (pollination) 85
hooked fruits 90
hormone rooting powder 110
hormones 175, 177–8, 205, **212–15**
  excretion 159
  transport in blood 145
'hormone' weed-killer 109
hornwort (*Ceratophyllum*) 284–5
horse chestnut (bud) 67
host 250, 314
houseflies (and disease) 315
human population 295–7
humerus 11, 182–4, 186, 188
humidity 353
  in transpiration 73, 77
humus 255–6
hunting 264–5
hybrid 233
hybrid wheat 234
Hydra 302, 350
hydration 17
hydrochloric acid in stomach 128–9, 135
hydrogencarbonate indicator 49, 52, 363
hydrogen peroxide 20
hydroponics 54
hyphae 321–4
hypocotyl 93–4, 108
hypothermia 169
hypothesis 45
hypothesis testing 31

IAA (indole acetic acid) 108, 110
identical twins 176
identification key 308
ileum 127, 130
image 196–7
  formation 197
immunity 145, 147, 318
implantation 174
incisors 189–90
incomplete dominance 232
incomplete metamorphosis 337
incubation (birds) 343
incus 199–200
indole acetic acid (IAA) 108, 110
induced abortion 177
industrialization 264
infection (defence) 146–7
inflorescence 81–3
influenza 315
ingestion 125, 127
inhaling 152–3
inherited characteristics 235
inhibition 208
inner ear 200–201
inoculation 147, 318
insecticide 266–7, 271
insectivore 301
insect pollination 84–6
insects 301–2, 307, 335, **336–8**
insulin 213–14, 225, 314
intercellular space 71
intercostal muscles 152–3
interdependence 242
interferon 318
intermediate neurone 208
internal environment 134, 145
internal fertilization 343–4, 349
internal respiration 24, 155
International Whaling Commission (IWC) 280
internode 58
inter-specific competition 282
intra-cellular enzymes 19
intra-specific competition 282
intra-uterine device 180
invertebrates 333–9
in-vitro fertilization 180
involuntary muscle 185–6
iodine
  in diet 116
  in starch test 122–3, 363
ions 17
iris (eye) 196, 198–9, 203
iron 178
  in diet 116
  in haemoglobin 137–8
  storage 133
irritability 352
islets (pancreas) 213–14
IUD (intra-uterine device) 180
IWC (International Whaling Commission) 280

jaw 183, 189–91
jellyfish 302
joints 184
jugular vein 142

keratomalacia 117
key (identification) 308
kidneys 159–62, 164
kidney transplant 162–3
kidney tubule 3, 10
kilojoules 118–19
kingdom 300–306
knee cap 183
knee jerk 207–8

labour (childbirth) 176
lacrimal duct 196
lactation 119
lacteal 130–31
lactic acid 25
ladybird 302
lamina 59
laparoscopy 180
large intestine 131
larva (insect) 337
larynx 127–8
lateral bud 58, 66, 100–102
lateral line 340
lateral root 58, 62, 66, 94, 96
leaching 248
lead-free petrol 276
lead pollution 273–4, 276
leaf 58
  absorption of chemicals 76
  evaporation from 71–3, 78
  fall 73
  gaseous exchange 79
  photosynthesis 48
  structure 59–61
  transpiration 71–3, 78
leaf blade 59
leaf stalk 59, 62
leech 288, 302
leg (skeleton) 183–4, 187
leguminosae 84
leguminous plants 247–9
Lemna (duckweed) 284–5
lens (eye) 195–8
lenticels 76
leucine 15, 115–16
lever effect (limb) 188
LH (luteinizing hormone) 180, 214
ligaments 184
light
  effect on germination 99
  penetration in water 283, 289
  plant responses 106–8, 110
light intensity
  and photosynthesis 50–52
  and transpiration 73, 77
  and vision 198–9
lightning 248
light-sensitive cells 197
lignin 63
limbs (skeleton) 183–4, 187–8
lime 257
lime water 27, 30, 363
limiting factors 50–51
Limnea (pond snail) 334
line transect 290
lipase 129
lipids **15**, 115
  digestion 129
lipoproteins 15
liver 127, 129, 131, **132–3**, 134
  and excretion 159
  in homeostasis 164
liver fluke 302
liverworts 304, 306–8
lizard 303, 343
load 188
lobster 335
locomotion 187
locust 336
long-day plants 105–6
longitudinal muscle 125–6, 128, 130
longitudinal section **3**–5, 62
Lumbricus terrestris (earthworm) 333
lung cancer 155, 157
lung capacity 154
lungs 127, 141, **151–4**, 342
  as excretory organs 159
  in homeostasis 164
lupin
  flower structure 81–2
  fruit 87
  pollination 84
  seed dispersal 90–91
luteinizing hormone (LH) 180, 214
lymph 143
lymphatic 143
lymphatic duct 144
lymphatic system 131, 144–5
lymph nodes 144
lymphocyte 137–8, 145, 147, 318
lysine 115–16
lysozyme 196

magnesium 52, 54
magnification 5
maize (genetics) 232–3
malaria 328
malarial parasite 328
'male' fern 304, 331
malleus 199
malpighian layer 165–7
maltase 129
Malthus 238
maltose 16, 18, 126–7, 129
mammals 301, 303, 307, **344**
  and seed dispersal 90
mammary glands 177, 344
manometer 28
manure 249, 256
marrow (bone) 138
marsh zone 285
mating 171, 173
mayfly larva 287–8
meat (in diet) 114, 116, 118
median fins 340
medulla
  adrenal gland 213
  brain 210–11
  kidney 160–61
meiosis **223**–4, 235
Meissner's corpuscle 166, 193
membrane
  artificial 37, 39
  cell, 3, **4**–5, 37–8
  nuclear 5–6
  selectively permeable 36–7, 39
menopause 178
menstrual cycle 178–9, 214
menstrual period 178–9
menstruation 178–9
mercury pollution 272
mesenteric artery 142
mesophyll 59, 61
metabolic water 348
metabolism 26, 119, 348
metamorphosis 337–8
micro-organisms
  (*see also* viruses, bacteria, protista and fungi)
  in soil 262
micropyle 87, 92–3
microscopes 2–5
microvilli 34, 130
mid-brain 210–11
middle ear 199–200
midrib 59
milk 177–8, 344
  in diet 116–17, 119
milk teeth 190–91
millipedes 258–9, 302, 335
Minamata bay 272
mineral elements
  in plant nutrition 53–4
mineral particles
  in soil 254–5, 260
mineral salts 249
  in the diet 116
  in plant nutrition 53–5
  in soil 256
miscarriage 156, 175, 177
mites 258–9, 302
mitochondria 3, 5–**6**
mitosis **220–21**, 222–4
models 43
molars 189–90
molluscs 301–2, 307, **334**
monera 300
monocotyledons 92, 305–7
  seed germination 95
monoculture 265–6
monosaccharides 16
mortality rate 292
mosquito larva 286–8
moss 304, 306–8, **330–31**
  leaf 12
motor fibre 208–9
motor impulse 206
motor neurones 206, 208
moulds 323, 325
moulting (insect) 336
mouth 127
mouth parts (insects) 337
movable joint 184
movement 184, 186, 352
Mucor 323
mucus 125, 129, 151, 155–6, 173, 179
mud (sampling) 290
multicellular organisms 300, 307
multi-polar neurone 206, 212
muscle 10–11, 125, 139, 140, 142, **185–8**, 206–8
  cells 10–11
  contraction 26, 30, 186–8
  fibre 185–6, 195
mushroom 324, 326
mussel 302, 334
mutation 226–7, 235, 315
mutualism 251
mycelium 321–4
mycorrhiza 257–8
myriapods 339
myxomatosis 244

nasal cavity 127, 194
nastic responses 105
national park 278
natural immunity 147
natural selection 237–8
Nature Conservancy Council 277
nature reserves 277

nectar 81, 84–6
nectary 80–81, 84, 86
negative feed-back 214
Neisseria 317
nematodes 301
  in soil 258
nephron 160–61
nerve 11, 206–7
  cell 9, 10, 206
  endings 190–91, 193–4, 166
  fibre 9, 206–8, 212
  impulse 35, 202, 206, **207**–8, 212
nervous system 11, 205, **206–11**
nets (field-work) 289
neural arch 182
neurones 206
newt 303, 342
niche 282
nitrates 52–4, 247–9, 267–8
nitrifying bacteria 247–8, 314
nitrogen 52–4, 247–9
nitrogen cycle 247–8
nitrogen fixation 248
nitrogen-fixing bacteria 247–9, 251
nitrogenous excretion 348
nitrogenous waste 159
  transport in blood 145
nitrogen oxide 272–3, 275–6
node 58
Notonecta (water boatman) 287
NPK fertilizer 53, 249
nuclear membrane 5–6
nuclear pores 6
nucleic acid **17**, 312
nucleolus 5–**6**
nucleotide 17, 25
nucleus 3–**4**, 5–9
nutrition 250
Nymphaea (water lily) 284–5

objective lens 2
octopus 302, 334
oesophagus 127
oestrogen 175, 178–9, 214–15
oil pollution 272
oils 15, 115
omnivores 250
operculum 340–41
optic lobe 211
optic nerve 196–8
optimum 19
oral groove 327–8
orchid (pollination) 85
order 301
organelle 3, 5–7
organic matter
  in soil 255–6, 260
organism 11
organs 10–11
osmo-regulation 162, 328
osmosis 36–42
  experiments 40–42
  in guard cell 60
osmotic gradient 75
ossicle 199–200
osteomalacia 117
outer ear 199–200
oval window 199–200
ovary
  human **171**–3, 178, 212–15
  plant 80, **81**–9
over-cooling 168–9
over-grazing 270–71
over-heating 168
overweight 120
oviduct 171–80
ovulation 172, 178–80, 214
ovule 80, **81**, 86–9
ovum 170–74, 176, 223
oxidation 24
oxygen
  diffusion into cells 34
  dissolved in water 268, 283
  in exhaled air 157
  in germination 97, 99
  from photosynthesis 45–9, 51–2
  in respiration 24–6, 28–9
  transport in blood 138, 145
  transport in plants 76
  uptake in lungs 154
oxygenated blood 138–9, 141, 145, 175
oxygen debt 25
oxy-haemoglobin **138**, 145, 154
ozone 273

'paçe-maker' 140
Pacinian corpuscle 193–4
pain 193, 202, 216
paired fins 340
palate 127, 189
palisade cell 4, 48, 50, 59, 61
palisade mesophyll 59, 61
pancreas 127, 129, 212–13
pancreatic amylase 129
pancreatic juice 129
paracetamol 216
'parachute' fruits and seeds 90
Paramecium 327–8
  population 292–4
parasite 250, 347
parasitic fungi 322
parasitic protozoa 328
parathyroid 116
parental care 343, 349
  in humans 177–8
partially permeable membrane 36, **37**, 39
passive smoking 156
pasteurization 121
pathogen 314
pectoral fin 340
pectoral girdle 182
pelvic fin 340
pelvic girdle 176, 182–4, 187
pelvis
  hip girdle 183, 187
  kidney 160–61
penicillin 226, 258, 325
Penicillium 258
penis 171–3
peppered moth 238
'pepper-pot' dispersal 89–90
pepsin 128–9, 134–5
pepsinogen 129
peptide 15, 127–9
peptide bond 15
peripheral nervous system 206
period (menstrual) 178–9
periodontal disease 191–2
periodontitis 192
periphyton 284, 288
peristalsis 125–6, 128
permanent teeth 190–91
permeability (soil) 261
persistent insecticide 267
pest control 295
pesticide 266, 271, 295
  systemic 76
petals **80**–89
petroleum 245–6
pH
  effect on enzymes 19, 22
  and freshwater 289
  of soil 257, 262–3
phagocyte 137–8, 147, 318
phagocytosis 34–5
phantom midge larva (*Chaoborus*) 287
pharynx 127
phenotype 229–30
phloem 9, 62–4, 74–5
phosphate 25, 52–5, 224–5, 267–8
phosphorus 52–3, 249
  in diet 116
photomicrographs 2
photoperiodism 105–6
photorespiration 51
photosynthesis 6, **44–52**, 61, 95, 245–6, 252
  experiments 45–7, 49, 51–2
  rate of 50–52
phototropism 106–8, 110
phylloquinone 117
phylum 301–2, 333
phytoplankton 243, 284, 329
pia mater 210
pickling 121
PIDCP (vitamin C test) 123, 363
pill (contraceptive) 180, 215
pinna (ear) 199, 344
pith 62–4, 68
pituitary gland 211–12, **214**–15
placenta 174–6
planarian 302
plankton 243, 286, 329
plankton net 289–90
Planorbis (pond snail) 334–5
plant cells **4**–5, 8, 12
  osmotic effects 38, 41
plant community 284–5
plant growth substances 108–9
plant kingdom 300, 304–6
plaque 191–2
plasma 137, **139**, 143, 145, 161
plasma proteins 133, 139
Plasmodium (Malarial parasite) 328
plasmolysis 41–2
plastid 5
platelets 137, **138**, 146
pleural fluid 152–3
pleural membrane 152–3
plumule 92–5
poikilothermic 340
pollen 80, 83–7
  grain 224
  sac 81, **84**–5
  tube **87**, 91
pollination **83**, 86
  of grasses 85
  by insects 84
  of lupin 84
  of wallflower 84
pollution
  of air 272–3, 275–6
  control 275–6
  of water 268, 271, 276–7
polypeptide 15, 19
polypody 304, 331–2
poly-saccharide 16
polyunsaturates 120
pond animals 334–5
pond ecology 284–90
pond skater 286
poppy, seed dispersal 89–90
population 264–5, **281**–2, **292–7**
  dynamics 294
  fluctuations 293–4
  growth 292–3, 295–6
  patterns 296
  of world 122, 180
pores
  nucleus 5–6
  skin 166–7
  stoma 9–10
pore space 254–5
posture 195, 201
potassium 53–4, 60, 249
  in diet 116
potassium nitrate 53–4
potato 102, 249
potato blight 249
potometer 77
predation 294
predator 242
pregnancy 174–6
  dietary needs 119
premolars 189
preservation (food) 121
pressure-sensory endings 193
primary consumer 243, 252
primate 301
producer 243, 246, 250, 252
progesterone 175, 178–9, 214–15
'pro-legs' 338
proprioceptor 195
prostate gland 171–2
protease 19, 128–9
protection (skeleton) 184
proteinase 128
proteins **14–15**, 19, 225
  in diet 114–15, 118–19
  digestion 127–9
  test for 122–3
protista 300, 307, **327**–8, 348
protophyta 300, 327
protoplasm 4
protozoa 300, 308, 327
  in soil 263
'Pruteen' 314
ptyalin 127
puberty 178
puff-ball 323
pulmonary artery and vein 139, 141–2, 152, 154
pulmonary cavity (snail) 334–5
pulmonary circulation 140–41
pulp (tooth) 190
pulse (arterial) 141–2
pupa 337–8
pupil 196, 198–9
pyloric sphincter 127–9
pyramid
  biomass 243, 252
  energy 243
  numbers 243
  population 296
pyrogallic acid 97
pyruvic acid 25

radiation 227
radicle 92–4, 96
  growth 69, 97
  response to stimuli 106, 109–10
radioactive waste 272
radius 11, 183–4, 186
radula 334–5
ragworm 302
ram's horn snail 334
Ranunculus aquatilis (water crowfoot) 284–5
Ranunculus uitans (river crowfoot) 284
rate
  of diffusion 34
  of enzyme reactions 19
  of photosynthesis 50–52
  of respiration 29
receptacle **81**, 88–9
receptor 193, 353
recessive 228–9
recessive back-cross 231
reclamation 278
rectum 127, 131
recycling 245–9, 278–9
red cells 116, **137**–8, 146, 154, 239
reed (*Phragmites communis*) 285
reedmace (*Typha latifolia*) 285
reed swamp 285
refined sugar, and tooth decay 191
reflex action 207–9
reflex arc 207–9
refraction 195–7
refrigeration 121, 311
rejection (transplants) 163
relay neurone 208–9
renal artery and vein 141, 160–61
renal capsule 160–61
renal tubules 160–61
replication 221–2

reproduction 342, 351
bacteria 311
bird 343
fish 341
flowering plants 80–89
frog 342
human 170–80
mammal 344
reptile 342
virus 313
reproduction rate 292
reproductive organs 214
human 171–2
plant 80–81
reproductive system
human 171–2
reptile 301, 303, 307, **343**
resistance
to disease 318
to drugs 226–7
respiration 6, **24–30**, 48–9, 52, 246, 346–7
experiments 26–30
rate 29
respiratory surface 155, 347
respiratory system 155
respirometer 28–9
retina 196–8
retinol 117
rhinovirus 315
rhizoid 330–31
rhizomes 100–102, 331
rhizosphere 258
rib cage 11
riboflavin 117
ribose 17, 25
ribose-nucleic acid 6, 17
ribosome 3, 5–6
ribs 182–4
in breathing 152–3
rice (in diet) 115–17
rickets 117
river crowfoot (*Ranunculus fluitans*) 284
RNA (ribose-nucleic acid) 6, **17**, 311–13
rodent 301
rods (eye) 197
root (plant) 58, 227
function 75
growth 65–6, 68–9
responses 106–10
structure 64–5
uptake of water 75
root (tooth) 189–91
root cap 64–5, 75
root hairs 64–6, **75**, 94
rooting 'hormone' 104
root nodules 247–9, 251
root pressure **74**–5, 79
root system 58, 66
rootstock 100–101
roughage 117
rubella 176–7
runner (stolon) 100–101
'rust' disease 323
rye grass 83

Sabin (polio vaccine) 147
'saddle' (earthworm) 334
saliva 21–2, 127, 134–5
salivary amylase 127, 134
salivary gland 127
Salk (polio vaccine) 147
salmonella 315–17
salt (taste) 184, 203–4
salting (food) 121
salts 17
in the diet 116
excretion 159–61
in soil 53, 256
transport in plants 74
uptake by roots 76
sampling techniques 289
sand particles 254–6, 260
saprophyte 245, 250, 322
saprophytic fungi 322
saturated fatty acids 119
scale leaf 101–2
scales 343
scanning electron microscope 6
scapula (shoulder blade) 183
scavenger 242
scent (flowers) 85–6
scion 103–4
sclera 196, 198
scrotum 171–2
scurvy 117
sea anemone 302
seal 303
search techniques 290
seaweeds 304, 330
sebaceous gland 166–7
secondary consumer 243–4, 252
secondary sexual characteristics 178
secretin 129
section 3–4, 7
freeze etched 6
sedative 216
seeds 80–81
dispersal 89–90
dormant 99
formation 87–9
germination 93–9
structure 92–3, 95
seed-bearing plants 305–6, 308
seed coat 87, 92
seed dispersal 89–90
segment 333, 335–6, 338–9
segmentation 333, 335–9
selection 238–9
selection pressure 238
selectively permeable membrane 36
selective reabsorption 160, 162
selective weed-killer 109
self-pollination 83
semen 172–3
semicircular canal 200–201
semi-lunar valve 139–40
seminal vesicle 171–2
'semi-permeable' membrane 36
sensation 202, 207
sense organs 193–4
senses 193–204
sensitivity 352
plants 105
skin 165
sensory cell 193–4, 201–2
sensory fibre 208–9
sensory impulse 202, 206
sensory nerve endings 193–4, 200
sensory neurones 206
sepal 80, **81**–4, 88–9
serum 147–8
sewage 267, 271
disposal 277, 317
sludge 277
sex chromosome 232–3
sex determination 232–3
sex hormones 178
sexually transmitted diseases 317
sexual reproduction 236, 349, 350–51
human 170–80
plants 80–91
sheath (contraceptive) 179
shell (molluscs) 334
shivering 168
shoot 2, 58
response to stimuli 106–8
short-day plants 106
shoulder blade 182–4, 186
shoulder girdle 182
shoulder joint 183–4
shrubs 305, 307
sickle-cell anaemia 226, 232, 239
sickle-cell trait 232
sieve plate 9, 63–4
sieve tube 59, 62–3, **64**
sight 196–9, 211
sigmoid curve 292–3, 352
silicon oxide 255
silt particle 254–5, 260
single-cell protein 314, 325
single factor inheritance 228–31
Site of Special Scientific Interest (SSSI) 278
skeletal muscle 185–6
skeletal system 11
skeleton 182–3
functions 184
skin 165–8
excretory function 159
skin senses 193, 203
skull 11, 182–4
'sleep' movements (plants) 105
slime capsule 310
slow-worm 303
sludge 277
slug 302, 334
small intestine 11, 127, 129, 130–31
smell
receptor 194
sense 194–5, 211
'smog' 272, 275
smoking
and bronchitis 156
and heart attacks 149, 156
and lung cancer 155, 157
and pregnancy 175
smooth muscle 185–6
snail 302, 334
snake 303, 343
soda-lime 27–9, 46
sodium (in diet) 116
sodium hydrogencarbonate 129
soft palate 127–8
soil 254–8, 260–61
soil erosion 269–71
soil-less culture 54
somatic cell 221
sori 331–2
sound 199
sour taste 194, 203–4
specialization
cells 7, 9–10
protista 328
species 301, 306
specificity
of antibodies 147
of enzymes 19
of sense organs 202
spermatazoa 170
sperm duct 171–2
sperm production 172
sperms 170–73, 223
sphincter 186
bladder 160–61
pyloric 128–9
spider 302, 335, 339
spinal column 182
spinal cord 11, 182, 184, 206, 208–9, **210**–11
spinal nerve 206, 208–9
spinal reflex 208
spine (spinal column) 182
spiracle (insect) 338
Spirogyra 329–30
spirometer 154
spleen 144–5
spongy mesophyll 59, 61
spontaneous abortion 177
sporangia
ferns 332
Mucor 323
spore capsule 330–31
spores 330, 332
bacteria 311
fungi 322–4, 326
springtails 258–9, 286
squash preparation 227
SSSI (Site of Special Scientific Interest) 278
stamens 80, **81**–6, 88–9
stapes 199–200
starch **16**, 20–22, 102, 114–15
digestion 126, 129, 134–5
digestion by fungi 325–6
in seeds 94, 96
test for 122–3
starch phosphorylase 21
stearic acid 15
stem 3, 58
growth 66, 69
strength 64, 68
structure 62–4
stem tuber 102
sterilization (bacterial) 319
sternum 153, 183
steroids 15
stickleback 340–41
stigma 80, **81**–9
stimulant 215
stimulus 105–6, 193–4, 202, 352–3
stock (graft) 103–4
stolon 100–101
stoma (stomata) 9–10, 48, 59, **60–61**, 68, 71–3, 78–9, 284
stomach 127–9
stomatal pore 60–61
storage (of food) 131–2, 347
in plants 52
strawberry
flowers and fruits 88–9
runners 100–101
Streptococcus 315
Streptomyces 258
streptomycin 258, 325
stress 213
and heart disease 150
stretch receptor 195, 199, 208
striated muscle 185
structural proteins 14
style (flower) **81**–3, 87–9
subclavian artery and vein 142
submerged plants 284–5
subsoil 254
substrate 19, 284
succession 285–6
suckling 177–8, 344
sucrase 19
sucrose 16, 19, 74, 114, 119
and tooth decay 191
sugar **15–16**, 114–15, 119
test for 122–3
and tooth decay 191
sulphates 52–4
sulphur 52–4
sulphur dioxide 272–3, 275–6
sunburn 165
sunlight 245, 251–2
in photosynthesis 45–8, 50
effect on skin 165
sun-tan 165
superphosphate 53
support (skeleton) 184
surface area 34
leaves 50
lungs 155–6
intestine 130
respiratory organs 347
root system 65
soil particles 255
and volume 329
surface feeder 286
surface film 283, 286
surface tension 283
surface waters (pond) 284
survival value 238, 353
suspensory ligament 196, 198

swallowing 127–8
swamp carr 286
sweat 168
  duct 166–7
  gland 166–8
sweating 160, 168
sweep net 289
sweet taste 194, 203–4
swim bladder 340
sycamore (fruit) 90
symbiont 250
symbiosis 250–51, 258
sympathetic nervous system 213
synapse 207–9
synovial fluid 184
synovial joint 184
syphilis 317
syrup (food preservation) 121
system 11
systemic circulation 140–41
systemic pesticide 76

tadpole 342
tail fin 340
tapeworm 250, 302
tap root 58, 62, **66**
target organs 145, 212
tartar 191
tartrazine 122
taste (sense) 194, 203–4
taste-buds 194
tear glands 196
tears 196
teeth 116, 189–92
temperature
  of body 167–9
  changes in water 283
  and enzyme reactions 19, 21
  and germination 98–9
  measurement in water 298
  and photosynthesis 50–51
  and transpiration 73
temperature control 168–9
tendon 184, 186
tentacle 334–5
terminal bud 58, 66–7, 101
tertiary consumer 243–4, 252
testa 87, 92–4
test cross 231
testis 171–2, 212, 214
testosterone 178, 214
'test-tube babies' 180
thallus 330
thermal capacity 14
thigh bone (femur) 182–3
thorax
  human 152–3
  insect 335–8
thrombus 149
thymine 225
thymus 145
thyroid gland 116, **212**, 215
thyroid-stimulating hormone (TSH) 214–15
thyroxine 116, **212**–15
tibia 11, 183, 187–8, 208
tiger moth 238
tinea 315, 322
tissue 10–11
tissue culture 12
tissue fluid **142**–3, 145, 163–4
tissue respiration 24, 155
TMV (tobacco mosaic virus) 312
toad 303, 342
toadstool 245, 322
tobacco mosaic virus (TMV) 312
tocopherol 117
toes 183
tolerance (drugs) 216
tomato 239
  flower and fruit 88
tongue 127, 194, 203–4
toothache 191
topsoil 254
touch
  sense 166
  sensitivity 203
  sensory endings 193
toxins 121, 147
toxoid 147
trace elements 52
trachea 127, 151–3
training 210
tranquillizer 216
transducer 193
transect 290
translocation 74
transmission of disease 314–17
transmitter substance 207
transpiration **71–3**, 78
  control 73
  functions 72
  rate 73, 76–7
transpiration stream 71
transport
  by circulatory system 145
  food in plants 74–5
  gases in plants 76
  salts in plants 74
  water in plants 71–3, 77–8
transport systems 347–8
transverse section **3**–5, 62
trees 305, 307
Treponema 317
triceps 186–7, 208
tricuspid valve 139–40
triglyceride 15
tripeptide 15
tristearin 15
trophic level 243
tropical forests 269
tropisms 105–10, 353
  advantages 107
  mechanism 108
trout 341
true-breeding 229
'true-legs' 338
trypsin 129
trypsinogen 129
tryptophan 116
TSH (thyroid-stimulating hormone) 214–15
tuber 102
tuberculosis 316
Tubifex 288
tubule (kidney) 3, 10
turgid 38
turgor 40, 42, 60
turgor pressure 38
twins 176
Typha latifolia (reedmace) 285
typhoid 315

ulna 11, 183–4, 186
ultra-filtration 160–61
ultra-violet light 165, 227, 273
umbilical cord 174–6
unicellular organisms 300, 327
unisexual 80, 83
unsaturated fatty acids 119–20, 150
unstriated muscle 185
urbanization 264
urea 133, 145, 159–61, 348
ureter 160
urethra 160–61, 171–2
uric acid 159, 161, 348
urine 159, 161–2
uterus **171**, 173–6, 178–9, 215
utriculus 200–201

vaccination 147
vaccine 147
vacuole 4, **5**–6, 8–9, 12, 38
vagina 171, 173, 176
valine 15, 115–16, 225–6
valves
  heart 139–40
  lymphatics 144
  veins 143
variables 98
variation 235–8
variegated leaf 45
vascular bundle 59, 62, 77
  function 63–4
vascular plants 306, 331
vasectomy 180
vaso-constriction 168
vaso-dilation 168
vegetative propagation 100–102
  advantages 102
vein
  human 11, 141–2, **143**
  leaf 59, 62, 64
vena cava 139, 141–2
venereal disease 317
venous system 142
ventilation 347
  lungs 152–3, 155
ventral fin 340
ventral root 208–9
ventricle
  heart 139–41
  brain 211
venules 142–3
vertebra 182–4
vertebral column 182–4, 187, 340, 343
vertebrates 301, 303, 307–8, **340–45**
vessel 59, 62, **63–4**, 68
vibrissae 344
villi (villus) 129–31
virus diseases 315–16
viruses 311–16
vision 196–9
'Visking' tubing 39
vitamins 17, 116–17
  absorption 131
  storage 133
  test for 123
vitreous humour 196
viviparity 343
voluntary action 210
voluntary muscle 185–6
Volvox 306
Vorticella 327
vulva 171, 176

Wallace 237
wallflower
  flower structure 80–81
  pollination 84
  seed dispersal 90
washing powders 314
wasp 302
water 14
  in diet 117–18
  and disease transmission 317
  as an environment 283
  excretion 159–62
  in germination 94, 97–8
  in photosynthesis 44–5, 47–8, 50
  physical properties 282
  from respiration 24
  in soil 254–7, 260–61
  in transpiration 71–3, 78
  transport in plants 71–3, 75, 78
  uptake by plants 77
  uptake by roots 75
Water Act 276
water balance 162
water beetle 287–8, 337
water boatman 287–8
water crowfoot 284–5
water cultures 53–4
water cycle 248–9
water flea 286
water lily 284–5
water louse (*Asellus*) 287
water pollution 268, 271, 276–7
water potential **37**-8, 60
water sac 174, 176
water-soluble vitamins 117
weaning 178
weed-killer 109
weight (human) 120
western diets 119
whales 280
wheat 249
  genetics 234
  seed germination 95
white cells 9, 137–8, 146
white matter 208, 210
wilting 38, 72–3
wind dispersal (seeds) 89–90
windpipe 127, 151
wind pollination 85–6
winged fruits 90
wings (insect) 336–8
'wisdom' tooth 189–90
withdrawal symptoms 216
womb 171
woodland carr 286
woodlouse 302
world food 122
world population 180, 296
World Wildlife Fund 279–80
worm cast 334
worms 333
wrist bones 183
WWF (World Wildlife fund) 279–80

X-chromosome 232–3
X-rays 227
xylem 64, 73, 75, 78

Y-chromosome 232–3
yeast 6, 24, 29–30, **324–5**, 326
  population 294
yolk sac 341

zona pellucida 172–4
zonation 286
zooplankton 243, 329
zygote 170–71, 349